technical
DRAWING
for ENGINEERING
communication   SEVENTH EDITION

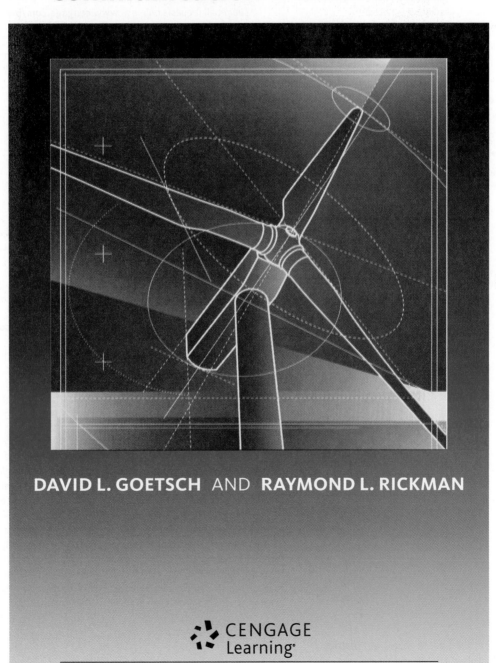

DAVID L. GOETSCH AND RAYMOND L. RICKMAN

✦ CENGAGE
Learning·

Australia • Brazil • Japan • Korea • Mexico • Singapore • Spain • United Kingdom • United States

Technical Drawing for Engineering Communication, 7e
David L. Goetsch and Raymond L. Rickman

SVP, GM Skills & Global Product Management:
Dawn Gerrain

Product Manager: Daniel Johnson

Senior Director, Development:
Marah Bellegarde

Senior Product Development Manager:
Larry Main

Content Developer: Richard Hall

Product Assistant: Andrew Ouimet

Vice President, Marketing Services:
Jennifer Ann Baker

Marketing Manager: Scott Chrysler

Senior Production Director: Wendy Troeger

Production Director: Andrew Crouth

Senior Content Project Manager:
Kenneth McGrath

Art Director: Bethany Casey

Technology Project Manager: Joe Pliss

Cover image(s): ©instamatics/iStock Vectors/
Getty Images, ©iStockPhoto.com/2ndpic

Library of Congress Control Number: 2014940177

ISBN: 978-1-285-17301-6

Cengage Learning
20 Channel Center Street
Boston, MA 02210
USA

Cengage Learning is a leading provider of customized learning solutions with office locations around the globe, including Singapore, the United Kingdom, Australia, Mexico, Brazil, and Japan. Locate your local office at:
www.cengage.com/global

Cengage Learning products are represented in Canada by Nelson Education, Ltd.

To learn more about Cengage Learning, visit **www.cengage.com**

Purchase any of our products at your local college store or at our preferred online store **www.cengagebrain.com**

Notice to the Reader
Publisher does not warrant or guarantee any of the products described herein or perform any independent analysis in connection with any of the product information contained herein. Publisher does not assume, and expressly disclaims, any obligation to obtain and include information other than that provided to it by the manufacturer. The reader is expressly warned to consider and adopt all safety precautions that might be indicated by the activities described herein and to avoid all potential hazards. By following the instructions contained herein, the reader willingly assumes all risks in connection with such instructions. The publisher makes no representations or warranties of any kind, including but not limited to, the warranties of fitness for particular purpose or merchantability, nor are any such representations implied with respect to the material set forth herein, and the publisher takes no responsibility with respect to such material. The publisher shall not be liable for any special, consequential, or exemplary damages resulting, in whole or part, from the readers' use of, or reliance upon, this material.

Printed in the United States of America
Print Number: 01   Print Year: 2014

brief contents

contents

preface

Purposes

Technical Drawing for Engineering Communication is intended for use in such courses as basic and advanced drafting, engineering graphics, descriptive geometry, mechanical drafting, machine drafting, tool and die design and drafting, and manufacturing drafting. It is appropriate for those courses offered in comprehensive high schools, area vocational schools, technical schools, community colleges, trade and technical schools, and at the freshman and sophomore levels in universities.

Prerequisites

There are no prerequisites. The text begins at the most basic level and moves step by step to the advanced levels.

It is as well-suited for students who have had no previous experience with technical drawing as it is for students with a great deal of prior experience.

how to use this book

Technical Drawing for Engineering Communication is a comprehensive teaching and learning tool that contains several special features to promote the student's development and to make learning easier. Students and instructors can make use of the following features:

Career Profiles located in the Section Opener focus on the occupations of specific individuals. These profiles were chosen as representative of the types of jobs that students could acquire after completing an education in drafting. The career profiles also relate directly to the content covered in the section.

CAREER PROFILE

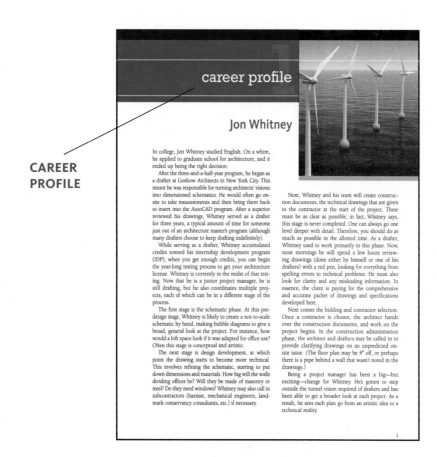

career profile

Jon Whitney

In college, Jon Whitney studied English. On a whim, he applied to graduate school for architecture, and it ended up being the right decision.

After the three-and-a-half-year program, he began as a drafter at Goshow Architects in New York City. This meant he was responsible for turning architects' visions into dimensioned schematics. He would often go on-site to take measurements and then bring them back to insert into the AutoCAD program. After a superior reviewed his drawings, Whitney served as a drafter for three years, a typical amount of time for someone just out of an architecture master's program (although many drafters choose to keep drafting indefinitely).

While serving as a drafter, Whitney accumulated credits toward his internship development program (IDP); when you get enough credits, you can begin the year-long testing process to get your architecture license. Whitney is currently in the midst of that testing. Now that he is a junior project manager, he is still drafting, but he also coordinates multiple projects, each of which can be in a different stage of the process.

The first stage is the schematic phase. At this pre-design stage, Whitney is likely to create a not-to-scale schematic by hand, making bubble diagrams to give a broad, general look at the project. For instance, how would a loft space look if it was adapted for office use? Often this stage is conceptual and artistic.

The next stage is design development, at which point the drawing starts to become more technical. This involves refining the schematic, starting to put down dimensions and materials. How big will the walls dividing offices be? Will they be made of masonry or steel? Do they need windows? Whitney may also call in subcontractors (hazmat, mechanical engineers, landmark conservancy consultants, etc.) if necessary.

Next, Whitney and his team will create construction documents, the technical drawings that are given to the contractor at the start of the project. These must be as clear as possible; in fact, Whitney says, this stage is never completed. One can always go one level deeper with detail. Therefore, you should do as much as possible in the allotted time. As a drafter, Whitney used to work primarily in this phase. Now, most mornings he will spend a few hours reviewing drawings (done either by himself or one of his drafters) with a red pen, looking for everything from spelling errors to technical problems. He must also look for clarity and any misleading information. In essence, the client is paying for the comprehensive and accurate packet of drawings and specifications developed here.

Next comes the bidding and contractor selection. Once a contractor is chosen, the architect hands over the construction documents, and work on the project begins. In the construction administration phase, the architect and drafters may be called in to provide clarifying drawings on an unpredicted on-site issue. (The floor plan may be 4″ off, or perhaps there is a pipe behind a wall that wasn't noted in the drawings.)

Being a project manager has been a big—but exciting—change for Whitney. He's gotten to step outside the tunnel vision required of drafters and has been able to get a broader look at each project. As a result, he sees each plan go from an artistic idea to a technical reality.

1

Chapter Objectives at the beginning of each chapter identify the skills and knowledge that students will acquire from reading the material. When students have finished the chapter, they should review the objectives to ensure that they have met each one.

The **Chapter Outline** lists the title of each major topic covered in the chapter. This provides a preview of the content coverage.

Key Terms, listed at the beginning of each chapter, are important words and phrases that students will encounter as they study the chapter. Each key term is highlighted in *italics* at the first significant use in the text and is included in the glossary or index at the back of the book.

Industry Application is a boxed article that explains how the skills and knowledge discussed in the chapter may be applied to a real-world job-related setting. A variety of skills will be covered, including math, science, communications, and computers.

A chapter **Summary** provides a recapitulation of the key topics covered in the chapter. This enables students to reassess their comprehension of the material before proceeding to the end-of-chapter questions and problems.

CHAPTER OUTLINE

KEY TERMS

CHAPTER OBJECTIVES

INDUSTRY APPLICATION

Review Questions are written to ensure that students have adequately read the chapter and that they understand the material. These questions prepare students for class tests given by the instructor and for the certification test offered by the American Design and Drafting Association. (Visit the American Design and Drafting Association website at **www.adda.org** to learn about the certification programs and ADDA membership.) These questions do not require a computer or drafting materials.

Chapter Problems contain drawing projects that allow students to test their drafting skills. Advanced problems are marked by **icons**. These special icons have been placed adjacent to the advanced drawing projects to indicate the engineering and design field the drawing pertains to. These icons call out a specific drafting field or discipline.

ADVANCED
ICON

SUMMARY

REVIEW
QUESTIONS

CHAPTER
PROBLEMS

ICONS

Architectural	Structural	Mechanical
Architectural	*Structural*	*Mechanical*
Pipe Drafting	Civil	Electrical
Pipe Drafting	*Civil*	*Electrical*

Content Overview

Chapters 1–18 are to be used to help students develop the design and drafting skills that are fundamental to *all* drafting fields. Instructors are encouraged to use all of these chapters to build a solid footing of design and drafting knowledge for students.

Chapters 19–21 allow students to develop advanced knowledge and skills beyond the fundamentals.

Instructors are encouraged to use these chapters to help their students develop an in-depth understanding of these discipline areas: Welding; Modern Manufacturing: Materials, Processes, and Automation; and Drafting Applications: Pipe, Architectural, and Civil Engineering.

Chapter 21, The Design Process and Advanced Concepts, is important in preparing the student for entry into the job force. The steps of the design process are defined, and the reader is taught how to be creative in analyzing and solving problems. Modern design processes such as DFM, DFA, rapid prototyping, and reverse engineering are addressed.

New Features in the Seventh Edition

The Seventh Edition is based on the concept that "less is more." After polling a wide cross section of instructors who use this textbook, the authors eliminated material reviewers thought was superfluous. This was done so that new material could be added where appropriate without increasing the size of the book and, in turn, the cost. A summary of changes for the Seventh Edition follows:

- The name of the book was changed to reflect the movement in the field away from the traditional name and concept of "drafting." Because technicians in this field prepare drawings and other forms of documentation to communicate engineering concepts and information, the newly adopted title of the book for the Seventh Edition is *Technical Drawing for Engineering Communication*.

- Drawings and photographs were updated throughout the book as needed to reflect the most modern technologies and practices.

- Chapter 1, Employability Skills, is a new chapter that covers the following topics: employability skills defined, importance of employability skills, specific employability skills needed by drafting and design technicians, and job-seeking skills.

- Chapter 2, Drafting Tools: Conventional, CAD, and Solid Modeling, was completely rewritten to present the types of tools drafting and design technicians typically use, ranging from conventional tools that are still used to CAD hardware and software to solid modeling software.

- Old Chapter 10, Dimensioning and Notation, became new Chapter 6 at the request of reviewers to put it in a more logical order for teaching and learning. In addition, the chapter was updated to reflect changes that grew out of the latest revision to ANSI Y14.5.

- Chapter 12, Geometric Dimensioning and Tolerancing, was completely rewritten to bring it in line with the latest revisions to ANSI Y14.5.

- Old Chapter 21, Drafting Applications, was moved to the website and is available there for instructors who use it.

- Old Chapter 22, The Design Process, became new Chapter 21, The Design Process and Advanced Topics. The following new material was added to this chapter: rapid prototyping, parametric design, ISO 9000, and reverse engineering.

Tested and Proven Features

- An enhanced Instructor Companion Website contains electronic instructional material.

- Chapter 21, The Design Process and Advanced Topics, covers modern design practices and standards.

- Discipline-specific icons highlight the drawing problems, enabling the reader to quickly choose problems pertaining to his or her drawing field.

- Review questions are formatted as multiple-choice and true/false for rapid testing and grading.

- New drawing problems are contained in the primary drawing chapters.

- Boxed articles covering real-world applications exist in all chapters. This reinforces the relevance of the chapter content to today's job environment.

- New math problems are contained in selected chapters for additional practice.

- Step-by-step explanations of drawing procedures and techniques are provided.

- The book is written in language students will understand; technical terms are defined as they are used.

- Unique black and rust color format depicts isometric views more clearly than "flat" black-and-white drawings.

- Text and illustrations are located in *direct* relationship to each other wherever possible.

- Real-world techniques are highlighted in the Industry Application boxed articles.

- Although the emphasis is on mechanical drafting, other pertinent drafting subjects are included for a comprehensive, well-rounded approach to technical drawing.

- The book contains in-depth drafting applications in architectural, structural, civil, and piping drafting.

The Learning Package

The complete Instructor Companion Website supplements package was developed to achieve two goals:

1. To assist students in learning the essential information needed to prepare for the exciting field of drafting.

2. To assist instructors in planning and implementing their instructional programs for the most efficient use of time and other resources.

The *Technical Drawing for Engineering Communication* package was created as an integrated whole. Supplements are linked to and integrated with the text to create a comprehensive supplement package that supports students and instructors, beginning or veteran. The package includes:

Instructor Companion Website—The ICW is an educational resource that creates a truly electronic classroom. The website contains tools and instructional resources that enrich your classroom and make your preparation time shorter. The elements of the ICW link directly to the text and tie together to provide a unified instructional system. With the ICW, you can spend your time teaching, not preparing to teach.

Features contained on the Instructor Companion Website include:

- **Syllabus:** Lesson plans created by chapter. You have the option of using these lesson plans with your own course information.

- **Chapter Hints:** Objectives and teaching hints that provide the basis for a lecture outline that helps you to present concepts and material. Key points and concepts can be graphically highlighted for student retention.

- **Answers to Review Questions:** These solutions enable you to grade and evaluate end-of-chapter tests.

- **PowerPoint® Presentation:** These slides provide the basis for a lecture outline that helps you to present concepts and material. Key points and concepts can be graphically highlighted for student retention.

- **Computerized Test Bank:** Over 800 questions of varying levels of difficulty are provided in true/false and multiple-choice formats so you can assess student comprehension.

Cengage Learning Testing (CLT)

Powered by Cognero, CLT is a flexible, online system that allows you to:

- Author, edit, and manage test bank content from multiple Cengage Learning solutions.

- Create multiple test versions in an instant.

- Deliver tests from your LMS, your classroom, or wherever you want.

Contact Cengage Learning or your local sales representative to obtain an instructor account.

Accessing an Instructor Companion Website from SSO Front Door

1. Go to http://login.cengage.com and log in, using the instructor e-mail address and password.

2. Enter author, title, or ISBN in the Add a Title to Your Bookshelf search.

3. Click Add to My Bookshelf to add instructor resources.

4. At the Product page, click the Instructor Companion site link.

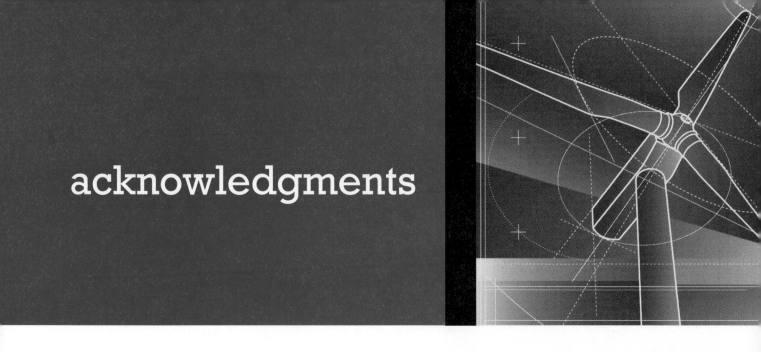

acknowledgments

The authors would like to acknowledge the efforts of many people without whose assistance this project would not have been completed. We thank Deborah M. Goetsch for her assistance with photography.

The following individuals reviewed the manuscript and made valuable suggestions to the authors. The authors and the publisher greatly appreciated their contributions to this textbook.

Sharon Bagby, Texas State Technical College

Steven P. Bauer, The Pennsylvania State University, DuBois

Glenn Frostholm, Augusta Technical College

Jerry M. Gray, West Georgia Technical College

Carrie Miller, Middle Georgia Technical College

Alicia O'Kelley, Western Dakota Technical Institute

About the Authors

David L. Goetsch is Emeritus Vice President and Professor of Design, Drafting, Quality, and Safety at Northwest Florida State College in Niceville, Florida. His drafting and design program has won national acclaim for its pioneering efforts in the area of computer-aided drafting (CAD). In 1984, his school was selected as one of only 10 schools in the country to earn the distinguished U.S. Secretary of Education's Award for an outstanding Vocational Program. Professor Goetsch is a widely acclaimed teacher, author, and lecturer on the subject of drafting and design. He won Outstanding Teacher of the Year honors in 1976, 1981, 1982, 1983, and 1984. In 1986, he won the Florida Vocational Association's Rex Gaugh Award for outstanding contributions to technical education in Florida. In 2003, Dr. Goetsch was selected as the University of West Florida's Distinguished Alumnus. In 2007, he was named one of the University of West Florida's top 40 alumni in its first 40 years of operation. He entered education full time after a successful career in design and drafting in the private sector, where he spent more than eight years as a Senior Drafter and Designer for a subsidiary of Westinghouse Corporation. He was elected Florida's Outstanding Economic Development person in 1992, 1996, and 2007.

Raymond L. Rickman is chairman of the Manufacturing and Technology Department and Professor of Design and Drafting at Northwest Florida State College.

Professor Rickman has extensive experience in the private sector and the classroom and is a consultant on the subject of geometric dimensioning and tolerancing. He was a member of the ANSI Y14.5 technical committee and is the co-author of several career mathematics textbooks.

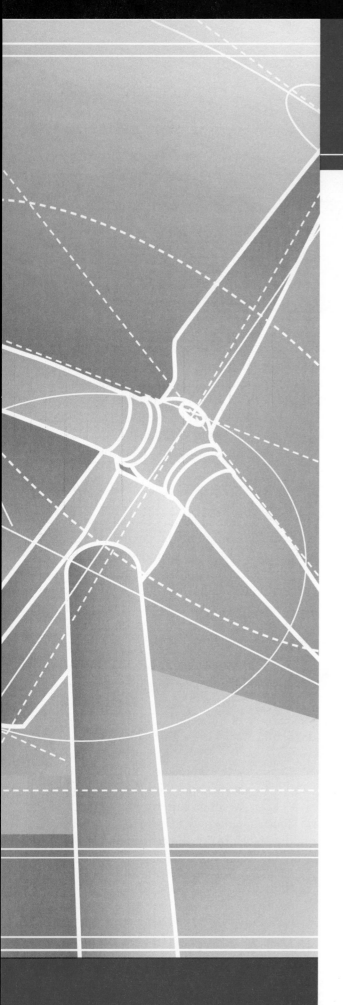

section 1

Basics

career profile

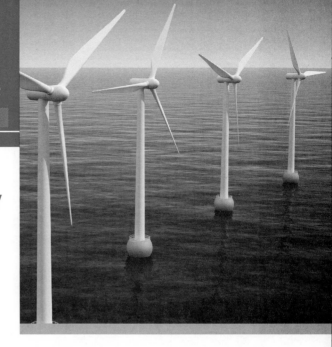

Jon Whitney

In college, Jon Whitney studied English. On a whim, he applied to graduate school for architecture, and it ended up being the right decision.

After the three-and-a-half-year program, he began as a drafter at Goshow Architects in New York City. This meant he was responsible for turning architects' visions into dimensioned schematics. He would often go on-site to take measurements and then bring them back to insert into the AutoCAD program. After a superior reviewed his drawings, Whitney served as a drafter for three years, a typical amount of time for someone just out of an architecture master's program (although many drafters choose to keep drafting indefinitely).

While serving as a drafter, Whitney accumulated credits toward his internship development program (IDP); when you get enough credits, you can begin the year-long testing process to get your architecture license. Whitney is currently in the midst of that testing. Now that he is a junior project manager, he is still drafting, but he also coordinates multiple projects, each of which can be in a different stage of the process.

The first stage is the schematic phase. At this pre-design stage, Whitney is likely to create a not-to-scale schematic by hand, making bubble diagrams to give a broad, general look at the project. For instance, how would a loft space look if it was adapted for office use? Often this stage is conceptual and artistic.

The next stage is design development, at which point the drawing starts to become more technical. This involves refining the schematic, starting to put down dimensions and materials. How big will the walls dividing offices be? Will they be made of masonry or steel? Do they need windows? Whitney may also call in subcontractors (hazmat, mechanical engineers, landmark conservancy consultants, etc.) if necessary.

Next, Whitney and his team will create construction documents, the technical drawings that are given to the contractor at the start of the project. These must be as clear as possible; in fact, Whitney says, this stage is never completed. One can always go one level deeper with detail. Therefore, you should do as much as possible in the allotted time. As a drafter, Whitney used to work primarily in this phase. Now, most mornings he will spend a few hours reviewing drawings (done either by himself or one of his drafters) with a red pen, looking for everything from spelling errors to technical problems. He must also look for clarity and any misleading information. In essence, the client is paying for the comprehensive and accurate packet of drawings and specifications developed here.

Next comes the bidding and contractor selection. Once a contractor is chosen, the architect hands over the construction documents, and work on the project begins. In the construction administration phase, the architect and drafters may be called in to provide clarifying drawings on an unpredicted on-site issue. (The floor plan may be 4″ off, or perhaps there is a pipe behind a wall that wasn't noted in the drawings.)

Being a project manager has been a big—but exciting—change for Whitney. He's gotten to step outside the tunnel vision required of drafters and has been able to get a broader look at each project. As a result, he sees each plan go from an artistic idea to a technical reality.

Introduction

Graphic Communication and Technical Drawing

OUTLINE

Graphic communication • Drawings described • Types of drawings • Types of technical drawings • Purpose of technical drawings • Applications of technical drawings • Regulation of technical drawings • What students of technical drawing, drafting, and CAD should learn • Technical drawing and quality/competitiveness • Summary • Review questions • Introduction problems

OBJECTIVES

Upon completion of this introduction, students should be able to do the following:

▶ Explain the concept of graphic communication.

▶ Define the term *drawing*.

▶ Differentiate between artistic and technical drawings.

▶ List and explain the types of technical drawings.

▶ Explain the purpose of technical drawings.

▶ Explain the different applications of technical drawings.

▶ Explain the concept of the regulation of technical drawings.

▶ Describe the role of design and drafting in promoting quality and competitiveness.

Graphic Communication

Graphic communication involves using visual material to relate ideas. Drawings, photographs, slides, transparencies, and sketches are all forms of graphic communication. Most children are able to draw before they are able to write. This is graphic communication. When one person sketches a rough map in giving another directions, this is graphic communication. Any medium that uses a graphic image to aid in conveying a message, instructions, or an idea is involved in graphic communication. One of the most widely used forms of graphic communication is the drawing.

Drawings Described

A *drawing* is a graphic representation of an idea, a concept, or an entity that actually or potentially exists in life. The drawing itself is: (1) a way of communicating all necessary information about an abstraction, such as an idea or a concept, or (2) a graphic representation of some real entity, such as a machine part, a house, or a tool, for example.

Drawing is one of the oldest forms of communication, dating back even farther than verbal communication. Cave dwellers painted drawings on the walls of their caves thousands of years before paper was invented. These crude drawings served as a means of communicating long before verbal communications had developed beyond the grunting stage. In later years, Egyptian hieroglyphics were a more advanced form of communicating through drawings.

The old adage "one picture is worth a thousand words" is still the basis of the need for technical drawings.

Types of Drawings

There are two basic types of drawings: artistic and technical. Some experts believe there are actually three types: the two mentioned and another type that combines these two. The third type is usually referred to as an illustration or rendering.

Artistic Drawings

Artistic drawings range in scope from the simplest line drawings to the most famous paintings. Regardless of their complexity or status, artistic drawings are used to express the feelings, beliefs, philosophies, or abstract ideas of the artist. This is why the layperson often finds it difficult to understand what is being communicated by a work of art.

In order to understand an artistic drawing, it is sometimes necessary to first understand the artist. Artists often take a subtle or abstract approach in communicating through their drawings. This gives rise to the various interpretations often associated with artistic drawings.

Technical Drawings

The technical drawing, on the other hand, is not subtle or abstract. It does not require an understanding of its creator, but only an understanding of technical drawings. A *technical drawing* is a means of clearly and concisely communicating all of the information necessary to transform an idea or a concept into reality. Therefore, a technical drawing often contains more than just a graphic representation of its subject. It also contains dimensions, notes, and specifications.

The mark of a good technical drawing is that it contains all of the information needed by individuals for converting the idea or concept into reality. The conversion process may involve manufacturing, assembly, construction, or fabrication. Regardless of the process involved, a good technical drawing allows the conversion process to proceed without having to ask designers or drafters for additional information or clarification.

Figures 1 and **2** contain samples of technical mechanical drawings that are used as guides by the people involved in various phases of manufacturing the represented parts. Notice that the drawings contain a graphic representation of the part, dimensions, material specifications, and notes.

Illustrations or Renderings

Illustrations or renderings are sometimes referred to as a third type of drawing because they are neither completely technical nor completely artistic; they combine elements of both, as shown in **Figures 3**, **4**, **5**, and **6**. They are technical in that they are drawn with mechanical instruments or on a computer-aided drafting system, and they contain some degree of technical information. However, they are also artistic in that they attempt to convey a mood; an attitude; a status; or other abstract, nontechnical feelings.

Types of Technical Drawings

Technical drawings are based on the fundamental principles of projection. A *projection* is a drawing or representation of an entity on an imaginary plane or planes. This projection plane serves the same purpose in technical drawing as is served by the movie screen in a theater.

FIGURE 1   Technical drawing (mechanical).

As shown in **Figure 7**, a projection involves four components: (1) the actual object that the drawing or projection represents, (2) the eye of the viewer looking at the object, (3) the imaginary projection plane (the viewer's drawing paper or the graphics display in a computer-aided drafting system), and (4) imaginary lines of sight called *projectors*.

Two broad types of projection, both with several subclassifications, are parallel projection and perspective (converging) projection.

Parallel Projection

Parallel projection is divided into the following three categories: orthographic, oblique, and axonometric projections.

Orthographic projections are drawn as multiview drawings that show flat representations of principal views of the subject (**Figure 8**). *Oblique projections* actually show the full size of one view and are of three varieties: *cabinet* (half-scale), *cavalier* (full-scale), and *general* (between half- and full-scale). **Figures 9** and **10** show cavalier and cabinet projections. *Axonometric projections* are three-dimensional drawings and are of three different varieties: isometric, dimetric, and trimetric, as shown in **Figures 11, 12, and 13.**

Perspective Projection

Perspective projections are drawings that attempt to replicate what the human eye actually sees when it views an object. That is why the projectors in a perspective drawing converge. There are three types of perspective projections: one-point, two-point, and three-point projections, as shown in **Figures 14, 15, and 16.**

Purpose of Technical Drawings

To appreciate the need for technical drawings, one must understand the design process. The design process is an orderly, systematic procedure used in accomplishing a needed design.

FIGURE 2   Technical drawing (mechanical).

FIGURE 3   Rendering.

FIGURE 4   Rendering.

Any product that is to be manufactured, fabricated, assembled, constructed, built, or subjected to any other type of conversion process must first be designed. For example, a house must be designed before it can be built. An automobile must be designed before it can be manufactured. A printed circuit board must be designed before it can be fabricated.

The Design Process

The *design process* is an organized, step-by-step procedure in which mathematical and scientific principles, coupled with experience, are brought to bear in order to solve a problem or meet a need. The design process has five steps. Traditionally, these steps have been:

FIGURE 5   Mechanical illustration (Courtesy Ken Elliott).

FIGURE 6   Mechanical illustration (Courtesy Ken Elliott).

(1) identification of the problem or need, (2) development of initial ideas for solving the problem, (3) selection of a proposed solution, (4) development and testing of models or prototypes, and development of working drawings (**Figure 17**).

The age of computers has altered the design process slightly for those companies that have converted to computer-aided drafting (CAD). For these companies, the expensive, time-consuming fourth step in the design process—the making and testing of actual models or prototypes—has been substantially altered as shown in **Figure 18**. This fourth step has been replaced with three-dimensional computer models that can be quickly and easily produced on a

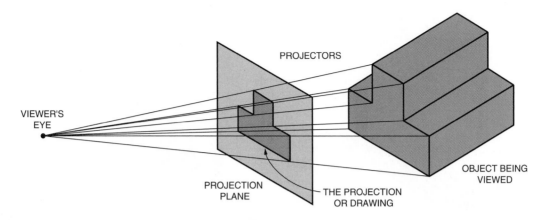

FIGURE 7   The projection plane.

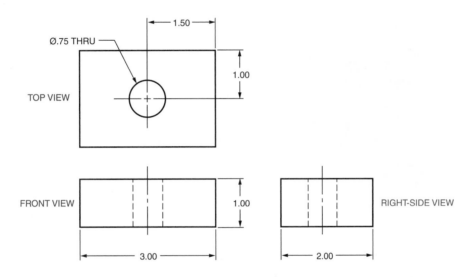

FIGURE 8   Orthographic multiview drawing.

FIGURE 9   Oblique projection (cabinet).

FIGURE 10   Oblique projection (cavalier).

CAD system using the database built up during the first three phases of the design process (**Figure 19**).

Whether in the traditional design process or the more modern computer version, working drawings are an integral part of the design process from start to finish. The purpose of technical drawings is to document the design process. Creating technical drawings to support the design process is called *drafting*. People who do drafting are known as *drafters*, *drafting technicians*, or CAD technicians (computer-aided drafting technicians). The words "draftsman" or "draughtsman" are no longer used.

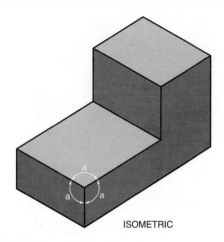

ISOMETRIC

FIGURE 11   Axonometric projection (isometric).

DIMETRIC

FIGURE 12   Axonometric projection (dimetric).

TRIMETRIC

FIGURE 13   Axonometric projection (trimetric).

In the first step of the design process, technical drawings are used to help clarify the problem or the need. The drawings may be old ones on file or new ones created for the purpose of clarification. In the second step, technical drawings—often in the form of sketches or preliminary drawings—are used to document the various ideas and concepts formed. In the third step, technical drawings—again, usually preliminary drawings—are used to communicate the proposed solution.

If the traditional fourth step in the design process is being used, preliminary drawings and

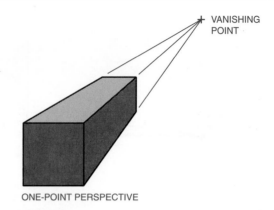

ONE-POINT PERSPECTIVE

FIGURE 14   One-point perspective projection.

TWO-POINT PERSPECTIVE

FIGURE 15   Two-point perspective projection.

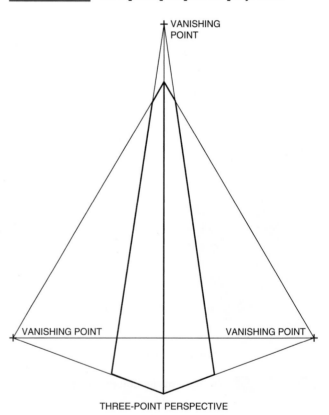

THREE-POINT PERSPECTIVE

FIGURE 16   Three-point perspective projection.

sketches from the first three steps will be used as guides in constructing models or prototypes for testing. If the more modern fourth step is being

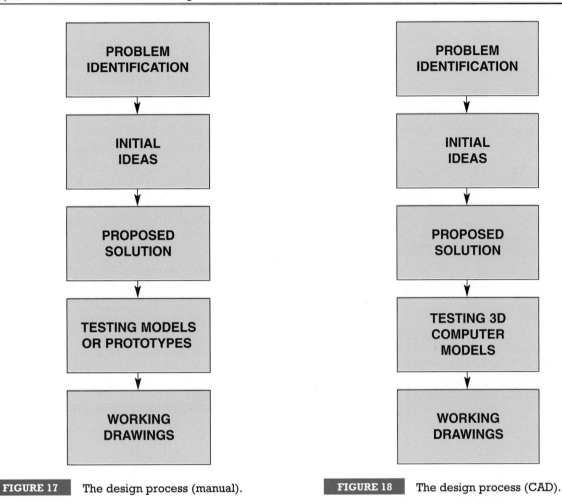

FIGURE 17   The design process (manual).

FIGURE 18   The design process (CAD).

used, the database built up during documentation of the first three steps can be used in developing three-dimensional computer models. In both cases, the final step is the development of complete working drawings for guiding individuals involved in the conversion process. **Figure 20** is a working drawing documenting the design of a simple mechanical part. The drawing was produced manually. **Figure 21** is the same drawing produced on a CAD system.

Applications of Technical Drawings

Technical drawings are used in many different applications. They are needed in any setting that involves design and in any subsequent form of the conversion process. The most common applications of technical drawings can be found in the fields of manufacturing, engineering, architecture, and construction, as well as in all of their various related fields.

Architects use technical drawings to document their designs of residential, commercial, and industrial

FIGURE 19   CAD image.

Simple mechanical drawing (manual).

Simple mechanical drawing (CAD).

buildings (**Figures 22** and **23**). Structural; electrical; and mechanical (heating, ventilating, air conditioning [HVAC], and plumbing) engineers who work with architects also use technical drawings to document those aspects of the design for which they are responsible (**Figures 24**, **25**, and **26**).

Surveyors and civil engineers use technical drawings to document such work as the layout of a new subdivision or the marking-off of the boundaries for a piece of property (**Figure 27**). Contractors and construction personnel use technical drawings as their blueprints in converting architectural and engineering designs into reality (**Figures 28** and **29**).

Technical drawings are equally important to engineers, designers, and various other individuals working in the manufacturing industry. Manufacturing engineers use technical drawings to document their designs. Technical drawings guide the collective

Technical drawing (architectural).

Technical drawing (architectural).

TYPICAL 6" WALL BASE CON FTG DETAIL

FIGURE 24   Technical drawing (structural).

FIGURE 26   Technical drawing (HVAC).

FIGURE 25   Technical drawing (electrical).

FIGURE 27   Technical drawing (civil).

efforts of individuals who are concerned with the same common goal (**Figures 30** and **31**).

Regulation of Technical Drawings

Technical drawing practices must be regulated because of the diversity of their applications. Just as the English language must have certain standard rules of grammar, the graphic language must have certain rules of practice. This is the only way to ensure that all people attempting to communicate using the graphic language are speaking the same language.

Standards of Practice

A number of different agencies have developed standards of practice for technical drawing. The most widely used standards of practice for technical drawing and drafting are those of the U.S. Department of Defense (DOD), the U.S. military (MIL), the American National Standards Institute (ANSI), and the American Society of Mechanical Engineers (ASME).

ANSI does not limit its activities to the standardization of technical drawing and drafting practices. In fact, this is just one of the many fields for which ANSI maintains a continuously updated set of standards.

Standards of interest to drafters, designers, checkers, engineers, and architects are contained in the "Y" series of ANSI standards. **Figure 32** contains a list of ANSI standards frequently used in technical drawing and drafting specifications.

FIGURE 28   Architectural/engineering drawing.

What Students of Technical Drawing, Drafting, and CAD Should Learn

Many people in the world of work use technical drawings in various forms. Engineers, designers, checkers, drafters, CAD technicians, and a long list of related occupations use technical drawings as an integral part of their jobs. Some of these people must be able to actually make drawings; others are required only to be able to read and interpret drawings; some must be able to do both.

What students of technical drawing, drafting, and CAD should learn depends on how they will use technical drawings in their jobs. Will they make them? Will they read and interpret them? This textbook is written for students in the fields of engineering, design, drafting, and architecture, among others, who must be able to make, read, and interpret technical drawings. These students should develop a wide range of knowledge and skills (**Figure 33**).

The learning required of technical drawing students can be divided into three categories: fundamental knowledge and skills, related knowledge, and advanced knowledge and skills.

In the fundamentals category, students of technical drawing, drafting, and CAD should develop knowledge and skills in the areas of drafting equipment; fundamental drafting techniques such as line work, lettering, scale use, and sketching; basic CAD system operation; geometric construction; multiview drawing; sectional views; descriptive geometry; auxiliary views; general dimensioning; and notation.

In the related knowledge category, students should develop a broad knowledge base in the areas of related math, welding, shop processes, and media and reproduction.

In the advanced category, students should develop knowledge and skills in the areas of development, geometric dimensioning and tolerancing, threads and fasteners, springs, cams, gears, machine design drawing, pictorial drawing, drafting shortcuts, and CAD/CAM technology and operations. The latter area represents a significant change in techniques used to create, maintain, update, and store technical drawings.

Figures 34 through **37** contain examples of several different kinds of technical drawings taken from the "real world" of drafting.

Architectural/engineering drawing.

FIGURE 29

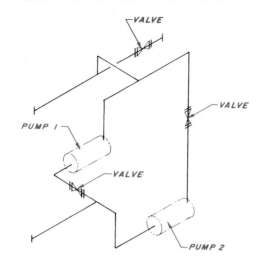

FIGURE 30   Isometric pipe drawing.

FIGURE 31   Single-line pipe drawing.

DECIMAL INCH SIZE AND FORMAT _____ Y14.1M

LINE CONVENTIONS AND LETTERING ___ Y14.2M

MULTIVIEW AND SECTION _____ Y14.3M

PICTORIAL DRAWING_____ Y14.4M

DIMENSIONS AND TOLERANCING _____ Y14.5M

SCREW THREADS_____ Y14.6

GEARS, SPLINES, AND SERRATIONS____ Y14.7.1

CASTINGS AND FORGINGS _____ Y14.8M

ENGINEERING DRAWING PRACTICES ___ Y14.100M

FIGURE 32   Sample list of drafting standards.

Technical Drawing and Quality/Competitiveness

Companies that design and manufacture products or design and build structures operate in a competitive environment. Surviving and succeeding in the intensely competitive modern marketplace requires companies to focus more attention than ever before

LEARNING CHECKLIST FOR STUDENTS OF TECHNICAL DRAWING

FUNDAMENTAL KNOWLEDGE AND SKILLS	RELATED KNOWLEDGE	ADVANCED KNOWLEDGE AND SKILLS
DRAFTING EQUIPMENT	NOTATION	DEVELOPMENT AND
FUNDAMENTAL DRAFTING TECHNIQUES	BASIC CAD SYSTEM OPERATION	INTERSECTIONS
SKETCHING	MATH	GEOMETRIC DIMENSIONING AND
GEOMETRIC CONSTRUCTION	WELDING	TOLERANCING
VISUALIZATION	MEDIA AND REPRODUCTION	THREADS AND FASTENERS
MULTIVIEW DRAWING	MANUFACTURING	SPRINGS
SECTION VIEWS	MATERIALS AND PROCESSES	CAMS
GRAPHICAL DESCRIPTIVE GEOMETRY		GEARS
AUXILIARY VIEWS		MACHINE DESIGN DRAWING
GENERAL DIMENSIONING		PICTORIAL DRAWING
		DRAFTING SHORTCUTS
		CAD TECHNOLOGY

FIGURE 33   Checklist for students of technical drawing.

on quality. Such concepts as flexible manufacturing, computer-integrated manufacturing (CIM), just-in-time manufacturing (JIT), ISO 9000, and statistical process control (SPC) are all attempts at improving quality and competitiveness.

As effective as these concepts may be at improving the quality of manufacturing processes, they cannot overcome a poor design or misinterpretations that result from poorly prepared technical drawings. The best way to ensure quality and the competitive edge it can give a company is to do the following: (1) design good quality from the outset and, (2) prepare technical drawings that accurately and properly communicate the design to those who will manufacture or build it.

Quality can be defined as the extent to which a product conforms to or exceeds the customer's expectations. In design and manufacturing, customer expectations are typically communicated in the form of specifications. Consequently, in this field, quality is often viewed as the extent to which specifications are met or exceeded. Companies that consistently meet or exceed customer expectations are more competitive.

In addition to developing high-quality designs, it is important to develop a design that is as easy as possible to manufacture. Often there will be several different design options that will meet or exceed customer expectations. When this is the case, the design chosen should be the one that is easiest to manufacture. This concept is known as *design for manufacturability (DFM)* and is covered in Chapter 17.

FIGURE 34   Typical rigid frame section of a metal building.

INDUSTRY APPLICATION

DESIGN FOR MANUFACTURABILITY

Metal Tech Manufacturing Company (MTM) had a reputation for high-quality products produced on time. Unfortunately, so did its principal competitor. Both companies kept their personnel and their processes up to date and working at peak performance levels. For years, neither company could gain a sustainable competitive advantage over the other—not until MTM decided to adopt *design for manufacturability (DFM)* as a competitiveness strategy.

MTM created DFM teams for all of its products. These teams were composed of design, drafting, and manufacturing personnel. Their charge was to develop designs that met the needs of customers but could be manufactured as efficiently and, in turn, as economically as possible. In other words, with the DFM approach, designers had to consider not just functionality but also manufacturability. A good product was no longer one that met the customer's needs in terms of function. It also had to be efficiently and economically manufactured.

The traditional attitude of "we designed it; making it is your problem" was replaced with "let's design it so that it works, and so that it can be efficiently produced." With manufacturing personnel involved in the design process, MTM's engineers and drafters found themselves being sent back to the drawing board frequently during the early stages of adopting DFM. Every time this happened, the design was simplified, and the manufacturing process was streamlined as a result. Soon DFM became second nature, and MTM saw its production costs fall by 32 percent. The company was able to pass the savings along to customers and, as a result, within a year of adopting DFM had almost doubled its sales.

FIGURE 35   Connection details.

Technical drawing plays a key role in determining the ultimate quality of a product. A good technical drawing is one that properly and conveniently communicates all of the information needed to transform a design into a product that meets or exceeds customer expectations. Anything less can detract from the quality of the product. The following examples illustrate this connection:

- An incorrect dimension on a technical drawing can cause the part to be produced at the wrong size. Each successive copy produced becomes waste.
- A dimension that is left off a technical drawing or just improperly placed can cause manufacturing personnel to interrupt their work to calculate the dimension or dimensions they need. The worst-case result is that they make a computation error and, consequently, produce a faulty part. The

best-case result is that the interruption decreases productivity and, in turn, competitiveness.

- An entry left off a parts list can cause purchasing personnel to order an insufficient amount of material and, as a result, interrupt the flow of the manufacturing process.
- An improperly drawn view can cause a part to be manufactured in the wrong configuration.

These are only a few examples of how technical drawings can affect the ultimate quality of a product. By doing so, they also affect the competitiveness of the company that manufactures the product. This book is designed to help students learn how to produce designs and technical drawings that make a positive contribution to quality and competitiveness.

FIGURE 36 Residential plot plan.

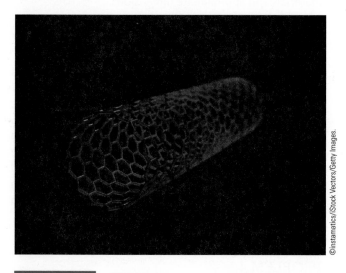

©instamatics/iStock Vectors/Getty Images.

FIGURE 37 CAD image.

SUMMARY

- Graphic communication involves using visual material to relate ideas. Drawings, sketches, slides, photographs, and transparencies are all forms of graphic communication.

- A drawing is a graphic representation of an idea, a concept, or an entity that actually or potentially exists in life.

- Artistic drawings are used to express the feelings, beliefs, philosophies, or abstract ideas of the artist. Technical drawings are used to clearly and concisely communicate all of the information necessary to transform an idea or concept into reality.

- Types of technical drawings include graphic, oblique, and axonometric projections (parallel projections) and one-, two-, and three-point perspective (perspective projections).

- The purpose of technical drawings is to document the design process.

- The principal applications of technical drawings are manufacturing, engineering, architecture, and construction.

- Technical drawings are regulated by such organizations as the U.S. Department of Defense, the U.S. military, the American National Standards Institute, and the American Society of Mechanical Engineers.

- Design and drafting can promote good-quality competitiveness by ensuring that high quality is designed in from the outset and ensuring that all drawings accurately communicate the design.

REVIEW QUESTIONS

Answer the following questions either true or false.

1. A drawing is a graphic representation of an idea, a concept, or an entity that actually or potentially exists in life.

2. The old adage "a picture is worth a thousand words" is still the basis of the need for technical drawings.

3. The two basic types of drawings are practical and technical.

4. Technical drawings can be subtle or abstract.

5. Artistic drawings are commonly used in the manufacturing field.

6. A good technical drawing is one that properly and conveniently communicates all the information needed to transform a design into a product that meets or exceeds customer expectations.

Answer the following questions by selecting the best answer.

1. Which of the following is **not** a component of a projection?
   a. Imaginary lines of sight called projectors
   b. Three-dimensional view

c. The eye of the viewer looking at the object
d. The imaginary projection plane

2. Which of the following is a subdivision of a parallel projection?
   a. Trimetric projection
   b. Cavalier projection
   c. Cabinet projection
   d. Oblique projection

3. Which of the following is **not** a type of perspective projection?
   a. One-point
   b. Two-point
   c. Three-point
   d. Four-point

4. Which of the following is **not** a step in the design process?
   a. Identify a problem or a need.
   b. Identify ideas to solve the problem.
   c. Obtain examples of several different kinds of technical drawings taken from the real world of drafting.
   d. Select a proposed solution.

5. Which of the following is **not** a field that uses technical drawings extensively?
   a. Manufacturing
   b. Retail
   c. Engineering
   d. Architecture and construction

6. Which of the following is **not** a U.S. organization that regulates technical drawing practices?
   a. American Society of Mechanical Engineers (ASME)
   b. U.S. Department of Defense (DOD)
   c. International Standards Organization (ISO)
   d. U.S. military (MIL)

INTRODUCTION PROBLEMS

To the student:

The majority of problems available for you in this text are related to technical drawing; however, additional problems are included that aim to do the following:

- Introduce you to the design process.
- Develop your critical thinking skills.
- Involve you in team activities.
- Exercise your imagination and creativity.
- Give you practice in using basic mathematics.
- Introduce you to some product design concepts.

You will need a scientific hand calculator with the usual functions: addition, subtraction, multiplication, division, trigonometric functions, logarithms, etc. Some problems may have more information than you need to set up the problem for solution. Some problems may have more than one solution, which is typical of real-life problems in the industry.

PROBLEM 1

In a team of three or four students, discuss how the design of products might differ from present designs if everyone had only a thumb and an index finger on each hand. Summarize your ideas in freehand sketches and report orally to the rest of the class.

PROBLEM 2

Figure 7 illustrates the concept of a projection plane (for third-angle projection). In a team of three or four students, discuss how you would go about preparing a device to illustrate this concept to a class of junior high school students. As an aid to your discussions, follow the steps in the design process described in Figure 17.

Prepare a freehand sketch of your final solution. Include materials and approximate dimensions.

PROBLEM 3

The weight of the flanged hub shown in Figures 20 and 21 needs to be calculated. The hub is steel with a density of 0.28 pounds per cubic inch. A suggestion for doing this type of calculation is to break up the object into convenient parts for doing the calculations. Separate the flanged hub into a hollow cylinder and a flange with one large hole and four small ones. Calculate the volume of metal and multiply by the density of steel. Use the format shown in the figure for Problem 4-30 (page 127) for your solution presentation.

PROBLEM 4

Optimization in product design usually means satisfying the design requirements for the lowest cost. For a box design, the lowest cost means using the least amount of material. Assume that a square box with an open top is to have a volume of 50 cubic inches. To visualize the optimum conditions for the box, prepare a plot of TOTAL AREA OF MATERIAL versus dimension D. That is, plot AREA on the vertical axis, dimension D on the horizontal axis.

Select a trial dimension H. Then, calculate the total area of the material. The suggested axes and calculation steps are given in the following figure:

V = VOLUME = 50 in^3
= D^2H
EXAMPLE CALCULATION:
LET H = 2 in
$50 = D^2(2)$
$D = \sqrt{25}$
D = 5 in
AREA = $D^2 + 4$ DH
= $(5)^2 + 4(5)(2)$
= 65 in^2

Try different values of H.
Calculate and plot A.
Construct a best-fit curve through the A points.
Lowest point of A curve determines H.

PROBLEM 5

A student was asked to study Figure 33 to compare what skills and knowledge the student already had to the topics listed in the checklist. One comment the student had was, "I believe that math should be on all three lists." What is your opinion regarding math and why? Discuss this question in a group of three or four and document your conclusions.

PROBLEMS

1

Employability Skills for Drafting and Design Technicians

CHAPTER OUTLINE

Employability skills defined • Importance of employability
skills • Employability skills needed by drafting and design
technicians • Job-seeking skills • Summary • Review questions

CHAPTER OBJECTIVES

Upon completion of this chapter, students should be able to do
the following:

▶ Define the term *employability skills*.

▶ Explain the importance of employability skills to drafting and
design technicians.

▶ List the most important employability skills for drafting and
design technicians.

▶ Demonstrate the skills necessary to secure employment in
drafting and design.

This chapter provides a comprehensive treatment of the concept of employability skills as they relate specifically to drafting and design technicians. Although many employability skills are transferable from one occupation to another, some skills apply specifically or are more critical in a drafting and design setting.

Employability Skills Defined

Employability skills are sometimes referred to as the *soft skills* or *people skills* needed in a drafting and design career. They are called soft skills to distinguish them from the *hard skills* (technical skills) required in drafting and design, such as geometric construction, spatial visualization, creating multiview drawings, creating sectional and auxiliary views, dimensioning and notation, CAD, three-dimensional modeling, and geometric dimensioning and tolerancing. They are called people skills to indicate that they relate to how you interact with people in a workplace setting. Employability skills are the nontechnical skills needed to secure a good drafting and design position after completing school, performing well in that position, and progressing up the career ladder over the course of your career.

Employability skills do not replace or supplant the technical skills needed to succeed in drafting and design; they augment them. Where technical skills will help you do a good job on drafting and design projects, employability skills will help you be a good employee and excel at interacting with people. Consequently, it is important to develop both the technical skills needed for a career in drafting and design as well as the employability skills (**Figure 1-1**). If two drafting and design technicians have equal technical skills, the one with superior employability skills is usually the one who will enjoy the more successful career. An example from the workplace illustrates why this is the case.

Susan and Dan (not their real names) both graduated from school on the same day with diplomas in drafting and design. They both made excellent grades and both had outstanding technical skills in drafting and design. But this is where their similarities ended. Susan had excellent employability skills. She had a strong work ethic, excellent communication skills, and worked well in teams.

Dan, on the other hand, did not like to interact with people, so his work in teams suffered. He was

TECHNICAL SKILLS + EMPLOYABILITY SKILLS

= SUCCESSFUL CAREER

FIGURE 1-1   Formula for building a successful career in drafting and design.

also a poor listener. Consequently, he sometimes failed to understand instructions, deadlines, and other important information that was critical to getting the job done right, on time, and within budget. He did excellent technical work but had to be kept away from customers and allowed to work on his own rather than in a team as much as possible.

Susan quickly climbed the career ladder in her profession, but Dan remained stuck near the bottom. After several years on the job, Susan had advanced from drafting and design technician to checker to project leader and to drafting and design supervisor. Dan, on the other hand, was still a drafting and design technician who had not advanced beyond receiving the normal salary increases that come with time on the job. Dan was considered a valuable employee but not a promotable employee because he failed to develop the employability skills that are so important to drafting and design professionals.

Importance of Employability Skills

To understand why employability skills are so important, you must first understand two other workplace concepts: 1) the human aspect of organizations, and 2) competition. Drafting and design technicians work in organizations such as engineering, architectural, construction, and manufacturing firms as well as in a variety of governmental agencies. All of these organizations, regardless of their specific areas of specialization, are staffed by people. People do the work of organizations, and they must work well together to get it done. In other words, to succeed in an organization, it's necessary to work effectively with people.

People don't want to work with colleagues who don't communicate well, lack integrity, won't accept responsibility for their work, don't work well in teams, or have negative attitudes—in other words, people who lack good employability skills. Such people just make the work more difficult for everyone, which, in turn, undermines the morale of the people who do the work. When this happens, just coming to work can become stressful, and the work itself can become drudgery. This is the human aspect of organizations (**Figure 1-2**).

The competition aspect of organizations is an extension of the human element. Private sector firms—engineering, architectural, construction, and manufacturing—compete against other similar firms for business. For example, if a professional football team decides to build a new stadium, many architectural firms will want to win the contract to design the form and function aspects of the stadium; many engineering firms will want to provide the civil,

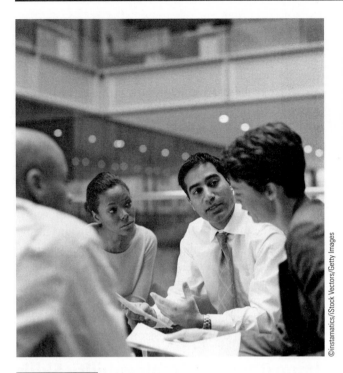

FIGURE 1-2   People in organizations work in teams.

FIGURE 1-3   Many of the organizations that employ drafting and design technicians must compete globally.

structural, electrical, and mechanical designs; and many construction firms will want to build the stadium. In today's global environment, there are likely to be architectural, engineering, and construction firms from all over the world competing to win their parts of the contract (**Figure 1-3**). Typically, the firm that can best properly fulfill the requirements of the stadium owner on time and within budget at the most competitive price will win the contract. The firm that can do this is the most productive and most efficient.

Competition means that to survive, organizations must perform at their best at all times and improve their performance continually. Peak performance and continual improvement cannot be achieved unless the people in the organization are able to work together cooperatively and collaboratively. People in organizations must be able to work effectively as teams or the organization in question will not be competitive. To work effectively as a team, an organization must have personnel with good employability skills.

Employability Skills Needed by Drafting and Design Technicians

There is wide agreement among employers concerning the specific employability skills they prefer drafting and design technicians to have. For ease of learning, those employability skills can be arranged in categories as shown in (**Figure 1-4**). The various

specific employability skills in each category shown in Figure 1-4 are explained in this section.

Personal Values

Employers want to know that they can trust drafting and design technicians and depend on them to work hard, work smart, and, when necessary, work long. Consequently, employers place a high priority on such personal values as honesty, integrity, dependability, and loyalty. They want employees who can be dependent upon to always tell the truth, who will do the right thing—even when doing so is difficult for some reason—and who can be trusted to look out for the best interests of the employer (loyalty).

Categories of Employability Skills for Drafting and Design

- Personal values
- Communication
- Critical thinking and problem solving
- Information management
- Attitude
- Responsibility/accountability
- Flexibility/adaptability
- Continual improvement
- Workplace safety
- Teamwork/project work
- Positive work ethic

FIGURE 1-4   The employability skills most important to drafting and design technicians fall into these categories.

Of course, the employer is responsible for establishing an environment in which honesty, integrity, dependability, and loyalty are not just expected but reciprocated and rewarded.

Communication

Employers want drafting and design technicians who are good communicators. This means first and foremost that they are good listeners, and also that they can convey information by speaking and writing in a clear, accurate, and comprehensive manner that is concise and easy to understand. Being a good communicator requires that drafting and design technicians know how and when to ask questions for clarification or to gain more complete information. In today's technology-driven workplace, being a good communicator also means effectively using all of the technological aids available to convey information (e.g., e-mail, texting, voice mail, social networking applications, etc.).

Critical Thinking and Problem Solving

Employers want drafting and design technicians who are critical thinkers and good problem solvers. These two employability skills go hand in hand and cannot be separated. Critical thinkers and problem solvers hear different points of view and analyze them according to facts rather than emotions or assumptions. They recognize bias in arguments made by others, identify the root causes of problems, creatively explore potential solutions, collect relevant information and apply it in making decisions, and think creatively to find optimum solutions to problems. They also recognize when others are making decisions based on assumptions rather than facts or, at least, the best information available at the time.

Information Management

Employers want drafting and design technicians who can locate, access, and collect relevant information from a variety of sources; organize the information; synthesize it to eliminate information that is not applicable, redundant, or inaccurate; analyze the information that is relevant; and apply it to the completion of drafting and design projects in ways that ensure the projects are done right, on time, and within budget.

Attitude

Employers want drafting and design technicians who have positive, can-do attitudes. Individuals with positive attitudes exhibit enthusiasm for their work; have a passion for getting the job done well, on time, and within budget; and are not discouraged by the roadblocks, problems, and challenges that inevitably arise on the job. Regardless of circumstances, people with positive attitudes take a *can-do* approach to their work. They focus on solutions rather than problems—on what they can do rather than what they can't do. And, at all times, no matter how rushed or difficult a project becomes, they convey—verbally and nonverbally—the message that "we can do this."

Responsibility and Accountability

Employers want drafting and design technicians who take responsibility for themselves, their jobs, their individual work, and their work in teams and who expect to be held accountable for the quality of their work. Finger pointers, excuse makers, and blamers do not help an organization outperform the competition. People who are responsible and accountable can be depended upon to arrive at work on time every day, work hard and smart every minute they are on the job, make the best use of their time, mutually support their team members, and in every situation act in ways that serve the best interests of the organization as long as those interests are ethical and legal.

Flexibility and Adaptability

Employers want drafting and design technicians who are flexible enough to work in teams or by themselves as the situation demands, and who are adaptable enough to work on different kinds of projects. People who can do only one thing or who must always do things in the same way inhibit the competitiveness of employers. Today's fast-paced global work environment requires personnel who are open to change and who can quickly adapt themselves to it. Drafting and design technicians who resist changes that will help improve their performance and that of the organization are an impediment to their employer's competitiveness.

Continual Improvement

Employers want drafting and design technicians who will work hard and smart to continually improve their knowledge and skills (hard and soft) in ways that improve their performance and, in turn, the performance of the organization. Global competition demands this concept of continual improvement. Organizations that compete with your employer strive every day to find ways to outperform it. Consequently, performance that was good enough yesterday may not be good enough today. To survive over the long term, organizations—and drafting and design technicians—have to get better and better at what they do all the time, forever.

Workplace Safety

Employers want drafting and design technicians who will work safely and follow established safety practices and procedures. Every year, workplace accidents drain organizations of both human resources and financial resources, which are needed to remain competitive. Workplace accidents and injuries dampen the morale of an organization; put good, talented personnel out of action; and cost the organization in medical, insurance, and legal expenses. Every dollar drained by workplace accidents is a dollar that could have been invested in making the organization more competitive. Consequently, it's important that drafting and design technicians know their employers' safety rules and follow them. Careless employees harm themselves, their teammates, the organization's property and equipment, and the organization's competitiveness.

Teamwork and Project Work

Employers want drafting and design technicians who can work well in teams and work together effectively on projects. This means being able to work well with people of diverse backgrounds, points of view, and perspectives. Drafting work is typically organized into projects that are assigned to teams. These teams might include a project manager, engineers, checkers, drafting and design technicians, manufacturing technicians (or construction personnel), and administrative support personnel. Consequently, it's important for drafting and design technicians to work well with others, set aside personal agendas for the good of the team, respect human diversity (individual differences in people), lead or support when appropriate, and prevent and resolve conflict. Conflict is best prevented by learning to disagree without being disagreeable and by being open to listening to differing points of view and opinions. Conflict is best resolved by asking a simple question: Which opinion, recommendation, or action best serves the mission and goals of the team? One additional capability that will serve you well in teams and projects is the ability to plan and carry out assigned project tasks from start to finish.

Positive Work Ethic

Employers want drafting and design technicians who have a positive work ethic. A positive work ethic encompasses many of the employability skills needed by drafting and design technicians. Individuals with a positive work ethic exhibit such characteristics as the following: 1) take pride in their work, 2) are willing to work hard, smart, and long, 3) maintain a positive

---

Work Ethic Checklist

People with a positive work ethic consistently do the following things that help them achieve success in careers such as drafting and design:

- ✔ Take pride in their work
- ✔ Work hard, work smart, and work long when necessary to get the job done
- ✔ Maintain a positive attitude toward their work and their team members
- ✔ Give their best effort on every job, project, and task
- ✔ Continually improve their technical and employability skills
- ✔ Strive to get their work done properly, on time, and within budget
- ✔ Make a point of being punctual in all aspects of their jobs

FIGURE 1-5   Work ethic checklist for drafting and design technicians.

attitude toward their work, those they work with, and their employer—even in difficult circumstances, 4) always give their best effort, 5) try to continually improve, 6) always strive to get their work done properly, on time, and within budget, and 7) are punctual in all aspects of their work (e.g., meeting deadlines, arriving at work, attending meetings) (**Figure 1-5**).

Job-Seeking Skills

Closely related to employability skills are job-seeking skills, which you need to secure a position in drafting and design so you can apply your technical skills and employability skills to build a successful career. Job-seeking skills include the following: 1) developing a résumé, 2) developing letters of introduction, 3) developing a list of references, 4) identifying job openings, and 5) interviewing (**Figure 1-6**). These skills are explained in this section.

Developing a Résumé

The résumé is a job-seeking tool that summarizes your qualifications in the field of drafting and design. Think of your résumé as a marketing pamphlet, and the product being marketed is you. A good résumé has a future-oriented tone, is brief and easily read, contains both drafting and design and employability

Checklist of Job-Seeking Skills

The following skills are necessary to secure a good position in drafting and design after completing school:

✔ Developing an effective résumé
✔ Developing letters of introduction
✔ Developing a list of references
✔ Identifying job openings
✔ Interviewing effectively

FIGURE 1-6 Drafting and design students must develop job-seeking skills.

skills, and is written in the terminology of the field in question (drafting and design). A résumé that has these characteristics will do the following:

- Summarize your career goal as it relates to a position in drafting and design
- Summarize your qualifications in a way that shows employers how you can help them
- Help you get your foot in the door for an interview

A résumé for a student about to graduate from a drafting and design program should contain the following elements: 1) your contact information, 2) career goal, 3) education summary, 4) summary of drafting and design skills, 5) summary of employability skills, and 6) summary of work experience (even if part-time and outside the field). **Figure 1-7** is an example of a well-written résumé for a student who just graduated from an architectural drafting program. **Figure 1-8** is a well-written résumé for a student who just graduated with a specialization in mechanical drafting.

Developing Letters of Introduction

Depending on what is called for in job announcements, you might need to write a cover letter (letter of introduction) to submit along with your résumé. The cover letter and résumé can be submitted electronically or the old-fashioned way. The same rules apply either way. The cover letter serves three specific purposes: 1) lets the employer know what specific job you are interested in, 2) points out specific details in your résumé that you would like to call the employer's attention to, and 3) helps ensure that your résumé is read by the employer rather than being tossed in a trash can.

Mark Jones

12 North Cedar Avenue

Northwoods, Florida 34672

850-865-9160   mjones@coss.com

Career Goal

To begin work in an architectural drafting and design position and rise over time to the level of drafting department manager.

Technical Skills

I have developed the following entry-level skills for architectural drafting and design:

- *Fundamentals.* Orthographic projection, isometric drawing, perspective drawing, and technical illustration.
- *Architectural drafting and design.* All applicable codes, space planning, design and layout of floor plans, development of elevations and three-dimensional illustrations, material selection, and development of specifications.
- *CAD skills.* Well-skilled in the latest CAD software for architectural applications.

Employability Skills

I have developed the following values and communication skills: honesty, integrity, dependability, loyalty, communication, critical thinking/problem solving, information management, positive attitude, responsibility/accountability, flexibility/adaptability, continual improvement, workplace safety, teamwork/project work, and a positive work ethic.

Education

- Diploma in Architectural Drafting and Design. Northwoods Technical institute, January XXXX.
- High school diploma. Northwoods High School, June XXXX.

FIGURE 1-7 Sample résumé for recent graduate from an architectural drafting program.

Sarah Rodriguez

1300 South Fallwoods Street

Tuskegee, Alabama 34672

334-960-9611   srodriguez@msp.com

Career Goal

To begin work in a mechanical drafting and design position and rise over time to the level of senior designer.

Technical Skills

I have developed the following entry-level skills for mechanical drafting and design:

- *Fundamentals*. Orthographic projection, auxiliary views, sectional views, descriptive geometry, dimensioning and notation, geometric dimensioning and tolerancing, and isometric drawing.
- *Mechanical drafting and design*. Patterns and developments, jigs and fixtures, springs, cams, gears, fasteners, assembly and detail drawings, parts lists, and specifications.
- *CAD skills*. Well-skilled in the latest CAD software for mechanical applications.

Employability Skills

I have developed the following values and communication skills: honesty, integrity, dependability, loyalty, communication, critical thinking/problem solving, information management, positive attitude, responsibility/accountability, flexibility/adaptability, continual improvement, workplace safety, teamwork/project work, and a positive work ethic.

Education

- Associate Degree in Mechanical Drafting and Design. Tuskegee Technical Community College, June XXXX.
- High school diploma. Tuskegee High School, June XXXX.

FIGURE 1-8   Sample résumé for a recent graduate with a specialization in mechanical drafting.

A cover letter should be brief and to the point—never more than one page—and it should contain three specific elements: 1) a reference to the job you are interested in, 2) references to any aspect of your résumé that relates specifically to the job in question, and 3) your up-to-date contact information. **Figure 1-9** is an example of a properly written cover letter.

Developing a List of References

If you get your foot in the door and are invited to fill out an application, or if the application is provided online and you submit it along with your résumé and cover letter, employers are going to want you to provide a list of references. References are people who can attest to your personal character, technical skills, and employability skills, preferably people of status and good standing in the community in question.

Those who can attest to your character are people such as personal friends, teachers, coaches, and pastors. The key is to identify people who can vouch for your good character and who will also have status in the eyes of the employer. For example, an individual who is known to the employer for some reason will make a stronger impression than one who is not. You do not want to use a reference that the employer would discount as an individual who would make a positive recommendation as a favor to you.

Those who can attest to your technical skills are the teachers in your drafting and design program or your supervisor if you happen to have experience in a drafting job. Those who can affirm your employability skills are supervisors from any type of position you have had—paid or volunteer—and teachers in your drafting and design program.

The key to ensuring a positive reference from people, regardless of who they are, is to ask their permission before listing them as a reference. Never put someone down as a reference on a job application without consulting him or her first. If individuals are willing to serve as references, tell them about the specific job in question and what you would like them to say about you, including what specific characteristics to emphasize. Do not leave reference recommendations to chance.

Identifying Job Openings

The Internet has simplified the process of identifying job openings. Several well-known Internet

Date: June XXXX

Mr. Joe Harris

Department Manager – Drafting and Design

TYBRON Industries, Inc.

601 Booth Avenue

Cantonment, Florida 34567

Re: Drafting Technician I opening

Dear Mr. Harris:

I am applying for the position advertised on your company's website: Drafting Technician I. My résumé is attached, and I can provide a list of references upon your request. I meet the technical qualifications listed in the job posting. In addition, I have the employability skills necessary to fit in well in your organization and to be an effective team member. I have a strong work ethic and am a fast learner.

I would appreciate an opportunity to interview for this position so that I can offer additional evidence of my ability to do an excellent job for your company.

Respectfully,

Jane Andrewson

1719 South Elm Street

Crestview, Florida 32547

850-786-7766

janea@tox.com

FIGURE 1-9   **Sample cover letter.**

sites are available for job seekers, but the two best-known and most popular sites are careerbuilder.com and monster.com. However, there are many others. Students are encouraged to conduct an Internet search on the topic "Job Openings in Drafting" to identify additional websites that might be helpful.

Interviewing

Employers who are interested in your résumé and application will invite you to come in for an interview. This is a critical make-or-break point in the job-seeking process. The interview is your opportunity to make a good impression on the employer. Consequently, it is important that you develop good interviewing skills. Preparation is the key to a successful interview. To prepare for an interview, you need to know the following: 1) how to pass the "fit" test, 2) how to dress properly for an interview, 3) how to send the right nonverbal cues during an interview, and 4) what types of questions you will be asked.

Employers are looking for employees who pass the "fit" test, meaning they will fit in well in the organization. Consequently, it's important to do your research before going to an interview. Go online and learn everything you can about the company, and use what you learn to prepare for the interview. For example, if an organization publishes its corporate values on its website, memorize those values and incorporate how you exemplify them in your answers to interview questions.

Regardless of how you like to dress on a day-to-day basis, it's important to dress appropriately for an interview. You do not dress for style, comfort, or to make a statement when interviewing. Rather, you should dress in a way that conveys the following nonverbal messages: 1) you respect the interviewer enough to take the time to "dress up," 2) you care enough to want to make a good impression, and 3) the interview is important to you. In addition, what you wear to an interview is better to be understated and conservative, not flashy. You want to wear clothing that subtly sends the right nonverbal messages, not clothing that makes it look like you are headed to a fashion show, cocktail party, or wedding.

The success or failure of an interview can be determined by the nonverbal cues you send to the interviewer. You send these cues by how you dress, which has already been covered, and also with your handshakes, posture, eye contact, facial expressions, and rate/tone of speech. A handshake should be firm but not bone-crushing and should be delivered while looking the other person in the eye and smiling. This type of handshake communicates warmth, professionalism, and confidence.

Whether standing or sitting, your posture should be erect—no slumping or slouching—to convey confidence and professionalism. Making eye contact is important as well. When talking to interviewers, look them directly in the eyes in a friendly, open manner. If others are present during an interview, spread your eye contact around when answering questions. Facial expressions are also important. A facial expression that conveys interest, attention, and friendliness is best. Do not allow your face to convey anger, frustration, confusion, impatience, arrogance, or deceitfulness.

If you are nervous during an interview—which you probably will be—your tendency will be to talk faster than normal. Consequently, you must discipline yourself to slow down and assume a relaxed and measured rate of speech during interviews. Also, your tone of voice should be confident, calm, and moderated. A loud and agitated tone of voice or one in which you mumble and cannot be heard will not serve your purposes. Speak calmly, conversationally, and at a pace that can be easily followed by the interviewer.

Finally, it's helpful to anticipate the types of questions you might be asked during an interview, although this can be difficult to do. You might think that all of the questions would pertain specifically to drafting and design, but this is rarely, if ever, the case. Some interviewers have pet questions they like to ask just to see what kind of response they will get. But typically you can expect to be asked the types of questions shown in (**Figure** 1-10).

Sample Interview Questions

Job seekers in drafting and design should be prepared to answer the following types of questions as well as specific technical questions during interviews. Although you can answer "yes" or "no" to some of these questions, you will be expected to expound briefly.

- Why do you want to work here?
- What do you know about our organization?
- Why do you think you are qualified for this job?
- Are you able to take directions from a supervisor in a positive manner?
- Do you find it difficult to work with people who are different from you?
- Can you disagree with others without being disagreeable?
- Did you have a job while you were in school?
- Are you willing to take on new and unfamiliar tasks?
- Are you punctual?
- Are you willing to work overtime if circumstances require it?
- Are you able to work well as a member of a team?
- Are you a good communicator? Listener?
- Where would you like to be in your career five years from now?
- Can you work well without supervision?
- If we offer you this job, when can you start?
- Are you able to interact comfortably with people you have never met (customers)?

FIGURE 1-10   These questions represent just a sampling of the types of questions asked during interviews.

SUMMARY

- Employability skills are sometimes called *people skills* or *soft skills* to distinguish them from the *hard skills* (technical skills) needed in drafting and design. Employability skills are the nontechnical skills needed to secure a good position in drafting and design after completing school, perform well in that position, and progress up the career ladder over the course of your career.

- Employability skills are important because people do the work of organizations, and they must work well together to get it done. Consequently, working effectively with people is important. Further, organizations are more competitive when their personnel work well together.

- Employability skills needed by drafting and design technicians include personal skills such as honesty, integrity, dependability, and loyalty; communication (speaking, writing, and listening); critical thinking and problem solving; information management; attitude; responsibility and accountability; flexibility and adaptability; continual improvement; workplace safety; teamwork and project work; and positive work ethic.

- Job-seeking skills are needed to secure a position in drafting and design. They include developing a résumé, developing letters of introduction, developing a list of references, identifying job openings, and interviewing.

REVIEW QUESTIONS

Mark the following statements as **either** *true* or *false*.

1. Employability skills can also be referred to as people skills or soft skills.

2. The term "hard skills" refers to such skills as communication and critical thinking.

3. One of the reasons employability skills are important is that they help people work together more effectively in organizations.

4. Competition demands that drafting and design technicians have good employability skills.

5. Teamwork is not an important employability skill because drafting and design technicians can work on their projects alone.

6. People with a positive work ethic take pride in their work and are willing to work hard, smart, and long to do a good job.

Answer the following questions by selecting the best answer.

1. Which of the following is considered an employability skill for drafting and design technicians?
   a. Using orthographic projection
   b. Creating auxiliary views of drawings
   c. Effective listening
   d. Using geometric dimensioning and tolerancing

2. Which of the following is a reason that employability skills are so important?
   a. Competition
   b. Three-dimensional modeling
   c. Improved sectional views
   d. Critical to proper notation

3. A summary of your qualifications relating to drafting and design is called what?
   a. Introduction
   b. Personal biography
   c. Résumé
   d. Qualifications letter

4. Which of the following is NOT a characteristic of a well-written résumé?
   a. Lengthy (usually more than five pages)
   b. Future-oriented
   c. Brief and easily read
   d. Written in the terminology of the field

5. Which of the following is an important purpose served by a cover letter?
   a. Summarizes all of your experience even if the jobs you have had are not in drafting and design
   b. Explains your career goal
   c. Lists your job references
   d. Tells the employer which specific job you are interested in

2

Drafting Tools: Conventional, CAD, and Solid Modeling

KEY TERMS

CAD software

Calculator

Center wheel bow compass

Circle template

Design intent

Design tools and methods

Drafting brush

Drafting dots or tape

Drafting machine or parallel bar

Dry cleaning pad

Ease of use

Eraser

Erasing shield

Features

Features and capabilities

File compatibility

Generic CAD software skills

Geometric parameters

Lead pointer

Mechanical pencils

Micrometer

Numeric parameters

Protractor

Solid modeling

Support available

Tools and methods

Triangles

Triangular scales

Vernier caliper

Whiteprinter

CHAPTER OUTLINE

Note to instructors and students • Conventional drafting tools • Triangular scales • Miscellaneous conventional tools • Paper sizes • Whiteprinter • Computer-aided drafting (CAD) • CAD software • Solid modeling • Summary • Review questions • Chapter 2 Problems

CHAPTER OBJECTIVES

Upon completion of this chapter, students should be able to do the following:

▶ Recognize the most frequently used conventional drafting tools.

▶ Demonstrate the proper use of the most frequently used conventional drafting tools.

▶ Explain the concept of computer-aided drafting (CAD).

▶ Explain the most important features of CAD software.

▶ Explain the concept of solid modeling.

Note to Instructors and Students

As the field of drafting and design has evolved over the years, the tools of the trade have also evolved. Computer-aided drafting (CAD) is now the norm in classrooms and the workplace. However, some instructors find it beneficial to teach certain basic concepts using conventional tools before transitioning students to CAD. To accommodate the teaching strategies of these instructors, this chapter covers the most frequently used conventional drafting tools before explaining the concepts of CAD and solid modeling. However, only the most frequently used conventional tools are included in this chapter.

Conventional Drafting Tools

Conventional drafting tools can still be valuable in helping beginning students learn basic drafting principles and techniques. **Figure 2-1** is a photograph of the most frequently used conventional drafting tools. **Figure 2-2** is a diagram of silhouettes that corresponds with Figure 2-1 and contains numbers for each conventional tool that match the numbers in the following list:

1. Center wheel bow compass
2. Triangles: 45-degree and 30–60-degree
3. Circle template
4. Irregular curves templates
5. Protractor
6. Mechanical pencils
7. Lead pointer
8. Erasing shield
9. Drafting brush
10. Dry cleaning pad
11. Lettering guide
12. Eraser
13. Calculator
14. Drafting dots or tape
15. Triangular scales: architect's, engineer's, and metric (left to right in Figure 2-1)
16. Drafting machine or parallel bar (not pictured in Figure 2-1)

Drafting Machines

A drafting machine is a device that attaches to the drafting table and replaces both the T-square and the parallel bar. There are two basic kinds of drafting machines. One is the arm type, and the other—the

FIGURE 2-1   Frequently used conventional drafting tools.

FIGURE 2-2   Diagram of conventional drafting tools numbered.

newer—is the track type. On both types, a round head holds two straightedges at right angles to each other. The head can be rotated to set the straightedges at any angle. Most machines are available with interchangeable straightedges marked with different scales along their edges.

Drafting machines replace straightedges, scales, triangles, and protractors. They increase accuracy and greatly reduce drafting time. A drafting machine is one of the few tools that can be purchased either as a right-handed or a left-handed instrument.

Most drafting machines have a protractor and a vernier, which permit readings to 5 minutes of an arc (**Figure 2-3**). Notice that zero on the protractor is in line with the zero on the vernier. The vernier is graduated in 5-minute increments from zero to 60 minutes. To read the vernier, first read the protractor (**Figure 2-4**). In this example, the zero

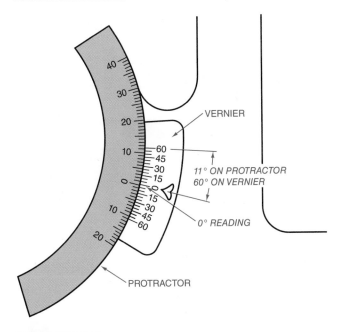

FIGURE 2-3   Drafting machine protractor and vernier.

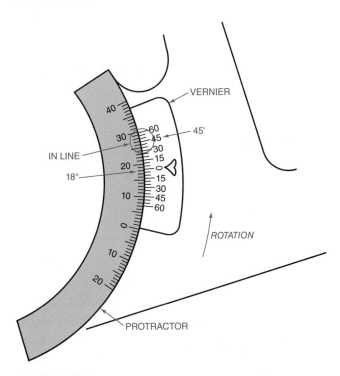

FIGURE 2-4   Reading the vernier.

on the vernier points between 18° and 19°. On the vernier, notice that the only line that lines up with a line on the protractor is the 45; thus, this is read as 18°45'. Some drafting machine heads simplify this process by adding a digital readout.

Drafting machine straightedges come in sizes of 9" (23 cm), 12" (30 cm), and 18" (45 cm), graduated or ungraduated, in both transparent plastic and aluminum scales.

A drafting machine, although a precision instrument, should be checked for accuracy at least once a week. The instructions for checking and adjusting a drafting machine are included with the manufacturer's information.

Compasses

There are several types of compasses, but the *bow compass* is the most frequently used type.

The bow type of compass (Figures 2-1 and 2-2, Number 1) is best for pencil drawings and tracings because it retains its setting even when strong pressure is applied to obtain dark lines. The adjusting screw is used to make fine adjustments. Technical pen adapters allow ink attachments to be used.

Compass leads should extend approximately 3/8" (0.9 cm). The metal point of the compass extends slightly more than the lead to compensate for the distance the point penetrates the paper (**Figure 2-5**). The lead is sharpened with a sandpaper paddle to produce clean, sharp lines. The flat side of the lead faces outward in order to produce circles of very small diameter. Sharpen with a paddle in the direction of the arrow, as shown, in order to keep the lead sharp longer.

To draw a circle with the bow compass, the compass is revolved between the thumb and the index finger (**Figure 2-6**). Pressure is applied downward on the metal point to prevent the compass from jumping out of the center hole.

Triangles

Two standard triangles are used by drafters. One is a 30–60-degree triangle, usually written as a 30°–60° triangle (**Figure 2-7**).

The other is a 45-degree triangle, written as a 45° triangle. The 45° triangle consists of two 45-degree angles and one 90-degree angle, Figure 2-7.

FIGURE 2-5   Bow compass lead and metal points.

REVOLVE COMPASS
BETWEEN THUMB
AND INDEX FINGER

15°

FIGURE 2-6   Drawing a circle with a bow compass.

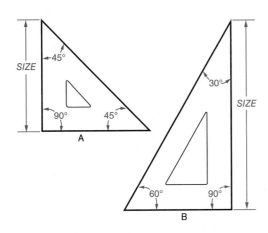

SIZE
45°
90°   45°
A

30°
SIZE
60°   90°
B

FIGURE 2-7   (A) 45-degree triangle, and
(B) 30–60-degree triangle.

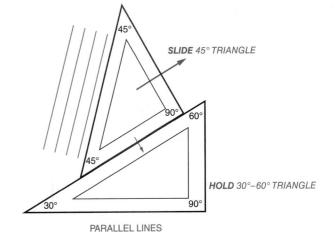

45°
SLIDE *45° TRIANGLE*
90°   60°
45°
HOLD *30°–60° TRIANGLE*
30°   90°
PARALLEL LINES

FIGURE 2-8   Drawing parallel angular lines.

The 30°–60° triangle contains a 30-degree angle, a 60-degree angle, and a 90-degree angle, Figure 2-7.

Triangles are made of plastic and come in a variety of sizes other than those mentioned. When lines are laid out, triangles are placed firmly against the upper edge of the straightedge. Pencils are placed against the left edge of the triangles, and lines are drawn upward, away from the straightedge. Parallel angular lines are made by moving the triangle to the right after each new line has been drawn (**Figure 2-8**).

Templates

A template is a thin, flat piece of plastic containing various cutout shapes. Figure 2-1 contains a circle template and two templates for drawing irregular curves (Numbers 3 and 4 in Figure 2-2). It is designed

to speed the work of the drafter and to make the finished drawing more accurate. Templates are available for drawing circles, ellipses, plumbing fixtures, bolts and nuts, screw threads, electronic symbols, springs, gears, and structural metals, to name just a few uses.

Templates come in many sizes to fit the scale being used on the drawing. A template should be used wherever possible to increase accuracy and speed. It is preferable to purchase templates that are stamped and not molded because molded templates become brittle in time and break.

Protractor

A protractor is used to measure and lay out angles, Figure 2-1 (Number 5 in Figure 2-2). It can be used in place of a drafting machine or an adjustable triangle.

To use the protractor, place the center point (located at the lower edge of the protractor) on the corner point of the angle. Align the base of the protractor along one side of the angle. The degrees are read along the semicircular edge.

Mechanical Pencils and Leads

Lead for a mechanical drafting pencil comes in 18 degrees of hardness, ranging from 9H, which is very hard, to 7B, which is very soft. For drafting purposes, the scale of hardness is as follows: 3H or 4H lead is recommended for layout work, and 2H, H, or HB leads are recommended for all other lines. Experiment with various leads to determine which ones give the best line thickness. This varies depending on the pressure applied to the point while drawing lines.

Mechanical pencils are sometimes called lead holders because they hold sticks of lead. The leads designed for lead holders come in the same range of hardness as those for regular pencils and are used for the same purposes. The main advantage is that they are more convenient to use. Leads are usually sharpened in a lead pointer. Electric lead pointers are fully automatic. A slight downward pressure of the lead starts the motor action (**Figure 2-9**). This machine produces a perfectly tapered point and eliminates all loose clinging graphite.

Erasers

Various kinds of erasers are available to a drafter. One of the most commonly used is a soft, white,

FIGURE 2-9   Lead pointer.

block-type eraser, Figure 2-1 and Number 12 in Figure 2-2. There are also electric erasers. If one develops good drawing habits, erasing can be kept to a minimum.

Erasing Shield

An erasing shield restricts the erasing area so that correctly drawn lines will not be disturbed during the erasing procedure. It is made from a thin, flat piece of metal with variously sized cutouts, Figure 2-1 and Number 8 in Figure 2-2. The shield is used by placing it over the line to be erased and erasing through the cutout.

Drafting Brush

The drafting brush is used to remove loose graphite and eraser crumbs from the drawing surface (Figure 2-1 and Number 10 in Figure 2-2). Do not brush off a drawing surface by hand because this tends to smudge the drawing. Drafting brushes come in various sizes from 10½″ to 14″ (26 cm to 35 cm). The bristles can be either horsehair or nylon, and they can be cleaned with warm, soapy water.

Dry Cleaning Powder

Dry cleaning powder is used to keep drawings clean, avoid smearing, and speed up the drafting process. Cleaning powder comes in cans or pads (Figure 2-1 and Number 10 in Figure 2-2) and is sprinkled over the drawing before starting. It is also used throughout the drawing process. It should be removed periodically as it soaks up dust and grime and replaced with new powder. It is imperative that all cleaning powder be removed from a drawing before placing it into a whiteprinter because the powder will stick to the roller. If good drawing habits are followed, dry cleaning powder becomes less necessary.

Triangular Scales

Various kinds of scales are used by drafters, Figure 2-1 and Number 15 in Figure 2-2. Several different scales are included on each instrument. Scales save the drafter the work of computing new measurements every time a drawing is made larger or smaller than the original.

Scales come open-divided and full-divided. A *full-divided scale* is one in which the units of measurement are subdivided throughout the length of the scale. An *open-divided scale* has its first unit of measurement subdivided, but the remaining units are open or free from subdivision.

Mechanical Engineer's Scale (Fraction)

Mechanical engineer's scales are divided into inches and parts to the inch. To lay out a full-size measurement, use the scale marked 16. This scale has each inch divided into 16 equal parts or divisions of 1/16″. It is used by placing the 0 on the point where measurement begins and stepping off the desired length (**Figure 2-10**).

To reduce a drawing by 50 percent, use the scale marked 1/2. The large 0 at the end of the first subdivided measurement lines up with the other unit measurements that are part of the same scale. The large numbers crossed out in (**Figure 2-11**) go with the 1/4 scale starting at the other end. These numbers are ignored while using the 1/2 scale. To lay out 1¾″ at the 1/2 scale, read full inches to the right of 0 and fractions to the left of 0.

The 1/4 scale is used in the same manner as the 1/2 scale. However, measurements of full inches are made to the left of 0 and fractions to the right because the 1/4 scale is located at the opposite end of the 1/2 scale (**Figure 2-12**).

Architect's Scale

The purpose of the architect's scale is primarily for drawing large buildings and structures, although the

FIGURE 2-10   Regular full-size scale.

FIGURE 2-11   Half-size scale.

FIGURE 2-12   Quarter-size scale.

FIGURE 2-13   Architect's scale (full-size).

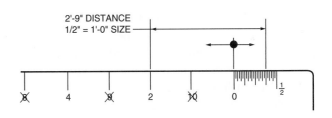

FIGURE 2-14   Architect's scale (half-size).

full-size scale is used frequently for drawing smaller objects. Because of this, the architect's scale is generally used for all types of measurements. It is designed to measure in feet, inches, and fractions of an inch. Measure full feet to the right of 0, inches and fractions of an inch to the left of 0. The numbers crossed out in (**Figure 2-13**) correspond to the 1/2 scale. They can be used, however, as 6″ because each falls halfway between full-foot divisions. Measurements from 0 are made in the opposite direction of the full scale because the 1/2 scale is located at the opposite end of the scale (**Figure 2-14**).

With the ten overlapping scales, a drawing may be made in various sizes from full size to 1/28 size. Since the architect's scale is showing divisions reciprocating feet and inches, a scale marked 1″ = 1′0″ makes a drawing with this scale 1/12 size.

EXAMPLES

3/4″ = 1′0″	would be 1/16 size	(3/4 × 1/12)
1/4″ = 1′0″	would be 1/48 size	(1/4 × 1/12)
1/2″ = 1′0″	would be 1/24 size	(1/2 × 1/12)
1″ = 1′0″	would be 1/12 size	(1 × 1/12)

Full size is measured on the full-size side of the architect's scale, where each division is 1/16.

Half size is measured by using the full-size side and dividing each measurement by 2.

Quarter size is measured by using the 3″ = 1′0″ scale. 3″ = 1′0″ would be 1/4 size (3 × 1/12 = 3/12 = 1/4).

Scale does not = size

Incorrect	1/4 scale ≠ 1.4 size
Correct	1/4 scale = 1/48 size

Mechanical Engineer's Scale (Decimal)

The mechanical engineer's scale is divided into inches and parts of an inch in decimal units. Each inch is divided into ten units or one tenth of an inch (.1). Each tenth is divided into five units where each is .02 hundredths of an inch. When using this scale, it is important to note that for accuracy of machined parts, measurements in thousandths of an inch are often required. When reading this scale, although .02 is often thought of as two hundredths of an inch, it is also 20 thousandths of an inch, written as .020. Another example would be one tenth. One tenth can be written as .1 in decimal form or 1/10 in fractional form. One tenth is one tenth of an inch written as .1. However, one tenth is also 10 hundredths of an inch or 100 thousandths of an inch. Mechanical engineers, machinists, and quality assurance inspectors generally speak in terms of thousandths of an inch when designing, manufacturing, and inspecting parts. Therefore, drafting technicians who use this scale should practice thinking in thousandths of an inch (**Figure 2-15**).

To read the mechanical engineer's scale (decimal), place the 0 at the point where the measurement begins and step off the desired length (**Figure 2-16**). The measurement in this example reads 1.660″. When viewing the scale, the correct measurement can be determined by breaking the measurement up into its individual components. The individual components of 1.660 are 1.000″, .6″, and .06″. The reading, therefore, is 1.660″, Figure 2-16. Additional examples are provided in **Figures 2-17** and **2-18**, where the scale readings are 2.320 and 3.040, respectively.

FIGURE 2-15   Mechanical engineer's scale.

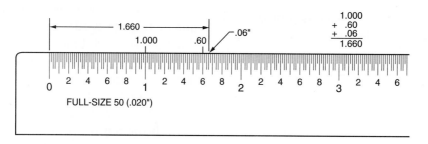

FIGURE 2-16   Reading the mechanical engineer's scale.

FIGURE 2-17   Mechanical engineer's scale.

FIGURE 2-18   Mechanical engineer's scale.

Civil Engineer's Scale

A civil engineer's scale is also called a *decimal-inch scale*. The number 10, located in the corner of the scale in **Figure 2-19**, indicates that each graduation is equal to 1/10 of an inch or .1″. Measurements are read directly from the scale. The number 20, located in the corner of the scale shown in **Figure 2-20**, indicates that it is 1/20 of an inch.

Using the same scale for civil drafting, one inch equals two hundred feet (**Figure 2-21**), and one inch equals one hundred feet (**Figure 2-22**).

Metric Scales

The metric system uses the meter (m) as its basic unit of linear measurement. A meter is 3.281 feet long or about 3⅜″ longer than a yardstick. Its multiples or parts are expressed by adding prefixes. These prefixes represent equal steps of 1,000 parts. The prefix for a thousand (1,000) is kilo; the prefix for a thousandth (1/1,000) is milli. One thousand meters (1,000 m), therefore, equals one kilometer. One thousandth (1/1,000 m) equals one millimeter (1.0 mm). Consider the following comparisons of metric measurements to American standard measurements:

- One millimeter (1.0 mm) = 0.001 meter = .03937 inch
- One thousand millimeters (1,000 mm) = 1.0 meter (1.0 m) = 3.281 feet
- One thousand meters (1,000 m) = 1.0 kilometer (1.0 km) = 3,281.0 feet

A metric scale is used if the millimeter is the unit of linear measurement. Common ratios of metric scales include: 1:100, 1:50, 1:20, 1:40, 1:80, and 1:33⅓ (**Figure 2-23**).

FIGURE 2-19   Civil engineer's scale.

FIGURE 2-20   Civil engineer's scale (half-size).

FIGURE 2-21   Civil engineer's scale (half-size).

FIGURE 2-22   Civil engineer's scale.

How to Read a Micrometer Graduated in Thousandths of an Inch (.001″)

Closely related to scales is the micrometer, a tool used to make precision measurements of objects. A micrometer consists of a highly accurate ground screw or spindle that is rotated in a fixed nut, thus opening or closing the distance between two measuring faces on the ends of the anvil and spindle (**Figure 2-24**). A piece of work is measured by placing it between the anvil and spindle faces and rotating the spindle by means of the thimble until the anvil and spindle both contact the work. The desired work dimension is then found from the micrometer reading indicated by the graduations on the sleeve and thimble, as described in the following paragraphs.

Since the pitch of the screw thread on the spindle is 1/40″ or 40 threads per inch in micrometers that are graduated to measure in inches, one complete revolution of the thimble advances the spindle face toward or away from the anvil face precisely 1/40 or .025″.

The reading line on the sleeve is divided into 40 equal parts by vertical lines that correspond to the number of threads on the spindle. Therefore, each vertical line designates 1/40 or .025″, and every fourth line, which is longer than the others, designates hundreds of thousandths. For example: the line marked "1" represents .100″, the line marked "2" represents .200″, the line marked "3" represents .300″, and so forth.

The beveled edge of the thimble is divided into 25 equal parts, with each line representing .001″

FIGURE 2-23   Metric scale.

FIGURE 2-24   How to read a micrometer.

indicated by the line on the thimble that coincides with the reading line on the sleeve.

EXAMPLE (See Figure 2-24):

The 1″ line on the sleeve is visible, representing .100″

Three additional lines are visible, each representing .025″. Thus, 3 × .025″ = .075″

The third line on the thimble coincides with the reading line on the sleeve, each line representing .001″. Thus, 3 × .001″ = .003″

The micrometer reading is

100″ + .075″ + .003″ = .178″

An easy way to remember how to read a micrometer is to think of the various units as if you were making change from a ten-dollar bill. Count the figures on the sleeve as dollars, the vertical lines on the sleeve as quarters, and the divisions on the thimble as cents. Add up your change and put a decimal point instead of a dollar sign in front of the figures.

Micrometers come in both English and metric graduations. They are manufactured with an English size range of 1 inch through 60 inches and a metric size range of 25 millimeters to 1,500 millimeters. The micrometer is a very sensitive device and must be treated with extreme care.

and every line numbered consecutively. Rotating the thimble from one of these lines to the next moves the spindle longitudinally 1/25 of .025″ or .001″, rotating two divisions represents .002″, and so forth. Twenty-five divisions indicate a complete revolution: .025″ or 1/40 of an inch.

To read the micrometer in thousandths, multiply the number of vertical divisions visible on the sleeve by .025″, and to this add the number of thousandths

Vernier Caliper

Closely related to scales and the micrometer for making precise measurements is the vernier caliper. Vernier calipers can measure both the outside and the inside of an object (**Figures 2-25** and **2-26**). Use the bottom scale when measuring an outside size. Use the top scale when measuring an inside size.

USE BOTTOM SCALE AND VERNIER FOR *OUTSIDE* MEASUREMENTS

OUTSIDE MEASUREMENT

FIGURE 2-25   Caliper—outside measurement.

USE TOP SCALE AND VERNIER FOR *INSIDE* MEASUREMENTS

INSIDE MEASUREMENT

FIGURE 2-26   Caliper—inside measurement.

Miscellaneous Conventional Tools

In addition to the conventional drafting tools explained so far, there is the engineering calculator. Drafting and design students should have an engineering calculator for performing the various math functions that go with the occupation. There are also other drafting tools that were once commonly used and are still found in some drafting settings, including technical inking pens and mechanical lettering sets (scriber, tracing pin, and lettering template). Students may come across these additional conventional tools, so they are explained in this section.

Technical Pens

The key to successful inking is a good technical pen (**Figure 2-27**).

Technical pen points are manufactured of stainless steel, tungsten, or jewels. The stainless steel point is chromium-plated for use on tracing paper or vellum. Tungsten points are long-wearing for use on abrasive, coated plotting film or triacetate. Jewel points are used on a plotter that has a controlled pen force.

Pen points are available in thirteen standard sizes of varying widths. For general drafting inking, numbers .45/1 and .70/2½ are recommended.

Mechanical Lettering Sets

Lettering sets come in a variety of sizes and templates. All sets contain a scriber, various pen sizes, and templates. Scribers are used with pens and templates to form letters.

Scribers may be adjusted to form vertical or slanted letters of several sizes from a single guide by simply unlocking the screw underneath the scriber and extending the arms (**Figure 2-28**).

Tracing Pin

Better, more expensive scribers use a double tracing pin (**Figure 2-29**). The blunt end is used for single-stroke lettering templates or very large templates that have wide

RESERVOIR PEN

INK CONTAINER   PEN BODY   CLEANING NEEDLE   POINT SECTION

LOCK RING   SPACER RING   NEEDLE RETAINER   COVER OR CAP

FIGURE 2-27   Caliper—internal parts of a technical inking pen.

FIGURE 2-28   Scribers are adjustable.

FIGURE 2-29   Double tracing pen.

grooves. The sharp end is used for very small lettering templates, double-stroke letters, or script-type lettering using a fine groove. Most tracing pins have a sharp point, but some do not. Always screw the cap back on the unused end after turning the tracing pin to the desired tip. Be careful with the points because they will break if dropped and can cause a painful injury if mishandled.

Standard Template

Learning to form mechanical letters requires a great deal of practice. **Figure 2-30** shows a template having three sets of uppercase and lowercase letters. Practice forming each size letter and number until they can be made rapidly and neatly. Use a very light, delicate touch so as not to damage the template, scriber, or pen.

INCHES		MILLIMETERS	
SIZE	DIMENSIONS	SIZE	DIMENSIONS
A	8 1/2 x 11 9 x 12	A-4	210 x 297
B	11 x 17 12 x 18	A-3	297 x 420
C	17 x 22 18 x 24	A-2	420 x 594
D	22 x 34 24 x 36	A-1	594 x 841
E	34 x 44 36 x 48	A-0	841 x 1189

FIGURE 2-31   Paper sizes.

Paper Sizes

The two standard paper sizes are classified as "architectural" and "engineering" sizes. Engineering, also defined as American National Standards Institute (ANSI) sizes, begin with A-size ($8\frac{1}{2}'' \times 11''$). Architectural sizes begin with A-size that is $9'' \times 12''$. The basic standard metric size for paper is A-4 (210×297 millimeters) (**Figure 2-31**). Examples of paper folded to A-size are shown in **Figure 2-32**.

Whiteprinter

Many types of whiteprinters are available for use in drafting rooms. A *whiteprinter* reproduces a drawing through a chemical process. Most of these machines work on the same basic principle (**Figure 2-33**). A bright light passes through the translucent original drawing and onto a coated whiteprint paper. The light breaks down the coating on the whiteprint paper, but wherever lines have been drawn on the original drawing, no light strikes the coated sheet. Then, the whiteprint paper is passed through ammonia vapor for developing. This chemical developing causes the unexposed areas—those that were shaded by lines on the original—to turn blue or black.

Most whiteprinters have controls to regulate the speed and flow of the developing chemical. Each type of machine requires different settings and has different controls. Before operating any whiteprinter, read all of the manufacturer's instructions.

FIGURE 2-30   Lettering template.

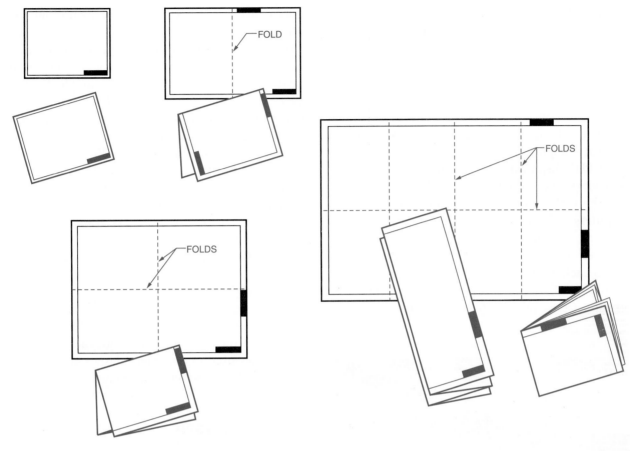

FIGURE 2-32   Paper folded to A-size.

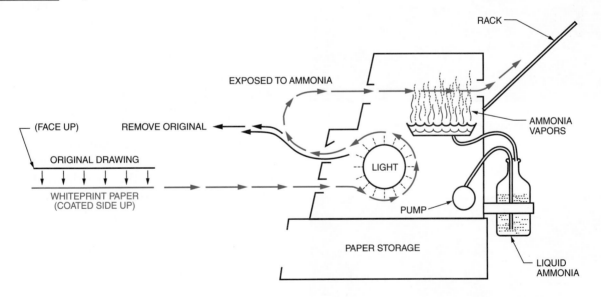

FIGURE 2-33   Whiteprinter process.

Today, whiteprinters are rarely seen in drafting rooms. The technology of choice is now a wide-format digital plain paper printer or inkjet plotter. Most digital printers can be equipped with scanners to facilitate both copying of hard-copy drawings as well as scanning of drawings for archiving or electronic distribution.

Computer-Aided Drafting (CAD)

Drafting done in any manner involves creating the drawings and other materials (bills of material, parts lists, specifications) that document designs, plans, surveys, and other ideas that must be conveyed from one individual or group to another. Computer-aided drafting (CAD) is just what the title implies: using a computer and various peripheral devices to aid in developing the documentation of designs, plans, surveys, or any other idea that must be conveyed from one individual or group to another. Whereas the tools of conventional drafting are those shown in Figure 2-1, the tools of CAD are those found in a typical computer system (e.g., central processing unit [CPU], display devices, input devices, data storage devices, and output devices) (**Figure 2-34**).

In the early days of their development, desktop computers did not have the processing power necessary to support CAD software. However, those days are long in the past. Today, even the smallest devices such as laptop and notebook computers can support CAD software, including solid modeling software. Consequently, CAD—a concept that was once considered a luxury—is now the normal way of doing the tasks associated with drafting and design.

Computer hardware changes rapidly and is always undergoing upgrading, improvement, and development. As hardware changes—for example, becoming ever smaller but more powerful—so do CAD system configurations. A typical computer system that might be used in a drafting and design setting looks similar to the one in **Figure 2-35** at the moment. But this will change quickly, as will the next generation of computer hardware and capabilities and the next and the next. For example, "cloud" technology has already revolutionized how computer files are stored, maintained, and retrieved.

Now that computer technology has developed to the point that two-dimensional CAD, three-dimensional CAD, and solid modeling are normal applications, the key to the future development of CAD is primarily in ongoing software development. Even more than the ever-changing hardware devices and capabilities used to document design ideas, software is the drafting technician's principal tool.

FIGURE 2-34   Typical computer hardware.

CAD Software

CAD software is readily available in a variety of packages and capabilities. Some of the more widely used CAD software packages at the moment include AutoCAD, DesignCAD 3D, TurboCAD, Punch! ViaCAD, ZWCAD, SmartDraw, Alibre Design, progeCAD, and DeltaCAD. These are just a few examples of the many CAD software packages that are available. Just as there are numerous brand-name makers of conventional drafting tools, there are many different developers of CAD software. Frankly, CAD software packages come and go as each new generation of software eclipses the capabilities of the next. Consequently, one of the biggest challenges facing drafting and design training programs is staying current with the latest CAD software developments.

Fortunately, CAD software is like automobiles in that although different cars have different options, capabilities, and prices, they are all driven in pretty much the same manner. Likewise, students who become proficient in using one CAD software package should be able to easily and quickly adapt to another package. Because this is the case, students in drafting and design programs should 1) become as proficient as possible in using whatever CAD software is available in their programs, and 2) learn the generic concepts that all CAD software is based on.

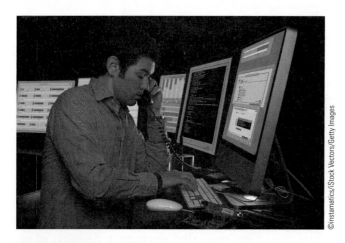

FIGURE 2-35   Typical computer workstation.

Generic CAD Software Skills

To understand what is meant by "generic CAD software skills," think of driving a car. No matter what kind of car it is, you will need to know how to put the car in the appropriate gear, steer, accelerate, and brake. In the same way, no matter what software package you plan to use, you will need to know how to do the following:

- Give commands for basic geometric shapes, text, and symbols (e.g., points, lines, arcs, circles, triangles, rectangles, polygons, splines, solids, surfaces, cross-hatching, symbols, dimensions, notes, text, etc.).
- Give commands to control how the drawing on the display screen is viewed (e.g., zoom in, zoom out, rotate, etc.).
- Give commands for modifying a drawing (e.g., editing, correcting, deleting, grouping, mirroring, etc.).

In the current state of CAD software development, commands in each of these three broad areas are given by selecting the appropriate command from a menu or submenu.

Regardless of the specific CAD software in question, students will need to learn how to perform such commonly used CAD-related tasks as using: 1) custom design palettes, 2) drag-and-drop functions, 3) data import functions, 4) text editing, 5) command history, 6) measurement commands, 7) macros, 8) cross-hatching, 9) textures, 10) snap tools, 11) colors, and 12) transparency options. Those who use CAD software for making pictorial illustrations will also have to learn to use such capabilities as lighting effects, photorealistic rendering and, in some cases, animation.

Selecting the Best CAD Software for a Given Application

Today's students of drafting and design are tomorrow's drafting technicians. Consequently, there will come a time when students can expect to be asked to help make decisions about choosing the next CAD software for their employers. It is important to understand the factors that should be considered when selecting from among various CAD software packages. Primary among these factors are the following:

- *Features and capabilities.* A software package's features and capabilities determine its usability (ease of use and convenience). They also determine what it can do and how well. Typical CAD software features and capabilities include drag-and-drop, command history, cross-hatching, textures, customizing

of tools, macros, and so on. It is important to know what specific features and capabilities the organization must have and which it would just like to have. Often, trade-offs between what you would like to have and what you must have are required when selecting CAD software.

- *Tools and methods.* It is important to select CAD software that is versatile—that is, software that contains both two- and three-dimensional tools. Ideally, the software will be capable of two-dimensional drafting, three-dimensional drafting, and modeling. Tools such as transparency, layering, lighting, snapping, and texturing are also valuable.
- *Compatibility of files.* Current computer capabilities allow for many different types of files (e.g., PDF, JPEG, VMF, PNG, WPG, DWG, and DXF). Consequently, it is important to know the types of files that will be used—in addition to DWG and DXF—in your setting. The highest possible level of compatibility is desired. Few things limit the usefulness of a CAD system more than incompatibility of files.
- *Support available.* No matter how good one becomes in using a given CAD software package, there will be problems that drafting technicians cannot solve, problems that can cause work to grind to a halt at the most inopportune times. Consequently, ongoing, after-purchase support is critical. The most important type of support to look for is built-in "help" elements that allow drafting technicians to seek help in real time without exiting the projects they are working on. In addition, online and telephone help should be readily and conveniently available.

Solid Modeling

As computer hardware and software capabilities evolved, solid modeling was the natural next step up from two- and three-dimensional CAD. The better solid modeling software packages are parametric feature–based. Parameters are the values that determine the shape or geometry of an object. A solid model is a solid representation of the object in question in that it not only represents length, width, and depth but also volume. A solid model is as much like the actual object as it is possible to be in an electronic replication. For example, the machined and welded part in **Figure 2-36** appears on the display screen as it would in real life and has the same features and characteristics.

The parameters for a solid model can be numeric in nature (*numeric* parameters), such as line lengths

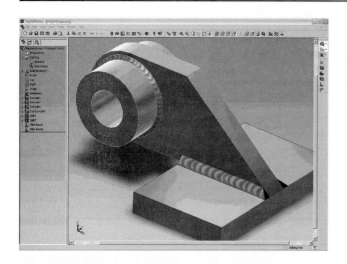

FIGURE 2-36   Solid model of a machined and welded part (Courtesy of SolidWorks Corporation).

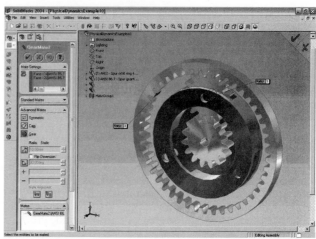

FIGURE 2-37   Solid model of a machined gear (Courtesy of SolidWorks Corporation).

or circle diameters, or they can be geometric in nature (*geometric* parameters), such as horizontal, vertical, concentric, parallel, tangent, or perpendicular. One of the benefits of solid modeling is that since parameters can be related to each other, they can automatically revise and update the object when a drafting technician changes a given parameter.

The software determines how to respond to changes in parameters through a concept known as *design intent*. For example, if a hole that is to be drilled in the center of a part is supposed to stay the same size and stay in the center of the part no matter what other changes might be made to the part, this is *design intent*. Drafting technicians can specify that intent and the software will follow it. *Features* in solid modeling software are the building blocks of the part. Features are the shapes and operations that will be used to electronically make the part in question.

The most common way to make a solid model is to begin with a two-dimensional sketch or drawing of the object in question. A sketch is made up of geometric entities such as points, lines, arcs, cones, and splines. The size and location of these geometric entities is explained by adding dimensions to the two-dimensional sketch or drawing. Relationships of features in the model are defined using such attributes as tangency, parallelism, perpendicularity, concentricity, etc. The geometry of a model is determined by the dimensions and relationships of the features. Examine the solid model of the machined gear in (**Figure 2-37**). The larger outside gear and the smaller inside gear would share the same center point (concentricity) no matter how many times the diameter of each gear might be revised.

Potential Benefits of Solid Modeling

Two-dimensional CAD software has revolutionized how the work of engineering, architecture, and drafting/design is done. But the benefits of two-dimensional software are felt primarily in the engineering and drafting departments. Solid modeling, on the other hand, has the potential to revolutionize the entire production process from bidding to purchasing to design to drafting to manufacturing to marketing. Good solid modeling software can generate, store, and share all of the information about a given product needed by all of the departments involved in the chain of production.

In fact, solid modeling software will not be worth the cost if its use is limited to just design and drafting. A solid model can provide the information needed for requests for proposals (RFPs) and requests for quotes (RFQs), engineering analysis, development of virtual prototypes, computer-numerical control (CNC) machine tool paths, purchase orders, marketing materials, and even invoicing. However, using solid modeling from one end of the production process to the other requires careful planning. It can also require hardware upgrades and will definitely require extensive training for all personnel in all departments that will use the software.

SUMMARY

- The most frequently used conventional drafting instruments include the following: triangles, scales, compasses, drafting brush, mechanical pencils, protractors, templates, erasing shield, eraser, lead pointer, drafting tape or dots, calculator, and dry cleaning pad.

- The most frequently used scales are the engineer's, architect's, and metric scales.

- Precision measuring instruments that technical drawing students should be able to use include the micrometer and vernier caliper.

- Standard paper sizes in inches are as follows: A = 8½ × 11, B = 11 × 17, C = 17 × 22, D = 22 × 34, and E = 34 × 44. Standard sizes in millimeters are as follows: A-4 = 210 × 297, A-3 = 297 × 420, A-2 = 420 × 594, A-1 = 594 × 841, and A-0 = 841 × 1,189.

- In the whiteprinting process, an original drawing is placed on top of coated whiteprint paper (coated side up). The two are exposed to intense light that burns off all of the coating except that covered by the pencil or ink lines of the original drawing. The original drawing is then removed, and the whiteprint paper is exposed to ammonia vapors. These vapors turn the remaining coating a deep purplish-blue color, creating a print of the original.

REVIEW QUESTIONS

Mark the following statements as either true or false.

1. Dry cleaning powder is used after the completion of a drawing to clean up any specks or smearing.

2. For drafting purposes, a 3H or 4H lead is recommended for layout work, and 2H, H, or HB leads are recommended for all other lines.

3. A drafting brush is used to remove loose graphite and eraser crumbs from the drawing surface.

4. An easy way to remember how to read a micrometer is to compare it to making change for a $1 bill.

5. An open-divided scale is one in which the units of measurement are subdivided throughout the length of the scale.

6. The tools of CAD are similar to those found in a typical computer system.

7. The key to the future development of CAD is in computer hardware development.

8. Students who learn one CAD software package should be able to quickly adapt to another.

9. When selecting a CAD software package, it is important to consider the type and amount of support available.

10. Versatility in CAD software is not an important selection criterion.

Answer the following questions by selecting the best answer.

1. Which of the following is a type of drafting triangle?
   a. 30°–60°
   b. Isosceles triangle
   c. Beam triangle
   d. Compass triangle

2. Which of the following drafting tools is commonly used to draw horizontal lines?
   a. Vernier
   b. Drafting machine
   c. Protractor
   d. Engineering scale

3. Which of the following paper sizes are commonly used in the industry today to produce engineering drawings?
   a. 8.5 × 11
   b. 9 × 12
   c. 22 × 34
   d. Both b and c

4. Which of the following is **not** true regarding the vernier on a drafting machine?
   a. Permits readings to 5 minutes of an arc
   b. Is graduated in increments from 0 to 60 minutes
   c. Takes the place of the arm-type drafting machine
   d. The zero on the protractor is in line with the zero on the vernier
   e. None of the above

5. Which of the following is **not** true regarding the whiteprinter?
   a. Most of them work on different basic principles.
   b. It reproduces a drawing through a chemical process.
   c. The chemical used is ammonia.
   d. The chemical developing causes the exposed areas to turn black or blue.

6. Which of the following would be considered a CAD tool?
   a. Data storage device
   b. Vernier protractor
   c. Compass
   d. Whiteprinter

7. With CAD, what is the drafting technician's principal tool?
   a. Computer hardware
   b. Engineering scales
   c. Software
   d. Templates

8. Which of the following is **not** considered a basic geometric shape in CAD?
   a. Line
   b. Point
   c. Cross-hatching
   d. Rotate

9. Which of the following is **not** considered a modifying function in CAD?
   a. Dimensioning
   b. Mirroring
   c. Zoom in
   d. Notation

10. The most important type of support to look for with CAD software is which of the following?
    a. Face-to-face
    b. 24-hour telephone service
    c. Well-written operator's manual
    d. Built-in "help" feature

CHAPTER 2 PROBLEMS

PROBLEM 2-1

On an A-size sheet of paper, lay out lines to scale as indicated:
Scale 16 (Figure 2-10)
   5-$\frac{7}{16}$"
   3-$\frac{5}{8}$"
Scale 1/2 (Figure 2-11)
   9-$\frac{1}{2}$"
   11-$\frac{3}{8}$"
Scale 1/4 (Figure 2-12)
   11-$\frac{3}{8}$"
   7-$\frac{7}{8}$"
Scale 1 (Figure 2-13)
   3"–8-$\frac{1}{2}$"
   4"–1-$\frac{3}{4}$"
Scale 1/2 (Figure 2-14)
   5"–9-$\frac{3}{4}$"
   7"–8-$\frac{1}{4}$"
Scale 10 (Figure 2-15)
   3-$\frac{3}{8}$"
   5-$\frac{7}{16}$"
Scale mm (Figure 2-19)
   42.5 mm
   37.7 mm

PROBLEM 2-2

Assume that a whiteprinter can process paper at 25 feet per minute (fpm). Approximately how many seconds would the printer take to print a D-size drawing the long way? (Dimension B. See Figure 2-31.) Note that the paper passes through the printer two times while traveling a total of 26".

PROBLEM 2-3

A rich uncle says to you that he will contribute to your education by giving you a check equal to a stack of one-dollar bills equal to your height. How large should the check be? A micrometer shown in Figure 2-22 would be helpful. What else could you do to determine the thickness of a one-dollar bill?

PROBLEM 2-4

At a certain time of day, a drafter used a standard steel tape to measure the shadow of a 1.8-m-tall person.
   The shadow measured approximately 0.9 m. Next, the drafter measured the shadow of a nearby water tower and found it to be approximately 7.7 m. Determine the approximate height of the water tower by using two methods.
1. Use a proportional equation.
2. Use trigonometry.

PROBLEM 2-5

Read the micrometer shown to determine the outside dimension of a typical drafting pencil.

PROBLEM 2-6

Prepare sketches to show how you would use a 30°–60° triangle with a 45° triangle to construct lines at the following angles to the horizontal: 15°, 75°, 105°, and 165°.
   Refer to Figure 2-8.

Problems 2-7 through 2-12

Carefully measure each line in inches, or in millimeters if metric is indicated. Neatly enter your answers on a sheet of paper. For extra practice, measure each line full size as given, half size as given, quarter size as given, or ten times scale as assigned by the instructor.

PROBLEM 2-7

FULL SIZE

1)
2)
3)
4)
5)
6)
7)
8)
9)
10)

PROBLEM 2-8

HALF SIZE

1)
2)
3)
4)
5)
6)
7)
8)
9)
10)

PROBLEM 2-9

3/4 SCALE

1)
2)
3)
4)
5)
6)
7)
8)
9)
10)

PROBLEM 2-10

METRIC (FULL SIZE) MILLIMETERS

1)
2)
3)
4)
5)
6)
7)
8)
9)
10)

PROBLEM 2-11

METRIC (HALF SIZE) MILLIMETERS

1)
2)
3)
4)
5)
6)
7)
8)
9)
10)

PROBLEM 2-12

METRIC (QUARTER SIZE) MILLIMETERS

1)
2)
3)
4)
5)
6)
7)
8)
9)
10)

Problems 2-13 through 2-28

Construct each object by using the given dimensions. Use consistent thick, black object lines throughout. Keep all corners tight and sharp. Where indicated, draw thin, black center lines.

PROBLEM 2-13

3.00
(76)

.25 (6)

.25
(6)

3.00
(76)

PROBLEM 2-14

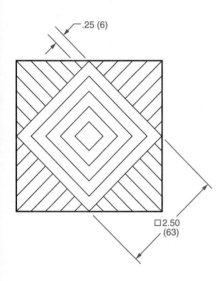

.25 (6)

□2.50
(63)

PROBLEM 2-15

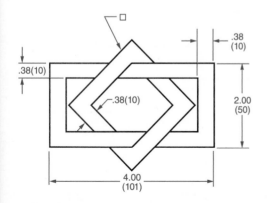

□

.38
(10)

.38(10)

.38(10)

2.00
(50)

4.00
(101)

PROBLEM 2-16

.38
(10)   TYPICAL

4.18
(106)

4.18
(106)

PROBLEM 2-17

.25 (6)
(TYPICAL)

4.25
(108)

PROBLEM 2-18

.25 (6)
TYPICAL
SPACE

.25
(6)
TYPICAL

2.50
(63)

2.50
(63)

5.25
(133)

PROBLEM 2-19

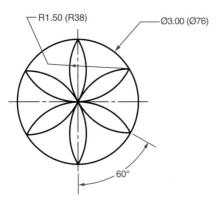

R1.50 (R38)   Ø3.00 (Ø76)

60°

PROBLEM 2-22

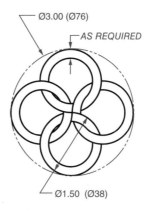

Ø3.00 (Ø76)

AS REQUIRED

Ø1.50 (Ø38)

PROBLEM 2-20

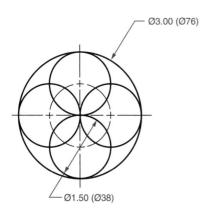

Ø3.00 (Ø76)

Ø1.50 (Ø38)

PROBLEM 2-23

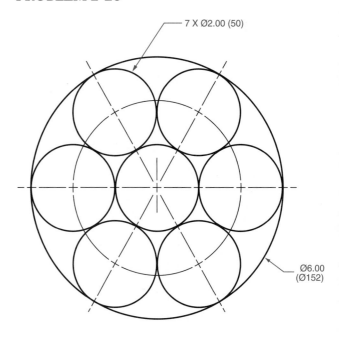

7 X Ø2.00 (50)

Ø6.00 (Ø152)

PROBLEM 2-21

Ø3.00 (Ø76)

Ø2.50 (Ø63)

Ø1.50 (Ø38)

AS REQUIRED

PROBLEM 2-24

Ø1.12 (Ø28)

Ø2.50 (Ø63)

Ø1.50 (Ø38)

PROBLEM 2-25

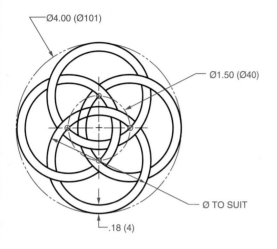

Ø4.00 (Ø101)

Ø1.50 (Ø40)

Ø TO SUIT

.18 (4)

PROBLEM 2-26

Ø5.50 (Ø139)

ALL SPACES = .25 (6)

PROBLEM 2-27

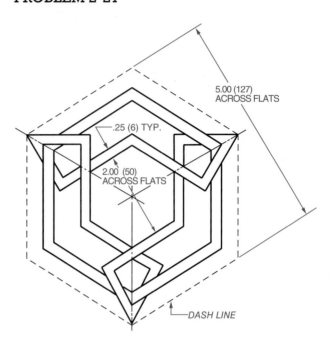

5.00 (127)
ACROSS FLATS

.25 (6) TYP.

2.00 (50)
ACROSS FLATS

DASH LINE

PROBLEM 2-28

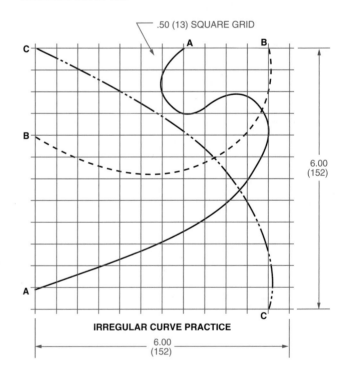

.50 (13) SQUARE GRID

C   A   B

B

A

C

6.00
(152)

IRREGULAR CURVE PRACTICE

6.00
(152)

3

Sketching and Lettering for Engineering Communication

CHAPTER OBJECTIVES

Upon completion of this chapter, students should be able to do the following:

▶ Explain the concept of talking sketching, including the two kinds.

▶ List the various styles of freehand lettering and the characteristics of good lettering.

▶ Explain the techniques one must know in order to do freehand lettering.

▶ Illustrate the various types of lines used on technical drawings.

▶ Illustrate the four types of sketches.

▶ Explain what materials are needed in order to make sketches.

▶ Demonstrate the most commonly used sketching techniques.

KEY TERMS

Axonometric sketching

Boardroom sketching

Center line

Conversational sketching

Cutting plane line

Dimension line

Extension line

Hidden line

Isometric (sketching)

Line work

Oblique sketching

Orthographic sketching

Perspective sketching

Phantom line

Text

Visible line

Note to Instructors and Students

Although modern CAD technologies have radically altered the way that design documentation is prepared, there is still a need for engineers, designers, and CAD technicians to develop their skills in the areas of sketching, hand lettering, and line work. Sketches remain an important form of communication between engineers, designers, and the CAD technicians who will convert their ideas into finished drawings and other types of documentation. Consequently, the authors continue to offer this chapter and recommend that instructors use it to help students develop the sketching, lettering, and line work skills they will need in the workplace.

Talking Sketching

Two types of sketching, which are often neglected in technical drawing texts but nevertheless complement the skills of a successful drafter, designer, or engineer, are conversational sketching and boardroom sketching.

Conversational Sketching

Conversational sketching usually occurs between two or more individuals huddled around a drafting board or a cafeteria table. The one doing the talking combines several types of sketches in one drawing as the idea is talked through. The first part of the sketch is most likely an orthographic view, the easiest projection to draw rapidly. Then, as the talking progresses, a pictorial is used to give the observers a feeling of depth. For example, a designer communicating an idea for an AM/FM receiver would begin with the orthographic view (**Figure 3-1**), and finish with a pictorial (**Figure 3-2**). A drafter, designer, or engineer sketches rapidly while talking and doesn't have time to develop a single specific kind of sketching, such as orthographic views, an isometric, an oblique, or a two-point perspective.

Boardroom Sketching

Communication with more individuals than a drafting board can accommodate requires a more visible

FIGURE 3-2   Talking sketch completed with pictorial.

surface for the larger group. The drawing surface for *boardroom sketching* sometimes is a chalkboard, but more often is a pastel-colored porcelainized enamel, a glass panel, or a dry-erase surface. All of these surfaces require a special marker and can be erased with a cloth.

Good-quality talking sketches, like good-quality sketching discussed earlier in this chapter, require practice to develop visible lines and especially to develop lettering large enough to be read by those sitting at the back of the room.

Note that sketching and lettering on overhead projectors is another medium for talking sketching. Practice to make sure that all those present can read all the lettering and see the lines.

Freehand Lettering

Text is an important part of a technical drawing. Not all information required on technical drawings can be communicated graphically, the most obvious being dimensions. *Text* on technical drawings consists of dimensions, notes, legends, and other data that are best conveyed using alphanumeric characters (**Figure 3-3**).

Several different ways are used to create text on technical drawings. The traditional method is by freehand lettering. Other methods include such mechanical lettering techniques as scriber templates, typewritten notation, and typed lettering generated by computer-aided drafting systems. This chapter focuses on freehand lettering. Other methods are described elsewhere in this text.

Lettering Styles

There are numerous different lettering styles or fonts (**Figure 3-4**). The standard style for freehand lettering on technical drawings, as established in American National Standards document Y14.2M-1992, is

FIGURE 3-1   Talking sketch (orthographic view).

FIGURE 3-3   Examples of text on a technical drawing.

THIS IS CALFLISCH FONT

THIS IS MATHEMATICAL FONT

THIS IS CHICAGO FONT

THIS IS LITHOS FONT

THIS IS ROSEWOOD FONT

THIS IS CURLZ FONT

THIS IS SABON FONT

THIS IS BRUSH FONT

FIGURE 3-4   Sample lettering fonts.

SINGLE - STROKE GOTHIC LETTERING SAMPLE

FIGURE 3-5   Single-stroke Gothic lettering.

single-stroke Gothic lettering. Vertical, single-stroke Gothic letters are the most universally used of the various styles available to drafters (**Figure 3-5**).

Some modifications of the standard Gothic configuration of letters are often made without actually changing from the Gothic style of lettering. One way is through the use of uppercase and lowercase letters (**Figure 3-6**), but this is seldom acceptable on technical drawings. Another method is to condense or extend the letters (**Figure 3-7**).

The most common way of modifying Gothic letters is to incline them slightly to the right

UPPERCASE GOTHIC
lowercase Gothic

FIGURE 3-6   Uppercase and lowercase Gothic lettering.

EXTENDED VARIATION

CONDENSED VARIATION

FIGURE 3-7   Extended and condensed variations of Gothic lettering.

SAMPLE OF INCLINED LETTERING

FIGURE 3-8   Inclined Gothic lettering.

(**Figure 3-8**). Inclined lettering is easier to make, as it lends itself to a natural direction of wrist action. The correct angle of the inclined elements is a two-unit incline to the right for each five units of letter height or at 68° using a letter guide. Errors are not as detectable with inclined letters as they are with vertical elements. Because the inclined elements are longer, they are easier to read. However, inclined lettering is not universally accepted, and caution must be exercised not to conflict with customary drafting styles. A backhanded or left-leaning inclination is never an acceptable modification.

Characteristics of Good Lettering

Good freehand lettering, regardless of whether it is uppercase or lowercase, condensed or extended, or vertical or inclined, must have certain characteristics. These requisites include neatness, uniformity, stability, proper spacing, and speed.

Neat lettering is important so that the information being conveyed can be easily read. Few things detract from the appearance and quality of a technical drawing more than sloppy lettering (**Figure 3-9**).

For uniformity, all letters should be the same in height, proportion, and inclination. A necessary tactic for maintaining uniformity is the use of guidelines (**Figure 3-10**). The customary heights of characters in technical drawing are 1/8″ (3 mm) for

FIGURE 3-9   Sloppy lettering is difficult to read.

FIGURE 3-10   Guidelines improve uniformity.

FIGURE 3-11   Top-heavy letters are not balanced.

SPACING WITHIN WORDS SHOULD BE CLOSE.

SPACING BETWEEN WORDS SHOULD BE FAR.

FIGURE 3-12   The proper spacing of letters and words.

regular text and 3/16″ (4.5 mm) for headings and titles.

The proper stability or balance of letters is an important characteristic in freehand lettering. Each letter should appear balanced and firmly positioned to the human eye. Top-heavy letters are not balanced because they appear about to topple over (**Figure 3-11**).

The proper spacing of letters and words is important, and it takes a lot of practice to accomplish. A good rule of thumb to follow in terms of spacing is to use close spacing within words and far spacing between words (**Figure 3-12**). The proper positions of letters relative to one another in words is accomplished by spacing the letters in the word equally in the area, not by trying to equalize the spacing between letters. This becomes automatic if the drafter concentrates on the word being lettered, not on each letter. Another rule of thumb for spacing is to allow the width of one round letter, such as O, C, Q, or G, between words. **Figure 3-13** illustrates how this type of spacing can make the lettering much easier to read.

THIS IS A PROPERLY SPACED SENTENCE.

THISISANIMPROPERLYSPACEDSENTENCE.

FIGURE 3-13   Spacing between words is important.

In the modern drafting room, because time is money, speed in freehand lettering is critical. Typically, freehand lettering is one of the slowest, most time-consuming tasks drafters must perform. It takes many hours of practice to develop freehand lettering that is neat, uniform, balanced, properly spaced, and fast. Some drafters never reach this goal. Those who do reach it through constant practice coupled with continual efforts to improve.

Freehand Lettering Techniques

Freehand lettering techniques are learned by knowing what grades of lead to use, how to make the basic lettering strokes, and how to use guidelines, along with constantly practicing and trying to improve (**Figure 3-14**).

All letters and numbers are created using six basic strokes (**Figure 3-15**). The first stroke is a single stroke made downward and to the right at approximately 45°. The second stroke is made downward and to the left at approximately 45°. The third stroke is vertical and is made from top to bottom. Stroke number four is horizontal and is made from left to right. The fifth stroke is a half-circular stroke to the left made from top to bottom. The sixth stroke is a half-circular stroke to the right made from top to bottom. All alphanumeric characters can be created using combinations of these six strokes. **Figure 3-16** shows how these strokes are used for making selected characters.

Lettering Guidelines

Guidelines are a critical part of freehand lettering. Uniformity, neatness, and stability cannot be achieved without using guidelines.

A lettering guide is used to make guidelines, as shown in **Figure 3-17**. One type of lettering guide, Figure 3-17, can be used to lay out the vertical, horizontal, and inclined guidelines, as well as text spacing, for good freehand lettering. The lettering guide has a fixed rectangular part and a round disk that can be rotated inside that fixed part. The rectangular part has two scales on it, and the disk has two scales on it. In addition to the four scales, there are five rows of holes, Figure 3-17. The scale on the far left hole

FIGURE 3-14   Good lettering takes practice.

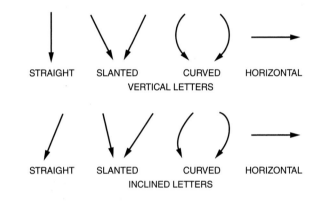

FIGURE 3-15   Basic strokes used for lettering.

pattern is present to 1/8″ spacing. It can be used to create lines with 1/8″ spacing between them on any multiple of 1/8″ up to 2″ spacing.

The scale to the immediate right of the far left scale is a metric scale. This scale is on the rotating disk and ranges from 2 to 10 mm with three odd values of 3.08, 6.1, and 9.75 mm. These values correspond closely to the inch equivalents of 1/8″, 1/4″, and 3/8″, respectively. This scale is used in conjunction with the mark that has an "M" above it, Figure 3-17. When the disk is rotated and the number on the scale lines up with the mark that has the "M" above it, the spacing between the holes that are connected by the line is equal to the value selected. **Figure 3-18** shows the 6.1 lined up with the mark below the "M." The spacing between the holes connected by the line is 6.1 mm.

The hole pattern below the 3/5 notch and the 2/3 mark is used in conjunction with the scale located at the bottom of the disk. The scale graduation is in 32nds of an inch and ranges from 3/32 to 10/32 of an inch, Figure 3-18. When the disk is rotated and the number on the scale lines up with the mark on the bottom of the rectangular part, the

FIGURE 3-16   Forming uppercase Gothic letters and numerals—vertical style.

FIGURE 3-17   Settings on the lettering guide.

spacing between the holes that are connected by the line is equal to the value shown. Figure 3-18 shows the 6 lined up with the mark. The spacing between the holes connected by the line is 6/32 or 3/16″.

To use the lettering guide, the base of the guide is placed in contact with the parallel bar or other straightedge. The pencil is inserted into the hole and used to pull the guide across the paper while maintaining contact with the straightedge. A line is lightly drawn by the pencil as the guide is pulled across the paper. The pencil is then inserted into another hole, connected by the

line to the first hole, and an additional line is created as the guide is pulled across the paper in the opposite direction. The spacing between the guidelines created is equal to the value selected, Figures 3-10 and 3-18.

Line Work

Line work is the generic term given to all of the various techniques used in creating the graphic data on technical drawings. Mechanical line work is made using either mechanical pencils or technical pens.

FIGURE 3-18   Metric settings on the lettering guide.

Devices such as parallel bars, drafting machines, triangles, scales, and numerous other tools are used to guide the line-making.

Characteristics of Lines

Twelve types of lines are used in manual drafting. Each has its own individual characteristics. The *visible line* is solid, thick, and dark. It is used to represent the outline of the features that are visible in a given view of an object. The *hidden line* is a thin line formed by a series of short dashes separated by even shorter breaks. It is used to indicate features that are not visible in a given view of an object. *Dimension lines* and *extension lines* are solid, thin lines of approximately the same width as hidden lines. Dimension lines may be broken for the dimension and should have arrowheads for terminations. Extension lines start with a 1/16″ break away from the visible lines of the object and represent an extension of the visible features of the object to accommodate dimension lines. Dimension lines are normally drawn perpendicular to extension lines with the arrowhead terminating at the extension line.

The *center line* is broken with one short dash in its center that is repeated approximately every 1.25″. It is the same width as the hidden, dimension, and extension lines. It is used to show the center of circular features or to represent symmetry. The *phantom line* is a solid, thin line that resembles a center line, except that it has two short dashes that are repeated approximately every two inches. It is used to represent a moveable feature in its different positions, a repeated detail, or a repeated pattern. *Viewing plane*

lines and *cutting plane lines* are thick like the visible line and consist of two styles. One type resembles a hidden line, and the other type resembles a phantom line. The viewing plane line is used to show where a viewing plane is being placed, as well as to indicate the direction in which the resulting section is to be viewed. The *crosshatching line*, also known as the *section line* or *section lining*, is used in sectional views of drawings. It is a thin, solid line used in sectional views to indicate material that touches the cutting plane as it slices through the object.

Freehand break lines and *mechanical break lines* are both used to show when an object is broken so as to save drawing space. The freehand break line is also used to show interior features in sectional drawings and is a solid, thick line produced without the aid of a straightedge. The mechanical break line is a solid, thin line with at least one zigzag in it. It is created with the use of a straightedge. The *stitch line* is a solid, thin line used to indicate where there is sewing or stitching. All lines used on technical drawings should closely match those in **Figure 3-19**.

Horizontal and Vertical Lines

Horizontal lines are formed by pressing the straightedge (T-square, parallel bar, drafting machine scale, and so forth) against the worksheet with one hand and moving the pencil with the other. Uniformity of line widths and weights can be achieved by holding the pencil at approximately 60° from the drawing surface, maintaining an even pressure downward, and slowly revolving the pencil axially as it moves

───────────────────		VISIBLE LINE
THICK	(H, F, or HB)	
– – – – – – – – – – –		HIDDEN LINE
THIN	(2H or H)	
───────────────────		CROSSHATCHING LINE
THIN	(2H or H)	
── ─ ── ─ ── ─ ──		CENTER LINE
THIN	(2H or H)	
◄───────────────►		DIMENSION AND EXTENSION LINES
THIN	(2H or H)	
──────────────╲──►		LEADER LINE
THIN	(2H or H)	
── ─ ── ─ ── ─ ──		CUTTING PLANE LINE
THICK	(H, F, or HB)	
∿∿∿∿∿∿∿∿∿∿		FREEHAND BREAK LINE
THICK	(H, F, or HB)	
──────╱╲╱────╱╲╱──		MECHANICAL BREAK LINE
THIN	(2H or H)	
── ─ ── ─ ── ─ ──		PHANTOM LINE
THIN	(2H or H)	
– – – – – – – – – – –		STITCH LINE
THIN	(2H or H)	

FIGURE 3-19   Line types used on technical drawings.

FIGURE 3-20   Maintaining uniformity of lines.

across the drawing surface (**Figure 3-20**). This keeps the lead tip symmetrical.

Vertical lines are created according to the same principles, except that the drafter's hand moves upward rather than from left to right. The angle of inclination, the amount of pressure, and the rotating motion are the same as they are for horizontal lines.

Angular Lines

Many modern devices are available to assist drafters in making angular lines. These include protractors, adjustable triangles, and adjustable arms on drafting machines. However, most angular lines can be created simply by using the standard 30°–60° and 45° triangles alone and in various combinations (**Figures 3-21, 3-22, 3-23,** and **3-24**). These standard tools create angles of 15°, 30°, 45°, 60°, and 75°.

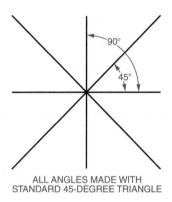

ALL ANGLES MADE WITH
STANDARD 45-DEGREE TRIANGLE

FIGURE 3-21   Making angular lines with the 45° triangle.

ALL ANGLES MADE WITH
STANDARD 30°–60° TRIANGLE

FIGURE 3-22   Making angular lines with the 30°–60° triangle.

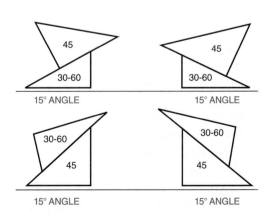

FIGURE 3-23   Making 15° angles.

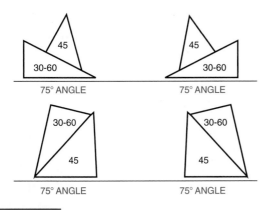

FIGURE 3-24   Making 75° angles.

Parallel Lines

Parallel lines can be created in a number of different ways. Vertical (and horizontal) parallel lines are made by simply moving the straightedge the required distance and making each successive line (**Figure 3-25**).

Parallel lines at angles can be created by using the 30°–60° and 45° triangles in combination much the same as they are used for making angular lines. When one uses triangles to create angular lines, the first line is created at the desired angle. Aligning one edge of a triangle to the line, register any side of the second triangle against one of the nonaligned edges of the first triangle. Holding the second triangle to prevent it from moving and sliding the first triangle along the engaged

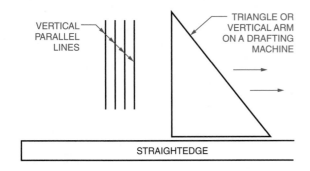

FIGURE 3-25   Making vertical parallel lines.

FIGURE 3-26   Creating successive parallel lines.

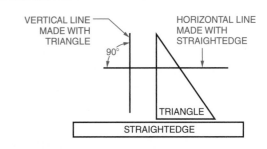

FIGURE 3-27   Making perpendicular lines.

FIGURE 3-28   Creating a line perpendicular to a nonvertical or nonhorizontal line.

edge of the second triangle will reposition the originally aligned edge to any desired parallel position. Successive parallel lines are created in the same way (**Figure 3-26**).

Perpendicular Lines

Drawing perpendicular lines can be accomplished in a manner similar to drawing parallel lines. Horizontal and vertical perpendicular lines can be created by using a straightedge and a triangle (**Figure 3-27**).

Creating a line perpendicular to a nonhorizontal or nonvertical line is accomplished by using triangles in conjunction with a straightedge (**Figure 3-28**). Line 1 in this figure is drawn first. Then, the 45° triangle is slid along the 30°–60° triangle, and the perpendicular line is created with the opposite side of the 45° triangle.

Sketching

Even in the world of high technology and computers, sketching is still one of the most important skills for drafters and designers. Sketching is one of the first steps in communicating ideas for a design, and it is used in every step thereafter. It is common practice for designers to prepare sketches that are turned over to drafters for conversion to finished working drawings. **Figure 3-29** is an example of a typical design sketch.

Sketching Lines

The lines used in creating sketches closely correspond to those used in creating technical drawings, except, of course, that they are not as sharp and crisp.

Figure 3-30 illustrates the various types of lines used in making sketches.

The basic line types are visible line, hidden line, center line, dimension line, sectioning line, extension line, and cutting plane line. These lines represent the various lines available for creating sketches. The character of each line, as illustrated in **Figure 3-31** should be closely adhered to when making sketches.

Types of Sketches

The types of sketches correspond to the types of technical drawings. There are four types of sketching: orthographic, axonometric, oblique, and perspective (**Figures 3-32, 3-33, 3-34,** and **3-35**).

Orthographic sketching relates to flat, graphic facsimiles of a subject showing no depth. Six principal

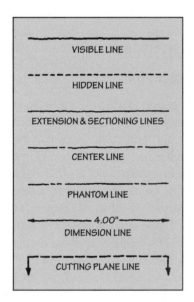

FIGURE 3-29 Typical design sketch.

FIGURE 3-31 Sample design sketch.

FIGURE 3-30 Lines used in sketching.

FIGURE 3-32 Orthographic sketch.

FIGURE 3-33 Axonometric sketch.

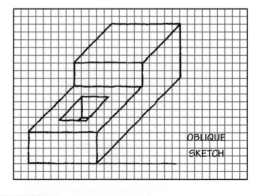

FIGURE 3-34 Oblique sketch.

views of a subject may be incorporated in an ortho-graphic sketch: top, front, bottom, rear, right side, and left side (**Figure 3-36**). The views selected for use in a sketch depend on the nature of the subject and the judgment of the sketcher.

Axonometric sketching may be one of three types: isometric, dimetric, or trimetric (**Figure 3-37**). The type most frequently used is isometric, in which length and width lines recede at 30° to the horizontal, and height lines are vertical (**Figure 3-38**). In sketching, the use of these terms is academic, as they relate to propor-tional scales and angle positions of height, width, and depth, which are only estimated in sketching.

Oblique sketching involves a combination of a flat, orthographic front surface with depth lines receding at a selected angle of 30°, 60°, or 45° (usually 45°) (**Figure 3-39**).

FIGURE 3-35   Perspective sketch.

FIGURE 3-38   Isometric sketch.

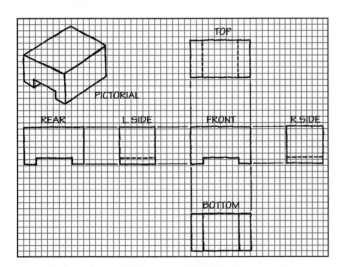

FIGURE 3-36   Six principal views in orthographic sketching.

FIGURE 3-39   Oblique sketch.

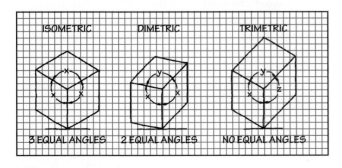

FIGURE 3-37   Three types of axonometric sketches.

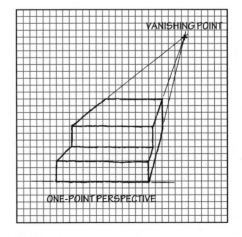

FIGURE 3-40   One-point perspective sketch.

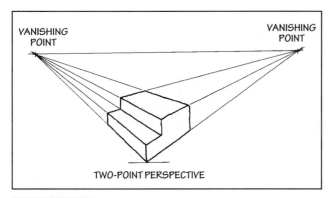

FIGURE 3-41   Two-point perspective sketch.

Perspective sketching involves creating a graphic facsimile of the subject. Consequently, depth lines must recede to a hypothetical vanishing point (or points) (**Figures 3-40** and **3-41**). In fact, all pictorial sketches naturally tend to assume characteristics of perspective sketches as a result of how the eye views the apparent relative proportions of objects. This is not necessarily undesirable.

Sketching Materials

An advantage of sketching is that it requires very few material aids. Whereas drafters must have a complete collection of tools, equipment, and materials in order to do working drawings, sketching requires nothing more than a pencil and a piece of paper. It is not uncommon for a sketch to be drawn on a paper napkin during a hurried luncheon meeting.

Sketching done in an office environment requires three basic materials: pencil, media (paper or graph paper), and an eraser. Graph paper simplifies the sketching process considerably, especially for students just learning, and it should be used freely.

Sketching Techniques

Sketches, as with drawings, consist of straight and curved lines. With practice, drafters can become skilled in creating neat, sharp, clear examples—straight or curved—of all the various line types introduced previously. When sketching, the following general rules apply.

1. Hold the pencil firmly, but not so tightly as to create tension or hand fatigue.
2. Grip the pencil approximately one inch to one-and-a-half inches up from the point.
3. Maintain a comfortable angle between the pencil and the sketching strokes.
4. Draw horizontal lines from left to right using short, slightly overlapping strokes.
5. Draw vertical lines from top to bottom using short, slightly overlapping strokes.
6. Draw curved lines using short, slightly overlapping strokes.

In addition to these general rules, some specific techniques are used in making the various line types for sketching.

SKETCHING ABILITY LEADS TO PROMOTION

"Don't they teach sketching in school any more? I can't find one drafting technician in this department who can make a decent sketch." Dan Johnson, chief drafter for Precision Machining, Inc. (PMI), had been frustrated when he made this statement. The problem he had faced at the time was simple. He needed to send a drafting technician into the field to make sketches of several parts for a warehouse full of machines that were to be refurbished and retrofitted. PMI had a contract to machine the parts. Unfortunately, the machines were old enough that technical drawings of the parts were no longer available.

Johnson's only option was to send a drafting technician to the warehouse armed with a sketch pad and micrometers. The parts would be measured, sketches would be made, and the drafting department would have the information needed to create a new technical drawing package. It was after seven technicians in a row had told Johnson that they either never learned sketching or had forgotten how that the chief drafter had vented his frustration. Fortunately for Johnson, Maria Sims overheard him. Sims was PMI's newest junior drafting technician, having graduated from Clark County College of Technology just a month earlier. Sims told Johnson that not only could she make comprehensive, accurate, readable sketches but that she could read a micrometer, too. She was given the assignment on the spot, and her sketching skills served her well.

Johnson was so impressed with his newest junior drafting technician that he not only promoted her, he asked her to give the entire department a seminar on sketching.

Sketching Straight Lines and Curves

Making straight lines on graph paper is a simple process of guiding the pencil using the existing lines. If graph paper is not available, pencil dots can be positioned to plot the path of the line (**Figure 3-42**). In this figure, the sketcher enters a series of pencil dots on the paper that provide a basic outline of the shape of the object. Then, with a series of short, slightly overlapping strokes, the pencil dots are connected (**Figures 3-43, 3-44,** and **3-45**). This technique is also used for curved lines (**Figure 3-46**).

Sketching Circles

Figure 3-47 illustrates a series of six steps that can be used for sketching a circle. Vertical and horizontal center lines are sketched, which positions the center of the circle (Step 1), and the radial d istances of the desired circle size are marked on each of these lines, equidistant from the center (Step 2). A square is drawn symmetrically around the center, with the sides located at the radial line marks (Step 3). On the diagonals of the square (Step 4), the radial distances are again marked

FIGURE 3-46   Using dots in sketching curved lines.

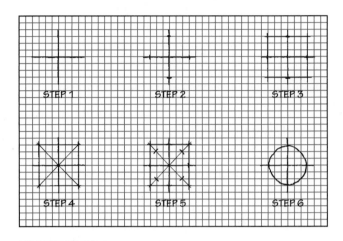

FIGURE 3-47   Sketching a circle.

off from the center (Step 5). This provides four positions for the circumference to pass through at the sides of the square and four more positions on the diagonals of the square. The right half of the circle is sketched in from top to bottom using short, slightly overlapping strokes, and then the left half is sketched in the same manner (Step 6).

Sketching Ellipses

A similar technique is used for sketching ellipses on an orthographic view, except that the square becomes a rectangle (**Figure 3-48**). Ellipses are oriented on the pictorial of the object being sketched as shown in **Figure 3-49**.

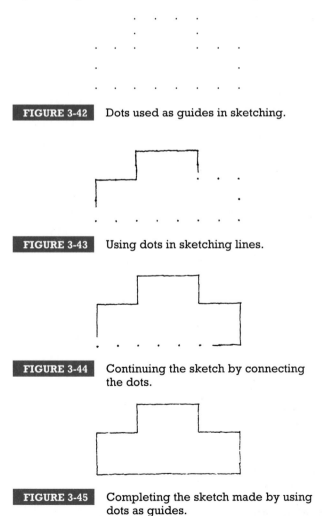

FIGURE 3-42   Dots used as guides in sketching.

FIGURE 3-43   Using dots in sketching lines.

FIGURE 3-44   Continuing the sketch by connecting the dots.

FIGURE 3-45   Completing the sketch made by using dots as guides.

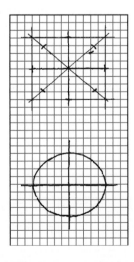

FIGURE 3-48   Sketching an ellipse.

FIGURE 3-49   Orienting an ellipse on an object.

Proportion in Sketching

Sketches are not done to scale, but it is important that they be made proportionately accurate. All of the various components of a sketch should be kept in proportion to those of the actual object. This technique takes a great deal of practice to master.

Some methods for achieving proportion recommend using a pencil or a strip of paper as a simulated scale. These techniques are not only unrealistic in terms of the real world, they defeat the very purpose of sketching. A skilled sketcher must learn to maintain proportion without the use of tools and aids. The best device for accomplishing proportion in sketching is the human eye. With practice, the drafter can become proficient in maintaining proportion without the use of extraneous, time-consuming devices. The following general rules relating to proportion will also help.

STEP 1   In sketching, use graph paper whenever possible.

STEP 2   Examine the object to be sketched and mentally break it into its component parts.

STEP 3   Beginning with the largest components (width and height), estimate the proportion, such as the width is 4/3 times the height or 5/2 times the height, and so on.

STEP 4   Lay out the largest component according to the proportions decided upon in Step 3. Use construction line squares and rectangles to block in irregularly shaped components (**Figure 3-50**).

Repeat Steps 3 and 4 until the entire object is finished.

FIGURE 3-50   Blocking in components.

Orthographic Sketching

Orthographic sketching may involve sketching any combination of the six principal views of the subject. The top, front, and right-side views are normally selected for representing an object in an orthographic sketch. However, these views are not always appropriate. The sketcher must learn to choose the most appropriate views. These are the views that show the most detail and the fewest hidden lines. A good rule of thumb to use in selecting views is to select the views that would give you all of the information you would need if you had to make the object yourself.

Once the views have been selected, the orthographic sketch may be laid out using the techniques set forth earlier in this chapter. To ensure that the sketched views align, the entire sketch should be blocked in before adding details (**Figure 3-51**). Once the layout is blocked in, the details can be added one view at a time (**Figures 3-52, 3-53,** and **3-54**).

Axonometric Sketching

As mentioned previously, there are three types of axonometric projection: isometric, dimetric, and

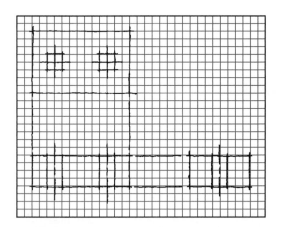

FIGURE 3-51   Blocking in an orthographic sketch.

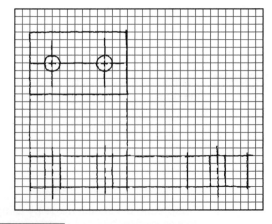

FIGURE 3-52   Completing the top view.

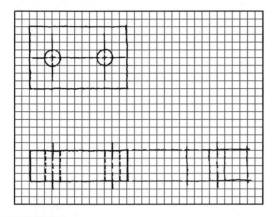

FIGURE 3-53   Completing the front view.

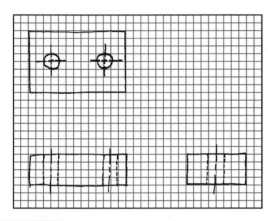

FIGURE 3-54   Completing the sketch.

trimetric. Isometric projection is used in sketching. Dimetric and trimetric projection have little application in sketching due to the difficulty in proportioning scale values of depth, width, and height. Isometric views have the same scaling value in all three directions, eliminating the need to vary proportions among the three directions. In an isometric sketch, height lines are vertical, and width and depth lines recede at approximately 30° and 150° (180°–30°) from the horizontal.

The first step in creating an isometric sketch is to lay out the isometric axis (**Figure 3-55**). All normal lines will be parallel to one of the axis lines. The next step is to block in the object by using construction lines (**Figure 3-56**). Five steps in the development of an isometric sketch are shown in **Figures 3-57, 3-58, 3-59, 3-60,** and **3-61**.

Oblique Sketching

Oblique sketching involves laying out the front view of an object and showing the depth lines receding at an angle (usually 45°) from the horizontal. Oblique sketching is particularly useful for dealing with an

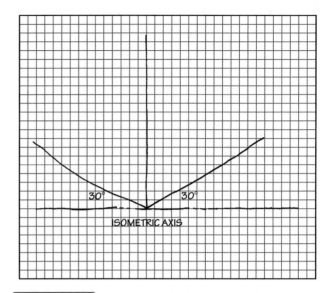

FIGURE 3-55   The isometric axis in sketching.

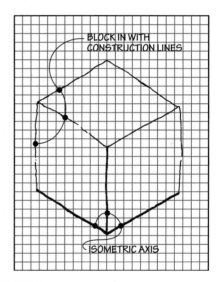

FIGURE 3-56   Blocking in the object.

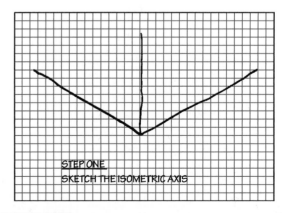

FIGURE 3-57   Step 1 in making an isometric sketch.

object having round components. Oblique sketching allows round components to be drawn round rather than elliptical.

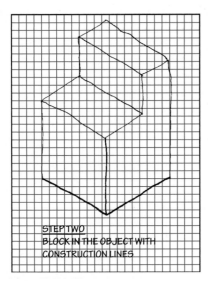

FIGURE 3-58   Step 2 in making an isometric sketch.

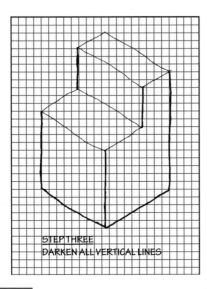

FIGURE 3-59   Step 3 in making an isometric sketch.

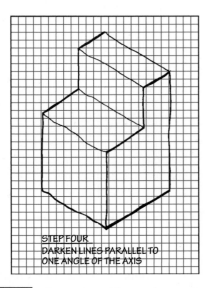

FIGURE 3-60   Step 4 in making an isometric sketch.

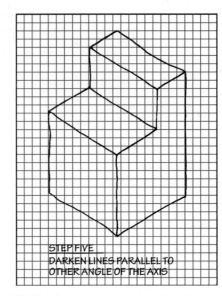

FIGURE 3-61   Step 5 in making an isometric sketch.

FIGURE 3-62   Step 1 in making an oblique sketch.

FIGURE 3-63   Step 2 in making an oblique sketch.

Using the blocking-in method, the flat front surface of the object is laid out. The depth is then blocked in using parallel lines, and the sketch is completed by outlining the exposed profile of the rear surface. **Figures 3-62, 3-63,** and **3-64** illustrate three steps in creating an oblique sketch.

Perspective Sketching

Perspective sketching closely approximates how the human eye actually sees an object. Two common

FIGURE 3-64   Step 3 in making an oblique sketch.

types of perspective sketches are one-point and two-point perspectives.

A one-point perspective sketch is similar to an oblique sketch, except that depth lines recede to a vanishing point instead of receding parallel to one another (**Figure 3-65**). In constructing a one-point perspective, the following procedures apply.

STEP 1   Lay out the flat front surface of the object by using the blocking-in method (**Figure 3-66**).

STEP 2   Select and mark a single vanishing point. Project all points on the front surface back to the vanishing point (**Figure 3-67**).

STEP 3   Estimate the depth of the object and mark it off on all line projectors (**Figure 3-68**).

STEP 4   Complete the sketch by outlining the exposed profile of the rear surface (**Figure 3-69**).

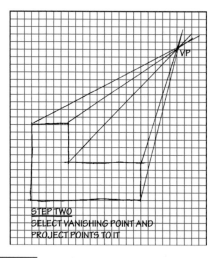

FIGURE 3-67   Step 2 in making a one-point perspective sketch.

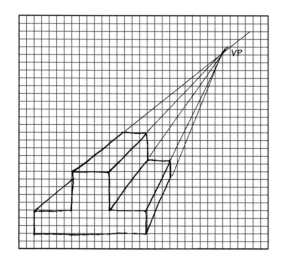

FIGURE 3-65   One-point perspective sketch.

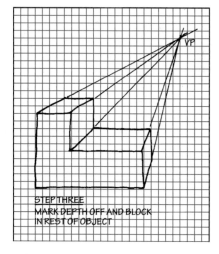

FIGURE 3-68   Step 3 in making a one-point perspective sketch.

FIGURE 3-66   Step 1 in making a one-point perspective sketch.

FIGURE 3-69   Step 4 in making a one-point perspective sketch.

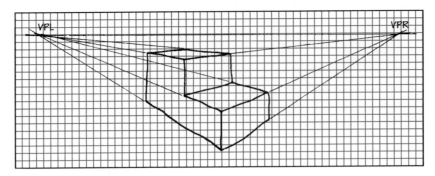

FIGURE 3-70   Two-point perspective sketch.

A two-point perspective resembles an isometric sketch, except that width and depth lines recede to the left and right vanishing points rather than receding in parallel (**Figure 3-70**).

In constructing a two-point perspective, the following procedures apply.

STEP 1   Lay out the two-point perspective frame, which consists of the vertical height line, the vanishing point left, the vanishing point right, and the receding lines (all estimated locations) (**Figure 3-71**). The horizon line

when positioned below the view provides a view of the bottom of the object, and when positioned above it shows the top. The vanishing points must be on the horizon.

STEP 2   Block in the object, estimating the length and width for proportion (**Figure 3-72**).

STEP 3   Lay out the details, lightly giving special attention to proportion (**Figure 3-73**).

STEP 4   Complete the two-point perspective sketch (**Figure 3-74**).

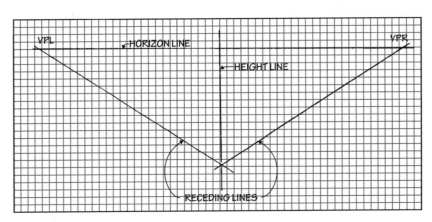

FIGURE 3-71   Laying out the two-point perspective frame.

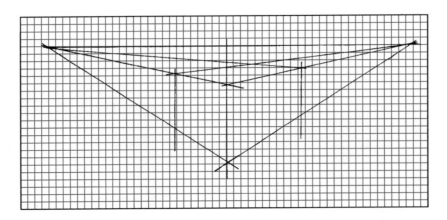

FIGURE 3-72   Blocking in the object.

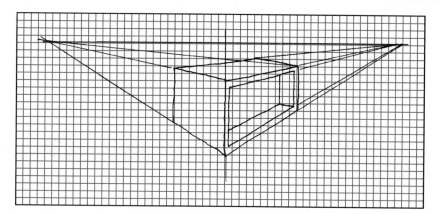

FIGURE 3-73   Laying out the details.

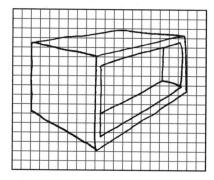

FIGURE 3-74   Completing the two-point perspective sketch.

SUMMARY

- Talking sketching is of two types: conversational and boardroom sketching. Conversational sketches are developed as two or three individuals discuss an idea. Boardroom sketches are developed on a marker board, chalkboard, or flip chart while a group discusses an idea.

- Freehand lettering may be done in several styles, including Block, Fast, Futura, Leroy, Old English, Times, and Helvetica. Single-stroke Gothic lettering may be done in upper- or lowercase, extended or condensed, and vertical or inclined styles. The characteristics of good lettering are neatness, uniformity, stability, proper spacing, and speed.

- Students should be able to make and demonstrate the proper use of the following types of lines: visible, hidden, crosshatching, center, dimension, extension, leader, cutting plane, freehand break, mechanical break, phantom, and stitch.

- Students should be able to illustrate the following types of sketches: orthographic, axonometric, oblique, and perspective.

- The materials needed for making sketches are pencils, paper or graph paper, and an eraser.

- Sketching techniques with which students should be proficient are making straight lines, curves, circles, and ellipses. Students should also be proficient in maintaining proper proportion when sketching.

REVIEW QUESTIONS

Mark the following statements as either true or false.

1. The standard style of freehand lettering on technical drawings is vertical, single-stroke Arial.

2. The correct slope of slanted lettering is a two-unit incline to the right for each five units of letter height.

3. Three strokes are required to make the letter B.

4. In regard to space left between words, it is a rule of thumb to leave the width of one round letter such as Q, C, or O.

5. Uniformity, neatness, and stability can all be achieved without using guidelines.

6. A setting of 8 on a lettering guide will produce letters 1/4″ high.

7. *Line work* is a term used to describe one specific technique used to create graphic data on technical drawings.

Answer the following questions by selecting the best answer.

1. Which of the following is **not** a characteristic of good freehand lettering?
   a. Neatness
   b. Speed

PROBLEMS

c. Uniformity

d. Incline

2. Which of the following is **not** one of the basic line types used on technical drawings?
   a. Angular, perpendicular, nonhorizontal, nonvertical
   b. Visible, hidden, crosshatch
   c. Cutting plane, freehand break, mechanical
   d. Center, dimension, extension, leader

3. Which of the following is an advantage of sketching compared to mechanical drawing?
   a. It requires few material aids.
   b. It can be mastered without much practice.
   c. It can be done with just a pencil and a piece of paper.
   d. Both a and c.

4. Which axonometric projection is preferred for sketching?
   a. Isometric
   b. Dimetric
   c. Trimetric
   d. Quadmetric

5. Which kind of projection exhibits a circle as an elliptic shape?
   a. Mercator
   b. Isometric
   c. Multiview
   d. Auxiliary

6. How many principal views are there in an orthographic projection?
   a. Two
   b. Four
   c. Six
   d. Eight

7. Guidelines are a critical part of freehand lettering because:
   a. They improve uniformity.
   b. They improve neatness.
   c. They improve stability.
   d. All of the above.

8. What is the customary height of characters in a technical drawing?
   a. 1/16″ for regular text and 3/16″ for headings and titles
   b. 1/8″ for regular text and 3/16″ for headings and titles
   c. 1/8″ for regular text and 1/2″ for headings and titles
   d. 1/4″ for regular text and 1/2″ for headings and titles

9. Which of the following is a type of axonometric sketch?
   a. Perspective
   b. Orthographic
   c. Oblique
   d. Trimetric

CHAPTER 3 PROBLEMS

Problems 3-1 through 3-9 are intended to give beginning drafters experiences in visualizing that objects are made of convenient shapes and features. Visualizing these shapes and features will help you to completely specify an object's dimensions. Another objective of these nine problems is to encourage you to use approximations, when appropriate, to obtain ballpark answers before diving into detailed calculations.

Problem 3-22 will be used here as an example to illustrate the requirements for the nine problems. Assume that you have been asked to calculate the weight of the object in Problem 3-22, which is made of aluminum with a density of 0.097 pounds per cubic inch. To make a ballpark calculation, consider the base to be $1″ \times 2″ \times 4″$ or a volume of approximately 8 cubic inches and the vertical portion $1/2″ \times 1\frac{1}{2}″ \times 4″$ or 3 cubic inches, for a total of $8 + 3 = 11$ cubic inches, which multiplied by the density equals approximately 1.0 pound. The correct volume is calculated by deducting the semicircular 1″ radius hole and deducting some volume for the rounded corners of 1/2″ radius. The correct volume is found to be 9.84 cubic inches. Multiply by the density to obtain 0.954 pounds for the correct weight.

Unless otherwise specified, use the format shown in Problem 4-37 to present your solutions for the following problems.

PROBLEM 3-1
Assume that the object in Problem 3-17 is aluminum. Calculate its weight.

PROBLEM 3-2
Assume that the object in Problem 3-18 is aluminum. Calculate its weight.

PROBLEM 3-3

Assume that the object in Problem 3-21 is aluminum. Calculate its weight.

PROBLEM 3-4

Assume that the object in Problem 3-23 is steel with a density of 0.28 pounds per cubic inch. Calculate its weight.

PROBLEM 3-5

Assume that the object in Problem 3-28 is brass with a density of 0.32 pounds per cubic inch. Calculate its weight.

PROBLEM 3-6

Assume that the object in Problem 3-29 is brass with a density of 0.32 pounds per cubic inch. Calculate its weight.

PROBLEM 3-7

Assume that the object in Problem 3-31 is steel with a density of 0.28 pounds per cubic inch. Calculate its weight.

PROBLEM 3-8

Assume that the object in Problem 3-33 is steel with a density of 7,765 kilograms per cubic meter. Calculate its weight in kilograms. Note that 1 kilogram equals 2.2 pounds. Also, the squares in Problem 3-33 are given as 10 mm or .010 m.

PROBLEM 3-9

Assume that the object in Problem 3-25 is cast iron with a density of 6,933 kilograms per cubic meter. Calculate its weight in kilograms.

Problems 3-10 through 3-13

*The following problems are intended to give the beginning drafter practice in making neat uppercase, vertical, and Gothic-style letters and numbers. On an A-size sheet of vellum, carefully lay out light .12 spaces as shown. Use .50 borders all around and **be sure to skip a space** between lines of letters or numbers. Practice letters, numbers, words, sentences, and paragraphs, as given.*

PROBLEM 3-10

.12
(3)

A
B
C
D
.09
(2)
E
F
G
H
I
J
K
L
M
N
O
P
Q
R
S
T
U
V
W
X
Y
Z

1
2
3
4
5
6
7
8
9
0

PROBLEM 3-11

.12
(3)

YOUR FIRST NAME

YOUR LAST NAME

.09
(2)
DATE (EXAMPLE, 8 AUG 88)

YOUR SCHOOL/COLLEGE NAME

YOUR TOWN

YOUR STATE

YOUR MAIL ZIP CODE

YOUR TELEPHONE NUMBER

PROBLEM 3-12

.12
(3)

GOOD LETTERING TAKES PRACTICE

.09
(2)

IMPROVEMENT IN LETTERING REQUIRES EFFORT

PATIENCE AND HARD WORK EQUALS GOOD LETTERING

SINGLE-STROKE, UPPERCASE, VERTICAL GOTHIC
LETTERING IS USED IN THE MECHANICAL DRAFTING FIELD

PROBLEM 3-13

LETTERING IS EXTREMELY IMPORTANT. IT MUST BE NEAT,
LEGIBLE, CLEAR, AND IN PROPORTION AND MUST BE DONE
IN A REASONABLE AMOUNT OF TIME

.12
(3)

.09
(2)

Problems 3-14 through 3-33

*These problems are intended to give the beginning drafter
practice in sketching. On an A-size sheet of vellum, neatly
sketch the top, front, and right-side views of each pictorial
figure shown. The grid is provided for proportion only. Try
to keep the object in scale as you sketch; keep all parallel
lines parallel, and sketch as neatly as possible. Do not use
any drawing instruments. Add all lines, including hidden
lines and center lines.*

*Next, sketch each object from your three-view drawings
into isometric pictorial views.*

PROBLEM 3-14

.50 (10) SQUARES

PROBLEM 3-15

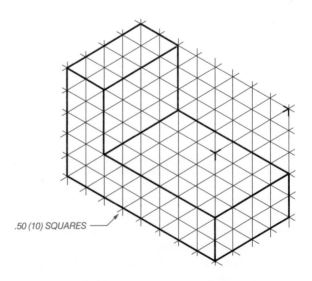

.50 (10) SQUARES

PROBLEM 3-16

.50 (10) SQUARES

PROBLEM 3-19

.50 (10) SQUARES

PROBLEM 3-17

.50 (10) SQUARES

PROBLEM 3-20

.50 (10) SQUARES

PROBLEM 3-18

.50 (10) SQUARES

PROBLEM 3-21

.50 (10) SQUARES

PROBLEM 3-22

.50 (10) SQUARES

PROBLEM 3-23

.50 (10) SQUARES

PROBLEM 3-24

.50 (10) SQUARES

PROBLEM 3-25

.50 (10) SQUARES

PROBLEM 3-26

.50 (10) SQUARES

PROBLEM 3-27

.50 (10) SQUARES

PROBLEM 3-28

.50 (10) SQUARES

PROBLEM 3-31

.50 (10) SQUARES

PROBLEM 3-29

.50 (10) SQUARES

PROBLEM 3-32

.50 (10) SQUARES

PROBLEM 3-30

.50 (10) SQUARES

PROBLEM 3-33

.50 (10) SQUARES

Problems 3-34 through 3-37

These few problems are intended to provide future drafters, designers, and engineers with experiences in developing an appreciation for sketching from memory, for recording preliminary design ideas, and for orally communicating technical ideas while sketching.

PROBLEM 3-34

Visit a facility and mentally (no notes or rough sketches) study an object, such as a machine component, an object or a tool in a shop, a part of an automobile, or a piece of sports equipment with a goal in mind that upon returning to class you will sketch the object(s) from memory. Then, prepare orthographic or isometric-type sketches.

*The skill emphasized here is that if you observe and study an object as though you were required to sketch it later, you will **see more** than if you just casually look at it. Develop this skill of collecting technical ideas for designing.*

PROBLEM 3-35

Sketch orthographic views and an isometric of a representative object to be used as your nameplate for a desk to match your career interests. For example: a bridge could represent a career in construction or civil engineering; a pressurized tank, for welding; an airplane, for flying; a computer, for computer science; and so on.

PROBLEM 3-36

Sketch orthographic views of your design of one of the following.

a. a modern chair
b. a storage cabinet for:
   1. wardrobe
   2. drafting tools
   3. typical home-use garden tools
c. a basic outline of a sports car
d. a basic outline of an electric car
e. a cabinet for trophies
f. a two-person skateboard
g. an outdoor logo for your school
h. a toolbox for:
   1. car maintenance tools
   2. woodworking tools
   3. hobbies (fishing, models, etc.)

PROBLEM 3-37

Visit one of the locations listed and select a device, not too complicated, to sketch. Make a rough sketch (for your use only) for later use in a drafting class. Be prepared to sketch the device while describing its attributes (shape, size, function, material, aesthetic properties, etc.) to either another student or at the board to the class.

a. metal shop
b. wood shop
c. electronics shop
d. hardware store
e. store
f. garage
g. kitchen
h. computer store

4

Geometric Construction

KEY TERMS

Acute angle	Hypocycloid
Altitude	Involute
Angle	Isosceles triangle
Bisect	Lead
Central angle	Line
Chord	Obtuse angle
Circle	Obtuse angle triangle
Circumference	Ogee curve
Concentric circle	Parallelogram
Cone	Point
Conic section	Pole
Cylindrical helix	Polygon
Diameter	Polyhedron
Eccentric circle	Prism
Epicycloid	Pyramid
Equilateral triangle	Quadrant
Helix	Quadrilateral
Hemisphere	Radius

(continued)

CHAPTER OUTLINE

Geometric nomenclature • Prism • Pyramid and cone • Sphere • Elemental construction principles • Polygon construction • Circular construction • Supplementary construction • Summary • Review questions • Chapter 4 problems • CAD instructions

CHAPTER OBJECTIVES

Upon completion of this chapter, students should be able to do the following:

▶ Define the most frequently used terms in geometric nomenclature.

▶ Properly apply the elemental principles of geometric construction.

▶ Demonstrate the proper procedures for polygon construction.

▶ Demonstrate the proper procedures for circular construction.

▶ Demonstrate the proper procedures for supplementary construction, including the following: spiral; helix; involute of a line, triangle, square, and circle; and cycloidal curve.

To be truly proficient in the layout of both simple and complex drawings, the drafter must know and fully understand the many geometric construction methods used. These methods are illustrated in this chapter and are basically simple principles of pure geometry. These simple principles are used to develop a drawing with complete accuracy, and in the fastest time possible, without wasted motion or any guesswork. Applying these geometric construction principles gives drawings a finished, professional appearance.

In laying out the various geometric constructions, it is important to use a very sharp, 4H lead and to be extremely accurate at all times. Always draw light construction lines that can hardly be seen when held at arm's length. These light construction lines should not be erased, as this takes up valuable drawing time, and they can also be reused to check layout work if necessary.

Geometric Nomenclature

Points in Space

A *point* is an exact location in space or on a drawing surface (**Figure 4-1**). A point is actually represented on the drawing by a crisscross at its exact location. The exact point in (drawing) space is where the two lines of the crisscross intersect. When a point is located on an existing line, a light, short, dashed line or crossbar is placed on the line at the location of the exact point. Never represent a point on a drawing by a dot, except for sketching locations. This is not accurate enough, and it is considered to be poor drafting practice.

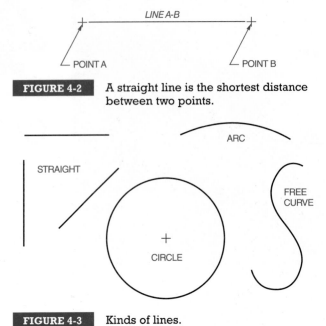

FIGURE 4-2   A straight line is the shortest distance between two points.

FIGURE 4-3   Kinds of lines.

Line

A straight line is the shortest distance between two points (**Figure 4-2**). It can be drawn in any direction. A *line* can be straight, an arc, a circle, or a free curve, as illustrated in **Figure 4-3**. If a line is indefinite and the ends are not fixed in length, the actual length is a matter of convenience. If the end-points of a line are important, they must be marked by means of small, mechanically drawn crossbars, as described by a point in space.

Straight lines and curved lines are considered to be parallel if the shortest distance between them remains constant. The symbol used for parallel lines is //. Lines that are tangent and at 90° are considered perpendicular. The symbol for perpendicular lines is ⊥ (singular) (**Figure 4-4**), and ⊥'s (plural). The symbol for an angle is ∠ (singular) and ∠'s (plural). To draw an angle, use the drafting machine, a triangle, or a protractor. For extra accuracy, use the vernier on the drafting machine or a vernier protractor.

Angle

An *angle* is formed by the intersection of two lines. There are three major kinds of angles: right angles, acute angles, and obtuse angles (**Figure 4-5**). The

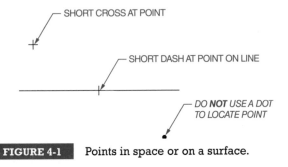

FIGURE 4-1   Points in space or on a surface.

FIGURE 4-4   Perpendicular lines.

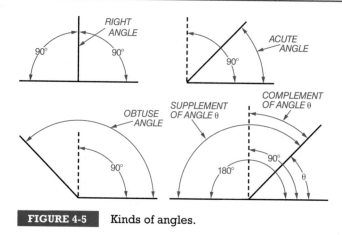

FIGURE 4-5   Kinds of angles.

right angle is an angle of 90°. An *acute angle* is an angle at less than 90°, and an *obtuse angle* is an angle at more than 90°. Note that a straight line is 180°. Figure 4-5 also illustrates the complementary and supplementary angles of a given angle.

There are 360 degrees (360°) in a full circle. Each degree is divided into 60 minutes (60′). Each minute is divided into 60 seconds (60″). Example: 48°28′38″ is read as 48 degrees, 28 minutes, and 38 seconds. To convert minutes and seconds to decimal degrees, divide minutes by 60 and seconds by 3,600.

EXAMPLE:

$21°18′27″ = 21° + (18.60)° + (27.3600)° = 21.3075°$
To convert decimal degrees to degrees, minutes, and seconds, multiply the degree decimal by 60 to obtain minutes and the minute decimal by 60 to obtain seconds.

EXAMPLE:

$77.365° = 77° + (.365 × 60)′ = 77°21′.9′$
$77°21′.9′ = 77°21′ + (.9 × 60)″ = 77°21′54″$
A vernier may be used to measure and read off minutes and seconds of a degree. The vernier scale is discussed in Chapter 2.

Triangle

A *triangle* is a closed plane figure with three straight sides and three interior angles. The sum of the three internal angles is always exactly 180°, half of the 360° in a full circle. **Figure 4-6** shows the various kinds of triangles: a right triangle, an equilateral triangle, an isosceles triangle, a scalene triangle, and an obtuse angle triangle.

A *right triangle* is a triangle having a right angle or an angle of 90°. The two sides forming the right angle are called legs, and the third side (the longest) is the hypotenuse. Any triangle inscribed in a semicircle is a right triangle (**Figure 4-7**).

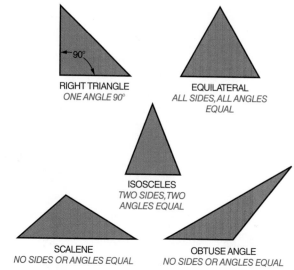

FIGURE 4-6   Kinds of triangles.

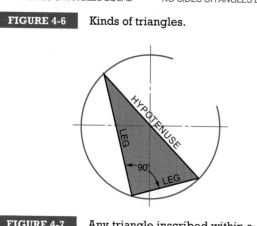

FIGURE 4-7   Any triangle inscribed within a semicircle is a right triangle.

An *equilateral triangle,* as its name implies, is a triangle with all sides of equal length. All of its interior angles are also equal. An *isosceles triangle* has two sides of equal length and two equal interior angles. A *scalene triangle* has no equal sides or angles. An *obtuse angle triangle* is a triangle having an obtuse angle greater than 90°, with no equal sides.

Polygon

A *polygon* is a closed plane figure with three or more straight sides (**Figure 4-8**). More specifically, shown in the figure are regular polygons, meaning that all sides are equal in each of these examples. The most important of these polygons as they relate to drafting are probably the triangle with three sides, the square with four sides, the hexagon with six sides, and the octagon with eight sides.

Quadrilateral

A *quadrilateral* is a plane figure bounded by four straight sides. When opposite sides are parallel, the quadrilateral is also considered to be a *parallelogram* (**Figure 4-9**).

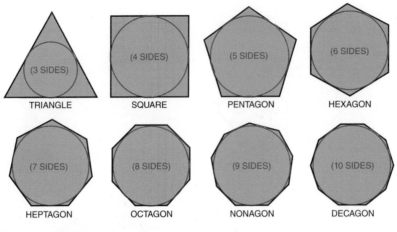

FIGURE 4-8   Examples of polygons.

FIGURE 4-9   Quadrilaterals.

Circle

A *circle* is a closed curve with all points on the circle at the same distance from the center point. The major components of a circle are the diameter, the radius, and the circumference (**Figure 4-10**).

The *diameter* of a circle is the straight distance from one outside curved surface through the center point to the opposite outside curved surface. The diameter of a circle is twice the size of the radius.

The *radius* of a circle is the distance from the center point to the outside curved surface. The radius is half the diameter and is used to set the compass when drawing a diameter.

The *circumference* of a circle is the distance around the outer surface of the circle. To calculate the circumference of a circle, multiply the value of π (use the approximation 3.1416) by the diameter. A chart similar to the one found in Appendix A may be used.

Other important parts of a circle are the central angle, the sector, the quadrant, the chord, and the segment (**Figure 4-11**).

FIGURE 4-11   Other parts of a circle.

FIGURE 4-10   The major components of a circle.

FIGURE 4-12   Concentric and eccentric circles.

TETRAHEDRON   HEXAHEDRON   OCTAHEDRON   DODECAHEDRON   ICOSAHEDRON

REGULAR SOLIDS

FIGURE 4-13   Polyhedrons (solids).

A *central angle* is an angle formed by two radial lines from the center of the circle.

A *sector* is the area of a circle lying between two radial lines and the circumference.

A *quadrant* is a sector with a central angle of 90° and usually with one of the radial lines oriented horizontally.

A *chord* is any straight line whose opposite ends terminate on the circumference of the circle. (A diameter is a chord passing through the center of the circle.)

A *segment* is the smaller portion of a circle separated by a chord.

Concentric circles are two or more circles with a common center point (**Figure 4-12**).

Eccentric circles are two or more circles without a common center point, Figure 4-12.

A *semicircle* is half of a circle.

Polyhedron

A *polyhedron* is a solid object bounded by plane surfaces. Each surface is called a face. If the faces are equal, regular polygons, the solid figure is a regular polyhedron (**Figure 4-13**).

Prism

A *prism* is a solid having ends that are parallel, matched polygons and sides that are parallelograms (**Figure 4-14**). This definition also applies to round or circular objects, such as a cylinder. When the

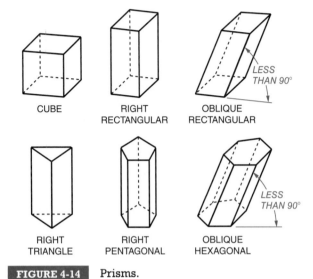

CUBE   RIGHT   OBLIQUE
            RECTANGULAR   RECTANGULAR

RIGHT   RIGHT   OBLIQUE
TRIANGLE   PENTAGONAL   HEXAGONAL

FIGURE 4-14   Prisms.

polygon on one end of a prism is not parallel to the other end, it is said to be *truncated*. The *altitude* of a prism is the perpendicular distance between its end polygons (or bases).

Pyramid and Cone

A *pyramid* is a polyhedron having a polygon as its base. Three or more triangles form its lateral sides, which meet at a common vertex (**Figure 4-15**). A *cone* is a pyramid with a central axis and an infinite number of sides that form a continuous curved lateral surface. When the vertex of a pyramid or cone has

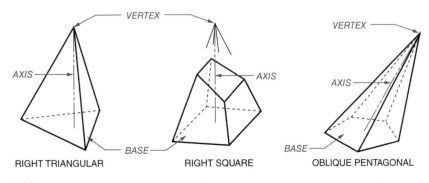

RIGHT TRIANGULAR   RIGHT SQUARE   OBLIQUE PENTAGONAL

FIGURE 4-15   Pyramids.

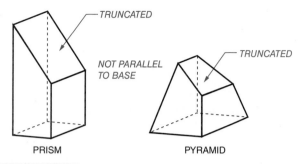

FIGURE 4-16   Truncated prism and pyramid.

FIGURE 4-17   Sphere.

been removed by a plane that intersects all the lateral sides (which forms a new polygon), the pyramid or cone is said to be truncated (**Figure 4-16**).

Sphere

A *sphere* is a closed surface, every point on which is equidistant from a common point or center (**Figure 4-17**). If a sphere is cut into two equal parts, the parts are called *hemispheres*. *Poles* are two reference measuring positions on the surface of the sphere on opposite sides of its center.

Elemental Construction Principles

The remaining portion of this chapter is devoted to illustrating step by step the many geometric construction principles used by the drafter to develop various geometric forms. It is important that each step be fully understood and followed. As the beginning drafter uses these geometric construction principles, various shortcuts will become evident, thus reducing the drawing time and increasing accuracy even more. At the end of the chapter, each of these techniques is incorporated or used in some way to complete the problems.

How to Bisect a Line

To *bisect* a line means to divide it in half or to find its midpoint. In the given process, a line will also be constructed at the exact center point of exactly 90°.

GIVEN: Line A-B (**Figure 4-18A**).

STEP 1   Set the compass approximately two-thirds of the length of line A-B, and swing an arc from point A (**Figure 4-18B**).

STEP 2   Using the same compass setting, swing an arc from point B (**Figure 4-18C**).

STEP 3   At the two intersections of these arcs, locate points X and Y (**Figure 4-18D**).

STEP 4   Draw a straight line connecting point X with point Y. Where this line intersects line A-B is the bisect of line A-B (**Figure 4-18E**). Line X-Y is also perpendicular to line A-B at the center point.

How to Bisect an Angle

To bisect an angle means to divide it in half or to cut it into two equal angles.

GIVEN: Angle ABC (**Figure 4-19A**).

STEP 1   Set the compass at any convenient radius, and swing an arc from point B (**Figure 4-19B**).

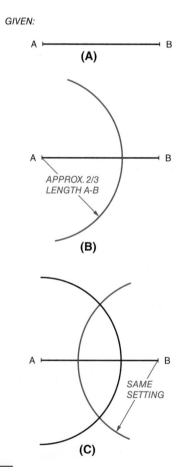

FIGURE 4-18   (A) How to bisect a line; (B) Step 1; (C) Step 2;

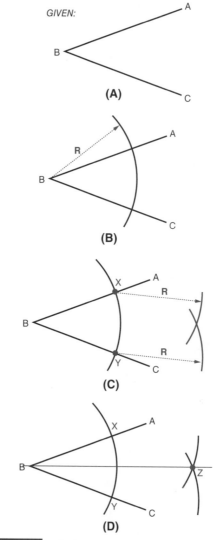

FIGURE 4-18   (*Continued*) (D) Step 3; (E) Step 4.

STEP 2   Locate points X and Y on the legs of the angle, and swing two arcs of the same length from points X and Y, respectively (**Figure 4-19C**).

STEP 3   Where these arcs intersect, locate point Z. Draw a straight line from B to Z. This line will bisect angle ABC and establish two equal angles: ABZ and ZBC (**Figure 4-19D**).

FIGURE 4-19   (A) How to bisect an angle; (B) Step 1; (C) Step 2; (D) Step 3.

HOW TO DRAW AN ARC OR CIRCLE (RADIUS) THROUGH THREE GIVEN POINTS

GIVEN: Three points in space at random: A, B, and C (**Figure 4-20A**).

STEP 1   With straight lines, lightly connect points A to B and B to C (**Figure 4-20B**).

STEP 2   Using the method outlined for bisecting a line, bisect lines A-B and B-C (**Figure 4-20C**).

STEP 3   Locate point X where the two extended bisectors meet. Point X is the exact center of the arc or circle (**Figure 4-20D**).

STEP 4   Place the point of the compass on point X; adjust the lead to any of the points A, B, or C (they are the same distance); and swing the circle. If all work is done correctly, the arc or circle should pass through each point, as shown in (**Figure 4-20E**).

HOW TO DRAW A LINE PARALLEL TO A STRAIGHT LINE AT A GIVEN DISTANCE

GIVEN: Line A-B and a required distance to the parallel line (**Figure 4-21A**).

STEP 1   Set the compass at the required distance to the parallel line. Place the point of the compass at any location on the given line, and swing a light arc whose radius is the required distance (**Figure 4-21B**).

STEP 2   Adjust the straightedge of either a drafting machine or an adjustable triangle so that it lines up with line A-B and slide the straight edge up or down to the extreme high point, which is the tangent point of the, arc; then, draw the parallel line (**Figure 4-21C**).

NOTE: The distance between parallel lines is measured on any line that is perpendicular to both.

GIVEN:

(A)

(B)

(C)

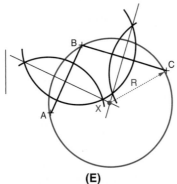

INTERSECTION POINT
(CENTER OF ARC OR CIRCLE)

(D)

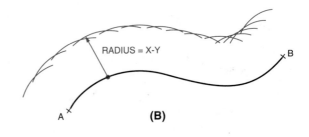

(E)

FIGURE 4-20   (A) How to draw an arc or circle (radius)
through three given points; (B) Step 1;
(C) Step 2; (D) Step 3; (E) Step 4.

HOW TO DRAW A LINE PARALLEL TO A CURVED LINE AT A GIVEN DISTANCE

GIVEN: Curved line A-B and a required distance to
the parallel line (**Figure 4-22A**).

STEP 1   Set the compass at the required distance to the
parallel line. Starting from either end of the

GIVEN:

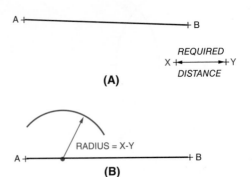

(A)

RADIUS = X-Y

(B)

HIGH POINT OF ARC

PARALLEL LINE

(C)

FIGURE 4-21   (A) How to draw a line parallel to
a straight line at a given distance;
(B) Step 1; (C) Step 2.

GIVEN:

REQUIRED
X ├──►├ Y
DISTANCE

(A)

RADIUS = X-Y

(B)

HIGH POINTS OF ARC

PARALLEL LINES

(C)

FIGURE 4-22   (A) How to draw a line parallel to a
curved line at a given distance;
(B) Step 1; (C) Step 2.

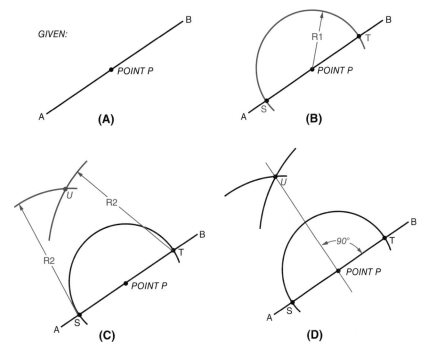

FIGURE 4-23   (A) How to draw a perpendicular to a line from a point on the line (Method 1); (B) Step 1; (C) Step 2; (D) Step 3.

curved line, place the point of the compass on the given line, and swing a series of light arcs along the given line (**Figure 4-22B**).

STEP 2   Using an irregular curve, draw a line along the extreme high points of the arcs (**Figure 4-22C**).

HOW TO DRAW A PERPENDICULAR TO A LINE FROM A POINT ON THE LINE (METHOD 1)

GIVEN: Line A-B with point P on the same line (**Figure 4-23A**).

STEP 1   Using P as the center, make two arcs of equal radius or one continuous arc (R1) to intercept line A-B on either side of point P at points S and T (**Figure 4-23B**).

STEP 2   Swing larger but equal arcs (R2) from each of points S and T to cross each other at point U (**Figure 4-23C**).

STEP 3   A line from U to P is perpendicular to line A-B at point P, (**Figure 4-23D**).

HOW TO DRAW A PERPENDICULAR TO A LINE FROM A POINT ON THE LINE (METHOD 2)

GIVEN: Line A-B with point P on the line (**Figure 4-24A**).

STEP 1   Swing an arc of any convenient radius whose center O is at any convenient location not on

line A-B but positioned to make the arc cross line A-B at points P and Q (**Figure 4-24B**).

STEP 2   A line from point Q through center O intercepts the opposite side of the arc at point R (**Figure 4-24C**).

STEP 3   Line R-P is perpendicular to line A-B. (A right triangle has been inscribed in a semicircle.) See **Figure 4-24D**.

HOW TO DRAW A PERPENDICULAR TO A LINE FROM A POINT NOT ON THE LINE

GIVEN: Line A-B and point P (**Figure 4-25A**).

STEP 1   Using P as a center, swing an arc (R1) to intercept line A-B at points G and H (**Figure 4-25B**).

STEP 2   Swing larger-but-equal-length arcs (R2) from each of the points G and H to intercept each other at point J (**Figure 4-25C**).

STEP 3   Line P-J is perpendicular to line A-B (**Figure 4-25D**).

HOW TO DIVIDE A LINE INTO EQUAL PARTS

GIVEN: Line A-B (**Figure 4-26A**).

STEP 1   Draw a line 90° from either end of the given line. A 90° line from point B is illustrated in **Figure 4-26B**, but either end or either direction will work as well.

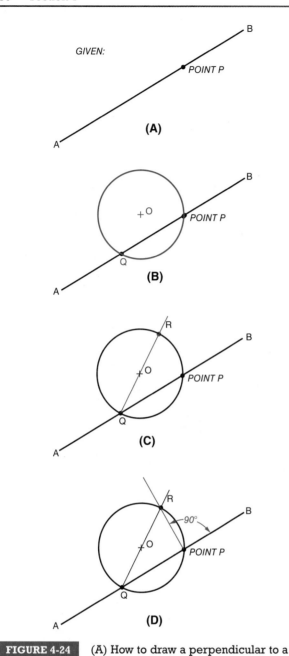

FIGURE 4-24   (A) How to draw a perpendicular to a
line at a point (Method 2); (B) Step 1;
(C) Step 2; (D) Step 3.

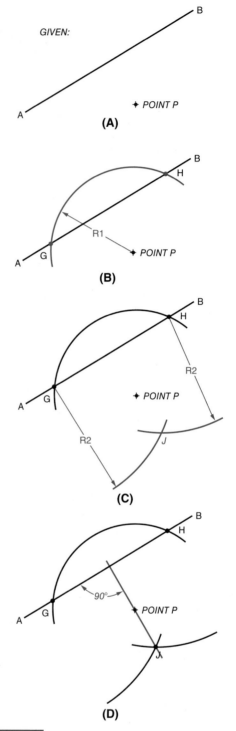

FIGURE 4-25   (A) How to draw a perpendicular to a
line from a point not on the line;
(B) Step 1; (C) Step 2; (D) Step 3.

STEP 2   Place a scale with its zero at point A of the
given line. If three equal parts are required,
pivot the scale until the 3″ measurement (or
any length representing three equal units of
measure: 30, 60, and 90 mm, for example)
is on the perpendicular line drawn in Step 1.
Place a short dash at these points. The example
in **Figure 4-26C** shows these dashes at the 1″
and 2″ marks.

STEP 3   Project lines downward from these points,
and add short hash marks where these projected lines cross the original given line A-B.

This divides line A-B into three equal parts.
Check all work, comparing your final step
with **Figure 4-26D**. An example of equal
spacing, where equally spaced holes are
required with a given length, is illustrated in
Figure 4-26C.

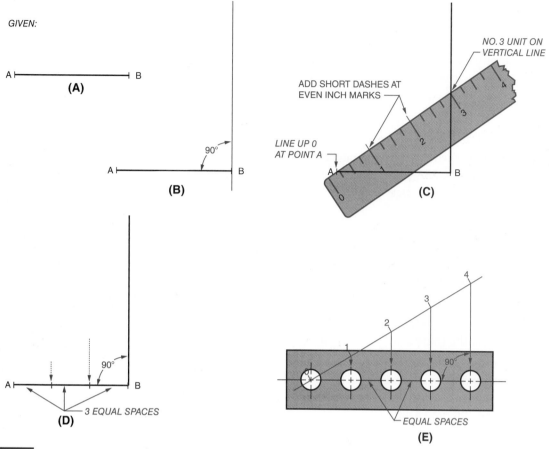

(A) How to divide a line into equal parts; (B) Step 1; (C) Step 2; (D) Step 3; (E) An example of equal spacing within a given length.

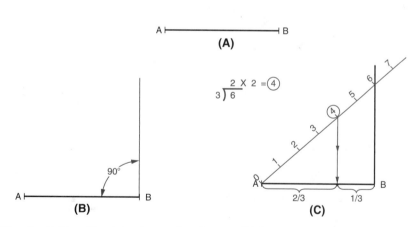

(A) How to divide a line into proportional parts; (B) Step 1; (C) Step 2.

HOW TO DIVIDE A LINE INTO PROPORTIONAL PARTS

GIVEN: Line A-B (**Figure 4-27A**). *Problem:* Locate point X at two-thirds of the distance from point A to point B.

STEP 1   Draw a line 90° from either end of the line. A 90° line from point B is illustrated in **Figure 4-27B**.

STEP 2   Place a scale with its zero on point A of the given line. Because a two-thirds proportion is required, pivot the scale until any multiple of three units of measure intersects the perpendicular line drawn in Step 1. In this example, 6 is used, representing three two-unit increments. The two-thirds position of this length is two two-unit increments, or the 4 position, where a hash mark is made, as shown in **Figure 4-27C**. Projecting this point downward to line A-B, it becomes point X, which is two-thirds the overall distance from point A.

HOW TO TRANSFER AN ANGLE

GIVEN: An angle formed by two straight lines, 0A-0B (**Figure 4-28A**), and one location of where the transferred angle begins (point 0′) (**Figure 4-28B**).

STEP 1   Refer to **Figure 4-28C**. Draw an arc through both legs of a given angle (R1) and then duplicate this radius at the transferred angle location (**Figure 4-28D**).

STEP 2   Transfer the chord length between the two angle legs at the intersection of the arc (R2) to the arc at the transfer angle location.

STEP 3   A line from the arc center to the intersection of the arc and chord length forms the second line, forming an angle equal to the original (**Figures 4-28E** and **4-28F**).

HOW TO TRANSFER AN ODD SHAPE

GIVEN: Triangle ABC (**Figure 4-29A**).

STEP 1   Letter or number the various corners and point locations of the odd shape in counterclockwise order around its perimeter. In this example, place the compass point at point A of the original shape and extend the lead to point B. Refer back to **Figure 4-29A**. Swing a light arc at the new desired location (**Figure 4-29B**). Letter the center point as A′, and add letter B′ at any convenient location on the arc. It is a good habit to lightly letter each point as you proceed.

STEP 2   Place the compass point at letter B of the original shape (**Figure 4-29C**), and extend the compass lead to letter C of the original shape.

STEP 3   Transfer this distance, B-C, to the layout (**Figure 4-29D**).

STEP 4   Going back to the original object, place the compass point at letter A (**Figure 4-29E**), and extend the compass lead to letter C.

STEP 5   Transfer the distance A-C, as illustrated in **Figure 4-29F**. Locate and letter each point.

STEP 6   Connect points A′, B′, and C′ with light, straight lines. This completes the transfer of

FIGURE 4-28   (A) How to transfer an angle; (B) Given: Point 0′; (C) Step 1; (D) Transferred angle; (E) Step 2; (F) Step 3.

the object (**Figure 4-29G**). Recheck all work and, if correct, darken lines to the correct line weight.

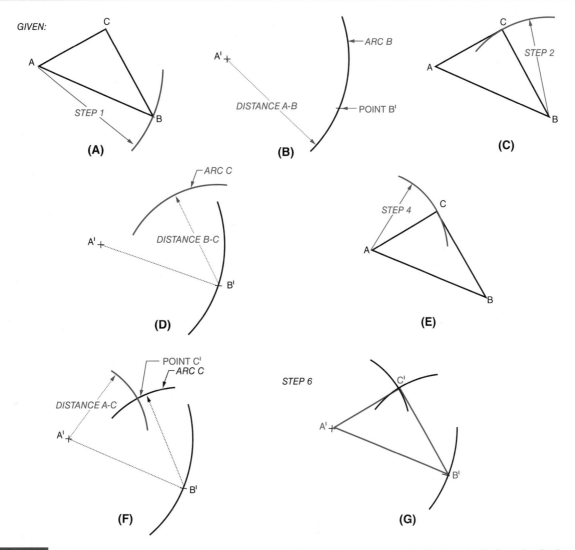

FIGURE 4-29   (A) How to transfer an odd shape; (B) Step 1; (C) Step 2; (D) Step 3; (E) Step 4; (F) Step 5; (G) Step 6.

HOW TO TRANSFER COMPLEX SHAPES

A complex shape can be transferred in exactly the same way by reducing the shape into simple triangles and transferring each triangle by using the aforementioned method.

GIVEN: An odd shape, A, B, C, D, E, F, G (**Figure 4-30A**). Letter or number the various corners and point locations of the odd shape in clockwise order around the perimeter, Figure 4-30A. Use the longest line or any convenient line as a starting point. Line A-B is chosen here as the example.

STEP 1 Lightly divide the shape into triangle divisions, using the baseline if possible. Transfer each triangle in the manner described in Figures 4-29A through 4-29G. Suggested triangles to be used in example Figure 4-30A are ABC, ABD, ABE, ABF, and ABG (**Figure 4-30B**).

STEP 2 This completes the transfer. Check all work and, if correct, darken in lines to correct line thickness. See (**Figure 4-30C**).

HOW TO PROPORTIONATELY ENLARGE OR REDUCE A SHAPE

GIVEN: A rectangle (**Figure 4-31A**). *Problem*: To enlarge or reduce its size proportionately.

STEP 1 Draw a line from corner to corner diagonally and extend it, as shown in **Figure 4-31B**.

STEP 2 The rectangle is enlarged to any size proportionately if the vertical and horizontal sides are located from the extended diagonal line (**Figure 4-31C**).

STEP 3 The rectangle is reduced proportionately if the vertical and horizontal lines are located on the unextended diagonal line (**Figure 4-31D**).

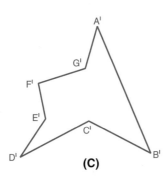

FIGURE 4-30   (A) How to transfer complex shapes; (B) Step 1; (C) Step 2.

Polygon Construction

HOW TO DRAW A TRIANGLE WITH KNOWN LENGTHS OF SIDES

GIVEN: Lengths 1, 2, and 3 (**Figure 4-32A**).

STEP 1 Draw the longest-length line—in this example, length 3—with end-points A and B (**Figure 4-32B**). Swing an arc (R1) from point A, whose radius is either length 1 or length 2; in this example, length 1.

STEP 2 Using the radius length not used in Step 1, swing an arc (R2) from point B to intercept the arc swung from point A at point C (**Figure 4-32C**).

STEP 3 Connect A to C and B to C to complete the triangle (**Figure 4-32D**).

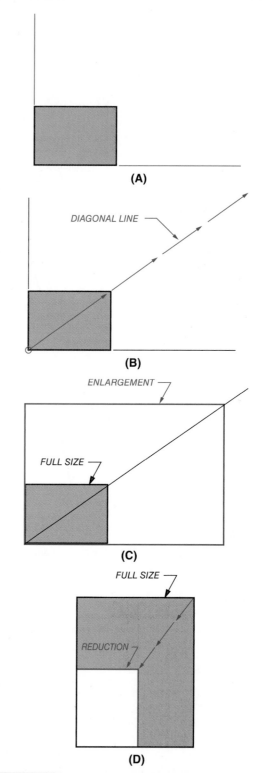

FIGURE 4-31   (A) How to proportionately enlarge or reduce a shape; (B) Step 1; (C) Step 2; (D) Step 3.

HOW TO DRAW AN EQUILATERAL TRIANGLE

GIVEN: A baseline and the given length of each side (**Figure 4-33A**).

STEP 1 Either adjust the drafting machine angle to 60° or use a 30°–60° triangle and project

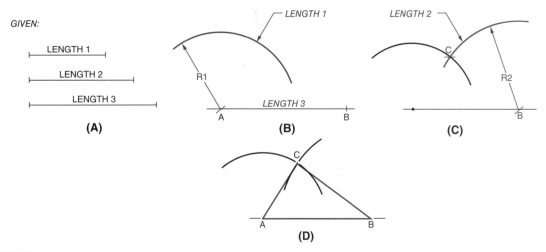

FIGURE 4-32   (A) How to draw a triangle with known lengths of sides; (B) Step 1; (C) Step 2; (D) Step 3.

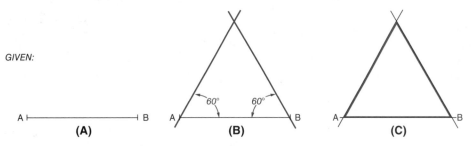

FIGURE 4-33   (A) How to draw an equilateral triangle; (B) Step 1; (C) Step 2.

from points A and B at 60° using light lines (**Figure 4-33B**). Allow light construction lines to cross as shown; do not erase them.

STEP 2   Check to see that there are three equal sides and, if so, darken in the actual triangle by using the correct line thickness (**Figure 4-33C**). Care should be exercised in constructing the sharp corners. Again, do not erase the light construction lines.

HOW TO DRAW A SQUARE

GIVEN: The location of the center and the required distance across the "flats" of a square (**Figure 4-34A**).

STEP 1   Lightly draw a circle with a diameter equal to the distance around the flats of the square. Set the compass at half the required diameter (**Figure 4-34B**).

STEP 2   Using a triangle, lightly complete the square by constructing tangent lines to the circle. Allow the light construction lines to project from the square, as shown, without erasing them (**Figure 4-34C**).

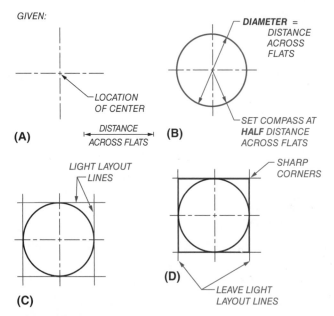

FIGURE 4-34   (A) How to draw a square; (B) Step 1; (C) Step 2; (D) Step 3.

STEP 3   Check to see that there are four equal sides and, if so, darken in the actual square by using the correct line thickness (**Figure 4-34D**). Care should be exercised in constructing the sharp corners. Again, do **not** erase the light construction lines.

A CLASSROOM DEBATE: WHY SHOULD WE LEARN THIS?*

The drafting class at Franklin Grove Community College was halfway through its lesson on geometric construction, and some of the students were becoming frustrated. "This is the most boring, tedious work I've ever done," said one student. "Why should we learn this when the computer will do it for us anyway?" Overhearing this comment, the instructor, John Diamond, asked the class to stop working and gather around.

This is what he told the class.

"I know that geometric constructions can be tedious and boring. I didn't enjoy learning how to make them either when I was a student, but I'm glad I did. I began teaching technical drawing around the time that CAD was in its infancy. Software developers were having problems back then trying to build programs that could create good technical drawings. One problem was that the personal computer hardware of the time had neither the memory nor the processing power to create complex technical drawings. Of course, time and technology would solve this problem."

"Another problem was that computer programmers didn't understand geometric constructions or other technical drawing techniques well enough to write programs that would do what drafting technicians and engineers needed them to do. What they needed was someone who really understood geometric constructions to advise the computer programmers. That's where I became involved. Because I understood geometric constructions to a sufficient level of detail to explain them to computer programmers, I was hired as a consultant. Not only did I make extra money, but I was involved in the original development of one of today's most successful CAD programs."

"You see, it's always better to understand the *why* behind the *how*. Not learning geometric constructions before learning CAD is like not learning mathematical computations before using a calculator. The student who depends on the computer instead of learning fundamentals such as geometric constructions will be like the auto mechanic who can only change parts because he does not know how to diagnose the problem."

*The names in this article have been changed to protect privacy.

HOW TO DRAW A PENTAGON (5 SIDES)

GIVEN: The location of the pentagon center and the diameter that will circumscribe the pentagon (**Figure 4-35A**).

STEP 1 Locate point A at the top-center of the circle and, using a drafting machine, position an angle of 72° (360/5) from the horizontal (or 18° from the vertical) through point A to locate point B where the angle crosses the circumference of the circle (**Figure 4-35B**).

STEP 2 Draw a horizontal line from point B to locate point C on the circumference of the circle on the opposite side (**Figure 4-35C**).

STEP 3 Set the compass at the distance from point B to point C, and swing this distance from the points, as illustrated in **Figure 4-35D**, to locate points X and Y.

STEP 4 Lightly connect the points. Check to see that there are five equal sides and, if correct, darken in the actual pentagon, taking care to construct five sharp corners (**Figure 4-35E**).

HOW TO DRAW A HEXAGON (6 SIDES)

GIVEN: The location of the required center and the required distance across the flats of a hexagon (**Figure 4-36A**).

STEP 1 Lightly draw a circle with a diameter equal to the distance across the flats of the hexagon. Set the compass at half the required diameter (**Figure 4-36B**).

STEP 2 Draw two horizontal lines tangent to the curve or two vertical lines if the hexagon is to be oriented at 90° to the illustrated position (**Figure 4-36C**).

STEP 3 Using a 30°–60° triangle, lightly complete the hexagon by constructing tangent lines to the circle (**Figure 4-36D**). Allow the light construction lines to extend as shown; do not erase them.

STEP 4 Check to see that there are six equal sides and, if so, darken in the actual hexagon by using the correct line thickness and taking care to construct six sharp corners (**Figure 4-36E**). Again do **not** erase the light construction lines.

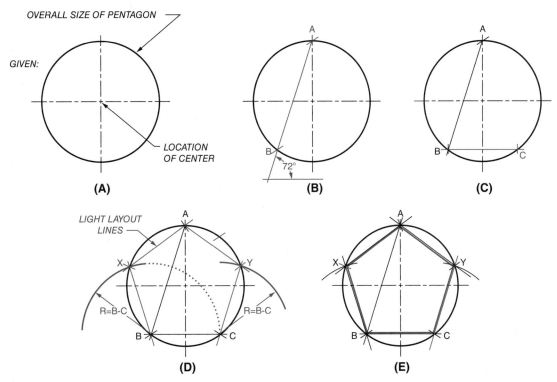

FIGURE 4-35   (A) How to draw a pentagon; (B) Step 1; (C) Step 2; (D) Step 3; (E) Step 4.

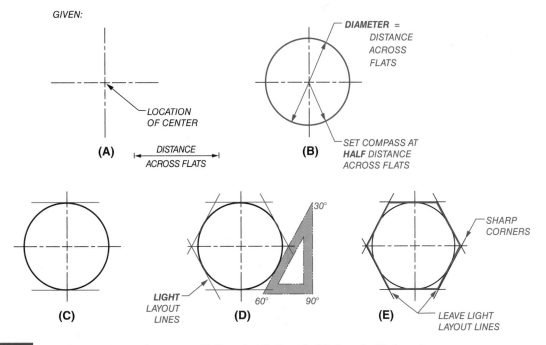

FIGURE 4-36   (A) How to draw a hexagon; (B) Step 1; (C) Step 2; (D) Step 3; (E) Step 4.

HOW TO DRAW AN OCTAGON (8 SIDES)

GIVEN: The location of the required center and the required distance across the flats of an octagon (**Figure 4-37A**).

STEP 1   Lightly draw a circle with a diameter equal to the distance across the flats of the octagon.

Set the compass at half the required diameter (**Figure 4-37B**).

STEP 2   Lightly draw two horizontal lines and two vertical lines tangent to the circle, as illustrated in **Figure 4-37C**.

(A)

(B)

(C)

(D)

(E)

FIGURE 4-37   (A) How to draw an octagon; (B) Step 1;
(C) Step 2; (D) Step 3; (E) Step 4.

STEP 3   Using a 45° triangle, lightly complete the octagon by constructing tangent lines to the circle (**Figure 4-37D**). Allow the light lines to extend.

STEP 4   Check that there are eight equal sides and, if so, darken in the actual octagon by using the correct line thickness and taking care to construct eight sharp corners (**Figure 4-37E**). Again, do **not** erase the light construction lines.

Circular Construction

HOW TO LOCATE THE CENTER OF A GIVEN CIRCLE

GIVEN: A circle without a center point (**Figure 4-38A**).

STEP 1   Using the drafting machine or T-square, draw a horizontal line across the circle approximately halfway between the **estimated** center of the given circle and the uppermost point on the circumference. Label the endpoints of the chord thus formed as A and B (**Figure 4-38B**).

STEP 2   Draw perpendicular lines (90°) downward from points A and B. Locate points C and D where these two lines pass through the circle (**Figure 4-38C**).

STEP 3   Carefully draw a straight line from point A to point D and from point C to point B. Where these lines cross is the exact center of the given circle. Place a compass point on the center point; adjust the lead to the edge of the circle, and swing an arc to check that the center is accurate (**Figure 4-38D**).

Alternatively, the intersection of perpendicular bisectors of any two nonparallel chords serves to locate the center. This is a modification of the previous construction for passing a circle through any three nonaligned points.

Tangent Points

A *tangent point* is the exact location or point where one line stops and another line begins. Tangent also means to "touch." As an example, a tangent point is the exact point where a curved line stops and a straight line begins (**Figure 4-39**).

How to Locate Tangent Points

The tangent point is a point 90° from the straight line to the swing point of the curved line. Place a light hash mark at each tangent point. It is a good habit to always find all tangent points on the object in all views before darkening in the drawing.

GIVEN:

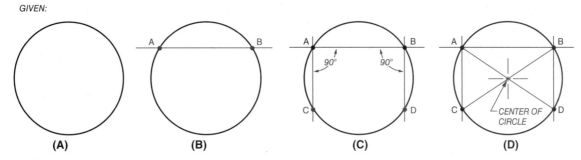

FIGURE 4-38   (A) How to locate the center of a given circle; (B) Step 1; (C) Step 2; (D) Step 3.

FIGURE 4-39   Tangent points.

The tangent points for a right-angle bend are illustrated in **Figure 4-40A**. The tangent points for an acute angle bend are illustrated in (**Figure 4-40B**). The tangent points for an obtuse angle bend are illustrated in **Figure 4-40C**.

In each preceding example, a light line is constructed 90° from the straight line to the exact swing point of the radius. Find all the tangent points before darkening in the final work. Always darken in all compass work first, followed by the straight lines.

The tangent point between two arcs or circles is the exact point where one arc or circle ends and the next arc or circle begins. The tangent point could also be where one arc or circle touches another arc or circle (**Figure 4-40D**).

The tangent point is found by drawing a straight line from the swing point of the first arc or circle to the swing point of the second arc or circle (**Figure 4-40E**). Place a short, light dash at the exact tangent point.

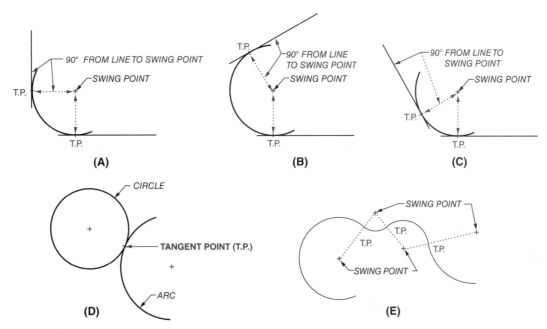

FIGURE 4-40   How to locate tangent points (A) On a right angle; (B) On an acute angle; (C) On an obtuse angle; (D) Between arcs or circles; (E) Tangent points between arcs or circles.

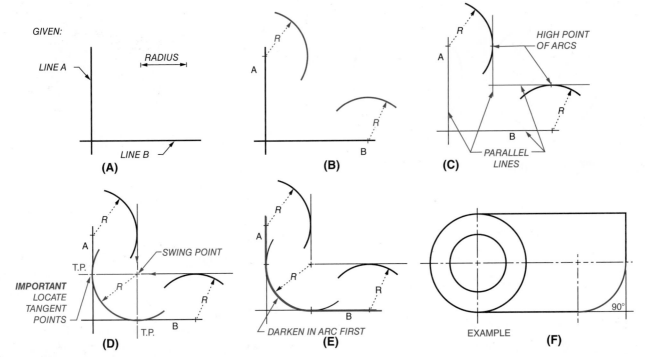

FIGURE 4-41 (A) How to construct an arc tangent to a right angle; (B) Step 1; (C) Step 2; (D) Step 3; (E) Step 4; (F) Example of an arc tangent to two lines at a right angle.

HOW TO CONSTRUCT AN ARC TANGENT TO A RIGHT ANGLE (90°)

GIVEN: A right angle (90°), lines A and B, and a required radius (**Figure 4-41A**).

STEP 1   Set the compass at the required radius and, out of the way, swing a radius from line A and one from line B, as illustrated in **Figure 4-41B**.

STEP 2   From the extreme high points of each radius, construct a light line parallel to line A and another line parallel to line B (**Figure 4-41C**).

STEP 3   Where these lines intersect is the exact location of the required swing point. Set the compass point on the swing point and lightly construct the required radius (**Figure 4-41D**). Allow the radius swing to extend past the required area, as shown in the figure. It is important to locate all tangent points (T.P.) before darkening in.

STEP 4   Check all work and darken in the radius by using the correct line thickness. Darken in connecting straight lines as required. Always construct compass work first, followed by straight lines. Leave all light construction lines (**Figure 4-41E**).

An example of an arc tangent to two lines at a right angle (90°) is illustrated in **Figure 4-41F**.

HOW TO CONSTRUCT AN ARC TANGENT TO AN ACUTE ANGLE (LESS THAN 90°)

GIVEN: An acute angle, lines A and B, and a required radius (**Figure 4-42A**). Follow the same procedure as outlined in the preceding example.

STEP 1   Set the compass at the required radius and, out of the way, swing a radius from line A and one from line B, as illustrated in **Figure 4-42B**.

STEP 2   From the extreme high points of each radius, construct a light line parallel to line A and another line parallel to line B (**Figure 4-42C**).

STEP 3   Where these lines intersect is the exact location of the required swing point. Set the compass point on the swing point and lightly construct the required radius (**Figure 4-42D**). Allow the radius swing to extend past the required area, as shown in Figure 4-42D.

STEP 4   Check all work and darken in the radius using the correct line thickness. Darken in connecting straight lines as required. Always construct compass work first, followed by straight lines. Leave all light construction lines (**Figure 4-42E**).

An example of an arc tangent to two lines at an acute angle (22°) is illustrated in **Figure 4-42F**.

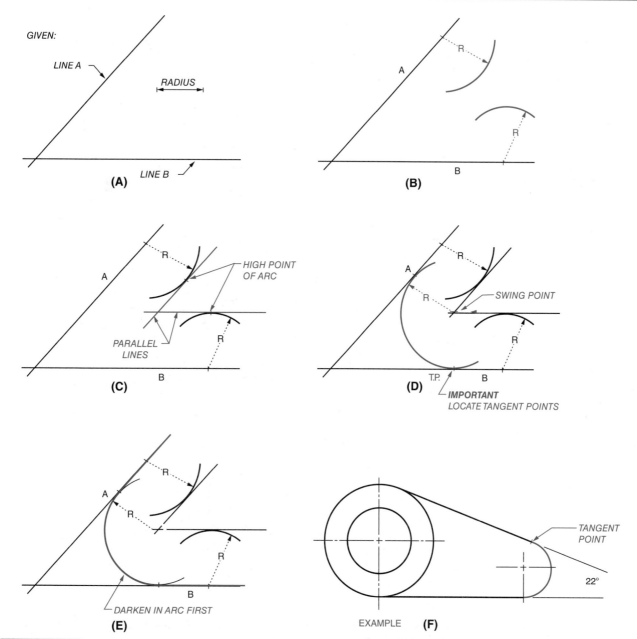

FIGURE 4-42 (A) How to construct an arc tangent to an acute angle; (B) Step 1; (C) Step 2; (D) Step 3; (E) Step 4; (F) Example of an arc tangent to two lines at an acute angle.

HOW TO CONSTRUCT AN ARC TANGENT TO AN OBTUSE ANGLE (MORE THAN 90°)

GIVEN: An obtuse angle between lines A and B and a required radius (**Figure 4-43A**). Follow the same procedure as outlined in the two preceding examples.

STEP 1 Set the compass at the required radius and, out of the way, swing a radius from line A and one from line B, as illustrated in **Figure 4-43B**.

STEP 2 From the extreme high points of each radius, construct a light line parallel to line A and one parallel to line B (**Figure 4-43C**).

STEP 3 Where these lines intersect is the exact location of the required swing point. Set the compass point on the swing point and lightly construct the required radius. Allow the radius swing to extend past the required area, as shown in **Figure 4-43D**.

STEP 4 Check all work and darken in the radius by using the correct line thickness. Darken in connecting straight lines as required. Always construct compass work first, followed by straight lines. Leave all light construction lines, as shown in **Figure 4-43E**.

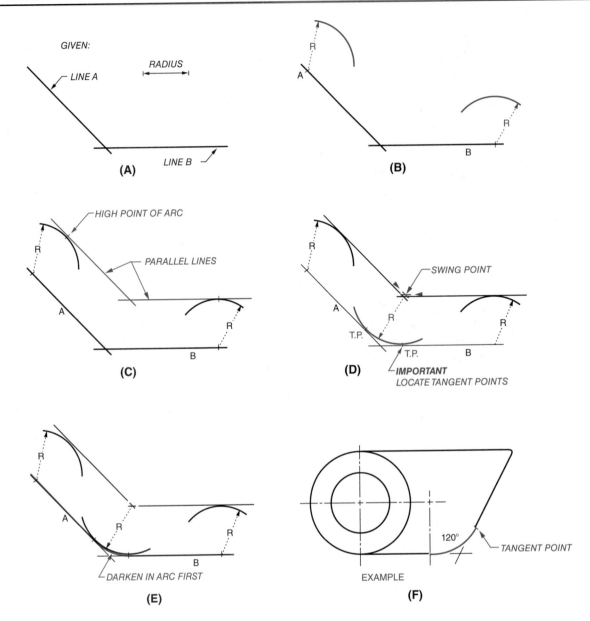

FIGURE 4-43   (A) How to construct an arc tangent to an obtuse angle; (B) Step 1; (C) Step 2; (D) Step 3; (E) Step 4; (F) Example of an arc tangent to two lines at an obtuse angle.

An example of an arc tangent to two lines at an obtuse angle (120°) is illustrated in **Figure 4-43F**.

HOW TO CONSTRUCT AN ARC TANGENT TO A STRAIGHT LINE AND A CURVE

GIVEN: Straight line A, arc B with a center point, and a required radius (**Figure 4-44A**).

STEP 1   Set the compass at the required radius and, out of the way, swing a radius from the given arc B and one from the given straight line A (**Figure 4-44B**).

STEP 2   From the extreme high points of each radius, construct a light straight line parallel to line A, and construct a radius outside the given arc B equal to the required radius, as illustrated in **Figure 4-44C**.

STEP 3   Where these lines intersect is the exact location of the required swing point. Set the compass point on the swing point and lightly construct the required radius (**Figure 4-44D**). Allow the radius swing to extend past the required area as shown. Locate all tangent points (T.P.) before darkening in.

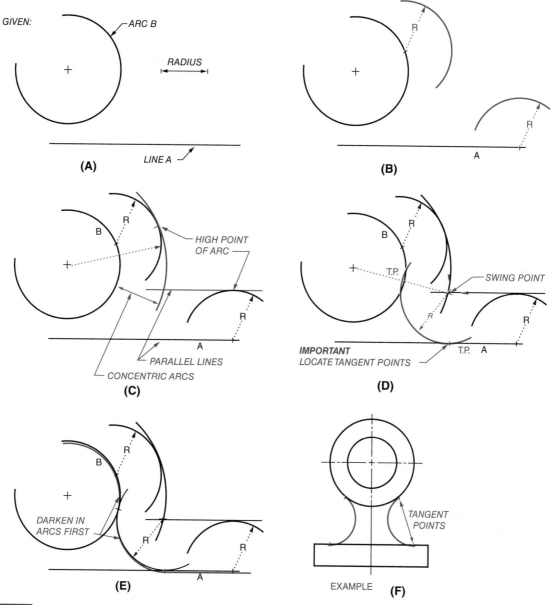

FIGURE 4-44   (A) How to construct an arc tangent to a straight line and a curve; (B) Step 1; (C) Step 2; (D) Step 3; (E) Step 4; (F) Example of an arc tangent to a straight line and a curve.

STEP 4   Check all work and darken in the radius by using the correct line thickness. Darken in the arcs first and the straight line last, (**Figure 4-44E**).

An example of an arc tangent to a straight line and a curve is illustrated in **Figure 4-44F**.

HOW TO CONSTRUCT AN ARC TANGENT TO TWO RADII OR DIAMETERS

GIVEN: Diameter A and arc B with center points located and the required radius (**Figure 4-45A**).

STEP 1   Set the compass at the required radius and, out of the way, swing a radius of the required length from a point on the circumference of the given diameter A. Out of the way, swing a required radius from a point on the circumference of a given arc B (**Figure 4-45B**).

STEP 2   From the extreme high points of each radius, construct a light radius outside of the given radii A and B, as illustrated in **Figure 4-45C**.

STEP 3   Where these arcs intersect is the exact location of the required swing point. Set the compass point on the swing point and lightly

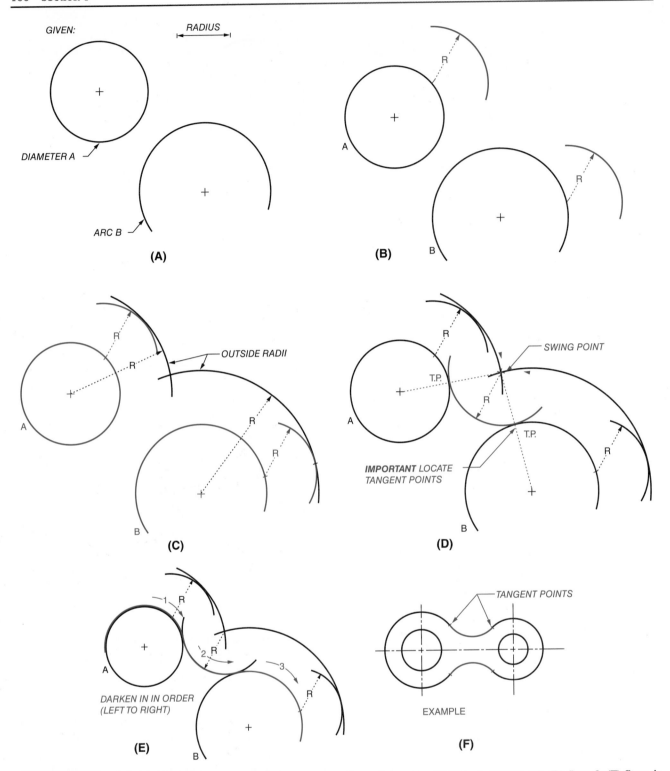

FIGURE 4-45   (A) How to construct an arc tangent to two radii or diameters; (B) Step 1; (C) Step 2; (D) Step 3; (E) Step 4; (F) Example of an arc tangent to two radii.

construct the required radius (**Figure 4-45D**). Allow the radius swing to extend past the required area as shown. Before darkening in, it is important to locate all tangent points (T.P.), Figure 4-45D.

STEP 4   Check all work and darken in the radii by using the correct line thickness. Darken

in the arcs or radii in consecutive order from left to right or from right to left, thus constructing a smooth connecting line having no apparent change in direction (**Figure 4-45E**).

An example of an arc tangent to two radii is illustrated in **Figure 4-45F**.

FIGURE 4-46   (A) How to draw an ogee curve; (B) Step 1; (C) Step 2; (D) Step 3; (E) Step 4; (F) Step 5; (G) Step 6.

How to Draw an Ogee Curve

An *ogee curve* is used to join two parallel lines. It forms a gentle curve that reverses itself in a neat, symmetrical geometric form.

GIVEN: Parallel lines A-B and C-D (**Figure 4-46A**).

STEP 1   Draw a straight line connecting the space between the parallel lines—in this example, from point B to point C (**Figure 4-46B**).

STEP 2   Make a perpendicular bisector to line B-C to establish point X (**Figure 4-46C**).

STEP 3   Make perpendicular bisectors to the lines B-X and X-C (**Figure 4-46D**).

STEP 4   Draw a perpendicular from line A-B at point B to intersect the perpendicular bisector of B-X, which locates the first required swing center. Draw a perpendicular from line C-D at point C to intersect the perpendicular bisector of C-X, which locates the second required swing center (**Figure 4-46E**).

STEP 5 Place the compass point on the first swing point, adjust the compass lead to point B, and swing an arc from B to X. Place the compass point on the second swing point and swing an arc from X to C (**Figure 4-46F**). This completes the ogee curve.

NOTE: Point X is the tangent point between arcs. Check and, if correct, darken in all work.

An example of an ogee curve is illustrated in Figure 4-46G.

Conic Sections

A *conic section* is a section cut by a plane passing through a cone. These sections are bounded by various kinds of shapes. Depending on where the section is cut, the shape can be a triangle, a circle, an ellipse, a parabola, or a hyperbola (**Figures 4-47A** through **4-47E**).

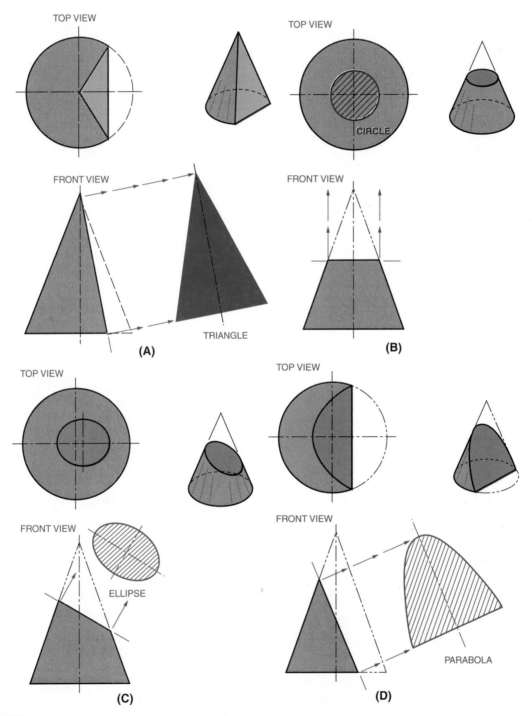

FIGURE 4-47   (A) Triangle section; (B) Circle section; (C) Ellipse section; (D) Parabola section;

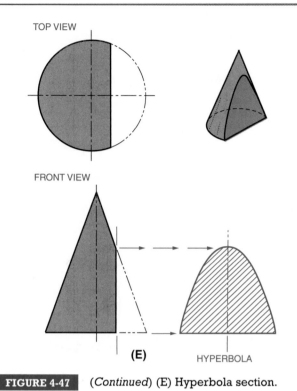

TOP VIEW

FRONT VIEW

(E)

HYPERBOLA

FIGURE 4-47   (*Continued*) (E) **Hyperbola section.**

Triangle

Figure 4-47A. This shape results when a plane passes through the apex of the cone.

Circle

Figure 4-47B. This shape results when a plane passes through the cone, parallel with the base and perpendicular to the axis.

Ellipse

Figure 4-47C. This shape results when a plane passes through the cone inclined to the axis.

Parabola

Figure 4-47D. This shape results when a plane passes through the cone parallel to one element.

Hyperbola

Figure 4-47E. This shape results when a plane passes through the cone parallel with the axis of the cone.

HOW TO DRAW AN ELLIPSE (CONCENTRIC CIRCLE METHOD)

GIVEN: The location of the center point, the major diameter, and the minor diameter (**Figure 4-48A**).

STEP 1 Lightly draw one circle equal to the major diameter and another circle equal to the minor diameter (**Figure 4-48B**).

STEP 2 Divide both circles into 12 equal divisions by passing lines through the center at every 30°. Number points 1 through 5 in clockwise consecutive order on the major diameter, positioning the 3 location on the right horizontal axis. Number points 7 through 11 in clockwise consecutive order, positioning the 9 location on the left horizontal axis. Point 6 is on the lower vertical axis at the minor diameter, and point 12 is on the upper vertical axis at the minor diameter (**Figure 4-48C**). If the ellipse were to be positioned at 90° from this example, the point locations would be rotated accordingly.

STEP 3 Project down from point 1 on the major diameter and to the right from point 1 on the minor diameter. Where these lines intersect is the exact location where point 1 will be on the ellipse (**Figure 4-48D**).

STEP 4 Project down from point 2 on the major diameter and to the right from point 2 on the minor diameter. Where these lines intersect is the exact location where point 2 will be on the ellipse (**Figure 4-48E**).

STEP 5 This completes the first quadrant of the ellipse. Continue around the circle to locate points 4, 5, 6, 7, 8, 9, 10, and 11 in the same manner, except in reverse (**Figure 4-48F**).

STEP 6 Lightly draw straight lines connecting points 1 through 12 in order to get a general idea of the ellipse outline (**Figure 4-48G**).

STEP 7 Darken in the ellipse by using an irregular curve. Carefully connect all the points with a smooth, continuous line of the correct thickness (**Figure 4-48H**). It is sometimes helpful to divide into 10° or 15° spaces the two ends where the ellipse curves the fastest in order to have more points around the extreme ends. In this example, this is done between points 2 and 4 and 8 and 10.

Examples of ellipses are illustrated in **Figure 4-48**, showing how a rotated circle would look at 0°, 15°, 45°, 60°, and 90°.

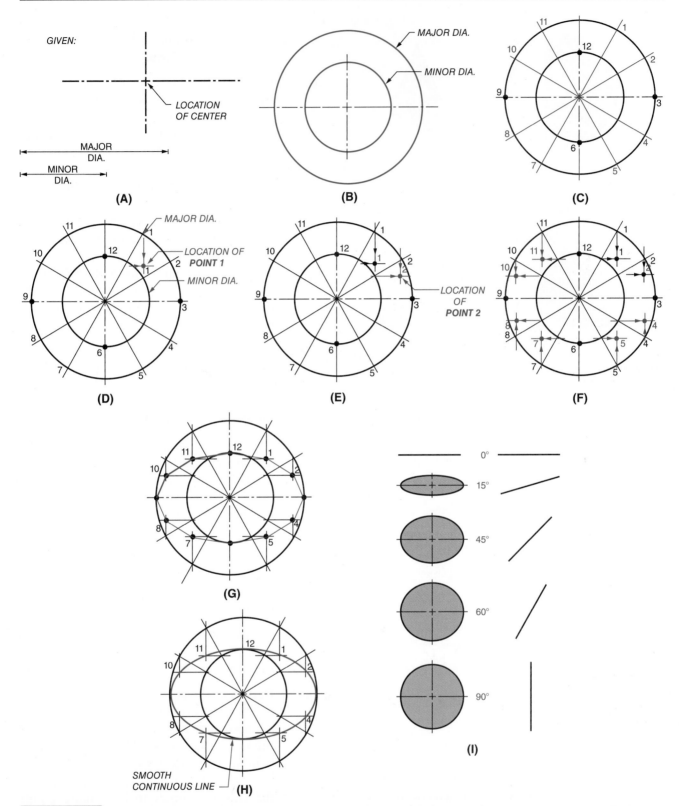

FIGURE 4-48   (A) How to draw an ellipse (concentric method); (B) Step 1; (C) Step 2; (D) Step 3; (E) Step 4; (F) Step 5; (G) Step 6; (H) Step 7; (I) Example of ellipses at various degrees.

HOW TO DRAW A TRAMMEL ELLIPSE

GIVEN: A major or minor axis of a required ellipse and any located point through which the ellipse must pass (**Figure 4-49A**). If a major axis is given, mark the end-points A and B. If the minor axis is given, as shown in the figure, mark the end-points C and D. Mark as P the known location of where the ellipse must pass.

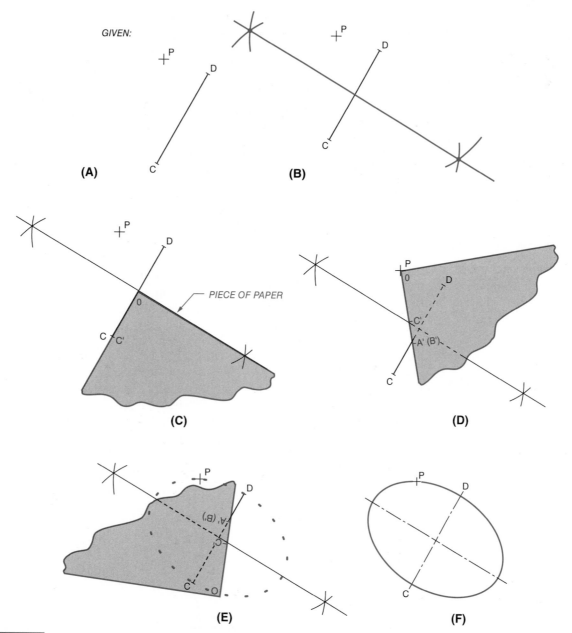

FIGURE 4-49   (A) Trammel ellipse; (B) Step 1; (C) Step 2; (D) Step 3; (E) Step 4; (F) Step 5.

STEP 1   Draw a perpendicular bisector of the given axis to locate the unknown axis position, crossing the known axis at a point that will become 0 (**Figure 4-49B**).

STEP 2   Mark an end corner of a separate strip of paper as point 0, and mark along the edge of the strip one-half of the distance A-B or C-D, whichever is known. Mark this point on the strip as A' or C' to correspond to the known axis (**Figure 4-49C**). Position the strip so that point 0 lies on the known point through which the ellipse must pass.

Position the other strip mark on the axis that is not represented by the half-length on the strip at the position (of two possible positions) that is nearest to the ellipse center.

STEP 3   With the strip positioned as specified in Step 2, mark the strip where it crosses the other axis. The length of end point 0 to this new position represents one-half of the unknown axis length, and it should correspondingly be labeled as A' or C' (**Figure 4-49D**).

STEP 4   Using the two marked strip locations, reposition the strip so that the mark representing one-half of the major axis always lies on the minor axis and the mark representing one-half of the minor axis always lies on the major axis. Repeat as needed to establish enough points for constructing the ellipse (**Figure 4-49E**).

STEP 5   Using a French curve, pass a smooth line through each of these established points (**Figure 4-49F**).

HOW TO DRAW A FOCI ELLIPSE

GIVEN: Major and minor axes of a required ellipse. The end-points of the major axis are labeled A and B, the end-points of the minor axis

are labeled C and D, and the ellipse center is labeled 0 (**Figure 4-50A**).

STEP 1   Swing a radius of one-half of the major axis (0A) from the end point of the minor axis (C or D) to locate focal points E and F at the points where the arc crosses the major axis (**Figure 4-50B**).

STEP 2   Select a random array of points on the major axis between one focal point and the ellipse center. (Space those at the ends more closely.) Number these in consecutive order (**Figure 4-50C**).

STEP 3   Swing an arc whose radius is the distance from point B to number 1. Swing this distance from focal point F. Intersect this arc with an arc whose radius is the distance from point A to the same numbered point—

FIGURE 4-50   (A) How to draw a foci ellipse; (B) Step 1; (C) Step 2; (D) Step 3; (E) Step 4; (F) Step 5; (G) Step 6.

GIVEN:

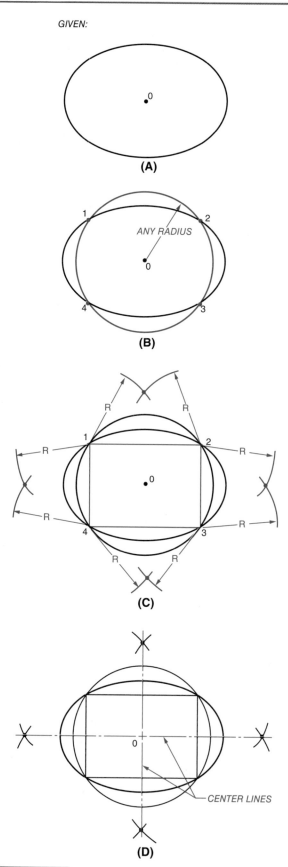

(A)

ANY RADIUS

(B)

(C)

CENTER LINES

(D)

FIGURE 4-51   (A) How to locate the major and minor axes of a given ellipse; (B) Step 1; (C) Step 2; (D) Step 3.

number 1 in this example—but whose center is at focal point E (**Figure 4-50D**).

STEP 4   Repeat Step 3, but reverse the focal point positions of the arc centers (**Figure 4-50E**).

STEP 5   Repeat Steps 3 and 4 for each point selected in Step 2 (**Figure 4-50D**).

STEP 6   Connect the points into a smooth curve by using a French curve (**Figure 4-50E**).

HOW TO LOCATE THE MAJOR AND MINOR AXES OF A GIVEN ELLIPSE WITH A LOCATED CENTER

GIVEN: An ellipse, with center 0 located (**Figure 4-51A**).

STEP 1   From ellipse center 0, draw a circle at a radius that allows intersecting the ellipse at any four locations. Label these points in successive order 1, 2, 3, and 4, moving clockwise around the ellipse center (**Figure 4-51B**).

STEP 2   Draw a rectangle by connecting each of the labeled points in successive order. As shown in **Figure 4-51C**, the major and minor axes are parallel to these sides, found by drawing perpendicular bisectors of two consecutive sides.

STEP 3   Draw the major and minor axes parallel to sides 1-2 and 2-3 through point 0 (**Figure 4-51D**).

HOW TO DRAW A TANGENT TO AN ELLIPSE AT A POINT ON THE ELLIPSE

GIVEN: An ellipse and a point P on its perimeter (**Figure 4-52A**). If not provided, locate major axis A-B, minor axis C-D, and focal points E and F.

STEP 1   Draw E-P and F-P (**Figure 4-52B**).

STEP 2   Bisect angle EPF (**Figure 4-52C**).

STEP 3   Draw a perpendicular to the angle bisector of angle EPF at point P. This is the required tangent (**Figure 4-52D**).

HOW TO DRAW A TANGENT TO AN ELLIPSE FROM A DISTANT POINT

GIVEN: An ellipse and a distant point P (**Figure 4-53A**).

STEP 1   If not provided, locate major axis A-B and focal points E and F (**Figure 4-53B**).

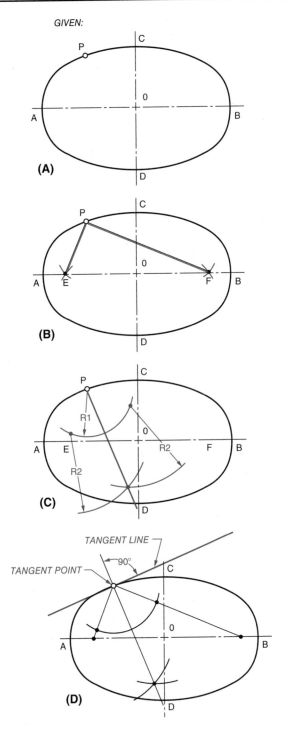

FIGURE 4-52   (A) How to draw a tangent to an ellipse at a point on the ellipse; (B) Step 1; (C) Step 2; (D) Step 3.

STEP 2   Swing an arc R1 from point P to pass through either focal point. F is selected for this example (**Figure 4-53C**).

STEP 3   Swing an arc whose radius is equal to the major axis from the other focal point (E in this example). Intersect the arc from Step 2 at points R and S (**Figure 4-53D**).

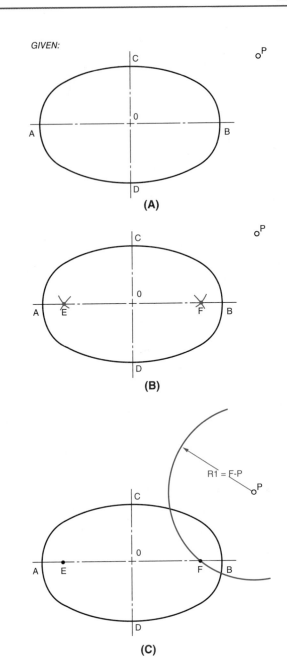

FIGURE 4-53   (A) How to draw a tangent to an ellipse from a distant point; (B) Step 1; (C) Step 2;

STEP 4   Extend the lines from E to each of the points R and S to cross the ellipse at U and V (**Figure 4-53E**).

STEP 5   The two possible points of tangency of the two possible lines from point P are U and V. Select and draw either, or both (**Figure 4-53F**).

HOW TO DRAW A PARABOLA (METHOD 1)

GIVEN: The front view and plan view of cone ABC with cutting plane X-Y (**Figure 4-54A**).

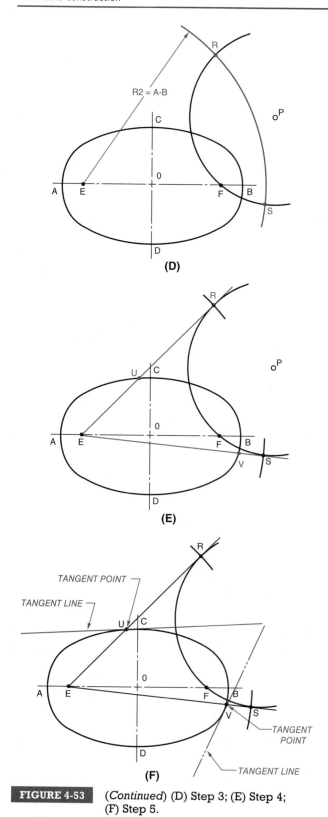

(D)

R2 = A-B

FIGURE 4-53 (*Continued*) (D) Step 3; (E) Step 4;
(F) Step 5.

(E)

(F)

TANGENT POINT

TANGENT LINE

TANGENT POINT

TANGENT LINE

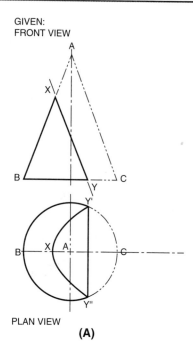

GIVEN:
FRONT VIEW

PLAN VIEW

(A)

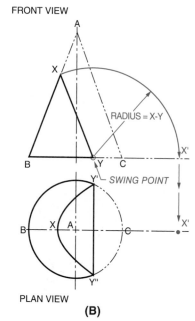

FRONT VIEW

RADIUS = X-Y

SWING POINT

PLAN VIEW

(B)

FIGURE 4-54 (A) How to draw a parabola (Method 1);
(B) Step 1;

STEP 1   Locate points Y′ and Y″ in the plan view.
In the front view, place the point of the
compass on point Y, and position the lead
on point X. Swing point X to the extended
baseline and down into the plan view, as
illustrated in **Figure 4-54B**.

STEP 2   Locate a point along line X-Y (in this
example, point D) and, using point Y
as a swing point, swing point D to the
extended baseline and down into the plan
view. In the front view, reset the compass
to a distance equal to the horizontal distance from the cone axis to the outer edge
(identified as length 1), as illustrated in
Figure 4-54C. Transfer this distance to the
plan view. Project point D down into the

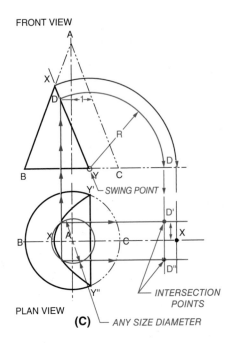

FRONT VIEW

PLAN VIEW

(C)

SWING POINT

INTERSECTION
POINTS

ANY SIZE DIAMETER

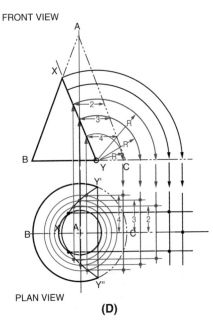

FRONT VIEW

PLAN VIEW

(D)

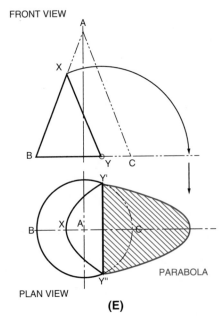

FRONT VIEW

PLAN VIEW

PARABOLA

(E)

FIGURE 4-54 *(Continued)* (C) Step 2; (D) Step 3; (E) Step 4.

plan view and, where point D intersects the arc, locate points D′ and D″. Project these points to the right, intersecting with point D from above at the actual intersection points D′ and D″.

STEP 3 Choose other points along line X-Y in the front view and project them over and down as described in Step 2 (**Figure 4-54D**).

STEP 4 Using an irregular curve, connect all points. This completes the parabola (**Figure 4-54E**).

HOW TO DRAW A PARABOLA (METHOD 2)

GIVEN: Required rise A-B and required span A-C. *Problem:* Construct a parabola within the required rise and span (**Figure 4-55A**).

STEP 1 Divide half the span distance A-0 into any number of equal parts. (In this example, 4 equal parts are used.) See **Figure 4-55B**. Divide half the rise A-B into equal parts amounting to the square of the equal parts in Step 1 (in this example, $4^2 = 16$ equal parts).

FIGURE 4-55   (A) How to draw a parabola (Method 2);
(B) Step 1; (C) Step 2; (D) Step 3.

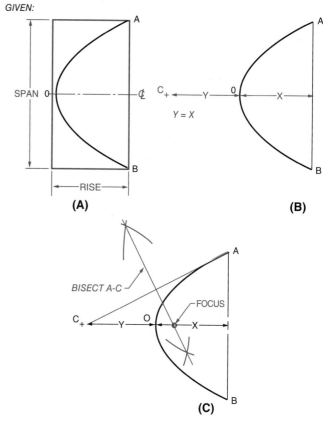

FIGURE 4-56   (A) How to find the focus point of a
parabola; (B) Step 1; (C) Step 2.

STEP 2   From line A-0, each point on the parabola
is offset by a number of units equal to the
square of the numbers of units from point 0
(**Figure 4-55C**). For example, point 1 projects
1 unit, point 2 projects 4 units, point 3 projects 9 units, and so forth.

STEP 3   Using an irregular curve, connect the
points to form a smooth parabolic curve
(**Figure 4-55D**).

HOW TO FIND THE FOCUS POINT OF A PARABOLA

GIVEN: A parabolic curve with points A, 0, and B
(**Figure 4-56A**). *Problem:* Find the focus
point of parabola A0B.

STEP 1   Draw a line from point A to point B. Continue
the center line to a point equal to length X,
distance Y, to find point C (**Figure 4-56B**).

STEP 2   Draw a line from point C to point A and
bisect it. The intersection of the bisect line
and the axis is the focus of the parabola
(**Figure 4-56C**).

HOW TO DRAW A HYPERBOLA (METHOD 1)

GIVEN: The front view and plan view of cone ABC
with cutting plane X-Y (**Figure 4-57A**).

STEP 1   Locate points Y′ and Y″ in the plan view.
In the front view, place the point of the compass on point Y, and set the lead to point
X. Swing point X to the extended baseline
and down to the plan view, as illustrated in
(**Figure 4-57B**).

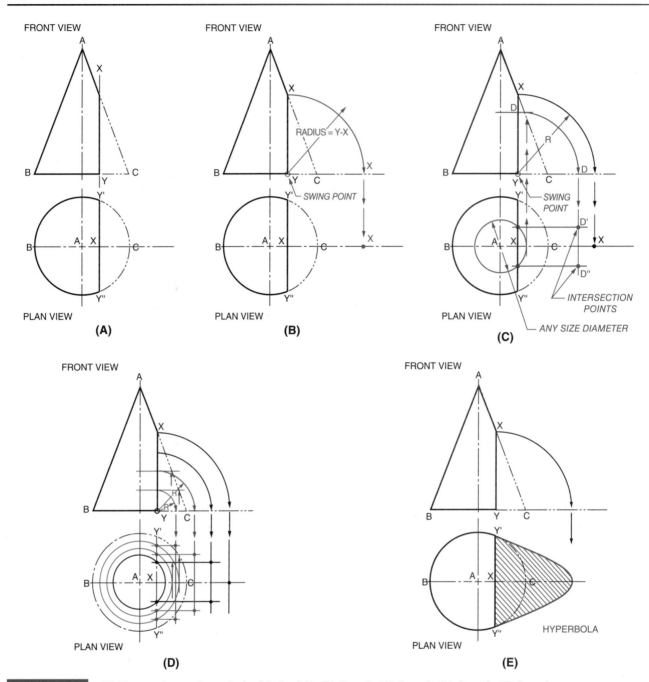

FRONT VIEW

FIGURE 4-57   (A) How to draw a hyperbola (Method 1); (B) Step 1; (C) Step 2; (D) Step 3; (E) Step 4.

STEP 2   Locate a point along line X-Y (in this example, point D). Using Y as a swing point, swing point D to the extended baseline and down into the plan view. In the front view, reset the compass to a distance equal to the horizontal distance from the cone axis to the outer edge (length 1), as illustrated in **Figure 4-57C**. Transfer the arc to the plan view; where the arc intersects with the cutting plane line X-Y, project to the right, and where these points intersect with point D from above is the actual location of points D′ and D″.

STEP 3   Choose other points along line X-Y in the front view, and project them over and down as described in Step 2 (**Figure 4-57D**).

STEP 4   Using an irregular curve, connect all points. This completes the hyperbola (**Figure 4-57E**).

HOW TO DRAW A HYPERBOLA (METHOD 2)

GIVEN: Coordinates, a square whose sides equal the transverse axis, and points A and B (**Figure 4-58A**).

FIGURE 4-58   (A) How to draw a hyperbola (Method 2); (B) Step 1; (C) Step 2; (D) Step 3; (E) Step 4.

STEP 1   With the center at the intersection of the diagonal lines from opposite corners of the square, place the compass lead at the corner of the given square, and swing arcs to intersect the horizontal line through the center. This locates focus A and focus B. Progressing outward along the horizontal line, mark off equal spaces of arbitrary length from the focus points (**Figure 4-58B**).

STEP 2   Set the compass at dimension X (A-1), and swing an arc from focus A. Set the compass at dimension Y (B-1), and swing an arc from focus B (**Figure 4-58C**).

STEP 3   Follow the same procedure for as many points as are required (in this example,

6 points). Set the compass at distance A-2 and swing from focus A. Set the compass at distance B-2 and swing from focus B, and so on (**Figure 4-58D**).

STEP 4   Using an irregular curve, carefully complete the hyperbola curve (**Figure 4-58E**).

HOW TO JOIN TWO POINTS BY A PARABOLIC CURVE

GIVEN: Points X and Y with 0 an assumed point of tangency (**Figures 4-59A** through **4-59C**).

STEP 1   Divide line X-0 into an equal number of parts; divide 0-Y into the same number of equal parts (**Figures 4-59D** through **4-59F**).

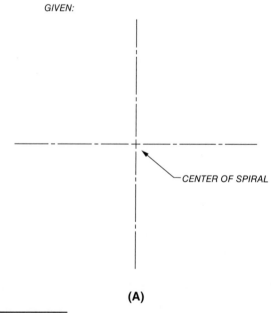

FIGURE 4-59   (A) How to join two points by a parabolic curve (given: 90° angle); (B) Given: Less than 90° angle; (C) Given: More than 90° angle; (D) Step 1; (E) Step 1; (F) Step 1; (G) Step 2; (H) Step 2; (I) Step 2.

STEP 2   Connect corresponding points (**Figures 4-59G** through **4-59 I**). Using an irregular curve, draw the parabolic curve as a smooth, flowing curve.

Supplementary Construction

HOW TO DRAW A SPIRAL

GIVEN: Crossed axes (**Figure 4-60A**).

STEP 1   Divide the circle into equal angles (in this example, 30°). Set the compass at a required radius, and draw various diameters to suit, evenly spaced, as illustrated in **Figure 4-60B**.

STEP 2   Starting any place along the angles, step over one angle and up one diameter until the required spiral is completed (**Figure 4-60C**).

FIGURE 4-60   (A) How to draw a spiral;

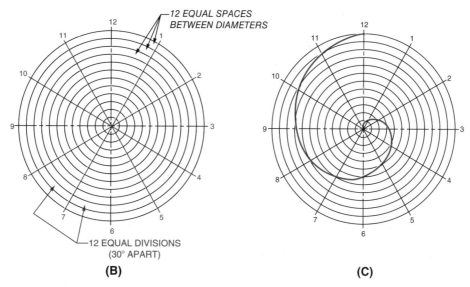

FIGURE 4-60   (*Continued*) (B) Step 1; (C) Step 2.

How to Draw a Helix

A *helix,* a form of spiral, is used in screw threads, worm gears, and spiral stairways, to mention but a few uses. A helix is generated by moving a point around and along the surface of a cylinder with uniform angular velocity about its axis. A *cylindrical helix* is simply known as a helix, and the distance measured parallel to the axis traversed by a point in one revolution is called the *lead.*

GIVEN: The top and front views of a cylinder with a required lead (**Figure 4-61A**).

STEP 1   Divide the top view into an equal number of spaces (12 in this example). Draw a line equal to one revolution and/or circumference, project the lead distance from the end, and label all points per **Figure 4-61B**.

STEP 2   Divide the circumference into the same amount of equal spaces, and project each

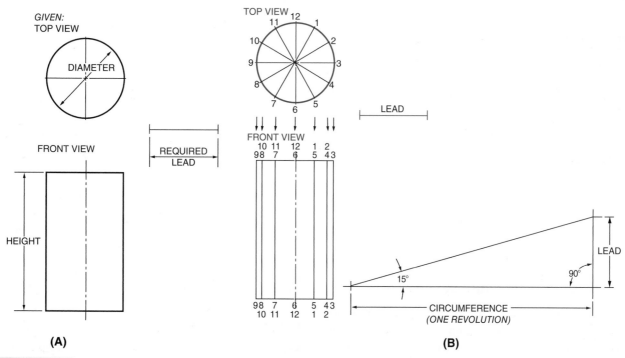

FIGURE 4-61   (A) How to draw a helix; (B) Given views;

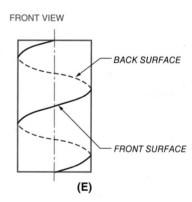

FIGURE 4-61   (*Continued*) (C) Steps 1 and 2; (D) and (E).

from the inclined line to the corresponding point projected from the front view (**Figure 4-61C**).

STEP 3   Connect all points of the helix, and draw as hidden lines the lines that would disappear from view on an actual helix form (**Figure 4-61D**). Notice that a right-hand helix is illustrated. To construct a left-hand helix, simply project all the points in the opposite direction (**Figure 4-61E**).

How to Draw an Involute of a Line

The curved path of a point on a string as it unwinds from a line, a triangle, a square, or a circle is an *involute*. The involute is used to construct involute gears. The involute forms the face and part of the flank of the teeth of the gear.

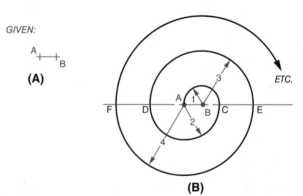

FIGURE 4-62   (A) How to draw an involute of a line; (B) Step 1.

GIVEN: Line A-B (**Figure 4-62A**).

STEP 1   Extend line A-B, as illustrated in **Figure 4-62B**. Set the compass at length A-B, and using point B as the swing point, swing semicircle

A-C. With the compass set at length A-C, and using point A as the swing point, swing semicircle A-D. Continue in this manner, alternating between points A and B until the required involute is completed, as shown in the figure.

HOW TO DRAW AN INVOLUTE OF A TRIANGLE

GIVEN: Triangle ABC (**Figure 4-63A**).

STEP 1 Extend straight lines from the triangle, as illustrated in **Figure 4-63B**. Set the compass at length A-C, and using point A as the swing point, swing semicircle A-C. With the compass set at length B-D, and using point B as the swing point, swing semicircle B-E. Continue similarly around the triangle until the required involute is completed, as shown in the figure.

HOW TO DRAW AN INVOLUTE OF A SQUARE

GIVEN: Square ABCD (**Figure 4-64A**).

STEP 1 Extend straight lines from the square, as illustrated in **Figure 4-64B**. Set the compass at length A-B, and using point A as the swing point, swing semicircle A-E. With the compass set at length B-E, and using point B as the swing point, swing semicircle B-F. Continue in the same manner around the square until the required involute is completed, as shown in the figure.

HOW TO DRAW AN INVOLUTE OF A CIRCLE

GIVEN: Circle A (**Figure 4-65A**).

STEP 1 Divide the circle into a number of equal parts and number each point clockwise around the circle (in this example, 12 equal parts). Project

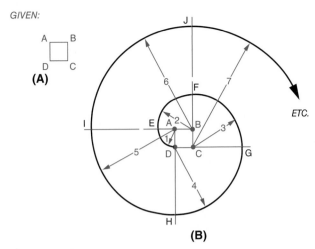

FIGURE 4-64   (A) How to draw an involute of a square; (B) Step 1.

a line perpendicular to the radius at each point in a clockwise direction (**Figure 4-65B**).

STEP 2 Set the compass at length 1-2, and using point 2 as the swing point, swing semicircle 1-2. With the compass set at length 1-2 plus 2-3, and using point 3 as the swing point, swing

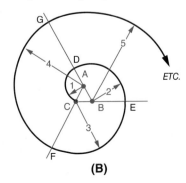

FIGURE 4-63   (A) How to draw an involute of a triangle; (B) Step 1.

FIGURE 4-65   (A) How to draw an involute of a circle; (B) Step 1; (C) Step 2.

semicircle 1-3. Continue in the same manner around the circle until the required involute is completed, as shown in **Figure 4-65C**.

HOW TO DRAW A CYCLOIDAL CURVE

GIVEN: A required span of a cycloid.

STEP 1   As the span represents the rolling distance of one revolution of a diameter, divide the span length by π to find the required diameter. Divide the span into an equal number of divisions. Twelve is a convenient number (**Figure 4-66A**).

STEP 2   Draw the rolling diameter tangent to the given span at point 0/12. Divide this diameter into the same number of equal spaces as in the Step 1 span division. Consecutively number the radial end-points of these divisions to make them meet their corresponding number of the span divisions as the diameter rolls along the span (**Figure 4-66B**).

STEP 3   At each span division, draw the rolling diameter tangent to the span in the rolled position, with the division numbers of span and diameter matching at their contact point. Locate the point 0/12 position on the diameter at each of these locations, marked with a small cross (**Figure 4-66C**).

STEP 4   Connect all the point 1 diameter division positions with a smooth curve to complete the cycloidal curve (**Figure 4-66D**).

NOTE: *If the given span is a concave arc, a hypocycloid will be drawn* (**Figure 4-67**). *If the span is convex, an epicycloid will be drawn* (**Figure 4-68**).

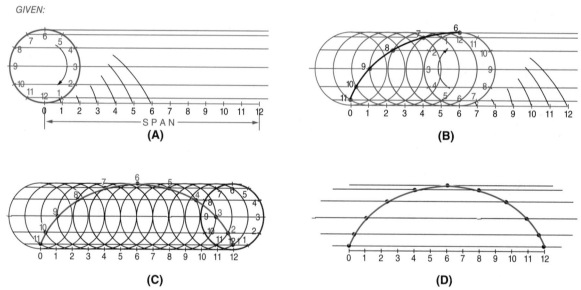

FIGURE 4-66   (A) How to draw a cycloidal curve Step 1; (B) Step 2; (C) Step 3; (D) Step 4.

FIGURE 4-67   A hypocycloid.

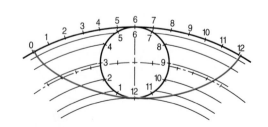

FIGURE 4-68   An epicycloid.

SUMMARY

- Geometric nomenclature consists primarily of the following concepts: points in space, line, angle, triangle, polygon, quadrilateral, circle, polyhedron, prism, pyramid, cone, and sphere.

- Elements and polygon construction principles consist primarily of how to do the following: bisect a line, bisect an angle, draw an arc or circle through three points, draw a line parallel to a straight line at a given distance, draw a line perpendicular to a line at a point, draw a line parallel to a curved line at a given distance, draw a perpendicular to a line from a point not on the line, divide a line into equal parts, divide a line into proportional parts, transfer an angle, transfer an odd shape, transfer complex shapes, proportionately enlarge or reduce a shape, draw a triangle with sides of known length, draw a square, draw a pentagon, and draw a hexagon.

- Circular construction consists primarily of knowing how to do the following: locate the center of a given circle; construct an arc tangent to a right angle, an acute angle, an obtuse angle, a straight line, a curve, and two radii or diameters; draw an ogee curve; draw an ellipse using the concentric circle method; draw a trammel ellipse; draw a foci ellipse; locate the major and minor axes of a given ellipse with a located center; draw a tangent to an ellipse at a point on the ellipse and from a distance point; draw a parabola; find the focus point of a parabola; draw a hyperbola; and join two points by a parabolic curve.

- Supplementary constructions require knowing how to do the following: draw a spiral; draw a helix; draw an involute of a line, triangle, square, and circle; and draw a cycloidal curve.

REVIEW QUESTIONS

Mark the following statements as either true or false.

1. A right angle is an angle of 90°.

2. An acute angle is an angle at more than 90°.

3. An obtuse angle is an angle at more than 90°.

4. There are six types of polygons.

5. Geometric construction procedures are not particularly important to the drafter.

6. Lines that are tangent and at 90° are considered perpendicular.

Answer the following questions by selecting the best answer.

1. Which of the following is not a part of a circle?
   a. Angle
   b. Center
   c. Sector
   d. Isosceles

2. Which of the following is true regarding a line?
   a. The shortest distance between two points
   b. Can be drawn in any direction
   c. Can be straight, an arc, a circle, or a free curve
   d. All of the above

3. Which of the following is true regarding the sum of three internal angles of a triangle?
   a. Always exactly 180°
   b. Always half of the 360° in a full circle
   c. Equal to half the sum of the three straight sides
   d. Both a and b

4. The circumference of a circle can be calculated by:
   a. Multiplying π by the diameter of the circle
   b. Dividing the diameter of the circle by π
   c. Multiplying π by 2 times the radius
   d. Both a and c

5. Which of the following is not a kind of triangle?
   a. Equilateral
   b. Obtuse
   c. Quadrilateral
   d. Scalene

6. A point in space is represented on a drawing by:
   a. A crisscross at its exact location
   b. A dot at its exact location
   c. A crossbar on an existing line
   d. Both a and c

7. Which of the following is **not** true regarding concentric and eccentric circles?
   a. Concentric circles have a common center point.
   b. Eccentric circles have no common center point.
   c. Concentric circles have no common center point.
   d. Two or more circles with a common center point are called concentric.

8. Which of the following is **not** true regarding an angle?
   a. There are three major types of angles.
   b. There are four major types of angles.
   c. An obtuse angle is one of the major types of angles.
   d. An angle is formed by two intersecting lines.

CHAPTER 4 PROBLEMS

Conventional Instructions

The following problems are intended to give the beginning student practice in the many geometric construction techniques used to develop drawings. Accuracy, line work, neatness, speed, and centering are stressed. It is recommended that dimensions not be added to problems at this time. The beginning student should practice using drafting instruments and correct line thicknesses and should concentrate on developing good drafting habits.

The steps to follow in laying out all drawings throughout this book are:

STEP 1 All geometric construction work should be made very accurately using a sharp 4H lead.

STEP 2 All geometric construction work must be laid out very lightly.

STEP 3 Do not erase construction lines. If constructed lightly, they will not be seen on the whiteprint copy.

STEP 4 Make a rough sketch of each problem before beginning to calculate the overall shape.

STEP 5 Lightly draw each problem completely first.

STEP 6 Locate all tangent points as you proceed. Make short, light dashes at each tangent point.

STEP 7 Try to center the problem in the work area.

STEP 8 Check each dimension for accuracy.

STEP 9 Darken in the drawing by using the correct line thickness and the following steps:

▸ Locate and draw all center lines with a thin black line.

▸ Darken in all diameters by using a compass.

▸ Darken in all radii by using either a compass or a circle template.

▸ Darken in all horizontal lines, either from right to left or from left to right.

▸ Recheck all dimensions.

▸ Check all lines for correct thickness.

▸ Fill out the title block by using light guidelines and neat lettering.

CAD INSTRUCTIONS

Students who plan to complete the following problems on a CAD system should begin with these general instructions.

1. Analyze the problem carefully.
2. Plan your drawing sequence. In what order will you develop the drawing?
3. Read the specific instructions for the problem and begin.

Structural ## PROBLEM 4-1

Triangular spaces are utilized for rigidity in frames and trusses. The scissors truss shown is typical of trusses premade for house construction. Using the dimensions and slopes given, lay out an approximate truss and determine approximate lengths of the 2 × 4s. Assume that a lumber store near you has 2 × 4s in stock in 8-, 10-, 12-, and 16-foot lengths. How many and what lengths of 2 × 4s are needed? (Note that the actual dimensions of standard 2 × 4s are 1½″ × 3½″.) Approximately what percentage of the lumber is wasted?

NOTE: SPECIAL GUSSET PLATES (NOT SHOWN) ARE USED AS JOINTS

Civil ## PROBLEM 4-2

The given drawing shows preliminary measurements to determine the distance across the river from line AB to the outcropping of rock on the other side. Using the information given, calculate the shortest distance from the 1,000'-long line to the rocks.

∠A 86° 44' 49"
∠A 83° 40' 25"

PROBLEM 4-3

The following X,Y coordinates in feet for determining the border of an oddly shaped parcel of property are as follows: A(0,0), B(110,-30), C(190,0), D(130,40), E(130,110), F(0,60). You are given the task of determining the area of the parcel. Suggestion: Lay out the parcel to scale, and divide the parcel into convenient areas for calculating.

Structural   ## PROBLEM 4-4

The rectangular frame shown needs a diagonal to ensure the frame's rigidity. The frame is made up of 3 × 3 × 1/4 structural steel angles and 1/4″ steel gusset plates. Thirty-five frames are to be made. Your task is to lay out the frame to the dimensions given and determine the length of the diagonal based on information given in the drawing. Then, assume that the standard length of available structural angles is 20′, and determine how many 20′ lengths are required for the 35 frames. What percentage of the structural angles will probably be wasted?

numbers on the lines are elevations in feet, and A is 100′ from B in a horizontal direction. You are requested to determine the angle and percentage slope of the terrain between A and B.

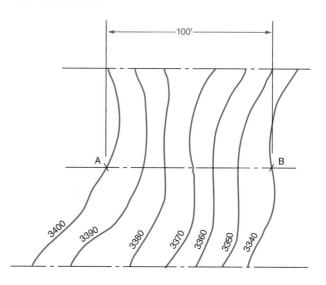

Mechanical   ## PROBLEM 4-6

Calculate the volume of the notched tubular object shown by utilizing equations for the areas of segments of a circle. Then, calculate the weight of the object by multiplying the volume by the density of the material.

STEEL 0.28 lb/in³

STEEL FRAME

Civil   ## PROBLEM 4-5

A ski slope designer made the sketch shown, which shows contour lines for a particular portion of a ski run. The

PROBLEM 4-7

Construct an ellipse using the concentric circle method shown in Figure 4-48. Minor diameter = 25 mm. Major diameter = 40 mm.

PROBLEM 4-8

Divide the work area into four equal spaces. In the upper left-hand space, draw an inclined line 3.25″ long. Bisect this line. Show all construction lines lightly.

In the upper right-hand space, draw an acute angle with intersecting lines, each approximately 2.75″ long. Bisect this angle. Show all light construction lines.

PROBLEMS

In the lower left-hand space, draw an inclined line 2.88″ long. Divide this line into five spaces. Show all light construction lines.

In the lower right-hand space, draw an inclined line 3.125″ long. Divide this line into three proportional parts 2, 3, and 4. Show all light construction lines.

PROBLEM 4-9

Divide the work area into four equal spaces. In the center of the upper left-hand space, draw an equilateral triangle having sides of 2.0. Bisect the interior angles. Show all light construction lines.

In the center of the upper right-hand space, draw a square having sides of 2.0. Show all light construction lines.

In the center of the lower left-hand space, draw a hexagon having the distance of 2.0 across the flats. Show all light construction lines.

In the center of the lower right-hand space, draw an octagon having the distance of 2.0 across the flats. Show all light construction lines.

PROBLEM 4-10

Divide the work area into four equal spaces. In the upper left-hand space, draw two lines intersecting at 60°, and draw an arc with a .88 radius tangent to the two lines. Show all light construction lines.

In the upper right-hand space, draw two lines intersecting at 120°, and draw an arc with a .625 radius tangent to the two lines. Show all light construction lines.

In the lower left-hand space, draw two lines intersecting at 90°, and draw an arc with a 1.25 radius tangent to the two lines. Show all light construction lines.

In the lower right-hand space, draw an ellipse with a major diameter of 3.00 and a minor diameter of 1.75.

PROBLEM 4-11

Divide the work area into four equal spaces. In the upper left-hand space, center a line 10 mm long and label it line A-B. Construct a straight line involute with five arcs.

In the upper right-hand space, center a triangle with sides equal to .25 and label the triangle ABC. Construct a triangular involute with five arcs.

In the lower left-hand space, center a square with sides equal to 10 mm and label the square ABCD. Construct a square involute with five arcs.

In the lower right-hand space, center a circle with a diameter of 1.00. Construct a circle involute with as many arcs as space will allow.

PROBLEM 4-12

Divide the work area into four equal spaces. In the center of the upper left-hand space, draw a spiral with 30° spaces, at 4 mm increments, for one 360° rotation. In the center of the upper right-hand space, draw two lines at right angles intersecting at point 0. One line is to be 3.5″ long and the other 2.5″ long. Draw a parabolic curve between these two lines.

In the center of the lower left-hand space, draw a rectangle 2.0″ × 4.50″ in size. Label the 2.0″ side as the rise and the 4.50″ as the span. Construct a parabola (Method 2) and locate the focus point.

In the center of the lower right-hand space, draw a line 80 mm long, and label its end-points A and B. Locate a point on line A-B at five-eighths of the line's length from point A. Label the point X.

Problems 4-13 through 4-26

Using the art for Problems 4-13 through 4-26, center each object within the work area using the correct line thickness.

Do not erase light construction lines.

Locate and draw a light, short dash at all tangent points.

Do not dimension objects.

Mechanical ## PROBLEM 4-13

Mechanical PROBLEM 4-14

Mechanical PROBLEM 4-15

Mechanical PROBLEM 4-16

Mechanical PROBLEM 4-17

PROBLEMS

Mechanical **PROBLEM 4-18**

—1/4 ELLIPSE FROM POINT A TO POINT B

—1/2 ELLIPSE FROM POINT C TO POINT D

Ø.63

ø3.0

1,75

.50

R.63

5.0

Mechanical **PROBLEM 4-19**

ELLIPSE
MINOR DIA = 100
MAJOR DIA = 150

45°

R6 TYP.

12

ø44

8

ø50   R38

4

6

12

38

METRIC

Mechanical **PROBLEM 4-21**

R15
(TYP.)

44

R14

2X ø12

57 ø38

25

R32

R32

R12

30°

R62

R12 (TYP.)

R22 (TYP.)

METRIC

Mechanical **PROBLEM 4-20**

45°

30°

R3.0   R3.12

3Xø1.25

R.25
(TYP.)

2Xø.56

R1.26

R1.63

.63

HEX .68 ACROSS FLATS

Mechanical **PROBLEM 4-22**

R.62

R.25

15°

R1.38

R.63

Ø.75

R2.0

R1.31

30°

Mechanical **PROBLEM 4-23**

ELLIPSE
ø1.38 MINOR DIA.
ø1.75 MAJOR DIA.

R.25 (TYP.)
OGEE CURVE
R1.18
.31 .62
2.0
.75
.38
.76
R1.18
A
B
ELLIPSE
ø1.25 MINOR DIA.
ø1.63 MAJOR DIA.
4.12

Mechanical **PROBLEM 4-24**

15 (TYP.)
82
R56
HEX 14 (ACROSS FLATS)
68
B
OGEE CURVE FROM POINT A TO B
A
5X ø12, EVENLY SPACED
METRIC

Mechanical **PROBLEM 4-25**

ø.68
ø.53
R.63
R.5
R1.25
ø3
R.43
R.38
30°
30°
2X ø.38
15°
30°
R2.5
ø4.0
R.63
R.43
30°
R.56
R.56
6X ø.50—EVENLY SPACED
4.18
ALL UNMARKED RADII, R.38

PROBLEMS

PROBLEM 4-26

45°

R.88 (TYP.)
R.38 (TYP.)
Ø.81
R.88

ø4.18
R3.0
30°
R.25
R1.25
R1.75
1.38
R1.63
R1.13
R1.98
30°
R.88
3X R.25
R1.25
R.41
.75

ALL UNMARKED RADII R .63

Problems 4-27 through 4-32

On an A-size sheet of vellum, carefully construct each of the assigned figures. Use thin, black center lines and thick, black object lines. Leave all light construction work. Do not add dimensions.

PROBLEM 4-27

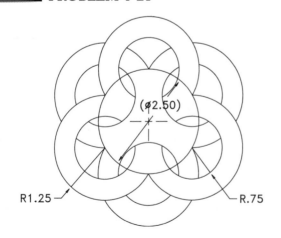

(ø2.50)

R1.25
R.75

PROBLEM 4-28

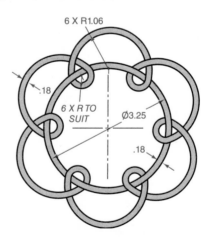

6 X R1.06

.18

6 X R TO
SUIT
Ø3.25

.18

Mechanical PROBLEM 4-29

METRIC

Mechanical PROBLEM 4-30

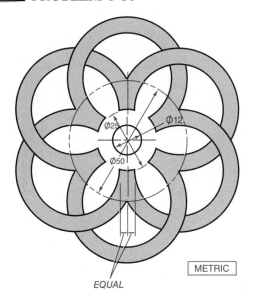

METRIC

EQUAL

Mechanical PROBLEM 4-31

Mechanical PROBLEM 4-32

METRIC

PROBLEM 4-33

Design and lay out to scale a clock face to match your interest(s) in technology.

PROBLEM 4-34

Find a logo for one of the following. Prepare a drawing of it.

    a. an automobile
    b. a computer company
    c. an airline
    d. a product of your choice
    e. an organization

PROBLEM 4-35

Refer to the figures indicated in this chapter and do the following:

    a. Figure 4-22—Draw a curve parallel to a curve similar to the lower edge of the object for Problem 4-9.
    b. Figure 4-30—Reproduce the drawing for Figure 4-30 and transfer it.
    c. Figure 4-46—Change the spacing of lines A-B and C-D and draw an ogee curve to join them.
    d. Figure 4-61—Draw the helix shown.
    e. Figure 4-65—Follow the steps and draw an involute curve.

    NOTE: The involute curve is the basis for the profiles of spur gears and helical gears.

    f. Figure 4-68—Construct the epicycloid shown.

Mechanical ## PROBLEM 4-36

The golden section shown is attributed to the Greeks who used it as a guide to create pleasing proportions for architecture, sculpture, and pottery. Three nested golden sections can be formed by following the construction pattern shown. Note that the ratio of the sides of each golden rectangle is 1.618 to 1.

$(2.618)/(1.618) = 1.618$; and $(4.236)/(2.618) = 1.618$

(For interest, check the ratios of the sides of credit cards and some food packaging.) Refer to the following figure and reconstruct the three golden rectangles. Then, lay out some shelves for a living room for books and hi-fi equipment. Include vertical dividers that incorporate the golden section proportions.

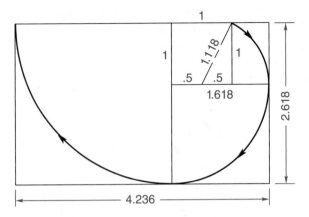

Mechanical ## PROBLEM 4-37

The freehand sketch given shows a format for a typical classroom engineering problem statement and solution. On a sheet of 8½ × 11″ (A-size) paper, redo the statement (GIVEN and FIND) and the solution. Use a straightedge for the sketch and diagrams and good lettering style for the statement and solutions.

PROBLEM 4-38

You have been asked to help an individual lay out a tentative plan for a service garage with parking facilities. The commercially zoned lot measures 25 m × 35 m. Sketch No. 1 shows a preferred arrangement of the service-pits building with parking. Lay out the lot using the given information. Use a scale to make the best use of an A-size drawing.

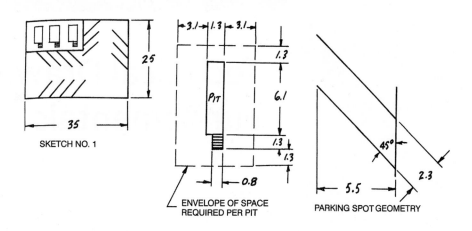

SKETCH NO. 1

ENVELOPE OF SPACE
REQUIRED PER PIT

PARKING SPOT GEOMETRY

PROBLEMS

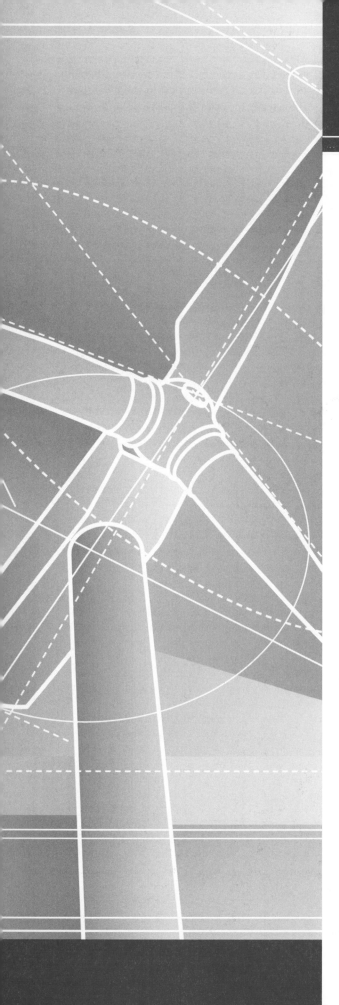

section 2

Technical Drawing Fundamentals

Alexander Torres

Graduating with a master's degree in fine arts, Alexander Torres thought he might become a painter—but then reality (and bills!) set in. Having studied technical drawing in high school and then gone on to get a degree in engineering and drafting, he knew that he *did* have experience in a more lucrative field.

Using the skills he learned in college and high school, he started doing illustrations and flowcharts for companies like Lockheed Martin. After a few years, he found his way into magazine work, which required both his drafting skills and his artistry.

Today, as a technical illustrator for *TK* magazine, he often has to create drawings that communicate the ideas explained in an article. The function of these schematics is part artistic (to illustrate the point: He recently used drafting software to draw a candy bar to metaphorically show that when working with a motherboard, you need to break it into its components) and part functional (the art must often explain complex parts of computers in ways that mere text cannot).

On this day, he has just been handed an article about circuit boards. His first move in planning the drawing: Read the abstract. He will then go over the concept with the editors to find out what they are looking for—do they want a dynamic look (in which case he might present an orthographic projection) or true representation? If the latter, what focal point should he take? In drawing something like this, he tries to look at the real thing. If he doesn't have it accessible at home, he will search the Internet. These will help him to visualize the drawing that he is about to do. Then, Torres does a pencil sketch, organizing the information so that it will work in a rectilinear format.

He will then plug in numbers into Infinity, a computer drafting system that allows him to build in 3D—a benefit because then he can move the point of view of the camera as needed. Sometimes he may decide that a fish-eye view will help the object jump off the page; other times he must stick with the realistic focal point. Either way, the artist in him tries to ensure that the final product will not be merely technical and accurate but also an element that draws the reader into the magazine.

5

Spatial Visualization and Multiview Drawings

CHAPTER OBJECTIVES

Upon completion of this chapter, students should be able to do the following:

▶ Explain the concept of spatial visualization.

▶ Demonstrate proficiency in the use of spatial visualization.

▶ Demonstrate proficiency in the use of orthographic projection.

▶ Demonstrate proficiency in all five steps of planning a drawing.

▶ Demonstrate proficiency in centering a drawing.

▶ Demonstrate proficiency in making proper use of technical drawing conventions.

▶ Demonstrate proficiency in applying visualization techniques.

▶ Explain the difference between third- and first-angle projections.

KEY TERMS

Blind hole

Conventional break

Counterbored hole

Countersunk hole

Fillets

Fold line

Foreshortening

Front view

Hidden lines

Orthographic projection

Perspective projection

Principal viewing planes

Right-side view

Rounds

Runouts

Spatial visualization

Spotfaced hole

Tapered hole

Through hole

Top view

Drafters, designers, and engineers follow a design process to solve technical product problems and use technical drawings to document and develop their ideas. The steps in the process leading to the technical drawings, primarily working drawings, were noted in the general comments in the Introduction. In Chapter 2, the drawing tools used to prepare the drawings were discussed, followed by another chapter that emphasized the first format drafters, designers, and engineers use to record their ideas and sketching techniques. This chapter focuses on the ever-present challenge to a design person to prepare multiview drawings, to scale, of technical, three-dimensional concepts on a two-dimensional surface so that the users of the drawings will clearly understand, in three dimensions, what the design person had in mind and have sufficient information to accurately produce the concept.

Providing accurate shape descriptions of objects by drawing methods requires that three-dimensional object information be presented in a flat, two-dimensional drawing space. This is usually done by drawing images of the object from multiple directions. Commonly shared methods and interpretations are essential for all who make or use such drawings. Technical drawings seldom use more than lines to outline an object's features. Visual qualities, such as color and texture, are more accurately specified by written requirements.

Viewing an object by eye creates a depth distortion. This phenomenon is recreated in a field of drawing called *perspective projection* (**Figure 5-1**). This distortion occurs because the visual rays used to create the image in the viewer's eye are not parallel. This distortion, while useful in providing the illusion of distance, results in a loss of image accuracy and fails to provide the information needed for most detailed technical communication. In order to develop multiview drawings, you must be proficient in the concept of spatial visualization.

Spatial Visualization

We see the world in three dimensions. When we look at an object, it has height, length, and depth. But when working in engineering or drafting, we sometimes have to communicate our ideas in a two-dimensional format. Making two-dimensional representations of three-dimensional objects or looking at a two-dimensional representation and visualizing the object in three dimensions requires a skill known as *spatial visualization.*

Spatial visualization is the act of (1) forming an accurate mental picture of how a three-dimensional object will look when represented in a two-dimensional format, or (2) forming an accurate mental picture of how a three-dimensional object will look based on the information given in a two-dimensional representation. Some people are innately talented at spatial visualization and some are not. However, with practice, most people can become proficient at spatial visualization.

Two-Dimensional Spatial Visualization

To gain a better understanding of the concept of spatial visualization, try the following experiment:

1. Pick up your copy of this book and look at it as a three-dimensional object (see **Figure 5-2**).

2. Now orient the book so that your line of sight is perpendicular to the front cover. You should see no depth—just a flat surface. This is part of the concept of orthographic projection used when constructing multiple views of an object. Orthographic projection is covered later in this chapter.

FIGURE 5-1   Perspective projection.

FIGURE 5-2 Three-dimensional view of this book.

3. Repeat Step 2 for all sides of the book. You should see the following sides: front, back, top, bottom, right side, and left side (see **Figure 5-3**).

4. At this point, do not concern yourself with the orientation of the six views of the book shown in Figure 5-3.

Orienting views is important, but it is covered later in this chapter. For now, you need only concentrate on visualizing what the views should look like in a two-dimensional format, and you should understand that any three-dimensional object has these six views when applying the concept of orthographic projection. Even a plain globe has six views (all six would look the same).

Try this experiment with several other objects that are conveniently available, such as a chair, soda can, computer mouse, coin, or any other object that is

handy. But this time, try to visualize each view without actually turning the object around and looking at it. Based on what you can see in three dimensions, what would the front, back, right side, left side, top, and bottom of the object look like in two dimensions? Make a sketch of each view of each object. This is a convenient experiment that can be conducted almost anywhere. Continue the experiment for a week with various objects in your environment, and you will find yourself becoming skilled at two-dimensional spatial visualization.

Three-Dimensional Spatial Visualization

Visualizing what a three-dimensional object will look like when drawn in two dimensions is half of the spatial visualization challenge. The other half involves using a two-dimensional representation of an object as the basis for visualizing what the actual object will look like in three dimensions. For example, look at the two-dimensional drawing in **Figure 5-4**. How would this object look in three dimensions?

Looking at the front and back views, you can see that the object in question is shaped like an L. By looking at the right and left sides, you can discern how wide the object is. The top and bottom views show in which directions the legs of the L point. With this information and a little practice, you can determine that the three-dimensional object in question would look like **Figure 5-5**.

The ability to apply spatial visualization is important to engineers and drafting technicians. For most people, developing this skill takes time, patience, and practice—lots of practice. Fortunately, spatial visualization is a concept that can be practiced anywhere. Until your spatial visualization skills are well-developed, practice it on objects in your environment.

FIGURE 5-3 Orthographic views of this book.

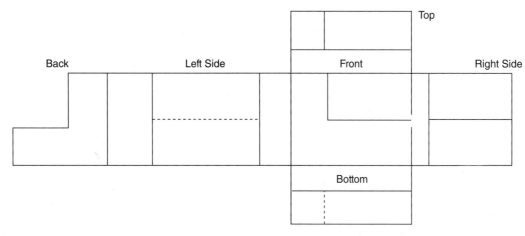

FIGURE 5-4   Multiple views of an object.

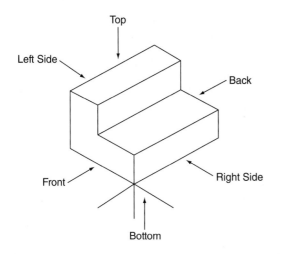

FIGURE 5-5   Three-dimensional view of an object.

Solid modeling software (Chapter 11) is making spatial visualization easier for engineers and drafting technicians by doing much of the work for them. But you cannot develop a three-dimensional solid model every time you need to visualize an object. Consequently, it is important to develop your skills in applying spatial visualization manually.

Orthographic Projection

Orthographic projection is the most accurate term for shape description wherein an undistorted image of the object appears in a flat, transparent, but imaginary projection plane. A projection plane may be thought of as an imaginary pane of glass with an object underneath it or behind it (**Figure 5-6A**). Viewed from the side, the pane of glass appears as a line (**Figure 5-6B**). Assume that light beams emitting from the surfaces of the object are projected to the

projection plane and that each light beam from each exposed surface is directed toward the viewing plane (**Figures 5-6C** and **5-6D**). The image shown in Figure 5-6A traces the path of each light beam as it intersects the viewing plane. This image is called a projection, and it illustrates the features of the exposed surfaces of the object.

Normal Surfaces

The surface images of the figures just described do not represent the size and shape of their corresponding real surfaces. Rectangular surfaces appear as parallelograms. The lengths of edges are shorter in the image than their actual lengths (a phenomenon called *foreshortening*). However, if the viewing plane were positioned parallel to several of the object's surfaces, as shown in **Figure 5-7A**, then the image appearing in that viewing plane would provide the actual size and shape of those surfaces, as shown in **Figure 5-7B**. A surface that is parallel to a viewing plane is said to be "normal" to the viewing plane.

A simple object, such as the cannonball in **Figure 5-8**, needs only one view to describe its true size and shape. An object such as the flat gasket in **Figure 5-9** needs only one view and a callout stating its required thickness. The callout can be noted on the drawing, as illustrated, or listed in the title block under "material." **Figure 5-10A** shows an object with a viewing plane placed in a normal position relative to some surfaces of the object, resulting in the orthographic views shown in **Figure 5-10B**. The sizes and shapes of the object's normal surfaces are shown in perfect outline in the orthographic view, but the viewer still cannot determine other features of the object. For example, **Figures 5-10C** through

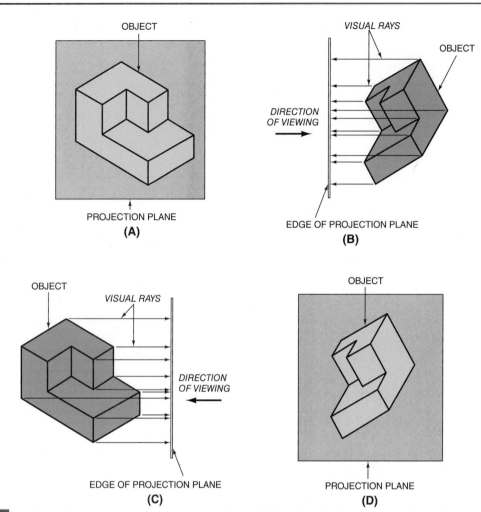

FIGURE 5-6   (A) Imaginary windowpane in front of object; (B) Side view of windowpane; (C) and (D) Reversed projection.

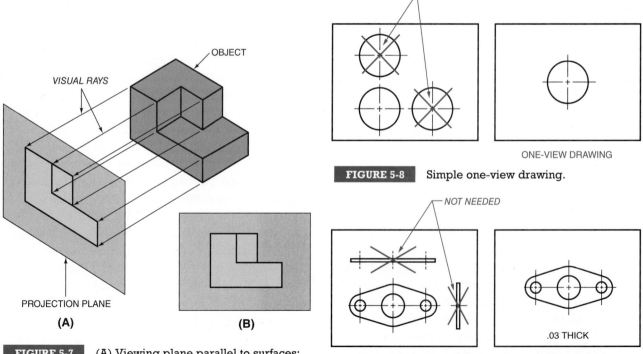

FIGURE 5-7   (A) Viewing plane parallel to surfaces; (B) Viewing plane provides actual size and shape of surfaces.

FIGURE 5-8   Simple one-view drawing.

FIGURE 5-9   Simple one-view drawing.

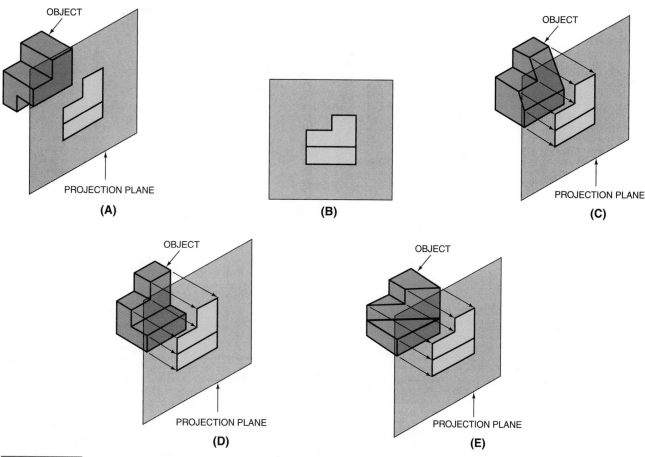

FIGURE 5-10   (A) Viewing plane in a normal position; (B) Orthographic view of the object; (C) Different object with the same orthographic view; (D) Another object with the same orthographic view; (E) Another object with the same orthographic view.

5-10E are different objects that provide exactly the same projected view. Furthermore, not all the surfaces of the projected views of Figures 5-10C and 5-10E are shown in their real size and shape, as they are not normal to the viewing plane. There is no way to fully distinguish these without additional information.

Two Orthographic Views

In order to present the images of each viewing plane in a flat drawing area, one of the viewing planes must be repositioned to lie in the same plane of the drawing as the other. The procedure used to do this is to create a *fold line* where the imaginary perpendicular-orthographic-viewing planes meet each other along the straightedge of the intersection, as shown in **Figure 5-11A**. Figure 5-11A shows the object in Figure 5-10A with a second viewing plane in position. The fold line acts as a hinge as the two intersecting viewing planes are swung into the same plane (**Figure 5-11B**). The resulting orthographic views are shown in **Figure 5-11C**. Each view contains the image of a 90° rotated view or projection of the other. In this procedure, the features and surfaces

are always aligned to each other and thus can be located.

Conversely, the same two orthographic views are shown in **Figures 5-11D** through **5-11F** in haphazard relationship to each other, putting them each in error and making it impossible to determine surface positions or the real shape of the object.

Labeling Two Views

Figures 5-12A and **5-12B** show the positioning of two adjoining orthographic viewing planes to describe the object shown in Figure 5-10C. The orthographic view that presents the most characteristic shape of the object is usually selected and identified as the *front view*. Once this view is selected, the orthographic view that is projected to the right of the front view is identified as the *right-side view*. The front view is often referred to as the front elevation, and the right-side view is referred to as the right profile.

Figure 5-13A shows viewing planes with view images that accurately describe Figure 5-10D. But finished orthographic drawings should not include

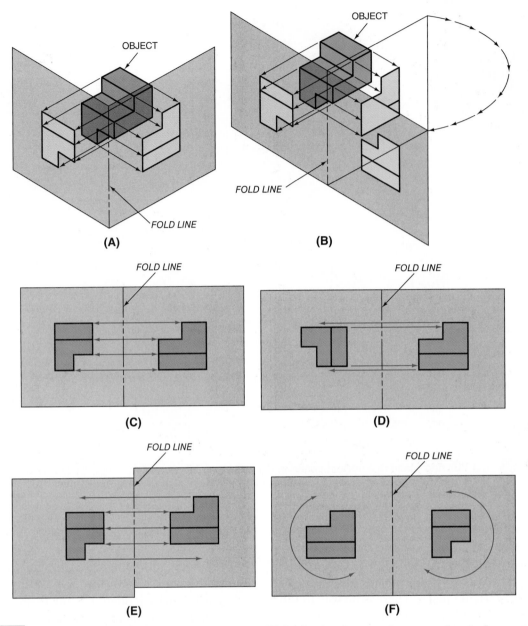

(A)   **(B)**

(C)   **(D)**

(E)   **(F)**

FIGURE 5-11 (A) Fold line; (B) Fold line acts as a hinge; (C) Projecting features from one view to the next view; (D), (E), and (F) Incorrect positioning of views.

(A)   **(B)**

FIGURE 5-12 (A) Positioning of two adjoining orthographic viewing planes; (B) Two adjoining orthographic viewing planes flattened out.

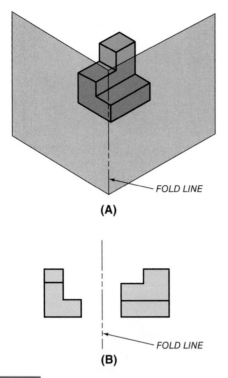

(A)

(B)

FIGURE 5-13   (A) Viewing planes and images;
(B) Orthographic drawing omitting
viewing planes.

the borders of viewing planes. For example, **Figure 5-13B** shows the views without the borders. Viewing planes were earlier identified as imaginary; therefore, the fold line is imaginary also. Often, fold lines are needed for view constructions, as illustrated later in this chapter and future chapters, but in general, fold lines are removed on technical drawings.

The distance of the image from the fold line identifies the object's location from the viewing plane. The drafter decides what this distance will be in order to make the best use of the space on the drawing paper and to provide room for dimensions and notes (see **Figure 5-14**).

Multiple Orthographic Views

Figure 5-15A shows two orthographic views that do not completely describe the object. Multiple

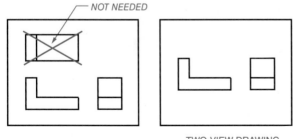

TWO-VIEW DRAWING

FIGURE 5-14   Only two views are required.

interpretations of the objects are possible, causing the viewer to be misled by inadequate information. Correctly interpreted, the object appears as shown in **Figure 5-15B**. Incorrect interpretations are shown in **Figures 5-15C** through **5-15E**.

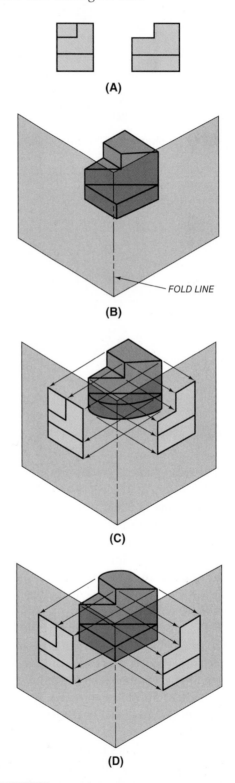

FIGURE 5-15   (A) Orthographic views of an object;
(B) Correct interpretation; (C) and
(D) Incorrect interpretations;

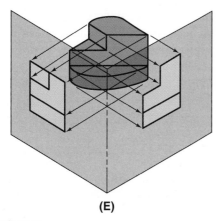

FIGURE 5-15   (*Continued*) (E) Incorrect interpretation.

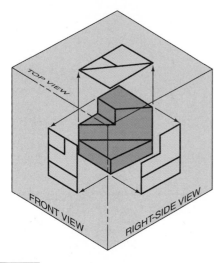

FIGURE 5-16   Third viewing plane added.

A third viewing plane is commonly used to ensure adequate definition of the represented object. The third viewing plane is usually above the object and perpendicular to the other two. It is called the *top view*. The top view is sometimes called the *plan view* in construction-related drawings. This third view is shown in **Figure 5-16**.

Select the views to best present the object's shape and make the most efficient use of the drawing area (**Figure 5-17A**), or to maintain the orientation of the object that is most easily understood by the reader, as illustrated in **Figure 5-17B**. Recall that the view selected should be an attempt to show the most characteristic shape of an object. For example, seeing a building drawn sideways makes reading the drawing very awkward.

The three viewing planes in Figure 5-16 are called *principal viewing planes,* as they are all perpendicular to one another. Orient the front view first. Three other principal planes complete the "box"; they make up the sides of a transparent box that completely surrounds the object (**Figure 5-18A**). Figure 5-18B

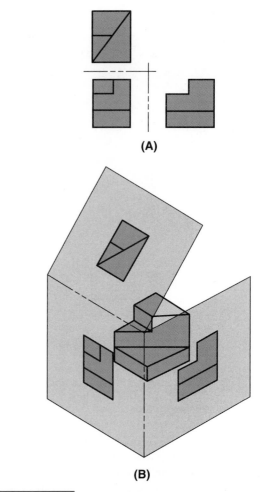

FIGURE 5-17   (A) Orthographic view of an object; (B) Three principal planes.

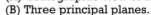

FIGURE 5-18   (A) Transparent box;

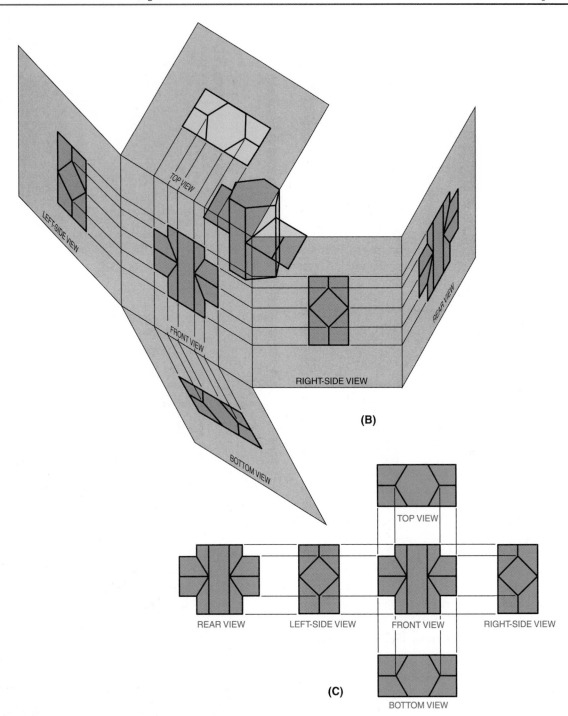

(B)

(C)

FIGURE 5-18   (*Continued*) (B) Fold lines selection; (C) Resulting multiple views.

shows the possible fold line selections that could be used with orthographic views on the principal planes. **Figure 5-18C** shows the six principal views.

Object Description Requirements

The six views of Figure 5-18C are not all necessary to provide sufficient information; three are sufficient for this object. The common arrangement of top, front, and right-side views is shown in **Figure 5-19**.

The top view aligns directly above the front view, and the right-side view aligns directly to the right of the front view. Both the top view and the right-side view images are the same distance behind the front view projection plane. See dimension b in **Figure 5-20**.

Dimension Transfer Methods

Figure 5-19 shows a method of transferring dimensions between the top and right-side views. This

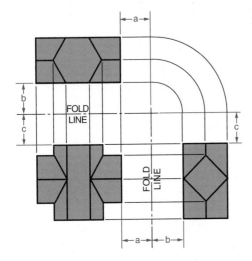

FIGURE 5-19   Transferring dimensions—
45° projection method.

FIGURE 5-20   Transferring dimensions—compass
method.

method, used only with principal views, employs a line at 45° to the fold lines of the principal planes.

Use a compass for a different method shown in Figure 5-20.

Figure 5-21 shows the transfer method, which is usually the most accurate. It is the method used in descriptive geometry discussed in Chapter 9. Draw the front and top views first and arbitrarily position a fold line between them. Recall that the fold line must be perpendicular to the projection lines between views. Construct a second fold line perpendicular to the projection lines that will go to the right to be used for the right-side view. This second fold line can be at any distance from the front view. Once the fold line is drawn, the distances from it to the right-side view must agree with distances to the top view (dimensions a, b, and c). Transfer these dimensions with a scale or dividers. If the front and right-side views are drawn first, then construct a fold line for the top view in a similar manner.

FIGURE 5-21   Transferring dimensions—transfer
method.

Hidden Lines

Figure 5-22 is an isometric view of an object. The orthographic top, front, and right-side views are illustrated in **Figure 5-23**. *Hidden lines* represent

FIGURE 5-22   Isometric view of an object.

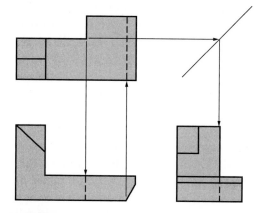

FIGURE 5-23   Orthographic views of the same object.

feature outlines—for example, intersections of plane surfaces, real but invisible—and are used in each viewing plane where appropriate.

A hidden line may be covered by a visible (solid) object line in a particular view. This concept is shown in **Figures 5-24A** and **5-24B**. Figure 5-24B shows surface B aligned behind the solid line of the intersection of plane A with the inclined plane. As expected in orthographic projection, there are many hidden lines covered by solid lines. Mentally perceiving where the hidden lines lie in orthographic views requires visualization.

Figure 5-25 shows typical hidden line length and spacing, approximately 1/8″ to 1/4″ long spaced approximately 1/32″ to 1/16″ (3.2 mm to 6.4 mm and .8 mm to 1.6 mm, respectively). The lengths and spaces can vary depending on the feature and the space available on the drawing, but consistency is encouraged. Hidden lines are narrower than solid lines. Use the accepted drafting practices of hidden line applications shown in **Figure 5-26**.

Center Lines

Center lines are used primarily as locations of circular, cylindrical, or spherical features. They can also be used to specify locations of other principal symmetries as well, such as the center line of a ship or a rocket. Crossed perpendicular center lines indicate

(A)

OBJECT LINE COINCIDES WITH HIDDEN LINE

(B)

FIGURE 5-24   (A) Isometric view of an object; (B) Object line takes precedence over hidden line.

DASH   SPACE

APPROX. $\frac{1}{8}$ APPROX. $\frac{1}{32}$

FIGURE 5-25   Average hidden line construction.

DO NOT TOUCH

CORNERS TOUCH

ALL CORNERS TOUCH

SPACE

STAGGER

TOUCH

FIGURE 5-26   Hidden line drafting practices.

a two-coordinate location of where an axis exists. A single center line indicates the path of an axis. Identify all axes, on all views, of circular features except for noncritical radii, Figure 5-26.

Surface Categories

Recall that normal surfaces are parallel to one of the principal projection planes of the imaginary transparent box surrounding the object. In **Figure 5-27**, these surfaces are identified as the top, front, and right side. The viewing planes containing these images are identified as horizontal, frontal, and profile projection planes, respectively. A normal surface's true size and shape appear in one of the principal views (**Figure 5-28**), but the surface appears as an edge view (a line) in the other principal views.

Surfaces A, B, and C, Figures 5-27 and 5-28, are inclined surfaces. They are not parallel to any one of the principal viewing planes, but each inclined surface is perpendicular to one of them. For example, inclined plane A is perpendicular to the frontal projection plane. Moreover, a surface that is perpendicular to a viewing plane will appear as an edge or line when projected on that plane. Surface A appears

smaller than its true size in both the top and right-side views, Figure 5-28, because of foreshortening. Inclined surfaces B and C are configured similarly to surface A, with B perpendicular to the profile projection plane, and C is perpendicular to the horizontal projection plane. Surface D is not perpendicular to any of the principal projection planes and is therefore categorized as an oblique plane. It appears foreshortened in all of the principal projection planes.

The transparent box of principal projection planes that surround cylindrical surfaces uses the cylindrical axis as the line of orientation (**Figure 5-29**). If the cylindrical axis is made perpendicular to one of the principal viewing planes, then the cylindrical surface appears as a circle or an arc. Figure 5-29 shows a cylinder whose axis is perpendicular to the top horizontal principal projection plane, and the circular image that results in the top view is shown in **Figure 5-30**. The curved surface of the cylinder cannot be classified as normal, inclined, or oblique. Plane surface A in Figure 5-30 is inclined to the horizontal viewing plane and shows as an ellipse in the right-side view. If surface A were at 45° to the horizontal viewing plane, a circle would appear as surface A in the right-side view.

FIGURE 5-27   Surface categories.

FIGURE 5-28   Orthographic views.

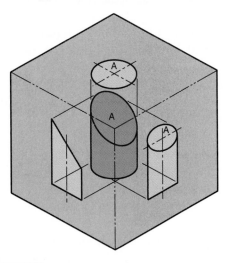

FIGURE 5-29   Object inside transparent box.

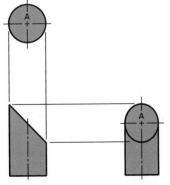

FIGURE 5-30   Circular image results in this top view.

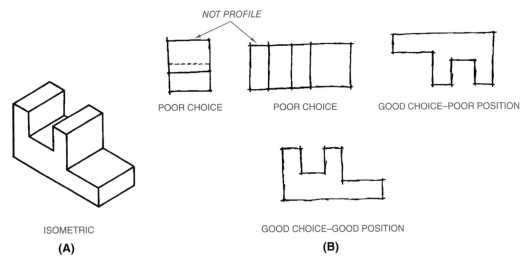

FIGURE 5-31   (A) Isometric view of an object; (B) Sketches of object to determine front view.

Planning the Drawing

Before beginning a three-view drawing of any object, make preliminary sketches of the object. Sketches of the object in **Figure 5-31A** are shown in **Figure 5-31B**. The sketches were used as an aid in selecting the best front view and its best position. Sketches are helpful in selecting the required views and their positions.

HOW TO PREPARE A THREE-VIEW DRAWING

STEP 1   Visualize the object to be sure you have a good mental picture of it.

STEP 2   Decide which view to use as the front view by sketching the object in several positions, Figure 5-31B. Keep in mind the following:

▸ The front view is the most important view; viewers see this one first.

▸ The front view should show the most basic profile; for example, use a side view of a vehicle with the wheels on the ground.

▸ The front view should appear stable; place the heavy part at the bottom.

▸ The front view should be positioned to minimize the number of hidden lines in all views; this enhances visualization.

STEP 3   Decide how many views are needed to completely illustrate the object.

STEP 4   Decide, keeping in mind the guidelines in Step 2, in which position to place the front view. Study **Figures 5-32** through **5-35** to understand why Figure 5-35 is the best position for the object from Figure 5-31A.

FIGURE 5-32   Poor front view—too many hidden lines.

FIGURE 5-33   Poor front view—unstable, too many hidden lines.

FIGURE 5-34   Poor front view—unstable.

FIGURE 5-35   Best position—front view.

STEP 5   Decide on the best placement of the views in the work area, usually centered. Work neatly.

Positioning the Views within the Work Area

The example shown on the left of **Figure 5-36** is centered, but poor judgment was exercised in using the space in the drawing. Space is wasted on both sides, and little room is available at the top and bottom for dimensions and notes. An inch (25.4 mm) of space near the borders is desirable. The example shown on the right in Figure 5-36 is balanced and well-centered. Try to keep open spaces evenly distributed around views. A systematic method for centering views is discussed next, after some comments on a general sketching procedure for centering views.

Sketching Procedure

Sketch efficiently by doing it freehand, quickly, and only to approximate scale on a sheet of practice drawing paper. First, select the front view, using the guidelines previously discussed, and sketch a light outline of the object in approximately the best position. Next, sketch outlines of the top and right-side views. Locate center lines of important features. Finally, darken the outline and center lines.

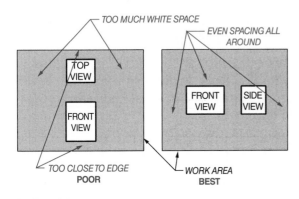

FIGURE 5-36   Positioning the views within the work area.

Centering the Drawing

Figures 5-37 and **5-38** show a simple object and a procedure used to center a sketch of the object within a specified work area. The simple object in Figure 5-37 is shown in a dimensioned isometric drawing. The views of the object in both figures are to be sketched freehand. Use an 8½″ × 11″ practice sheet, and follow the procedure given. There is no need to use a scale, but a drawing sheet with fade-out lines would be helpful.

FIGURE 5-37   Centering views horizontally.

FIGURE 5-38   Centering views vertically.

Calculate the total horizontal distance needed for the front view plus the right-side view by adding the front view width of 4.0 plus 2.0 for the depth of the right-side view for a total of 6.0″. To center these views, subtract 7.0 from 11.0 to get 4.0. Divide this into thirds for the spacing at each end and between views (D = 1.66″) (see Figure 5-37).

Calculate the total vertical distance needed for the front view plus the top view by adding the front view height of 2.5 plus 2.0 for the depth of the object for a total of 4.5″. To center these views, subtract from 8.5 to get 4.0, and divide this into thirds for the upper and lower and between-view spacing (D = 1.33″) (see Figure 5-38).

Try an alternate approach for centering a drawing. Lay out to scale on a practice sheet, with drawing instruments, outlines of the front view and the right-side view with a 1.0″ space between the two views. Next, lay out the top view with a 1.0″ space between the front and top views. Next, place a good drawing sheet over the practice sheet and move it around until the drawing is basically centered. Finally, trace the drawing onto the good sheet.

Numbering for Making Drawings

Sometimes a multiview drawing of an object, to be prepared from an isometric drawing, is so complicated that the person making the drawing is not quite sure how some part of a view will look. These more difficult drawings are easier to visualize if points of the various features are numbered and then systematically analyzed. Think of the corners of the features as lines connecting points. For instance, think of each line as connecting two points in space (**Figure 5-39**). If the ends of a line can be found, all that needs to be done to complete the line is to connect the ends. Once the ends have been found and numbered in one view, the same can be done in other views by simple projection. **Figure 5-40** is an isometric view of a simple object, which is simple for discussion purposes. Each end of each line has been assigned a number.

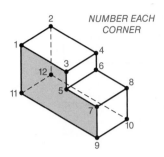

FIGURE 5-40   Given: An isometric view of an object.

Use the following steps for practice in numbering and analyzing drawings.

STEP 1   Lightly draw the basic shape outline of the required views and the 45° projection line (**Figure 5-41**).

STEP 2   Assign the numbers from the isometric sketch to each corner of the view.

STEP 3   Project these points up to the top view, and number them as shown in (**Figure 5-42**). Start with the number seen first, followed by a comma or slash, and then the number hidden.

STEP 4   Project these points from the front view to the right-side view, and project the same

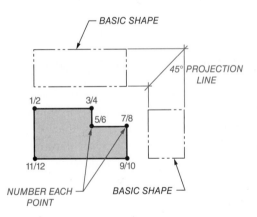

FIGURE 5-41   Number one view.

FIGURE 5-42   Project numbers into the top view.

FIGURE 5-39   A line is made up of two points in space.

FIGURE 5-43   Project numbers into the right-side view.

numbered points from the top view via the 45° line. The exact position of each point is located where the projection lines cross; number them (**Figure 5-43**). Projecting and locating points can be done with alternative methods, as shown in Figures 5-20 and 5-21.

NOTE: If a point can be located in any two of the three views, it can be projected into the third view.

STEP 5   Connect points together—for example, point 1 to point 2, 3 to 4, and so on—until all the line connections are made. Check the planes made by the lines. Refer to the isometric view, Figure 5-40, as needed.

COMMENT: These steps describe how to make a complicated drawing. Read and understand a complicated drawing already made by first locating and numbering points on an object in at least three views. Connect the numbers and systematically study the object, line by line, plane by plane.

Technical Drawing Conventions

Assume now that preliminary sketching has been done to locate the best positions for views of an object for a multiview drawing and that the views are going to be drawn, to scale, with the drawing tools discussed in Chapter 2. The following topics are important for multiview, technical drawings.

- Rounds and Fillets
- Runouts
- Treatment of Intersecting Surfaces
- Curve Plotting
- Cylindrical Intersections
- Incomplete Views
- Aligned Features
- How to Represent Holes
- Conventional Breaks

The following discussions on recommended techniques for preparing multiview technical drawings all emphasize the same message: effective communication of three-dimensional concepts—that is, to enhance the visualization process for the use of the drawings.

Rounds and Fillets

Metal casting processes require an object's material to flow into cavities and corners in a mold. Sharp corners are not only undesirable on a casting but are also difficult to cast, so *rounds* are specified for outside corners and *fillets* for inside corners. Cast objects tend to crack due to the strain placed on the metal during cooling. Fillets distribute the strain and alleviate the tendency to crack. Rounds and fillets improve an object's ability to withstand cyclic fatigue loading and also enhance the appearance of an object (**Figures 5-44** and **5-45**).

Runouts

Runouts are curved surfaces formed where a flat and a curved surface meet, Figure 5-45. To find the exact intersection where a runout occurs, locate the tangent points of the curved surfaces (**Figure 5-46**).

FIGURE 5-44   Rounds and fillets.

FIGURE 5-45   Runouts.

FIGURE 5-46   Locate the tangent points.

Project these tangent points into the next appropriate view and add the runouts, as shown in **Figure 5-47**. Note the direction of each runout.

Treatment of Intersecting Surfaces

Figures 5-48 through **5-51** show how to illustrate various intersecting surfaces. In these examples, rounds, fillets, and runouts have been omitted to simplify the drawings. In actual practice, rounds, fillets, and runouts would have been added. In Figure 5-48, the flat surface meets the cylinder sharply (see top view), making a visible edge necessary in the front view. The meeting point in the top view is the exact location of the object line in the front view and right-side view.

In Figure 5-49, the flat surface blends into the cylinder, stopping at the point of tangency. In the top view, the tangent point is located at the center line and blends in at that location in the front view.

In Figure 5-50, the surface blends into the cylinder, stopping at the point of tangency, similar to Figure 5-49 except on an angle. The tangent point is projected into the front view and right-side view.

FIGURE 5-47   Complete runouts at tangent points.

FIGURE 5-48   Intersecting surfaces.

FIGURE 5-49   Intersecting surfaces.

FIGURE 5-50   Intersecting surfaces.

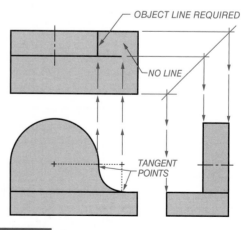

FIGURE 5-51   Intersecting surfaces.

In Figure 5-51, the radius blends into the base, stopping at the point of tangency. In the top view, a line is drawn extending to the tangent point of the arc in the front view.

Curve Plotting

Curved surfaces in multiview drawings need to be projected into other views. Follow the steps listed to plot a curved surface.

STEP 1 Lightly complete the basic outline of each view (**Figure 5-52**). Locate the center of an arc, point X.

STEP 2 Draw the arc and divide it into equal spaces (**Figure 5-53**). Use 30° spacing for smaller arcs but 10° spacing for larger ones. If accuracy is not needed, use larger spacing to save time.

STEP 3 Number the points in order clockwise, Figure 5-53.

STEP 4 Project these points up to the 45° axis line and to the left to the top view (**Figure 5-54**).

STEP 5 Project from the right-side view to the slanted surface in the front view and locate each point (**Figure 5-55**).

STEP 6 Project these points up from the front view to the top view.

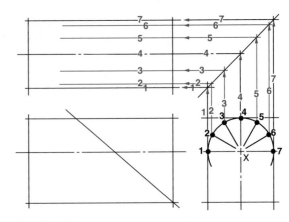

FIGURE 5-54 Curve plotting (Step 4).

FIGURE 5-55 Curve plotting (Step 5).

FIGURE 5-52 Curve plotting (Step 1).

FIGURE 5-56 Curve plotting (completed).

STEP 7 Number the points 1 through 7 in the top view, where the numbered lines from the right-side view intersect the numbered lines from the front view (**Figure 5-56**).

NOTE: *The method shown in Figure 5-21 could be used for projecting and locating curves.*

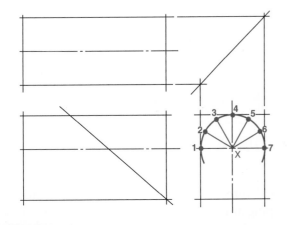

FIGURE 5-53 Curve plotting (Steps 2 and 3).

Cylindrical Intersections

If there is a considerable difference in the diameters of each of the intersecting cylinders, then use a straight line for the intersection (**Figures 5-57A** and **5-57B**). If there is little difference between the intersecting cylindrical diameters, then use a simple arc, constructed or approximated through three points (**Figures 5-58A** and **5-58B**).

In sheet metal construction, locate the actual edge of intersection, as shown in **Figures 5-59** and **5-60**. Figure 5-59 shows line elements on each of two intersecting cylindrical surfaces. The line elements are parallel to their respective cylindrical axes and to a common (frontal) viewing plane. For example, line elements number 1 on each cylinder are the same distance from the frontal viewing plane. Also, line elements numbers 2, 3, and 4 are each the same distance from the frontal plane, respectively. Use this relationship of the elements to construct a curved line of intersection between the large cylinder and the smaller cylinder shown in Figure 5-59.

STEP 1   Refer to **Figure 5-60A** and locate line elements 1, 2, 3, and 4 on the smaller cylinder in the right-side view. Project them into the front view.

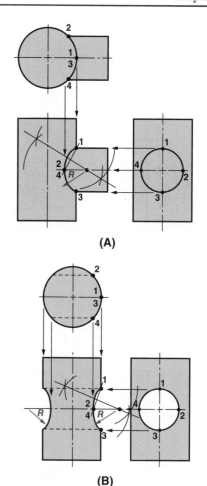

(A)

(B)

FIGURE 5-58   (A) Cylindrical intersections (arc); (B) Cylindrical intersections (arc).

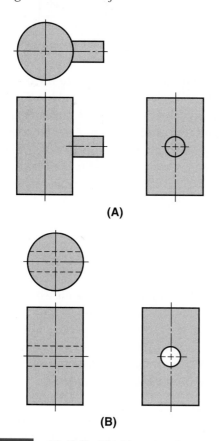

(A)

(B)

FIGURE 5-57   (A) Cylindrical intersections (straight line); (B) Cylindrical intersections (straight line).

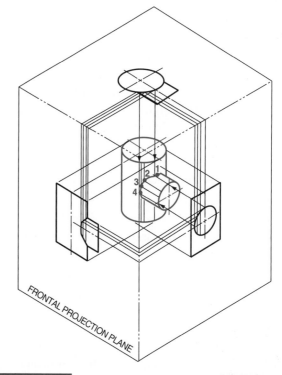

FIGURE 5-59   Two intersecting cylindrical surfaces.

(A)

(B)

FIGURE 5-60   (A) Step 1; (B) Step 2.

STEP 2 Locate a fold line between the front and right-side views (**Figure 5-60B**), at a distance A from the center line of the large cylinder. Locate a fold line at the distance A from the top view.

Transfer the distances labeled 1 through 4 in the right-side view to the top view as shown.

Next, because each corresponding line element from each cylinder is parallel to the front viewing plane and the same distance from it, they must intersect. Fit a curve through the points of intersection, Figure 5-60B. In a like manner, complete the lower portion of the intersection.

Incomplete Views

A drawing can be absolutely correct but confusing, as in **Figure 5-61**. The left- and right-side views are correct,

CORRECT BUT POOR PRACTICE

FIGURE 5-61   Overlapping views—poor practice.

INCORRECT BUT BETTER PRACTICE

FIGURE 5-62   Omitting some details—better practice.

but they are difficult to read due to the overlapping of the details on both ends of the bracket. Omit some of the features in order to make a view clearer and easier to understand (**Figure 5-62**). For example, the right-side view shows only the right-side details. Omitting some of the details in a view for clarity is a standard practice used in technical drawing. Sometimes a note is added such as, "only right side shown."

Aligned Features

Show features at their actual distance from an axis or center line rather than in the projected or foreshortened position. **Figure 5-63A** is an isometric view of a symmetrical part. **Figure 5-63B** indicates how this symmetrical part appears confusing in conventional projection (top illustration) but gives a better indication of the part's symmetry when projected from the aligned position (lower illustration).

How to Represent Holes

Since products contain various kinds of holes, be able to draw each kind and, therefore, be able to identify different kinds of holes when reading drawings.

(A)

CONVENTIONAL PROJECTION
(AS SEEN)

FEATURE ALIGNMENT

ALIGNED PROJECTION

(B)

FIGURE 5-63   (A) Isometric view of a symmetrical part; (B) Aligned projection.

FIGURE 5-64   Through hole and blind hole.

FIGURE 5-65   Tapered hole.

Plain Holes

A *through hole* goes completely through an object; a *blind hole* is machined to a specific depth. Through holes and blind holes constitute the majority of plain holes in products (**Figure 5-64**). The drawing at the left in Figure 5-64 shows a top view and a front view of a through hole.

Holes are usually drilled. A drilled hole is not as accurate as a reamed hole. For example, the tolerance for a 1″ diameter drilled hole might be ±.002″, but ±.0002″ for a 1″ diameter reamed hole. A hole, specified in a callout to be reamed, is first drilled to a size approximately .015″ smaller in diameter than the size

specified and then reamed to the callout specifications. The drafter, designer, or engineer specifies the reamed hole required; the machinist selects the proper drill.

A blind hole is drilled to a specific depth, Figure 5-64.

Twist drills, as shown in the figure, have a conical point of approximately 118° for general-purpose drilling. The full depth of the hole is called out for the cylindrical portion of the hole only.

Tapered Holes

A *tapered hole* (**Figure 5-65**), must first be drilled .015″ smaller in diameter than the smallest diameter of the tapered hole and then reamed.

Countersunk Holes

A *countersunk hole* is usually a plain through hole with an upper portion enlarged conically (**Figure 5-66**), most often to 82° to accommodate standard, flat-head machine screws. A machinist makes the countersink to a depth that produces a specified diameter slightly larger than the head of the flat-head screw.

Counterbored Holes

A *counterbored hole* is usually a plain through hole with an upper portion enlarged cylindrically (**Figure 5-67**), to a specified diameter and depth. A callout would specify the through hole to be slightly larger than the fastener and the counterbore to be slightly larger than the head of the fastener and slightly deeper than the height of the head.

Spotfaced Holes

A *spotfaced hole* is essentially a plain through hole with a shallow counterbore (**Figure 5-68**), deep enough to provide a flat surface for a head of a bolt, machine screw, or a nut for a threaded member.

Conventional Breaks

Use a *conventional break* for an effective way to draw long objects of constant cross section, such as a pipe, tubing, structural members, or shafting, on a relatively small sheet of drawing paper. A long pipe, drawn to 1/4 size, is shown in **Figure 5-69A** but redrawn to full size, with a conventional break, in **Figure 5-69B**. This presents a much clearer understanding of the pipe or other member, especially if there are holes to be located and dimensioned at either end.

Kinds of Conventional Breaks

Three kinds of breaks are commonly used in technical drawings: the S, the Z, and the freehand. Follow the steps in **Figure 5-70** for constructing an S break for a shaft or tube. The S break may be drawn with instruments or done freehand. Use **Figure 5-71A** as a guide for constructing a Z break, usually used for long, wide parts of constant cross section. The Z part is usually done freehand.

The freehand break (**Figure 5-71B**), is only one example of the use of the freehand break line.

Visualization

Visualization is the process of re-creating a three-dimensional image of an object in a person's mind, using the evidence and clues provided by orthographic

COUNTERSUNK HOLE (V)

FIGURE 5-66   Countersunk hole.

COUNTERBORED HOLE (⊔)

FIGURE 5-67   Counterbored hole.

SPOTFACED (SF)

FIGURE 5-68   Spotfaced hole.

FIGURE 5-69   Conventional breaks.

FIGURE 5-70   S break.

THICK BLACK LINE

THIN BLACK LINES

Ⓐ

2.0

24.0

.08

THICK BLACK LINES
(FREEHAND)

Ⓑ

.43

18.0   .125

FIGURE 5-71   Z break and freehand break.

drawings. This process has been the focus of much investigation, often used for studying a person's ability to perceive (such as in intelligence tests). The goal of reading an orthographic drawing is to accurately visualize information about the relative positions of an object's surfaces and geometric features.

There is some evidence that an active what-if imagination is a key ingredient in the visualization process. For example, a solid line circle in a view causes a what-if visualizer to seek further evidence. If an adjoining view has a solid line rectangle, bisected by a center line (axis) that aligns with the circle's center, a solid cylinder is indicated (**Figure 5-72**). If the aligned rectangle has hidden lines, a hole is indicated (**Figure 5-73**). A single orthographic view seldom is capable

of providing three-dimensional evidence, and the reader must consider two or more views simultaneously in seeking an understanding of feature outlines.

Engineering drawings all follow the same procedures, and none are any more or less difficult to read than another. However, the quantity of geometrical information varies considerably among drawings, and the time required to interpret the information varies accordingly.

Visualization practice using simple drawings is required by most individuals to gain experience in mental three-dimensional image construction. Even experienced designers often use modeling clay to form a three-dimensional object from orthographic views. Some build models out of cardboard and glue. Some do quasi-isometric drawings of the overall object, based on the orthographic views, and then systematically try to draw in lines, planes, and solids until all parts of the puzzle fit into a three-dimensional concept. Chapter 18 has excellent guidelines for creating pictorial drawings from orthographic views.

First-Angle Projection

While the United States, Great Britain, and Canada use the third-angle projection multiview drawing system, most of the rest of the industrial world uses first-angle projection.

The main difference between the two systems is the position of the projection plane. Recall the following sentence from the beginning of this chapter: "A projection plane may be thought of as an imaginary pane of glass with an object underneath it or

FIGURE 5-72   Visualization of an object.

FIGURE 5-73   Visualization of an object.

FRONT VIEW   LEFT SIDE

TOP VIEW

FIGURE 5-74   First-angle projection.

(F.V.)   (R.S.V.)   (R.S.V.)   (F.V.)

THIRD-ANGLE PROJECTION   **FIRST-ANGLE PROJECTION**
(A)   **(B)**

FIGURE 5-75   (A) Third-angle projection symbol;
(B) First-angle projection symbol.

behind it, Figure 5-6A." That is, it is observer first, then projection plane, and last, the object. For first-angle projection, it is observer first, then the object, and last, the projection plane.

First-angle projection, like third-angle projection, starts with the front view as the most important view.

Refer to **Figure 5-74** and imagine that the observer is at the left of the front view and looking toward the object, the letter E. What the observer sees, looking at the object, is on the projection plane behind the object—that is, observer-object-plane. Next, imagine that the observer is above the front view and looking toward the object. What the observer sees is the top view, which is placed below the object on the drawing.

Due to the increase in the international exchange of parts and drawings, the projection method used should be indicated on the drawing, with the standard symbol, or the name of the country issuing the drawing. The standard symbol is usually placed on a drawing in or near the title block. **Figure 5-75A** shows the standard symbol used to denote third-angle projection. **Figure 5-75B** shows first-angle projection. Note that in Figure 5-75A, the observer is assumed to be at the right of the front view (F.V.) and what the observer sees is shown in the right-side view (R.S.V.). In Figure 5-75B, the observer is also assumed to be at the right of the object and what the observer sees is shown in a right-side view, but the projection plane is behind the object.

CONFUSION OVER THE ANGLE OF PROJECTION

Mary Anderson had never seen anything like it. Her company, Precision Metal Components (PMC), had recently teamed with a German company on a large contract to manufacture parts for automobile engines.

The plan was for the German partner to provide a set of existing drawings that PMC personnel would revise as necessary depending on design changes.

Anderson had expected metric dimensioning, so that aspect of the drawings came as no surprise. What she hadn't expected was the arrangement of views on the drawings. "Look at these drawings!" said Anderson. "The top view is where the front view should be, and the left side is where the right side should be." In every respect, the drawings were well-drawn. In fact, they were some of the best drawings Anderson had ever seen, except, that is, for the arrangement of views. Anderson couldn't understand how someone could produce a set of drawings that could be so perfect in every way but one. It didn't make sense.

As she pondered the question, Anderson remembered a different angle of projection she had learned in school. Taking her old technical drawing book off the shelf, Anderson blew away the accumulated dust and turned to the chapter on orthographic projection. She soon had her answer. While the United States uses third-angle projection, the method Anderson and her team are accustomed to, Germany uses first-angle projection.

Anderson made copies of the pertinent pages for her team members. Smiling, Anderson commented to no one in particular, "We are going to need a little training before we revise this drawing package."

SUMMARY

- Spatial visualization is the act of forming an accurate mental picture of how a three-dimensional object will look when represented in a two-dimensional format or forming an accurate mental picture of how a three-dimensional object will look based on the information given in a two-dimensional representation.

- Students studying orthographic projection techniques should master the following: projecting normal surfaces, projecting two orthographic views, projecting multiple orthographic views, deciding how many views and which views are needed to adequately describe an object, how to transfer dimensions between views, proper use of hidden lines, proper use of center lines, and projection of inclined surfaces.

- Planning the development of a typical drawing requires five steps: (1) visualize the object, (2) select which view will be the front view, (3) decide how many views are needed, (4) decide where to place the front view, and (5) decide on the best placement of the other views.

- Properly centering a drawing within a specialized work area requires the use of simple mathematical calculations. The amount of space occupied by the various views to be drawn (vertically and horizontally) is subtracted from the space available. The remaining space is then divided proportionally.

- Students should learn to make proper use of the following technical drawing conventions: rounds and fillets, runouts, intersecting surfaces, curve plotting, cylindrical intersections, incomplete views, aligned features, holes, and conventional breaks.

- Students should learn how to mentally visualize an object in orthographic views and how to visualize from orthographic views to the object.

- First-angle projection is used in the industrialized countries of the world except the United States, Great Britain, and Canada. With first-angle projection, you have the viewer first, then the object, and last, the projection plane.

REVIEW QUESTIONS

Mark the following statements as either true or false.

1. Before beginning a three-view drawing, it is a good practice to make preliminary sketches of the object.

2. The work area on a drawing is the size of the piece of paper you are working with.

3. An orthographic projection is the most accurate description of shape wherein an undistorted image of the object appears in the flat, transparent, but imaginary projection plane.

4. Runout is the formation of a curved surface where a flat surface and a curved surface meet.

5. Fillets are interior rounded intersections between two surfaces.

6. The usual depth (call-off) for a counterbored hole is **slightly** deeper than the height of the head of the fastener.

7. The lightly constructed 45° angle projection line is used to represent visual rays.

8. A micrometer is used in dimensioning a blind hole.

9. Three views should be used to describe an object.

10. All drawings should be lightly numbered.

Answer the following questions by selecting the best answer.

1. Which of the following is **not** a frequent use of center lines?
   a. The location of a circular or cylindrical feature
   b. The location of an intersection of inclined planes
   c. The location of a spherical feature
   d. To specify the location of other principal symmetries

2. The difference between an inclined plane and an oblique plane is:
   a. Position in the drawing.
   b. Different projection angle.
   c. Inclined planes are used in third-angle projection only.
   d. Inclined planes are not parallel to any normal surface but are perpendicular to at least one. Oblique planes are neither parallel nor perpendicular.

3. Which of the following is **true** regarding ways to draw a cylindrical intersection?
   a. If there is a considerable difference in the dimensions, use a straight line for the intersection.
   b. If there is little difference in the dimensions, use a simple arc.
   c. Both a and b.
   d. Neither a nor b.

4. Where is the third-angle projection multiview system used?
   a. United States
   b. Great Britain
   c. Canada
   d. All of the above

5. What is the main difference between first-angle and third-angle projection?
   a. The standard symbols
   b. The time required to interpret the information
   c. The position of the projection plane
   d. None of the above

6. To illustrate a countersunk hole, what angle is used to draw the countersunk portion?
   a. 81°
   b. 82°
   c. 83°
   d. 84°
   e. None of the above

CHAPTER 5 PROBLEMS

The following problems are intended to give the beginning student experience in visualizing objects that are made of convenient shapes and features. Visualizing these shapes and features will help you to completely dimension an object.

Problem 3-22 will be used here as an example to illustrate the requirements for the problems.

Assume that you have been asked to calculate the weight of the object in Problem 3-22, which is made of aluminum with a density of 0.097 pounds per cubic inch.

You calculate the volume to be 9.84 cubic inches, which when multiplied by 0.097 pounds per cubic inch equals 0.954 pounds. Unless otherwise specified, use the format shown in Problem 4-37 to present your solutions to the following problems.

PROBLEM 5-1

Assume that the object in Problem 5-8 is cast iron with a density of 0.25 pounds per cubic inch. Calculate the weight of the object in pounds.

PROBLEM 5-2

Assume that the object in Problem 5-14 is steel with a density of 7765 kilograms per cubic meter. Calculate the weight of the object in kilograms.

PROBLEM 5-3

Assume that the object in Problem 5-18 is aluminum with a density of 0.097 pounds per cubic inch. Calculate the weight of the object in pounds.

PROBLEM 5-4

Assume that the object in Problem 5-27 is aluminum with a density of 0.097 pounds per cubic inch. Calculate the weight of the object in pounds.

PROBLEM 5-5

Assume that the object in Problem 5-41 is steel with a density of 7765 kilograms per cubic meter. Calculate the weight of the object in kilograms.

PROBLEM 5-6

Assume that the object shown here is steel with a density of 0.28 pounds per cubic inch. Calculate the weight of the object in pounds.

STEEL SHAFT

CONVENTIONAL INSTRUCTIONS

The following problems are intended to give the beginning student practice in visualizing the multiview system, choosing and sketching the required view, using drafting instruments, and using correct line thickness.

As these are beginning problems, no dimensions will be used at this time.

The steps to follow in laying out all drawings throughout this book are:

STEP 1 Study the problem carefully.

STEP 2 Choose the view with the most detail as the front view.

STEP 3 Position the front view so that there will be the least number of hidden lines in the other views.

STEP 4 Make a sketch of all required views.

STEP 5 Center the required views within the work area with a 1″ (25 mm) space between each view.

PROBLEMS

STEP 6   Use light projection lines. Do **not** erase them.

STEP 7   Lightly complete all views.

STEP 8   Check to see that all views are centered within the work area.

STEP 9   Check to see that there is a 1″ (25 mm) space between all views.

STEP 10   Carefully check all dimensions in all views.

STEP 11   Darken in all views by using the correct line thickness.

STEP 12   Recheck all work and, if correct, neatly fill out the title block by using light guidelines and neat lettering.

CAD INSTRUCTIONS

Students who plan to complete the following problems on a CAD system should begin with these general instructions.

1. Analyze the problem carefully.

2. Plan your drawing sequence. In what order will you develop the drawing?

3. Read the specific instructions for the problem and begin.

Problems 5-7 through 5-17

Construct a three-view drawing of each object using the listed steps.

PROBLEM 5-7

PROBLEM 5-8

PROBLEM 5-9

PROBLEM 5-10

PROBLEM 5-11

PROBLEM 5-12

PROBLEM 5-13

PROBLEM 5-14

PROBLEM 5-15

PROBLEM 5-16

PROBLEM 5-17

METRIC

PROBLEM 5-18

Construct a three-view drawing of this object. Project points 1, 2, and 3 into the right-side view and into the top view. Complete all views by using the listed steps.

PROBLEM 5-19

Construct a three-view drawing of this object. Project points 1 through 8 into the top view and into the front view. Complete all views by using the listed steps.

PROBLEM 5-20

Construct a three-view drawing of this object. Project points 1 through 6 into the front view and into the right-side view. Complete all views by using the listed steps.

PROBLEM 5-21

Construct a three-view drawing of this object. Project points 1 through 7 into the right-side view and into the front view. Complete all views by using the listed steps.

PROBLEM 5-22

Construct a three-view drawing of this object. Project points 1 through 7 into the right-side view and into the top view. Complete all views by using the listed steps.

PROBLEM 5-23

Construct a three-view drawing of this object. Locate and number 12 points around the 1.5 diameter hole. Project points 1 through 12 into the front view and into the right-side view. Complete all views by using the listed steps.

Problems 5-24 through 5-31

Construct a three-view drawing of each object using the listed steps.

PROBLEM 5-24

PROBLEMS

PROBLEM 5-25

PROBLEM 5-26

PROBLEM 5-27

PROBLEM 5-28

PROBLEM 5-29

PROBLEM 5-30

PROBLEM 5-31

METRIC

PROBLEM 5-32

Construct a four-view drawing of this object (front, top, right-side, and left-side views). Use the listed steps.

Problems 5-33 through 5-42

Construct a three-view drawing of each object using the listed steps.

PROBLEM 5-33

PROBLEM 5-34

PROBLEM 5-35

PROBLEM 5-36

PROBLEM 5-37

PROBLEM 5-38

PROBLEM 5-39

PROBLEM 5-40

PROBLEM 5-41

METRIC

PROBLEM 5-42

ALL UNMARKED RADII, R.38

Problems 5-43 through 5-56

Choose the front view and construct a three-view drawing of the objects in Problems 5-43 through 5-56 using the listed steps.

PROBLEM 5-43

ALL UNMARKED RADII, R.125

PROBLEMS

PROBLEM 5-44

45 90

R100

2X Ø16 THRU

R25 R37

76

28 12

64

152

12

38

25 50

ALL UNMARKED RADII = R4

METRIC

PROBLEM 5-46

38

46

24

2X Ø10 THRU
(IN LINE)

15

25

4X R15

25 15

2X Ø12 THRU

65

15

R12

15

15

48

15

100

15

ALL UNMARKED
RADII = R5

15

R

50

METRIC

PROBLEM 5-45

50

14 22

Ø 48 THRU

180

96

50

26

48

50

24

10

30

12

6 38

24 56

2XR12

52

METRIC

104

24 100

ALL UNMARKED RADII = R8

PROBLEM 5-47

1.00

.50

or

.50

2X Ø.38 THRU

1.62

15°

Ø1.06 THRU

R1.03

1.00

.56

R.50

.44 .75

1.00

.44

1.50

3.75

1.00

1.00

.50 2.00

PROBLEM 5-48

PROBLEM 5-49

MATERIAL THICKNESS .125
INTERNAL RADII .0125

PROBLEM 5-50

12 30
54
12
170
30
13
(9) 8 15 (104)
6
Ø26
Ø6 ⬇ 14
4X Ø6 THRU
EVENLY SPACED ON A
Ø46 B.C.
Ø60
Ø26
Ø22
Ø26
Ø18 50
32
3
Ø12
SAW KERF
1 X 45°
CHAMFER
2

METRIC

PROBLEM 5-51

20
80
60
45°
45°
30
10
28
120
60
Ø10 THRU
25
30
80
15
25
10
15
40
10
30
(40)
10
15
Ø20 THRU
⌴ Ø30 ⬇ 4

METRIC

PROBLEM 5-52

R20
25
34
70
45°
(TYP) 46
20
120
50
2X Ø16 THRU
30
20
5
12
(36)
12
60

ALL UNMARKED RADII = R4

METRIC

PROBLEM 5-53

PROBLEM 5-54

PROBLEM 5-55

METRIC

ALL UNMARKED
RADII = R3

PROBLEM 5-56

PROBLEM 5-57

For practice in reading multiview drawings that have already been drawn, study the pattern of the sample responses related to the following numbered multiview drawing, and then complete the surface analyses as indicated.

Sample Responses

1.	Surface 25-26	in R is 5-6-10-9	in T
2.	Surface 16-17	in F is 29-30	in R
3.	Surface 16-17	in F is 2-3-6-10-11-7	in T

Complete the Responses

1.	Surface 13-14-15-16	in F is _____	in T
2.	Surface 24-25-26-27-28	in R is _____	in T
3.	Surface 29-25-26-27-30	in R is _____	in T
4.	Surface 8-7-12-11	in T is _____	in F
5.	Surface 13-14	in F is _____	in T
6.	Surface 29-25-24	in R is _____	in F
7.	Surface 11-7-2	in T is _____	in F
8.	Surface 11-7-2	in T is _____	in R
9.	Surface 32-30-27-28	in R is _____	in T
10.	Surface 6-5-9-10	in T is _____	in F

PROBLEM 5-58

Six possible T views, all correctly drawn, are shown for the one F view. Draw an A view for each one of the T views.

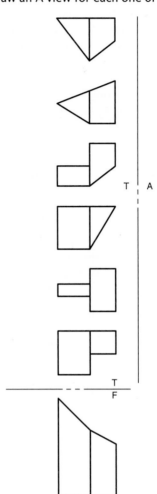

PROBLEM 5-59

The steel end cover shown has eight through holes 9/16″ in diameter for 1/2″ diameter fillister head cap screws. (See Appendix A.) Change the callout for the holes to be counterbored. (Refer to Chapter 10 for callouts.) The heads of the cap screws will not protrude above the surface of the cover. (This change is for safety reasons; some counterbored holes are specified for aesthetic reasons.) Prepare a two-view drawing of the cover with the new callout and the dimensions. Show one hole in cross section in a separate view.

PROBLEM 5-60

Visit a lab or shop listed and do a field sketch (rough sketch) of a piece of apparatus (not too complicated). Later, use the sketch to prepare a multiview drawing of the apparatus. Show approximate dimensions and note general materials.

 a. Chemistry lab
 b. Physics lab
 c. Metal shop
 d. Wood shop
 e. Electrical/electronics shop
 f. Computer lab

PROBLEM 5-61

Visit a track and field area and a gymnasium and do a field sketch (rough sketch) of one of the following. Later, use the sketch to prepare a multiview drawing of the item sketched. Give approximate dimensions and note general materials.

 a. Hurdle
 b. Starting block

 c. Javelin
 d. High-jump post
 e. Horse
 f. Springboard (used with a horse)
 g. Parallel bars
 h. Other

PROBLEM 5-62

Visit a weight room and do a field sketch (rough sketch) of a portion of a complicated assembly or all of a relatively simple assembly. Later, use the sketch to prepare a multiview drawing of the item sketched. Give approximate dimensions and note general materials.

PROBLEM 5-63

Do a conceptual design of a contemporary desk lamp for a student's desk. That is, do the general outline of the lamp only, not the insides. Prepare a multiview drawing of the lamp and indicate approximate sizes.

PROBLEM 5-64

Design a ladder for use around a typical home. Use wood and fasten the rungs to the side rails with glue, unless otherwise directed. Prepare a multiview drawing of the ladder, to scale, with approximate dimensions. (A Z break may be useful here.)

PROBLEM 5-65

Utilize the following anthropometric data and design one of the objects listed. Use the dimensions given or use your own body dimensions as a guide. Prepare a multiview drawing of the design, with approximate dimensions and general materials.

 • A door (standard height 6′8″ to 7′; doorknob or handle, 38″ above floor)
 • A working counter for the center of a kitchen (standard height 36″ to 40″)
 • A bar stool (standard height 27″ to 30″)
 • A simple workbench (standard height 36″)
 • A computer workplace (25″ to 26″ knee clearance)

PROBLEM 5-66

Design a simple metal grill for use over an open fire, collapsible and small enough to fit in a backpack. Prepare a multiview drawing of the grill. Include approximate dimensions.

PROBLEM 5-67

Design, for a junior high school science class, an inexpensive, easy-to-assemble, portable, demonstration apparatus for illustrating the change of supporting force, F, (in pounds) in the diagonal member BC as the angle θ is changed, while the weight W is constant.

Refer to the frame outline shown. Prepare a multiview drawing of the design, with approximate dimensions, and note the materials used. (A scale may need to be designed; for example, use a spring with a pointer attached, calibrated for different forces pulling on the spring.)

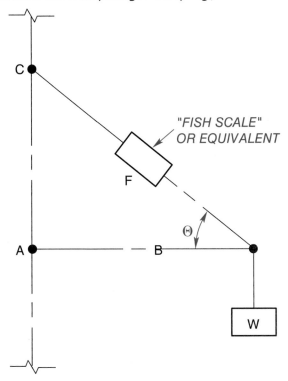

PROBLEM 5-68

Design an easy-to-assemble apparatus for a junior high school science class. It should be portable and inexpensive to demonstrate two facets of a pendulum. First, demonstrate how the period (a complete cycle back and forth) of a pendulum changes with a change in length. Second, demonstrate how the period doesn't change with a change in weight for any one length.

Prepare a multiview drawing of the design, indicate approximate dimensions, and note the materials in general. The equation for the period T of a pendulum is $T = 2\pi \times$ square root of (length/g). The g in the equation is the gravity constant. If the length is in meters and the gravity constant equals 9.81 meters/second squared, then the period will be in seconds. For example, a swing (a type of pendulum) with a length of 12 meters and your best friend in the swing would have a period of approximately 6.9 seconds. (This would be about a 40′ long swing.)

PROBLEM 5-69

The drawing shown is for a special aluminum circular sealing plate, used on an experimental apparatus. The plate has a groove for an O-ring seal and threaded holes for a pressure transducer. A similar plate is needed for another project, but the outside diameter is to be changed to 200 mm, and the threaded holes on the 41 mm bolt circle are to be drilled and tapped for a metric fastener M8 × 1.25. The depth of the threaded portion of holes varies from 1 times the diameter of the fastener if the material is steel to 2 times the fastener diameter for aluminum. Prepare a similar drawing for the circular plate and use your own judgment for the changes in the dimensions. Note that the title block has been omitted for convenience; however, the third-angle projection symbol is shown. Note: Problem 5-69 is a "real-world" drawing. It does not conform to standard drafting conventions in all cases.

6

Dimensioning and Notation

CHAPTER OBJECTIVES

Upon completion of this chapter, students should be able to do the following:

▶ Explain the three main dimensioning systems used in the United States: metric, decimal-inch, and fractional dimensioning.

▶ Demonstrate proficiency in properly using the following dimension components: extrusion lines, dimension lines, leader lines, and arrowheads.

▶ Demonstrate the steps used in laying out dimensions for an object.

▶ Demonstrate proper application of the dimensioning techniques for the following: dimensioning symbols, dimensioning chords, arcs, angles, radii, curved surfaces, offsets, irregular curves, contours, multiple radii, normal objects by offset, spheres, round holes, making simple hole callouts, applying drill size tolerances, dimensioning hole locations for single and multiple holes, and locating holes about a bolt center.

(continued)

KEY TERMS

Allowance

Arrowhead

Attributes

Chamfer

Clearance fit

Decimal-inch dimensioning

Detail notes

Dimensioning

Dimension line

Extension line

Fractional dimensioning

General notes

Interference fit

Knurl

Leader line

Least material condition (LMC)

Manufacturing part family

Maximum material condition (MMC)

Metric dimensioning

Nominal dimension

Notation

Opitz system

Part family

Tolerance

Transition fit

Verification

Written notes

(CHAPTER OBJECTIVES continued)

► Demonstrate proficiency at applying the following dimensioning techniques: locating holes on center lines and concentric arcs, multiple holes along the same center line, repetitive features, callouts for tapered holes, callouts for countersunk holes, callouts for counterbored holes, dimensioning a cylinder and a square, double dimensioning, reference dimensioning, dimensioning slots and rounded ends, not-to-scale dimensions, nominal dimensions, dimensioning external chambers, internal chambers, necks, undercuts, knurls and keyways, staggered dimensions, and dimensioning flat tapers, round tapers, threads, pads, bosses, sheet metal bends, sectional views, pyramids, and cones.

► Demonstrate proficiency at applying the following dimensioning techniques: dimensioning concentric and nonconcentric shafts, rectangular coordinate dimensioning, dimensioning holes on a bolt center diameter, finish marks, X-Y-Z coordinates, tabular dimensioning, tabular drawing, tolerancing, shaft limits, hole limits, allowance, clearance, interference fit, transition fit, size limits, design size, maximum and minimum sizes, location limits, design location, maximum and minimum location, calculating fits, matching parts, and standard fits.

► Demonstrate proficiency in the proper application of notation.

► Explain the concept of group technology as it relates to design and drafting.

One of the most fundamental drafting tasks is to meet the requirements of the engineering definition of the part while providing for the most economical production process and interchangeability considerations. All of this is accomplished by the use of proper dimensions and notation on drawings. *Dimensioning* is the process whereby size and location data for the subject of a technical drawing are provided. *Notation* is the process whereby needed information not covered by dimensions is placed on a technical drawing.

It is critical that drafters, designers, and engineers be proficient in standard dimensioning practices. The most widely accepted dimensioning standard is the American Society of Mechanical Engineers document Y14.5M-2009. Similar standards are produced by the International Standards Organization (ISO). However, unless otherwise specified, ASME Y14.5M-2009 is the standard used for guiding dimensioning practices.

Modern dimensioning practices described in ASME Y14.5M-2009 apply in most instances in which interchangeability of parts is a major consideration. The concept dictates that parts produced from a drawing at one manufacturing site should be interchangeable with those produced at another manufacturing site. Automotive parts are an excellent example of production for interchangeability. Some parts are manufactured in America, some in Europe, and some in Japan, but they must all fit together in one car during assembly. Although interchangeability is not a factor with all parts that are produced, the drafter should still use the basic dimensioning principles of ASME Y14.5M-2009. This is particularly important when the parts will be produced using such ever-increasing automated, semi-automated, or integrated processes as numerical control, computer-aided manufacturing (CAM), or computer-integrated manufacturing (CIM).

Dimensioning Systems

Three dimensioning systems are used on technical drawings in the United States: metric dimensioning, decimal-inch dimensioning, and fractional dimensioning. Certain rules of practice with which drafters should be familiar pertain to each of these dimensioning systems.

Metric Dimensioning

The standard metric unit of measurement for use on technical drawings is the millimeter (.001 meter) or .039 inch. **Figure 6-1** is a chart of various metric units of measurement of less than a meter that shows where the millimeter fits in.

When using *metric dimensioning,* several general rules should be observed. When a dimension is less than 1 millimeter, a zero must be placed to the left of the decimal point (**Figure 6-2A**). When a metric dimension is a whole number, neither the zero nor the decimal is required (**Figure 6-2B**). When a metric dimension consists of a whole number and a decimal portion of another millimeter, it is written as follows: whole number first, decimal point second and, finally, the decimal part of the number. The decimal

UNIT	MULTIPLE OF A METER
METER	1.0
DECIMETER	0.1
CENTIMETER	0.01
MILLIMETER	0.001

FIGURE 6-1   Metric linear measurements.

FIGURE 6-2   Metric dimensioning.

FIGURE 6-3   No commas in metric dimensions.

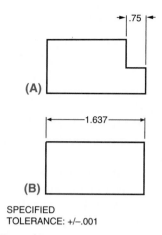

SPECIFIED TOLERANCE: +/−.001

FIGURE 6-4   Decimal-inch dimensioning.

fraction are required. A zero is not required preceding the decimal point (**Figure 6-4A**). The number of places beyond the decimal point that a decimal-inch dimension is carried is determined by the specified tolerance for the part in question (**Figure 6-4B**). In this figure, a tolerance of .002 (±.001), three places to the right of the decimal, is specified. Consequently, the dimension of 1.637 is carried out three places to the right of the decimal.

There are no specified sizes for decimal points, but they should be made dark enough and large enough to be seen and to reproduce through any normal reproduction process (diazo, photocopy, or microfilm).

Fractional Dimensioning

Fractional dimensioning is not frequently used on mechanical technical drawings. Its primary use is on architectural and structural engineering drawings. However, since it is occasionally still used on mechanical drawings, drafters should be familiar with this system. When you are using fractional dimensions on mechanical drawings, several rules should be observed. The line separating the numerator and denominator of a fraction should be a horizontal line, not an inclined line (**Figure 6-5**). Full-inch dimensions should be a minimum of one-eighth inch in height. The combined height of the numerator, denominator, and horizontal line of a fraction should be one-quarter inch (**Figure 6-6**), for A-, B-, and C-size drawings and a minimum of five-sixteenths inch for D-, E-, and F-sizes. Also, be sure to leave a

$$\frac{1}{8}$$ $$\frac{1}{\diagup 8}$$

CORRECT   INCORRECT

FIGURE 6-5   Horizontal line is the correct method.

part of the number is not followed by a zero in metric dimensioning (**Figure 6-2C**). Individual digits in metric dimensions are not separated by commas or spaces (**Figure 6-3**). Drawings prepared with metric dimensions are identified with the word "metric" contained in a small rectangle below the part.

Decimal-Inch Dimensioning

Decimal-inch dimensioning is frequently used in the dimensioning of technical drawings. It is a much less cumbersome system for mechanical drawings than is the fractional system, and it is still used more than the metric system. When using the decimal-inch dimensioning system, several rules should be observed. If a dimension is less than one inch, only a decimal point and the numbers of the decimal

FIGURE 6-6   Proportions for fractions.

visible gap between the numerator, denominator, and horizontal line of the fraction.

Orienting Dimensions

Regardless of whether you are dimensioning in the metric, decimal-inch, or fractional dimensioning system, there are two subsystems for orienting dimensions. These orientation subsystems are known as the aligned system and the unidirectional system. The aligned dimensioning system is illustrated in **Figure 6-7**. In this system, the dimensions are aligned with the dimension line so that they can be read either from the bottom of the drawing sheet or from the right-hand side. Notice in Figure 6-7 the area that has been marked off to be avoided if possible. Dimensions written in this area are difficult to read from either the bottom or the right-hand side of the drawing sheet and, therefore, should be avoided. The unidirectional dimensioning system is illustrated in **Figure 6-8**. In this system, all dimensions are written so that they can be read from the bottom of the drawing sheet. This means that the guidelines used in entering the dimensions should always be parallel with the bottom of the page.

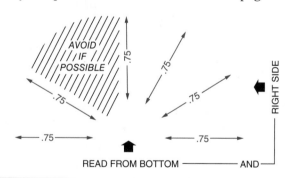

FIGURE 6-7   Aligned dimensioning system.

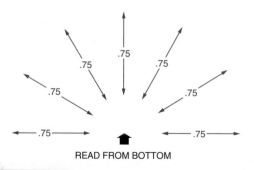

FIGURE 6-8   Unidirectional dimensioning system.

Dimension Components

Several components are common to all dimensioning systems. These include extension lines, dimension lines, arrowheads, leader lines, and the actual numbers or dimensions. Drafters and engineers should be knowledgeable in the proper use of these components.

Extension Lines

An *extension line* is a thin, solid line that extends from the object in question or from some feature of the object. Several rules should be observed when placing extension lines on technical drawings. There should be a small but visible gap (approximately .06″) between the object or object feature and the beginning of an extension line (**Figure 6-9A**). Extension lines should extend uniformly beyond dimension lines for a distance of approximately one-eighth inch (**Figure 6-9B**). Extension lines that originate on the object, such as center lines, may cross visible lines with no gap required (**Figure 6-10**). **Figure 6-11** is a fully dimensioned mechanical drawing that summarizes the basic rules drafters and engineers should remember about the spacing of extension lines, dimension lines, and dimensions. Notice in this figure that when center lines become extension lines, there is

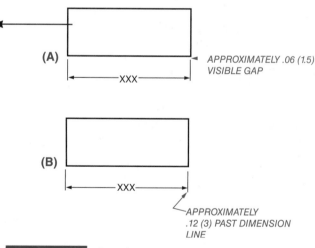

FIGURE 6-9   Drawing extension lines.

FIGURE 6-10   Center lines as extension lines.

FIGURE 6-11   Spacing for extension lines, dimension lines, and dimensions.

REQUIRED DIMENSIONS BETWEEN VIEWS	SUGGESTED DISTANCE BETWEEN VIEWS	
1	1.25	30
2	1.62	39
3	2.00	48
4	2.38	57
5	2.75	66
6 MAX.	3.12	75

no gap required as they cross object lines. Notice also that leader lines are not broken when they cross object lines. Note the visible gap (approximately .06″) between the object itself and the beginning of extension lines. Note the .12″ or approximately one-eighth inch overhang of the extension line beyond dimension lines. Suggested spacing of dimension lines and distances between views are summarized in the chart in Figure 6-11.

Dimension Lines

A *dimension line* is a thin, solid line used to indicate graphically the linear distance being dimensioned. Dimension lines are normally broken for placement

of the dimension (**Figure 6-12A**). If a horizontal dimension line is not broken, the dimension is placed above the dimension line with guidelines parallel to it (**Figure 6-12B**). **Figure 6-13** illustrates the

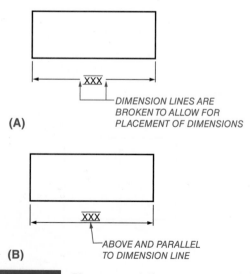

FIGURE 6-12   Placement of dimensions.

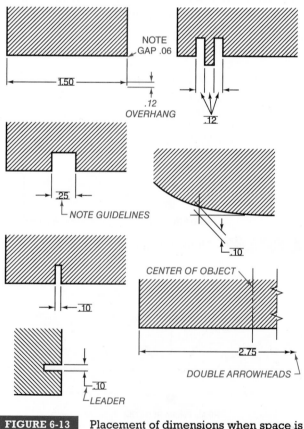

FIGURE 6-13   Placement of dimensions when space is limited.

proper methods to be used for locating dimensions when space is limited on a drawing. *Notice from this illustration that one dimension can be written, and then leaders can be used to point to where the dimension applies.* A dimension may be written within the extension lines with dimension lines extending inward rather than outward. Dimension lines may be bent at a 90° angle, and extension lines, when necessary, can be drawn at an oblique angle. Double arrowheads can be used on a dimension line when a long or short break line has been used.

When dimensioning multiple features of an object, dimensions should be aligned uniformly rather than staggered or randomly scattered about the object,

(**Figure 6-14**). **Figure 6-15** illustrates another rule for the proper placement of dimensions. When dimensioning several views of the same object, always place the dimensions between the views, and always include the overall size, width, height, and depth of the object.

Dimension lines are drawn parallel to the direction of measurement. Sufficient distance between the object and the dimension lines, and between successive dimension lines, is important so that cramped and crowded dimensions do not result. First dimension lines should be at least .38″ from the object (**Figure 6-16A**). Successive dimension lines should be at least .25″ apart. If using metric dimensions, the first dimension lines should be at least 10 mm away from the object. Successive lines should have at least 6 mm between them (**Figure 6-16B**). **Figure 6-17** illustrates the methods to be used when successive dimensions are called for in close areas. Notice how the dimension lines and corresponding dimensions are offset, and notice also the gap left in the extension lines to accommodate crossover of dimension lines.

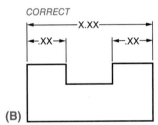

FIGURE 6-14 Proper placement of dimensions when dimensioning multiple features.

FIGURE 6-16 Successive dimension lines.

FIGURE 6-15 Dimensioning between views.

FIGURE 6-17 Successive dimension lines in close areas.

When the shape of an object requires a series of parallel lines, the breaks and dimensions should be slightly staggered to make it easier to read the dimensions (**Figure 6-18**). **Figure 6-19** illustrates the proper way of stacking a successive group of dimension lines. Drafters and engineers should always place the shortest dimensions closest to the objects, and always keep dimensions outside and away from the object, as illustrated in Figure 6-19.

Leader Lines

A *leader line* is a thin line that begins horizontally, breaks at an angle, and terminates in an arrowhead or, on occasion, a dot (**Figure 6-20**). Leaders that terminate at an edge or at some other specific point on a drawing should have an arrowhead. Leaders that

terminate inside an object on a flat surface should end in a dot. The preferred angle of a leader line is from a minimum of 30° to a maximum of 60° (**Figure 6-21**). The horizontal portion of a leader should be from .125″ to .25″ in length, Figure 6-21. To avoid confusion, leader lines should not be drawn parallel to extension lines or dimension lines. When placing a dimension and/or note at the end of a leader line, it is easier to letter from the leader than into the leader (**Figure 6-22**).

When using a leader line to direct a dimension to its appropriate feature on a drawing, one dimension for each leader is the preferred method (**Figure 6-23**). More than one leader line extending from the same dimension can create a confusing situation that is difficult to interpret. Leader lines pointing to the

FIGURE 6-18   Staggering dimensions.

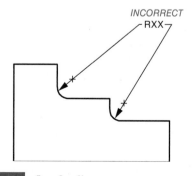

INCORRECT   CORRECT

RULES:
ALWAYS PLACE SHORTEST DIMENSIONS
CLOSEST TO THE OBJECT
ALWAYS KEEP DIMENSIONS OUTSIDE AND AWAY
FROM THE OBJECT

FIGURE 6-19   Stacking dimension lines.

FIGURE 6-20   Leader lines.

FIGURE 6-21   Angle of a leader line.

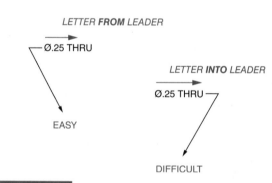

FIGURE 6-22   Lettering and leaders.

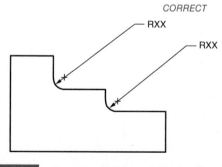

FIGURE 6-23   One dimension per leader line is preferred.

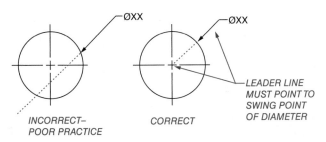

FIGURE 6-24   Leader lines point at the center of a circle.

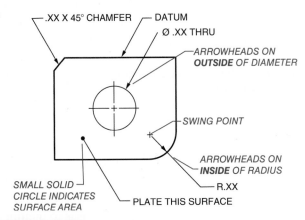

FIGURE 6-25   Proper use of leader lines.

center of a circle should be directed toward but not extended into the circle center (**Figure 6-24**). **Figure 6-25** illustrates the proper use of leader lines in a variety of dimensioning situations.

Arrowheads

An *arrowhead* is the most commonly used termination symbol for dimension and leader lines. Arrowheads should be approximately three times as long as they are wide. They should be large enough to be seen but small enough that they do not detract from the appearance of the drawing. A commonly accepted length for arrowheads is .12″. Arrowheads may be slightly larger or smaller than this but, regardless of their size, they should be uniform throughout a drawing (**Figure 6-26**). Although the same standard

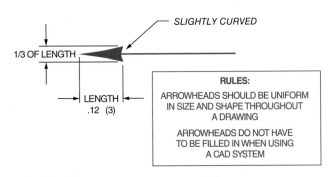

FIGURE 6-26   Proper size and shape of arrowheads.

FIGURE 6-27   Drawing arrowheads.

applies to drawings prepared on a CAD system, arrowheads do not have to be filled in on drawings prepared using automated processes. Some CAD systems fill in arrowheads and some do not. Both open and filled-in arrowheads have been considered acceptable since the advent of CAD.

Figure 6-27 illustrates the three steps used in drawing arrowheads.

Laying Out Dimensions

Figure 6-28 illustrates the four steps in laying out dimensions for an object. First, place light layout lines on the drawing sheet. Second, draw in light guidelines for the dimensions. Third, darken your extension lines, dimension lines, and dimensions. Fourth, add arrowheads.

FIGURE 6-28   Laying out dimensions.

Steps in Dimensioning

There are two basic steps in dimensioning objects, regardless of the type of object.

STEP 1   Apply the size dimensions. These are dimensions that indicate the overall size of the object and the various features that make up the object.

STEP 2   Apply the locational dimensions. Locational dimensions are dimensions that locate various features of an object from some specified datum or surface. **Figure 6-29** gives examples of size and location dimensions.

In order to properly dimension an object, drafters and engineers must be able to mentally break the object down into component parts and subelements. **Figure 6-30** illustrates graphically how this is done. On the left side of the figure is a two-dimensional representation of the object containing a top and front view. On the right side of the figure is a three-dimensional geometric breakdown of the object. This three-dimensional breakdown shows the drafter and engineer what geometric shapes must be sized using dimensions and which must be located using dimensions.

Specific Dimensioning Techniques

All of the information on dimensioning so far has been of a general nature. The following sections deal with techniques used for applying dimensions to specific situations that are recurrent in drafting. With a thorough knowledge of the general information presented earlier, and the specific information presented in these sections, drafters and engineers will be able to dimension any situation confronted on technical drawings.

Dimensioning Symbols

There are a number of symbols used in different dimensioning situations with which drafters and engineers should be familiar. These symbols are summarized in **Figure 6-31**. Review these symbols carefully and commit them and their uses to memory before proceeding.

Dimensioning Chords, Arcs, and Angles

Chords, arcs, and angles are dimensioned in a similar manner. When dimensioning a chord, the dimension line should be perpendicular and the extension lines parallel to the chord. When dimensioning an arc, the dimension line runs concurrent with the arc

FIGURE 6-29   Size and location dimensions.

FIGURE 6-30   Geometric breakdown of an object.

CF	CONTINUOUS FEATURE
CR	CONTROLLED RADIUS
R	RADIUS
Ø	DIAMETER
X	NUMBER OF TIMES–PLACES
▼	DEEP OR DEPTH
∨	COUNTERSINK
⊔	COUNTERBORE
⌊SF⌋	SPOTFACE
□	SQUARE
( )	REFERENCE DIMENSION
.25	NOT-TO-SCALE DIMENSION
◿	SLOPE
▷	CONICAL TAPER
⌒.25	ARC LENGTH
SR	SPHERICAL RADIUS
SØ	SPHERICAL DIAMETER
.25	BASIC DIMENSION
ST	STATISTICAL TOLERANCE

FIGURE 6-31   Dimensioning symbols.

curve, but the extension lines are either vertical or horizontal. An arc symbol is placed above the dimension. When dimensioning an angle, the extension lines extend from the sides forming the angle, and the dimension line forms an arc. These methods are illustrated in (**Figure 6-32**). **Figures 6-33** and **6-34**

FIGURE 6-32   Dimensioning chords, arcs, and angles.

FIGURE 6-33   Dimensioning angles.

FIGURE 6-34   Dimensioning angles.

contain additional information relative to dimensioning angles.

Figure 6-33 shows that angles are normally written in degrees, minutes, and seconds. The symbols used to depict degrees, minutes, and seconds are also shown in this figure. Angular measurements may also be stated in decimal form. This is particularly advantageous when they must be entered into an electronic digital calculator. The key to converting angular measurements to decimal form is in knowing that each degree contains 60 minutes, and each minute contains 60 seconds. Therefore, converting a measurement stated in degrees, minutes, and seconds into decimal form is a two-step process. Consider the example of the angular measurement 2 degrees, 15 minutes, and 30 seconds. This measurement would be converted to decimal form as follows:

STEP 1   The seconds are converted to decimal form by dividing by 60. Thirty seconds divided by 60 equals .50. The 15 minutes are added to this so that the minutes are expressed in decimal form or 15.50.

STEP 2   The minutes stated in decimal form are converted to decimal degrees by dividing by 60. The number 15.50 divided by 60 equals .25833. The 2 degrees are added to this number to have the measurement stated in decimal form or 2.2583 degrees.

Figure 6-34 illustrates how angles can be dimensioned for size and location. The size dimension gives the overall size of the angular cut. However, it does not locate the angular cut in the object. This is done by using a locational dimension off of the center line of the object. **Figure 6-35** shows how normal linear dimensions can be used for effecting the proper angle. When this method is used in the example on the left, all surfaces are dimensioned except the angular surface. When this method is used, the angular surface is defined by default. In the method on the right, three linear dimensions and one angular dimension define the angular surface.

FIGURE 6-35   Using normal linear dimensions for effecting the proper angle.

Dimensioning a Radius

Dimension lines used to specify a radius have one arrowhead, normally at the arc end. An arrowhead should not be used at the center of the arc. Where space permits, the arrowhead should be placed between the arc and the arc center with the arrowhead touching the arc. If space permits, the dimension is placed between the arc center and the arrowhead. If space is not available, the dimension may be placed outside the arc by extending the dimension line into a leader. The arc center of a radius is denoted with a small cross. **Figure 6-36** illustrates the preferred method for dimensioning a radius.

Dimensioning a Foreshortening Radius

On occasion, the center of an arc radius will fall outside the drawing or will be so far removed from the drawing as to interfere with other views. When this is the case, the radius dimension line should be foreshortened and the radius center located using coordinate dimensions. This is done by relocating the arc center and placing a zigzag in the radius dimension line, as shown in **Figure 6-37**. When this method is used, it is important that the arc center actually lay upon the real center line of the arc.

Dimensioning Curved Surfaces

Curved surfaces containing two or more arcs are dimensioned by showing the radii of all arcs and

FIGURE 6-37   Dimensioning a foreshortening radius.

locating the arc centers using coordinate dimensions. Other radii may be located using their points of tangency. This method is illustrated in **Figure 6-38**.

Dimensioning Offsets

Offsets are dimensioned from the points of intersection of the tangents along one side of the object (**Figure 6-39**). The distance from one end of the offset to the intersection is specified with a coordinate dimension. The distance from the other end to the

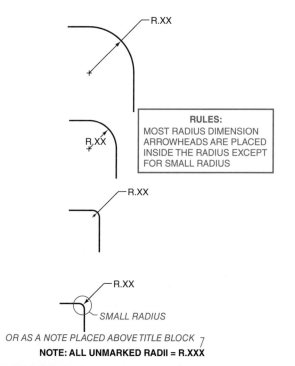

FIGURE 6-36   Dimensioning a radius.

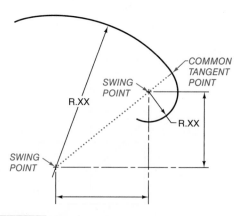

FIGURE 6-38   Dimensioning curved surfaces.

FIGURE 6-39   Dimensioning offsets.

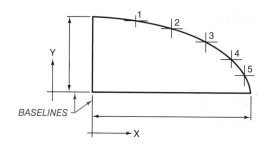

STATION	1	2	3	4	5
X	1.12	2.12	3.00	3.62	4.00
Y	1.75	1.50	1.12	.75	.31

FIGURE 6-41   Dimensioning contours not defined as arcs.

offset is also specified with a coordinate dimension, as shown in Figure 6-39.

Dimensioning Irregular Curves

You have already seen in Figure 6-38 how to dimension curved surfaces using arc centers, radius dimensions, and tangent points. Irregular curves may also be dimensioned using coordinate dimensions from a specified datum. When this is the case, the coordinate dimensions extend from a common baseline to specified points along the curve (**Figure 6-40**).

Dimensioning Contours not Defined as Arcs

Contours not defined as arcs can be dimensioned by indicating X–Y coordinates at points along the surface of the contour. Each of these points, sometimes referred to as stations, is numbered. The X and Y coordinates for each station are tabulated and placed in table form under the drawing (**Figure 6-41**).

Dimensioning Multiple Radii

When dimensioning an object that requires several radii, arcs should be dimensioned showing the radius in a view that gives the true shape of the curve. The dimension lines for a radius should be drawn as a radial line at an angle rather than horizontally or vertically. Only one arrowhead is used. The R

precedes the dimensional value when dimensioning in the decimal-inch or metric system. This method is illustrated in **Figures 6-42** through **6-44**. Notice in Figure 6-44 that where a radius is dimensioned

FIGURE 6-42   Dimensioning multiple radii.

FIGURE 6-40   Dimensioning irregular curves.

FIGURE 6-43   Dimensioning multiple radii.

FIGURE 6-44   Dimensioning multiple radii.

in a view that does not show the true radius, a note should be used to indicate what the true radius is.

Dimensioning by Offset (Round Objects)

Another way to dimension a round object is the off-set method. In this method, dimension lines are used as extension lines. Dimension lines are distributed across the object perpendicular to the center line of the object and spaced using coordinate dimensions (**Figure 6-45**).

Dimensioning by Offset (Flat Objects)

Flat objects may also be dimensioned using the offset method. Again, dimension lines on the object become extension lines. Dimension lines are spaced across the object perpendicular to the center line of the datum and extended to become extension lines for the coordinate dimensions that space them (**Figure 6-46**).

Dimensioning Spheres

Figure 6-47 illustrates the proper method for dimensioning spheres. When the diameter method is used, a leader points to the center of the sphere, and the

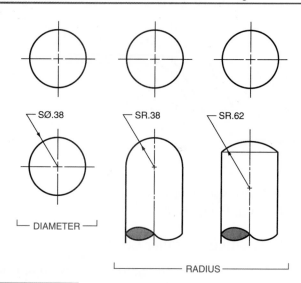

FIGURE 6-47   Dimensioning spheres.

diameter note is preceded with a capital S to indicate that it is a spherical diameter. When the radius method is used, a dimension extends from the arc center to the arc and is extended on with a leader, and the note is preceded by a capital SR to indicate a spherical radius.

Dimensioning Round Holes

Round holes are dimensioned in the view in which they appear as circles (**Figure 6-48**). Holes may be dimensioned using a leader that points toward the

FIGURE 6-45   Dimensioning by offset (round object).

FIGURE 6-46   Dimensioning by offset (flat object).

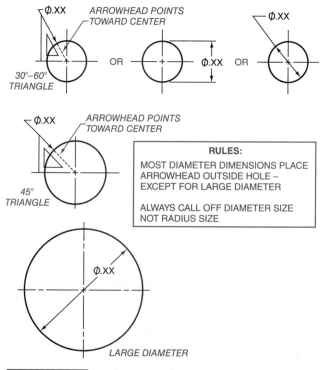

FIGURE 6-48   Dimensioning round holes.

center of the hole in which the note gives the diameter, or extension lines may be drawn from the circle with a dimension that also indicates the diameter. Larger circles are dimensioned with a dimension line drawn across the circle through its center at an angle with the diameter dimension shown. Except for very large holes, the arrowhead and the dimensional value are placed outside the hole. It is important when dimensioning holes to call off the diameter, not the radius.

Simple Hole Callouts

Drafters and engineers need to know how to apply simple callouts to both through holes and blind holes. A through hole is one that passes all the way through the object. A blind hole is one that cuts into but does not pass through the object. Both types of holes, and the callout used for each, are illustrated in **Figure 6-49**. Machine holes are generally drilled to make the rough hole and then reamed to refine the hole. Figure 6-49 shows the difference between a drill and a ream. No hole is only reamed. A hole must be drilled before it can be reamed.

A through-hole callout has a leader line extending toward the center of the hole in the view in which the hole appears as a circle. The note attached to the leader gives the diameter of the hole and the word "thru" to indicate that the hole passes through an object. Blind-hole callouts are similar to through-hole callouts, except that the depth symbol is followed by the actual depth of the hole.

FIGURE 6-49   Simple hole callouts.

Drill Size Tolerancing

Holes are not drilled to the exact size specified on a drawing. This is because there are several factors that mitigate against a perfectly sized hole. The accuracy of the drill, the tolerance level of the machine, and the qualifications of the machinist all have an impact on the actual size of a hole once it's drilled. It is accepted in manufacturing that no hole, even with the added accuracy provided by computer-aided manufacturing, is going to be drilled exactly the size specified. Therefore, drafters and engineers need to know how much variation to expect in a hole so that they can decide what limits to give the hole and whether the actual drilled hole will give the fit required in a given situation.

Drill size tolerance charts have been developed to assist drafters and engineers in determining the expected upper and lower limits of a drilled hole. Such charts are contained in Appendix A as Table 16. Turn to this table and you will notice that the standard drill size is given a number/letter, a fractional designation, a decimal designation, and a metric designation. To the right-hand side of the table, the tolerance for each drill size is given in decimals. The left-hand column is the plus tolerance, and the right-hand column is the minus. To use this table, apply the following steps:

1. Find the letter or number of the drill in question on the left-hand side of the table.

2. Find the corresponding size for the drill in decimal form.

3. Find the plus tolerance for that size drill and add it to the decimal drill size. This will give you the upper limit for the hole size.

4. Find the minus tolerance for that drill size and subtract it from the decimal drill size. This will give you the lower limit for the hole.

5. Write the upper and lower limits with the smaller number on top separated from the larger number by a horizontal line. For example, an H-size drill has a decimal diameter of 0.2660. To find the upper limits for a hole using this drill size, add the plus tolerance of 0.0064 to this decimal drill size to get 0.2724 as the upper limit for the hole. To find the lower limit, subtract the minus tolerance of .002 from the decimal drill size to get a lower limit of 0.2640.

Dimensioning Hole Locations

In addition to dimensioning the size of a hole, drafters and engineers must also dimension the locations of holes. **Figure 6-50** illustrates the proper methods

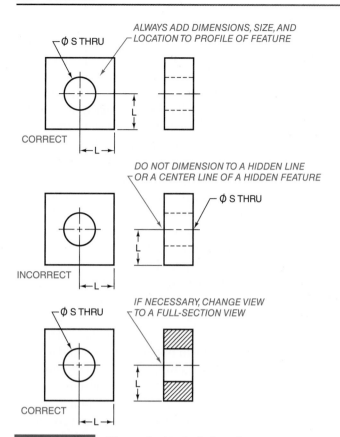

ALWAYS ADD DIMENSIONS, SIZE, AND LOCATION TO PROFILE OF FEATURE

Ø S THRU

CORRECT

DO NOT DIMENSION TO A HIDDEN LINE OR A CENTER LINE OF A HIDDEN FEATURE

Ø S THRU

INCORRECT

Ø S THRU

IF NECESSARY, CHANGE VIEW TO A FULL-SECTION VIEW

CORRECT

FIGURE 6-50   Dimensioning hole locations.

for dimensioning the location of a hole. The preferred method is to dimension the location of the hole in the view where the hole appears as a circle. It is common practice to dimension from a reference side of the object to the center line of the hole. One should never dimension to a hidden line, however. If it is necessary to show one of the locational dimensions in a hidden view, convert the view into a sectional view.

Dimensioning Hole Locations (More than One Hole)

Figure 6-51 illustrates the proper method of dimensioning the locations of more than one hole within

Ø SIZE

LOCATION

LOCATION

Ø SIZE

THE MOST IMPORTANT FACTOR IS HOLE LOCATIONS RELATIVE TO EACH OTHER

FIGURE 6-51   Dimensioning hole locations (more than one hole).

an object. When there is more than one hole in an object, what is important is the relationship of the holes to each other. Because of this, the dimensions applied show the distance horizontally and vertically between the center lines of the holes.

Locating Holes About a Bolt Center

A common dimensioning situation is to have holes spaced about a bolt center. When this is the case, the holes may be dimensioned using coordinate dimensions or angular dimensions (**Figure 6-52**). For the greatest accuracy, coordinate dimensions can be given from the center line of the bolt's center and center-to-center dimensions of the holes. This method is illustrated in the top half of Figure 6-52. Notice that the diameter of the bolt center is specified as a reference dimension. Reference dimensions are given only for informational purposes. They are not intended to be measured or to be used in dictating manufacturing operations. Another method that can be used for locating holes about a bolt center is illustrated in the bottom half of Figure 6-52.

With this method, the diameter of the bolt circle is shown, and a note specifying the number of holes, their diameters, and the indication "equally spaced" is given. If the holes are unequally spaced, the same method can be used except that the note "equally spaced" is dropped, and an angular measurement with reference to one of the center lines is indicated.

3X Ø

(Ø BC.)

REF. DIMENSION

3X R

-OR-

3X R

Ø

3X Ø EQUALLY SPACED ON A Ø BC.

FIGURE 6-52   Locating holes about a bolt center.

Locating Holes on Center Lines and Concentric Arcs

When locating holes on common center lines, the center lines become baselines from which coordinate measurements can be made, as illustrated in the top half of **Figure 6-53**. All three holes in the object in this figure share the same horizontal center line. Consequently, the only dimensions required are dimensions from center to center of the circles. This is accomplished by providing a coordinate dimension from the center of a large circle to the center of one of the smaller circles and the overall coordinate dimension between the centers of the two small circles, as illustrated. When centers are located on concentric arcs, the polar system of dimensioning can be used, as illustrated in the bottom half of Figure 6-53. When this is the case, the radial distances from the point of concurrency and the angular measurements between the holes are used to locate the centers of the holes. The distance from the point of concurrency to the center of each hole is indicated by the radius, as shown.

Dimensioning Multiple Holes along the Same Center Line

Figure 6-54 shows the proper method used for dimensioning multiple holes along the same center line. The method shown at the top of Figure 6-54 is known as "chain dimensioning." The center line

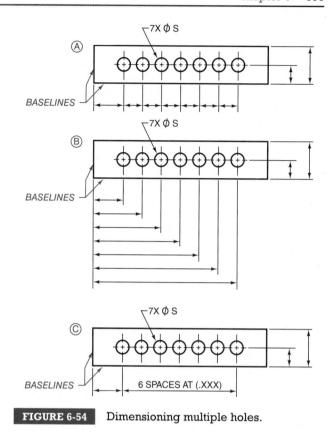

FIGURE 6-54   Dimensioning multiple holes.

of the first hole is dimensioned from a reference baseline, which is normally one side of the object. Then, each successive center line is dimensioned from the immediate preceding center line. A problem with this type of dimensioning is that if one of the features in the chain is improperly located, all successive features will be improperly located. As a result, another method of dimensioning multiple holes or features is often used in place of chain dimensioning. This dimensioning concept is known as "datum" or "baseline" dimensioning. As shown in Figure 6-54B, each successive center line is dimensioned from a baseline or datum. In this way, any one of the multiple features can be improperly located by the machinist without affecting the others. When this happens, the error can sometimes be corrected.

Another method that can be used for dimensioning multiple features within an object is illustrated at the bottom of Figure 6-54. With this method, the center line of the first feature is dimensioned from a given datum, usually the edge of the object. Remaining features are dimensioned using a note that indicates the number of spaces at a given distance. This method does not have the disadvantage of needing an accurate placement of the first feature in order for each successive feature to be accurately located.

FIGURE 6-53   Locating holes on center lines and concentric arcs.

Dimensioning Continuous Features

Many times, features have interruptions in them. When dimensioning such a feature, an extension line is used to indicate that the specified dimension applies across the interruptions. If an extension line is not used, then it is implied that the specified dimension applies across the interruptions. The *continuous feature symbol* is the symbolic means of indicating that the specified dimension or tolerance applies across interruptions in a feature so that it acts as a single continuous feature when considering the requirements for the feature. When using the continuous feature symbol, the use of the extension line is not needed (**Figure 6-55**).

Dimensioning Repetitive Features

Figure 6-55 illustrates the proper methods for dimensioning an object that contains repetitive features. Repetitive features may be located on a drawing using a note that specifies the number of times a given dimension is repeated and giving one particular dimension and the total length as a reference. When dealing with repetitive features, it is necessary to use both size and locational dimensions. Note the dimensions that have been given in Figure 6-55. Some are size dimensions and some are location dimensions.

Callouts for Tapered Holes

Occasionally the design of a manufactured part will require a tapered hole. **Figure 6-56** illustrates the proper method for annotating a tapered hole and shows an example of a 4° tapered bit. Callouts for tapered holes are similar to those for regular holes

L = LOCATION DIMENSION
S = SIZE DIMENSION

FIGURE 6-55   Dimensioning repetitive features.

FIGURE 6-56   Callouts for tapered holes.

except that the note must also contain the taper symbol and the degree and the length of taper. The callout for the tapered hole in Figure 6-56 means that the subject hole has a .50 diameter at the top. It is drilled through and tapered at 4° for the full length of the hole.

Callouts for Countersunk Holes

Countersinking is a machining process to form a conical head so that a fastener can fit flush with the top of a part. The callout for a countersunk hole should contain the diameter of the countersink hole, which is the maximum diameter on the surface, and the angle of the countersink. The proper method for calling out a countersunk hole is illustrated in **Figure 6-57**. First, a .50 diameter hole is drilled through the part. Then, using a countersink bit, the top of the hole is countersunk at 82° until the head of the countersink has a diameter of 1.12.

Callouts for Counterbored Holes

On occasion, a hole in a part is counterbored to allow the head of a bolt or another type of fastener to be recessed. The bottom of a counterbore is flat rather than angled as in a countersink. The callout for a counterbored hole shows the diameter of the drilled hole, the counterbore symbol, the diameter of the counterbore itself, and the depth of the counterbore. When a hole is to be counterbored, the hole itself is drilled first, and then the counterbore is applied using a special counterbore bit, as shown in **Figure 6-58**.

Callouts for Spotfaces

A spotface is a machining process used to finish the surface around the top of a hole so that a bolt or other type of threaded fastener can feed properly.

FIGURE 6-57   Callouts for countersunk holes.

FIGURE 6-58   Callouts for counterbored holes.

FIGURE 6-59   Callouts for spotfaces.

A spotface resembles a very shallow counterbore. The callout for a spotface contains the diameter of the actual hole and the diameter of the spotface. The counterbore symbol is also used for a spotface. The depth of a spotface is not normally shown. When a hole is to be spotfaced, the hole itself is drilled first, and then the spotface is applied with a special spotface bit, as illustrated in **Figure 6-59**.

Dimensioning a Cylinder

Figure 6-60 shows the proper way to dimension a cylinder. Notice from this figure that two methods can be used. In the first method, the longitudinal view and the end view of the cylinder are given. When this method is used, the length of the cylinder is shown and the diameter dimension is applied between the longitudinal view and the end view. In the second method, the end view is left off. Consequently, the diameter dimension is applied to the longitudinal view. When dimensioning a cylinder, always specify the diameter rather than the radius.

Dimensioning a Square

Figure 6-61 illustrates the proper method for dimensioning an object that is square in cross section. No center line on the object indicates that the object is square rather than round in cross section. This type of object requires a length dimension on the longitudinal view and a height dimension placed between the longitudinal view and the end view, preceded by the square symbol.

Double Dimensioning

Figure 6-62 illustrates the concept of double dimensioning. This is sometimes referred to as

FIGURE 6-60   Dimensioning a cylinder.

FIGURE 6-61   Dimensioning a square.

FIGURE 6-62   Double dimensioning.

FIGURE 6-63   Reference dimensioning.

"superfluous" dimensioning. Drafters and engineers should avoid double dimensioning. It is redundant and can lead to confusion.

Reference Dimensioning

Figure 6-63 illustrates the concept of reference dimensioning. In the object shown, the .62 dimension in parentheses is a reference dimension. Reference dimensions are not required. They are occasionally given for information purposes only. The old method used the letters REF following the dimension to indicate a reference dimension. The newer and accepted method is to place reference dimensions in parentheses. **Figures 6-64** and **6-65** further illustrate the concept of reference dimensions.

In Figure 6-64, the second .25 dimension in the front view of the object is a double dimension; it is superfluous. In Figure 6-65, the problem has been solved by placing the second dimension in parentheses, thereby making it a reference dimension. The drafter or engineer had two options in this case for solving the problem: omit the dimension altogether or make it a reference dimension.

FIGURE 6-64   Reference dimensions.

FIGURE 6-65   Reference dimensions.

FIGURE 6-66   Dimensioning internal slots.

Dimensioning Internal Slots

Internal slots represent a common manufacturing situation. Drafters and engineers approach internal slots as two partial holes separated by a space. There is a variety of methods for dimensioning internal slots. The preferred methods are illustrated in **Figure 6-66**. Notice from these examples that when the width of the slot is dimensioned, there is no need to indicate the size of the radius. To do so would be double dimensioning. All of the methods shown in Figure 6-66 are acceptable methods and can be substituted for each other based on the needs of the individual drawing and local preferences.

Figures 6-67 and **6-68** are examples of how to dimension internal slots for two other manufacturing situations. Note in Figure 6-67 that the distance from the concurrent center point to the center line of the slot is indicated with a radius dimension. In addition, the angle formed by the extension of the two-hole center lines to their intersection at the converging point must also be dimensioned as an angle. Figure 6-68 illustrates the proper method for dimensioning an internal slot when one of the holes is larger than the other. If a width dimension is provided, a radius value is not provided. If a width dimension is not provided, the radius dimension must be.

FIGURE 6-67   Dimensioning internal slots.

FIGURE 6-68   Dimensioning internal slots.

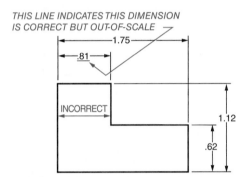

FIGURE 6-69   Dimensioning external rounded ends.

Dimensioning Rounded Ends

Figure 6-69 illustrates the proper method for dimensioning the three most common situations involving external rounded ends. Parts with fully rounded ends should be dimensioned with an overall dimension and the radius should be indicated but not dimensioned, as indicated at the top of Figure 6-69. Figures with partially rounded ends, such as the example in the middle of Figure 6-69, require a radius specification. For parts with rounded ends and holes, overall dimensions can be given as reference dimensions, as shown in the bottom example of Figure 6-69.

Dimensioning and Hidden Lines

There is only one rule for dimensioning and hidden lines: *Do not dimension to hidden lines.*

Figure 6-70 illustrates what drafters and engineers can do when faced with a situation where it is necessary to dimension hidden lines. When this is the case, and it cannot be avoided in any other way, convert the hidden-line view to a sectional view so that you are dimensioning to a solid line.

Not-to-Scale (NTS) Dimensions

On occasion, it is necessary to enter a dimension not to scale. When this is the case, the not-to-scale

FIGURE 6-70   Do not dimension to hidden lines.

FIGURE 6-71   Not-to-scale (NTS) dimensions.

dimension should be underlined with a solid straight line, as illustrated in **Figure 6-71**. The old method used a wavy line under the dimension. This is no longer an acceptable practice.

Nominal Dimensions

A *nominal dimension* is one that represents a general classification size given to a part or product. It may not, and usually does not, express the actual size of the object dimensioned. There is usually a small amount of difference between the actual size of a part and the nominal size, as illustrated in **Figure 6-72**.

FIGURE 6-72   Nominal dimensions.

Dimensioning External Chamfers

A *chamfer* is a beveled edge normally applied to cylindrical parts and a variety of different types of fasteners. Chamfers are used to eliminate rough edges that might break off or impede the assembly of parts. **Figure 6-73** illustrates three accepted procedures for dimensioning external chamfers. The illustration at the top shows how to dimension a chamfer using a leader and note. This is frequently done when the chamfer is a 45° chamfer. The middle illustration shows how to dimension larger chamfers at angles other than 45°. Notice that the length of the chamfer is dimensioned, as is the angle of the chamfer. The bottom illustration shows how to dimension a 45° chamfer that is very small. The dimension is called out and followed by the term "MAX." This tells the machinist that a 45° chamfer is to be applied but its length is to be no more than .010.

Dimensioning Internal Chamfers

Figure 6-74 illustrates two accepted methods for dimensioning internal chamfers. The example on the left illustrates how to call out a 45° chamfer. The example on the right illustrates how to call out a chamfer other than 45°. Notice that when the chamfer is other than 45°, the dimension of the top of the hole after the chamfer has been applied must be given, as must the angle of the chamfer.

Staggered Dimensions

Figure 6-75 illustrates the proper method for applying staggered dimensions. A succession of parallel dimensions can become confusing and difficult to read. As a result, the accepted practice is to stagger the series of parallel dimensions, as shown in the bottom half of Figure 6-75. Staggered dimensions are usually easier to read than aligned parallel dimensions.

Dimensioning Necks and Undercuts

Necks and undercuts are recessed cuts into cylindrical parts. They are generally used where cylinders or other geometric shapes come together. **Figure 6-76** illustrates the proper method for dimensioning necks and undercuts. Notice from the three examples on the left side of Figure 6-76 that necks and undercuts may be dimensioned using a leader and accompanying note. The three examples on the right side of Figure 6-76 show that necks and undercuts may also be dimensioned using a combination of notes, angular dimensions, and size dimensions.

FIGURE 6-73   Dimensioning external chamfers.

FIGURE 6-74   Dimensioning external chamfers.

FIGURE 6-75   Staggered dimensions.

FIGURE 6-76   Dimensioning necks and undercuts.

DP = DIAMETRICAL PITCH

FIGURE 6-78   Dimensioning knurls.

Head Fits for Undercuts and Chamfers

Figure 6-77 illustrates the proper method for showing head fits for parts with undercuts and shoulders. With an undercut, the head fits flush with the surface. With a shoulder, the head will only fit flush with the surface if the surface has been chamfered.

Dimensioning Knurls

Knurls are diamond-shaped or parallel patterns cut into cylindrical surfaces to improve gripping to improve the bonding between parts for permanent press fits and, sometimes, for decoration. The accepted method for dimensioning both parallel and diamond-shaped knurls is illustrated in **Figure 6-78**. Notice in this figure that when knurls are fully dimensioned on a drawing, there is no need to actually show them. The callouts for a knurl should include the type, pitch, and diameter. The initials DP in the knurl callouts stand for diametrical pitch. The most commonly used diametrical pitches for knurling are

64 DP, 96 DP, 128 DP, and 160 DP. In addition to the knurling callout, the length of the knurl and the diameter of the cylinder that will receive the knurls must be shown.

Dimensioning Keyways

Figure 6-79 illustrates the preferred methods for dimensioning both exterior and interior keyways. Notice that both size and location dimensions are required to completely dimension a keyway. The size dimension gives the overall width of the keyway, and the location dimension locates the keyway on the object in question from a specified datum such as the center line of the object. With an exterior keyway, the depth is shown using a dimension that runs from the bottom of the keyway to the bottom of the part in question and parallel to the center line or axis of the shaft. With an interior keyway, the depth of the keyway is specified using a dimension that runs from the top of the keyway to the bottom

FIGURE 6-77   Head fits for undercuts and chamfers.

FIGURE 6-79   Dimensioning keyways.

of the hole and parallel to the center line of the hole. **Figure 6-80** illustrates the preferred method for dimensioning a woodruff keyway. A woodruff keyway requires both size and location dimensions. The size dimensions specify the width and depth of the keyway. The location dimension from a specified datum such as a center line locates the keyway in the part, as shown in Figure 6-80. In addition, the location of the center of the keyway from one end of the shaft must also be shown. The diameter of the shaft and the woodruff cutter must also be dimensioned, as shown in Figure 6-80.

Dimensioning Flat Tapers

Figure 6-81 illustrates the preferred method for dimensioning flat tapers. The method on the left uses 2-size dimensions and a leader with a taper symbol attached. In addition, the taper ratio must be shown. The 5:1 taper ratio called out for the part on the left means that the part tapers one unit in the vertical direction for every five units traveled in the horizontal. The example on the right illustrates another acceptable method for dimensioning flat tapers. In this method, 3-size dimensions provide all of the information needed to manufacture the part with the proper taper.

Dimensioning Round Tapers

Figure 6-82 illustrates two acceptable methods for dimensioning round tapers. Round tapers can be dimensioned by providing the following dimensional data: (a) the diameter or width at both ends of the taper, (b) the length of the tapered feature, and (c) the rate of taper. The example on the right shows the diameter at both ends of the taper. The example on the left shows the length of the tapered feature and is in the taper ratio of 10:1, meaning that the part is tapered 1 unit for every 10 units parallel to the horizontal center line of the shaft.

Dimensioning Threads

Figure 6-83 shows the preferred methods for dimensioning threaded fasteners. Notice that it is not necessary to attempt to draw an actual representation of the thread. Rather, the schematic approach shown in the middle and bottom examples is preferred. The threads themselves are called out in a thread note attached to a leader. In addition, the nominal diameter of the shaft being threaded and the longitudinal

NOTE: DIAMETER OF CUTTER, DEPTH AND WIDTH OF WOODRUFF KEY SIZE IS FOUND IN APPENDIX A

FIGURE 6-80   Dimensioning a woodruff keyway.

FIGURE 6-81   Dimensioning flat tapers.

FIGURE 6-82   Dimensioning round tapers.

FIGURE 6-83   Dimensioning threads.

FIGURE 6-84   Dimensioning a boss and a pad.

FIGURE 6-86   Dimensioning sectional views.

distance of the threading must also be dimensioned, as shown in Figure 6-83. A chamfer callout should be shown with the leader line to indicate how the end of the threaded fastener is to be chamfered.

Dimensioning a Pad and a Boss

Figure 6-84 illustrates the preferred method for dimensioning a pad and a boss. A boss is a round projection from a part. A pad is a projection of any shape other than round. Notice from Figure 6-84 that the boss requires a diameter dimension, a dimension showing how far the boss projects out from the part, and a radius indicator. The pad requires 2-size dimensions: one for length and one for width. It also requires a projection dimension and a radius indicator.

Dimensioning Sheet Metal Bends

Figure 6-85 illustrates the different methods for dimensioning sheet metal bends. When dimensioning a sheet metal object, allowances must be made for the bends. The intersection of the surfaces adjacent to a bend is called the "mold line." This is the line that is used for dimensioning purposes. Notice in Figure 6-85 that each bend has an inside mold line and an outside mold line. Dimensions begin at the intersection of the outside mold line, as shown.

Dimensioning Sectional Views

Figure 6-86 illustrates the rule for dimensioning sectional views. The dimension should, whenever possible, be placed outside the object. Placing dimensions inside an object causes them to conflict with the section lines. When there is no other way to apply a dimension than to place it inside a sectioned area, remove enough of the section lining to permit the dimension to be easily read and the dimension lines to be easily seen, as shown in the right-hand example in Figure 6-86.

Dimensioning Pyramids

Figure 6-87 illustrates the acceptable method for dimensioning a pyramid. A pyramid is dimensioned by showing the height in the front view and the dimensions of the base and the center of the vertex in the top view. If the pyramid base is square, it is only necessary to show the base dimension on one side of the base in the top view and to indicate that the base is square by using the appropriate symbol.

Dimensioning a Cone

Figure 6-88 illustrates the preferred methods for dimensioning a cone. A cone is dimensioned by showing its altitude in the view where the cone appears as a

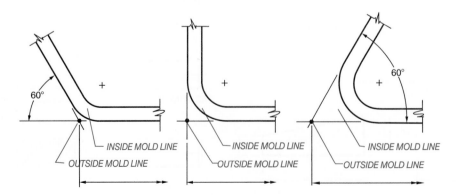

FIGURE 6-85   Dimensioning sheet metal bends.

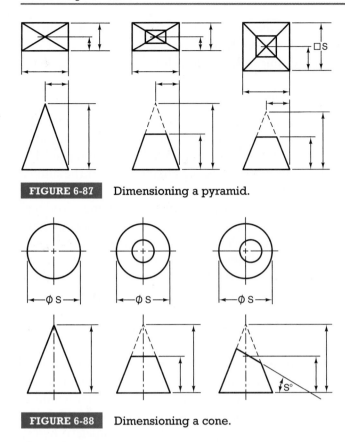

FIGURE 6-87   Dimensioning a pyramid.

FIGURE 6-88   Dimensioning a cone.

FIGURE 6-89   Dimensioning concentric and nonconcentric shafts.

triangle and showing the diameter of the base between the top view and front view. The frustum of a cone is dimensioned by giving the diameter between the top and front view and both heights in the front view where the cone would appear as a triangle. The example on the right-hand side of Figure 6-88 illustrates how to dimension a cone that is cut off at an angle. The diameter of the cone's base is shown between the top and front views. The height of the cone before it is cut and the height of the cone at the center line of the cone after the cut are shown in the front view where the cone would appear as a triangle. The angle of the cut is also dimensioned in this view.

Dimensioning Concentric and Nonconcentric Shafts

Figure 6-89 illustrates the various methods that can be used for dimensioning concentric and nonconcentric shafts and cylinders. The example at the top uses the chain dimensioning concept. The example in the center uses the datum or baseline dimensioning concept. The example at the bottom uses the datum dimensioning concept combined with a reference dimension.

Rectangular Coordinate Dimensioning

Figure 6-90 illustrates the proper methods for applying rectangular coordinate dimensioning. Rectangular

FIGURE 6-90   Rectangular coordinate dimensioning.

coordinate dimensioning is a system that locates features of the drawing relative to each other, individually or as a group, from a datum or series of datums. Round holes and other similar features are located by dimensioning distances to their center points. Linear dimensions in the rectangular coordinate dimension system specify distances from two or three mutually perpendicular planes that are used as datums. Coordinate dimensions should clearly indicate what features of the part being dimensioned establish these datums.

Dimensioning Holes on a Bolt Center Diameter

Figure 6-91 illustrates the proper methods used in dimensioning holes on a bolt center diameter. In the example on the left, the diameter of the bolt's circle is given, and one angle is shown to indicate the number of degrees between the center lines of the six circles. This is sufficient information for the machinist to properly drill the holes. In the example on the right, a note is used to accomplish the same thing. Notice that the note specifies the number and size of the holes. It explains that the holes are drilled through the object and equally spaced on a bolt center. Either of these methods may be used for dimensioning holes around a center. **Figure 6-92** shows two other methods that are used for dimensioning holes around a center. These methods are used when the holes around the bolt center are not equally spaced. The example on the left gives a dimension for the bolt center, a note containing the number of holes and their size, and angles turned from a

reference center line for locating the center lines of the other three holes. The example on the right substitutes linear dimensions for angular dimensions.

Finish Marks

Figure 6-93 illustrates three different systems for applying finish marks to drawings. The traditional system and the general system have been replaced by the new system. However, engineers should be familiar with all three systems so that they are prepared to interpret finish marks applied to drawings produced in years past. Note the sizes specified when applying finish marks. A finish mark in the new system, which is the type of finish mark that should be used on all drawings now, resembles a mechanically produced check mark that forms a 60° "v." Finish marks are used for specifying the type of surface finish to be applied to any machined surface. Finish marks should always appear where the surface in question appears as an edge view. Finish marks are not used when dimensioning notes are given for such machining processes as drilling and boring, as they would be redundant. Finish marks are also omitted when the part in question is made of machined stock. When all surfaces on a part are finished, a note stating "Finish all over" or the letters FAO may be substituted.

Surface imperfections of a part are usually so small that they may be barely visible to the human eye. Consequently, they are measured in microinches. **Figure 6-94** illustrates the concept of the microinch.

FIGURE 6-91   Dimensioning holes on a bolt center diameter.

FIGURE 6-93   Finish marks.

FIGURE 6-94   Microinches.

FIGURE 6-92   Dimensioning holes around a center.

Figure 6-95 illustrates the proper method for applying microinch callouts to finish marks. Notice that microinch callouts are not used with the traditional finish mark system.

Figure 6-96 illustrates the rule for applying finish marks to drawings. Note that finish marks are to be repeated in every view where the surface in question appears as an edge. Figure 6-97 is a table of typical

MICROINCH CALLOFF

NO NUMBER

63   63

TRADITIONAL   GENERAL   NEW
SYSTEM   SYSTEM   SYSTEM

OUTDATED

FIGURE 6-95 Microinch callouts.

125/   125/

125   125

RULE:
FINISH MARKS MUST BE REPEATED IN EVERY VIEW
WHERE THE SURFACE APPEARS AS AN EDGE –
EVEN AS A HIDDEN SURFACE

FIGURE 6-96 Applying finish marks to drawings.

MICROMETERS AA RATING	MICROINCHES AA RATING	APPLICATION
25.2/	1000/	Rough, low-grade surface resulting from sand casting, torch or saw cutting, chipping, or rough forging. Machine operations are not required because appearance is not objectionable. This surface, rarely specified, is suitable for unmachined clearance areas on rough construction items.
12.5/	500/	Rough, low-grade surface resulting from heavy cuts and coarse feeds in milling, turning, shaping, boring, and rough filing, disc grinding, and snagging. It is suitable for clearance areas on machinery, jigs, and fixtures. Sand casting or rough forging produces this surface.
6.3/	250/	Coarse production surfaces, for unimportant clearance and cleanup operations, resulting from coarse surface grind, rough file, disc grind, rapid feeds in turning, milling, shaping, drilling, boring, grinding, etc., where tool marks are not objectionable. The natural surfaces of forgings, permanent mold castings, extrusions, and rolled surfaces also produce this roughness. It can be produced economically and is used on parts where stress requirements, appearance, and conditions of operations and design permit.
3.2/	125/	The roughest surface recommended for parts subject to loads, vibration, and high stress. It is also permitted for bearing surfaces when motion is slow and loads light or infrequent. It is a medium commercial machine finish produced by relatively high speeds and fine feeds taking light cuts with sharp tools. It may be economically produced on lathes, milling machines, shapers, grinders, etc., or on permanent mold castings, die castings, extrusion, and rolled surfaces.
1.6/	63/	A good machine finish produced under controlled conditions using relatively high speeds and fine feeds to take light cuts with sharp cutters. It may be specified for close fits and used for all stressed parts, except fast rotating shafts, axles, and parts subject to severe vibration or extreme tension. It is satisfactory for bearing surfaces when motion is slow and loads light or infrequent. It may also be obtained on extrusions, rolled surfaces, die castings, and permanent mold castings when rigidly controlled.
0.8/	32/	A high-grade machine finish requiring close control when produced by lathes, shapers, milling machines, etc., but relatively easy to produce by centerless, cylindrical, or surface grinders. Also, extruding, rolling, or die casting may produce a comparable surface when rigidly controlled. This surface may be specified in parts where stress concentration is present. It is used for bearings when motion is not continuous and loads are light. When finer finishes are specified, production costs rise rapidly; therefore, such finishes must be analyzed carefully.
0.4/	16/	A high-quality surface produced by fine cylindrical grinding, emery buffing, coarse honing, or lapping, it is specified where smoothness is of primary importance, such as rapidly rotating shaft bearings, heavily loaded bearing, and extreme tension members.
0.2/	8/	A fine surface produced by honing, lapping, or buffing. It is specified where packings and rings must slide across the direction of the surface grain, maintaining or withstanding pressures, or for interior honed surfaces of hydraulic cylinders. It may also be required in precision gauges and instrument work, or sensitive value surfaces, or on rapidly rotating shafts and on bearings where lubrication is not dependable.
0.1/	4/	A costly refined surface produced by honing, lapping, and buffing. It is specified only when the design requirements make it mandatory. It is required in instrument work, gauge work, and where packing and rings must slide across the direction of surface grain such as on chrome-plated piston rods, etc, where lubrication is not dependable.
0.05/	2/	Costly refined surfaces produced only by the finest of modern honing, lapping, buffing, and superfinishing equipment. These surfaces may have a satin or highly polished appearance depending on the finishing operation and material. These surfaces are specified only when design requirements make it mandatory. They are specified on fine or sensitive instrument parts or other laboratory items, and certain gauge surfaces, such as precision gauge blocks.
0.025/	1/	

FIGURE 6-97 Typical surface roughness height applications.

surface roughness height applications. Micrometer ratings are listed in the left-most column of the table. The corresponding microinch ratings are listed in the next column. Applications are then explained horizontally beside each rating. Engineering students should familiarize themselves with these ratings and their corresponding applications.

X-Y-Z Coordinates

Figure 6-98 illustrates the concept of X-Y-Z coordinates. In order to be able to fully understand the various dimensioning systems, one must understand the concept of X-Y-Z coordinates. This is particularly true in the case of tabular dimensioning. Note from this illustration that X coordinates move from an origin in a horizontal direction, Y coordinates move from an origin in a vertical direction, and Z coordinates move from an origin in a direction indicating depth.

Tabular Dimensioning

Figure 6-99 illustrates the concept of tabular dimensioning. Tabular dimensioning is a system in which size and location dimensions are given in a table rather than on the drawing itself. Notice that the only dimensions applied to the drawing itself are the overall length, width, and depth dimensions. Each hole is given an alphanumeric designation. Holes with the same letter designation have the same diameter. Holes with the same letter designation but a different numerical suffix have the same diameter but a different depth. The holes are located in the part using X, Y, and Z coordinates from the specified baselines. In certain cases such as the one illustrated in Figure 6-99, tabular dimensioning is a more convenient, less confusing approach.

HOLE	Ø	X	Y	Z	NOTE
A1	.25	.30	1.75	THRU	–
A2		1.38	1.06	.38	–
B1		.30	.94	THRU	–
B2	.18	3.25	1.75	THRU	–
B3		3.25	1.50	THRU	–
C1		.88	1.68	.38	–
C2	.38	.88	.62	.25	–
C3		1.12	.38	.38	–
D1		2.12	1.81	THRU	–
D2	.08	3.00	.62	.38	–
D3		3.62	.62	.38	–

FIGURE 6-99   Tabular dimensioning.

Tabular Drawing

When parts have similar features but different dimensions, the dimensions can be given in tabular form to decrease the amount of drawing that must be done. With a tabular drawing, one drawing of the part is made with all appropriate dimension lines applied. However, instead of dimension notations, numerical notations are substituted. Then, a table of dimensions is prepared showing the part numbers for all of the parts that the drawing applies to and the dimensions for each letter notation (**Figure 6-100**).

FIGURE 6-98   X-Y-Z coordinates.

A85254-3	4.50	1.62	Ø.38	3/8-16 UNC-2A	1.50
A85254-2	4.88	2.50	Ø.38	3/8-16 UNC-2A	2.00
A85254-1	2.50	1.50	Ø.25	1/4-20 UNC-2A	.88
PART NUMBER	A	B	C	D	E

REGULAR DRAWING NUMBER

PLAN NO 85254

FIGURE 6-100   Tabular drawing.

Tolerancing

Even the best machinist operating the most accurate machine tool is not able to produce a part as precisely and accurately as specified on a drawing. For this reason, it is necessary to apply tolerances when dimensioning parts. A *tolerance* is the total amount of variation permitted from the design size of a part. A basic rule of thumb when applying tolerances is to make them as large as possible and still be able to produce a usable part. Tolerances may be expressed as limits, as shown in **Figures 6-101** and **6-102**, or as the design size followed by the tolerance (12.382 ± 0.003). The design size of a part is also known as the nominal size. This is the dimension from which the limits are calculated. To calculate the upper limit of a toleranced dimension, add the allowable tolerance to the nominal size. To calculate the lower limits for such dimensions, subtract the allowable tolerance from the nominal size, Figures 6-101 and 6-102. There are several definitions with which drafters and engineers should be familiar before applying tolerances on drawings. The most important of these are:

ACTUAL SIZE

Also called the produced size, this is the actual size of the part after it is produced.

.XXX = + .010

.510 UPPER LIMIT
.490 LOWER LIMIT

FIGURE 6-101   Shaft limits.

.XXX = ± .010

.535 UPPER LIMIT
.515 LOWER LIMIT

FIGURE 6-102   Hole limits.

ALLOWANCE

The amount of room between two main parts, usually considered the tightest possible fit between the two parts, is called the *allowance.*

BILATERAL TOLERANCE

A bilateral tolerance is one in which variation from the nominal dimension is allowed in both the plus and the minus directions (.500 ± .002).

DATUM

A theoretically perfect surface, plane, axis, center plane, or point from which dimensions for related features are established.

LEAST MATERIAL CONDITION (LMC)

The condition that exists when the least material is available for the features being dimensioned. *Least material condition (LMC)* is the lower limit for an external feature and the upper limit for an internal feature.

LIMITS

The largest and smallest possible size in which a part may be produced and still be usable. The limits are calculated by applying the specified tolerance to the nominal dimension.

MAXIMUM MATERIAL CONDITION (MMC)

The condition of a part in which the most material is present. *Maximum material condition (MMC)* is the upper limit for an external feature and the lower limit for an internal feature.

NOMINAL DIMENSION

This is the dimension from which the limits are calculated. By applying the specified tolerance to the nominal dimension, one can calculate the upper and lower limits.

TOLERANCE

The total permissible variation in size for a part. The tolerance is the difference between the upper and lower limits for a part.

UNILATERAL TOLERANCE

A tolerance that may vary in only one direction, either in the plus or minus direction but not both:

$$.500^{+.000}_{-.002} \qquad .250^{+.002}_{-.000}$$

In addition to these tolerance-related definitions, drafters and engineers should also be familiar with the types of fits that are used with mating parts. The following definitions relate specifically to fits between mating parts.

FIT

Fit refers to the amount of tightness or looseness between mating parts. There are three types of fits: clearance, interference, and transition.

CLEARANCE FIT

A clearance fit is one in which there is space remaining after mating parts have been assembled.

INTERFERENCE FIT

The type of fit that occurs when there is a negative allowance, or in other words, when the shaft is larger than the collar with which it must mate.

TRANSITION FIT

The type of fit in which there may be either a clearance or an interference fit after mating parts have been assembled. This occurs when the smallest shaft size within the allowable tolerance will fit within the largest hole size, resulting in a clearance fit, and when the largest shaft size within the allowable tolerance will interfere with the smallest hole within an allowable tolerance, creating an interference fit.

Shaft Limits

Figure 6-101 illustrates the concept of shaft limits. In this figure, the shaft has a nominal dimension of .500. The tolerance specified is ±.010. To determine the upper limit for the shaft, the tolerance of .010 is added to the nominal dimension of .500, resulting in an upper limit of .510. To determine the lower limit of the shaft, the tolerance of .010 is subtracted from the nominal dimension of .500, resulting in a lower limit of .490.

Hole Limits

Figure 6-102 illustrates the concept of hole limits. The hole through the part shown has a nominal diameter of .525. The specified tolerance is ±.010. To determine the upper limit for the hole, the tolerance of .010 is added to the nominal dimension of .525, resulting in an upper limit of .535. The lower limit is determined by subtracting the tolerance of .010 from the nominal dimension of .525, resulting in a lower limit of .515.

Allowance

Figure 6-103 illustrates the concept of allowance or the tightest fit between mating parts. Part 1 is a shaft that, when assembled, will mate with Part 2. To determine the allowance between Part 1 and Part 2, the maximum material condition of the shaft (the largest allowable size) is subtracted from the maximum material condition of the hole (the smallest size hole). The maximum material condition of the shaft is .510. The maximum material condition of the hole is .515. Therefore, the allowance between Part 1 and Part 2 is .005.

Clearance Fit

Figure 6-104 illustrates the concept of *clearance fit*. The shaft, Part 1, is to mate with the collar, Part 2, during assembly. To determine the clearance between Part 1 and Part 2, the least material condition of the shaft is subtracted from the least material condition of the hole. The least material condition of the shaft is .490. The least material condition of the hole is .535. Therefore, the clearance between Part 1 and Part 2 is .045.

Interference Fit

Figure 6-105 illustrates the concept of *interference fit*. The smallest shaft in this example is larger than the largest corresponding hole. To determine the amount of interference, you can calculate the allowance and the

.515 MMC HOLE
.510 MMC SHAFT
.005 **ALLOWANCE** *(TIGHTEST FIT)*

FIGURE 6-103   Allowance.

.535 LMC HOLE
.490 LMC SHAFT
.045 **CLEARANCE** *(LOOSEST FIT)*

FIGURE 6-104   Clearance.

.501 MMC SHAFT
.494 MMC HOLE
(–).007 **ALLOWANCE** *(TIGHTEST FIT)*

.500 MMC SHAFT
.496 MMC HOLE
(–).004 **CLEARANCE** *(LOOSEST FIT)*

FIGURE 6-105   Interference fit.

clearance. The allowance, in the case of an interference fit, represents the maximum amount of interference. The clearance, in an interference fit, represents the smallest amount of interference. You can see from the calculations in Figure 6-105 that the maximum amount of interference during assembly will be .007, or the allowance. The minimum amount of interference will be .004, or the clearance.

Transition Fit

Figure 6-106 illustrates the concept of *transition fit*. In this example, there can be both a clearance fit and an interference fit. The largest shaft in this example would interfere with the smallest hole. However, the smallest shaft will fit through the largest hole. This condition is illustrated in the calculations.

Size Limits

Figure 6-107 illustrates the concept of size limits. The cylindrical part shown in this figure will be usable if produced at any size between the upper and lower length limits and at any diameter between the upper and lower diameter limits, inclusive. Applying size limits such as those illustrated in Figure 6-107

.500 MMC SHAFT
.499 MMC HOLE
(–).001 **ALLOWANCE** *(TIGHTEST FIT)*

.502 LMC HOLE
.496 LMC SHAFT
(+).007 **CLEARANCE** *(LOOSEST FIT)*

FIGURE 6-106   Transition fit.

FIGURE 6-107   Size limits.

cuts down on waste, requires less time to produce the part and, therefore, cuts the cost of production and the corresponding costs of the part.

Design Size

Figure 6-108 illustrates the concept of design size. Design size is another way of stating nominal size. The design size represents the ideal the drafter or engineer is seeking. Ideally, the produced size will match the design size. However, since this is not practical, tolerances are normally applied.

Maximum and Minimum Sizes

Figure 6-109 illustrates the concept of maximum and minimum sizes. These are the sizes between which the actual length and diameter of the produced part must fall. After all machining processes have been completed, the inspection department will check the part. If the actual produced size falls within the established maximum and minimum sizes (inclusive), the part will be passed as usable. If it is smaller in any dimension than the minimum size, the part will be scrapped as waste. If it is larger in any dimension than the maximum size, it might be salvaged through additional machining.

FIGURE 6-108   Design size.

FIGURE 6-109   Maximum and minimum sizes.

Location Limits

Just as limits are applied to size dimensions, they are also applied to location dimensions, and for the same reasons. Machinists cannot locate features on a part precisely to the design dimension. Consequently, limits must be applied to location dimensions.

Figure 6-110 illustrates how limits are applied to location dimensions. The part shown in Figure 6-110 will be accepted if the central line of the hole falls within the limits shown.

Design Location

Figure 6-111 illustrates the concept of design location. The design location is the location the drafter or engineer actually wants. It is the theoretically ideal location of the feature. The dimensions shown on this figure are the design dimensions, sometimes referred to as nominal dimensions. The specified tolerance is applied to these nominal dimensions to determine the upper and lower limits for locating the feature.

Maximum and Minimum Locations

Figure 6-112 illustrates the concept of maximum and minimum locations. By applying the specified tolerance to the nominal dimensions, the maximum and minimum locations of the center line of the feature in this figure can be determined. The tolerances applied to the nominal dimension give the machinist latitude in both the X and Y directions. This establishes a square tolerance zone within which the actual center point must fall. So long as the center of the feature

FIGURE 6-112   Maximum and minimum locations.

falls anywhere within the prescribed tolerance zone, the parts will be accepted as usable.

Calculating Fits

Figure 6-113 illustrates the mathematics involved in calculating fits. Take special note of the rule that states "All dimensions should be given as large as possible without interfering with the function of the part in order to keep costs down—close dimensions are expensive." The tighter the fit, the more difficult the part is to manufacture. The more difficult the part is to manufacture, the more it costs to manufacture. Therefore, an engineer and a drafter should never assign a tighter fit than is absolutely necessary for the part to properly function.

Matching Parts

Figure 6-114 illustrates the concept of matching parts. During assembly, Part A must fit over Part B. You can see from this drawing that the hole in Part A must be large enough to accommodate the size and location tolerances of the pins in Part B. The parts must be able to fit if the holes in Part A are bored and located on the low side of the tolerance and the pins in Part B are made and located on the high side of the tolerance. Obviously, determining the dimensions

FIGURE 6-113   Calculating fits.

FIGURE 6-110   Location limits.

FIGURE 6-111   Design location.

.515 1.515 2.515 3.515 4.015

.515
1.515
2.515
3.515
4.015

PART A

PART B

TOLERANCE +/- .015

FIGURE 6-114   Matching parts.

to be used for matching parts takes a good deal of careful planning. It is critical that features on mating parts be located from the same datum, as shown in this figure. Some additional factors to consider when planning the dimensions for matching parts are:

1. The length of time the parts are to be engaged
2. The speed at which mating parts will move after assembly
3. The load of the parts after assembly
4. The type of lubrication used if lubrication is necessary
5. The average temperature at which parts will operate
6. The level of humidity during operation
7. The materials of which parts are made
8. The required functional lifespan of the part
9. The capabilities of machine tools and the people that operate them in the manufacturing plant
10. Overall costs of the part

Standard Fits

Figure 6-115 shows the different types of standard fits. The American National Standards Institute document governing fits is ANSI B4.1.

STANDARD FITS	
RC	RUNNING AND SLIDING FITS
LC	CLEARANCE LOCATIONAL FITS
LT	TRANSITION LOCATIONAL FITS
LN	INTERFERENCE LOCATIONAL FITS
FN	FORCE AND SHRINK FITS

FIGURE 6-115   Standard fits.

This standard covers the five types of fits listed in Figure 6-115. Each of these fits is explained in this section.

RUNNING OR SLIDING CLEARANCE FITS (RC)

Running or sliding fits are used to provide a similar running or sliding performance, with an allowance for suitable lubrication, for all sizes within the specified range. **Figure 6-116** is a table of American National Standard Running and Sliding Fits (ANSI B4.1). From this table you can see that there are nine classes of running and sliding fits: Classes RC1, RC2, RC3, RC4, RC5, RC6, RC7, RC8, and RC9.

Class RC1 is used with close sliding fits for accurately locating parts that must fit together without any play. Class RC2 is used for sliding fits where accurate location is important, but where greater maximum clearance that is provided by Class RC1 fits is necessary. Parts manufactured to this fit do not run freely, but they do move and turn easily. Class RC3 is used for precision running fits. This class ensures the closest possible fit that can be expected to run freely. This fit is used for precision work that will operate at slow speeds under light pressures, but it is not suitable where significant temperature differences are likely to occur.

Class RC4 is used for close running fits on accurate machinery with medium surface speeds and pressure when accurate location and very little play is desired. Classes RC5 and RC6 are used for medium running fits on accurate machinery with higher running speeds and/or heavier pressures. Class RC7 is used for free running fits where accuracy is less important and/or where significant temperature variations are likely. Classes RC8 and RC9 are used for loose running fits when large tolerances are necessary, together with an allowance on the external parts.

CLEARANCE LOCATIONAL FITS (LC)

Clearance locational fits are used with parts that are stationary but must be able to be freely assembled or disassembled. **Figure 6-117** is a table of American National Standard Clearance Locational Fits from ANSI 4.1.

TRANSITION LOCATIONAL FITS (LT)

Transition locational fits fall on a continuum between clearance and interference fits and are used in applications where locational accuracy is important but a small amount of clearance or interference is allowable. **Figure 6-118** is a table of American

Nominal Size Range, Inches Over To	Class RC 1			Class RC 2			Class RC 3			Class RC 4		
	Clearance*	Hole H5	Shaft g4	Clearance*	Hole H6	Shaft g5	Clearance*	Hole H7	Shaft f6	Clearance*	Hole H8	Shaft f7
	Values shown below are in thousandths of an inch											
0–0.12	0.1 / 0.45	+0.2 / 0	−0.1 / −0.25	0.1 / 0.55	+0.25 / 0	−0.1 / −0.3	0.3 / 0.95	+0.4 / 0	−0.3 / −0.55	0.3 / 1.3	+0.6 / 0	−0.3 / −0.7
0.12–0.24	0.15 / 0.5	+0.2 / 0	−0.15 / −0.3	0.15 / 0.65	+0.3 / 0	−0.15 / −0.35	0.4 / 1.12	+0.5 / 0	−0.4 / −0.7	0.4 / 1.6	+0.7 / 0	−0.4 / −0.9
0.24–0.40	0.2 / 0.6	+0.25 / 0	−0.2 / −0.35	0.2 / 0.85	+0.4 / 0	−0.2 / −0.45	0.5 / 1.5	+0.6 / 0	−0.5 / −0.9	0.5 / 2.0	+0.9 / 0	−0.5 / −1.1
0.40–0.71	0.25 / 0.75	+0.3 / 0	−0.25 / −0.45	0.25 / 0.95	+0.4 / 0	−0.25 / −0.55	0.6 / 1.7	+0.7 / 0	−0.6 / −1.0	0.6 / 2.3	+1.0 / 0	−0.6 / −1.3
0.71–1.19	0.3 / 0.95	+0.4 / 0	−0.3 / −0.55	0.3 / 1.2	+0.5 / 0	−0.3 / −0.7	0.8 / 2.1	+0.8 / 0	−0.8 / −1.3	0.8 / 2.8	+1.2 / 0	−0.8 / −1.6
1.19–1.97	0.4 / 1.1	+0.4 / 0	−0.4 / −0.7	0.4 / 1.4	+0.6 / 0	−0.4 / −0.8	1.0 / 2.6	+1.0 / 0	−1.0 / −1.6	1.0 / 3.6	+1.6 / 0	−1.0 / −2.0
1.97–3.15	0.4 / 1.2	+0.5 / 0	−0.4 / −0.7	0.4 / 1.6	+0.7 / 0	−0.4 / −0.9	1.2 / 3.1	+1.2 / 0	−1.2 / −1.9	1.2 / 4.2	+1.8 / 0	−1.2 / −2.4
3.15–4.73	0.5 / 1.5	+0.6 / 0	−0.5 / −0.9	0.5 / 2.0	+0.9 / 0	−0.5 / −1.1	1.4 / 3.7	+1.4 / 0	−1.4 / −2.3	1.4 / 5.0	+2.2 / 0	−1.4 / −2.8
4.73–7.09	0.6 / 1.8	+0.7 / 0	−0.6 / −1.1	0.6 / 2.3	+1.0 / 0	−0.6 / −1.3	1.6 / 4.2	+1.6 / 0	−1.6 / −2.6	1.6 / 5.7	+2.5 / 0	−1.6 / −3.2
7.09–9.85	0.6 / 2.0	+0.8 / 0	−0.6 / −1.2	0.6 / 2.6	+1.2 / 0	−0.6 / −1.4	2.0 / 5.0	+1.8 / 0	−2.0 / −3.2	2.0 / 6.6	+2.8 / 0	−2.0 / −3.8
9.85–12.41	0.8 / 2.3	+0.9 / 0	−0.8 / −1.4	0.8 / 2.9	+1.2 / 0	−0.8 / −1.7	2.5 / 5.7	+2.0 / 0	−2.5 / −3.7	2.5 / 7.5	+3.0 / 0	−2.5 / −4.5
12.41–15.75	1.0 / 2.7	+1.0 / 0	−1.0 / −1.7	1.0 / 3.4	+1.4 / 0	−1.0 / −2.0	3.0 / 6.6	+2.2 / 0	−3.0 / −4.4	3.0 / 8.7	+3.5 / 0	−3.0 / −5.2
15.75–19.69	1.2 / 3.0	+1.0 / 0	−1.2 / −2.0	1.2 / 3.8	+1.6 / 0	−1.2 / −2.2	4.0 / 8.1	+2.5 / 0	−4.0 / −5.6	4.0 / 10.5	+4.0 / 0	−4.0 / −6.5

See footnotes at end of table.

Nominal Size Range, Inches Over To	Class RC 5			Class RC 6			Class RC 7			Class RC 8			Class RC 9		
	Clearance*	Hole H8	Shaft e7	Clearance*	Hole H9	Shaft e8	Clearance*	Hole H9	Shaft d8	Clearance*	Hole H10	Shaft c9	Clearance*	Hole H11	Shaft
	Values shown below are in thousandths of an inch														
0–0.12	0.6 / 1.6	+0.6 / 0	−0.6 / −1.0	0.6 / 2.2	+1.0 / 0	−0.6 / −1.2	1.0 / 2.6	+1.0 / 0	−1.0 / −1.6	2.5 / 5.1	+1.6 / 0	−2.5 / −3.5	4.0 / 8.1	+2.5 / 0	−4.0 / −5.6
0.12–0.24	0.8 / 2.0	+0.7 / 0	−0.8 / −1.3	0.8 / 2.7	+1.2 / 0	−0.8 / −1.5	1.2 / 3.1	+1.2 / 0	−1.2 / −1.9	2.8 / 5.8	+1.8 / 0	−2.8 / −4.0	4.5 / 9.0	+3.0 / 0	−4.5 / −6.0
0.24–0.40	1.0 / 2.5	+0.9 / 0	−1.0 / −1.6	1.0 / 3.3	+1.4 / 0	−1.0 / −1.9	1.6 / 3.9	+1.4 / 0	−1.6 / −2.5	3.0 / 6.6	+2.2 / 0	−3.0 / −4.4	5.0 / 10.7	+3.5 / 0	−5.0 / −7.2
0.40–0.71	1.2 / 2.9	+1.0 / 0	−1.2 / −1.9	1.2 / 3.8	+1.6 / 0	−1.2 / −2.2	2.0 / 4.6	+1.6 / 0	−2.0 / −3.0	3.5 / 7.9	+2.8 / 0	−3.5 / −5.1	6.0 / 12.8	+4.0 / 0	−6.0 / −8.8
0.71–1.19	1.6 / 3.6	+1.2 / 0	−1.6 / −2.4	1.6 / 4.8	+2.0 / 0	−1.6 / −2.8	2.5 / 5.7	+2.0 / 0	−2.5 / −3.7	4.5 / 10.0	+3.5 / 0	−4.5 / −6.5	7.0 / 15.5	+5.0 / 0	−7.0 / −10.5
1.19–1.97	2.0 / 4.6	+1.6 / 0	−2.0 / −3.0	2.0 / 6.1	+2.5 / 0	−2.0 / −3.6	3.0 / 7.1	+2.5 / 0	−3.0 / −4.6	5.0 / 11.5	+4.0 / 0	−5.0 / −7.5	8.0 / 18.0	+6.0 / 0	−8.0 / −12.0
1.97–3.15	2.5 / 5.5	+1.8 / 0	−2.5 / −3.7	2.5 / 7.3	+3.0 / 0	−2.5 / −4.3	4.0 / 8.8	+3.0 / 0	−4.0 / −5.8	6.0 / 13.5	+4.5 / 0	−6.0 / −9.0	9.0 / 20.5	+7.0 / 0	−9.0 / −13.5
3.15–4.73	3.0 / 6.6	+2.2 / 0	−3.0 / −4.4	3.0 / 8.7	+3.5 / 0	−3.0 / −5.2	5.0 / 10.7	+3.5 / 0	−5.0 / −7.2	7.0 / 15.5	+5.0 / 0	−7.0 / −10.5	10.0 / 24.0	+9.0 / 0	−10.0 / −15.0
4.73–7.09	3.5 / 7.6	+2.5 / 0	−3.5 / −5.1	3.5 / 10.0	+4.0 / 0	−3.5 / −6.0	6.0 / 12.5	+4.0 / 0	−6.0 / −8.5	8.0 / 18.0	+6.0 / 0	−8.0 / −12.0	12.0 / 28.0	+10.0 / 0	−12.0 / −18.0
7.09–9.85	4.0 / 8.6	+2.8 / 0	−4.0 / −5.8	4.0 / 11.3	+4.5 / 0	−4.0 / −6.8	7.0 / 14.3	+4.5 / 0	−7.0 / −9.8	10.0 / 21.5	+7.0 / 0	−10.0 / −14.5	15.0 / 34.0	+12.0 / 0	−15.0 / −22.0
9.85–12.41	5.0 / 10.0	+3.0 / 0	−5.0 / −7.0	5.0 / 13.0	+5.0 / 0	−5.0 / −8.0	8.0 / 16.0	+5.0 / 0	−8.0 / −11.0	12.0 / 25.0	+8.0 / 0	−12.0 / −17.0	18.0 / 38.0	+12.0 / 0	−18.0 / −26.0
12.41–15.75	6.0 / 11.7	+3.5 / 0	−6.0 / −8.2	6.0 / 15.5	+6.0 / 0	−6.0 / −9.5	10.0 / 19.5	+6.0 / 0	−10.0 / −13.5	14.0 / 29.0	+9.0 / 0	−14.0 / −20.0	22.0 / 45.0	+14.0 / 0	−22.0 / −31.0
15.75–19.69	8.0 / 14.5	+4.0 / 0	−8.0 / −10.5	8.0 / 18.0	+6.0 / 0	−8.0 / −12.0	12.0 / 22.0	+6.0 / 0	−12.0 / −16.0	16.0 / 32.0	+10.0 / 0	−16.0 / −22.0	25.0 / 51.0	+16.0 / 0	−25.0 / −35.0

All data above heavy lines are in accord with ABC agreements. Symbols H5, g4, etc. are hole and shaft designations in ABC system. Limits for sizes above 19.69 inches are also given in the ANSI Standard.
* Pairs of values shown represent minimum and maximum amounts of clearance resulting from application of standard tolerance limits.

FIGURE 6-116   ANSI standard running and sliding fits (Courtesy ASME).

Nominal Size Range, Inches Over To	Class LC 1 Clearance*	Class LC 1 Hole H6	Class LC 1 Shaft h5	Class LC 2 Clearance*	Class LC 2 Hole H7	Class LC 2 Shaft h6	Class LC 3 Clearance*	Class LC 3 Hole H8	Class LC 3 Shaft h7	Class LC 4 Clearance*	Class LC 4 Hole H10	Class LC 4 Shaft h9	Class LC 5 Clearance*	Class LC 5 Hole H7	Class LC 5 Shaft g6
						Values shown below are in thousandths of an inch									
0–0.12	0 / 0.45	+0.25 / 0	0 / −0.2	0 / 0.65	+0.4 / 0	0 / −0.25	0 / 1	+0.6 / 0	0 / −0.4	0 / 2.6	+1.6 / 0	0 / −1.0	0.1 / 0.75	+0.4 / 0	−0.1 / −0.35
0.12–0.24	0 / 0.5	+0.3 / 0	0 / −0.2	0 / 0.8	+0.5 / 0	0 / −0.3	0 / 1.2	+0.7 / 0	0 / −0.5	0 / 3.0	+1.8 / 0	0 / −1.2	0.15 / 0.95	+0.5 / 0	−0.15 / −0.45
0.24–0.40	0 / 0.65	+0.4 / 0	0 / −0.25	0 / 1.0	+0.6 / 0	0 / −0.4	0 / 1.5	+0.9 / 0	0 / −0.6	0 / 3.6	+2.2 / 0	0 / −1.4	0.2 / 1.2	+0.6 / 0	−0.2 / −0.6
0.40–0.71	0 / 0.7	+0.4 / 0	0 / −0.3	0 / 1.1	+0.7 / 0	0 / −0.4	0 / 1.7	+1.0 / 0	0 / −0.7	0 / 4.4	+2.8 / 0	0 / −1.6	0.25 / 1.35	+0.7 / 0	−0.25 / −0.65
0.71–1.19	0 / 0.9	+0.5 / 0	0 / −0.4	0 / 1.3	+0.8 / 0	0 / −0.5	0 / 2	+1.2 / 0	0 / −0.8	0 / 5.5	+3.5 / 0	0 / −2.0	0.3 / 1.6	+0.8 / 0	−0.3 / −0.8
1.19–1.97	0 / 1.0	+0.6 / 0	0 / −0.4	0 / 1.6	+1.0 / 0	0 / −0.6	0 / 2.6	+1.6 / 0	0 / −1	0 / 6.5	+4.0 / 0	0 / −2.5	0.4 / 2.0	+1.0 / 0	−0.4 / −1.0
1.97–3.15	0 / 1.2	+0.7 / 0	0 / −0.5	0 / 1.9	+1.2 / 0	0 / −0.7	0 / 3	+1.8 / 0	0 / −1.2	0 / 7.5	+4.5 / 0	0 / −3	0.4 / 2.3	+1.2 / 0	−0.4 / −1.1
3.15–4.73	0 / 1.5	+0.9 / 0	0 / −0.6	0 / 2.3	+1.4 / 0	0 / −0.9	0 / 3.6	+2.2 / 0	0 / −1.4	0 / 8.5	+5.0 / 0	0 / −3.5	0.5 / 2.8	+1.4 / 0	−0.5 / −1.4
4.73–7.09	0 / 1.7	+1.0 / 0	0 / −0.7	0 / 2.6	+1.6 / 0	0 / −1.0	0 / 4.1	+2.5 / 0	0 / −1.6	0 / 10.0	+6.0 / 0	0 / −4	0.6 / 3.2	+1.6 / 0	−0.6 / −1.6
7.09–9.85	0 / 2.0	+1.2 / 0	0 / −0.8	0 / 3.0	+1.8 / 0	0 / −1.2	0 / 4.6	+2.8 / 0	0 / −1.8	0 / 11.5	+7.0 / 0	0 / −4.5	0.6 / 3.6	+1.8 / 0	−0.6 / −1.8
9.85–12.41	0 / 2.1	+1.2 / 0	0 / −0.9	0 / 3.2	+2.0 / 0	0 / −1.2	0 / 5	+3.0 / 0	0 / −2.0	0 / 13.0	+8.0 / 0	0 / −5	0.7 / 3.9	+2.0 / 0	−0.7 / −1.9
12.41–15.75	0 / 2.4	+1.4 / 0	0 / −1.0	0 / 3.6	+2.2 / 0	0 / −1.4	0 / 5.7	+3.5 / 0	0 / −2.2	0 / 15.0	+9.0 / 0	0 / −6	0.7 / 4.3	+2.2 / 0	−0.7 / −2.1
15.75–19.69	0 / 2.6	+1.6 / 0	0 / −1.0	0 / 4.1	+2.5 / 0	0 / −1.6	0 / 6.5	+4 / 0	0 / −2.5	0 / 16.0	+10.0 / 0	0 / −6	0.8 / 4.9	+2.5 / 0	−0.8 / −2.4

See footnotes at end of table.

Nominal Size Range, Inches Over To	Class LC 6 Clearance*	Class LC 6 Hole H9	Class LC 6 Shaft f8	Class LC 7 Clearance*	Class LC 7 Hole H10	Class LC 7 Shaft e9	Class LC 8 Clearance*	Class LC 8 Hole H10	Class LC 8 Shaft d9	Class LC 9 Clearance*	Class LC 9 Hole H11	Class LC 9 Shaft c10	Class LC 10 Clearance*	Class LC 10 Hole H12	Class LC 10 Shaft	Class LC 11 Clearance*	Class LC 11 Hole H13	Class LC 11 Shaft
								Values shown below are in thousandths of an inch										
0–0.12	0.3 / 1.9	+1.0 / 0	−0.3 / −0.9	0.6 / 3.2	+1.6 / 0	−0.6 / −1.6	1.0 / 2.0	+1.6 / 0	−1.0 / −2.0	2.5 / 6.6	+2.5 / 0	−2.5 / −4.1	4 / 12	+4 / 0	−4 / −8	5 / 17	+6 / 0	−5 / −11
0.12–0.24	0.4 / 2.3	+1.2 / 0	−0.4 / −1.1	0.8 / 3.8	+1.8 / 0	−0.8 / −2.0	1.2 / 4.2	+1.8 / 0	−1.2 / −2.4	2.8 / 7.6	+3.0 / 0	−2.8 / −4.6	4.5 / 14.5	+5 / 0	−4.5 / −9.5	6 / 20	+7 / 0	−6 / −13
0.24–0.40	0.5 / 2.8	+1.4 / 0	−0.5 / −1.4	1.0 / 4.6	+2.2 / 0	−1.0 / −2.4	1.6 / 5.2	+2.2 / 0	−1.6 / −3.0	3.0 / 8.7	+3.5 / 0	−3.0 / −5.2	5 / 17	+6 / 0	−5 / −11	7 / 25	+9 / 0	−7 / −16
0.40–0.71	0.6 / 3.2	+1.6 / 0	−0.6 / −1.6	1.2 / 5.6	+2.8 / 0	−1.2 / −2.8	2.0 / 6.4	+2.8 / 0	−2.0 / −3.6	3.5 / 10.3	+4.0 / 0	−3.5 / −6.3	6 / 20	+7 / 0	−6 / −13	8 / 28	+10 / 0	−8 / −18
0.71–1.19	0.8 / 4.0	+2.0 / 0	−0.8 / −2.0	1.6 / 7.1	+3.5 / 0	−1.6 / −3.6	2.5 / 8.0	+3.5 / 0	−2.5 / −4.5	4.5 / 13.0	+5.0 / 0	−4.5 / −8.0	7 / 23	+8 / 0	−7 / −15	10 / 34	+12 / 0	−10 / −22
1.19–1.97	1.0 / 5.1	+2.5 / 0	−1.0 / −2.6	2.0 / 8.5	+4.0 / 0	−2.0 / −4.5	3.0 / 9.5	+4.0 / 0	−3.0 / −5.5	5.0 / 15.0	+6 / 0	−5.0 / −9.0	8 / 28	+10 / 0	−8 / −18	12 / 44	+16 / 0	−12 / −28
1.97–3.15	1.2 / 6.0	+3.0 / 0	−1.0 / −3.0	2.5 / 10.0	+4.5 / 0	−2.5 / −5.5	4.0 / 11.5	+4.5 / 0	−4.0 / −7.0	6.0 / 17.5	+7 / 0	−6.0 / −10.5	10 / 34	+12 / 0	−10 / −22	14 / 50	+18 / 0	−14 / −32
3.15–4.73	1.4 / 7.1	+3.5 / 0	−1.4 / −3.6	3.0 / 11.5	+5.0 / 0	−3.0 / −6.5	5.0 / 13.5	+5.0 / 0	−5.0 / −8.5	7 / 21	+9 / 0	−7 / −12	11 / 39	+14 / 0	−11 / −25	16 / 60	+22 / 0	−16 / −38
4.73–7.09	1.6 / 8.1	+4.0 / 0	−1.6 / −4.1	3.5 / 13.5	+6.0 / 0	−3.5 / −7.5	6 / 16	+6 / 0	−6 / −10	8 / 24	+10 / 0	−8 / −14	12 / 44	+16 / 0	−12 / −28	18 / 68	+25 / 0	−18 / −43
7.09–9.85	2.0 / 9.3	+4.5 / 0	−2.0 / −4.8	4.0 / 15.5	+7.0 / 0	−4.0 / −8.5	7 / 18.5	+7 / 0	−7 / −11.5	10 / 29	+12 / 0	−10 / −17	16 / 52	+18 / 0	−16 / −34	22 / 78	+28 / 0	−22 / −50
9.85–12.41	2.2 / 10.2	+5.0 / 0	−2.2 / −5.2	4.5 / 17.5	+8.0 / 0	−4.5 / −9.5	7 / 20	+8 / 0	−7 / −12	12 / 32	+12 / 0	−12 / −20	20 / 60	+20 / 0	−20 / −40	28 / 88	+30 / 0	−28 / −58
12.41–15.75	2.5 / 12.0	+6.0 / 0	−2.5 / −6.0	5.0 / 20.0	+9.0 / 0	−5 / −11	8 / 23	+9 / 0	−8 / −14	14 / 37	+14 / 0	−14 / −23	22 / 66	+22 / 0	−22 / −44	30 / 100	+35 / 0	−30 / −65
15.75–19.69	2.8 / 12.8	+6.0 / 0	−2.8 / −6.8	5.0 / 21.0	+10.0 / 0	−5 / −11	9 / 25	+10 / 0	−9 / −15	16 / 42	+16 / 0	−16 / −26	25 / 75	+25 / 0	−25 / −50	35 / 115	+40 / 0	−35 / −75

All data above heavy lines are in accordance with American-British-Canadian (ABC) agreements. Symbols H6, H7, s6, etc. are hole and shaft designations in ABC system. Limits for sizes above 19.69 inches are not covered by ABC agreements but are given in the ANSI Standard.

* Pairs of values shown represent minimum and maximum amounts of interference resulting from application of standard tolerance limits.

FIGURE 6-117   ANSI standard clearance locational fits (Courtesy ASME).

Nominal Size Range, Inches Over To	Class LT 1			Class LT 2			Class LT 3			Class LT 4			Class LT 5			Class LT 6		
	Fit*	Std. Tolerance Limits		Fit*	Std. Tolerance Limits		Fit*	Std. Tolerance Limits		Fit*	Std. Tolerance Limits		Fit*	Std. Tolerance Limits		Fit*	Std. Tolerance Limits	
		Hole H7	Shaft js6		Hole H8	Shaft js7		Hole H7	Shaft k6		Hole H8	Shaft k7		Hole H7	Shaft n6		Hole H7	Shaft n7
							Values shown below are in thousandths of an inch											
0–0.12	−0.12 +0.52	+0.4 0	+0.12 −0.12	−0.2 +0.8	+0.6 0	+0.2 −0.2							−0.5 +0.15	+0.4 0	+0.5 +0.25	−0.65 +0.15	+0.4 0	+0.65 +0.25
0.12–0.24	−0.15 +0.65	+0.5 0	+0.15 −0.15	−0.25 +0.95	+0.7 0	+0.25 −0.25							−0.6 +0.2	+0.5 0	+0.6 +0.3	−0.8 +0.2	+0.5 0	+0.8 +0.3
0.24–0.40	−0.2 +0.8	+0.6 0	+0.2 −0.2	−0.3 +1.2	+0.9 0	+0.3 −0.3	−0.5 +0.5	+0.6 0	+0.5 +0.1	−0.7 +0.8	+0.9 0	+0.7 +0.1	−0.8 +0.2	+0.6 0	+0.8 +0.4	−1.0 +0.2	+0.6 0	+1.0 +0.4
0.40–0.71	−0.2 +0.9	+0.7 0	+0.2 −0.2	−0.35 +1.35	+1.0 0	+0.35 −0.35	−0.5 +0.6	+0.7 0	+0.5 +0.1	−0.8 +0.9	+1.0 0	+0.8 +0.1	−0.9 +0.2	+0.7 0	+0.9 +0.5	−1.2 +0.2	+0.7 0	+1.2 +0.5
0.71–1.19	−0.25 +1.05	+0.8 0	+0.25 −0.25	−0.4 +1.6	+1.2 0	+0.4 −0.4	−0.6 +0.7	+0.8 0	+0.6 +0.1	−0.9 +1.1	+1.2 0	+0.9 +0.1	−1.1 +0.2	+0.8 0	+1.1 +0.6	−1.4 +0.2	+0.8 0	+1.4 +0.6
1.19–1.97	−0.3 +1.3	+1.0 0	+0.3 −0.3	−0.5 +2.1	+1.6 0	+0.5 −0.5	−0.7 +0.9	+1.0 0	+0.7 +0.1	−1.1 +1.5	+1.6 0	+1.1 +0.1	−1.3 +0.3	+1.0 0	+1.3 +0.7	−1.7 +0.3	+1.0 0	+1.7 +0.7
1.97–3.15	−0.3 +1.5	+1.2 0	+0.3 −0.3	−0.6 +2.4	+1.8 0	+0.6 −0.6	−0.8 +1.1	+1.2 0	+0.8 +0.1	−1.3 +1.7	+1.8 0	+1.3 +0.1	−1.5 +0.4	+1.2 0	+1.5 +0.8	−2.0 +0.4	+1.2 0	+2.0 +0.8
3.15–4.73	−0.4 +1.8	+1.4 0	+0.4 −0.4	−0.7 +2.9	+2.2 0	+0.7 −0.7	−1.0 +1.3	+1.4 0	+1.0 +0.1	−1.5 +2.1	+2.2 0	+1.5 +0.1	−1.9 +0.4	+1.4 0	+1.9 +1.0	−2.4 +0.4	+1.4 0	+2.4 +1.0
4.73–7.09	−0.5 +2.1	+1.6 0	+0.5 −0.5	−0.8 +3.3	+2.5 0	+0.8 −0.8	−1.1 +1.5	+1.6 0	+1.1 +0.1	−1.7 +2.4	+2.5 0	+1.7 +0.1	−2.2 +0.4	+1.6 0	+2.2 +1.2	−2.8 +0.4	+1.6 0	+2.8 +1.2
7.09–9.85	−0.6 +2.4	+1.8 0	+0.6 −0.6	−0.9 +3.7	+2.8 0	+0.9 −0.9	−1.4 +1.6	+1.8 0	+1.4 +0.2	−2.0 +2.6	+2.8 0	+2.0 +0.2	−2.6 +0.4	+1.8 0	+2.6 +1.4	−3.2 +0.4	+1.8 0	+3.2 +1.4
9.85–12.41	−0.6 +2.6	+2.0 0	+0.6 −0.6	−1.0 +4.0	+3.0 0	+1.0 −1.0	−1.4 +1.8	+2.0 0	+1.4 +0.2	−2.2 +2.8	+3.0 0	+2.2 +0.2	−2.6 +0.6	+2.0 0	+2.6 +1.4	−3.4 +0.6	+2.0 0	+3.4 +1.4
12.41–15.75	−0.7 +2.9	+2.2 0	+0.7 −0.7	−1.0 +4.5	+3.5 0	+1.0 −1.0	−1.6 +2.0	+2.2 0	+1.6 +0.2	−2.4 +3.3	+3.5 0	+2.4 +0.2	−3.0 +0.6	+2.2 0	+3.0 +1.6	−3.8 +0.6	+2.2 0	+3.8 +1.6
15.75–19.69	−0.8 +3.3	+2.5 0	+0.8 −0.8	−1.2 +5.2	+4.0 0	+1.2 −1.2	−1.8 +2.3	+2.5 0	+1.8 +0.2	−2.7 +3.8	+4.0 0	+2.7 +0.2	−3.4 +0.7	+2.5 0	+3.4 +1.8	−4.3 +0.7	+2.5 0	+4.3 +1.8

All data above heavy lines are in accord with ABC agreements. Symbols H7, js6, etc. are hole and shaft designations in ABC system.
* Pairs of values shown represent maximum amount of interference (−) and maximum amount of clearance (+) resulting from application of standard tolerance limits.

FIGURE 6-118   ANSI Standard Transition Locational Fits (Courtesy ASME).

National Standard Transition Locational Fits from ANSI B4.1.

INTERFERENCE LOCATIONAL FITS (LN)

Interference locational fits are used when locational accuracy is very important and for parts requiring rigidity and alignment but no special requirements for bore pressure. Interference locational fits are not used for parts that are designed to transmit frictional loads to other parts through the tightness of the fits. These types of applications are covered by force fits. **Figure 6-119** is a table of American National Standard Interference Locational Fits from ANSI B4.1.

FORCE AND SHRINK FITS (FN)

Force or shrink fits are a special type of interference fit used when maintenance of constant bore pressure for all sizes within the specified range is important. The amount of interference varies with the diameter, and the difference between minimum and maximum values is small to ensure maintenance of the resulting pressures within reasonable limits. **Figure 6-120** is a table of American National Standard Force and Shrink Fits from ANSI B4.1.

Using Fit Tables

Figures 6-116 through 6-120 are fit tables taken from ANSI B4.1. It is important for engineers and drafters to know how to use tables such as these. Look at Figure 6-116. The nine classes explained earlier are arranged across the top of the table. Under each class column, there are three other columns: one giving the clearance, one giving the standard tolerance limits for the hole, and one giving the standard tolerance limits for the shaft. On the extreme left-hand side of the chart, the nominal size range in inches is given. To use this table, as well as the others contained in this chapter, select the appropriate class across the top of the table and locate the nominal size at the extreme left-hand side of the table. Moving across the row that contains the appropriate nominal size range to the class column you have selected, you will find the tolerance limits you need. These tolerance limits are added to or subtracted from the basic size to obtain maximum and minimum sizes of mating parts. Each tolerance is preceded by either a plus or a minus to indicate whether it is added to or subtracted from the basic size. Values must be changed to inches before adding and subtracting.

Tolerance limits given in body of table are added or subtracted to basic size (as indicated by + or − sign) to obtain maximum and minimum sizes of mating parts.

Nominal Size Range, Inches	Class LN 1			Class LN 2			Class LN 3		
	Limits of Inter-ference	Standard Limits		Limits of Inter-ference	Standard Limits		Limits of Inter-ference	Standard Limits	
		Hole H6	Shaft n5		Hole H7	Shaft p6		Hole H7	Shaft r6
Over To	Values shown below are given in thousandths of an inch								
0–0.12	0 0.45	+0.25 0	+0.45 +0.25	0 0.65	+0.4 0	+0.65 +0.4	0.1 0.75	+0.4 0	+0.75 +0.5
0.12–0.24	0 0.5	+0.3 0	+0.5 +0.3	0 0.8	+0.5 0	+0.8 +0.5	0.1 0.9	+0.5 0	+0.9 +0.6
0.24–0.40	0 0.65	+0.4 0	+0.65 +0.4	0 1.0	+0.6 0	+1.0 +0.6	0.2 1.2	+0.6 0	+1.2 +0.8
0.40–0.71	0 0.8	+0.4 0	+0.8 +0.4	0 1.1	+0.7 0	+1.1 +0.7	0.3 1.4	+0.7 0	+1.4 +1.0
0.71–1.19	0 1.0	+0.5 0	+1.0 +0.5	0 1.3	+0.8 0	+1.3 +0.8	0.4 1.7	+0.8 0	+1.7 +1.2
1.19–1.97	0 1.1	+0.6 0	+1.1 +0.6	0 1.6	+1.0 0	+1.6 +1.0	0.4 2.0	+1.0 0	+2.0 +1.4
1.97–3.15	0.1 1.3	+0.7 0	+1.3 +0.8	0.2 2.1	+1.2 0	+2.1 +1.4	0.4 2.3	+1.2 0	+2.3 +1.6
3.15–4.73	0.1 1.6	+0.9 0	+1.6 +1.0	0.2 2.5	+1.4 0	+2.5 +1.6	0.6 2.9	+1.4 0	+2.9 +2.0
4.73–7.09	0.2 1.9	+1.0 0	+1.9 +1.2	0.2 2.8	+1.6 0	+2.8 +1.8	0.9 3.5	+1.6 0	+3.5 +2.5
7.09–9.85	0.2 2.2	+1.2 0	+2.2 +1.4	0.2 3.2	+1.8 0	+3.2 +2.0	1.2 4.2	+1.8 0	+4.2 +3.0
9.85–12.41	0.2 2.3	+1.2 0	+2.3 +1.4	0.2 3.4	+2.0 0	+3.4 +2.2	1.5 4.7	+2.0 0	+4.7 +3.5
12.41–15.75	0.2 2.6	+1.4 0	+2.6 +1.6	0.3 3.9	+2.2 0	+3.9 +2.5	2.3 5.9	+2.2 0	+5.9 +4.5
15.75–19.69	0.2 2.8	+1.6 0	+2.8 +1.8	0.3 4.4	+2.5 0	+4.4 +2.8	2.5 6.6	+2.5 0	+6.6 +5.0

All data in this table are in accordance with American-British-Canadian (ABC) agreements. Limits for sizes above 19.69 inches are not covered by ABC agreements but are given in the ANSI Standard.
Symbols H7, p6, etc. are hole and shaft designations in ABC system.
* Pairs of values shown represent minimum and maximum amounts of interference resulting from application of standard tolerance limits.

FIGURE 6-119   ANSI Standard Interference Locational Fits (Courtesy ASME).

EXAMPLE:

Using fit tables:
Values are limits in thousandths of an inch.
Basic size: .0156
Nominal size range: .0–0.12
Limits of clearance: 0.1 = 000.1 = .0001

0.45 = 000.45 = .00045

Standard limits: (Move decimal three places to the left to obtain values in inches.)

Hole: +0.2 = 000.2   = +.0002
         −0   = −0
Shaft: −0.1 = −000.1   = −.0001

       −0.25 = −000.25   = −.00035

Values shown below are in thousandths of an inch

Nominal Size Range, Inches Over To	Class FN 1 Inter-ference*	Hole H6	Shaft	Class FN 2 Inter-ference*	Hole H7	Shaft s6	Class FN 3 Inter-ference*	Hole H7	Shaft t6	Class FN 4 Inter-ference*	Hole H7	Shaft u6	Class FN 5 Inter-ference*	Hole H8	Shaft x7
0–0.12	0.05	+0.25	+0.5	0.2	+0.4	+0.85				0.3	+0.4	+0.95	0.3	+0.6	+1.3
	0.5	0	+0.3	0.85	0	+0.6				0.95	0	+0.7	1.3	0	+0.9
0.12–0.24	0.1	+0.3	+0.6	0.2	+0.5	+1.0				0.4	+0.5	+1.2	0.5	+0.7	+1.7
	0.6	0	+0.4	1.0	0	+0.7				1.2	0	+0.9	1.7	0	+1.2
0.24–0.40	0.1	+0.4	+0.75	0.4	+0.6	+1.4				0.6	+0.6	+1.6	0.5	+0.9	+2.0
	0.75	0	+0.5	1.4	0	+1.0				1.6	0	+1.2	2.0	0	+1.4
0.40–0.56	0.1	+0.4	+0.8	0.5	+0.7	+1.6				0.7	+0.7	+1.8	0.6	+1.0	+2.3
	0.8	0	+0.5	1.6	0	+1.2				1.8	0	+1.4	2.3	0	+1.6
0.56–0.71	0.2	+0.4	+0.9	0.5	+0.7	+1.6				0.7	+0.7	+1.8	0.8	+1.0	+2.5
	0.9	0	+0.6	1.6	0	+1.2				1.8	0	+1.4	2.5	0	+1.8
0.71–0.95	0.2	+0.5	+1.1	0.6	+0.8	+1.9				0.8	+0.8	+2.1	1.0	+1.2	+3.0
	1.1	0	+0.7	1.9	0	+1.4				2.1	0	+1.6	3.0	0	+2.2
0.95–1.19	0.3	+0.5	+1.2	0.6	+0.8	+1.9	0.8	+0.8	+2.1	1.0	+0.8	+2.3	1.3	+1.2	+3.3
	1.2	0	+0.8	1.9	0	+1.4	2.1	0	+1.6	2.3	0	+1.8	3.3	0	+2.5
1.19–1.58	0.3	+0.6	+1.3	0.8	+1.0	+2.4	1.0	+1.0	+2.6	1.5	+1.0	+3.1	1.4	+1.6	+4.0
	1.3	0	+0.9	2.4	0	+1.8	2.6	0	+2.0	3.1	0	+2.5	4.0	0	+3.0
1.58–1.97	0.4	+0.6	+1.4	0.8	+1.0	+2.4	1.2	+1.0	+2.8	1.8	+1.0	+3.4	2.4	+1.6	+5.0
	1.4	0	+1.0	2.4	0	+1.8	2.8	0	+2.2	3.4	0	+2.8	5.0	0	+4.0
1.97–2.56	0.6	+0.7	+1.8	0.8	+1.2	+2.7	1.3	+1.2	+3.2	2.3	+1.2	+4.2	3.2	+1.8	+6.2
	1.8	0	+1.3	2.7	0	+2.0	3.2	0	+2.5	4.2	0	+3.5	6.2	0	+5.0
2.56–3.15	0.7	+0.7	+1.9	1.0	+1.2	+2.9	1.8	+1.2	+3.7	2.8	+1.2	+4.7	4.2	+1.8	+7.2
	1.9	0	+1.4	2.9	0	+2.2	3.7	0	+3.0	4.7	0	+4.0	7.2	0	+6.0
3.15–3.94	0.9	+0.9	+2.4	1.4	+1.4	+3.7	2.1	+1.4	+4.4	3.6	+1.4	+5.9	4.8	+2.2	+8.4
	2.4	0	+1.8	3.7	0	+2.8	4.4	0	+3.5	5.9	0	+5.0	8.4	0	+7.0
3.94–4.73	1.1	+0.9	+2.6	1.6	+1.4	+3.9	2.6	+1.4	+4.9	4.6	+1.4	+6.9	5.8	+2.2	+9.4
	2.6	0	+2.0	3.9	0	+3.0	4.9	0	+4.0	6.9	0	+6.0	9.4	0	+8.0

See footnotes at end of table.

Values shown below are in thousandths of an inch

Nominal Size Range, Inches Over To	Class FN 1 Inter-ference*	Hole H6	Shaft	Class FN 2 Inter-ference*	Hole H7	Shaft s6	Class FN 3 Inter-ference*	Hole H7	Shaft t6	Class FN 4 Inter-ference*	Hole H7	Shaft u6	Class FN 5 Inter-ference*	Hole H8	Shaft x7
4.73–5.52	1.2	+1.0	+2.9	1.9	+1.6	+4.5	3.4	+1.6	+6.0	5.4	+1.6	+8.0	7.5	+2.5	+11.6
	2.9	0	+2.2	4.5	0	+3.5	6.0	0	+5.0	8.0	0	+7.0	11.6	0	+10.0
5.52–6.30	1.5	+1.0	+3.2	2.4	+1.6	+5.0	3.4	+1.6	+6.0	5.4	+1.6	+8.0	9.5	+2.5	+13.6
	3.2	0	+2.5	5.0	0	+4.0	6.0	0	+5.0	8.0	0	+7.0	13.6	0	+12.0
6.30–7.09	1.8	+1.0	+3.5	2.9	+1.6	+5.5	4.4	+1.6	+7.0	6.4	+1.6	+9.0	9.5	+2.5	+13.6
	3.5	0	+2.8	5.5	0	+4.5	7.0	0	+6.0	9.0	0	+8.0	13.6	0	+12.0
7.09–7.88	1.8	+1.2	+3.8	3.2	+1.8	+6.2	5.2	+1.8	+8.2	7.2	+1.8	+10.2	11.2	+2.8	+15.8
	3.8	0	+3.0	6.2	0	+5.0	8.2	0	+7.0	10.2	0	+9.0	15.8	0	+14.0
7.88–8.86	2.3	+1.2	+4.3	3.2	+1.8	+6.2	5.2	+1.8	+8.2	8.2	+1.8	+11.2	13.2	+2.8	+17.8
	4.3	0	+3.5	6.2	0	+5.0	8.2	0	+7.0	11.2	0	+10.0	17.8	0	+16.0
8.86–9.85	2.3	+1.2	+4.3	4.2	+1.8	+7.2	6.2	+1.8	+9.2	10.2	+1.8	+13.2	13.2	+2.8	+17.8
	4.3	0	+3.5	7.2	0	+6.0	9.2	0	+8.0	13.2	0	+12.0	17.8	0	+16.0
9.85–11.03	2.8	+1.2	+4.9	4.0	+2.0	+7.2	7.0	+2.0	+10.2	10.0	+2.0	+13.2	15.0	+3.0	+20.0
	4.9	0	+4.0	7.2	0	+6.0	10.2	0	+9.0	13.2	0	+12.0	20.0	0	+18.0
11.03–12.41	2.8	+1.2	+4.9	5.0	+2.0	+8.2	7.0	+2.0	+10.2	12.0	+2.0	+15.2	17.0	+3.0	+22.0
	4.9	0	+4.0	8.2	0	+7.0	10.2	0	+9.0	15.2	0	+14.0	22.0	0	+20.0
12.41–13.98	3.1	+1.4	+5.5	5.8	+2.2	+9.4	7.8	+2.2	+11.4	13.8	+2.2	+17.4	18.5	+3.5	+24.2
	5.5	0	+4.5	9.4	0	+8.0	11.4	0	+10.0	17.4	0	+16.0	24.2	0	+22.0
13.98–15.75	3.6	+1.4	+6.1	5.8	+2.2	+9.4	9.8	+2.2	+13.4	15.8	+2.2	+19.4	21.5	+3.5	+27.2
	6.1	0	+5.0	9.4	0	+8.0	13.4	0	+12.0	19.4	0	+18.0	27.2	0	+25.0
15.75–17.72	4.4	+1.6	+7.0	6.5	+2.5	+10.6	9.5	+2.5	+13.6	17.5	+2.5	+21.6	24.0	+4.0	+30.5
	7.0	0	+6.0	10.6	0	+9.0	13.6	0	+12.0	21.6	0	+20.0	30.5	0	+28.0
17.72–19.69	4.4	+1.6	+7.0	7.5	+2.5	+11.6	11.5	+2.5	+15.6	19.5	+2.5	+23.6	26.0	+4.0	+32.5
	7.0	0	+6.0	11.6	0	+10.0	15.6	0	+14.0	23.6	0	+22.0	32.5	0	+30.0

All data above heavy lines are in accordance with American-British-Canadian (ABC) agreements. Symbols H6, H7, s6, etc. are hole and shaft designations in ABC system. Limits for sizes above 19.69 inches are not covered by ABC agreements but are given in the ANSI standard.

* Pairs of values shown represent minimum and maximum amounts of interference resulting from application of standard tolerance limits.

FIGURE 6-120 ANSI Standard Force and Shrink Fits (Courtesy ASME).

Calculating for tolerance dimensions:

Hole: .0156   basic hole size
         +.0002   standard limits in thousands
          .0158   upper limit for hole

          .0156   basic hole size
           −0   standard limits in thousands
          .0156   lower limit for hole

Shaft: .0156   basic hole size
         −.0001   standard limits for shaft
          .0155   upper limit for shaft

          .0156   basic hole size
         −.00025   standard limit for shaft
          .01535   lower limit for shaft

Tolerance dimensions for basic size of .0156:

Hole: .01580   Shaft: .01550
          .01560   .01535

To check calculations:

Subtract: .01580   upper limit of hole
           −.01535   lower limit of shaft
            .00045   limit of clearance
            .01560   lower limit of hole
           −.01550   upper limit of shaft
            .0001   limit of clearance

Limit of Clearance:

.0001   = .1 from chart

.00045 = .45 from chart

Manufacturing Precision

Figure 6-121 is a table showing the levels of tolerances engineers and drafters can expect from various common manufacturing processes. It is important for engineers and drafters to know what to expect from the manufacturing processes that will be used in converting their design into an actual working part. It will serve no purpose to specify a tolerance greater than that which can be expected of a given manufacturing process.

Summary of Dimensioning Rules

This section summarizes most of the dimensioning rules with which engineers and drafters should be familiar. This section can be used as a checklist for ensuring that drawings you do throughout this book are properly dimensioned. Use this summary as a checklist until you are so familiar with these rules that the checklist is no longer necessary.

1. Give all dimensions necessary to accurately communicate the design to manufacturing personnel, but only those dimensions necessary.

2. Apply dimensions in such a way that they cannot be misinterpreted.

3. Show dimensions between surfaces or points that have a functional relationship.

Standard Tolerance Grades								
5	6	7	8	9	10	11	12	13
	Cylindrical Grinding							
	Surface Grinding							
	Diamond Turning							
	Diamond Boring							
	Broaching							
		Reaming						
			Turning					
			Boring					
					Milling			
					Planning and Shaping			
					Drilling			

FIGURE 6-121   Levels of tolerances for common manufacturing processes.

4. Do not dimension to or from rough surfaces. Always attempt to dimension from a finished surface.

5. Give complete dimensions so that manufacturing personnel do not have to interrupt their work to calculate or otherwise determine a dimension they need.

6. Apply dimensions in the views where the shape of the part is shown most accurately.

7. Apply dimensions in those views where the part being dimensioned appears true size and true shape.

8. Do not dimension to hidden lines. If it becomes necessary to dimension to hidden lines, change the view to a sectional view.

9. Attempt to place dimensions that apply to two adjacent views between the views.

10. Avoid crossing extension lines with dimension lines.

11. Clearly dimension and annotate all features. Never expect manufacturing personnel to assume anything.

12. Do not apply a complete chain of detail dimensions. Omit the last one or indicate that the last dimension in the chain is a reference dimension.

13. Avoid placing dimensions on the part itself.

14. Always place dimension lines uniformly throughout the drawing.

15. Never use a line that constitutes a portion of the drawing of a part as a dimension line.

16. Avoid crossing dimension lines wherever possible.

17. Extension lines that cross other extension lines or object lines should not be broken.

18. A center line may be used as an extension line.

19. Center lines should not extend from view to view.

20. Leader lines should point to the center of a hole or circular feature.

21. Leader lines should be straight and broken at precise angles. They should never curve.

22. Leader lines should begin approximately at the center of a note height and extend either from the beginning or end of the note.

23. If it becomes necessary to letter within a section-lined area, erase enough of the section lines to leave a clear space so that the dimension can be easily read.

24. Fraction bars should be horizontal rather than angled.

25. The numerator and denominator of a fraction should not touch the horizontal fraction bar.

26. Finish marks should be applied on views in which the surface appears as an edge.

27. Finish marks are not necessary on parts made from rolled stock.

28. If a part is finished all over, finish marks are not used. Substitute the letters FAO or apply a note that says "Finish All Over."

29. Dimension a cylinder by showing the diameter and length on the rectangular view or by showing the diameter as a diagonal dimension in the circular view.

30. Always specify drill sizes in decimals.

31. A metric or decimal-inch diameter dimension should be preceded by the diameter symbol.

32. A metric or decimal-inch radius dimension should be preceded by a capital R.

33. Dimensions that are not to scale should be underlined with a straight line.

34. Decimal dimensioning is preferred for mechanical and manufacturing-related drawings.

Notation

A good drawing may be defined as one that contains all of the information required by the various design and manufacturing people who will use it in producing the subject part. Most of this information can be conveyed graphically using standard dimensioning practices. However, it is not uncommon to encounter a situation in which all of the needed information cannot be communicated graphically. In these cases, notes are used to communicate or clarify the designer's intent.

Notes are brief, carefully worded statements placed on drawings to convey information not covered or not adequately explained using graphics (**Figure 6-122**). Notes should be clearly worded so as to allow only one correct interpretation.

There are no ANSI standards specifically governing the use of notes on technical drawings. However, several rules of a general nature should be observed. These rules apply to both general notes and the more specific detail notes.

NOTE: FAO TO 63 MICROINCHES

NOTE: ALL UNMARKED RADII = R.25

FIGURE 6-122   Notes on drawings improve communication.

Rules for Applying Notes on Drawings

Notes may be lettered freehand, entered using a keyboard in a computer-aided drafting (CAD) system, or added through any one of several mechanical lettering processes. Sample notes are included on the drawings in **Figure 6-123**. In any case, regardless of how they are put on the drawing, notes should be oriented horizontally on the drafting sheet (**Figure 6-124**). General notes should be located directly above the title block. When using manual processes to apply general notes, the first note is placed directly above the title block, the second is placed on top of it, and so on up the line. This allows notes to be added as needed without renumbering. However, when using a CAD system, this is not necessary since one of the advantages of CAD is that notes can be renumbered and

rearranged automatically. Detail notes should be located as close as possible to the detail they are describing and connected to it by a leader line.

Notes should be applied to drawings after all graphics have been completed. This prevents notes and graphics from overlapping and minimizes other technique problems such as smudging the worksheet and frequent erasing of notes as changes are required on a drawing.

Two basic types of notes are used on technical drawings: general notes and detail notes. They serve different purposes and, therefore, must be examined separately.

General Notes

General notes are broad items of information that have a job- or project-wide application rather than relating to just one single element of a project or a part. They

FIGURE 6-123   Sample note lettered mechanically.

FIGURE 6-124   Notes are placed horizontally.

are usually placed immediately above the title block on the drawing sheet and numbered sequentially.

Information placed in general notes includes such characteristics of a product as finish specifications; standard sizes of fillets, rounds, and radii; heat treatment specifications; cleaning instructions; general tolerancing data; hardness testing instructions; and stamping specifications.

Detail Notes

Detail notes are specific notes that pertain to one particular element or characteristic of a part. They are placed as close as possible to the characteristic to which they apply and are connected using a leader line (**Figure 6-125**).

Detail notes should not be placed on views (**Figure 6-126**). The only exception to this rule is in cases where a great deal of open space exists on a view, but very little around it (**Figure 6-127**). Notes should never be superimposed over other data such as dimensions, lines, or symbols (**Figure 6-128**).

Common sense is the best rule to follow in applying detail notes. Since detail notes are used to more completely communicate or to clarify intent, they should be placed as close as possible to the element to which they pertain and in such a way as to be easily read.

FIGURE 6-127   Acceptable placement of a detail note when there is space.

NOTE: *DO NOT DIMENSION TO HIDDEN LINES OR CENTER LINES OF HIDDEN LINES*

FIGURE 6-128   Superimposing a note is poor practice.

FIGURE 6-125   Detail note connected with a leader line.

FIGURE 6-126   Poor placement of a detail note.

Writing Notes

The written word lends itself to interpretations that may vary. Thus, *written notes* used on drawings must convey the exact intent of the designer. Consequently, drafters must be especially careful in the preparation of notes.

The first step is to place in a legend all abbreviations used in the notes (**Figure 6-129**). This ensures that all readers of the notes interpret the abbreviations consistently and correctly.

Another technique that will limit misinterpretation of notes is punctuation or indentation. Notes containing more than one sentence should be properly punctuated so that readers know where one sentence leaves off and the next one begins (**Figure 6-130**). Indentation is used when the length of a note requires more than one line. The second line is indented so that readers know it is part of the first line (**Figure 6-131**).

Another technique is particularly useful for beginning drafters. It is called *verification*. Notes to be placed on drawings are first jotted down, legibly,

LEGEND

1. ASSY = ASSEMBLY
2. FAO = FINISH ALL OVER
3. MS = MACHINE STEEL
4. FAB = FABRICATE
5. CARB = CARBURIZE
6. DVTL = DOVETAIL
7. HT TR = HEAT TREAT
8. KST = KEYSEAT
9. TPR = TAPER
10. SF = SPOTFACED

FIGURE 6-129   Abbreviations are explained in a legend.

NOTE

POWER BRUSH ALL SURFACES. APPLY THREE COATS PAINT. STAMP PART WITH NUMBER 01929612.

FIGURE 6-130   Proper punctuation of notes is important.

NOTE

ALL DIMENSIONS AND CONDITIONS SHALL BE FIELD CHECKED AND VERIFIED BY PROPER TRADES

FIGURE 6-131   Indenting for clarity.

GENERAL NOTES ← LETTERS .25 HIGH

ALL FILLETS AND ROUNDS .125 DIA. ← LETTERS .125–
ALL SURFACES MUST BE FREE OF BURRS. ← .1875 HIGH

FIGURE 6-132   Proper lettering height is important.

NOTES

1. SPACING BETWEEN LINES OF THE SAME NOTE SHOULD BE .0625.

2. SPACING BETWEEN SEPARATE NOTES SHOULD BE .1250.

FIGURE 6-133   Spacing of notes.

NOTES

NOTES SHOULD BE LIMITED TO 4 TO 6 WORDS PER LINE

FOUR TO SIX WORDS

FIGURE 6-134   Number of words per line limitations are important.

NOTES

AVOID LONG LINE IN A NOTE FOLLOWED BY ONE SHORT LINE

FIGURE 6-135   Lines in notes should be approximately equal.

on a print of the drawing. Another drafter, preferably one with experience, is then asked to read the notes to ensure that they are not open to multiple interpretations.

Note Specifications

A number of specifications should be observed when writing notes. The most widely accepted lettering style for notes is uppercase (all capital) block Gothic letters in a vertical format. Some companies accept uppercase block Gothic lettering in an inclined or slanted format. The proper lettering height is one-quarter inch for titles such as GENERAL NOTES, and one-eighth inch to three-sixteenths inch for actual notes (**Figure 6-132**).

Spacing between successive lines of the same note should be approximately one-sixteenth inch. Spacing between separate notes should be approximately one-eighth inch (**Figure 6-133**).

There are no hard and fast rules governing the length of a line of notation, but from four to six words per line is widely accepted (**Figure 6-134**).

When a note string will contain more than six words, it should be divided into more than one line. The number of words in the multiple-line note should be divided so as to balance the finished note. One long line followed by a drastically shorter line is bad form (**Figure 6-135**).

On occasion, the last word in a note string will have to be hyphenated. There are rules governing the dividing of words according to syllables. If the proper point of division is not obvious from the makeup of the word, consult a dictionary. Do not hyphenate abbreviated words.

Sample Notes

The following is a list of sample notes of the types frequently used on technical drawings:

- All fillets and rounds R1/8
- 45° chamfer all edges

- Surfaces A & B parallel
- Stock .125 thick
- Heat treat
- All internal radii R.0625
- FAO
- Ream for 1/4 dowels
- All bends R5/32
- CBORE from bottom
- Identical bolts—both sides
- 3/32 × 1/16 oil groove
- 87 CSK 3/4 diameter—4 holes
- Rounds R.125 unless otherwise specified
- 4 holes equally spaced
- 1/32 × 45° chamfer
- #7 Drill .75 deep
- Machine steel—4 required
- Power brush all exterior surfaces
- Sandblast before painting

Notice that these sample notes are very brief, concise, and to the point. Words such as "a," "an," "the," and "are" are used only sparingly. The notes are not always complete sentences in terms of proper grammar, but they are complete thoughts in terms of communication. Notice also the use of abbreviations such as FAO (finish all over), CBORE (counterbore), and CSK (countersink). Abbreviations can be used frequently to cut down on the length of notes and the time required to letter them. However, when used, abbreviations should always be placed in a legend to ensure consistency of interpretation.

Group Technology

Group technology is a manufacturing concept in which similar parts are grouped together in part groups or families. Parts may be alike in two ways:

1. In their design characteristics
2. In the manufacturing processes required to produce them

By grouping similar parts into families, manufacturing personnel can improve efficiency. Such improvements are the result of advantages gained in such areas as setup time, standardization of processes, and scheduling.

Group technology can also improve the productivity of design personnel by decreasing the amount of work and time involved in designing a new part. Chances are a new part will be similar to an existing part in a given family. When this is the case, the new part can be developed by simply modifying the design of the existing part. Design modifications tend to require less time and work than new design. This is especially true in the age of CAD. The advantages of group technology are dealt with at greater length later in this chapter.

Historical Development

Ever since the Industrial Revolution, manufacturing and engineering personnel have been searching for ways to optimize manufacturing processes. There have been numerous developments over the years since the Industrial Revolution mechanized production. Mass production and interchangeability of parts in the 1800s were major steps forward in optimizing manufacturing.

However, even with mass production and assembly lines, most manufacturing is done in small batches ranging from one workpiece to 2000 or 3000. In fact, even today, over 70% of manufacturing involves batches of less than 3000 workpieces. Historically, less has been done to optimize small-batch production than has been done for assembly line work.

There have been attempts to standardize a variety of design tasks and some working queuing and sequencing in manufacturing, but until recent years, design and manufacturing in small-batch settings has been somewhat random.

The underlying problem that has historically prevented significant improvements to small-batch manufacturing is that any solution must apply broadly to general production processes and principles rather than to a specific product. This is a difficult problem because the various workpieces in a small batch can be so random and different.

When manufacturing entered the age of automation and computerization, such developments as scheduling software, sequencing software, and materials requirements planning (MRP) systems became available to improve the production of both small and large batches. Even these developments have not optimized the production of small-batch manufacturing lots. The problem has become even more critical because, since the end of World War II, the trend has been toward smaller batches.

In recent years, the problems of small-batch production have finally begun to receive the attention necessary to bring about improvements. A major step is the ongoing development of group technology.

Part Families

It has already been stated that parts may be similar in design (size and shape) and/or in the manufacturing processes used to produce them. A group of such parts is called a *part family*. It is possible for parts in the same family to be very similar in design yet radically different in the area of production requirements. The opposite may also be true.

Figure 6-136 contains examples of two parts from the same family. These parts were placed in the same family based on design characteristics. They have exactly the same shape and size. However, you will notice that they differ in the area of production processes.

Part 1, after it is drilled, will go to a painting station for two coats of primer. Its dimensions must be held to a tolerance of plus or minus .125″. Part 2, after it is drilled, will go to a finishing station for sanding and buffing. Its dimensions require more restrictive tolerances of plus or minus .003″. The parts differ in material. The material for Part 1 is cold-rolled steel; Part 2 is aluminum.

Figure 6-137 contains examples of two parts from the same family. Although the design characteristics of these two parts are drastically different (i.e., different sizes and shapes), a close examination will reveal that they are similar in the area of production processes. Part 1 is made of stainless steel. Its dimensions must be held to a tolerance of plus or minus .002″, and three holes must be drilled through it. Part 2, in spite of the differences in its size and shape, has exactly the same manufacturing characteristics.

PART I

MATERIAL: STAINLESS
TOLERANCES: ±0.002

PART 2

MATERIAL: STAINLESS
TOLERANCES: ±0.002

FIGURE 6-137   Two parts from the same family.

The parts shown in Figure 6-136, because of their similar design characteristics, were grouped in the same family. Such a family is referred to as a design part family. Those in Figure 6-137 were grouped together because of similar manufacturing characteristics. Such a family is referred to as a *manufacturing part family*. The characteristics used in classifying parts are referred to as attributes.

By grouping parts into families, manufacturing personnel can cut down significantly on the amount of materials handled and movement wasted in producing them. This is because manufacturing machines can be correspondingly grouped into specialized work cells instead of the traditional arrangement of machines according to function (i.e., mills together, lathes together, drills together, etc.).

Each work can be specially configured to produce a given family of parts. When this is done, the number of setups, the amount of materials handling, the length of lead time, and the amount of in-process inventory are all reduced.

Grouping Parts into Families

Part grouping is not as simple a process as one might think. You already know the criteria used: design similarities and manufacturing similarities. But how does the actual grouping take place?

PART I   0.75 DIA. HOLE

MATERIAL: COLD ROLLED STEEL
TOLERANCES: ±0.0125″
FINISH: TWO COATS PRIMER

PART 2   0.75 DIA. HOLE

MATERIAL: ALUMINUM
TOLERANCES: ±0.003″
FINISH: SAND AND BUFF

FIGURE 6-136   Two parts from the same family.

There are three methods that can be used for grouping parts into families (**Figure 6-138**).

1. Sight inspection
2. Route sheet inspection
3. Parts classification and coding

All three methods require the expertise of experienced manufacturing personnel.

Sight inspection is the simplest, least sophisticated method. It involves looking at parts, photos of parts, or drawings of parts. Through such an examination, experienced personnel can identify similar characteristics and group the parts accordingly. This is the easiest approach, especially for grouping parts by design attributes, but it is also the least accurate of the three methods.

The second method involves inspecting the routing sheets used to route the parts through the various operations to be performed. This can be an effective way to group parts into manufacturing part families, provided the routing sheets are correct. If they are, this method is more accurate than the sight inspection approach. This method is sometimes referred to as the PFA or production flow analysis method.

The most widely used method for grouping parts is the third method: parts classification and coding. This is also the most sophisticated, most difficult, and most time-consuming method. Parts classification and coding is complex enough to require a more in-depth treatment than the other two methods.

Parts Classification and Coding

Parts classification and coding is a method in which the various design and/or manufacturing characteristics of a part are identified, listed, and assigned a code number. Recall that these characteristics are referred to as *attributes*. This is a general approach used in classifying and coding parts. There are many different systems that have been developed for actually carrying out the process, none of which has emerged as the standard.

The many different classification and coding systems that have been developed all fall into one of three groups (**Figure 6-139**).

1. Design attribute group
2. Manufacturing attribute group
3. Combined attribute group

Students of CAD/CAM should be familiar with how systems are grouped.

DESIGN ATTRIBUTE GROUP

Classification and coding systems that use design attributes as the qualifying criteria fall into this group. Commonly used design attributes include dimensions, tolerances, shape, finish, and material.

MANUFACTURING ATTRIBUTE GROUP

Classification and coding systems that use manufacturing attributes as the qualifying criteria fall

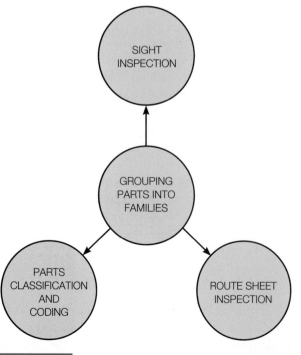

FIGURE 6-138 Methods of grouping parts into families.

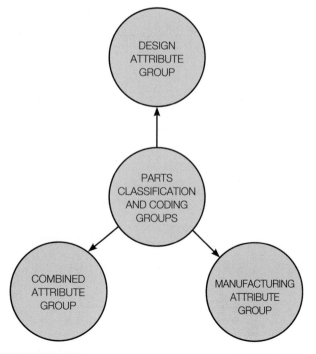

FIGURE 6-139 Parts classification and coding groups.

into this group. Commonly used manufacturing attributes include production processes, operational sequence, production time, tools required, fixtures required, and batch size.

COMBINED ATTRIBUTE GROUP

There are advantages in using design attributes and advantages in using manufacturing attributes. Systems that fall into the design attribute group are particularly advantageous if the goal is design retrieval. Those in the manufacturing group are better if the goal is any of a number of production-related functions. However, there is a need for systems that combine the best characteristics of both. Such systems are both design and manufacturing attributes.

SAMPLE PARTS CLASSIFICATION AND CODING SYSTEM

Some companies develop their own parts classification and coding systems. But this can be an expensive and time-consuming approach. The more widely used approach is to purchase a commercially prepared system. There are several such systems available. However, the most widely used of these is the Opitz system. It is a good example of a parts classification and coding system and how one works.

Opitz System

Classification and coding systems use alphanumeric symbols to represent the various attributes of a part. One of the many classification and coding systems is the Opitz system. This system uses characters in 13 places to code the attributes of parts and, hence, to classify them. These digit places are represented as follows:

12345 6789 ABCD

The first five digits (12345) code the major design attributes of a part. The next four digits (6789) are for coding manufacturing-related attributes and are called the supplementary code. The letters (ABCD) code the production operation and sequence.

The alphanumeric characters shown represent places. For example, the actual numeral used in each place can be 0–9. The numeral used in the 1 place indicates the length-to-diameter ratio of the part. The numeral used in the 2 place indicates the external shape of the part. The numeral used in the 3 place indicates the internal shape of the part. The numeral used in the 4 place indicates the type of surface machining. The numeral used in the 5 place indicates gear teeth and auxiliary holes. With such a system, a part might be coded as follows:

20801

The 2 means the part has a certain length-to-diameter ratio. The first 0 means the part has no external shape elements. The 8 means the part has an internal thread. The second 0 means no surface machining is required. The 1 means the part is axial, not on pitch.

Figure 6-140 is a chart that shows the basic overall structure of the Opitz system for parts classification and coding. **Figure 6-141** is a chart used

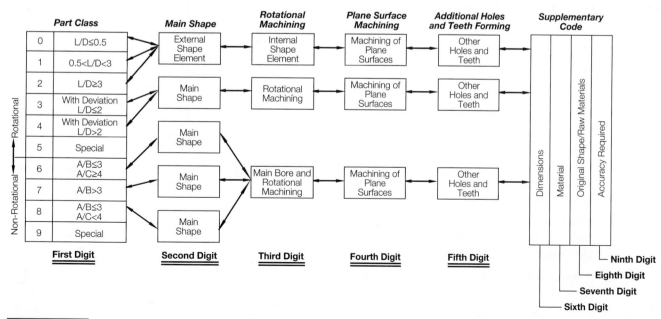

FIGURE 6-140   Basic structure of the Opitz system.

	PART CLASS	MAIN SHAPE EXTERNAL		ROTATIONAL MACHINING INTERNAL		PLANE SURFACE MACHINING	ADDITIONAL HOLES AND TEETH
	L/D<0.5		Smooth, no shape elements		No hole, no breakthrough	No surface machining	No auxiliary hole
1	0.5<L/D<3	Stepped to one end or smooth	No shape elements	Smooth or stepped to one end	No shape elements	Surface plane and/or curved in one direction, external	Axial, not on pitch circle diameter
2	L/D≤3		Thread		Thread	External plane surface related by graduation around a circle	Axial on pitch circle diameter
3	—		Functional groove		Functional groove	External groove and/or slot	Radial, not on pitch circle diameter
4	—	Stepped to both ends	No shape elements	Stepped to both ends	No shape elements	External spline (polygon)	Axial and/or radial and/or other direction
5	—		Thread		Thread	External plane surface and/or slot, external spline	Axial and/or radial on PCD and/or other directions
6	—		Functional groove		Functional groove	Internal plane surface and/or slot	Spur gear teeth
7	—		Functional cone		Functional cone	Internal spline (polygon)	Bevel gear teeth
8	—		Operating thread		Operating thread	Internal and external polygon, groove and/or slot	Other gear teeth
	First Digit	**Second Digit**		**Third Digit**		**Fourth Digit**	**Fifth Digit**

(Left margin: Rotational ↕ Non-Rotational)
(Fourth Digit column side label: No gear teeth / With gear teeth)

FIGURE 6-141   Assigning a code using the Opitz system.

for assigning an Opitz code to rotational parts. **Figure 6-142** is a drawing of an actual part that has been assigned an Opitz code of 15400. This code was arrived at as follows:

STEP 1   The total length of the part is divided by the overall diameter:

$$\frac{1.75(L)}{1.25(D)} = 1.40$$

Since 1.40 is greater than .5 but less than 3, the first digit in the code is 1.

STEP 2   An overall description of the external shape of the part would read ". . . a rotational part that is stepped on both ends with one stepped end threaded." Consequently, the most appropriate second digit in the code is 5.

STEP 3   A description of the internal shape of the part would read ". . . a through hole." Consequently, the third digit in the code is 4.

FIGURE 6-142   Sample part that can be coded using the Opitz system.

STEP 4   Upon examination, it can be seen that no surface machining is required. Therefore, the fourth digit in the code is 0.

STEP 5   It can also be seen that no auxiliary holes or gear teeth are required. Therefore, the fifth digit in the code is also 0.

GROUP TECHNOLOGY AND GLOBAL COMPETITION

"We need to adopt the group technology concept," said Juan Martinez. Martinez is the chief designer for Magna-Tech Manufacturing, a leading producer of metal components used in the manufacture of automobiles.

In recent years, Magna-Tech's competition has become both global and fierce. In order to survive and prosper, the company needs to optimize the performance of every employee and every process from design and drafting to manufacturing.

As the design and drafting department's representative on the company's management team, Martinez is expected to propose strategies for cutting costs and improving productivity, and he is expected to propose strategies to help Magna-Tech become more competitive on a continual basis. The company's CEO has made it clear to all of his managers that he is looking for strategies that will give Magna-Tech a sustainable competitive advantage in the global marketplace. This is why Juan Martinez scheduled the current departmental meeting. He needs input from his design and drafting personnel.

"I want us to brainstorm this morning," said Martinez. "Before the next manager's meeting, I need a list of benefits our company can gain by adopting group technology."

"The first benefit is that group technology will cut down on the amount of time needed to design a part," said Margaret Flagstone. "Right now we practically start over every time we have a new part we have already designed. With group technology, we could place the new part in a part family. Once we know what family it fits in, much of the design is already done."

"Excellent point," said Martinez. "I'll tell you what we might do. What if I write four or five benefits on the flip chart and see if you agree with them? Then when we have our top five benefits we can divide ourselves into five small groups and explore each benefit in depth. I'd like to have some concrete examples to offer at the manager's meeting." With that, Martinez began writing benefits on a flip chart.

Advantages and Disadvantages of Group Technology

There are several advantages that can be realized through the application of group technology. These advantages include:

1. Improved design
2. Enhanced standardization
3. Reduced materials handling
4. Simplified production scheduling
5. Improved quality control

Improved Design

Group technology allows designers to use their time more efficiently and productively by decreasing the amount of new design work required each time a part is to be designed. When a new part is needed, its various attributes can be listed. Then, an existing part with as many of these attributes as possible can be identified and retrieved. The only new design required is that which relates to attributes of the new part not contained in the existing part. Because this characteristic of group technology tends to promote design standardization, additional design benefits accrue.

Enhanced Standardization

Parts are classified into groups according to their similarities. The more similarities, the better. Consequently, enhanced standardization is promoted by group technology. As you have already seen, design factors account for part of the enhanced standardization. Setups and tooling also become more standardized because similar parts require similar setups and similar tooling.

Reduced Materials Handling

One of the major elements of group technology is the arrangement of machines into specialized work cells, each cell producing a given family of parts. Traditionally, machines are arranged by function (i.e., lathes together in one group, mills together in another, etc.). A functional arrangement causes excessive movement of parts from machine to machine and back. Specialized work cells reduce such movement to a minimum and, in turn, reduce materials handling.

Simplified Production Scheduling

With machines grouped into specialized work cells and arrangement of parts into families, there is less work center scheduling, requiring less scheduling overall. Group technology involves less sophisticated scheduling.

Improved Quality Control

Quality control is improved through group technology because each work cell is responsible for a specific family of parts. This does two things, both of which tend to improve the quality of parts: (1) it promotes pride in the work among employees assigned to each work cell, and (2) it makes production errors and problems easier to trace to a specific source. In traditional manufacturing shops, each work center performs certain tasks, but no one center is responsible for an entire part. Such an approach does not promote quality control.

In spite of these advantages, there are several disadvantages associated with group technology. Transitioning from machines in a traditional shop into specialized work cells can cause a major disruption. Such disruptions translate quickly into monetary losses. As a consequence, some managers are reluctant to pursue group technology.

Another problem with group technology is that parts classification coding is a time-consuming, difficult, and expensive process, especially in the initial stages. This causes some managers to balk at the expense.

SUMMARY

- The three dimensioning systems used on technical drawings in the United States are metric, decimal-inch, and fractional. In all of these systems, dimensions are oriented to be read from the bottom or right-hand side of the drawing sheet.
- The components that are common to all dimensioning systems are extension lines, dimension lines, arrowheads, leader lines, and dimensions.
- The four steps in laying out dimensions are as follows: (1) place eight layout lines on the drawing sheet, (2) lightly draw in guidelines for the dimensions, (3) darken the extension lines and dimension lines and add dimensions, and (4) add arrowheads.
- Tolerancing-related concepts with which students should be familiar are actual size allowance, bilateral tolerance, datum, least material condition, limits, maximum material condition, nominal dimension, tolerance, unilateral tolerance, fit, clearance fit, interference fit, and transition fit.
- Not all information needed to convey a design can be given graphically. In such cases, notes are used to provide the additional information. Notes are brief, carefully worded statements placed on drawings to convey information not adequately explained graphically.
- Group technology is a manufacturing concept in which similar parts are grouped together in part families. Such parts may be alike in two ways: (1) in their design characteristics and (2) in the manufacturing processes required to produce them. Advantages of group technology include improved design, enhanced standardization, reduced materials handling, simplified production scheduling, and improved quality control.

REVIEW QUESTIONS

Mark the following statements as either true or false.

1. The most frequently used dimensioning standard is fractional dimensioning.
2. The proper way to indicate on a drawing that a dimension is NTS is to put a wavy line under the dimension.
3. An extension line should extend beyond the dimension line approximately one-eighth inch.
4. The key word to remember when making arrowheads, regardless of the size, is "uniform."
5. Unidirectional dimensioning is aligning all dimensions with the dimension line.
6. Notes are brief, carefully worded statements placed on drawings to convey information not covered in the graphic.
7. Notes should be oriented on a drafting sheet horizontally.
8. The term *verification* as it relates to notes is printing a list of your notes and asking an experienced drafter to ensure that they are not open to multiple interpretations.
9. Group technology is a concept by which similar parts are grouped together in part groups or families.
10. Attributes are characteristics used to group parts.
11. General notes are specific and pertain to a particular element or characteristic of a part.
12. The proper height for lettering note titles is 1/4".

13. Parts are grouped into families by their design characteristics or the manufacturing process required to produce them.

Answer the following questions by selecting the best answer.

1. Of the dimensioning systems used, which one is most frequently used on mechanical technical drawings?
   a. Metric dimensioning
   b. Decimal-inch dimensioning
   c. Fractional dimensioning
   d. Proportional dimensioning

2. Which is the correct rule when labeling for a metric dimension that is less than one millimeter?
   a. A decimal is not required
   b. Neither a zero nor a decimal is required
   c. Place a zero to the right of the decimal
   d. None of the above

3. Which is the correct way to show the line separating the numerator and denominator of a fraction?
   a. /   c. \
   b. —   d. All of the above

4. Which is *not* a component of all dimensioning systems?
   a. Arrowheads   c. Leader lines
   b. Extension lines   d. Orientation lines

5. The recommended length/width ratio of arrowheads is that they are:
   a. Twice as long as they are wide
   b. Three times as long as they are wide
   c. Twice as wide as they are long
   d. Three times as wide as they are long

6. Which is the proper location for general notes?
   a. In the lower left-hand corner
   b. In the upper left-hand corner
   c. Above the title block
   d. Below the title block

7. Which of the following is not true regarding notation?
   a. The number of words in a multiple-line note should be divided to balance the finished note.
   b. Four to six words per line is widely accepted.
   c. There are hard and fast rules regarding notation.
   d. Hyphenation, by syllables, is accepted.

8. Which of the following is **not** a method for grouping parts into families?
   a. Size and shape
   b. Sight inspection
   c. Route sheet inspection
   d. Parts classification and coding

9. Which of the following is **not** a commonly applied design attribute?
   a. Dimensions   c. Tolerances
   b. Weight   d. Shape

10. Which of the following is **not** a commonly applied manufacturing attribute?
    a. Production processes
    b. Operation sequence
    c. Sight inspection
    d. Tools required

11. Which of the following is **not** an advantage of group technology?
    a. Improved design
    b. Reduced materials handling
    c. Easier parts classification coding
    d. Simplified production scheduling

CHAPTER 6 PROBLEMS

The following problems are intended to give beginning drafters and engineers practice in applying the various dimensioning techniques required on technical drawings in the industry. Students should follow the general instructions that apply to all problems and any additional instructions provided for special problems. All problems should be dimensioned in strict accordance with the rules set forth in this chapter.

Conventional Instructions

1. Study each problem carefully.
2. Make a complete sketch of the problem, applying all dimensions on the sketch. Note on the sketch the proper spacing between views, between the object and first dimension lines, and between successive dimension lines.
3. Compare your fully dimensioned sketch against all applicable sections of this chapter and "The Summary of Dimensioning Rules."

   NOTE: *These instructions do not apply to Problems 6-47 through 6-50.*

CAD INSTRUCTIONS

Students who plan to complete the following problems on a CAD system should begin with these general instructions.

1. Analyze the problem carefully.
2. Review the various commands you will use to develop the drawing.
3. Read the specific instructions for the problem and begin.

Problems 6-1 through 6-24

1. *Use the grid to determine all dimensions.*
2. *Redraw the views shown and fully dimension your drawing.*

PROBLEM 6-1

PROBLEM 6-2

PROBLEM 6-3

PROBLEM 6-4

PROBLEM 6-5

PROBLEM 6-6

PROBLEM 6-7

PROBLEM 6-8

PROBLEM 6-9

PROBLEMS

PROBLEM 6-10

PROBLEM 6-11

PROBLEM 6-12

PROBLEM 6-13

PROBLEM 6-14

PROBLEM 6-15

PROBLEM 6-16

PROBLEM 6-17

PROBLEM 6-18

PROBLEM 6-19

PROBLEM 6-20

PROBLEM 6-21

PROBLEM 6-23

PROBLEM 6-22

PROBLEM 6-24

Problems 6-25 through 6-46

1. *Convert the isometric drawings provided into orthographic drawings, showing as many views and/or sections as necessary to communicate the design.*
2. *Fully dimension your drawing.*

PROBLEM 6-25

PROBLEM 6-26

PROBLEM 6-27

METRIC

PROBLEM 6-28

ALL UNMARKED RADII = R.09

PROBLEM 6-29

Ø14 THRU
⌴ Ø26 ⍈ 8

16

100

8

Ø44

16

MIN RADII

8

2X Ø64

16X Ø6 ⍈ 10 (8 EACH LEVEL)
EVENLY SPACED (45°)

METRIC

PROBLEM 6-30

20
10
15
50
15
15
16
15
60°
12
40
26
16
25
115
14
6
22
15
6
15
20
58
2X Ø6 ⍈ 8
15
18
40
10

ALL UNMARKED RADII = R2

METRIC

PROBLEM 6-31

40

20

Ø30 ⍈ 62

RC–4
Ø10 THRU

2X R10

35

100

35

50

Ø46

(8)

70

8

METRIC

PROBLEM 6-32

Ø52

STANDARD SQUARE KEY
FOR A Ø28 SHAFT

RC – 6
Ø28 THRU

4 RIBS–
EVENLY SPACED

10
(TYP.)

Ø100

64

14

METRIC

PROBLEM 6-33

4X Ø.25 (FINE THREAD) EVENLY SPACED
FN–4
Ø1.12 THRU
1.05
Ø2.25 B.C.
Ø4.12 O.D.
.18
.56
Ø3.50 I.D.
.18
.36

PROBLEM 6-34

2X R10
12
25
R10
46
Ø30 THRU
90
54
2
4
METRIC
Ø50
STANDARD SQUARE KEY FOR A Ø30 SHAFT

PROBLEM 6-35

(.75)
38
56
18
1.25
3.12
R.38
RC–5
Ø.50 THRU
⌴ Ø1.25 ⌵.35
.31 (TYP.)
20°
1.06
ALL ⊠UNMARKED RADII = R.06
Ø1.75

PROBLEM 6-36

26
R14
R6
44
2X R12
88
36
2X R3
8
32
20
8
8
10
2X Ø6 THRU
⌴ Ø14 ⌵3
10 12
28
40
52
METRIC

PROBLEM 6-37

Ø38 COARSE THD.
1.38
Ø1.56
LC–8
Ø.88 THRU
2.62
(38)
.44
8°
TYP.
1.13
Ø2.00
RC–6
Ø1.12 THRU
2.50
45°
.50
.38
38 .38
8°
TYP.
3.12
1.13
30°
STANDARD SQUARE KEY FOR A Ø1.12 SHAFT
RC–4
Ø.56 THRU
1.06
.040
.080
1.38
ALL FILLETS/ROUNDS = R.06

PROBLEM 6-38

52
78
66
15
14
50
26
16
23
21
30
20
44

BASE

Ø46

LN–3
2X Ø10 THRU
CYLINDER

FN-4
Ø30 THRU

2X Ø6 THRU
(IN LINE)

1
2
13
13
3
R7

METRIC

NOTE: CYLINDER IS CENTERED ON THE BASE

PROBLEM 6-39

20
12

4X 2 WIDE X DEEP NOTCH
EVENLY SPACED

+ 6X R30

RC–3
Ø34
Ø60
Ø82

6X R12

RC–5
6X Ø8 THRU
EVENLY SPACED ON A
Ø100 B.C.

ALL UNMARKED RADII = R2   METRIC

PROBLEM 6-40

R.62

Ø.625/.620
THRU,
TWO IN LINE

.50
.50
2.00
.50
.50
.38

1.18

3.12

1.25

Ø1.25

Ø.38 THRU
⌴ Ø.75 ⦂ .25

ALL UNMARKED RADII = R.06

PROBLEM 6-41

2X Ø.25 THRU
(BOTH ENDS)

2X R.32
2X R.75

.09, BOTH ENDS

Ø2.50

6.38

2X Ø1.25
TOP/BOTTOM

Ø.66

RC – 4
Ø1.75
THRU

1.12

3.19

.83

2.00
.25

Ø22 THRU

4.50

(1.50)

RC – 6
2X Ø.875
THRU

ALL FILLETS/ROUNDS =
R.09

PROBLEM 6-42

PROBLEM 6-43

PROBLEM 6-44

PROBLEM 6-45

PROBLEM 6-46

PROBLEM 6-47

Convert Problem 6-47 into a tabular drawing.

PROBLEM 6-48

Convert Problem 6-48 into a standard coordinate dimensioning system drawing.

HOLE	SIZE
I	.750/.752
H	.120
G	2.562/2.565
F	.510/.512
E	.375/.380
D	.320
C	.188
B	.218/.220
A	.500

MATERIAL: STEEL .12 THICK

PROBLEM 6-49

Convert Problem 6-49 into a tabular drawing.

PROBLEM 6-50

Draw parts AO608-5 and AO6o8-8 full size.

ALL UNMARKED RADII = R.12

PART NO.	A	B	C	D	E	F	G	H	I	J
A0608-8	4.12	3.12	3.50	.50	3.00	.75	1.50	2.00	2	.50
A0608-7	3.38	1.75	2.50	.62	2.50	.50	–	–	–	–
A0608-6	3.50	1.50	2.50	.44	1.50	.50	1.00	.38	–	–
A0608-5	4.10	1.12	2.62	.25	1.25	.38	.75	.50	6	38
A0608-4	5.25	3.12	4.00	.75	2.25	1.00	2.12	.75	8	.62
A0608-3	3.25	1.50	2.38	.44	1.50	.50	1.00	.38	–	–
A0608-2	4.00	3.00	3.50	.62	1.00	.75	1.50	.62	3	.50
A0608-1	3.38	1.75	2.50	.50	2.00	.62	1.12	.25	4	.31

PROBLEMS 6-51 AND 6-52

Carefully fill in all blank lines.

PROBLEM 6-51

1 _____ NOMINAL SIZE
2 _____ MMC OF HOLE
3 _____ MMC OF SHAFT
4 _____ ALLOWANCE
5 _____ CLEARANCE

6 _____ NOMINAL SIZE
7 _____ MMC OF HOLE
8 _____ MMC OF SHAFT
9 _____ ALLOWANCE
10 _____ CLEARANCE

11 _____ NOMINAL SIZE
12 _____ MMC OF SLOT
13 _____ MMC OF KEY
14 _____ ALLOWANCE
15 _____ CLEARANCE

16 _____ NOMINAL SIZE
17 _____ MMC OF SLOT
18 _____ MMC OF KEY
19 _____ ALLOWANCE
20 _____ CLEARANCE

PROBLEM 6-52

PROB	NOM SIZE	BASIC SIZE	DIMENSION OF HOLE	DIMENSION OF SHAFT	HOLE TOLERANCE	SHAFT TOLERANCE	ALLOWANCE	CLEARANCE
1			.625 / .630	—		.005	+.002	
2	1.25		—	—	.001	.002	+.001	
3			.312 / .314	.316 / .314				
4	.75		—	—	.003	.002	+.001	
5			.812 / .813	.815 / .814				
6			22.2 / 22.3	22.5 / 22.4				
7	20		—	—	0.4	0.3	+0.2	
8			9.6 / 9.4	9.8 / 9.4				
9	30		—	—	0.2	0.4	+0.2	
10			15.6 / 15.4	—		0.6	+0.4	

(Rows 1–6: INCH; Rows 7–10: METRIC)

PROBLEM 6-53

Refer to the mounting guide drawing, No. 392450, and answer the following questions:

1. What is dimension A?
2. What tolerance is expected for two-place dimensions?
3. What are the dimension limits (maximum and minimum) of dimension B?
4. What size tap hole is specified for the 5/8 threaded hole?
5. What thread series is indicated for the 5/8 threaded hole?
6. When was the drawing last revised?
7. What are the dimension limits (maximum and minimum) of dimension C?
8. What is the maximum dimension of F?
9. For one assembly of the mounting guide, two stud bolts that are supposed to go through the two holes D are found to be spaced too far apart. The two stud bolts have a center-to-center distance of 1.625" and are .373" in diameter. The .422" D holes are 1.563" apart. One remedy for this type of problem is to enlarge the mating holes. Therefore, what is the minimum diameter for the two holes D that would permit the assembly?

7

Sectional Views

CHAPTER OUTLINE

Sectional views • Cutting-plane line • Direction of sight • Section lining • Multisection views • Kinds of sections • Sections through ribs or webs • Holes, ribs and webs, and spokes and keyways • Aligned sections • Fasteners and shafts in section • Intersections in section • Summary • Review questions • Chapter 7 problems • CAD instructions

CHAPTER OBJECTIVES

Upon completion of this chapter, students should be able to do the following:

▶ Describe the concept of a sectional view.

▶ Define what is meant by a cutting plane.

▶ Illustrate section lining.

▶ Prepare the following kinds of sections: full, offset, half, broken-out, revolved, removed, auxiliary, thinwall, and assembly.

▶ Demonstrate how to create the following in sections: ribs, webs, holes, spokes, fasteners, shafts, and keyways.

▶ Explain aligned sectioning.

▶ Show intersections in sections.

KEY TERMS

Aligning of features

Assembly section

Auxiliary section

Balloon

Broken-out section

Cutting-plane line

Full section

Half section

Offset section

Parts list

Removed section

Revolved section (Rotated section)

Section lining

Sectional view

Thinwall section

Sectional Views

A *sectional view* communicates more information about objects with complicated internal features than the conventional multiview drawing method can with hidden lines. The multiview drawing method, discussed in Chapter 5, is an excellent way to illustrate various external features of an object; however, complicated internal features are difficult to illustrate with hidden lines only. Sectional views solve the problem.

Sectional views are somewhat similar to x-ray images and very similar to magnetic resonance imaging in that each one shows internal details of an object not otherwise available to the observer.

A sectional view is a view of the surfaces that would be seen if an object could be cut open. The view is commonly referred to as a cross section or, simply, section (**Figure 7-1**).

Hidden lines are omitted in sectional views, unless it is absolutely necessary to illustrate some unique feature.

This hidden-line convention and other sectioning conventions described in ASME Y14.3M-1994 are part of the language of technical drawing. Discussions of the conventions follow.

Cutting-Plane Line

The *cutting-plane line* indicates the path that an imaginary cutting plane follows to slice through an object. Think of the cutting-plane line as a saw blade that is used to cut through the object. The cutting-plane line is represented by a thick black dashed line, as shown in **Figure 7-2**. The line on top is the newer cutting-plane line; sizes are approximated and spaced by eye. If the cutting-plane line passes through the center of the object, and it is very obvious where the cutting plane is located, it can be omitted. This is the drafter's decision.

Direction of Sight

The drafter must indicate the direction in which the object is to be viewed after it is sliced or cut through. This is accomplished by adding a short leader and arrowhead to the ends of the cutting-plane line (**Figure 7-3**). These arrowheads indicate the direction of sight. Letters are usually added to the ends of the

SCREW-HEX HD CAP
1/4-28 UNF, X 3/4 (19) LG
2 REQUIRED

SCREW-HEX HD CAP
1/4-28, UNF X $1\frac{1}{8}$ (28) LG
6 REQUIRED

KEY-SQ (NOM SIZE)
1/8 (3) X 1/8 (3) X 3/4 (20) LG

SCREW-HEX HD CAP
1/4-28 UNF, X 7/8 (22) LG
6 REQUIRED

PACKING MATERIAL

FIGURE 7-1   Sectional view.

FIGURE 7-2   Cutting-plane line.

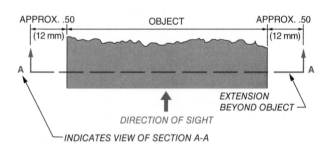

FIGURE 7-3   The direction of sight.

cutting-plane line to indicate exactly what cutting plane is used. The cutting plane extends past the object by approximately 1/2″, as shown.

Section Lining

Section lining shows where the object is sliced or cut by the cutting-plane line. Section lining is represented by thin black lines drawn at 45° to the horizon, unless there is some specific reason for using a different angle. Section lining is spaced by eye from 1/16″ (1.5 mm) to 1/4″ (6 mm) apart, depending on the overall size of the object. The average spacing used for most drawings is .13″ (3 mm) (**Figure 7-4**). Section lines must be of uniform thickness (thin black) and evenly spaced by eye.

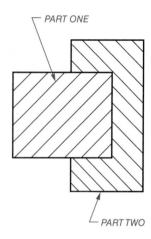

FIGURE 7-5   Two parts with section lining.

If a cutting plane passes through two parts, each part has section lines using a 45° angle or other principal angle. These section lines should not be aligned in the same direction and should not be joined together at the common visible edge (**Figure 7-5**). If the cutting plane passes through more than two parts, the section lining of each part must be drawn at a different angle. When an angle other than 45° is used, the angle should be 30° or 60°. Section lining should **not** be parallel with the sides of the object to be section lined (**Figure 7-6**).

In past years, section lining used various symbols to indicate the type of material used to make the object. These symbols used only general-type identifications such as cast iron, steel, brass, aluminum, and so forth. (Refer ahead to Figure 7-33 for examples.) Today, because there are so many different kinds of material, section lining symbols have been eliminated, and a single, all-purpose section lining is used (as illustrated in Figure 7-4). Specific information as to the type of material is given in the title block under "material," Figure 7-34.

FIGURE 7-4   Section lining.

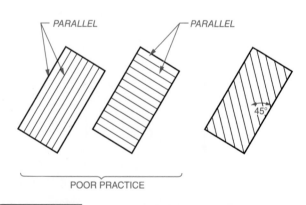

FIGURE 7-6   Section lining angle.

FIGURE 7-7   Multisection views.

Multisection Views

When an object is very complicated and cut in more than one place, each cutting-plane line must be labeled starting with section A-A, followed by B-B, and so forth (**Figure 7-7**).

Kinds of Sections

Nine kinds of sections are used today in the industry.

- Full section
- Offset section
- Half section
- Broken-out section
- Revolved section (rotated section)
- Removed section
- Auxiliary section
- Thinwall section
- Assembly section

Some sections are made up of a combination of the nine kinds of sections. Each is explained in full detail in the following paragraphs.

Full Section

A *full section* is simply a section of one of the regular multiviews that is sliced or cut completely in two.

See the given problem (**Figure 7-8**), a regular three-view drawing of an object.

Determine which view contains many hard-to-understand hidden lines. In this example, it is the front view. Add a cutting plane to either the top view or the right-side view. In this example, the top view is chosen. Indicate how the front view is to be viewed—that is, the direction of sight. After determining where the object is to be sliced or cut and viewed, change the front view into section A-A (**Figure 7-9**).

Think of the object as a pictorial drawing (**Figure 7-10**). An imaginary cutting-plane line is passed through the object (**Figure 7-11**). The front portion is

FIGURE 7-8   Given: Regular three views of an object.

FIGURE 7-9 Section A-A added.

FIGURE 7-10 Pictorial view of the object.

FIGURE 7-11 Imaginary cutting plane added.

FIGURE 7-12 Pictorial view of the full section.

removed, and the remaining section is viewed by the direction of sight (**Figure 7-12**).

Notice that section lining is applied only to the area the imaginary cutting plane passed through. The back side of the hole and the back sides of the notches are **not** section lined.

Offset Section

Many times, important features do not fall in a straight line as they do in a full section. These important features can be illustrated in an *offset section* by bending or offsetting the cutting-plane line. An offset section is very similar to a full section, except that the cutting-plane line is not straight (**Figure 7-13**).

Note that the features of the countersunk holes A, projection B with its counterbore, and groove C with a shoulder are not aligned with one another. The cutting-plane line is added, and changes of direction (staggers) are formed by right angles to pass through

FIGURE 7-13 Offset section.

these features. An offset cutting-plane line A-A is added to the top view, and the material behind the cutting plane is viewed in section A-A (**Figure 7-14**). The front view is changed into an offset section, similar to a full-sectional view. The actual bends of the cutting-plane lines are omitted in the offset section (**Figure 7-15**). With the use of a sectional view, another view often may be omitted. In this example, the right-side view could have been omitted, as it adds nothing to the drawing and takes extra time to draw.

Half Section

In a *half section,* the object is cut only halfway through and a quarter section is removed. See the given problem (**Figure 7-16**). showing two views of an object. A cutting plane is added to the front view, with only one arrowhead to indicate the viewing direction. Also, a quarter section is removed and, in this example, the right side is sectioned accordingly

FIGURE 7-15   Bends omitted from section view.

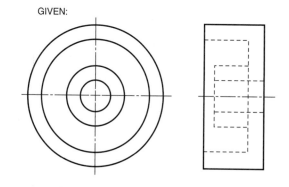

FIGURE 7-16   Given: Regular two views of an object.

(**Figure 7-17**). A pictorial view of this half section is illustrated in **Figure 7-18**. The visible half of the object that is not removed shows the exterior of the object, and the removed half shows the interior of the object. The half of the object not sectioned can be drawn as it would normally be drawn, with the appropriate hidden lines.

FIGURE 7-14   Pictorial view of the offset section.

FIGURE 7-17   Half section.

FIGURE 7-18   Pictorial view of the half section.

Half sections are best used when the object is symmetrical—that is, the same shape and size on both sides of the cutting-plane line. A half-section view is capable of illustrating both the inside and the outside of an object in the same view. In this example, the top half of the right side illustrates the interior; the bottom half illustrates the exterior. A center line is used to separate the two halves of the half section (refer back to Figure 7-17). A solid line would indicate the presence of a real edge, which would be false information.

Broken-out Section

Sometimes, only a small area needs to be sectioned in order to make a particular feature or features easier to understand. In this case, a *broken-out section* is used (**Figure 7-19**). As drawn, the top section is somewhat confusing and could create a question. To clarify this area, a portion is removed (**Figure 7-20**). The finished drawing would be drawn as illustrated in **Figure 7-21**. The broken line is put in freehand and is drawn as a visible thick line. The actual cutting-plane line is usually omitted.

FIGURE 7-19   Given: Regular two views of an object.

FIGURE 7-20   Pictorial view of the broken-out section.

Revolved Section (Rotated Section)

A *revolved section,* sometimes referred to as a *rotated section,* is used to illustrate the cross section of ribs, webs, bars, arms, spokes, or other similar features of an object. **Figure 7-22** is a two-view drawing of an arm. The cross-sectional shape of the center portion of the arm is not defined. In drafting, no feature should remain questionable, and a section through the center portion of the arm would provide the complete information.

FIGURE 7-21   Broken-out section.

FIGURE 7-22   Given: Regular two views of an object.

A revolved section is made by assuming a cutting plane perpendicular to the axis of the feature of the object to be described (**Figure 7-23**). Note that the rotation point occurs at the cutting-plane location and, theoretically, will be rotated 90°. Rotate the imaginary cutting-plane line about the rotation point of the object (**Figure 7-24**). Notice that dimension X is transferred from the top view to the sectional view of the feature—in this example, the front view.

FIGURE 7-23   Revolved section.

FIGURE 7-24   Pictorial view of revolved section.

Dimension Y in the top view is also transferred to the front view. The section is now drawn in place. The finished drawing is illustrated in **Figure 7-25**. Note that the break lines in the front view are on each side of the sectional view and are put in freehand.

FIGURE 7-25   Revolved section view.

WORKPLACE SKILLS

Betty Konrad was a CAD technician employed by a company that produced household appliances such as ovens, dishwashers, clothes washers, and dryers. The company planned to launch an advertising campaign to boost sales and expand its market base. The company needed an effective means to communicate to the public just how their appliances worked.

Betty was assigned the task of organizing a group of two CAD technicians and two illustrators to prepare informational drawings to show how the appliances performed and what they were made of. To organize the team and to get them looking forward to working together, Betty called a meeting and tossed out several questions. "What do you think the users want to know about the appliances?" "What types of drawings would best show how they work and how they are assembled?" The group concluded that the most prevalent technical drawings showing the "insides" of appliances were pictorials with broken-out sections (Figure 7-20). Furthermore, the group felt that the users wanted to know how easy an appliance was to use and its capacity, such as what size turkey fit in an oven and how large loads could be in the washer and dryer.

Betty assigned two appliances to each CAD technician for the creation of isometric broken-out sections. The illustrators were to use this output to prepare colorful presentations. The illustrators would also include capacities.

As the team launched their efforts, Betty prepared a planning chart so each member would have a visual reminder of the team's schedule.

The revolved section is not used as much today as it was in past years. Revolved sections tend to be confusing, and they often create problems for the people who must interpret the drawings. Today, we are encouraged to use a removed section instead of a revolved or rotated section.

Removed Section

A *removed section* is very similar to a rotated section except that, as the name implies, it is drawn removed or away from the regular views (**Figure 7-26**). The removed section, as with the revolved section, is also used to illustrate the cross section of ribs, webs, bars, arms, spokes, or other similar features of an object.

A removed section is made by assuming that a cutting plane, perpendicular to the axis of the feature of the object, is added through the area that is to be sectioned. (Refer back to Figures 7-23 and 7-24.) Transfer dimensions X and Y to the removed views, exactly as was done in the rotated section. Height features, such as dimensions A and B, are transferred from the front view in this example.

Note that a removed section must identify the cutting-plane line from which it was taken. In the sectional view, do not draw features other than the actual section.

Removed sections are labeled section A-A, section B-B, and so forth, corresponding to the letters at the ends of the cutting-plane line. The sections are usually placed on the drawing in alphabetical order from left to right or from top to bottom, away from the regular views.

Sometimes a removed section is simply drawn on a center line that is extended from the object (**Figure 7-27**). A removed section can be drawn to an enlarged scale if necessary to illustrate and/or dimension a small feature. The scale of the removed section must be indicated directly below the sectional view (**Figure 7-28**).

In the field of mechanical drafting, the removed section should be drawn on the same page as the

FIGURE 7-26 Removed section.

FIGURE 7-27   Removed section view.

FIGURE 7-28   Enlarged removed section.

SECTION A-A

FIGURE 7-29   Auxiliary section.

FIGURE 7-30   Thinwall section.

FIGURE 7-31   Space between thinwall sections.

regular views. If there is not room enough on the same page and the removed section is drawn on another page, a page number cross-reference must be given as to where the removed section may be found. The page where the removed section is located must refer back to the page from which the section is taken. For example, section A-A on sheet 2 of 4.

Auxiliary Section

If a sectional view of an object is intended to illustrate the true size and shape of an object's boundary, the cutting-plane path must be perpendicular to the axis or surfaces of the object. An *auxiliary section* is projected in the same way as any normal auxiliary view (**Figure 7-29**).

Thinwall Section

Any very thin object that is drawn in section, such as sheet metal, a gasket, or a shim, should be filled-in solid black, as it is impossible to show the actual section lining. This is called a *thinwall section* (**Figure 7-30**). If several thin pieces that are filled-in solid black are touching one another, a small white

space is left between the solid thinwall sections (**Figure 7-31**).

Assembly Section

When a sectional drawing is made up of two or more parts, it is called an *assembly section* (**Figure 7-32**). An assembly section can be a full section (as it is in this example), an offset section, a half section, or a combination of the various kinds of sectional views. The assembly section shows how the various parts go together.

Each one of the section linings in Figure 7-32 is the accepted standard described in the beginning

FIGURE 7-32 Assembly section.

FIGURE 7-33 Symbols for section lining.

of the chapter. Nevertheless, sometimes you will be given drawings using lining symbols to represent different materials, as illustrated in **Figure 7-33**. The symbols shown represent the more commonly used engineering metals and concrete. Symbols are also available for rubber, plastic, cork, insulation, slate, earth, rock, sand, and wood.

Each part in the assembly must be labeled with a name, part or plan number, and the quantity required for one complete assembly. If the assembly section does not have many parts, this information is added by a note alongside each part. If the assembly has many parts, and there is not enough room to prevent the drawing from appearing cluttered, each individual part may be identified by a number within

a circle called a *balloon*. The balloon callout system is used in this example. A table must be added to the drawing, listing the name, part or plan number, quantity required for one complete assembly, and a cross-reference to the corresponding balloon number. This is called a *parts list*. The exact form of the list varies from company to company. **Figure 7-34** is an example of a parts list used with the balloon system of callouts. Notice that entries are sometimes listed in reverse (bottom to top) order, as illustrated.

Occasionally, drafters deviate from conventional standards for clarity, even though the deviation may violate a basic standard. The first part of this chapter covered conventional standards for sectioning; the latter part will focus on deviations for clarity.

INDUSTRY APPLICATION

CLEARING UP THE CONFUSION

The drafting team was dealing with a complex, irregularly shaped machine part. To adequately provide all of the information needed to manufacture the part, the team had to draw all six orthographic views. This did not completely solve the problem. "This part just has too many hidden lines," said the team leader. "I don't think I could make this part from these drawings. With all of the hidden lines, it's hard to determine what is required. Any suggestions?"

"Why don't we cut some sections?" asked one of the team members.

"I've been out of school too long," said the team leader. "Where would you cut them and how many?"

"As I see it," said the team member, "we need to cut a full section to show the right side, and another for the left side. Then we need an offset section from the direction of the front view and one broken-out section." Various team members began sketching as they talked. Soon the types and numbers of sections needed had been determined, and the drafting team set to work revising the drawings. The final result was a set of drawings that gave manufacturing personnel the information they needed to make the part without difficulty.

CORRESPONDS TO BALLOON
NUMBERS IN FIGURE 7-32

PART NO.	NO. REQD.	DESCRIPTION	MATERIAL	NOTES
6	1	DRIVE WASHER	1018 STEEL	
5	1	BUSHING	LEADED BRONZE	
4	1	CRANKSHAFT	AISI 1040 STEEL	
3	1	CRANKCASE	ALUMINUM ALLOY	
2	1	CONNECTING ROD	ALUMINUM ALLOY 7075T6	
1				(NO. NOT USED)

USUAL CALLOUT INFORMATION

FIGURE 7-34   Examples of a parts list.

Sections through Ribs or Webs

True projection of a sectioned view often produces incorrect impressions of the actual shape of the object. **Figure 7-35** gives a front view and a right-side view. Its pictorial view would look like (**Figure 7-36**). A full section A-A would appear as it does in **Figure 7-37**. This is a true projection of section A-A, as the cutting-plane line passes through the rib.

However, such a sectional view gives an incorrect impression of the object's actual shape, and it is poor drawing practice. It misleads the viewer into thinking the object is actually shaped as it is shown in **Figure 7-38**. The conventional practice used to illustrate this section is to draw the section view as

GIVEN:

FIGURE 7-35   Given: Two-view drawing of an object.

PICTORIAL VIEW

FIGURE 7-36   Pictorial view of an object.

SECTION A-A

TRUE PROJECTION OF SECTION

FIGURE 7-37   True projection of an object.

MISLEADING

PICTORIAL VIEW

FIGURE 7-38   True projection can be misleading.

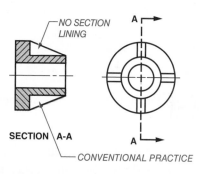

NO SECTION LINING

SECTION A-A

CONVENTIONAL PRACTICE

FIGURE 7-39   Conventional practice—web or rib not sectioned.

illustrated in **Figure 7-39**, which is not a true projection. Note that the web or rib is not section lined.

Some companies use another method to compensate for this problem. It is somewhat of a middle ground or a combination of true projection and correct representation (**Figure 7-40**). This is called

HIDDEN LINE

SECTION A-A

ALTERNATIVE PRACTICE

FIGURE 7-40   Alternate conventional practice.

FIGURE 7-41   Example of true projection.

FIGURE 7-42   Example of conventional practice.

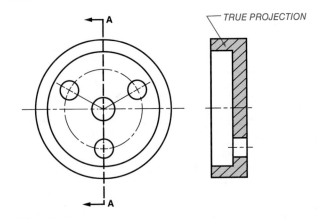

FIGURE 7-43   Holes using true projection.

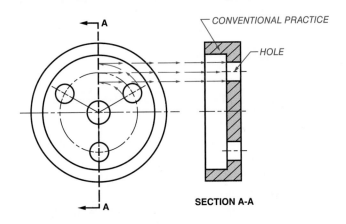

FIGURE 7-44   Holes using conventional practice (aligning of features).

alternate section lining. Section lining, over the rib or web section, is drawn by using every **other** section line, and the actual shape is indicated by hidden lines. However, most companies do not use alternate section lining.

Another example of a cutting plane passing through a rib or web is shown in **Figure 7-41**. This example is a true projection, but it is poor drafting practice because it gives the impression that the center portion is a thick, solid mass. **Figure 7-42** is drawn incorrectly, but it does not give the false impression of the object's center portion.

This is the conventional practice used.

Holes, Ribs and Webs, and Spokes and Keyways

Holes located around a bolt circle are sometimes not aligned with the cutting-plane line (**Figure 7-43**). The cutting plane passes through only one hole. This is a true projection of the object but poor drafting practice. In actual practice, the top hole is theoretically revolved to the cutting-plane line and projected

to the sectional view (**Figure 7-44**). This practice is called *aligning of features*.

Ribs and Webs

Ribs or webs sometimes do not align with the cutting-plane line (**Figure 7-45**). The cutting plane passes through only one web and only one hole. This is a true projection of the object but poor drafting practice. In actual practice, one of the webs is theoretically

FIGURE 7-45   Rib or web using true projection.

revolved up to the cutting-plane line and projected to the sectional view (**Figure 7-46**). Notice that the bottom hole is unaffected and is projected normally. This is another example of aligning of features.

Spokes and Keyways

Spokes and keyways and other important features sometimes do not align with the cutting-plane line

FIGURE 7-46   Rib or web using conventional practice.

FIGURE 7-47   Spokes and keyway using true projection.

FIGURE 7-48   Spokes and keyway using conventional practice.

(**Figure 7-47**). The cutting-plane line passes through only one spoke and misses the keyway completely. This is also a true projection of the object but poor drafting practice. In conventional practice, spoke B is revolved to the cutting-plane line and projected to the sectional view (**Figure 7-48**). The keyway is also projected as illustrated. This is another example of aligning of features.

Aligned Sections

Arms and other similar features are revolved to alignment in the cutting plane, as were spokes in the preceding section. This procedure is used if the cutting-plane line cannot align completely with the object, as illustrated in **Figure 7-49**.

The arm or feature is now revolved to the imaginary cutting plane and projected down to the sectional view (**Figure 7-50**). The actual cutting-plane line is bent and drawn through the arm or feature

FIGURE 7-49   Two-view drawing.

FIGURE 7-50   Aligned section.

and then revolved to a straight, aligned vertical position. Notice that section lining is **not** applied to the arm, and it is also omitted from the web area.

Fasteners and Shafts in Section

If a cutting plane passes lengthwise through any kind of fastener or shaft, the fastener or shaft is **not** sectioned (**Figure 7-51**). Section lining of a fastener or shaft would have no interior detail; thus, it would serve no purpose and only add confusion to the drawing. The round-head machine screw, the hex-head cap screw with nut, and the rivet are not sectioned. The other objects in the figure, such as fasteners, ball bearing rollers, and so forth, are also not sectioned.

If a cutting-plane line passes perpendicularly through the axis of a fastener or shaft, section lining is added to the fastener or shaft (**Figure 7-52**). The end view has section lining added as shown.

Intersections in Section

Where an intersection of a small or relatively unimportant feature is cut by a cutting-plane line, it is not drawn as a true projection (**Figure 7-53**). Because a

true projection takes drafting time, it is preferred that it be disregarded and the feature drawn by using conventional practice, as shown in **Figure 7-54**. This procedure is much quicker and more easily understood.

FIGURE 7-53 Intersection using true projection.

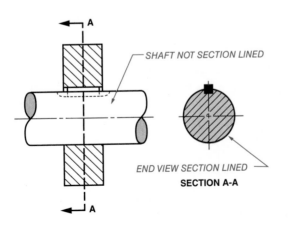

FIGURE 7-52 Shaft sectioned in end view only.

FIGURE 7-54 Intersection using conventional practice.

SUMMARY

- A sectional view is a view of the surfaces of an object that would be seen if an object could be cut open.

- A cutting-plane line indicates the path that an imaginary cutting plane follows to slide through an object.

- The direction of sight is shown by arrows on the cutting-plane line. It indicates the direction of sight for viewing the completed sectional view.

- Section lining shows where the object is sliced or cut by the cutting-plane line. It indicates the surface that is touched by the cutting plane.

- When an object is complicated and cut in more than one place, each cutting-plane line must be labeled starting with section A-A, B-B, C-C, and so on.

- The kinds of sections students can expect to draw are as follows: full, offset, half, broken-out, revolved, removed, auxiliary, thinwall, and assembly.

- A section cut through a rib or a web can result in a misleading representation if true projection is applied. Consequently, in conventional practice, section lines are not drawn in the rib or web.

- When cutting sections through objects that have holes, ribs, webs, spokes, and/or keyways that are not obliqued, a cutting plane can pass through one feature and miss another. This problem is solved by revolving the missed feature over to the cutting plane. This same technique is used in creating obliqued sections of objects with arms or other features that do not line up with a cutting plane.

- If a cutting plane passes lengthwise through any type of fastener or shaft, the fastener or shaft is not section lined.

REVIEW QUESTIONS

Mark the following statements as either true or false.

1. True projection of a sectional view always produces the actual shape of the object.

2. Section lining should be parallel with the sides of the object to be section lined.

3. A revolved section is the same as a rotated section.

4. A removed section is very similar to a rotated section except that it is drawn away from the regular view.

5. Hidden lines are usually not used in a sectional view.

6. A removed section is sometimes drawn to larger scale to illustrate and/or dimension a small feature.

7. Half sections are best used when an object is **not** symmetrical.

8. Today, it is recommended to use the removed section rather than the revolved section.

9. If a cutting-plane line passes lengthwise through any kind of fastener or shaft, the fastener or shaft is not sectioned.

10. If a cutting-plane line passes perpendicularly through the axis of a fastener or shaft, the fastener or shaft is not sectioned.

Answer the following questions by selecting the best answer.

1. Which of the following is **not** a list of types of sectional views?
   a. Full, half, offset
   b. Front, back, profile
   c. Broken-out, revolved, removed
   d. Thinwall, auxiliary, assembly

2. Which of the following is **not** true regarding alternate section lining?
   a. It is a combination of true projection and correct representation.
   b. Over the rib or web section, every other section line is drawn.
   c. The rib or web section is not lined.
   d. The actual shape is indicated by hidden lines.

3. What kind of section view illustrates both the exterior and interior of the object?
   a. Half section
   b. Full section
   c. Broken-out
   d. Revolved

4. Which of the following is **not** true regarding an assembly section?
   a. When a sectional drawing is made up of two or more parts, it is called an assembly section.
   b. A table, called a parts list, may be added to the drawing.
   c. Each part number may be identified by a number within a circle called a balloon.
   d. The assembly section is always represented using a full section view.

5. What must be done if a removed section needs to be placed on a page other than the page on which the cutting-plane line is placed?
   a. The page where the removed section is located must refer back to the page from which it was taken.
   b. A page number cross-reference must be given to where the removed section may be found.
   c. Both a and b
   d. None of the above

6. What must be included for each part in an assembly section?
   a. Name of part
   b. Part number or plan number
   c. Quantity required to complete one assembly
   d. All of the above

CHAPTER 7 PROBLEMS

The following problems are intended to give the beginning drafter practice in using the various kinds of sectional views used in the industry. As these are beginning problems, no dimensions will be used at this time.

Conventional Instructions

The steps to follow in laying out all problems in this chapter are:

STEP 1   Study the problem carefully.

STEP 2   Choose the view with the most detail as the front view.

STEP 3   Position the front view so there will be the least amount of hidden lines in the other views.

STEP 4   Make a sketch of all required views.

STEP 5   Determine what should be drawn in section, what type of section should be used, and where to place the cutting-plane line.

STEP 6   Center the required views within the work area with a 1″ (25 mm) space between views.

STEP 7   Use light projection lines. Do **not** erase them.

STEP 8   Lightly complete all views.

STEP 9   Check to see that all views are centered within the work area.

STEP 10   Check to see that there is a 1″ (25 mm) space between all views.

STEP 11   Carefully check all dimensions in all views.

STEP 12   Darken in all views by using the correct line thickness.

STEP 13   Add a cutting-plane line and section lining as required.

STEP 14   Recheck all work and, if it is correct, neatly fill out the title block by using light guidelines and neat lettering.

CAD INSTRUCTIONS

Students who plan to complete the following problems on a CAD system should begin with these general instructions.

1. Analyze the problem carefully.
2. Plan your drawing sequence. In what order will you develop the drawing?
3. Read the specific instructions for the problem and begin.

Mechanical   **PROBLEM 7-1**

Center three views within the work area and make the front view a full section.

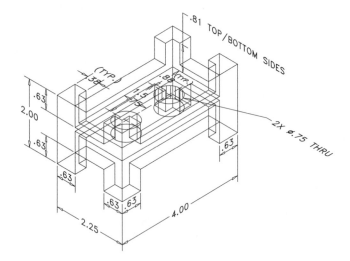

Mechanical **PROBLEM 7-2**

Center two views within the work area and make one view a full section. Use correct drafting practices for the ribs.

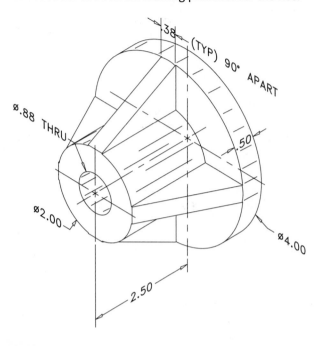

Mechanical **PROBLEM 7-3**

Center two views within the work area and make one view a full section. Use correct drafting practices for the holes.

Mechanical **PROBLEM 7-4**

Center the front view and top view within the work area. Make one view a full section.

Mechanical **PROBLEM 7-5**

Center two views within the work area and make one view a full section. Use correct drafting practices for the arms, horizontal hole and keyway.

PROBLEM 7-6

Center two views within the work area and make one view a full section. Use correct drafting practices for the keyway, ribs and holes.

Mechanical **PROBLEM 7-7**

Center three views within the work area and make one view an offset section. Be sure to include three major features.

Mechanical **PROBLEM 7-8**

Center three views within the work area, and make one view an offset section. Be sure to include three major features.

Mechanical **PROBLEM 7-9**

Center three views within the work area, and make one view an offset section. Be sure to include as many of the important features as possible.

Mechanical **PROBLEM 7-10**

Center three views within the work area, and make one view an offset section. Be sure to include as many of the important features as possible.

Mechanical **PROBLEM 7-11**

Center two views within the work area, and make one view an offset section. Be sure to include as many of the important features as possible.

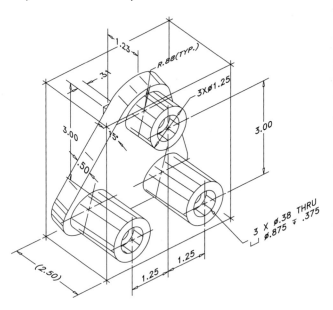

PROBLEMS

PROBLEMS

Mechanical **PROBLEM 7-12**

Center the front view and top view within the work area. Make one view a half section.

Mechanical **PROBLEM 7-14**

Center the two views within the work area and make one view a half section.

Mechanical **PROBLEM 7-13**

Center two views within the work area and make one view a half section.

Mechanical **PROBLEM 7-15**

Center two views within the work area and make one view a half section.

Mechanical **PROBLEM 7-16**

Center two views within the work area and make one view a full section.

ALL UNMARKED RADII ARE R.13

Mechanical **PROBLEM 7-17**

Center the required views within the work area and make one view a broken-out section to illustrate the complicated interior area.

PROBLEM 7-18

Center the required views within the work area and make one view a broken-out section as required.

 PROBLEM 7-19

Center three views within the work area and add removed sections A-A and B-B.

Mechanical ## PROBLEM 7-20

Center two views within the work area and add removed sections A-A, B-B, and C-C.

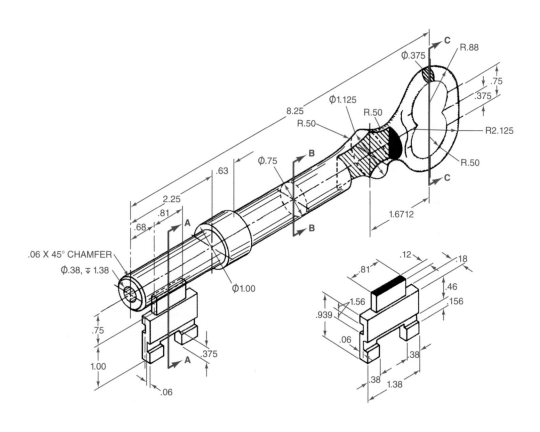

Mechanical ## PROBLEM 7-21

Center the required views within the work area and add the removed section as required.

Mechanical

PROBLEM 7-22

Center the required views within the work area and add the removed section as required.

CYLINDER WALL
THICKNESS .125

ALL UNMARKED RADII = R.06

Mechanical

PROBLEM 7-23

Center the required views within the work area and add removed sections A-A and B-B.

ALL UNMARKED RADII = R.06

Mechanical

PROBLEM 7-24

Center the required views within the work area and add removed section A-A.

METRIC

Mechanical **PROBLEM 7-25**

Center the four views within the work area. Make the top view section A-A and the right-side view section B-B.

ALL UNMARKED RADII = R.06

Mechanical **PROBLEM 7-26**

Center the required views within the work area and add removed sections A-A and B-B.

ALL UNMARKED RADII = R.06

Mechanical   **PROBLEM 7-27**

Center the required views within the work area and add removed sections A-A and B-B.

ALL UNMARKED RADII = R.06

Mechanical   **PROBLEM 7-28**

Center the front view, side view and removed sections A-A, B-B, and C-C within the work area.

Mechanical **PROBLEM 7-29**

Make a two-view assembly drawing of parts 1, 2, 3, and 4. Make one view a full-section assembly. Use correct section lining and all conventional drafting practices.

'ROUND' END SLIGHTLY

1.25

∅.20

PART NO. 1

1° TAPER

∅.20, THRU

R1.75

∅.31

R.0625

1.0   3.125   1.10   4.722   .25

10.0

PART NO. 2

∅.31, THRU

5° TAPER

R.12

∅.88

∅1.0

2X ∅.20 (IN LINE)

R.06

.06   1.0

1.69

PART NO. 3

4.88

R.125

.06 X 45° CHAMFER

45°   1.12

∅.63

1.12

∅.31, THRU
⌴ ∅1.12 X .25

2.44

2.38

2.12

1.94

1.63

1.0

5° TAPER

R.75

FLAT SURFACES

∅1.12

∅.88

.75

.75

∅19 THRU

∅1.25   ∅1.62

∅1.50

∅1.25   ∅1.50

PART NO. 4

Mechanical   ## PROBLEM 7-30

Make a two-view assembly drawing of parts 1, 2, and 3. Make one view a full-section assembly. Use correct section lining and all conventional drafting practices.

'ROUND' ENDS

$\phi \frac{.251}{.240}$

1.06

PART NO. 1
3 – REQ'D

3X $\phi \frac{.250}{.249}$ THRU, 120° APART

1.25

ϕ.625

(60°)

R.1144

ϕ1.12

ϕ1.68

(60°)

SHARP   SHARP

30°

(.63)

2.25

PART NO. 2

3X $\phi \frac{.250}{.249}$ THRU, 120° APART

SHARP

R1.8

ϕ2.0

(60°)

ϕ2.38

ϕ2.56

ϕ2.75

(60°)

.38 ← FRONT END

.75

3.25

PART NO. 3

Mechanical   ## PROBLEM 7-31

Make a two-view assembly drawing of parts 1, 2, and 3. Make one view a full-section assembly. Use correct section lining and all conventional drafting practices.

PART NO.1

PART NO.2

PART NO.3

PROBLEMS

Problems 7-32 through 7-37

Center required views within the work area. Leave a 1" or 25 mm space between views. Make one view into a section view to fully illustrate the object. Use either a full, half, offset, broken-out, revolved, or removed section. Do not add dimensions.

Mechanical **PROBLEM 7-32**

METRIC

Mechanical **PROBLEM 7-33**

ALL UNMARKED
RADII = R3

Mechanical **PROBLEM 7-34**

ALL FILLETS = R2

2 X 45° CHAMFER

12X Ø5 THRU
EVENLY SPACED ON A Ø80 B.C.

METRIC

Mechanical **PROBLEM 7-35**

8X Ø5 ⬇10
EVENLY SPACED
ON A Ø52 B.C.

Ø30 THRU

METRIC

PROBLEMS

Mechanical PROBLEM 7-36

Φ120
Φ102
Φ84
3
10
40
(Φ52)
(Φ32)
7
80
Φ100
7
Φ32
16
8
Φ20 THRU
Φ24 X 40 LONG OIL GROVE
Φ92 (INSIDE Φ)
Φ100 (OUTSIDE Φ)

ALL UNMARKED RADII = R2   METRIC

Mechanical PROBLEM 7-37

58 32
10
Φ132
Φ115
Φ100
Φ116
20
Φ86
Φ64
R4
Φ10 THRU
42
58   R3
100
Φ70
15
30
METRIC
Φ102
Φ50 THRU
4
8

ALL UNMARKED RADII = R5

PROBLEM 7-38

Refer to Problem 17-9 and use the full-section drawing for information on dimensions to determine the overall distance from the outside of part 1 on the left to the outside of part 1 on the right along the center line of the pulley. Prepare a sketch with dimensions for your answer.

PROBLEM 7-39

Refer to Problem 18-52 (assembly) and prepare an assembly section (front-view orientation) of the vise. Indicate approximate dimensions only.

PROBLEM 7-40

Refer to Problem 18-50 and prepare a full-section view of the object.

PROBLEM 7-41

Refer to Problem 6-21 and prepare an offset section view (profile orientation) of the object through the center line. Include a threaded hole.

Mechanical ## PROBLEM 7-42

Select an object having a nonuniform cross section like the scissor half shown and prepare removed section views to illustrate detail of the object. Suggested objects: hammer, pliers, bolt cutters, boat oar, snow ski, water ski, propeller (marine, airplane, contemporary wind machine, etc.), nozzle, handles, or spokes.

PROBLEM 7-43

Refer to Problem 6-12 and prepare an aligned section view of the object. Show the threaded holes and the keyway by using conventional practices.

PROBLEM 7-44

Refer to Problem 6-37 and prepare an aligned section view of the object.

PROBLEM 7-45

Refer to Problem 6-43. Prepare a conventional section view showing the ribs of the object.

PROBLEM 7-46

Refer to Problem 17-13 (top view). Prepare a section, drawn to scale.

Mechanical ## PROBLEM 7-47

Refer to the drawings shown. Prepare an offset section C-C, full size, of the distributor cap. Dimensions are not required.

PROBLEM 7-48

Select one of the following common devices, and prepare a section view(s) to best show inside details and how the device works for a nontechnical person. Simplify the details of the device for clarity. How-things-work books would be useful references. Indicate approximate sizes only.

a. clock

b. toaster

c. internal-combustion-engine cylinder

d. ski binding

e. electric drill

f. car window mechanism (manual or electrical)

g. electric motor application in an average home, including an automobile

NOTE: The average home with an automobile has approximately 35 electric motors.

PROBLEM 7-49

Find material on threaded inserts (usually a threaded collar made of steel, inserted in softer material such as aluminum). Prepare a section drawing showing an application. Indicate approximate dimensions.

PROBLEM 7-50

Select an object from one of the locations or categories listed and prepare a multiview drawing of your redesign ideas for the object. Use section views to replace traditional views where appropriate. (Refer to Figure 7-15.)

a. classroom

b. lab

c. shop

d. garage

e. automobile

f. hobby

g. sports equipment

PROBLEM 7-51

Use the partially completed drawings shown and do the following.

a. Assume that the beam used has a depth of 152 mm and a flange width of 84 mm. Estimate dimensions of the parts that make up dimension X, and show in a freehand sketch dimension X and your assumptions.

b. Make additional estimates: (1) of the thickness of the plates for the trolley wheels, and (2) of the distances from the trolley wheels to the lower bolted connection and to the hole. Prepare an offset section through the trolley wheels, through the bolted connection, and through the hole.

PROBLEMS

PROBLEM 7-52

Design a lever-type door handle for handicapped individuals who have prosthetics for hands and cannot use knob-type door openers. Use removed sections to show the details of the lever arm. Indicate approximate dimensions.

PROBLEM 7-53

Design a desk for a modern CAD station. Prepare a multiview drawing of the desk and show a cross-sectional view (profile view orientation) of the desk. A 26″ knee standard is a typical standard for computer desks. Indicate approximate dimensions.

PROBLEM 7-54

Name or describe the type of sectioning used in the figures and problems listed.
Figure 6-105, Figure 12-6, Figure 13-14, Figure 14-17, Problem 17-9, Figure 19-35, and Figure 20-13.

PROBLEM 7-55

Name or describe the type of sectioning used in the figures and problems listed.
Figure 13-31, Figure 16-30, Problem 17-10, and Figure 20-3.

PROBLEM 7-56

Name or describe the type of sectioning used in the figures and problems listed.
Problem 6-39, Figure 12-5, Figure 12-73, Problem 12-30, Figure 17-17, Problem 17-18, and Figure 20-8.

8

Auxiliary Views

CHAPTER OUTLINE

Auxiliary views defined • Secondary auxiliary views •
Partial views • Auxiliary section • Half auxiliary views •
Oblique surfaces, viewing planes, and auxiliary view
enlargements • Summary • Review questions •
Chapter 8 problems • CAD instructions

CHAPTER OBJECTIVES

Upon completion of this chapter, students should be able to do the
following:

▶ Define the concept of an auxiliary view and list its advantages.

▶ Describe and use the reference plane method to draw an auxiliary view.

▶ Describe and use the folding line method to draw an auxiliary view.

▶ Draw round and curved surfaces in auxiliary views.

▶ Illustrate secondary auxiliary views.

▶ Illustrate partial auxiliary views.

▶ Draw enlarged auxiliary views.

▶ Describe the use of a viewing plane for auxiliary views.

KEY TERMS

Auxiliary section

Auxiliary view

Enlargement of the
auxiliary view

Fold line

Front view auxiliary

Oblique surface

Partial auxiliary view

Primary auxiliary
view

Secondary auxiliary
view

Side view auxiliary

Top view auxiliary

True length

True size

Viewing plane

Auxiliary Views Defined

Many objects have inclined surfaces that are not always parallel to the regular planes of projection. For example, in **Figure 8-1**, the front view is correct as shown, but the top and right-side views do **not** correctly represent the inclined surface. To truly represent the inclined surface and to show its true shape, an auxiliary view must be drawn. An *auxiliary view* has a line of sight that is perpendicular to the inclined surface, as viewed looking directly at the inclined surface. Auxiliary views are always projected 90° from the inclined surface.

An auxiliary view serves three purposes:

- It illustrates the true size of a surface.
- It illustrates the true shape of a surface, including all true angles and/or arcs.
- It is used to project and complete other views.

An auxiliary view can be constructed from any of the regular views. An auxiliary view projected from the front view would appear as it does in Figure 8-1. This is referred to as a *front view auxiliary*. An auxiliary view projected from the top view would appear as it does in **Figure 8-2**. This is referred to as a *top view auxiliary*. An auxiliary view projected from the right-side view would appear as it does in **Figure 8-3**. This is referred to as a *side view auxiliary*. Note in each case that the auxiliary view is projected 90°

FIGURE 8-3   Side view auxiliary view.

from the inclined or slanted surface and is viewed from a line of sight 90° to the inclined or slanted surface or as viewed looking directly down upon the inclined surface.

Hidden Lines in an Auxiliary View

Hidden lines should be omitted in an auxiliary view, unless they are needed for clarity. This is the drafter's prerogative or decision.

How to Draw an Auxiliary View

GIVEN: The pictorial view of an object (**Figure 8-4**). Notice the inclined surface. Because the inclined surface is on the front view, this will be a front view auxiliary. The usual three views of an object (front view, top view, and right-side view) are shown in **Figure 8-5**.

STEP 1   Label all important points of the auxiliary view, as illustrated in **Figure 8-6A**.

STEP 2   Construct a reference line, which is also the edge view of a reference plane, in the right-side view. Always construct a reference line so that it is vertical and passes through as many points as possible. In this example, it passes through points a–d (**Figure 8-6B**).

FIGURE 8-1   Front view auxiliary view.

FIGURE 8-2   Top view auxiliary view.

FIGURE 8-4   Drawing an auxiliary view.

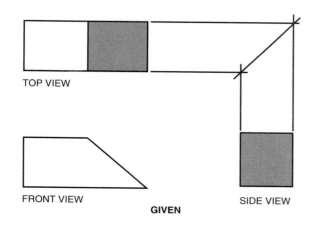

TOP VIEW

FRONT VIEW

SIDE VIEW

GIVEN

FIGURE 8-5 Given: Three views of an object.

STEP 3 Draw light projection lines 90° from the inclined surface and construct a reference line parallel to the inclined surface at any convenient distance, as shown in (**Figure 8-6C**). Label all important points established thus far. Notice points a and d are **on** the reference line.

STEP 4 In the right-side view, measure the distance each point is **from** the reference line. Project these distances back to the inclined surface of the front view and up to 90° from the inclined surface to the reference line above. Transfer each distance, and lightly label each point as illustrated in (**Figure 8-6D**).

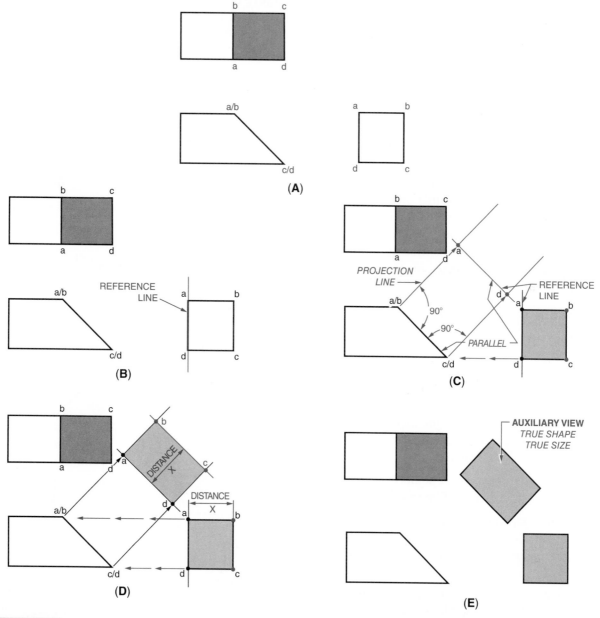

FIGURE 8-6 (A) Step 1; (B) Step 2; (C) Step 3; (D) Step 4; (E) Step 5 (finished drawing).

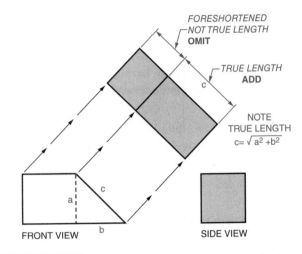

FIGURE 8-7   Draw only the inclined surface in the auxiliary view.

(**Figure 8-8C**). These distances are identical to the distances from a, b, c, and d to the fold line F/S and fold line F/T. Lightly connect a, b, c, and d bordering the inclined surface.

STEP 4   Recheck all work to ensure the following:

▸ That the projection lines are 90° from the inclined surface, which appears as an edge in the front view.

▸ That the fold line F/A is parallel to the inclined surface in the front view.

▸ That all distances have been transferred accurately.

Carefully darken in and complete the auxiliary view. The finished drawing will appear as it does in (**Figure 8-8D**). Figure 8-8D shows the good drafting practice of projecting only the auxiliary surface, the surface of interest.

STEP 5   Recheck all work. Be sure that:

▸ The projection is 90° from the inclined surface—in this example, the front view.

▸ The reference line is parallel to the inclined surface of the front view.

▸ All distances have been transferred accurately.

If correct, carefully darken in and complete all views. The final finished drawing will appear as it does in (**Figure 8-8E**). Notice that **only** the inclined surface is projected into the auxiliary view. Anything else would be foreshortened and, thus, not of true size or shape, and therefore of no use. Good drafting practice is to project **only** the surface of the inclined line (**Figure 8-7**).

How to Draw an Auxiliary View by Using a Fold Line

GIVEN: The pictorial view of an object, Figure 8-4, which is also shown in the usual three orthographic views—front, top, and side (Figure 8-5). Because the inclined surface is on the front view, we will use a front view auxiliary for this illustration.

STEP 1   Label all the important points of the auxiliary surface as shown in (**Figure 8-8A**).

STEP 2   Construct and label *fold lines*, which also represent the lines of intersection of adjacent orthographic projection planes (**Figure 8-8B**). (Refer to Figure 5-7.)

STEP 3   Draw light projection lines 90° from the inclined surface, and locate points a, b, c, and d in the auxiliary view from the fold line F/A

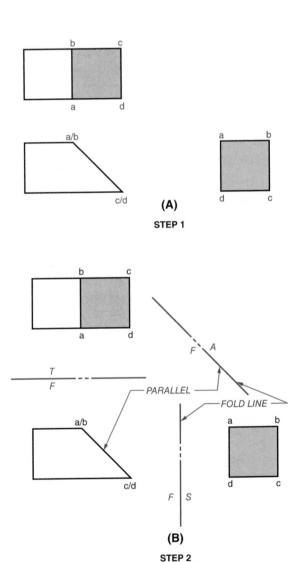

FIGURE 8-8   (A) Step 1; (B) Step 2;

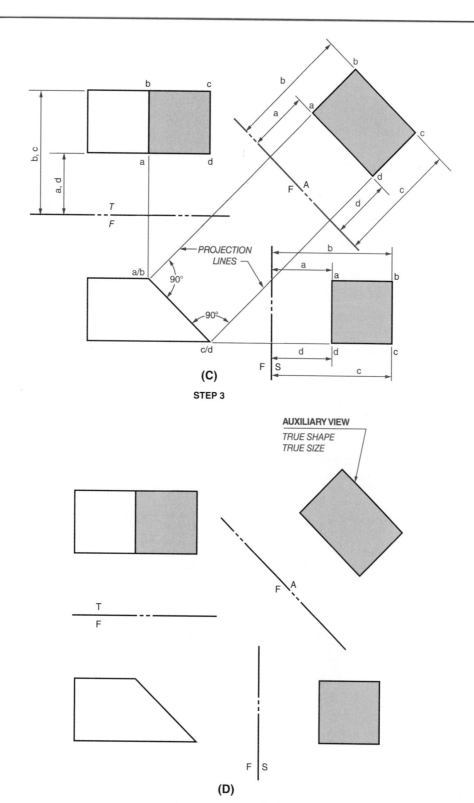

(C)

STEP 3

(D)

STEP 4 (FINISHED DRAWING)

FIGURE 8-8   (*Continued*) (C) Step 3; (D) Step 4 (finished drawing).

How to Project a Round Surface from an Inclined or Slanted Surface

GIVEN: The usual three views of an object—front view, right-side view, and unfinished top view—are shown in (**Figure 8-9**). (The usual 45° projection angle line will be omitted, as a slightly newer projection method will be used to complete the top view.) Refer also to the pictorial drawing of this object (**Figure 8-10**).

STEP 1 Divide the rounded view into equal spaces. In this example, using a 30° triangle, the right-side view is rounded and divided into 12 equal parts (**Figure 8-11A**). Letter each point clockwise, as shown in the figure.

STEP 2 Construct a vertical reference line in the right-side view so it passes through the center—in this example, through points a and g. (Always place the reference line through the center of any symmetrical object.) Construct a reference line in the top view that runs through the center, as illustrated in (**Figure 8-11B**).

STEP 3 Draw light projection lines from the 12 points in the right-side view to the inclined edge in the front view. Project these same 12 points

directly up to the top view from the inclined edge in the front view. Notice that points a and g are on the reference line (**Figure 8-11C**).

STEP 4 In the right-side view, measure the distance each point is from the reference line. Transfer each of these distances from the right-side view reference line to the top view reference line. Label each point lightly as each is found (**Figure 8-11D**).

STEP 5 Lightly connect all points and, if correct, darken in all views. This completes the top view (**Figure 8-11E**). Always darken in the compass and irregular curve layout work first. This completes the top view.

(A)

(B)

(C)

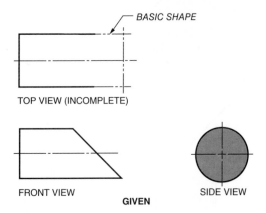

BASIC SHAPE

TOP VIEW (INCOMPLETE)

FRONT VIEW   SIDE VIEW

GIVEN

FIGURE 8-9   Projecting a round surface from an inclined surface.

FIGURE 8-10   Pictorial view of an object.

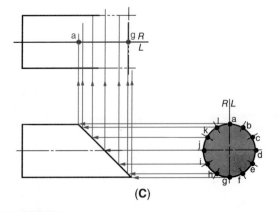

FIGURE 8-11   (A) Step 1; (B) Step 2; (C) Step 3;

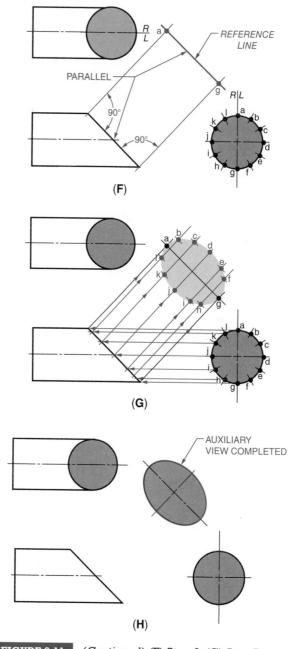

FIGURE 8-11   (*Continued*) (D) Step 4; (E) Step 5;

How to Draw an Auxiliary View of a Round Surface

STEP 6   Draw light projection lines 90° from the inclined surface of the front view. Construct the reference line parallel to the inclined surface at any convenient distance. Label the points that are on the reference line; in this example, a and g (**Figure 8-11F**).

STEP 7   Draw light projection lines from the 12 points in the right-side view to the inclined edge in the front view (Step 3). Project these same 12 points directly up to the auxiliary view from the inclined edge in the front view. Again, notice that points a and g are on the reference line (**Figure 8-11G**). In the right-side view, measure the distance each point is from the reference line. Transfer each of these distances from the right-side view reference line to the auxiliary view reference line. Label each point lightly as each is found.

STEP 8   Lightly connect all points and, if correct, darken in the auxiliary view. This completes the auxiliary view (**Figure 8-11H**). Notice that only

FIGURE 8-11   (*Continued*) (F) Step 6; (G) Step 7; (H) Step 8 (finished drawing).

the inclined surface has been projected into the auxiliary view.

How to Plot an Irregular Curved Surface

GIVEN: The usual three views: front view (incomplete), top view, and right-side view (**Figure 8-12**). This is an example of a top view auxiliary.

STEP 1   Add various points at random along the curved surface. Even spaces are not necessary, but try to choose points that pick up high and low

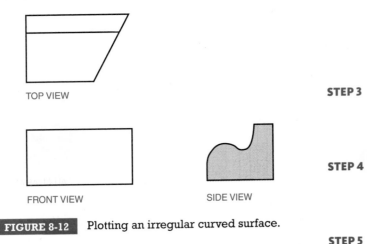

TOP VIEW

FRONT VIEW   SIDE VIEW

FIGURE 8-12   Plotting an irregular curved surface.

points along the line (**Figure 8-13A**). Always label the points in a clockwise direction.

STEP 2   From the right-side view, project these points up to the 45° projection line and over to the top view. Where these points intersect with the inclined surface, project directly down into the front view. Project these same points

from the right-side view to the front view. Where the points intersect is where the point actually is. Lightly locate and connect all lines and complete the front view (**Figure 8-13B**).

STEP 3   Establish a reference line in the right-side view. Construct this reference line so it passes through as many points as possible; in this example, points a and j (**Figure 8-13C**).

STEP 4   Add projection lines at 90° from the inclined surface. Add a reference line parallel to the inclined surface. Notice that points a and j again fall **on** the reference line (**Figure 8-13D**).

STEP 5   Project all points from the right-side view up and over to the inclined surface of the top view. Where the points intersect with the inclined surface, project 90° from the inclined surface. Transfer all distances from the reference line in the right-side view to the reference line in the auxiliary view. Lightly connect all points and, if correct, darken in all views (**Figure 8-13E**).

FIGURE 8-13   (A) Step 1; (B) Step 2; (C) Step 3;

FIGURE 8-14   Secondary auxiliary view.

FIGURE 8-13   (*Continued*) (D) Step 4; (E) Step 5 (finished drawing).

Secondary Auxiliary Views

Up to this point, primary auxiliary views have been dealt with. A *primary auxiliary view* can be projected from any of the regular views, as has been illustrated thus far.

Sometimes, a primary auxiliary view is not enough to fully illustrate an object; a secondary auxiliary view is needed. A *secondary auxiliary view* is projected directly from the auxiliary view. Many times, the auxiliary view and/or some of the other views cannot be fully completed until the secondary view has been completed first (**Figure 8-14**).

Partial Views

The use of a *partial auxiliary view* makes it possible to eliminate one or more of the regular views that, in turn, saves drafting time and cost. **Figure 8-15** is an example of a front view, top view, right-side view, and auxiliary view. Note that the auxiliary view is

FIGURE 8-15   Given: Regular three views and partial auxiliary view.

always drawn partial; only the inclined surface is drawn on the auxiliary view. By using a partial top and right-side view in conjunction with the particular auxiliary view, the drawing can be simplified without detracting from its clarity; in fact, in most cases, these partial views make the drawing easier to read (**Figure 8-16**).

FIGURE 8-16   Partial view and partial auxiliary views.

SAVING TIME AND CUTTING COSTS

"We've got to cut costs in this department," said Anna Valdez, CAD supervisor for Montgomery Engineering Company (MEC). MEC was undergoing a company-wide effort to cut costs, increase productivity, improve quality, and enhance competitiveness. The CAD department was feeling the pressure because MEC's latest contract consisted of more than 50 separate machine parts, all oddly shaped and complex. Valdez had called a departmental meeting to brainstorm ways to reduce drafting time on this contract. "We can save a lot of time by using partial auxiliary views where appropriate and by eliminating regular views when we do," suggested John Compton, a senior CAD technician. Compton laid a stack of drawings on the table and pointed out several places where full auxiliary views had been drawn when partials would have sufficed. "Not only could we be using partials," said Compton, pointing to the drawings for emphasis, "but the partial auxiliary views would eliminate the need for regular views in some cases."

"Good point, John," said Valdez. "Just this one suggestion would reduce drafting time by at least 30 percent. What other suggestions are there?"

"Look at these objects," said Tram Ly. "They are symmetrical. Why do we waste time drawing full auxiliary views when half views would be sufficient?"

"A good question," said Valdez. "Is anyone opposed to using half auxiliary views on symmetrical parts?" Hearing no comments, Valdez said, "Good. Let's do it." By the time the meeting finally ended, MEC's drafting department had identified several solid suggestions for reducing drafting time and cutting costs.

Auxiliary Section

An *auxiliary section,* as its name implies, is an auxiliary view in section. An auxiliary section is drawn exactly as is any removed sectional view, and it is projected in exactly the same way as any auxiliary view (**Figure 8-17**). All the usual auxiliary view rules apply, and generally only the surface cut by the cutting-plane line is drawn.

Half Auxiliary Views

If an auxiliary view is symmetrical, and space is limited, it is permitted to draw only half of the auxiliary view (**Figure 8-18**). Use of the half auxiliary view saves some time, but it should be used only as a last resort, as it could be confusing to those interpreting the drawing. Always draw the **near** half, as shown in the figure.

FIGURE 8-17   Auxiliary section views.

FIGURE 8-18   Half auxiliary view.

Oblique Surfaces, Viewing Planes, and Auxiliary View Enlargements

Oblique surfaces were described in Chapter 5 (Figure 5-27) as not being perpendicular to any one of the three principal projection planes. The Letter A Billboard in (**Figure 8-19A**) qualifies as an oblique surface. It is used in this discussion to illustrate three auxiliary views: finding an auxiliary view from an oblique surface, using a *viewing plane* to establish an auxiliary view, and enlarging an auxiliary view to show more information.

Finding the true size of an oblique surface requires the descriptive geometry procedure for finding a plane in true size. As the descriptive geometry technique (Figure 9-17) is thoroughly discussed in the next chapter, only a brief summary will be offered here. Any line in *true length* in the oblique surface may be used. Line b_1-b_2 in the Letter A Billboard shown in Figure 8-19A is used to illustrate the technique. Construct fold line T/F parallel to b_1-b_2 in the **front** view so b_1-b_2 will be in true length in the **top** view. Make fold line T/A_1 perpendicular to the direction

of line b_1-b_2 in the top view so b_1-b_2 will appear as a point in the A_1 view and the oblique surface will appear as an edge. Draw fold line A_1/A_2 parallel to the edge view of the oblique surface so that the oblique surface will appear in *true size* in the A_2 view. (Note that line b_1-b_2 will appear in true length.)

To illustrate the use of an auxiliary viewing plane, align viewing plane V-V in (**Figure 8-19B**) parallel to the edge view of the Letter A Billboard. View V-V will then appear in true size. Auxiliary view V-V could be placed at other locations on the drawing if more room is needed.

A two-to-one *enlargement of the auxiliary view* V-V is shown in Figure 8-19B. Construct enlargements when more detail is needed.

An additional type of enlargement drawing for better visualization is illustrated in Chapter 3, Figure 3-3. The example in Figure 3-3 is typical. Refer to the isolated drawing labeled **Detail B**. Also notice the full-section drawing of the compression housing. The partial drawing in the detail is an enlargement of the portion within the small circle located where the flange joins the tubular portion. The letter B appears in the circumference of the small circle.

FIGURE 8-19 (A) Oblique surface auxiliary view: fold line method; (B) Viewing plane and enlarged auxiliary view.

PROBLEMS

SUMMARY

- An auxiliary view is a view with its line of sight perpendicular to an inclined surface of an object. An auxiliary view serves three purposes: (1) it illustrates the true size of a surface, (2) it illustrates the true shape of a surface, and (3) it can project and complete other views.

- A secondary auxiliary view is one that is projected from a primary auxiliary view to more fully illustrate an object.

- A partial auxiliary view is used to save drafting time and costs by eliminating one or more regular views.

- An auxiliary section is an auxiliary view in sections. It is drawn in the same way as a removed section.

- A half auxiliary view can be used when an auxiliary view is symmetrical. When this is used, the near half is drawn.

REVIEW QUESTIONS

Mark the following statements as either true or false.

1. A partial auxiliary view makes it possible to eliminate one or more of the regular views.

2. Hidden lines in an auxiliary view should be omitted unless they are needed for clarity.

3. A primary auxiliary view is always projected from the front view.

4. A reference line should be drawn horizontally and pass through as many points as possible.

5. Use of the half auxiliary view is very common because it saves time.

6. When laying out a drawing with an auxiliary view, always choose the view with the most detail as the front view.

Answer the following questions by selecting the best answer.

1. Which of the following is **not** a purpose that an auxiliary view serves?
   a. To avoid having to create a removed sectional view
   b. To illustrate the true size of a surface
   c. To illustrate the true shape of a surface
   d. To project and complete other views

2. Which of the following is **not** a major kind of auxiliary view?
   a. Front view
   b. Top view
   c. Side view
   d. Bottom view

3. When we draw an auxiliary view, projection lines must be drawn at what angle from the inclined surface?
   a. 60°
   b. 45°
   c. 90°
   d. None of the above

4. Which of the following is **not** true regarding a secondary auxiliary view?
   a. It is projected directly from the auxiliary view.
   b. Many times, the secondary auxiliary view is completed first.
   c. A secondary auxiliary view is always created.
   d. None of the above.

5. Which of the following is **not** true regarding an auxiliary sectional view?
   a. It is an auxiliary view in sections.
   b. It is drawn exactly as is any removed sectional view.
   c. Generally, only the surface cut by the cutting-plane line is drawn.
   d. Not all of the auxiliary view rules apply.

CHAPTER 8 PROBLEMS

The following problems are intended to give the beginning drafter practice in sketching and laying out multiviews with an auxiliary view.

Conventional Instructions

The steps to follow in laying out any drawing with an auxiliary view are:

STEP 1 Study the problem carefully.

STEP 2 Choose the view with the most detail as the front view.

STEP 3 Position the front view so there will be the least number of hidden lines in the other views.

STEP 4 Determine from which view to project the auxiliary view.

STEP 5 Make a sketch of all views, including the auxiliary view.

STEP 6 Center the required views within the work area with approximately 1″ (25 mm) of space between the views. Adjust the regular views to accommodate the auxiliary view.

STEP 7 Use light projection lines. Do **not** erase them.

STEP 8 Lightly complete all views.

STEP 9 Check to see that all views are centered within the work area.

STEP 10 Carefully check all dimensions in all views.

STEP 11 Darken in all views by using the correct line thickness.

STEP 12 Recheck all work and, if it is correct, neatly fill out the title block by using light guidelines and neat lettering.

CAD INSTRUCTIONS

Students who plan to complete the following problems on a CAD system should begin with these general instructions.

1. Analyze the problem carefully.

2. Plan your drawing sequence. In what order will you develop the drawing?

3. Read the specific instructions for the problem and begin.

Problems 8-1 through 8-4

Draw the front view, top view, right-side view, and auxiliary view. Complete all views using the listed steps.

PROBLEM 8-1

PROBLEM 8-2

PROBLEM 8-3

PROBLEM 8-4

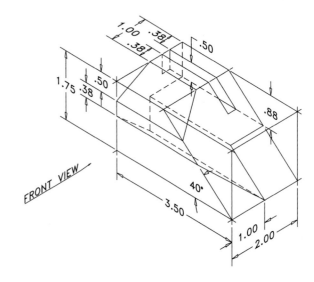

Problems 8-5 and 8-6

Draw the front view, top view, right-side view, and two auxiliary views to illustrate the true size and shape of the slanted surfaces. Complete all views by using the listed steps.

PROBLEM 8-5

PROBLEM 8-6

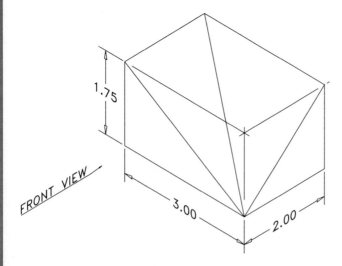

Problems 8-7 through 8-13

Draw the front view, top view, right-side view, and auxiliary view. Complete all views by using the listed steps.

PROBLEM 8-7

PROBLEM 8-8

PROBLEM 8-9

PROBLEM 8-10

PROBLEM 8-11

PROBLEM 8-12

PROBLEM 8-13

PROBLEM 8-14

Draw the front view, top view, right-side view, and *two* auxiliary views to illustrate the true size and shape of all surfaces. Complete all views by using the listed steps.

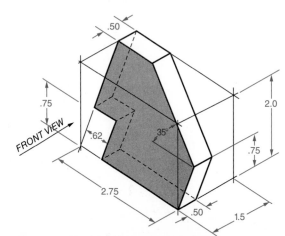

Problems 8-15 through 8-22

Draw the required views to fully illustrate each object.
Complete all views by using the listed steps.

PROBLEM 8-15

PROBLEM 8-16

PROBLEM 8-17

2.8146

R.31 (TYP.)

Ø.75 THRU
Ø1.25

2.0

1.0

60°

38

R.38

4.0

2.0   R.25 (TYP.)

FRONT VIEW

1.25   2.5

2X Ø.38 THRU

ALL UNMARKED RADII = R.06

PROBLEM 8-18

2X Ø.25, THRU
⌴ Ø.50 ⊽ .125

.25

.75

1.5

R.43 (TYP.)

60°

.38

1.0   1.5

1.0

R.38

ALL UNMARKED RADII = R.06

38

FRONT VIEW

R.88

PROBLEM 8-19

SECTION THRU WEB

ALL UNMARKED RADII ARE R.06

PROBLEM 8-20

PROBLEM 8-21

PROBLEM 8-22

ALL UNMARKED RADII = R.09

Problems 8-23 through 8-29

Make a finished drawing of selected problems as assigned by the instructor. Draw the given front and right-side views, add the top view with hidden lines if required, and add an auxiliary view. If assigned, design your own right-side view consistent with the given front view, and add a complete auxiliary view. Do not add dimensions.

PROBLEM 8-23

PROBLEM 8-24

PROBLEMS

PROBLEM 8-25

PROBLEM 8-26

PROBLEM 8-27

PROBLEM 8-28

PROBLEM 8-29

Problems 8-30 through 8-37

Draw the given views and add the required auxiliary view. Present the views within the work area so all views have a neat, centered appearance. Do not add dimensions.

PROBLEM 8-30

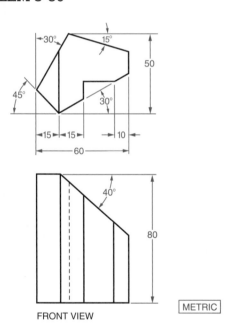

FRONT VIEW

PROBLEM 8-31

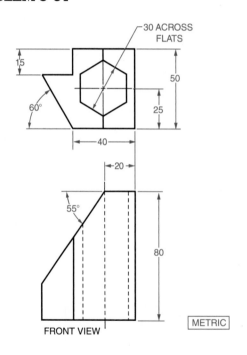

FRONT VIEW

PROBLEM 8-32

FRONT VIEW

PROBLEM 8-35

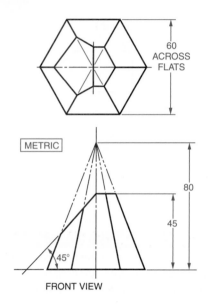

FRONT VIEW

PROBLEM 8-33

FRONT VIEW

PROBLEM 8-36

FRONT VIEW

PROBLEM 8-34

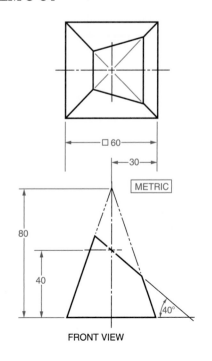

FRONT VIEW

PROBLEM 8-37

FRONT VIEW

PROBLEM 8-38

The partially completed building, in outline form, is to have the name of your city painted on one side of the roof, in letters to be read by individuals in aircraft as high as 5,000′ above the roof sign. A rule of thumb for good readability is that the letters on a sign, in a classroom, in an auditorium, or on a roof should be as high as approximately 1/250 of the distance the viewer is from the sign. Use an auxiliary view of the roof to lay out your city's name. Then, specify the length of the building to suit your sign. You may choose the font you wish to use.

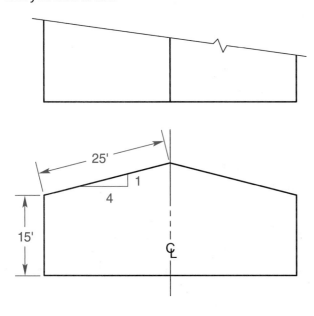

PROBLEM 8-39

The 750 mm outside diameter flange for the special 45° elbow shown is to have M24 × 3.0 in through holes on a 650 mm bolt circle. The recommended distance between fasteners (for strength and to provide space for wrenches) is 2.5–3.0 times the diameter of the fastener. Decide on the best, even number of bolts for the 650 mm bolt circle on this flange. Show your answer in an auxiliary view of the flange with the bolt holes.

PROBLEM 8-40

Panels A, B, and C, in the control station shown, satisfy anthropometric requirements. They are 30″ from the operator's head at location O, and the edge of the desk is 10″ from the operator. Panel B needs to be replaced with a new panel that has the five new instruments shown, located in a reasonable arrangement in the panel. Prepare an auxiliary view of panel B, and locate the five instruments, using your own judgment for the arrangement. Note approximate dimensions to their locations. Suggestion: Construct the auxiliary view of B, which will show the panel in true size. Cut out five rectangles of paper to represent the five instruments, and make trial arrangements of them on your auxiliary view. Select the best arrangement, in your opinion. Mark their locations and give approximate dimensions to their locations.

PROBLEM 8-41

The hopper shown needs a vibrator attached to the left side to shake down material to the conveyor below. Without the vibrator, some materials tend to hang up. Prepare an auxiliary view of the left side of the hopper to determine the approximate dimensions to locate a mounting plate that already has threaded holes, shown. (The plate is to be welded to the hopper.) The center of the plate should be approximately 1.9 m above the floor level so maintenance personnel can reach the vibrator. Also, locate the plate to either side of center so the maintenance person won't have to reach over the conveyor.

METRIC

9

Descriptive Geometry

CHAPTER OUTLINE

Descriptive geometry projection • Steps used in descriptive geometry projection • Notations • Fold lines • Bearings, slope, and grade • Applied descriptive geometry • Topographic topics • Vectors • Noncoplanar-concurrent force systems • Summary • Review questions • Chapter 9 problems • CAD instructions

CHAPTER OBJECTIVES

Upon completion of this chapter, students should be able to do the following:

▶ Explain the reasons for using the descriptive geometry method of projection.

▶ List the steps used to solve descriptive geometry problems.

▶ Demonstrate proficiency in the use of system of notations.

▶ Define the term *fold line*.

▶ Demonstrate proficiency at using fold lines.

▶ Demonstrate proficiency in determining the position of a line in space by using its bearing, slope, and grade.

▶ Demonstrate proficiency in plotting the boundaries of real estate.

▶ Demonstrate proficiency in applying the concepts of cut and fill.

▶ Demonstrate proficiency in applying the concepts of strike and dip.

▶ Demonstrate proficiency in the use of vectors.

Descriptive Geometry Projection

Primary auxiliary views are views that have been projected perpendicular from an inclined surface and viewed looking directly at that surface. As studied in Chapter 8, the auxiliary view shows the true size and true shape of the inclined surface and is sometimes used to complete other views. If a surface is oblique (that is, slanting in more than one direction and not 90° from one of the other views), the true shape and true size of the successive surface are based on auxiliary views.

Using the descriptive geometry method of projection not only gives true size and shape, but it also can be used to find intersections, true distances of lines in space, true angles between surfaces, and exact piercing points. Descriptive geometry graphically shows the solution to problems dealing with points, lines, and planes, and their relationship in space. In order to be able to apply descriptive geometry to various drafting problems, the drafter must know and understand the various basic steps involved. This chapter explains these basic steps. The basic theories covered in this chapter are:

- Projecting a line into other views
- Projecting points into other views
- Determining the true length of a line
- Determining the point view of a line
- Finding the true distance between a line and a point in space
- Determining the true distance between two lines in space
- Projecting a plane surface in space
- Developing an edge view of a plane surface in space
- Determining the true distance between a plane surface and a point in space
- Determining the true angle between plane surfaces in space

Steps Used in Descriptive Geometry Projection

All problems, regardless of their complexity, use the same procedures or steps. These steps ultimately must be done in the same order each time, although some intermediate steps may be selected or are optional. Like climbing a flight of stairs, each step must be executed, and only experience can give one the ability to perform multiple steps simultaneously. An example

FIGURE 9-1   Steps used to solve descriptive geometric problems.

of the sequential steps needed in descriptive geometry problems is as follows: In order to find an edge view of a plane (flat surface), we must locate a true-length line in that plane (Step 1); a point view of the true-length line must be projected, which yields an edge view of the plane (Step 2); see **Figure 9-1**. By projective means, it would be impossible to find the edge view of an oblique surface from normal views without performing Steps 1 and 2.

Notations

As we progress through sequential steps, it is important to keep track of all views and points in space. (Recall that a straight line is composed of two spatially located end-points.) To do this, we should use a system of *notations* or labeling. Each view and each point should be labeled to provide an accurate identity of each at all times. For purposes of this text, the following notations are used in this chapter.

Each point in space is called out in *lowercase letters*.

EXAMPLE:

a, b, c, d
Each *line* in space is identified by the two *lowercase end-points*.

EXAMPLE:

Line a-b
Each line in space is assumed to end at the indicated end-points, unless it is specified as an extended line (continues without end).

EXAMPLE:

Line p-q and extended line j-k
Each plane in space, such as a triangle, is called out in lowercase letters.

EXAMPLE:

Triangle abc
Each plane in space is assumed to be limited (stops at the indicated boundaries) unless otherwise defined as an *unlimited plane* (limitless extension beyond the indicated boundaries).

EXAMPLE:

Plane defg and unlimited plane mno
Each *viewing plane* is called out in *uppercase letters*.

EXAMPLE:

F = Frontal viewing plane (a principal plane).
T = Top (horizontal) viewing plane (a principal plane).
R = Right-side (profile) viewing plane (principal plane).
L = Left-side (profile) viewing plane (principal plane).
A(digit) = an auxiliary viewing plane, with the number of successive projections from a principal viewing plane.
B(digit) = same as A(digit), but using a different series of successive projections.

A combination of lowercase and uppercase letters is used to fully describe each point in space and the view in which it is located.

EXAMPLE:

aF would be a location as seen in the frontal viewing plane.
aT-bT would be line a-b as seen in the top (horizontal) viewing plane.
aR-bR-cR would be plane a-b-c as seen in the right-side (profile) viewing plane.

Fold Lines

A fold line is represented by a thin black line, similar to a phantom line. It indicates a 90° intersection of two viewing planes. Each fold line must be labeled using uppercase letters (**Figure 9-2**). In this example, the fold line is placed between the front view and the top view. This placement eliminates the usual 45°

FIGURE 9-2   Fold line.

FIGURE 9-3   Regular multiview drawing using 45° projection line.

projection line (**Figure 9-3**). It is important to add these notations in order to keep track of exactly which viewing plane is being executed.

How to Project a Line into Other Views

Fold lines are placed between successive orthographic views and labeled with uppercase letters (**Figure 9-4**). Any appropriate spacing is permissible, but the distance from the image to the frontal viewing plane fold line must be equal at both the top and right-side views. The spatial positioning of a line a-b (surrounded by an imaginary transparent box) is shown in **Figure 9-5**. Point a is closer to the frontal viewing plane than is point b, but it is farther below the top (horizontal) viewing plane. Represented on a normal layout drawing, it would be illustrated and labeled as shown in **Figure 9-6**.

The customary 45° projection line is not used in descriptive geometry projection. All points are projected along light projection lines drawn at 90° to the fold lines. All measurements are taken along these light projection lines and measured from the fold line to the point in space (**Figure 9-7**).

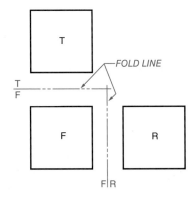

FIGURE 9-4   Regular multiview drawing using fold lines instead of 45° projection line.

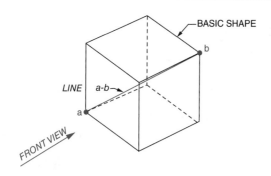

FIGURE 9-5   Spatial positioning of line a-b.

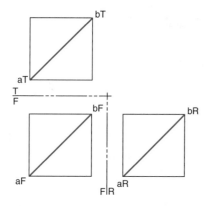

FIGURE 9-6   Line a-b in all three regular views.

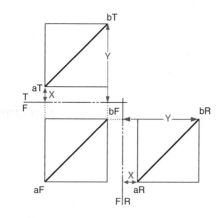

FIGURE 9-7   Skip a view when transferring distances.

An important rule to remember is to always skip a view between all measurements. In Figure 9-7, notice in the right-side view that dimension X (distance from F/R fold line to point aR) is projected into and through the front view and up to the top view and transferred into the top view (distance from F/T fold line to point aT). The front view was skipped.

Again, referring to the right-side view of Figure 9-7, dimension Y (distance from F/R fold line to point bR) is transferred into the top view (distance from F/T fold line to point bT), and the front view was skipped. In

each of the instances, the distance is measured as the perpendicular distance from the fold line to the point.

EXAMPLE:

To project a top view of a line from a given front and right-side view.

GIVEN: Line a-b in the front view and line a-b in the right-side view (**Figure 9-8**). Extend projection lines into the top view from the end-points in the front view, aF and bF. Find the distances X and Y from the line end-points in the right-side view aR and bR to the frontal viewing plane at fold line F/R, and transfer them into the top view. Label all points and fold lines in all views. Be sure that all projections are made at 90° from the relevant fold lines (**Figure 9-9**).

How to Locate a Point in Space (Right View)

A point in space is projected and measured in exactly the same way as a line in space, except that the point is a line with a single end point (**Figure 9-10A**).

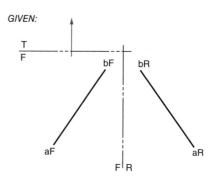

FIGURE 9-8   Locating line a-b in top view.

FIGURE 9-9   Line a-b in the top view using the fold line.

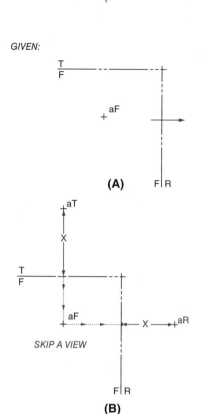

GIVEN:

(A)

(B)

FIGURE 9-10 (A) Locating a point in space (right view); (B) Step 1.

at any convenient distance, such as approximately 1/2″.

STEP 2 Label the fold line A for the auxiliary view.

STEP 3 Extend projection lines from the end-points of the line being projected into the auxiliary view. These must always be at 90° to the fold line.

STEP 4 Transfer the end point distances from the fold line in the second preceding view from the one being drawn to locate the corresponding end-points in the view being drawn.

EXAMPLE:
Refer to **Figure 9-11A**.

GIVEN: Line a-b in the front view, side view, and top view. The problem is to find the true length of line a-b. A true length can be projected from any of the three principal views by placing a fold line parallel to the line in any view. The right-side view is selected in this example.

STEP 1 Draw a fold line parallel to line a-b, and label it as shown in **Figure 9-11B**. Extend light projection lines at 90° to the fold line from points a and b into the auxiliary view.

STEP 2 Determine distances X and Y from the front view to the near-fold line and transfer them into the auxiliary view, as shown in **Figure 9-11C**. Label all points and fold lines. The result will be the true length of line a-b.

Use these steps to find the true length of any line.

How to Construct a Point View of a Line

To construct a point view of a line:

STEP 1 Find the true length of the line.

STEP 2 Draw a fold line perpendicular to the true-length line at any convenient distance from either end of the true-length line.

STEP 3 Label the fold line A-B (B indicates a secondary auxiliary view).

STEP 4 Extend a light projection line from the true-length line into the secondary auxiliary view.

STEP 5 Transfer the distance of the line end-points into the secondary auxiliary view (B) from the corresponding points in the second preceding view.

EXAMPLE:

To project the right-side view of a point from a given top and front view.

GIVEN: Point a in the front view and top view (Figure 9-10A). Project point aF from the front view into the right-side view. Point aR must lie on this projection line. Find X, which is the distance from fold line F/T to aT in the top view, and project it into and through the front view (**Figure 9-10B**). Project it over into the right-side view. Transfer distance X to find point a. Be sure to always label all points and fold lines in all views.

How to Find the True Length of a Line

Any line that is parallel to a fold line will appear in its true length in the next successive view adjoining that fold line.

To find the true length of any line:

STEP 1 Draw a fold line parallel to the line of which the true length is required. This can be done

EXAMPLE:
Refer to **Figure 9-12A**.

GIVEN: Line a-b in the front view, side view, and auxiliary view. The true length is located in

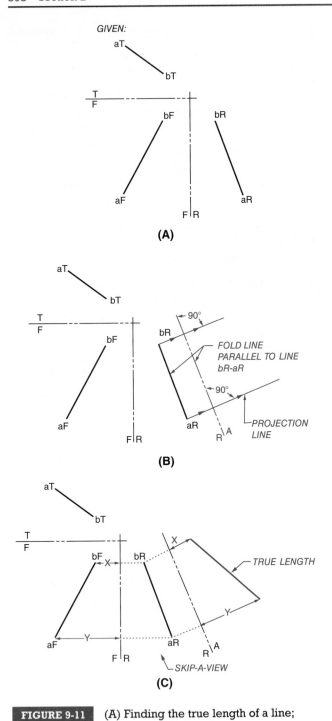

(A)

(B)

(C)

FIGURE 9-11   (A) Finding the true length of a line;
(B) Step 1; (C) Step 2.

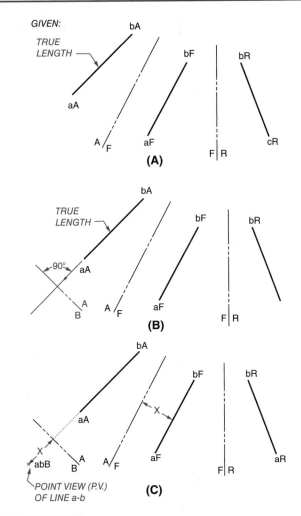

(A)

(B)

(C)

FIGURE 9-12   (A) Constructing a point view of a line;
(B) Step 1; (C) Step 2.

of the line's path. Extend a projection line aligned with the true-length line into this new secondary auxiliary view. The projection lines from both points a and b appear to align in this common projection line. Therefore, both points a and b must be located on this projection line in the secondary auxiliary view. In **Figure 9-12B**, the fold line A/B was added to the left of the true-length line a-b.

STEP 2   The front view is the second preceding view to the secondary auxiliary view being constructed. Therefore, distance X in the front view from the fold line A/F to line a/b is transferred to the secondary auxiliary view from the fold line A/B (**Figure 9-12C**). As points a and b in the front view are both at the same distance from the fold line A/F, and they both lie on the same projection line in the secondary auxiliary view, they will appear to coincide at the same

the auxiliary viewing plane (A), which is projected from the front view. (The true length could have been projected from any of the given views.)

STEP 1   At any convenient distance from either end, draw a fold line A/B perpendicular to the true length of line a-b. This will establish a viewing plane that is perpendicular to the direction location. This provides evidence that the end view of the line has been achieved.

How to Find the True Distance between a Line and a Point in Space

The true distance between a line and a point in space will be evident in a view that shows the end point of the line and that point, simultaneously.

STEP 1   Project an auxiliary view (viewing plane A) to find the true length of the line, and project the point into the same view.

STEP 2   Project a secondary auxiliary view (B) to find the point view of the line, and project the point into the same view.

STEP 3   The observable distance between the end view of the line and the point will be the actual distance between them. The location in the line that is nearest to the point is on the path that is perpendicular to the line from the point, but this is not discernible in the secondary auxiliary view.

EXAMPLE:
Refer to **Figure 9-13A.**

GIVEN: Line a-b and point c in the front view and right-side view.

STEP 1   Project the true length of line a-b by placing a fold line parallel to a given view of the line, and project point c into the auxiliary view. Recall that point locations are transferred from the second preceding view. Label all fold lines and all points (**Figure 9-13B**).

STEP 2   Draw a fold line perpendicular to the true length of line a-b and label the fold line. Project line a-b into the secondary auxiliary view (B) to find the point view of line a-b. Project point c along into the second auxiliary view (B). The actual distance between the point view of line a-b/B and point cB is evident (**Figure 9-13C**).

The location in line a-b that is nearest to point c lies on a path that is perpendicular to line ab and passes through c. Any path that is perpendicular to a line will appear perpendicular in a view where the line is true length. Therefore, the path can be drawn in the preceding view to locate point p on the line. Point p can be projected to its correct location on the original views of line a-b.

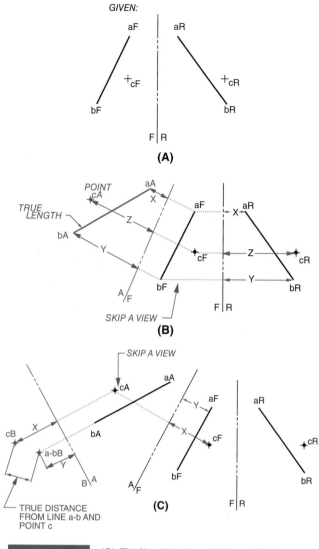

FIGURE 9-13   (A) Finding the true distance between a line and a point in space; (B) Step 1; (C) Step 2.

How to Find the True Distance between Two Parallel Lines

If two lines are actually parallel, they will appear parallel in all views. A view is needed to show the end view of both lines simultaneously, where the real distance between them will be apparent.

STEP 1   Find the true length of each of the two lines. Projecting a view that will provide the true length of a line will automatically provide the true length of any other parallel line that is projected into the same view.

STEP 2   Find the point view of each of the two lines. Projecting a view that will provide the end view of a line will automatically provide the end view of any other parallel line that is projected into the same view.

STEP 3 The true distance between the parallel lines is the straight-line path between their endpoints. There is no single location within the length of the lines where this occurs, as each location in a line has a corresponding closest point location on the other line, each on the path connecting them and perpendicular to their respective lines.

EXAMPLE:

Refer to **Figure 9-14A**.

GIVEN: The parallel lines a-b and c-d in a front view, a right-side view, and a top view.

STEP 1 Find the true lengths of line a-b and line c-d in auxiliary view A. The two parallel lines will also be parallel in the auxiliary, if they are actually parallel (**Figure 9-14B**).

STEP 2 Draw a fold line perpendicular to the true-length lines, a-b and c-d. Project the lines into the secondary auxiliary view (B) to find the point view of the two lines a-b and c-d. Measure the true distance between the point views of lines a-b/B and c-d/B (**Figure 9-14C**).

How to Find the True Distance between Two Nonparallel (or Skewed) Lines

If two lines are not parallel, they will appear non-parallel in at least one view. It is wise to first check two nonparallel (or skewed) lines to determine if they actually intersect, which would make the distance between them zero. This can be verified by projecting the apparent point of intersection of a view to the adjoining view. If the apparent points of intersection align, then the lines actually intersect.

To determine the true distance between two non-parallel lines:

STEP 1 Project the true length of either one of the two lines, and project the other line into that auxiliary view (A).

STEP 2 Find the point view of the true-length line, and project the other line into that secondary auxiliary view (B).

STEP 3 Measure the true distance between the point view line at a location that is perpendicular to the other line.

(A)

(B)

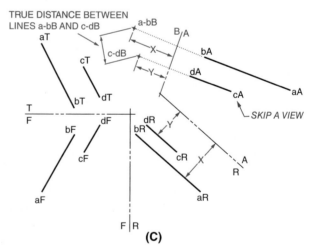

(C)

FIGURE 9-14   (A) Finding the true distance between two parallel lines; (B) Step 1; (C) Step 2.

EXAMPLE:

Refer to **Figure 9-15A**.

GIVEN: Nonparallel lines a-b and c-d in a front view and right-side view.

GIVEN:

(A)

(B)

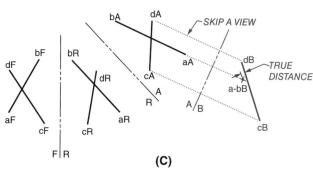

(C)

FIGURE 9-15   (A) Finding the true distance between two nonparallel (or skewed) lines; (B) Step 1; (C) Step 2.

STEP 1   Project the true length of line a-b in an auxiliary view (A), and project line c-d along into that auxiliary view (**Figure 9-15B**). Notice that c-d is not the true length.

STEP 2   Project the point view of line a-b into the secondary view (B), and project line c-d into this view. Measure the true distance between the point view of line a-b/B, perpendicular from line c-d/B (**Figure 9-15C**).

How to Project a Plane into Another View

A limited plane is located by points in space joined by straight lines. In order to transfer a plane, find at least three points in the plane, and project each point into the next view.

EXAMPLE:
Refer to **Figure 9-16A**.

GIVEN: Triangular plane abc shown in the top view and front view.

STEP 1   In the top view, find the distance from the fold line to points aT, bT, and cT (**Figure 9-16B**).

STEP 2   Project these points from the front view into the right-side view, and transfer the corresponding distances from the top view points to the top view fold line into the right-side view (**Figure 9-16C**). Construct straight lines between points a, b, and c. This transfers the plane surface.

How to Construct an Edge View of a Plane Surface

If the end view of a line that is in a plane is shown, then the edge view of that plane is shown also. Recall

(A)

(B)

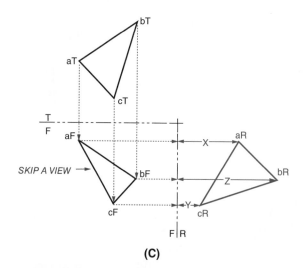

(C)

FIGURE 9-16   (A) Projecting a plane into another view; (B) Step 1; (C) Step 2.

that to find the end view of a line, a true length must be found first. When the boundaries of a plane are provided, projecting a view to find the true length of a selected boundary is a simple procedure. It is necessary only to locate a fold line parallel to the boundary to ensure its true length in the next projected view.

A shorter method of securing a true-length line in a plane is also used, if a true-length line is not present already. A line may be added on the plane in any view, arranged to be parallel to an adjoining fold line and to have its terminations at the plane boundaries. Projecting the added-line terminations to the corresponding boundaries in the adjoining view relocates the added line in that view. Moreover, it will be a true-length line, as it was made to be parallel to the fold line in the preceding view.

Whichever method is used to find a true-length line in the plane, all point locations within the plane should be projected to the view where the true-length line exists, if not there already. A fold line perpendicular to the true-length line will provide a direction for projecting the edge view of the plane. Projecting all the points in the plane will provide evidence of this, as they will align into a single, straight path.

EXAMPLE:
Refer to **Figure 9-17A.**

GIVEN: Plane surface abc in the top view and front view.

STEP 1 Construct a line parallel to fold line F/R and through one or more points, if possible. In this example, the line passes through point c. Label this newly constructed line cF/xF (**Figure 9-17B**).

STEP 2 Construct the true length of line cF/xF in the next view (**Figure 9-17C**).

STEP 3 Construct the point view of the true-length line cR/xR. Project the remaining points of the plane surface into this view. Also, label all points and fold lines. If done correctly, the result should be a straight line that passes through the point view. This straight line is the edge view of the plane (**Figure 9-17D**).

STEP 4 Construct a fold line A/B parallel to plane abc (shown as an edge in view A), and construct the plane surface abc in view B.

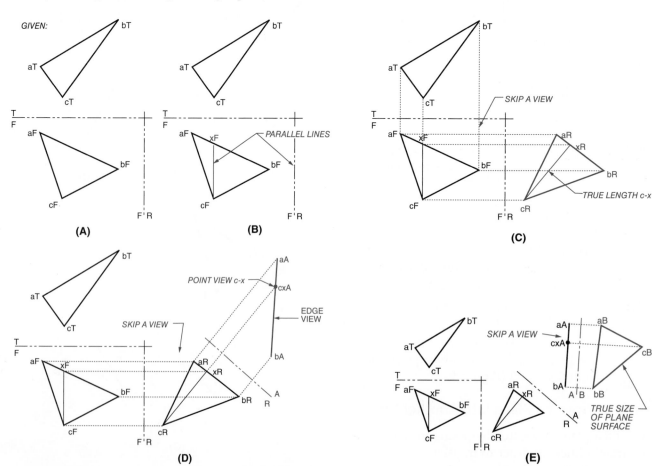

FIGURE 9-17   (A) Constructing an edge view of a plane surface; (B) Step 1; (C) Step 2; (D) Step 3; (E) Step 4.

Plane surface abc is now seen in true size, in view B (**Figure 9-17E**), because plane abc was seen as an edge in the adjacent view A.

How to Find the True Distance between a Plane Surface and a Point in Space

Finding the true distance between a given point and a plane in space requires a view that shows the edge view of the plane and the point in the same view. The distance from the point to the plane is the path that is perpendicular to the plane and passing through the point.

STEP 1   Add a line on the plane parallel to a fold line, if a true-length line is not already available in the plane.

STEP 2   Project the added line to an adjoining view, where it will be seen in true length.

STEP 3   Project a point view of the added true-length line.

STEP 4   Project the points of the plane into this view. The result should be a straight line that passes through the added line end view, which is the edge view. Project the point in space into this view. Measure the perpendicular distance from the plane's edge view to the point in space. This is the true distance between the plane surface and the point in space.

EXAMPLE:

Refer to **Figure 9-18A.**

GIVEN: Triangular plane abc and point X in the front view and right-side view.

STEP 1   Construct a line parallel to the fold line F/R and through one or more points. Point c is a convenient line termination. Label this newly constructed line cR/zR (**Figure 9-18B**).

STEP 2   Construct the true length of line cR/zR in the front view, and label the line cF/zF (**Figure 9-18C**). Bring point X into this view also. Label all points and fold lines.

STEP 3   Project the end view of the true-length line cF/zF in the auxiliary view (**Figure 9-18D**). Project the points of the plane surface into this view, and also point X, projecting 90° from the fold line. Label all points and fold lines. Measure the perpendicular distance from the edge view to the point in space. This is the true distance between the plane surface and the point X in space.

How to Find the True Angle between Two Planes (Dihedral Angle)

The edge of intersection between two planes is a line that is common to both planes. The end view of that line will provide the edge view of both planes simultaneously, and the angle between them will be evident.

To find the true angle between two surfaces:

STEP 1   Construct the true length of the intersection between the two surfaces, and project all other points of both planes into that auxiliary view.

STEP 2   Project the end view of the true-length edge of the intersection line, and project the points of both planes into this secondary auxiliary view. Label all points and fold lines. Measure the true angle between the two surfaces.

EXAMPLE:

Refer to **Figure 9-19A.**

GIVEN: Plane abc and plane abd that intersect one edge, a-b, in the front view, side view, and top view.

STEP 1   Project the true length of the edge of intersection a-b, and project all other points into this auxiliary view (A) (**Figure 9-19B**).

STEP 2   Project the point view of the true length of the edge of intersection a-b, and project all other points into this secondary auxiliary view (B) (**Figure 9-19C**). Measure the true angle between the two surfaces.

How to Determine the Visibility of Lines

To determine which line of an apparent intersection of two lines is closer to the viewer, the following steps are used.

STEP 1   From the exact crossover point of the lines, project to an adjoining view.

STEP 2   In the adjoining view, determine which of the lines is closest to the fold line between the views, on the projection line from the first view (or the first line that the projection line encounters on its path from the first view).

STEP 3   Whichever line is closest to the fold line at that point only is the line that is in front of the other line in the first view.

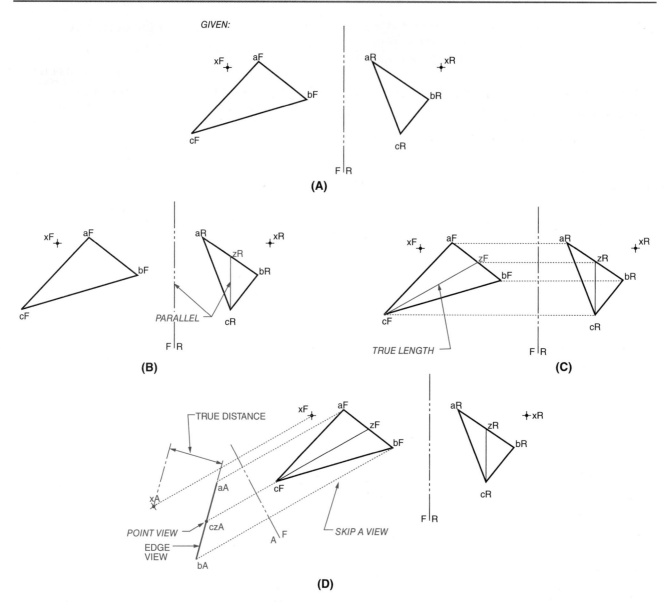

FIGURE 9-18   (A) Finding the true distance between a plane surface and a point in space; (B) Step 1; (C) Step 2; (D) Step 3.

EXAMPLE:

Refer to **Figure 9-20A**.

GIVEN: Lines a-b and c-d in the front view and top view.

STEP 1 At the exact crossover of lines a-b and c-d in the top view, project down through the fold line to the corresponding lines in the front view. At this exact point along the lines, line c-d is closer to the fold line and line a-b is farther away from the fold line; thus, in the top view, line c-d is in front of line a-b (**Figure 9-20B**).

STEP 2 See **Figure 9-20C**. At the exact crossover of lines a-b and c-d in the front view, project

up through the fold line to the corresponding lines in the top view. At this exact point along the lines, line a-b is closer to the fold line and line c-d is farther away from the fold line; thus, in the front view, line a-b is in front of line c-d.

STEP 3 The result is drawn to illustrate this crossover (**Figure 9-20D**).

How to Determine the Piercing Point by Inspection

The *piercing point* is the exact location of the intersection of a surface and a line. The exact piercing point is determined from the view where the surface

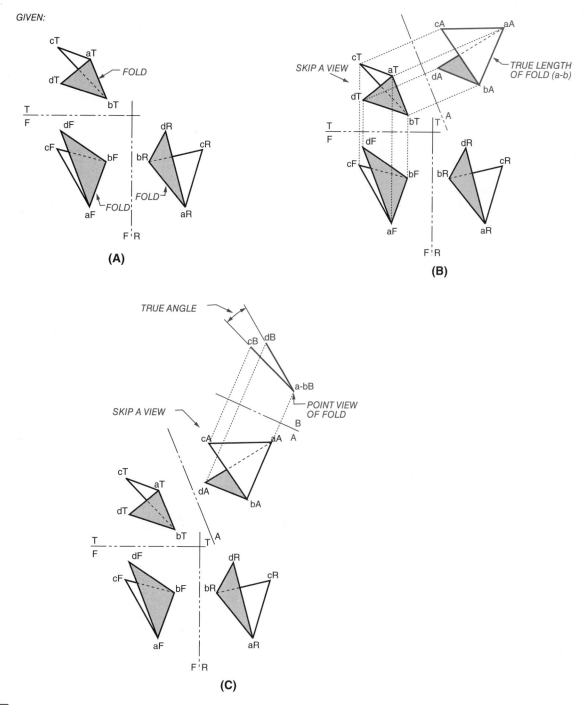

GIVEN:

(A)

(B)

(C)

FIGURE 9-19 (A) Finding the angle between two surfaces; (B) Step 1; (C) Step 2.

appears as an edge view. In **Figure 9-21A**, the top view illustrates the edge view of the surface. The piercing point is established in the top view and projected down into the front view (**Figure 9-21B**). In **Figure 9-22A**, the front view illustrates the edge view of the surface. The piercing point is established in the front view and projected up into the top view (**Figure 9-22B**).

How to Determine the Piercing Point by Construction

To determine the exact piercing point by construction:

STEP 1   Construct a true length of a line on the plane surface.

STEP 2   Project a point view of the true-length line and an edge view of the plane surface. The edge

FIGURE 9-20   (A) Determining the visibility of lines; (B) Step 1; (C) Step 2; (D) Step 3.

view of the plane lies in the path of the line in this view, indicating the piercing point location.

STEP 3   Transfer the piercing point back into the other views.

STEP 4   Determine the visibility of lines to find which part of the line is visible from the viewing direction. Use the fold line to determine visibility, if necessary.

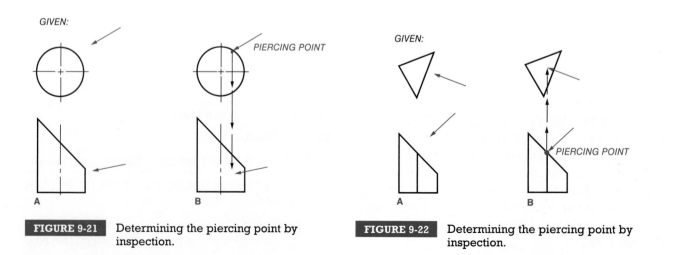

FIGURE 9-21   Determining the piercing point by inspection.

FIGURE 9-22   Determining the piercing point by inspection.

EXAMPLE:

Refer to **Figure 9-23A**.

GIVEN: Plane surface abc and line x-y in the top view and front view.

STEP 1 In the front view of plane abc, construct a line parallel to fold line bF-dF and find its true length in the top view (**Figure 9-23B**).

STEP 2 Project an auxiliary view to find the point view of the true-length line from the top view and the edge view of the plane. The line intersection with the edge view of the plane indicates the piercing point location (**Figure 9-23C**).

STEP 3 Project the piercing point back into other views (**Figure 9-23D**).

STEP 4 In each view that the piercing point is being projected to, the portion of the line that is visible can be determined by checking its preceding view (**Figure 9-23E**).

Notice in the auxiliary view that line x-y is closer to the fold line from the piercing point to point yA than is the edge view of the plane surface from the piercing point to point cA. Therefore, in the top view, line x-y is seen only from the piercing point to yT.

STEP 5 By viewing back into the top view, determine the visibility of lines of the plane surface and line x-y in the front view (**Figure 9-23F**). Notice in the top view that line x-y is closer to the fold line from the piercing point to point yT than is the edge view of the plane surface cT/bT; therefore, in the front view, line x-y is seen only from the piercing point to yF.

When the edge view does not appear in the regular views, it must be constructed using the preceding steps (**Figure 9-24A**). Once the edge view is constructed, the piercing point can easily be seen. The piercing

(A)

(B)

(C)

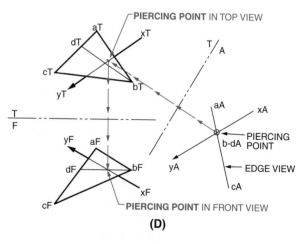

(D)

FIGURE 9-23 (A) Determining the piercing point by construction; (B) Step 1; (C) Step 2; (D) Step 3;

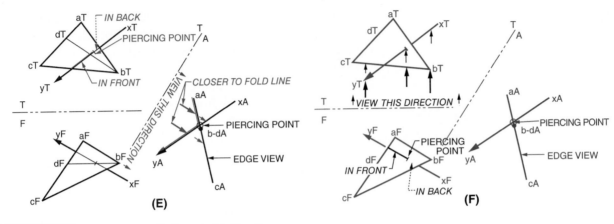

FIGURE 9-23   *(Continued)* (E) Step 4; (F) Step 5.

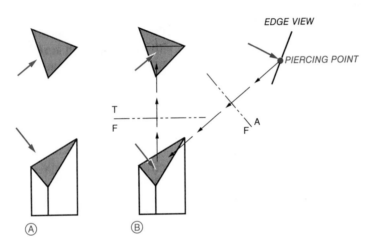

FIGURE 9-24   Determining the piercing point by construction.

point is now projected down into the front view and up into the top view (**Figure 9-24B**).

How to Determine the Piercing Point by Line Projection

An alternate method of determining the piercing point simply locates the path of a cutting plane that passes through the line, leaving a "scar" on the surface of the given plane. Since the line lies in the cutting plane, an intersection of the line and the scar on the given plane locates the piercing point. The end-points of the scar are found by aligning the cutting plane with the given line in one view, where it crosses the boundary of the given plane in two places. Projecting the cutting plane intersection to the corresponding boundaries of the given plane in the next view locates the scar end-points in that view, and it provides a view of the scar that the given line can cross at the piercing point.

EXAMPLE:
See **Figure 9-25A**.

GIVEN: Skewed surface abc and a line.

STEP 1   In the top view, lightly draw the piercing line and determine points x and y where they cross the boundary of plane abc. The piercing point is located somewhere between these points.

STEP 2   Project points x and y down into the front view to the corresponding edges of the plane abc (**Figure 9-25B**). Where line x-y crosses the given line is the exact piercing point.

STEP 3   Project the piercing point to the top view (**Figure 9-25C**). The visibility of the piercing line is determined by inspection.

Figure 9-26A illustrates a cylinder pierced by a line. Lightly draw the piercing line in the top view and find point x. Project point x down into the

GIVEN:

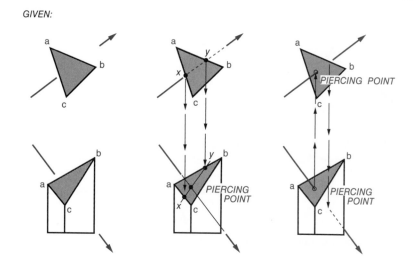

FIGURE 9-25   Determining piercing points by line projection (skewed surface).

GIVEN:

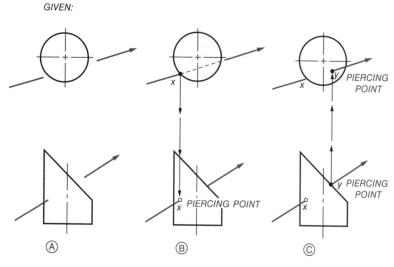

FIGURE 9-26   Determining piercing points by line projection (cylinder).

front view (**Figure 9-26B**), to find piercing point x. Because the piercing line exits the skewed surface, piercing point y is found on the edge view of the front view (**Figure 9-26C**). Project piercing point y up into the top view. Visibility of the piercing line is determined by inspection.

If the object is a cone, line segments must be located and drawn from the base of the cone up to the vertex of the cone.

GIVEN: A cone with a piercing line (**Figure 9-27A**).

Lightly draw the piercing line in the top view and find points x and y (**Figure 9-27B**). Project points x and y down to the base of the cone. Project light line segments from points x and y on the base up to the vertex of the cone. Where these lines cross the piercing line is the location of the two piercing

points. Visibility of the piercing line is determined by inspection (**Figure 9-27C**).

If the object is spherical, an imaginary flat surface on the sphere must be established along the piercing line.

GIVEN: A sphere with a piercing line (**Figure 9-28A**).

Lightly draw the piercing line in the top view. Where the piercing line intersects with the sphere is the location of the imaginary flat surface (**Figure 9-28B**). This also establishes the diameter of the flat surface. Transfer the imaginary flat surface to the front view. Where the piercing line intersects the imaginary flat surface are the exact piercing point locations. Once the piercing point locations are found, they are projected up into the top view (**Figure 9-28C**). The visibility of the piercing line is determined by inspection.

FIGURE 9-27   Determining piercing points by line projection (cone).

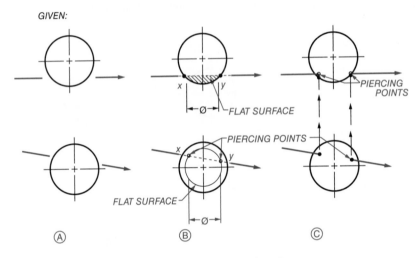

FIGURE 9-28   Determining piercing points by line projection (sphere).

How to Find the Intersection of Two Planes by Line Projection

The intersection of two planes can be determined by generating an edge view of one of the planes. Another method, somewhat simpler, is the line-projection method.

GIVEN: Two plane surfaces (**Figure 9-29A**).

In the top view, locate points x and y on edge ab of one of the planes. Project points x and y down into the front view, as illustrated in **Figure 9-29B**, and draw a light line from x-y in the front view. The piercing point is where line x-y crosses edge ab. Project points from edge ac of the same plane. Project these points down into the front view. Draw a line connecting these points; where they cross edge ac is the exact piercing point. Complete the views as illustrated in **Figure 9-29C**.

How to Find the Intersection of a Cylinder and a Plane Surface by Line Projection

The intersection of a cylinder and a plane surface can also be determined by generating an edge view of the plane surface. Another method is the line-projection method.

GIVEN: A cylinder and a plane surface (**Figure 9-30A**).

Divide the circle into equal parts; in this example, 12 equal parts. Extend these lines out to the edges of the plane surface (**Figure 9-30B**). The example illustrated uses points 1 o'clock and 7 o'clock to find points x and y on the edge of the plane surface. Project points x and y down into the front view. Draw a line from x-y in the front view. Points 1 o'clock and 7 o'clock are located somewhere on this line. To find their exact locations, project points 1 and 7 from the top view to the line in the front view.

GIVEN:

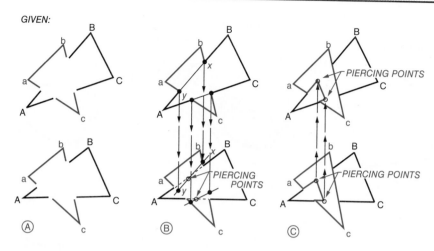

FIGURE 9-29   Determining the intersection of two planes.

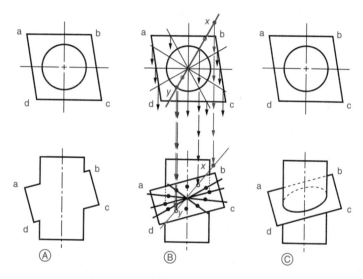

FIGURE 9-30   Determining the intersection of a cylinder and a plane surface.

Continue around the various points to locate each of the 12 points. Complete the drawing as shown in **Figure 9-30C**.

How to Find the Intersection of a Sphere and a Plane Surface

GIVEN: A sphere and a plane (**Figure 9-31A**).

Construct an edge view of the plane (**Figure 9-31B**). Divide the sphere into even spaces as illustrated. Think of each evenly spaced line as an imaginary sliced-off portion of the sphere. Draw each imaginary slice in the top view, and where the projection from the corresponding point on the edge view crosses is the piercing point. Complete the top view as illustrated. Once these points have been found in the top view, the line-projection method is used to transfer them to the front view (**Figure 9-31C**).

How to Find the Intersection of Two Prisms (Method 1)

GIVEN: The top, front, and right-side views of two intersecting prisms (**Figure 9-32**).

Label the various points as illustrated. Project each point to the 45° projection and into the next view, and where they intersect is the location of each point, as illustrated. The only point in question

GIVEN:

EVEN SPACES

EDGE VIEW

Ⓐ   Ⓑ   Ⓒ

FIGURE 9-31   Determining the intersection of a sphere and a plane surface.

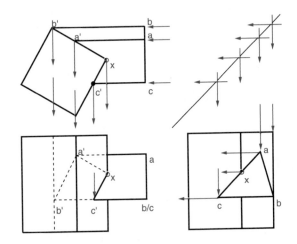

FIGURE 9-32   Finding the intersection of two prisms (Method 1).

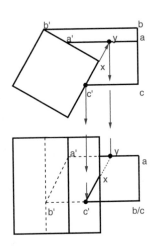

FIGURE 9-33   Finding the intersection of two prisms (Method 2).

is point x; using the three views, it can be easily found.

How to Find the Intersection of Two Prisms (Method 2)

With only two views, there is a problem locating point x in the front view. Refer to **Figure 9-33**. In the top view, extend line c'-x where it crosses line segment a-a'. Project point y down into the front view to line a-a'. Draw a line from c' to y to locate point x in the front view.

Bearings, Slope, and Grade

In some fields of drafting, the exact position of a line in space is described by its *bearing* and *slope* or by its bearing and *grade*.

Bearing of a Line

The bearing or *heading* of a line is the direction of that line as it is drawn in the *top view* of a drawing on a map. Using a map of a given area on the earth's surface, a view looking directly down would be an example of how a bearing of a line is used.

There are two methods used to call off a bearing. It is called off by either its *north/south deviation* or by its *azimuth bearing*.

North/South Deviation

A bearing of a line is measured in degrees with respect to the north or the south. It is customary to consider the north as being located from the top of the page and the south as being located from the bottom of the

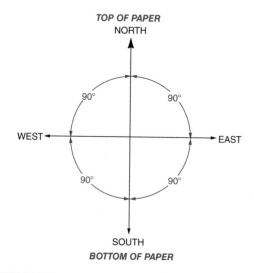

TOP OF PAPER
NORTH

FIGURE 9-34   North, south, east, west headings; north is usually located at the top of the page.

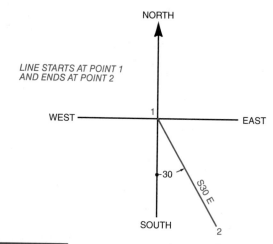

LINE STARTS AT POINT 1
AND ENDS AT POINT 2

FIGURE 9-36   A line with a heading of S 30° E.

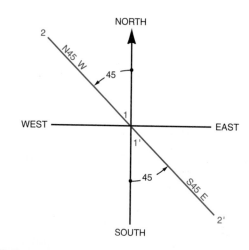

FIGURE 9-37   This line could be either N 45° W or S 45° W, depending on the starting point and the direction in which it is projected.

page, unless otherwise noted (**Figure 9-34**). Bearings are called off in angles less than 90° and are always given from the north or south, heading toward either the east or west (**Figure 9-35**). In this example, a line, 1-2, is projected 45° from the north and toward the west. It has a bearing of, and is called off as, N 45° W. A line, 1-2, is projected 30° from the south toward the east (**Figure 9-36**). It has a bearing of, and is called off as, S 30° E. It is important to know where the point of origin or the beginning of the line is, and in which direction the line is heading (**Figure 9-37**). Line 2-2′ could be called off as either N 45° W or as S 45° E, depending on its point of origin. If the line is projected from point 1 to point 2, the bearing would be N 45° W; if from 1′ to point 2′, the bearing would be S 45° E.

Azimuth Bearing

In using the azimuth bearing method, the total angle is used, going *clockwise* from the north. A given line, 1-2, is drawn at 45° from the north and heading toward the west (**Figure 9-38**). Using the azimuth method and measuring clockwise from the north, it has a bearing of 315° (90° plus 90° plus 90° plus 45°). It is usually understood that the north bearing is the starting point; thus, the N is usually omitted in verbally describing the bearing; on the drawing, however, it is best to use the N or North so as to avoid any confusion. Note that bearing is sometimes noted as heading.

Slope Angle

Using the slope angle is a way to describe the inclination of a line. The slope of a line is the angle, in degrees, that the line makes with a horizontal (level) plane

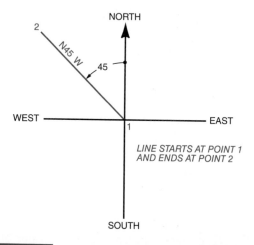

LINE STARTS AT POINT 1
AND ENDS AT POINT 2

FIGURE 9-35   A line with a heading of N 45° W; headings are always drawn in the top view.

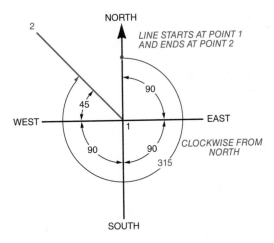

FIGURE 9-38 Azimuth bearing is indicated from the north clockwise to the line.

FIGURE 9-39 Slope angle is the angle in degrees from the horizon.

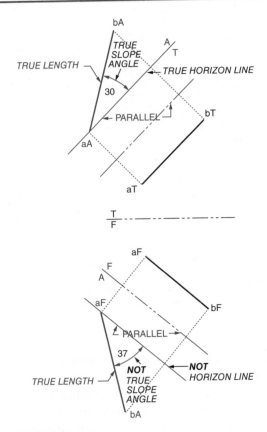

FIGURE 9-40 The true slope angle of a line can be seen only in a view that has a horizon line. It is noted in degrees.

(Figure 9-39). The line is heading from point 1 toward point 2, and it is rising; therefore, it is a plus (+) slope angle.

Note that the *true slope* of a line can be seen only in the view that has a horizon (level) line and in which it is drawn as a *true length* (Figure 9-40). In this example, the true length can be found from the front view and/or from the top view. However, projecting from the front view will **not** give a horizon (level) line—it is found only by projecting from the top view.

Grade Percent

Another way to describe the inclination of a line in relation to the horizon is by the ratio of the vertical rise (R) to the horizontal (H). It is expressed in percent and calculated by the simple formula:

$$\text{PERCENT OF GRADE} = \frac{\text{UNITS OF VERTICAL RISE (R)}}{\text{UNITS OF HORIZON (H)}} \times 100$$

Figure 9-41 illustrates a simple 45° angle line with 100 horizontal units and 100 vertical units. 100 divided by 100 times 100 equals a 100% grade. If the angle were 30°, the grade would be 57%

(Figure 9-42). The most common method of measuring the horizontal and vertical units is with an engineer's scale using multiples of 10. Percent of grade is another way to express the slope of a line

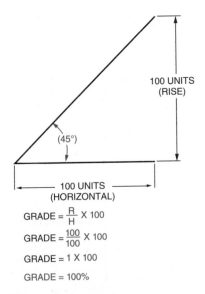

GRADE = $\frac{R}{H}$ X 100

GRADE = $\frac{100}{100}$ X 100

GRADE = 1 X 100

GRADE = 100%

FIGURE 9-41 Grade percent is the inclination of a line in relation to the horizon, noted in percent.

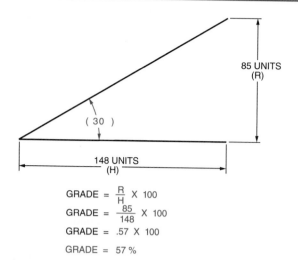

$$\text{GRADE} = \frac{R}{H} \times 100$$

$$\text{GRADE} = \frac{85}{148} \times 100$$

$$\text{GRADE} = .57 \times 100$$

$$\text{GRADE} = 57\%$$

FIGURE 9-42   Example of a line with a grade of 57%.

but, as in laying out grade percents, **the view that has both the true horizon and true length must be used.** Refer back to Figure 9-40.

Other Ways to Call Off the Slope of a Line

Various fields of drafting use slightly different methods to call off the specified inclination of lines. Each method is similar to that of calling off percent of slope and/or percent of grade in that each expresses the relationship of the line to the horizontal (level) plane. Three common methods are pitch, bevel, and batter.

PITCH

Pitch is usually used in the architectural field of drafting and is used to call off the slope of a roof. It is expressed as the ratio of vertical rise to 12″ or 12′ of horizontal span (**Figure 9-43**). In each horizontal 12″, the roof rises 6″.

BEVEL

In structural engineering, where steel members are used in the construction of the building, the slope of a member is called off by the *bevel* of the beam (**Figure 9-44**). It is very similar to that of the pitch used in architectural drafting.

FIGURE 9-43   Pitch is the slope of a roof and is a ratio of vertical rise for every 12″ or 12′ of horizontal span.

FIGURE 9-44   A bevel is used in structural engineering to indicate the slope of a structural member.

BATTER

In the field of civil engineering, the slope of the grade (earth) is used and is referred to as *batter*. It is called off as illustrated by **Figure 9-45**. This means that for each unit of rise (R), there are four horizontal (H) units.

Angles between Bearings or Headings

If you know the bearings of two lines, the angle between the lines can be determined (**Figure 9-46**). Given are lines 1-2 and 1-3, as illustrated. Both these lines fall within one 90° quadrant; therefore, to find the angle between the lines, subtract one bearing from the other (**Figure 9-47**). Fifteen degrees is subtracted from 80°, and the true angle between the bearings is 65°.

If the bearings fall within two or more quadrants, the actual angles in each quadrant must be added or subtracted, as illustrated in **Figures 9-48** through **9-51**.

FIGURE 9-45   Batter is used in the civil engineering field to indicate the slope of a grade.

FIGURE 9-46   Angle between two headings in one quadrant (90°).

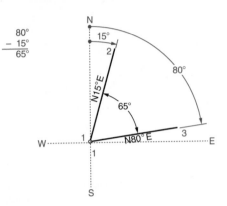

FIGURE 9-47 How to calculate the angle between headings in one quadrant.

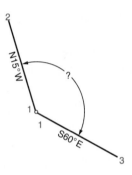

FIGURE 9-48 Angle between two headings in three quadrants.

FIGURE 9-49 How to calculate the angle between headings in three quadrants.

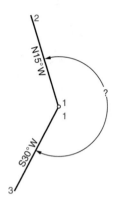

FIGURE 9-50 Angle between two headings in four quadrants.

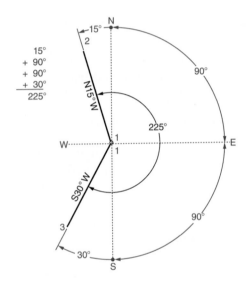

FIGURE 9-51 How to calculate the angle between headings in four quadrants.

How to Construct a Line with a Specified Bearing, Slope Angle, and Length

The bearing is always drawn in the top view, and the true length can be drawn if the line is to be exactly horizontal. Any line with an incline up (+) or down (−) from the horizon *must* be laid out using the following steps:

GIVEN: The starting point aT and aF of a line with a bearing of N 60° E, with a slope angle of +15°, and 230.0′ long (**Figure 9-52A**).

Because the line is **not** horizontal and has a slope angle of plus (+) 15°, the true length must be constructed before the heading length can be drawn in the top view.

STEP 1 In the top view, draw a line starting from aT, at 60° from the north and toward the east (**Figure 9-52B**). Temporarily draw the line longer than the 230.0′ required.

STEP 2 Find the true length of the line by adding a fold line, T/A, parallel to the bearing line aT. Project point a into the auxiliary view to find aA (**Figure 9-52C**), and refer to dimension X. Don't forget to skip a view.

From point aA in the auxiliary view, draw a light horizon line parallel to the fold line T/A. From point aA, construct a 15° angle toward the fold line as shown.

NOTE: *Draw the line toward the fold line for a rise (+) and away from the fold line for a drop (−) in elevation. Along this line measure the true length, 230.0',*

to establish point bA. Project bA back into the top view to find point bT. *Remember, this line is **not** the true length in the top view.*

STEP 3   Project point bT down into the front view to find point bF. Use the Y dimension to find point bF, but do not forget to skip a view (**Figure 9-52D**). Construct line aF/bF. Note

GIVEN:

LINE BEARING N 60° E
15° (PLUS) SLOPE
230.0 TRUE LENGTH

aF +

(A)

aF +

(B)

(C)

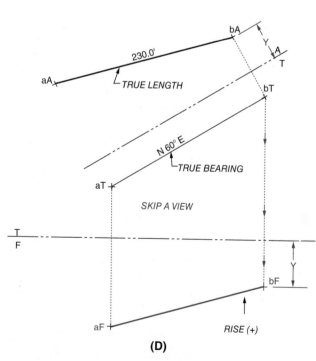

(D)

FIGURE 9-52   (A) Constructing a line with a specified bearing, slope angle, and length; (B) Step 1; (C) Step 2; (D) Step 3.

that the line rises from a to b in the front view, which indicates a plus (+) rise.

The true bearing is found in the top view, and the true length is found in the auxiliary view. Remember, true length must be found before the other views can be completed.

How to Construct a Line with a Specified Bearing, Percent of Grade, and Length

Approximately the same steps are used in constructing a line with a specified bearing, percent of grade, and length as was used in constructing a line with a specified bearing, slope angle, and length. The bearing is always drawn in the top view, and the true length of the line can be drawn if the line is to be *exactly* horizontal. Any line with an incline up (+) or down (−) from the horizon *must* be laid out using the following steps:

GIVEN: The starting point aT and aF of a line with a bearing of N 45° W, with a grade of (−) 40%, and 190.0 feet long (**Figure 9-53A**). Because the line is **not** horizontal and has a grade of minus (−) 40%, the true length must be constructed before the bearing length can be drawn in the top view.

STEP 1 In the top view, draw a line starting from aT, at 45° from the north and toward the west (**Figure 9-53B**). Temporarily draw the line actually longer than the 190.0′ required.

STEP 2 Find the true length of the line by adding a fold line, T/A, parallel to the bearing line aT. Project point a into the auxiliary view to find aA (**Figure 9-53C**), and refer to dimension X. Don't forget to skip a view.

From point aA in the auxiliary view, draw a light horizon line parallel to the fold line T/A. From point aA, and along the horizon line, measure 100 units. From this point, turn 90° away from the fold line and measure 40 units, and put a point there. Construct a line from aA to this new point, and extend it out to the 190.0′ true length to find point bA.

NOTE: Construct the triangle toward the fold line for a rise (+) and away from the fold line for a drop (−) in elevation. Project bA back into the top view to find point bT. Remember, this line is not the true length in the top view.

STEP 3 Project point bT down into the front view to find point bF. Use the Y dimension to find point bF, but do not forget to skip a view (**Figure 9-53D**). Construct the line aF/bF.

Note that the line falls from a to b in the front view, which indicates a minus (−) drop. The true bearing is found in the top view, and the true length is found in the auxiliary view. Remember, true length must be found before the other views can be completed.

Applied Descriptive Geometry

The topics in the remainder of this chapter are essentially applications of the descriptive geometry basics already presented. Some of the topics are contours and topographic maps, vectors, forces in trusses, and revolution.

Topographic Topics

Surveys

Compass directions with distances describe real estate boundaries in cities and counties for commercial and domestic property owners. The descriptions usually include references to governmental surveys, such as being located in a government lot, section, township, and range, and they have specific coordinates so a surveyor can locate the property and stake out the area. Architects and builders have drafters prepare scaled drawings of the areas for planning the locations of buildings and houses.

How to Plot the Boundary of Real Estate

EXAMPLE:

GIVEN: Compass bearings with distances in feet, outlining a real estate lot, shown in a rough sketch with north in the traditional direction toward the top of the sheet (**Figure 9-54A**).

STEP 1 Lay out the boundary to scale starting with the S17° W 59′, then the S31° W 47′, and next the S54°30′ E 170′ while utilizing the drawing area efficiently: for example, the direction north may not be toward the top of the sheet (**Figure 9-54B**). Note that the 71°30′ angle shown is an internal angle, and that the sum of all the internal angles equals (number of sides N−2)×180°. For the layout shown, (N−2)×180°=(5−2)×180°=540°. The 85°30′ angle is called a deflection angle; this is the smallest angle the surveyor must turn the transit to align with the next side. The sum of all the deflection angles must equal 360° for a complete traverse around the property.

GIVEN:

LINE BEARING N 45°W
40 % (MINUS) GRADE
190.0 TRUE LENGTH

FIGURE 9-53 (A) Constructing a line with a specified bearing, percent of grade, and length; (B) Step 1; (C) Step 2; (D) Step 3.

STEP 2 Complete the boundary with N26°25'104" next and N54°30' W 176' last. The last segment should close on the starting point (**Figure 9-54C**).

For excavation and landscaping planning, an architect may ask a drafter to draw in contour lines. This is the topic of the next section.

Contour Lines

A *contour line* on a plot or map is a line at a constant elevation on the surface of the earth, above sea level (**Figure 9-55A**). A *contour interval* is the difference in elevation between neighboring contour lines. It is the same value for any one drawing. For example,

50' is the interval in Figure 9-55A due to mountainous terrain. Intervals of 10' are common; architectural plots may have only 1' intervals on relatively level terrain for houses or buildings.

Usually, every fifth contour line is heavier and its elevation indicated. Lines close together represent steep terrain.

Profiles

Imagine a vertical plane cutting through a contoured map and being able to see the cross section at the cut. The cross section would show a profile of the terrain in an elevation view, as illustrated in **Figure 9-55B** for a straight path from point A to point B. The fold line

FIGURE 9-54   (A) Compass bearings and distances: Real estate lot; (B) Step 1; (C) Step 2.

is added as a reminder that all topographic maps are T views, and elevations, of course, occur in views projected off a T view.

Cuts and Fills

Highways passing through hilly terrain require that the earth of some hills be excavated and the earth used to fill in valleys—*cuts and fills* (**Figure 9-56**). The volume of the earth to be cut and filled needs to be determined prior to having an army of construction workers and their equipment arrive at the job site, for efficient use of labor and equipment. The volume of the cut determines how much fill can be

completed. Volumes of cuts and fills are determined from boundary maps.

How to Determine the Boundaries of Cuts and Fills from Topographic Maps with Contour Lines

EXAMPLE:

GIVEN: A portion of a topographic map with contours and the proposed location of a 60′ wide, level road, at an elevation of 500′ (**Figure 9-57A**). The fold line T/F is perpendicular

(A)

(B)

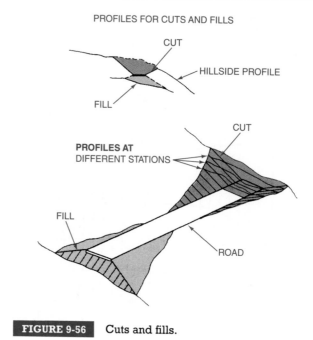

PROFILES FOR CUTS AND FILLS

FIGURE 9-56   Cuts and fills.

(A)

FIGURE 9-57   (A) Proposed road.

to the direction of the road, and the F view shows elevation planes at 10′ intervals; the topographic map and the elevations are drawn to the same scale.

STEP 1   Construct cutting planes at 45° to the horizontal in the elevation view F, to each side of

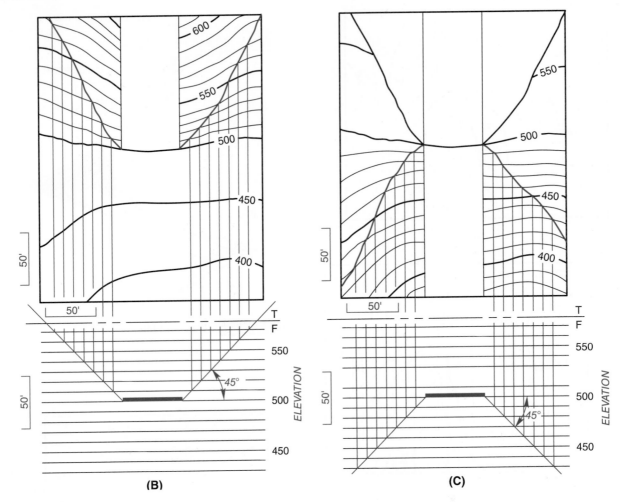

FIGURE 9-57   *(Continued)* (B) Step 1; (C) Step 2.

the proposed 60′ wide road (**Figure 9-57B**). Where the cutting planes intersect the elevation planes, mark the corresponding intersections on the map. Connect the intersections, freehand, to show the boundaries of the excavation cut, Figure 9-57B.

STEP 2   In a similar manner, construct cutting planes at 45° to represent the proposed earth fill, and mark the boundaries as before **Figure 9-57C**. Note that the angles of cuts and fills vary—for example, a commonly used fill has a slope of 1.5 horizontal to 1.0 vertical.

How to Determine the Volumes of Cuts and Fills (Graphical, Mechanical, and Electronic)

EXAMPLE:

GIVEN: The cut and fill boundaries from the preceding example shown in a simplified T view **Figure 9-58A**.

STEP 1   Start at the 500′ elevation. Divide the cut and the fill with vertical planes at 25′ intervals called stations (STA) (**Figure 9-58B**). Estimate the elevations of the points where the vertical planes at each station intersect the boundaries by reading the contour lines.

STEP 2   Construct scaled drawings of the profiles at the stations, and determine the area of each profile by using a graphical method for calculating geometric areas (**Figure 9-58C**). Tabulate the results.

Three methods used to calculate geometric areas of cross sections are graphical, mechanical, and electronic. The three are applied to a variety of problems, such as cuts and fills, cross sections of ship and boat hulls, and cross sections of layers of odd shapes for volume calculations.

A *graphical method* for approximating areas was used in Figure 9-58C. The method is to divide the area into convenient shapes, such as trapezoids and triangles, to scale appropriate lengths, and to calculate the areas.

A mechanical method utilizes a planimeter shown in the conceptual sketch (**Figure 9-58D**). Pole arm OA is confined to swivel about point O. Point P is a pointer the operator uses to trace the complete boundary, drawn to scale, causing an indicating

wheel to rotate and slide. The wheel's final position gives a reading of the area traversed.

An electronic method employs a computer and a digitizer (looks like a computer mouse). A cross-haired window at the bottom of the digitizer is positioned at small increments of distances around the boundary of an area (**Figure 9-58E**). At each increment along

EXAMPLE CALCULATION STA 3 + 00 CUT

③ *READ ELEVATIONS FROM FIG. 7-58B*

④ *DIVIDE AREA INTO TRAPEZOID AND TRIANGLE*

④F △

② *CONSTRUCT SLOPE OF CUT (OR FILL)*

① *LAYOUT ROAD*

⑤ *CALCULATE AREA*

$$A\,\text{STA 3 + 00 CUT} = \frac{(148)(10)}{2} + \frac{(148 + 60)(44)}{2}$$

$$= 740 + 4576 = \underline{5316\ \text{ft}^2} \leftarrow \text{AREA STA 3 + 00 CUT}$$

	STA	ft²
CUTS	5 + 00	10,412
	4 + 00	7,442
	3 + 00	5,316
	2 + 00	3,502
	1 + 00	900
	0	0
FILLS	1 + 00	2,070
	2 + 00	4,620
	3 + 00	6,900

(C)

FIGURE 9-58 (A) Cut and fill boundaries; (B) Step 1; (C) Step 2 (graphical method); (D) Mechanical method; (E) Electronic method.

the boundary, a key on the digitizer is clicked. This causes the X-Y coordinates to be stored in a computer. Finally, the computer uses the coordinates to calculate the digitized area.

STEP 3   Calculate the amount of earth to excavate (cut) between two stations (STA) by averaging the areas at the stations and multiplying by the 25' distance between the stations. For example, the area STA 2 + 00 of the cut = 3502 square feet; the area at STA 3 + 00 of the cut = 5,316 square feet; the volume between STAs 2 and 3 of the cut is (3502 + 5,316)/2 × (25) = 110,225 cubic feet. The volume in cubic yards (called yards in construction) is 110,225/27 = 4,082. Use this calculation procedure to determine the volume of earth between STAs 0 + 00 and 5 + 00 of the cut, Figure 9-58B. Get 20,709 yards available for fills. The fill requirement (using the same calculation procedure) for the volume between STAs 0 + 00 and 3 + 00 FILLS is only 9,390 yards. The remaining earth from the cut would be used to fill additional stations.

Strike and Dip

Strike and dip are terms used in mining to describe three attributes of ore veins (**Figure 9-59**). *Strike* is a compass direction of a horizontal line lying in the plane of an ore vein. *Dip* is the downward angle of the vein with respect to a horizontal plane at the site of the ore vein. Dip also includes the general compass direction of the downward slope.

The vein in Figure 9-59 would be described as N64°30' E, 35° NW—that is, the compass direction of the strike and the 35° dip downward toward the NW.

How to Determine the Strike and Dip of a Vein of Ore (or Stratum of Rock)

EXAMPLE:

GIVEN: The depths of three core-holes, a, b, and c, sunk over a discovered ore vein (**Figure 9-60A**).

STEP 1   Lay out the points a, b, and c in a T view to scale, Figure 9-60A. Since the points, which are at different depths, do not fall in a straight line, they describe a triangular plane lying in the vein (**Figure 9-60B**).

STEP 2   Find plane abc as an edge in view A by using fold line T/A (**Figure 9-60C**), and the technique described in Figures 9-17A through 9-17D in this chapter.

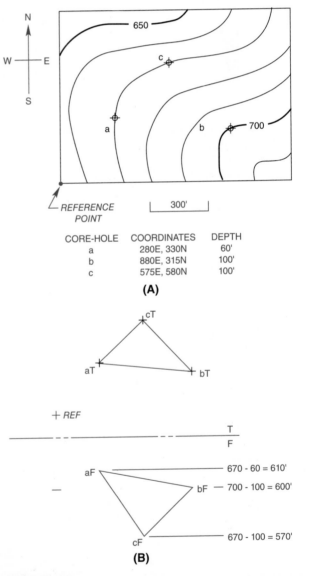

CORE-HOLE	COORDINATES	DEPTH
a	280E, 330N	60'
b	880E, 315N	100'
c	575E, 580N	100'

(A)

(B)

FIGURE 9-60   (A) Core-holes for vein location; (B) Step 1;.

FIGURE 9-59   Strike and dip attributes.

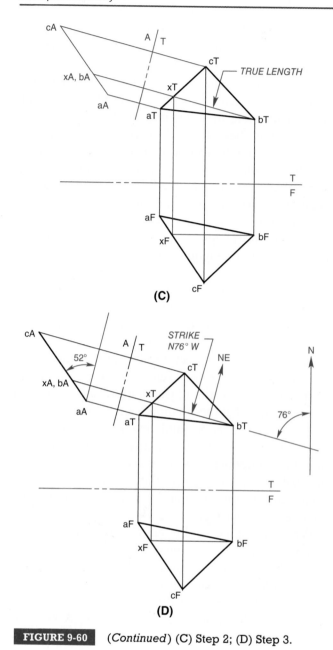

(C)

(D)

FIGURE 9-60   (*Continued*) (C) Step 2; (D) Step 3.

STEP 3   In view A, measure the dip angle, and in the T view, measure the strike bearing (**Figure 9-60D**). Note the general direction of the dip to complete the attributes of the ore vein—N76° W, 52° NE.

How to Determine the Strike and Dip of a Vein Given Two Lines in the Plane of the Vein and Intersecting

EXAMPLE:

GIVEN: Two apparent dip directions (not strikes) and angles of the dips as follows: N61° W, 40°

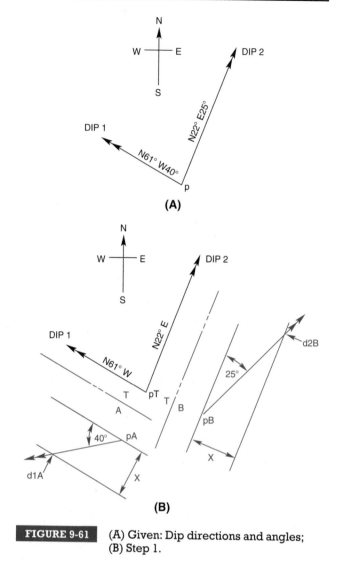

(A)

(B)

FIGURE 9-61   (A) Given: Dip directions and angles; (B) Step 1.

and N22° E, 25° both originating from the same point p, a bore-hole (**Figure 9-61A**).

STEP 1   Construct view A using fold line T/A parallel to DIP1 to obtain the true view of the dip. Construct view B using T/B parallel to DIP2 to obtain its true view (**Figure 9-61B**). From points pA and pB, lay out the same distance, X, downward. Label the marks d1A and d2B, respectively.

STEP 2   Project the points d1A and d2B into the T view and note that the line d1T-d2T is horizontal and, therefore, the strike of the ore vein (**Figure 9-61C**).

STEP 3   Project the triangle (p-d1-d2)T into a view C using fold line T/C perpendicular to the strike line, and measure the true dip angle (**Figure 9-61D**). The true strike and dip are N52° E, 43° NW.

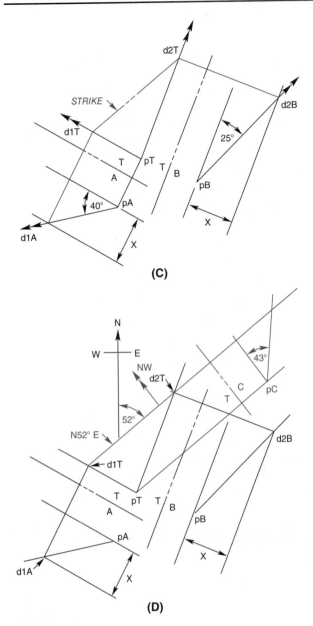

(C)

(D)

FIGURE 9-61 (*Continued*) (C) Step 2; (D) Step 3.

Vectors

A *vector,* as defined in engineering, technology, and mathematics, must have both magnitude and direction. The vector shown in **Figure 9-62**, drawn to scale and having an arrowhead at one end, could represent any one of the following: displacement (length and direction); force (pounds and direction); velocity

(distance per unit of time and direction); or acceleration (distance per unit of time squared and direction). As a comparison, direction only is a characteristic of the omnipresent straight line with an arrowhead at one end. A straight line with arrowheads at both ends in dimensioning indicates magnitude. These are not vectors, but they efficiently convey information in technical material, such as compass bearings, cursor positions, callouts, identifying features and, of course, dimensions on orthographic drawings (**Figure 9-63**). Magnitudes without directions are called *scalars,* such as temperature and pressure.

Vector equivalents of physical systems are simplifications used in engineering so that the vectors can be manipulated efficiently in mathematical formulas. For example, bulky objects or portions of a bridge deck are assumed to be loads concentrated at a point. Internal loads, although acting over an area, are assumed to be point loads acting at the centroid of the area. All parts of an object traveling at the same velocity are represented by one velocity vector acting at the center of mass of the object (**Figure 9-64**). Always keep in mind that these simplifications do not change the physical situation, and that point loadings are only approximate. Thus, graphical solutions employing these simplifications are as accurate as mathematical solutions. A number of the vector manipulations that can be done graphically are discussed in this chapter, for two main reasons—to demonstrate the efficacy of graphics and to emphasize that when you can visualize what is going on, you will have a better understanding of the analysis and the answer.

Vectors will be labeled with capital letters—e.g., F(AB) is a force F in the direction of, or in, member AB. An applied load would be W or F. A velocity vector would be V.

M ——————▶ N

1.5 feet——————1.5'
15 pounds ——————15 lbf
150 feet/second ——————150 fps
1.5 ft/sec/sec ——————1.5 ft/sec²

FIGURE 9-62 Vector.

FIGURE 9-63 Nonvectors.

FIGURE 9-64   Vectors used in engineering simplifications.

Vector Systems

Six possibilities exist for analyses using vectors: coplanar (concurrent, parallel, and nonparallel) and noncoplanar (concurrent, parallel, and nonparallel). Coplanar-concurrent and noncoplanar-concurrent are the categories emphasized in this chapter.

Coplanar-Concurrent Force Systems— The Parallelogram Law: Two Vectors

The *parallelogram law* basically states that two vectors, not parallel, drawn to scale, having the same units, and acting at the same point can be added or

subtracted, as shown in **Figure 9-65A**. Use vector A plus vector B to construct a parallelogram. The diagonal of the parallelogram vector (A + B) is called the resultant R, which is equivalent to the effect of the two vectors (**Figure 9-65B**). Vector A minus vector B can also be used to construct a parallelogram. The resultant vector (A–B) is the diagonal (**Figure 9-65C**).

If the vectors being added or subtracted are force vectors, the equal and opposite reaction to the resultant is defined as the *equilibrant*. The reaction must be present to satisfy a basic law of mechanics, which is that the sum of the forces acting on an object that is in equilibrium must add up to zero in any one

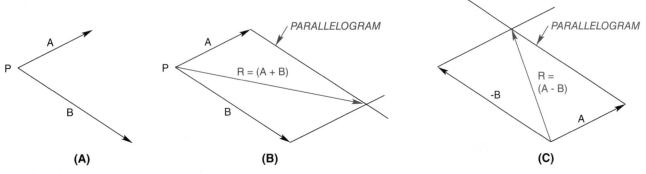

FIGURE 9-65   (A) Coplanar concurrent vectors; (B) Adding vectors: parallelogram law; (C) Subtracting vectors: parallelogram law.

direction—that is, $\Sigma F = 0$. A force may be defined as the action of one body on another, and it tends to change the state of rest or motion of the body acted on.

A vector force acting on a body can be considered to be moved along its line of action without changing its effect on the body.

EXAMPLE:
Refer to **Figure 9-66A**.

GIVEN: Force vectors P and Q acting on body B. Also, S and T acting on W.

STEP 1   Move vectors P and Q along their respective lines of action until they meet and are tail to tail (**Figure 9-66B**). Also, move S and T.

STEP 2   The resultant R of vectors P and Q shows the combined effect of P and Q on body B (**Figure 9-66C**). Also, the resultant U of vectors S and T shows the effect on body W.

How to Add More than Two Coplanar-Concurrent Vectors

Two graphical methods can be used to add more than two coplanar-concurrent vectors. One requires adding any two vectors first, and then using the

resultant to add to the third, etc. The second involves adding vectors "tail to head," in any order, to obtain a resultant.

EXAMPLE: METHOD 1
Refer to **Figure 9-67A**.

GIVEN: Vectors A, B, and C acting at point P.

STEP 1   Add vectors A and B using the parallelogram law to obtain resultant R1 (**Figure 9-67B**).

STEP 2   Add vectors R1 and vector C using the parallelogram law to obtain resultant R2 (**Figure 9-67C**). R2 is the resultant of the three vectors and would have the same effect as A, B, and C acting together at point P.

EXAMPLE: METHOD 2
Refer to **Figure 9-68A**.

GIVEN: Vectors A, B, and C acting at point P.

STEP 1   Move vector B parallel to its given position so that the tail of B is at the head of vector A (**Figure 9-68B**).

STEP 2   Locate vector C in a similar manner so that the tail of C is at the head of vector B (**Figure 9-68C**).

STEP 3   A vector from point P to the head of vector C is the resultant of the three vectors—that is, the resultant would have the same effect on point P as vectors A, B, and C (**Figure 9-68D**).

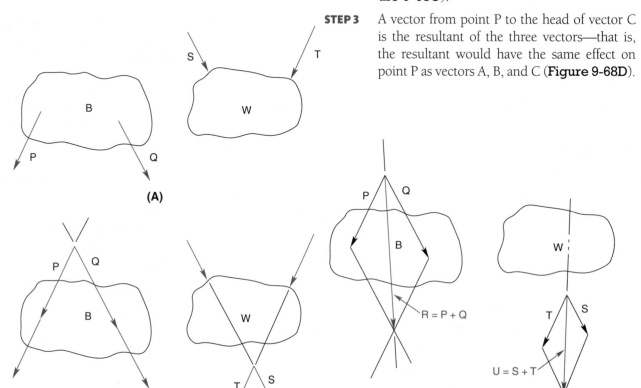

FIGURE 9-66   (A) Vector line of action; (B) Step 1; (C) Step 2.

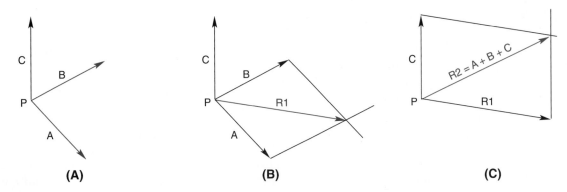

FIGURE 9-67 (A) Three coplanar concurrent vectors (Method 1); (B) Step 1; (C) Step 2.

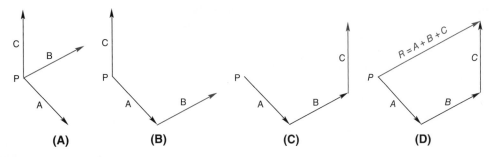

FIGURE 9-68 Four coplanar concurrent vectors (Method 2).

Resolution of Vectors

A given vector can be resolved into two coplanar vector components that act along two axes that intersect at the tail of the given vector.

How to Resolve a Vector onto Perpendicular Axes

EXAMPLE 1:

Refer to **Figure 9-69A.**

GIVEN: Vector A and axes X and Y intersecting at O.

STEP 1 From the tip of vector A, construct lines parallel to the two axes so that the lines intersect the axes X and Y at X(1) and Y(1) (**Figure 9-69B**).

STEP 2 The two vector components that replace vector A are OX(1) and OY(1) with arrowheads at X(1) and Y(1), respectively (**Figure 9-69C**).

EXAMPLE 2:

A three-dimensional mechanism problem in dynamics, but simplified to a statics problem at the instant shown. A further simplification presents the components as straight lines. The simplified mechanism and the velocity vector are drawn to scale. Refer to **Figure 9-70A.**

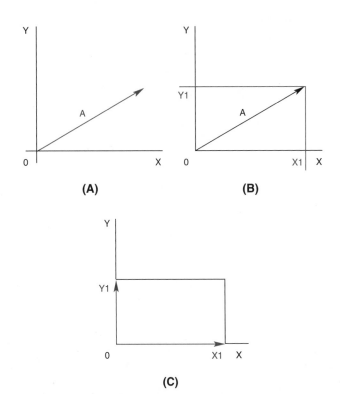

FIGURE 9-69 (A) Resolution on perpendicular axes; (B) Step 1; (C) Step 2.

GIVEN: Arm R confined to rotate about point A. Slider S pinned to C of crank BC. Crank BC confined to rotate about point B, 360°. (The apparent interference of arm R with crank BC doesn't occur in the three-dimensional mechanism.) Slider C is free to slide along arm R.

Slider C at the instant shown has a velocity V(C) m/sec = ω(BC) radians/sec × BC m. The direction of the velocity of slider S at the instant shown is perpendicular to the crank BC because slider C is attached to point C.

TO FIND: The angular velocity ω(ARM).

STEP 1 Resolve vector V(C), drawn to scale, into two components, one parallel to arm R and one perpendicular, V(perpendicular), to arm R (**Figure 9-70B**).

STEP 2 Divide V(perpendicular) m/sec by L m to obtain ω(arm R) radians/sec. (Note that radians/sec = $2 \times \pi \times$ RPS. RPS = revolutions per second.)

How to Resolve a Vector onto Two Axes NOT Perpendicular

EXAMPLE:

GIVEN: Vector V originating at o and axes m and n intersecting at o (**Figure 9-71A**).

STEP 1 From the tip of vector V, construct lines parallel to the two axes so that they intersect axes m and n at M(1) and N(1) (**Figure 9-71B**).

STEP 2 The two components that replace vector V are OM(1) and ON(1) with arrowheads at M(1) and N(1), respectively (**Figure 9-71C**).

Two-Force Members

Two-force members have forces at each end, equal in magnitude, in the opposite direction, and collinear. Therefore, they are in equilibrium. The weight of the member is considered to be negligible due to the usually much greater effect of the forces applied. See **Figure 9-72**.

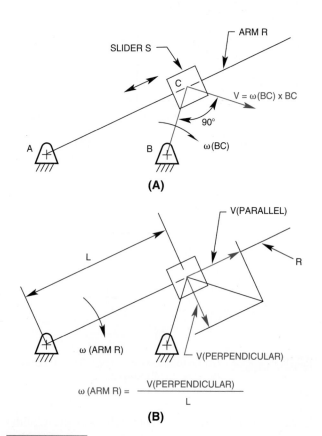

(A)

(B)

$$\omega \, (\text{ARM R}) = \frac{\text{V(PERPENDICULAR)}}{\text{L}}$$

(B)

FIGURE 9-70   (A) Quick-return mechanism;
(B) Steps 1 and 2.

FIGURE 9-71   (A) Resolving a vector onto two axes not perpendicular; (B) Step 1; (C) Step 2.

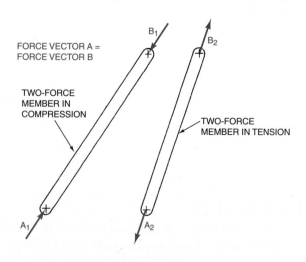

(WEIGHTS OF THE MEMBERS ARE USUALLY NEGLIGIBLE COMPARED TO THE EFFECTS OF THE APPLIED FORCES.)

FIGURE 9-72   Two-force members.

How to Determine the Forces in the Members of a Pinned Joint

EXAMPLE:

GIVEN: A space diagram showing weight W supported by a frame in equilibrium, made up of two-force members AB and BC, drawn to scale, and pinned to a wall (**Figure 9-73A**).

STEP 1 Isolate pinned joint B by "cutting" members AB and BC and redrawing the joint, parallel to the original drawing. An engineering expedient is to represent members of a frame (or truss) as straight lines. The lines of action of the internal forces F(AB) and F(BC) intersect at joint B. Their directions must be tentatively assumed by inspection. The direction of force F(AB) must keep the joint away from the wall. Force F(BC) keeps the joint from dropping. See **Figure 9-73B**.

STEP 2 Construct a vector force diagram of the forces at joint B. First, lay out the applied force W, from a point O, vertically downward, to scale. The two remaining forces F(AB) and F(BC) must add—tail to head—to W and end at the starting point O because joint B is in equilibrium. To determine the magnitudes of F(AB) and F(BC), construct a light line through the head of vector W parallel to the direction F(AB) and a light line through the tail of vector W parallel to the direction of F(BC) (**Figure 9-73C**). Where the two lines intersect determines the magnitudes of F(AB) and F(BC). The tail-to-head method of adding vectors determines their directions. The assumed directions are confirmed in the vector force diagram.

Note that only two unknown magnitudes of forces can be determined in these coplanar joint analyses. Their directions are determined in the space diagram.

Trusses

Trusses are frames that are usually two-dimensional and are made of structural shapes, such as angles, tubes, rounds, and channels. They are assembled in triangular patterns for rigidity. The members are considered to be two-force members, and the joints are assumed to be pinned even though welded, bolted, or riveted. Observe some features of the truss shown in **Figure 9-74A** and the simplified drawing of the truss in **Figure 9-74B**. Both are drawn to scale.

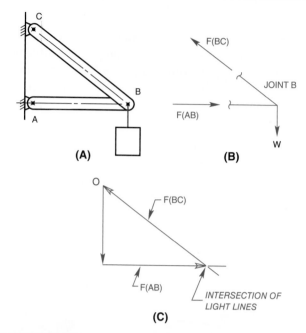

FIGURE 9-73   (A) Determining forces of a pinned joint; (B) Step 1; (C) Step 2.

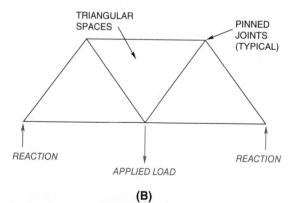

FIGURE 9-74   (A) Example truss; (B) Simplified drawing of truss.

Trusses seldom stand alone. They are used in pairs or in multiples. For example, some highway bridges and railroad bridges employ pairs. Multiples are used to support roofs (**Figure 9-75A**). In a coplanar analysis, each truss is assigned its share of the roadbed loading or the roof loads at the appropriate joints of the trusses (**Figure 9-75B**).

How to Determine the Forces in the Members of a Truss

EXAMPLE 1(A): METHOD OF JOINTS

GIVEN: Line drawing to scale of a symmetrical truss with one applied load and two equal reactions (**Figure 9-76A**).

STEP 1 Select a joint in the truss with only two unknowns. A and C meet this criterion. Use joint A and isolate it in a space diagram (**Figure 9-76B**). Assume the direction of F(AD) by inspection to "go into" the joint—that is, to be a compressive force. Assume F(AB) "goes away from" the joint and is thus a tensile force. Next, lay out a force diagram to scale, starting with the known force of 2500 N. Follow the procedure described in Step 2 of the preceding discussion, Figure 9-73C, where the light lines indicate the lines of action of the two unknown forces. Their intersection point determines the magnitudes. Scale the magnitudes and put arrowheads on the vectors so that the vectors add head to tail back to the tail of the 2500 N force, the starting point. Finally, make a small freehand sketch of the truss and circle the known members, AD and AB.

STEP 2 Joints B or D have only two unknowns. Select either one and isolate it in an equilibrium sketch. Use joint B (**Figure 9-76C**). Construct a force diagram starting with known information, the 5,000 N applied load, and the 7,500 N tensile force in AB. Construct the light lines for the forces in members BD and BC. Scale their magnitudes and add arrowheads so the forces add head to tail back to the tail of the 5,000 N force. On a new freehand sketch, circle the known members, AD, AB, BD, and BC.

STEP 3 The remaining member DC is known due to the symmetry of the truss, F(CD) = F(AD), and is a compressive force. If the truss or the loading were not symmetrical, joint C would be isolated and a force diagram prepared as shown (**Figure 9-76D**).

STEP 4 Prepare a summary of the forces on a sketch of the truss or in a table (**Figure 9-76E**).

UPPER STRUCTURE

SNOW

WIND

HIGHWAY OR RAILWAY

ROOF & MAINTENANCE PERSONNEL

(A)

UPPER STRUCTURE LOADS

NOT A JOINT

HIGHWAY OR RAILWAY & STRUCTURE LOADS

REACTION   REACTION

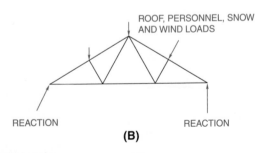

ROOF, PERSONNEL, SNOW AND WIND LOADS

REACTION   REACTION

(B)

FIGURE 9-75 (A) Truss applications; (B) Truss drawings for analyses.

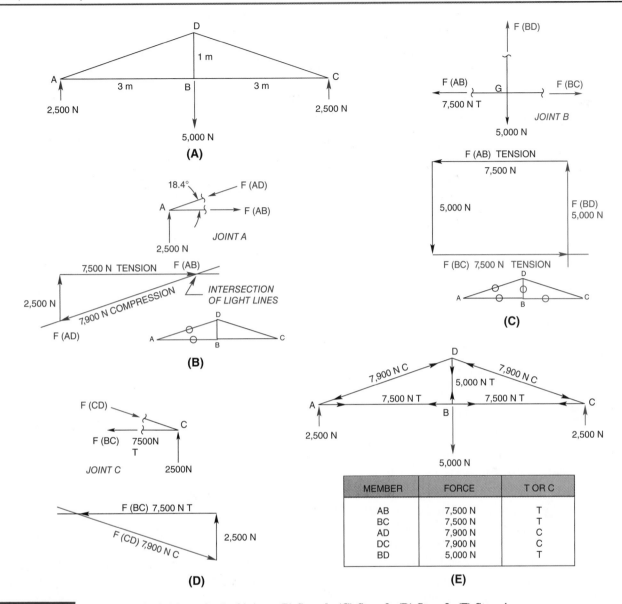

FIGURE 9-76   (A) Example 1 (a) method of joints; (B) Step 1; (C) Step 2; (D) Step 3; (E) Step 4.

EXAMPLE 1(B): METHOD: MAXWELL DIAGRAM WITH BOW'S NOTATION

GIVEN: The same truss used for the method of joints, but to be labeled with Bow's notation scheme (**Figure 9-77A**). Label the triangular spaces within the truss with numbers or letters. Create enclosed spaces between the applied forces and reaction forces by drawing a light envelope line surrounding the truss and passing through each applied force and reaction. Label each "space" with a letter.

STEP 1   Construct a force diagram from reading the external spaces in the envelope in a clockwise

direction (**Figure 9-77B**). Use a convenient scale to keep the force diagram on the drawing sheet and to use the space efficiently. Read spaces a to b for the left reaction, 2,500 N. Then, read spaces b to c for 2,500 N. Finally, read c to a for the 5,000 N force, which brings the force diagram back to the starting point. Thus, the truss is in equilibrium.

STEP 2   Select any joint with only two unknown members and read spaces clockwise around the joint (**Figure 9-77C**). Construct lines parallel to the two members crossed. For example, start at joint A and go a to b, b to 1, and 1 to a. Point 1 is located where the construction lines cross.

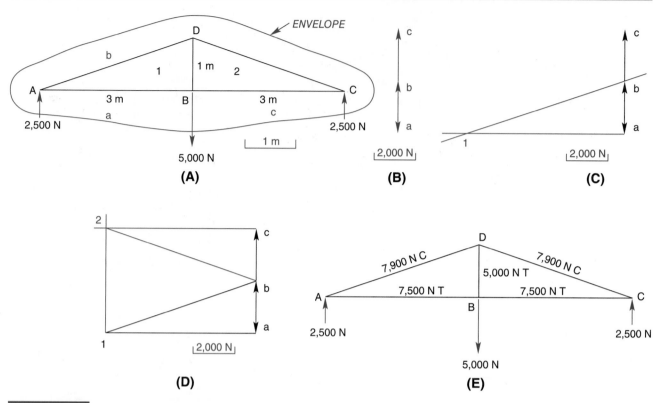

(A) Example 1 (b) Maxwell diagram with Bow's notation; (B) Step 1; (C) Step 2; (D) Step 3; (E) Step 4.

STEP 3   Select the next joint with only two unknown members, joint B or D. Around B go a to 1, 1 to 2, 2 to c, and c to a (**Figure 9-77D**). The two light lines for 1 to 2 and 2 to c locate point 2. To finish the force diagram for the truss, go around joint C or joint D. Choose joint C and go b to c, c to 2, and 2 to b.

STEP 4   Prepare a summary of the forces on a sketch of the truss (**Figure 9-77E**). Go clockwise around each joint of the truss to determine tensile and compressive forces. For example, reading clockwise around joint A, 1 to a in the force diagram goes in a direction away from A. Therefore, AB is in tension, and the magnitude is scaled to be 7,500 N. Going b to 1 goes into joint A, thus producing a compressive force of 7,900 N. In a similar clockwise pattern, determine the type and magnitude of the force for each member.

The next example focuses on a more complicated truss with asymmetrical loading.

EXAMPLE 2(A): METHOD OF JOINTS

GIVEN: A freehand sketch and a scaled line drawing of a symmetrical roof truss, but with asymmetrical loading due to anticipated wind and snow conditions (**Figure 9-78A**). The reactions at A and D were determined analytically using the fundamental equations of statics, $\Sigma F = 0$ and $\Sigma M = 0$. At joint A, the small freehand sketch shows a typical symbol for a pinned support; that is, it provides support vertically and horizontally.

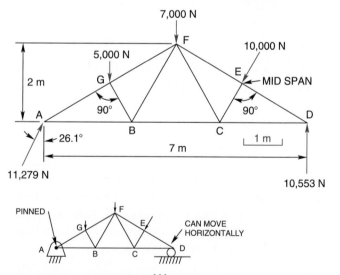

(A) Example 2 (a) method of joints 5,000 N A 11,279 N 26.1°;

The symbol at the right end indicates support in one direction only, normal to the surface of support.

STEP 1 Start at a joint with only two unknowns, A or D. Isolate joint A (**Figure 9-78B**) and construct a force diagram to find F(AB) and F(AG). (Use the technique discussed with respect to Figure 9-73C.) Then, to keep a handy record, circle the known members AB and AG in a miniature sketch of the truss. Circle additional known members as they are found.

STEP 2 Isolate joint G, the next joint with only two unknowns (**Figure 9-78C**). Construct a force diagram starting with known forces F(AG) and the 5,000 N applied force. Find F(GF) and F(GB), and circle those members in a miniature sketch.

STEP 3 Isolate joint B next (**Figure 9-78D**), and construct a force diagram starting with F(AB), then F(GB) to find F(BF) and F(BC). Note that the light line for F(BC) coincides with F(AB) because of the arbitrary clockwise selection of forces around joint B. However, any order of taking the forces around a joint can be used for the method of joints; the final magnitudes and directions would be identical. Circle the known members in a miniature sketch.

STEP 4 Isolate joint F (**Figure 9-78E**). Construct a force diagram by using any order of forces. A clockwise order was used for the figure, starting with known F(FB), then F(FG), followed by the 7,000 N applied force to determine F(FE) and F(FC). Add these to a miniature sketch.

Determine F(CE) by inspection to be 10,000 N. A force acting normal to a joint that has members at 90° to each other can only transmit force to the truss member along the same line as the force. Circle this member in the sketch.

STEP 5 Isolate the last joint D (**Figure 9-78F**) to determine F(DC) and F(DE). Observe that force F(DE) should equal F(EF) due to the 90° angles between member DE and CE, and between DE and the 10,000 N force. Circle the last two members to complete the analysis.

STEP 6 Prepare a table of the forces in the truss members (**Figure 9-78G**).

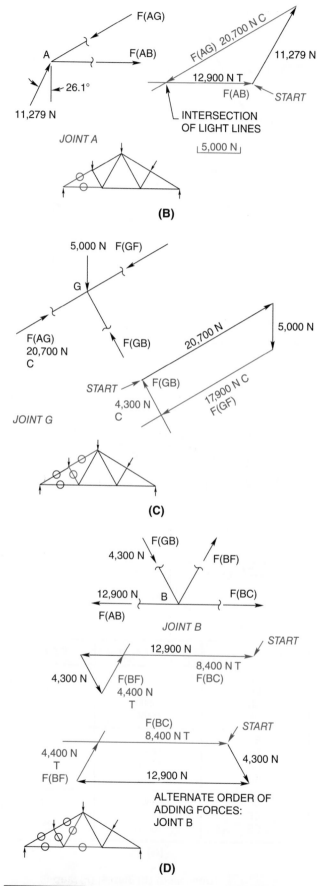

FIGURE 9-78   (*Continued*) (B) Step 1; (C) Step 2; (D) Step 3;

JOINT F

(E)

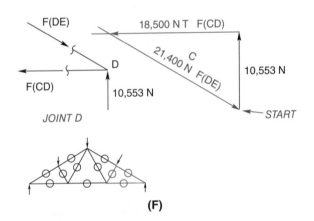

JOINT D

(F)

MEMBER	FORCE N	T OR C
AB	12900	T
BC	8400	T
CD	18500	T
DE	21400	C
EF	21400	C
FG	17900	C
GA	20700	C
BG	4300	C
BF	4400	T
CF	10200	T
CE	10000	C

(G)

FIGURE 9-78   (*Continued*) (E) Step 4; (F) Step 5; (G) Step 6.

EXAMPLE 2(B): METHOD: MAXWELL DIAGRAM WITH BOW'S NOTATION

GIVEN: The same symmetrical truss, drawn to scale, with asymmetrical loads, used in the preceding method of joints example (**Figure 9-79A**). Label the triangular spaces within the truss with numbers. Draw a light line envelope surrounding the truss and passing through each applied force and reaction. Label these spaces with letters.

STEP 1 Start a force diagram (to a scale to keep the diagram on the drawing sheet) by reading the external spaces in a clockwise direction, a to b, b to c, c to d, d to e, and e to a. Plot each applied force and reaction (**Figure 9-79B**).

Either clockwise or counterclockwise will work, but once a pattern is started, it must be used throughout the complete problem both for the Maxwell diagram and for determining whether a member is in tension or compression.

STEP 2 Start at a joint with only two unknowns, A or D. At joint D, go clockwise a to 1, 1 to e, and e to a to locate point 1 (**Figure 9-79C**).

STEP 3 Now joint E has only two unknowns, so continue d to e, e to 1, 1 to 2, and 2 to d to locate point 2 (**Figure 9-79D**).

STEP 4 For joint C, go a-3-2-1-a to locate point 3 (**Figure 9-79E**).

STEP 5 Joint F has only two unknowns, so go d-2-3-4-c-d to locate point 4 (**Figure 9-79F**).

STEP 6 Around point B, go a-5-4-3-a to locate point 5 (**Figure 9-79G**).

STEP 7 For the last joint, A, go a-b-5-a (**Figure 9-79H**). The closing line in direction 5-b should intersect point b on the Maxwell diagram for a correct solution.

STEP 8 Prepare a table of the forces in each member; scale the diagram for magnitudes and use the clockwise pattern to determine whether the force goes into the joint (compression) or goes away (tension) (**Figure 9-79I**). For example, around joint C for member CF, 3-2 goes away from the joint; thus, CF is in tension. Scale the magnitude to be 10,000 N. Around joint F, 2-3 goes away from joint F as expected because CF is a tension member.

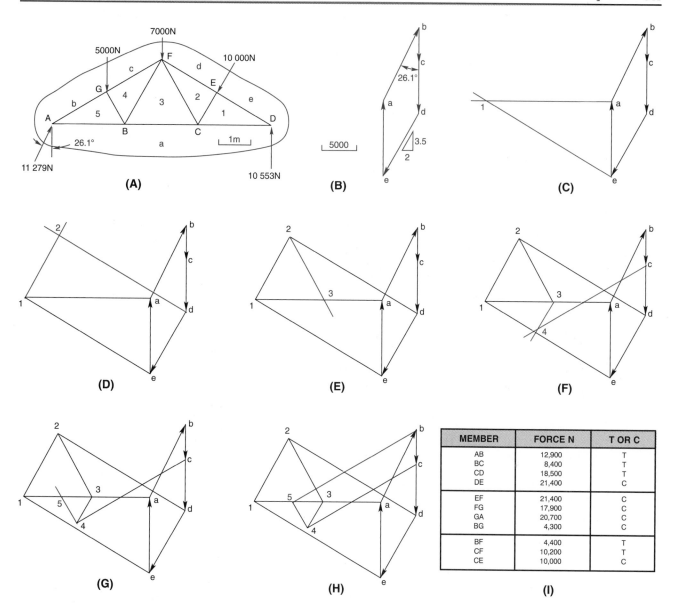

FIGURE 9-79   (A) Example 2 (b) Maxwell diagram with Bow's notation; (B) Step 1; (C) Step 2; (D) Step 3; (E) Step 4; (F) Step 5; (G) Step 6; (H) Step 7; (I) Step 8.

Noncoplanar-Concurrent Force Systems

How to Determine the Forces in the Members of a Noncoplanar-Concurrent Force System

A noncoplanar-concurrent force system could be described as a "tripod" with a force or forces applied at the apex of the tripod—that is, three two-force members, usually rigid, meeting at a point and supporting a load (**Figure 9-80**). Only three unknown forces may be present in order to solve for their magnitudes using basic statics; again the governing equation is $\Sigma F = 0$.

Furthermore, the vectors, representing the forces in the three members, plus the applied load(s) must add tail to head starting at the apex and ending at the apex because the system is in equilibrium, Figure 9-80.

Graphical solutions of three-member, noncoplanar-concurrent force systems incorporate two fundamental descriptive geometry techniques: (1) finding the piercing point of a line in a plane, and (2) finding the true length of a line. In a space diagram, the direction line of a force vector of one member pierces the plane made by the other two vectors. In a force diagram, the true lengths of the three vectors representing the forces in the members are found. The space diagram and the force diagram are drawn to scale and are usually superimposed for convenience.

FIGURE 9-80   Noncoplanar-concurrent force systems.

EXAMPLE 1:

GIVEN: A track beam for a hoist, attached to a wall, held up by two wire ropes, and supporting a load at the end, as illustrated in the simplified line drawing (**Figure 9-81A**). The line drawing of the beam and wire ropes is drawn to scale. The applied load of 40,000 N is drawn to scale.

STEP 1 Draw a direction line, originating at the tip of the applied load vector and parallel to any one of the three members (**Figure 9-81B**). Find the piercing point made by this line in the plane made by the other two members.

Select this first direction line for convenience in arriving at a solution. For example,

the plane made by (b-o)T and (c-o)T in view T shows as an edge in view F. Having the edge view expedites finding a piercing point. Therefore, construct a line parallel to the third member (a-o)F, starting the line at the tip of vector F, and note where the line pierces the (b-o-c)F plane, which actually extends beyond the apparent boundaries of (b-o-c)F. (Keep in mind that a willingness to "go beyond arbitrary boundaries" is a step toward being a creative individual.)

STEP 2 Find F(BO) and F(CO) now that F(AO) has been established. The force vectors parallel to members OB and OC, and lying in the plane of BOC extended, must add tail to head

DESCRIPTIVE GEOMETRY PAYS OFF

Du Vo Minh could remember really disliking descriptive geometry during his years as a student. Projecting lines into other views, determining the true length of a line, identifying the true angle between surfaces, and constructing force diagrams were not what Minh had thought drafting would be about. He wanted to design expensive mansions and unique commercial buildings. He wanted to draw the plans for majestic skyscrapers and towering parking decks. Minh's teacher had reassured him at the time that he would probably do all of these things after graduation, and that he would do them even better if he understood the proper application of descriptive geometry. At the time, the instructor's advice had been a bitter pill to swallow, but Minh had taken it and learned descriptive geometry.

Now, six years later, Minh silently thanks the instructor for insisting that he learn force diagrams and all of the other applications of descriptive geometry. The company he worked for, Commercial Construction, Inc. (CCI), had been awarded a contract to renovate the historic district of downtown Williamsboro, Tennessee, a prestigious multi-million dollar contract. But things were not going well, or at least they hadn't been.

CCI was using trusses in the renovation of several colonial-style buildings, and the trusses were failing.

Several had twisted under only partial loading, and truss members had actually snapped on one. The engineers who had designed the trusses were stumped; they couldn't find the problem. That's where Minh became a hero and earned a promotion.

Using the descriptive geometry concepts he had learned in school, Minh developed a force diagram for the trusses in question. He also created a member table showing the forces for each vector in the diagram.

By examining Minh's diagram and table, CCI's engineers were able to see clearly what they had missed mathematically.

The trusses were quickly redesigned, and CCI's construction crews resumed work. Du Vo Minh was promoted to senior drafter and assigned to the design team responsible for roof trusses.

to F(AO) and end at the origin of the force vector F (**Figure 9-81C**).

STEP 3 Determine the magnitudes of the three unknown forces and prepare a summary sketch (**Figure 9-81D**). Use the principle of revolution (to be discussed in the next section of this chapter) to determine the magnitudes. The sketch of isolated joint O with directions of the three forces found in the force diagram indicates whether each member is in tension or compression. Of course, cables supporting a load must be in tension.

EXAMPLE 2:

GIVEN: A tripod on a sloping terrain supporting a 12,000 N load (**Figure 9-82A**). The tripod and the applied load are drawn to scale.

STEP 1 Select a plane made by two of the members, to expedite the solution, and construct a view showing this plane as an edge (**Figure 9-82B**). Use members (a-o)F and (c-o)F to establish

the plane using auxiliary line (a-x)F, and project the plane (a-o-c)T into view A. Use the technique shown in Figure 9-17, discussed earlier in this chapter, for finding an edge view of a plane surface.

STEP 2 Find the piercing point in plane (a-o-c)A of the vector representing the force in member (o-b)A by constructing a direction line in view A, originating at the tip of the 12,000 N force and parallel to member (o-b)A. Label the piercing point pA and show F(OB) (**Figure 9-82C**). Project a line from point pA into view T, and locate pT by noting where extended line (o-b)T crosses the projected line from pA in view A.

STEP 3 In the T view, construct a line from the piercing point pT, parallel to (o-a)T, until it intersects (o-c)T. This establishes F(OA) and F(OC) (**Figure 9-82D**). Project the point of intersection of F(OA) with (o-c)T into view A and indicate both vectors F(OA) and F(OC) in view A.

FIGURE 9-81   (A) Determining the forces of a noncoplanar-concurrent force system—Example 1; (B) Step 1; (C) Step 2; (D) Step 3.

STEP 4   Determine the magnitudes of the three unknowns—F(OB), F(OA), and F(OC)—and list them in a table (**Figure 9-82E**). All three unknown forces could be determined by the true-length-of-a-line method. For example, F(OA) and F(OC) could be determined in an auxiliary view B by making a fold line A/B parallel to the edge view of plane (a-o-c)A. Then,

F(OA) and F(OC) would be in true size in view B. A different method, however, will be used for the three unknowns—the technique of revolution. For clarity, the technique of revolution will be shown for determining one force only, force F(OB). Revolution will be discussed in the next section.

COMMENT: An alternative method of finding the piercing point described in Step 2 would be to employ the line-projection method, Figure 9-25,

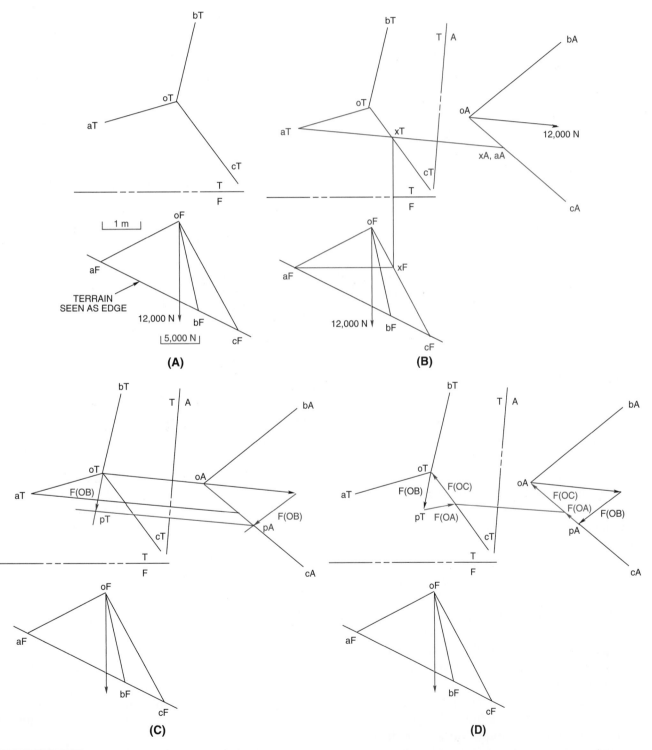

FIGURE 9-82   (A) Determining the forces of a noncoplanar-concurrent force system—Example 2; (B) Step 1; (C) Step 2; (D) Step 3.

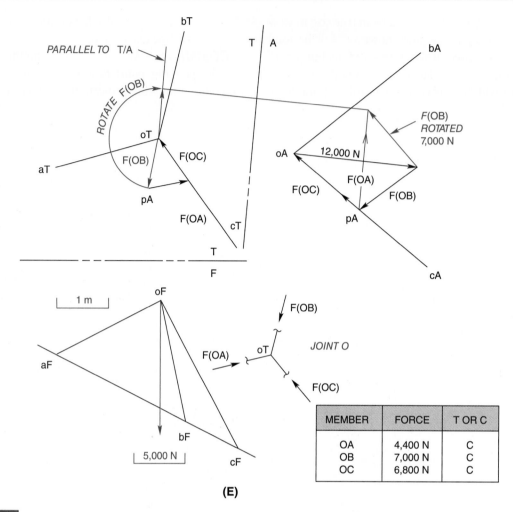

MEMBER	FORCE	T OR C
OA	4,400 N	C
OB	7,000 N	C
OC	6,800 N	C

(E)

FIGURE 9-82   (*Continued*) (E) Step 4.

*discussed earlier in this chapter. The method is shown in views F and T (**Figure 9-83**).*

Revolution

The revolution technique in graphics saves time and space. The technique can be used to solve many true-length and true-angle problems while requiring fewer views.

Consider the basic attributes of the technique by observing the straight member, a-b, rigidly attached to an axis m-n (which can rotate 360°), shown in the T and F views (**Figure 9-84A**). The true length of member a-b is seen twice in the F view as the axis makes one revolution (**Figure 9-84B**). The end bT of member a-b traces a circular path in the T view, and bF traces a straight line in the F view. The circle made by bT is in true size in the T view, and bF's path shows as an edge parallel to the fold line T/F in the F view.

The revolution technique works only if the axis, whether real or imaginary, shows as a point in one view and in true length in any adjacent view.

How to Determine the True Length of a Line

EXAMPLE:

GIVEN: The T and F views from Figure 9-11A are redrawn here (**Figure 9-85A**). In the T view, assume that an axis (m-n)T goes through bT, but it only shows as a point, mT, nT. Then, in the F view, axis mF-nF shows in true length.

STEP 1   In the T view, rotate member (a-b)T about axis mT, nT until (a-b)T is parallel to the fold line T/F (**Figure 9-85B**). Next, in the F view, move point aF parallel to the fold line T/F until it lines up with the projection line from aT's new position in the T view. Redraw (a-b) F in the revolved position where it is now in true length.

COMMENT: Revolution techniques also save time in laying out developments, as discussed in Chapter 10.

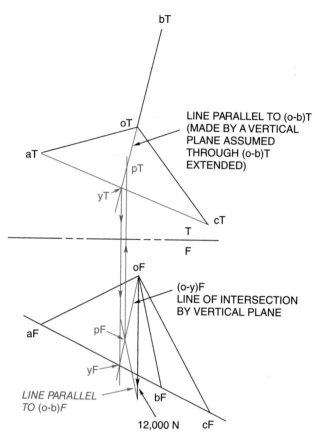

LINE PARALLEL TO (o-b)T
(MADE BY A VERTICAL PLANE ASSUMED THROUGH (o-b)T EXTENDED)

(o-y)F
LINE OF INTERSECTION BY VERTICAL PLANE

LINE PARALLEL TO (o-b)F

12,000 N

FIGURE 9-83   Piercing point by line-projection method.

CIRCLE MADE BY bT

TRUE LENGTH

TRUE LENGTH

PATH OF B SEEN AS AN EDGE

(A)   **(B)**

FIGURE 9-84   (A) Revolution attributes; (B) True length by revolution.

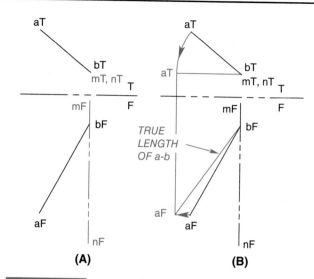

(A)   **(B)**

FIGURE 9-85   (A) Determining the true length of a line—Example 1; (B) Step 1.

How to Determine the True Angle a Line Makes with a Plane

EXAMPLE 1:

GIVEN: Plane (a-b-c-d)T in true size and line (k-j)T piercing the plane at point pT, in the T view (**Figure 9-86A**). In the F view, line (k-j)F pierces plane (a-b-c-d)F, which shows as an edge.

STEP 1   Construct an auxiliary axis m-n that shows as point mT, nT in the T view and goes through piercing point pT (**Figure 9-86B**). Construct the axis (m-n)F as a true-length line in the F view, perpendicular to the T/F fold line.

STEP 2   Rotate line (k-j)T about the axis, in the direction of the smallest angle, until line (k-j)T is parallel to the fold line T/F (**Figure 8-86C**). Construct projection lines from the new positions of kT and jT into the F view. Then, locate the new positions of kF and jF by moving them parallel to the fold line T/F. The point kF moves to the left to the projection line from kT, and point jF moves to the right to the projection from jT. Draw (k-j)F in its revolved position and observe that it is now in true length. Measure the true angle between the line (k-j)F and the plane (a-b-c-d)F.

In summary, the true angle a line makes with a plane must be seen in a view where the plane shows as an edge and the line is in true length.

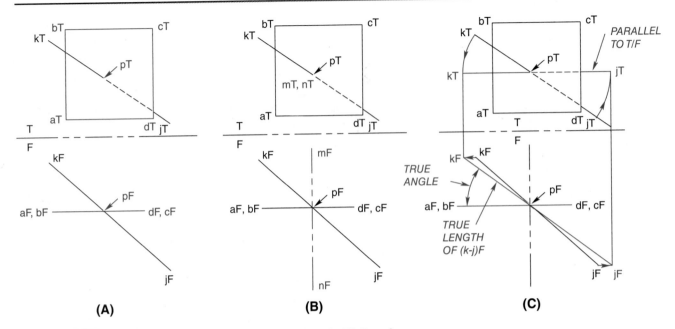

FIGURE 9-86   (A) Determining the true angle; (B) Step 1; (C) Step 2.

EXAMPLE 2:

GIVEN: Views F, T, and A from Figure 9-23F and redrawn here (**Figure 9-87A**).

This example shows that with only one more view, view B, the true angle between the piercing line (x-y)A and the plane (a-b-c)A can be found easily using revolution.

Construct a fold line A/B parallel to the edge view of plane (a-b-c)A shown in the A view (**Figure 9-87B**).

Project the plane and the piercing line into the B view. The plane (a-b-c)B will be in true size, but the line (x-y)B is not.

Therefore, in view A, construct a true-length axis (m-n)A going through the piercing point and perpendicular to plane (a-b-c)A. The axis will show as a point (m-n)B in view B, at the piercing point. Rotate the line (x-y)B about the axis in view B until the line is parallel to the fold line A/B. Then, in view A, move point xA parallel to fold line A/B to coincide with the new position of xB in view B. In view A, draw the line (x-y)A in its rotated position, where it now shows in true length. In view A, measure the true angle between the line (x-y)A and the edge view of plane (a-b-c)A.

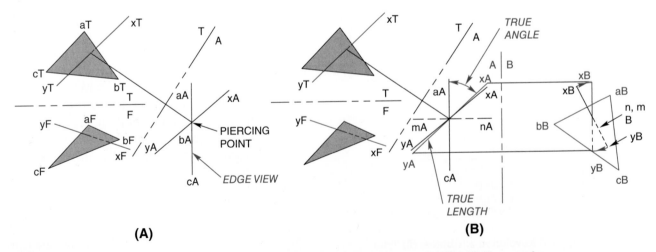

FIGURE 9-87   (A) Determining the true angle—Example 2 (refer to Figure 9-23F); (B) True angle.

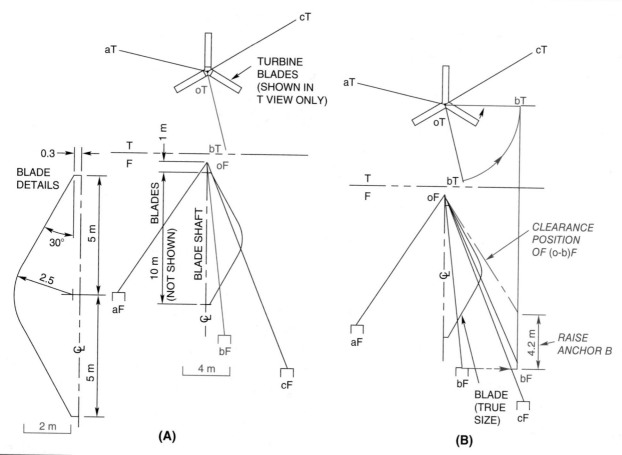

FIGURE 9-88 (A) Vertical wind turbine; (B) Step 1.

How to Determine Clearances Using Revolution

EXAMPLE:

GIVEN: The top and front views of a proposed installation of a vertical wind turbine and turbine blade details (**Figure 9-88A**). *Requirement:* Determine the vertical distance to raise anchor B so cable (o-b)F will clear any portion of the turbine blades by at least .3 meter. Cables (o-a)F and (o-c)F are assumed to meet this requirement, but they should be checked.

STEP 1 Rotate (o-b)T in the T view until it is parallel to the fold line T/F (**Figure 9-88B**). Note that the vertical axis of the turbine qualifies as the axis of revolution because it is in true length in the F view and shows as a point in the adjacent view, view T. Move point bF to the right in the F view as shown. Then, lay out the true size of the turbine blade in the F view.

Construct a straight line from oF, representing (o-b)F, which clears the turbine blade by

at least .3 meter. Find the intersection point of this straight line (o-b)F with a vertical line directly above anchor B.

The vertical distance down from the intersection point to the original elevation of anchor B is the required distance to raise anchor B.

SUMMARY

- Descriptive geometry shows graphically the solution to problems involving points, lines, and planes and their relationship in space.

- Basic theories of descriptive geometry include the following: projecting a line into other views; projecting points into other views; determining the true length of a line; determining the point view of a line; finding the true distance between a point and a line, a plane, or two lines in space; projecting a plane surface in space; developing an edge view of a plane surface in space; and determining the true angle between plane surfaces in space.

- The bearing or heading of a line is the direction of the line as it is drawn on a map in top view. There are two methods used for calling out a bearing: north/south deviation and azimuth bearing.

- The slope of a line is the angle, in degrees, that the line makes with a horizontal plane.

- The grade percent of a line describes the inclination of the line in relation to the horizon by the ratio of the vertical rise to the horizontal.

- The pitch of a line or surface describes the slope of a roof and is expressed as the ratio of vertical rise to 12″ or 12′ in horizontal span.

- A contour line on a plot or map is a line at a constant elevation of the earth, above sea level. Contour interval is the difference in elevation between neighboring contour lines.

- Strike is a compass direction of a horizontal line lying in the plane of an ore vein. Dip is the downward angle of the vein with respect to a horizontal plane at the site of the ore vein.

- A vector is a graphic representation of force, displacement, velocity, or acceleration. A vector has both direction and magnitude.

REVIEW QUESTIONS

Mark the following statements as either true or false.

1. Descriptive geometry is used by drafters, designers, and engineers because it graphically shows the solutions to problems dealing with points, lines, and planes and their relationship in space.

2. Notations are like footnotes.

3. Notations are important because they are used to keep track of all views and points in space.

4. The point view of a line describes the inclination of a line in relation to the horizon.

5. In order to find an edge view of a plane, a true-length line must be located in that plane.

6. A fold line is represented by a thin black line and must be labeled using lowercase letters.

7. The first step to determining the visibility of two intersecting lines (which of the two lines is closer) is to project the exact crossover point of the lines to an adjoining view.

8. The bearing and grade, or bearing and slope, refers to the exact position of a fold line.

9. A contour line on a plot or a map is a line at a constant elevation on the surface of the earth, above sea level.

10. The significance of contour lines being close together on a map is steep terrain.

11. The term cuts and fills refers to earth being excavated for highways.

12. A vector is a directed line segment which, in computer graphics, is defined by its two end-points.

13. Trusses are usually used in pairs or in multiples.

14. A truss is a frame that is usually three-dimensional and made of structural shapes.

15. Space diagrams and force diagrams are used in truss analysis to determine the forces in the members of a truss.

16. Methods of assembling truss joints include welding, riveting, and bolting.

17. When projecting a plane into another view, you must first find at least four points in the plane and project each point into the next view.

Answer the following questions by selecting the best answer.

1. Which of the following is **not** an acceptable method for determining the intersection of two planes?
   a. By inspection
   b. By generating an edge view of the plane surface
   c. By using the line-projection method
   d. Both b and c

2. Which of the following is **not** a step involved in determining the true distance between two parallel lines in space?
   a. Find the true length of each of the two lines.
   b. Find the point view of each of the two lines.
   c. Project a point view of the added true-length line.
   d. The true distance between the parallel lines is the straight-line path between their end-points.

3. Which of the following is **not** a step involved in determining the true distance between two nonparallel (or skewed) lines in space?
   a. Project the true length of either one of the two lines, and project the other line into that auxiliary view (A).
   b. Find the point view of the true-length line, and project the other line into that secondary auxiliary view (B).
   c. Measure the true distance between the point view line at a location that is perpendicular to the other line.
   d. Project the same points from the front view into the right-side view.

4. A point is called out in a given view by:
   a. Lowercase letters
   b. Uppercase letters

c. Either upper- or lowercase letters

d. All of the above

5. Which of the following is an important rule to remember when projecting from one view to another?

a. Always project the top view of a line from a given front.

b. The customary 45° projection line is used.

c. The fold line is always placed between the front V and the top V.

d. Always skip a view between all measurements.

6. Which of the following is **not** one of the basic steps involved in finding the true length of any straight line?

a. Draw a fold line parallel to the line for which the true length is required, and label it A for auxiliary.

b. Extend projection lines from the end-points of the line being projected into the auxiliary view.

c. Projection lines must always be parallel to the fold line.

d. Transfer the end point distances from the fold line in the second preceding view from the one being drawn to locate the corresponding end-points in the view being drawn.

7. Which of the following is **not** one of the basic steps used to find the point view of a line as projected from its true length?

a. Find the true length of the line.

b. Draw a fold line perpendicular to the true-length line at any convenient distance from either end of the true-length line and label it A-B.

c. Extend a light projection line from the true-length line into the secondary auxiliary view.

d. Measure the true distance between the point view line at a location perpendicular to the other line.

8. Which of the following is a use of vectors in engineering and technology?

a. Displacement   c. Velocity

b. Force   d. All of the above

9. Which of the following is **not** part of the parallelogram law?

a. Two vectors

b. Not parallel, drawn to scale

c. Parallel, drawn to scale

d. Having the same units and acting at the same point can be added or subtracted using a specific formula

10. Which of the following is the correct formula for determining the percent of grade?

a. Units of horizon ÷ units of vertical rise × 100

b. Units of vertical rise ÷ units of horizon × 100

c. Units of horizon × units of vertical rise ÷ 100

11. A piercing point is:

a. The exact location of an intersection of two lines

b. The exact location of an intersection of two surfaces

c. The exact location of an intersection of a plane and a surface line

d. None of the above

12. Given the bearing, slope, and length of a line, which of the following is **not** true when constructing the line in a drawing?

a. The bearing is always drawn in the top view.

b. The true length can be drawn if the line is to be exactly horizontal.

c. Any line with an incline up (+) or down (−) from the horizon must be laid out using specific steps.

d. If the line is horizontal, the true length must be constructed before the heading length can be drawn in the top view.

13. Which of the following is **not** true regarding trusses?

a. Trusses are seldom used in multiples.

b. Trusses seldom stand alone.

c. Trusses are frames that are usually two-dimensional.

d. Joints are assumed to be pinned, although they can be welded, bolted, or riveted.

CHAPTER 9 PROBLEMS

The following problems are intended to give the beginning drafter practice in using the various principles of descriptive geometry. Problems 1 through 13 deal with each of the various principles used to solve actual design problems. Problems 14 through 32 apply these principles to develop required views of various objects.

Conventional Instructions

The steps to follow in laying out problems 1 through 13 and 23 through 32 are:

1. On an 8½ × 11″ sheet of paper with a sharp 4-H lead, locate all lines, points, and fold lines per the given dimensions.

2. Complete the problem per the given instructions. Project in the direction of the large arrow.

PROBLEMS

The steps to follow in laying out problems 14 through 27 are:

STEP 1   Study the problem carefully.

STEP 2   Using the given front view, make a sketch of all required views.

STEP 3   Center the required views within the work area with a 1″ (25 mm) space.

STEP 4   Use light projection lines. Do **not** erase them.

STEP 5   Lightly complete all views.

STEP 6   Check to see that all views are centered within the work area.

STEP 7   Check that there is a 1″ (25 mm) space between all views.

STEP 8   Carefully check all dimensions in all views.

STEP 9   Darken in all views using the correct line thickness.

STEP 10 Recheck all work and, if correct, neatly fill out the title block using light guidelines and neat lettering.

CAD INSTRUCTIONS

Students who plan to complete the following problems on a CAD system should begin with these general instructions.

1. Analyze the problem carefully.

2. Plan your drawing sequence. In what order will you develop the drawing?

3. Read the specific instructions for the problem and begin.

PROBLEM 9-1

Locate line a-b and fold lines per the given dimensions. Locate line a-b in the right-side view. Label all points in all views.

PROBLEM 9-2

Locate line a-b and fold lines per the given dimensions. Locate line a-b in the top view. Locate point c on line a-b in all views. Label all points in all views.

PROBLEM 9-3

Locate line a-b and fold lines per the given dimensions. Locate line a-b in the right view. Find the true length of line a-b, projecting from the right side. Label all points in all views.

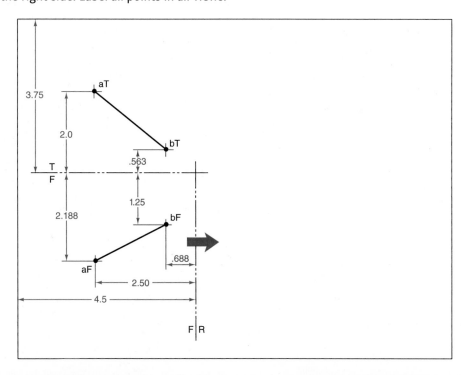

PROBLEMS

PROBLEM 9-4

Locate line a-b and fold lines per the given dimensions. Locate line a-b in the top view. Find the point view of line a-b, projecting from the top view. Label all points in all views.

PROBLEM 9-5

Locate line a-b, point c, and fold lines per the given dimensions. Find the true distance from line a-b to point c. Project from the right view. Label all points in all views.

PROBLEM 9-6

Locate parallel lines a-b and c-d and fold lines per the given dimensions. Find the true distance between lines a-b and c-d. Project from the front view. Label all points in all views.

PROBLEM 9-7

Locate lines a-b and c-d and fold lines per the given dimensions. Find the true distance between lines a-b and c-d. Project from the top view. Label all points in all views.

PROBLEM 9-8

Locate points a, b, and c and fold lines per the given dimensions. Connect points a, b, and c to form a plane. Project plane abc into the top view. Label all points in all views.

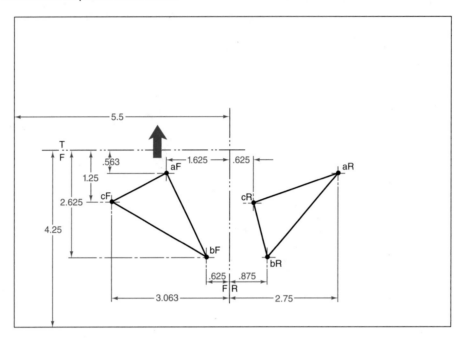

PROBLEM 9-9

Locate points a, b, c, and d, point x, and fold lines per the given dimensions. Connect points a, b, c, and d to form a plane. Project plane abcd into the right view; locate point x in all views. Label all points in all views.

PROBLEM 9-10

Locate points a, b, c, and fold lines per the given dimensions. Connect points a, b, and c to form a plane. Draw a line from point cF, parallel to fold line F-R; label this line cF-xF. Find the edge view of plane surface abc. Project from the right view. Label all points in all views.

PROBLEM 9-11

Locate points a, b, and c and fold lines per the given dimensions. Connect points a, b, and c to form a plane. Draw a line from point bF, parallel to fold line F-T; label this line bF-xF. Find the edge view of plane abc and project from the front view. Label all points in all views.

PROBLEM 9-12

Locate points a, b, and c, point x, and fold lines per the given dimensions. Connect points a, b, and c to form a plane. Find the true distance between plane surface abc and point x. Project from the front view. Label all points in all views.

PROBLEM 9-13

Locate points a, b, c, and d and fold lines per the given dimensions. Connect points a, b, c, and d to form a folded object. Find the true angle between the plane surfaces as viewed directly along the fold, a-b. Project from the right view. Label all points in all views.

Problems 9-14 through 9-22

Draw the front view, top view, right view, and auxiliary view per the listed steps.

PROBLEM 9-14

PROBLEM 9-15

PROBLEM 9-16

PROBLEM 9-17

PROBLEM 9-20

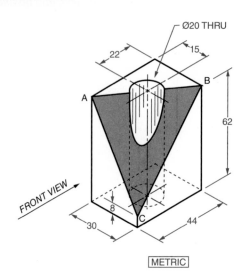

METRIC

PROBLEM 9-18

DEPTH OF GROOVES = .25

PROBLEM 9-21

PROBLEM 9-19

PROBLEM 9-22

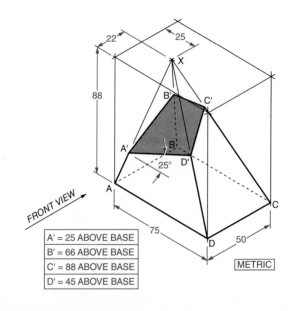

| A' = 25 ABOVE BASE |
| B' = 66 ABOVE BASE |
| C' = 88 ABOVE BASE |
| D' = 45 ABOVE BASE |

METRIC

PROBLEM 9-23

Draw the front view, top view, right view, and required auxiliary views per the listed steps. Find the true angle at abc as viewed along the fold.

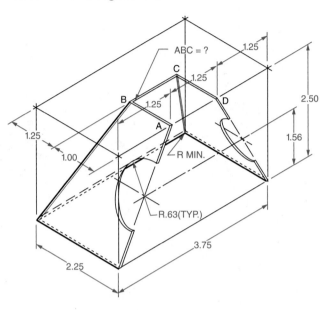

PROBLEMS 9-24 AND 9-25

Draw the front view, top view, right view, and auxiliary view per the listed steps.

PROBLEM 9-24

METRIC

PROBLEM 9-25

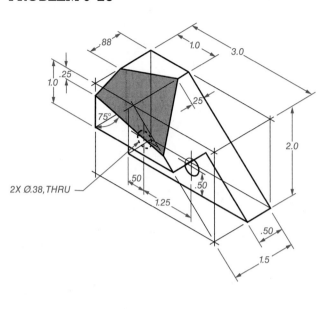

PROBLEMS 9-26 AND 9-27

Draw the front view, top view, right view, and required auxiliary views per the listed steps. Find the true angle between the various surfaces as viewed along the fold.

PROBLEM 9-26

ALL BENDS MIN. R

.062 THICK

PROBLEM 9-27

R.38
C
30°   45°
A
1.625
D
4.5
45°   1.25   30°   PARALLEL
.18
.38
.18   B
1.50   .38   .50
2X Ø.31, THRU
FRONT VIEW
R.38
ALL BENDS MIN. R
.2.0

PROBLEM 9-28

Complete the front view using the line-projection method.
Draw the correct visibility of all lines in the front view.

B
□1.5
C
5.25
A
□3.0
60°
30°
D
.63
T
F
1.0   4.0
2.50   B   2.13
3.56   C
A   3.25
3.00   D

PROBLEM 9-29

Complete the top and front views using the line-projection
method. Draw the correct visibility of all lines by inspection.

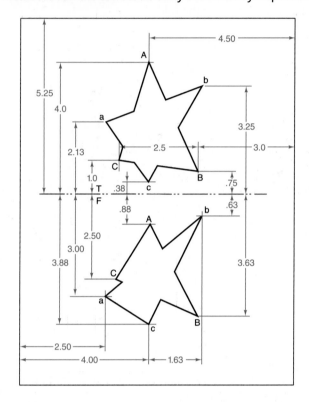

4.50
A
b
5.25
4.0
a
3.25
2.13   2.5   3.0
C
1.0   B
T   .38   c   .75
F
.88   .63
2.50   A   b
3.00
3.88   C
3.63
a
B
c
2.50
4.00   1.63

PROBLEM 9-30

Complete the front view by inspection and projection.
Draw the correct visibility of all lines.

3.25
2.0
5.0
.75   LINE
3.13   X-Y
.63   1.88
T
F
1.25   X   1.5
X
.88
3.50   .88
.56
3.25   2.75   Y
30°   F   A
2.88
Y

PROBLEM 9-31

Complete the front view using the line-projection method. Draw the correct visibility of all lines.

Lightly lay out and draw the two views and the fold line on an A-size sheet of vellum using the given dimensions. Complete all views using the correct line thickness—add all hidden lines.

PROBLEM 9-33

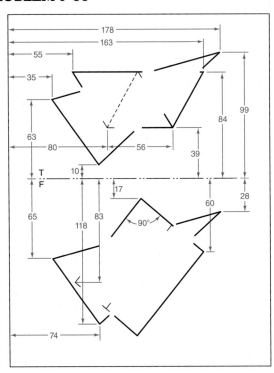

PROBLEM 9-32

Complete the top and front views using the line-projection method. Do not use the right-side view for solving the problem. Draw the correct visibility of all lines.

PROBLEM 9-34

PROBLEM 9-35

METRIC

PROBLEM 9-37

METRIC

PROBLEM 9-36

METRIC

PROBLEM 9-38

METRIC

PROBLEM 9-39

PROBLEM 9-41

PROBLEM 9-40

PROBLEM 9-42

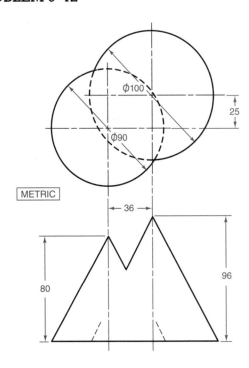

PROBLEM 9-43

Calculate the angles using the given bearings for A through F. Show all math work.

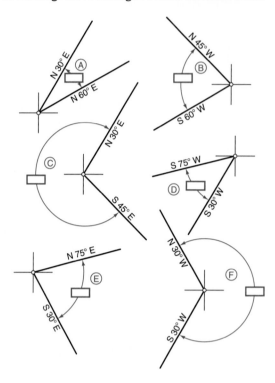

PROBLEM 9-44

On a B-size paper, draw the *top view* and *front view* of a straight roadway (a-b) 280.0 feet long, with an azimuth bearing of N 288° and a grade of −20%. Use a 1″ = 50.0′ scale. Show all light construction work.

GIVEN: Location of fold line F/T and location of points aT and aF.

PROBLEM 9-45

On a B-size paper, lay out the *top view* and *front view* of a radio tower. Use a 1" = 50.0' scale. Show all light construction work. Answer the following questions:

1. What is the grade percent for the guide wire a/b?
2. What is the slope angle of the guide wire?
3. What is the true length of the wire?

GIVEN: Location of the radio tower, guide wires, and fold line F/T.

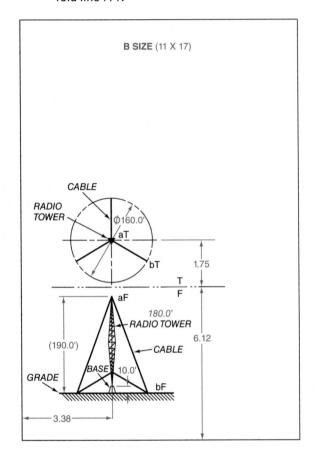

PROBLEM 9-46

On a B-size paper, draw an existing conveyor belt system with the given specifications (see specifications in box). The conveyor belt system is to be redesigned to go from point a directly to point d. Use a 1" = 50.0' scale. Show all light construction work. Answer the following questions:

1. What is the new redesigned bearing from points a to d?
2. What is the new redesigned true length?
3. What is the new grade percent?
4. How much shorter is the new design than the old design?

GIVEN: Location of grade, starting point a, and fold line F/T.

BELT	BEARING	TRUE LG.	GRADE %
a – b	N 30° W	180.0'	40%
b – c	N 80° W	140.0'	30%
c – d	S 45° W	145.0'	15%
ANSWER			
a – d			

B SIZE (11 X 17)

PROBLEM 9-47

On a B-size paper, construct two tunnels, a and b.
Tunnel a starts at point a and has a bearing of N 70° W. It
is 240.0 feet long and has a plus slope angle of 34°. Tunnel
b starts at point b and has a bearing of N 60° E. It is 200.0
feet long and has a plus slope angle of 19°. Use a 1″ = 50′
scale. Show all light construction work. Project from the
front view. Answer the following questions:

1. Do the tunnels ever intersect?
2. If not, how far apart are they?

GIVEN: Locations of tunnel entrances a and b and fold
line F/T.

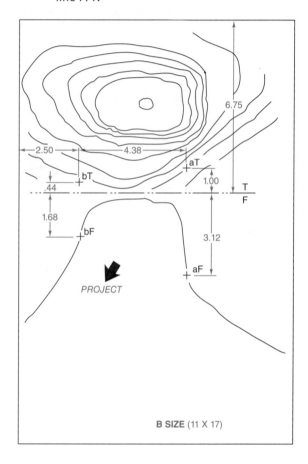

PROBLEM 9-48

On a B-size paper, construct two mining shafts into a
mountain from two different directions and elevations.
Both are cut straight into the mountain at different percents of grades. One shaft starts at point a, and the other
starts at point b. Shaft a has a bearing of S 60° W, with a
grade of + 35%, and is 165.0 feet long. Shaft b has a bearing of S 45° E, with a grade of + 40%, and is 176.0 feet
long. Use a 1″ = 50.0′ scale. Show all light construction
work. Answer the following questions:

1. Do the shafts intersect?
2. What is the true distance from the entrance of
shaft a to the entrance of shaft b?
3. What is the bearing and percent of grade from the
two entrances?

GIVEN: Locations of the shaft entrances a and b, fold
line F/T, and mountain.

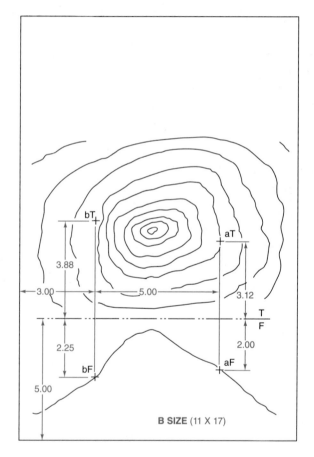

PROBLEM 9-49

A 40′ wide, level road, at elevation 150′, is to be run through a hilly terrain as shown on the contour map. Trace the map and use the accompanying scale to determine cut and fill boundaries for a slope of 1:1 for both the cuts and the fills.

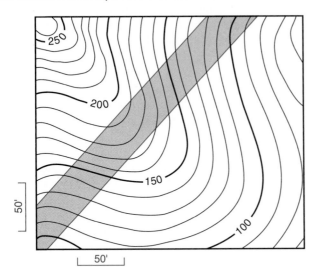

PROBLEM 9-50

Vehicle A is traveling due north at 150 kph (90 mph); vehicle B, due west at 110 kph (66 mph). Vehicle B is 40 kilometers north and 40 kilometers east of vehicle A at the instant shown. What is the shortest distance between the vehicles A and B when they pass each other? An example velocity vector diagram illustrates the consequences of A and B going at the same speed. What should vehicle B's speed be so that the shortest distance between A and B is 15 kilometers when they pass each other?

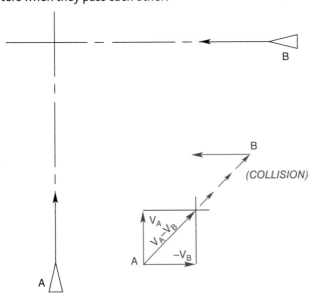

PROBLEM 9-51

An ore outcropping is shown on the topographic contour map. Points A and B are on the upper surface (called the hanging wall) of the vein. A vertical bore hole at point C indicates that the hanging wall is at 877′ elevation. The core hole extended at C indicates that the bottom surface of the ore vein (footwall) is at 841′ elevation. For the information given for the locations of points A, B, and C and the vein surfaces, determine the following:

1. The strike and dip of the ore vein.
2. The true thickness of the ore vein.

FROM 0		ELEV
A 435 E 610 N		980′
B 815 E 370 N		940′
C 605 E 195 N		955′

PROBLEM 9-52

The simplified drawing of the four-bar linkage shown (three links and the rigid support) has link A-B rotate 360° around pinned connection A. Link A-B's motion causes link C-D to oscillate about pinned connection D. At the instant shown, a velocity analysis has link C-D rotating CCW at 18.3 radians per second. Note that the velocity of point B of link A-B is V(B) = ω(AB) × (0.03 m) = 1.8 m per second. The component of V(B), which is coincident with member B-C, causes link B-C to move to the left with a velocity of V(BC) = 1.2 m per second as found graphically. Next, a component of V(BC) perpendicular to C-D causes C-D to rotate CCW about point D with an angular velocity of ω(CD) = (1.19)/0.065 = 18.3 radians per second. Determine the following:

1. The instantaneous angular velocities of ω(CD) when A-B is at 90°, 180°, and 270°.

2. At what values of ω will C-D have an angular velocity of zero? Why?

FOUR-BAR LINKAGE

PROBLEM 9-53

The cam and cam follower shown represent a camshaft cam and valve follower in an internal combustion engine. The cam rotates at a constant rate of 2,000 rpm, which is 209 radians per second (2000/60 = 33.33 RPS; and 33.33 × 2 × π = 209 radians per second). The cam has rotated 225° at the instant shown and is transferring a velocity of 120 inches per second to the follower as determined by a graphical analysis. The velocity of the point in contact T at this instant is V(T) = ω(cam) × 0.98 inch = (209 radians per second) × (0.98′) = 205″ per second. Then, the resolution of the 205″/sec gives 120″/sec to the follower, and the other component indicates a sliding velocity parallel to the bottom surface of the valve. Hence, good lubrication is required.

Lay out the cam from the dimensions given and graphically determine the velocity of the valve throughout one revolution, starting at θ = 0°. Record the results in a graph of V(follower) versus angle of rotation θ. *Suggestions*: (1) From θ = 180° to 360° determine V(follower) at 15° intervals. (2) Lay out the cam on one sheet of drawing paper, and use an overlay sheet of drawing paper for the intervals.

CAM DETAIL

PROBLEM 9-54

The laser beam a-b shines on the metal door shown. As the door swings open, the laser beam traces a path on the door's surface. Determine the path for the door opening 150° and show the path in view F.

PROBLEM 9-55

A basic toggle mechanism is used to generate a large force from a smaller force for punching holes in metal. For the position shown in the simplified mechanism, what force is generated at punch A, vertically downward? When the 40,000 N force is at B2, what force at A is generated vertically downward?

PROBLEM 9-56

For the coplanar-concurrent system shown, determine the forces in members AB and BC using the method of joints.

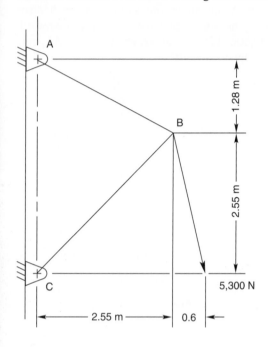

PROBLEM 9-57

Use the method of Maxwell diagram with Bow's notation to determine the forces in each member of the truss shown. Tabulate the magnitudes and whether the members are in tension or compression.

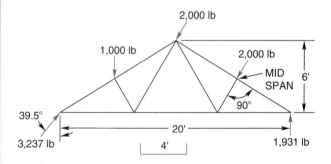

PROBLEM 9-58

Redraw the tripod and the 1,500 lb applied force starting at the lower left-hand area of a drawing sheet. Determine the force in each member by the method shown in this chapter, and use revolution to finally determine the magnitude of each force. *Suggestion*: Use members o-b and o-c to form a plane.

PROBLEM 9-59

From the given information, lay out the four noncoplanar-concurrent force vectors, and add them tail to head in any order (the order can differ in each view also) to determine the equivalent resultant of the four forces. Then, find the magnitude of the resultant by revolution. Also, determine the bearing and percent slope of the resultant. The coordinates of the four forces originating from point o are:

a. (10,5,5)
b. (4,–13,–10)
c. (–8,–8,8)
d. (–15,5,–10)

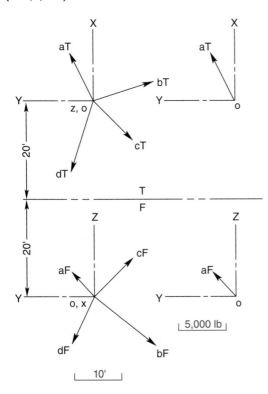

PROBLEM 9-60

A ham radio operator needs to know the lengths of three guy wires for his antenna installation and where to locate guy wire o-c. At the lower left corner of a sheet of drawing paper, construct a layout of the given roof information and solve the ham's guy wire problem using rotation. Guy wire o-c is to clear corner f of the chimney by one foot. Show all work and tabulate the answers.

PROBLEM 9-61

A tension spring a-b constrains a 30″ wide door. The spring shown in simplified schematic form has a spring constant of 5 lb/in.—i.e., 5 lb will stretch the spring 1″. The door shown in the closed position pivots about a hinge located at point c. At the position shown, the spring has an initial tension of 6 lb. What will be the total tension in the spring when the door is opened 90°? Solve the problem using revolution.

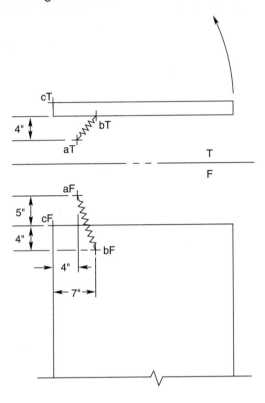

PROBLEM 9-62

The simplified drawing shows a personnel lift, powered by a hydraulic cylinder, DC. At different angles θ, the force F(DC) required by the cylinder is calculated by dividing the moment due to the 1,000 lb load (1,000 lb load × the shortest distance to the line of action of the 1,000 lb load from point A) by the shortest distance to the line of action of the cylinder's force F(DC). The shortest distances are called moment arms. Redraw the positions of the lift as specified in the table, locate the cylinder, measure moment arms, and do the calculations to finish the table.

	ANGLE	1,000 lb MOMENT	CYLINDER MOMENT	FORCE IN CYL.
ΣM_A = 0	0°	22,000 ft-lb = 3.1 x F(CYL)		7,097 lb
	15°			
	30°			
	40°	17,500 ft-lb = 2.3 x F(CYL)		7,609 lb
	50°			

NOTE: • ARMS AE AND BF ARE ALWAYS PARALLEL
• A, B, C, D, & E ARE PINNED JOINTS
• CHANGE IN LENGTH CD AS LIFT RISES IS STROKE OF CYLINDER
• A MOMENT ABOUT A POINT, DUE TO A FORCE, EQUALS THE PRODUCT, OF THE FORCE TIMES THE SHORTEST DISTANCE TO THAT POINT FROM THE LINE OF ACTION OF THE FORCE

$\Sigma M_A = 0$ AT 0° (1,000)(22.0) = (3.1)(FORCE IN CYLINDER) ft-lb
AT 40° (1,000)(17.5) = (2.3)(FORCE IN CYLINDER) ft-lb

PROBLEM 9-63

A floating golf-course green, supported by a concrete float, is moved by changing the lengths of four wire ropes powered by winches inside the float. From the position shown, the green is moved 130 yards west and 20 yards north. Use revolution to determine the change in length of each cable, A, B, C, and D. Tabulate the changes. (Ask your golfing friends if they could hit the green at the new position.)

10

Patterns and Developments

CHAPTER OUTLINE

Developments • Parallel line development • Radial line development • Triangulation development • True-length diagrams • Notches • Bends • Summary • Review questions • Chapter 10 problems • CAD instructions

CHAPTER OBJECTIVES

Upon completion of this chapter, students should be able to do the following:

▶ Define the term *development* and list the major kinds of developments.

▶ Demonstrate proficiency in making parallel line developments.

▶ Demonstrate proficiency in making radial line developments.

▶ Demonstrate proficiency in the application of triangulation developments.

▶ Demonstrate proficiency in constructing a true-length diagram.

▶ Explain when and how the two major types of notches are used.

▶ Demonstrate proficiency in performing bend calculations.

KEY TERMS

Bend allowance charts

Developed length

Development

Development surface

Double-curved surface

Flat-plane surface

Parallel line development

Parallel station lines

Plane surface

Radial line development

Seams

Single-curved surface

Stretchout line

Thickness gage

Triangulation

Triangulation development

True distance

True length

Warped surface

Weight gage

Developments

A *development* is the pattern or template of a shape that is laid out in a single flat plane in preparation for the bending or folding of a material to a required shape. Surface developments are used in many different industries. Some examples of objects requiring developments are cereal boxes, toolboxes, funnels, air-conditioning ducts, and simple mailboxes.

Three major kinds of surface developments are parallel line developments, radial line developments, and triangulation developments. *Parallel line developments* are used for objects having parallel fold lines. *Radial line developments* are those whose fold lines radiate from one point (**Figure 10-1**). *Triangulation development* is the development process of breaking up an object into a series of triangular plane surfaces (**Figure 10-2**). Each kind of development is explained here in full.

Surfaces

A *development surface* is the exterior and/or interior of the sheet material used to form an object. The various kinds of surfaces include plane surface, single-curved surface, and double-curved surface.

If any two points anywhere on a surface are connected to form a straight line, and that line rests upon the surface, it is a *plane surface*. If all points on a surface can be interconnected to form straight lines without

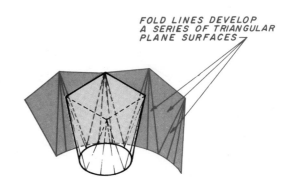

RADIAL LINE DEVELOPMENT

FIGURE 10-2   Triangulation development.

exception, it is a *flat-plane surface*. The top of a drawing board is an example of a flat-plane surface. A flat-plane surface can have three or more straight edges. Objects such as a cube or a pyramid are bounded by plane surfaces. If a surface can be unrolled to form a plane, it is a *single-curved surface*. A cylinder or a cone is an example of a single-curved surface.

A surface that cannot be developed because it is neither a plane surface nor a curved surface is a *warped surface*. An automobile fender is an example of a warped surface. This kind of surface is usually stamped or pressed into shape. An object fully formed by curved lines with no straight lines is a *double-curved surface*. A sphere is an example of a double-curved surface. This type of surface cannot be developed exactly by using flat patterns; only an approximate development can be made.

Laps and Seams

Extra material must be provided for laps and *seams*. Many kinds of seams are available (**Figure 10-3**). When making a choice, the drafter must take into account the thickness of the metal so that crowding of the metal at the joints is not a problem. Allowance must be made for the gluing, soldering, riveting, or welding processes for joints. The method of fastening joints together varies with the material and, accordingly, the elimination of rough or sharp edges. Tabs for the drawing problems at the end of this chapter should be designed to support the joining surfaces (**Figure 10-4**).

Design Practices

The drafter should lay out developments according to the dimensions of stock materials for economy and the best use of materials and labor. The stock material area required to cut a pattern should be kept to the smallest convenient size. It is good practice to put the seam at

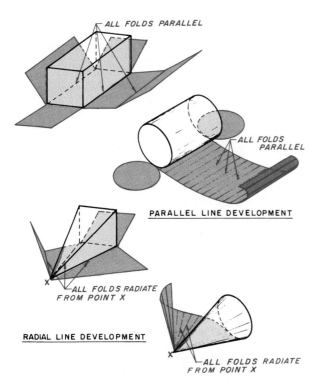

FIGURE 10-1   Parallel line development and radial line development.

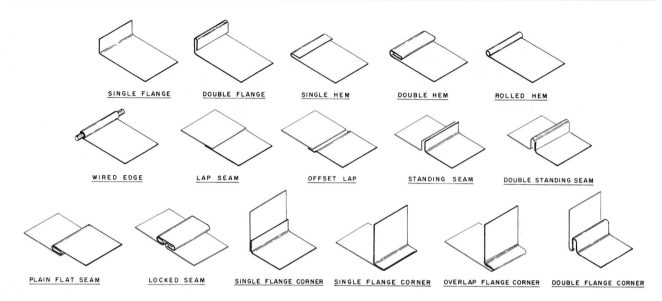

SINGLE FLANGE   DOUBLE FLANGE   SINGLE HEM   DOUBLE HEM   ROLLED HEM

WIRED EDGE   LAP SEAM   OFFSET LAP   STANDING SEAM   DOUBLE STANDING SEAM

PLAIN FLAT SEAM   LOCKED SEAM   SINGLE FLANGE CORNER   SINGLE FLANGE CORNER   OVERLAP FLANGE CORNER   DOUBLE FLANGE CORNER

FIGURE 10-3   Many types of seams are available.

FIGURE 10-4   Tabs should support the joining surfaces.

the shortest joint and to attach tops and bottoms along the longest possible seam or bend to reduce the length requiring soldering, riveting, or welding. It is assumed that the inside surface of the final object is the side that the pattern defines. (The important dimension sizes are usually the inside surfaces.) Fold or bend locations in the material are shown on the pattern with thin, solid lines and are locations that are also assumed to occur on the inside surface of the final object.

Thickness of Material

The actual thickness of sheet metal is specified by gage numbers. Each gage number indicates a particular thickness of material. **Figure 10-5** lists the gage number and gage sizes for sheet and plate steel. These are nominal thicknesses, subject to permissible tolerances. Today, there is considerable confusion when using the gage numbering system.

In 1893, Congress established the United States Standard Gage, and it was primarily a *weight gage* rather than a *thickness gage*. It was derived from the weight of wrought iron. At that time, the weight of wrought iron was calculated at 480 pounds per cubic foot; thus, a plate 12″ square and 1″ thick weighed 40 pounds. A No. 3 U.S. gage represents a wrought iron plate weighing 10 pounds per square foot. Therefore, if a weight per square foot 1″ thick is 40 pounds, the plate thickness for a No. 3 gage equals $10 \div 40 = 0.25″$, which is the original thickness equivalent for a No. 3 U.S. gage. Since this and all other gage numbers were based on the weight of wrought iron, they are not correct for steel. To add to the confusion, there is considerable variation in the gage thickness for different kinds of material. For example, a gage used for such nonferrous materials as brass and copper is sometimes also used to specify a thickness for steel or vice versa.

Today, to help eliminate the problems, the decimal or metric system of indicating gage size is now replacing the older gage size numbering system.

Parallel Line Development

In parallel line developments, all fold lines are parallel. (Refer back to Figure 10-1.)

A three-view drawing of a simple two-inch cube is shown in **Figure 10-6**. A cube is a specific

| SHEET METAL | | | | | |
| GAGE | ALUMINUM | | BRASS | | STEEL | |
	THICKNESS	WT./SQ. FT.	THICKNESS	WT./SQ. FT.	THICKNESS	WT./SQ. FT.
8	.1285	1.812	.1285	5.662	.1644	6.875
9	.1144	1.613	.1144	5.041	.1494	6.250
10	.1019	1.440	.1019	4.490	.1345	5.625
11	.0907	1.300	.0907	3.997	.1196	5.000
12	.0808	1.160	.0808	3.560	.1046	4.375
13	.0720	1.020	.0720	3.173	.0897	3.750
14	.0641	.907	.0641	2.825	.0747	3.125
15	.0571	.805	.0571	2.516	.0673	2.812
16	.0508	.720	.0508	2.238	.0598	2.500
17	.0453	.639	.0453	1.996	.0538	2.250
18	.0403	.580	.0403	1.776	.0478	2.000
19	.0359	.506	.0359	1.582	.0418	1.750
20	.0320	.461	.0320	1.410	.0359	1.500
21	.0285	.402	.0285	1.256	.0329	1.375
22	.0253	.364	.0253	1.119	.0299	1.250
23	.0226	.318	.0226	.996	.0269	1.125
24	.0201	.289	.0201	.886	.0239	1.000
25	.0179	.252	.0179	.789	.0209	.875
26	.0159	.224	.0159	.700	.0179	.750
27	.0142	.200	.0142	.626	.0164	.688
28	.0126	.178	.0126	.555	.0149	.625
29	.0113	.159	.0113	.498	.0135	.562
30	.0100	.141	.0100	.441	.0120	.500
31	.0089	.126	.0089	.392	.0105	.438
32	.0080	.112	.0080	.353	.0097	.406
33	.0071	.100	.0071	.313	.0090	.375
34	.0063	.089	.0063	.278	.0082	.344
35	.0056	.079	.0056	.247	.0075	.312
36	.0050	.071	.0050	.220	.0067	.281
37	.0045	.063	.0045	.198	.0064	.266
38	.0040	.056	.0040	.176	.0060	.250
39	.0035	.050	.0035	.154	–	–
40	.0031	.044	.0031	.137	–	–

FIGURE 10-5   Thickness of material.

FIGURE 10-6   Simple multiview drawing of a cube.

kind of prism. Notice that all the fold lines that form the lateral surfaces of a prism are parallel (**Figure 10-7A**). If the cube were to be unfolded, it would appear as it does in **Figures 10-7B** through **10-7D**. The finished development is shown in **Figure 10-7E**.

Notice that the cube was developed or laid out from a baseline, sometimes referred to as the *stretchout line*. Fold lines are always oriented 90° from the baseline, as illustrated. If the pattern were to be cut from a flat sheet of material, tabs would be needed to fasten the cube together. Tabs must be positioned so as to make each meet with an untabbed edge (**Figure 10-7F**). Notice that there are seven tabs and seven blank edges. Each tab folds to meet a particular blank edge.

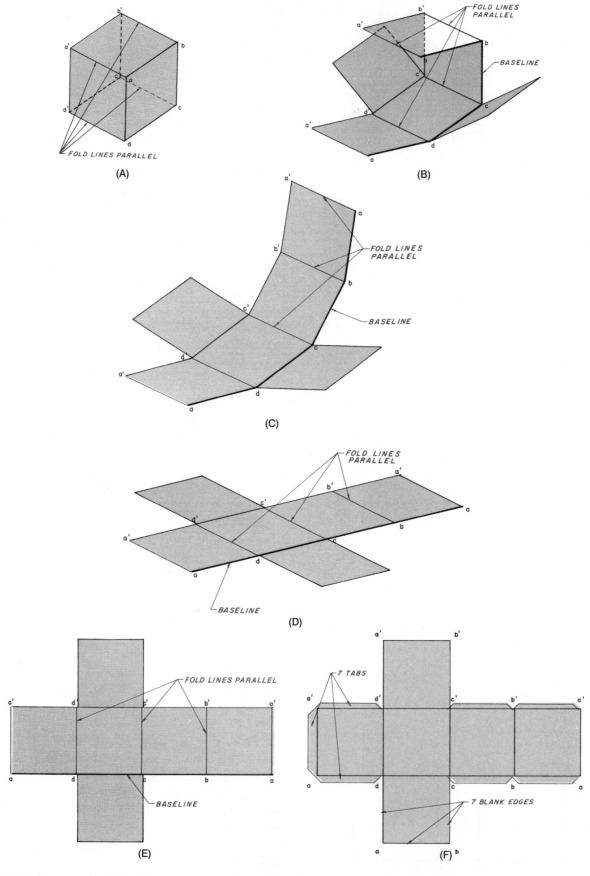

How to Develop a Truncated Prism Using Parallel Line Development

GIVEN: A prism with a front view, top view, and auxiliary view (**Figure 10-8A**).

STEP 1 Locate the baseline, which must always be 90° from the parallel fold lines. Label each fold line clockwise in alphabetical or numerical order. Always start with the shortest fold line (**Figure 10-8B**). Either fold line a-a′ or b-b′ would be eligible, and a-a′ is selected. Notice that the starting fold line is also the finishing fold line, and each is labeled at the same location. The finishing fold line is a-a′. All real (or true) lengths of the fold lines are seen in the front view, and the real distances between their locations are seen in the top view.

STEP 2 On a separate sheet of paper, construct a pattern baseline, allowing enough space for the complete development, top and bottom surfaces, and the required tabs. At the left end of the baseline, draw a line perpendicular to the baseline. This is the first fold line and is also a seam edge. Transfer the true length from fold line a-a′ (as measured from the three-view drawing in Figure 10-8A). Label this line a-a′ also, as illustrated in **Figure 10-8C**.

STEP 3 From the top view of Figure 10-8A, obtain the true distance from the first fold line location at a-a′ to the second fold line location at b-b′.

On the pattern baseline, transfer this distance, measuring from the first fold line a-a′, and draw the next fold line parallel to the first fold line. Label it b-b′ (**Figure 10-8D**).

STEP 4 Repeat the first three steps, alternating true fold line lengths and true distances between fold lines until all the true lengths and true distances are transferred. Each of the true lengths a-a′ through e-e′ is transferred from the front view, and the true distances between them are found from the top view of Figure 10-8A. Connect the points as illustrated in **Figure 10-8E**.

STEP 5 The edge selected for attachment of the top surface should be made to keep the rectangle size from which the whole pattern is cut as small as possible. Line a′b′ would be the best, except an acute angle cutout c′b′a′ is difficult to make. The fold line a′d′ is next best, as shown. From Figure 10-8A, transfer the true size and shape of the top surface (auxiliary view) and the bottom surface (**Figure 10-8F**).

STEP 6 Add tabs enough to fasten the edges. The locations should be selected so as not to enlarge the material area needed for the pattern cutout (**Figure 10-8G**). Check to verify that the number of tabs equals the number of blank edges in the same manner as is illustrated in Figure 10-7F. This completes the parallel line development of the object in Figure 10-8A.

(A)

(B)

FIGURE 10-8   (A) Development of a truncated prism using parallel line development; (B) Step 1;

FIGURE 10-8 (*Continued*) (C) Step 2; (D) Step 3; (E) Step 4; (F) Step 5; (G) Step 6 (completed development of the prism).

How to Develop a Truncated Cylinder Using Parallel Line Development

GIVEN: A cylinder with a front view, top view, and auxiliary view (**Figure 10-9A**). In developing a cylinder, the same procedure is used as when developing a prism. Because the cylinder has a rounded surface, line segments called station lines must be assumed or chosen on the lateral surface. This is explained in Step 1.

STEP 1 Locate a baseline that must always be 90° from the fold lines, sometimes referred to as *parallel station lines*. The bottom corner of the lateral surface is selected as a convenient location because it already exists. In the view where the cylinder appears as a diameter (the top view in this example), divide the diameter into equal divisions (**Figure 10-9B**). This example uses 12 equal divisions, but any

number of appropriate equal divisions would do. These divisions locate fold line positions, which are used only for construction and layout purposes. Starting with the shortest fold line, consecutively label each division—in this example, a through 1. Note that a-a′ marks the beginning and ending of the same line, which meets from opposite directions. The true "distances" between station lines can be approximated in this top view. Project each point into the next (front) view in order to find the true "lengths" of the fold lines. It should be re-emphasized that the fold lines must be 90° from the baseline.

STEP 2 On a separate sheet of paper, construct a pattern baseline, allowing enough space for the complete development, both ends, and the required tabs. At the left end of the pattern baseline, draw a line perpendicular to the baseline. This is the first fold line location

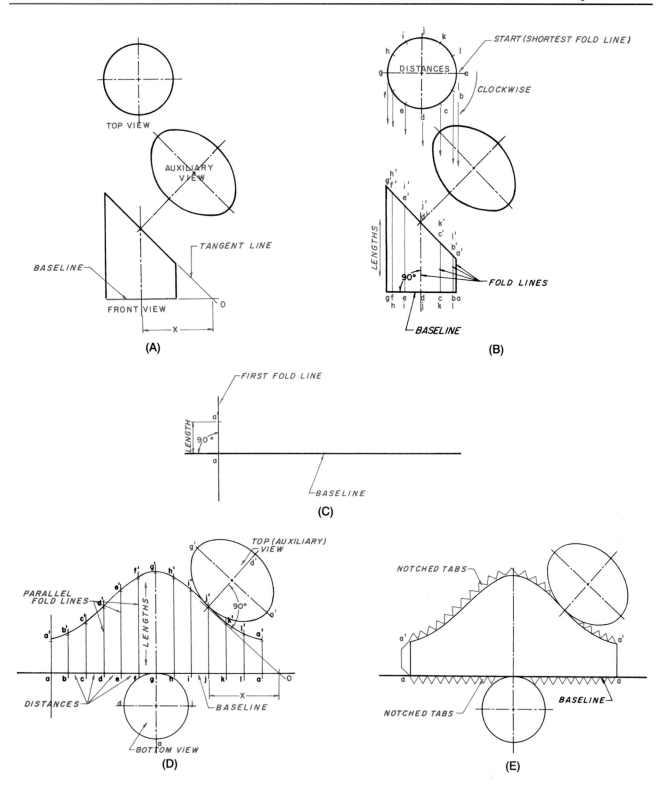

FIGURE 10-9 (A) Development of a truncated cylinder using parallel line development; (B) Step 1; (C) Step 2; (D) Steps 3 and 4; (E) Step 5 (completed development of the cylinder).

a-a'. Transfer the true length a-a' from the front view of Figure 10-9B. Label this line as illustrated in **Figure 10-9C**.

STEP 3   Two methods are used for finding the distances between station lines. Method A is

an approximate method that assumes the cylinder to be a prism, whereas Method B is mathematically correct.

METHOD A: From the top view of Figure 10-9B, obtain the direct chordal distance from the first station line a-a′ to the second fold line b-b′. On the pattern baseline, draw the second fold line parallel to the first station line a-a′ at the chordal distance found from the top view, and label it b-b′ (**Figure 10-9D**). Repeat these steps, alternating chordal lengths and true distances until all station lines have been located and drawn. Note that the last fold line length a-a′ is a duplicate of the first.

METHOD B: The required length to form the cylinder's circumference can be calculated using the following formula: circumference = cylinder diameter × π (π = 3.1416, approximately). The baseline is extended to this length and divided into the same number of divisions as was done at Step 1 in the top view of the cylinder, Figure 10-9B. The division can best be performed as a construction exercise, rather than mathematically. Refer back to Chapter 4, "Geometric Construction." Each of the 13 division locations (from 12 spaces) is the correct location of the required station lines and can be successively labeled.

STEP 4   Add the true size and shape of the bottom surface (bottom view) and the top surface (auxiliary view). The attachment locations of these surfaces are particularly difficult, as the mathematical point of tangency is not easily isolated on the pattern. The top surface location is chosen to allow the widest access of cutting to this point. The bottom surface can be added to any of the fold lines. In order to locate the top surface, the distance X from center to point 0 is transferred from Figure 10-9B to the development and projected 90°, as illustrated in Figure 10-9D.

STEP 5   Add the tabs necessary for edge attachments. In cylinders, a notched tab must be used in order to accommodate crowding by surface curvature. A tab must also be added at fold line a-a′ in order to attach edge a-a′ to the other edge a-a′; the top (auxiliary) and bottom fold to meet the notched tabs (**Figure 10-9E**)

Regardless of the shape, whether it be a cylinder or a prism, parallel line developments follow essentially the same procedure. **Figure 10-10** shows a prism truncated at both ends, which necessitates creating a new baseline other than a part edge to ensure that all fold lines (instead of station lines) are at 90° to the baseline. **Figure 10-11** is a cylindrical prism,

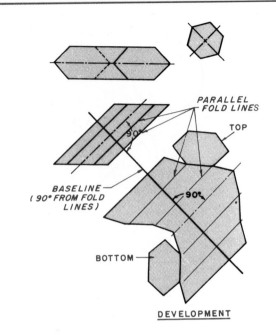

FIGURE 10-10   Creating a new baseline.

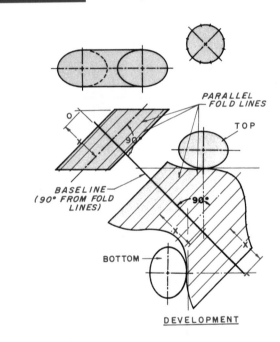

FIGURE 10-11   Using station lines instead of fold lines.

truncated at each end, which requires the selection of station lines instead of using existing fold lines. **Figure 10-12** shows a series of truncated cylinders, similar in construction to the preceding views, whose patterns are used to form an elbow.

Radial Line Development

Radial line development is different from parallel line development in that all fold lines or line segments radiate from one point. (Review Figure 10-1.)

Series of patterns used to form an elbow.

As in all patterns, true lengths and true distances must be used in laying out the developments.

How to Develop a Truncated Pyramid Using Radial Line Development

GIVEN: A two-view drawing and an auxiliary view of a pyramid with the top cut off at an angle (truncated) (**Figure 10-13A**) Label points clockwise starting with the shortest fold line. In this example, either a-1 or b-2 could be used. Line a-1 is selected, but the fold line location a-1 is also the meeting corner of edges a-1 and a-1. True distances between the end-points of the fold lines are evident in

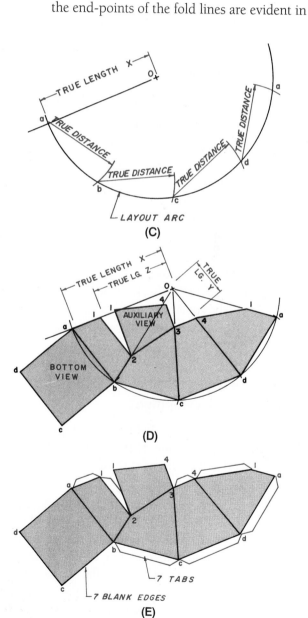

FIGURE 10-13   (A) Development of a pyramid using radial line development; (B) Step 1; (C) Step 2; (D) Step 3; (E) Step 4 (completed development of a pyramid).

the top view. For example, a-b, b-c, c-d, and d-a are true lengths.

STEP 1 To find the true length of the fold lines, they must be rotated to a position that is parallel to the frontal viewing plane (**Figure 10-13B**). In the top view, point a is rotated as shown and projected into the front view. This revolved line from a′ to 0 is its true length. Similarly, point c in the top view is rotated and projected into the front view, giving the true length of c′-0. The same procedure is used to find the true lengths 3′-0 and 1′-0. To construct a development of this object, continue with the following steps.

STEP 2 Using the true length from 0 to a (a′-0), swing the arc as shown in **Figure 10-13C**. On this arc, mark off as chord lengths the true distances between the fold line end-points a to b, b to c, c to d, and d to e, as transferred from the top view of Figure 10-13B. Label all points as shown. Connect the fold lines a to 0, b to 0, c to 0, d to 0, and a to 0.

STEP 3 Using the true lengths found in Figure 10-13B, locate points 1, 2, 3, and 4 on the respective fold lines 0-a, 0-b, 0-c, and 0-d. Note that 0-1 is the same length as both 0-2 and 0-1, and 0-3 is the same length as 0-4. Add the auxiliary and bottom views, transferring them directly from the original Figure 10-13A (**Figure 10-13D**).

STEP 4 Add the required tabs as illustrated in **Figure 10-13E**. Check to be sure the number of tabs equals the number of blank edges. In this example, there are seven tabs and seven blank spaces.

How to Develop a Truncated Cone Using Radial Line Development

As with cylindrical parallel line developments, station lines are line segments that must be positioned on the lateral surface in order to draw a development.

GIVEN: A two-view drawing of a round cone, with the top cut off at an angle (truncated), and having no fold lines (**Figure 10-14A**). It is necessary to first complete the given views and to draw a true view of the inclined flat surface. To project the flat surface to the top view, the front view is sliced at points A through G. These seven slices are projected into the top view and appear as circles, correspondingly labeled A through G. The point of intersection of the edge view with

each slice is projected into the top view. To construct the auxiliary view, project from the seven point intersections perpendicular from the edge view of the flat surface to any convenient distance for the auxiliary view. Draw and use the center line as a reference line, and transfer all point-to-center line distances from the top view to the auxiliary view; refer to reference distances X.

STEP 1 To construct a development of this object, station lines need to be established, and their true lengths and true distances must be determined. In the top view of **Figure 10-14B**, divide the base circle into 12 equal parts, and label each point. In this example, numbers 1 through 12 are arbitrarily used, positioned as illustrated. The station line 0-3 must have two identities because two edges will meet at this point. The true distances between station lines are located between their end-points on the cone base, seen in the top view. Project the 12 points down into the base of the front view. These are the end-points of the station lines in the front view. Draw a line from each of these end-points up to point 0.

STEP 2 There are two methods of drawing the pattern outline. Each begins with swinging an arc whose radius is the true length of the lateral distance from the apex to the base, or length 0-3 or 0-9. The compass radius can be set to this distance on Figure 10-14B. This distance can also be derived mathematically if the diameter of the base of the cone is known and the altitude (perpendicular distance from base to apex) is known. The formula is:

RADIUS = THE SQUARE ROOT OF [(½ CONE BASE DIAMETER)² + (ALTITUDE)²]

After swinging this arc, choose Method A, which approximates the cone as a pyramid, or Method B, which is mathematically correct.

METHOD A: In the top view of Figure 10-14B, the direct chordal distance between any two station line end-points—say, 5 and 6 (they are all equal)—is an approximation of the arc length between them. On the 0-3 radius just drawn, strike off this chordal distance between station line end-points repeatedly to equal the same number of spaces on the arc as the top view of the cone was divided into in Step 1, and ending at point 3 (**Figure 10-14C**).

Connect the station line end-points from point 0 to each of these arc intersections, and proceed to Step 3.

METHOD B: The central angle of the pattern is determined by a pattern arc length needed to equal the circumference of the base of the cone. The ratio of the central angle (A°) to a full circle 360° is the same as the circumference of the base of the cone is to the full circumference generated by the pattern's radius.

$$\frac{A°}{360°} = \frac{\text{CIRCUMFERENCE OF BASE OF CONE}}{\text{CIRCUMFERENCE OF PATTERN}}$$

$$\frac{A°}{360°} = \frac{\pi \times \text{CONE DIAMETER AT BASE}}{\pi \times \text{PATTERN DIAMETER}}$$

$$\frac{A°}{360°} = \frac{\text{CONE BASE DIAMETER}}{2 \times \text{PATTERN RADIUS}}$$

$$A° = \frac{180 \times \text{CONE BASE DIAMETER}}{\text{PATTERN RADIUS (A' − 0)}}$$

After the central pattern radius A° is found, the angle between successive station lines (a°) is found by dividing the same number of conic divisions done in Step 1 into A°. The

(A)

(B)

(C)

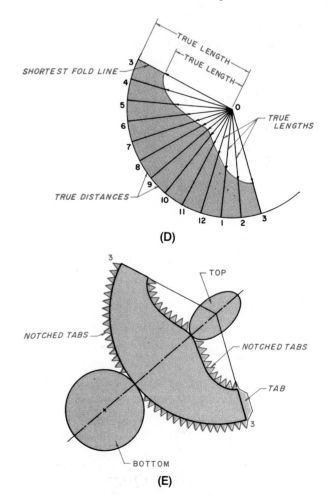

(D)

(E)

FIGURE 10-14   (A) Development of a truncated cone using radial line development; (B) Step 1; (C) Step 2; (D) Step 3; (E) Step 4 (completed development of a cone).

chordal lengths (D) of these divisions are found by using the formula:

$$D = 2 \times \text{PATTERN RADIUS} \times [\text{SIN} (\tfrac{1}{2} A°)]$$

This chordal distance can then be struck off on the pattern radius to form the same number of spaces as the conic divisions in Step 1 and labeled accordingly. See Figure 10-14C. Connect these station line end-points from point 0 to each of the intersections, and proceed to Step 3.

STEP 3 To draw the edge of the intersection that the cone's lateral surface makes with the upper flat surface, true lengths from the apex to this edge at each station line must be determined. Only 3′-0 and 9′-0 are visible as true lengths in the front view of Figure 10-14B. To find the other true lengths, each segment must be rotated to a position parallel to the frontal viewing plane; in this example, on line 3-0. Project the intersection of each station line at the truncated surface edge to line 3-0 on the cone profile in the front view (on line 3-0).

Starting with the shortest line segments of the cone, swing true-length arc 0-3′ on the corresponding pattern station lines, as illustrated in **Figure 10-14D**. Swing the apex-to-edge true length found for each station point at the corresponding station line on the pattern. Repeat for all 13 stations, 3 through 12 and back to 3. Label each point as illustrated.

Draw light line segments between each of the station line intersection positions found on the corresponding pattern station lines. This completes the true contour of the top edge, Figure 10-14D.

STEP 4 Add the top, bottom, and split tabs, as illustrated in **Figure 10-14E**.

Triangulation Development

Triangulation is the third major method used to lay out a surface development. *Triangulation* is a method of dividing a surface into a number of triangles and then transferring each triangle's true size and shape to the development. (Review Figure 10-2.) As with parallel line and radial line developments, *true lengths* and *true distances* must be used exclusively in pattern constructions.

True-Length Diagrams

Before any layouts can be started, true lengths and true distances of the object's boundary edges must be determined. A true-length diagram is usually used

to develop true lengths and true distances of these edges. A true-length diagram is often a quicker way to obtain the needed projections than descriptive geometry methods.

How to Develop a Transitional Piece Using Triangulation Development

GIVEN: **Figure 10-15A** describes a transitional piece using a top view, front view, and isometric view. Each of the object's corners is labeled as illustrated. Note that the true distances A to B, B to C, C to D, D to A, 1 to 2, 2 to 3, 3 to 4, and 4 to 1 can be measured directly from the top view. As drawn, the lengths A-1, B-1, B-2, C-2, C-3, D-3, D-4, and A-4 are not shown in their true lengths. A true-length diagram must be used to find these.

STEP 1 A true-length diagram is a combination of two views. In this example, it is a combination of the top view and the front view (**Figure 10-15B**). The height between points A and 1 is shown projected from the front view to the true-length diagram. The distance X between corresponding line end-points A and 1 is found in the top view and transferred to the true-length diagram. The illustrated diagonal line, drawn in the true-length diagram, is the true length of A-1. As all the heights and top view distances between end-points are the same for each of the lateral edges, then A-1 is also the same length as B-1, B-2, C-2, C-3, D-3, D-4, and A-4. These true lengths are needed to lay out the development.

STEP 2 Starting from line A-1 and using true lengths and true distances, reconstruct each triangle representing a surface of the object, as illustrated in **Figure 10-15C**. (Review Chapter 3 for aid in transferring a triangle.) Notice in this example that all true distances are transferred from the top view, and all true distances are transferred from the true-length diagram. A tab is added to join A-1 to A′-1′. This completes the transitional piece development as drawn in Figure 10-15A.

How to Develop a Transitional Piece with a Round End Using Triangular Development

A transitional piece that is square or rectangular at one end and round at the other is laid out using a procedure very similar to the one outlined previously. The only difference is that station lines must be positioned on the lateral surface of the round-to-corner transitions.

FIGURE 10-15   (A) Development of a transitional piece using triangulation development; (B) Step 1 (true-length diagram); (C) Step 2.

Figure 10-16A illustrates a two-view transitional piece with a round top end and a rectangular bottom end. In this example, the top end is divided into 12 equal spaces. Each point is numbered clockwise as illustrated. The bottom four corners making up the bottom rectangle are labeled A through D (**Figure 10-16B**). The true distances from A-B, B-C, C-D, and D-A′ are found in the top view. Station lines W, X, Y, and Z are not true lengths, but the true lengths must be determined with a true-length diagram. Note that in the true-length diagram, the true length of line W, from A′ to 0, is also the same length for

B-6, C-6, and D-12. Lines X, Y, and Z are also used four times at corresponding locations. Segment the object into triangles and develop the pattern by using true lengths and true distances to reconstruct each adjoining segment (**Figure 10-16C**).

How to Develop a Transitional Piece with No Fold Lines Using Triangular Development

Figure 10-17A shows a transitional piece with no evident fold lines or line segments. It is developed in the same manner as any other triangulation development.

FIGURE 10-16   (A) Development of a transitional piece with a round end using triangulation; (B) True-length diagram;
(C) Completed development of a transitional piece with a round end.

STEP 1   Divide the top and bottom surfaces into equal spaces, as illustrated in **Figure 10-17B**. In this example, the top edge is divided into 12 equal spaces, 0 through 12. The bottom edge is divided into the same number of equal spaces and lettered A through L to A.

STEP 2   Connect points 12 to A, 1 to B, 2 to C, and consecutively to points L to 11 with solid lines, as illustrated in **Figure 10-17C**. Dashed lines are added to segment the object into various triangles.

STEP 3   True distances from 1 to 2, 2 to 3, 3 to 4, and so on through 11 to 12, and the true distances from A to B, B to C through to L to A′ are found in the top view, Figure 10-17C. The other fold lines connecting the top and bottom are not true lengths and, therefore, true-length diagrams must be made, one for the solid lines and the other for the dashed lines (**Figure 10-17D**). The development is laid out by constructing connecting triangles using the lengths of each leg of each triangle (**Figure 10-17E**).

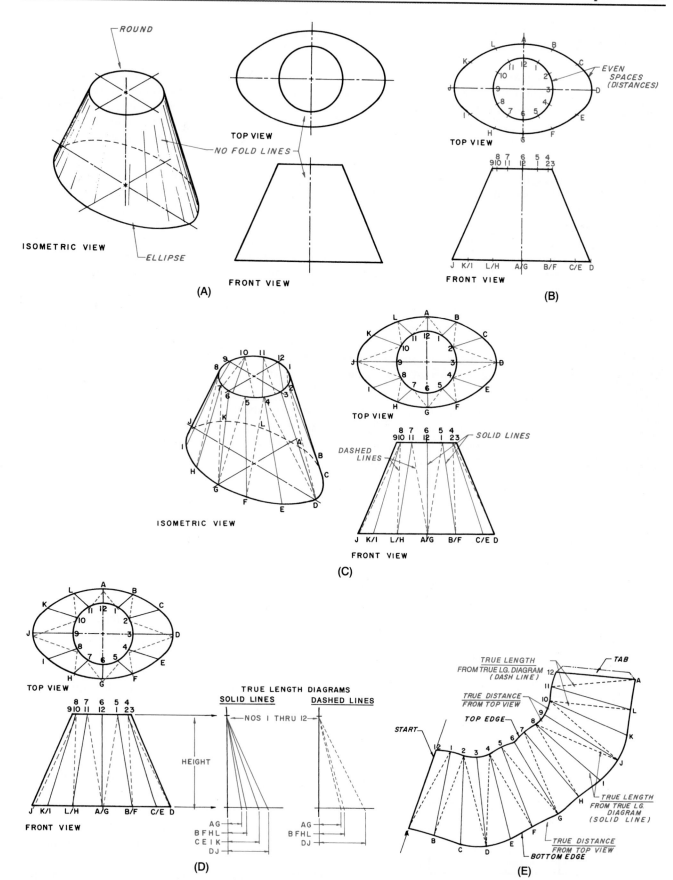

ISOMETRIC VIEW

TOP VIEW

FRONT VIEW

(A)

TOP VIEW

FRONT VIEW

(B)

ISOMETRIC VIEW

TOP VIEW

FRONT VIEW

(C)

TOP VIEW

FRONT VIEW

(D)

(E)

FIGURE 10-17   (A) Development of a transitional piece with no fold lines; (B) Step 1; (C) Step 2; (D) Step 3 (true-length diagram); (E) Completed development of a transitional piece.

A WIN-WIN-WIN PARTNERSHIP

Andrews Sheet Metal Fabrication (ASMF) is a successful family business that to date has supported three generations of the Andrews family. ASMF has been in business for more than 60 years. In the old days, all of the key management and technical positions were held by family members who learned their respective trades from grandfathers, fathers, older brothers, and uncles. But in recent years, ASMF's reputation for quality has resulted in rapid expansion. The company must now look outside the family for designers, drafters, and leading technicians.

This is causing a problem. John Andrews, the current CEO of ASMF, explains the situation as follows: "When we started looking outside of the family to fill technical positions—particularly design and drafting positions—we simply could not find anyone qualified. We received plenty of applications from people who had excellent mechanical drafting skills relating to precision metal parts, but most had never heard of sheet metal fabrication. No one knew how to perform bend calculations. Out of frustration, we started calling technical schools and community colleges as far away as 100 miles. Finally, after two weeks of telephone calls we came across a small technical college in a town 75 miles to the south that sounded promising. In fact, it sounded too good to be true.

"Fully expecting to be disappointed, I visited the college and examined the curriculum. To my great surprise, I could have hired any of the school's advanced students. They all knew patterns, developments, and bend calculations. I gave projects that included parallel line development, radial line development, triangulation development, and true-length diagrams. I talked with several students who actually pointed out errors in a set of hand calculations I had brought with me.

"I hired two students on the spot, and worked out an agreement to notify the head of the college's drafting program every time a position became available. We plan to have a long-term relationship with this college. In fact, I'm going to establish the ASMF scholarship for the sheet metal trades to encourage drafting students to study our field. Our new relationship with the college is a win-win-win partnership. ASMF wins, the college wins, and the students win."

Notches

Some developments require notches. Two major types of notches are usually used: a sharp V or a rounded V (**Figure 10-18**). The sharp V is used only if a minimal force is tending to part the material. The sharp point of the V will tear or crack under stress. The rounded V is used where parts would be under greater stress. The radius of the vertex of the rounded V should be at least twice the thickness of the metal used, and larger if possible.

Bends

When sheet metal is bent to form a corner, rib, or design, the outer surface stretches and the inner surface compresses. The length of material needed by each bend must be calculated and added to the straight, unbent portions of the pattern. The material needed by each bend is the length of the neutral axis within the material, where neither compression nor stretching occurs. This is calculated either by using formulas or various charts available for this purpose.

Bend allowance charts are included in Appendix A and are much faster to use than a formula. Two basic kinds of charts are used to calculate bend allowance, in both the English and metric systems. One type is used for 90° bends; the other for bends from 1° through 180°. The total length of a pattern is called the *developed length*. To determine the developed length, the stretched-out flat pattern must include all straight sides, plus the calculated bend allowances.

How to Find All Straight Sides of a 90° Bend

Figure 10-19A shows a simple 90° bend with an inside radius of .25, a sheet metal thickness of .125, and legs of 2.0 and 3.0. (The English system inch is

SHARP V NOTCH
USED FOR LOW STRESS

ROUNDED V NOTCH
USED FOR HIGH STRESS

FIGURE 10-18   Notches.

(A)

(B)

(C)

FIGURE 10-19 (A) Bend allowance of 90° bend; (B) Total of straight lengths; (C) Total length, including straight lengths and bend allowance.

used in this example.) Bend radii are always measured from the surface closest to the bend radius center.

STEP 1 Locate the tangent points at the ends of the straight sides.

STEP 2 Add the thickness of the sheet metal to the bend radius .25 (.125 + .25 = .375).

STEP 3 Subtract the sum of the sheet metal thickness .125 and the radius .25 from the 3.00 overall length of the object (3.00 − .375 = 2.625).

STEP 4 Subtract the sum of the sheet metal thickness .125 and the bend radius .25 from the 2.00 overall height of the object (2.00 − .375 = 1.625).

STEP 5 Add the two straight sides together (2.625 + 1.625 = 4.250). This is the total length of the straight sides of this object (**Figure 10-19B**).

How to Find the Bend Allowance of a 90° Bend

Review Figure 10-19A.

STEP 1 Note the metal thickness—in this example, .125—and the inside radius—in this example, .25.

STEP 2 Refer to the bend allowance chart in inches for 90° bends in Appendix A.

STEP 3 The left-hand column gives various metal thicknesses. Go down the left-hand column until the required size or the closest size is found. In this example, it is .125.

STEP 4 Along the top of the chart are listed various inside radii; go across the top of the chart to the required size or the closest radius—in this example, .25.

STEP 5 From the .125 number in the left-hand column, project across to the right; from the .25 number along the top of the chart, project down to where the two columns intersect. Given is the bend allowance for material .125 thick with a .25 radius. In this example, the bend allowance is .480 (**Figure 10-19C**).

STEP 6 The stretched-out dimension is found by adding the straight sides to the bend allowance; in this example, 4.250 + .480 = 4.730 (see Figure 10-19C).

How to Find the Total Straight-Side Length Adjoining a Bend Other than 90°

Figure 10-20A shows a simple 30° bend with an inside radius of 6.35 and a sheet metal thickness of 3.175 (the metric system is used in this example).

STEP 1 Locate the tangent points at the ends of the straight sides.

STEP 2 Determine the length of the straight sides and add them together (64.0 + 50.0 = 114.0). This is the total length of the straight sides of this object.

How to Find the Bend Allowance of Other than a 90° Bend

Review Figure 10-20A.

STEP 1 Note the metal thickness (in this example, 3.175) and the inside radius (in this case, R6.35).

STEP 2 Refer to the bend allowance chart in millimeters for 1° bends in Appendix A.

(A)

(B)

(C)

FIGURE 10-20   (A) Bend allowance of a bend other than 90°—total of straight lengths; (B) Total length of bend allowance; (C) Total length, including straight lengths and bend allowance.

STEP 3   The left-hand column gives various metal thicknesses. Go down the left-hand column until the required size or the closest size is found. In this example, it is 3.175.

STEP 4   Along the top of the chart are listed various inside radii. Go across the top of the chart to the required size or closest radius—in this example, 6.35.

STEP 5   From the 3.175 number in the left-hand column, project across to the right; from the 6.35 number along the top of the chart, project down to where the two columns intersect. Given is the factor used to calculate the bend allowance—in this example, .1356.

STEP 6   Multiply this factor times the actual degrees in the bend, from a straight 180° line (**Figure 10-20B**). This is the bend allowance (.1356 × 30° = 4.068).

STEP 7   Add the straight-side lengths to the bend allowance to get the total developed length: 114.0 + 4.068 = 118.068 (**Figure 10-20C**).

NOTE: A 30° bend dimension, as illustrated in **Figure 10-21A**, *is actually 150° (180° − 30° = 150°)* (**Figure 10-21B**). *The total bend must be calculated from a straight piece that is actually 180° before bending.*

(A)

(B)

FIGURE 9-21   (A) Total bend must be calculated from straight line dimension—given is 30°; (B) As dimensioned, 30° is actually a 150° bend from a straight line.

SUMMARY

- A development is the pattern or template of a shape that is laid out in a single flat plane in preparation for bending or folding into a required shape. The three types of developments are parallel, radial line, and triangulation.

- In parallel line development, all fold lines are parallel, and the object is laid out from a baseline known as the Stretchout line.

- In radial line development, all fold lines and line segments radiate from one point. True lengths and true distances must be used in laying out the developments.

- Triangulation is a method of dividing a surface into a number of triangles and then transferring each triangle's true size and shape to the development.

- When a development must be notched, two major types are used: sharp V and rounded V. Parts that will be subjected to greater levels of stress use the rounded V.

- When bending a sheet metal to form a corner, rib, or design, the outer surface stretches and the inner surface compresses. The length of material needed by each bend must be calculated and added to the straight, unbent portions of the pattern.

REVIEW QUESTIONS

Mark the following statements as either true or false.

1. A transitional piece is usually stamped or pressed into a particular shape.

2. A gage number represents the weight per square foot of a particular material.

3. The importance of developing the pattern or template with the inside surface up is that important dimension sizes and fold or bend locations are assumed to occur on the inside surface of the final object.

4. Tabs are used to fasten items together.

5. Bend allowance refers to the practice of calculating the length of material needed by each bend.

6. A true-length diagram often provides needed projections more quickly than descriptive geometry methods.

7. If a transitional piece does not have fold lines, it is developed in the same manner as any other triangulation development.

8. An automobile fender is an example of a single-curved surface.

Answer the following questions by selecting the best answer.

1. Which kind of development would be used to develop a pattern?
   a. Parallel line development
   b. Truncated line development
   c. Radial line development
   d. Triangulation line development

2. What are the two major kinds of notches used in a development?
   a. Sharp U or rounded U
   b. Sharp V or rounded U
   c. Rounded U or rounded V
   d. Rounded V or sharp V

3. In parallel line development, at which angle to the baseline must the fold lines be projected?
   a. 45°   c. 90°
   b. 60°   d. 180°

4. The gage system of calling out the thickness of a material is being phased out because:
   a. It is out of date.
   b. It causes considerable confusion.
   c. It is inaccurate.
   d. None of the above.

5. Which of the following elements **must** be located or laid out before a development can actually be started?
   a. True lengths
   b. True distances
   c. Station lines
   d. Both a and b

CHAPTER 10 PROBLEMS

The following problems are intended to give the beginning drafter practice in sketching, laying out auxiliary views if required, and drawing the stretched-out development of various objects. These problems will use one or more of the three standard methods of development: parallel line developments, radial line developments, and triangulation developments. The student will also practice calculating developed lengths by using various charts to determine full length before bending.

Conventional Instructions

The steps to follow in laying out any object that is to be developed are:

STEP 1   Study the problem carefully.

STEP 2   Determine which method or methods of development will be used (parallel line/ radial line/triangulation).

STEP 3 On a scrap sheet of paper, draw to scale the number of views required to lay out **all** true lengths and true distances. Add the end view and auxiliary views, in scale, if necessary. Draw with a **sharp** 4-H lead, and work as accurately as possible.

STEP 4 Label each point, if necessary, in order to keep track of progress.

STEP 5 Check all lengths and auxiliary views if included.

STEP 6 Starting from a baseline, lightly lay out the development. Use true lengths and true distances for all measurements. (Be sure to start the seam at the shortest fold line possible.)

STEP 7 Label each point, if necessary, in order to keep track of progress.

STEP 8 Use phantom lines to represent all fold lines.

STEP 9 Add tabs as required, and check to see that there are enough but without duplications.

STEP 10 Recheck all true lengths and true distances.

STEP 11 Darken in all lines.

CAD INSTRUCTIONS

Students who plan to complete the following problems on a CAD system should begin with these general instructions.

1. Analyze the problem carefully.
2. Plan your drawing sequence. In what order will you develop the drawing?
3. Read the specific instructions for the problem and begin.

Problems 10-1 through 10-18

Using the parallel line development method, develop each object starting from the given seam. Label each point clockwise; add ends and/or auxiliary views as necessary to develop a complete object. Add tabs .125 (3) × 45° to suit. Use phantom lines for all fold lines.
Extra assignment(s): *Cut out the developments and glue or tape them together to prove their accuracy.*

Mechanical   **PROBLEM 10-1**

Mechanical   **PROBLEM 10-2**

Mechanical   **PROBLEM 10-3**

Mechanical   PROBLEM 10-4

Mechanical   PROBLEM 10-5

Mechanical   PROBLEM 10-6

Mechanical   PROBLEM 10-7

METRIC

Mechanical   PROBLEM 10-8

Mechanical   PROBLEM 10-9

Mechanical **PROBLEM 10-10**

Mechanical **PROBLEM 10-11**

Mechanical **PROBLEM 10-12**

Mechanical **PROBLEM 10-13**

Mechanical **PROBLEM 10-14**

Mechanical **PROBLEM 10-15**

Mechanical PROBLEM 10-16

Mechanical PROBLEM 10-18

Mechanical PROBLEM 10-17

Problems 10-19 through 10-23

Using the parallel line development method, develop each object starting from the given seams. Add .125 (3) × 45° tabs as required to hold parts together. Design parts so there are as many identical parts as possible. Use phantom lines for all fold lines.
Extra assignment(s): *Cut out the developments and glue or tape them together to prove their accuracy.*

Mechanical | **PROBLEM 10-19**

Mechanical | **PROBLEM 10-20**

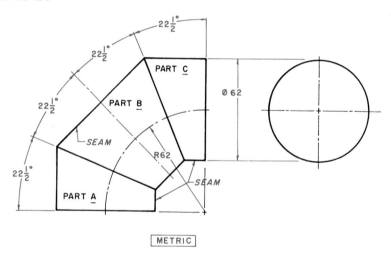

METRIC

Mechanical | **PROBLEM 10-21**

Mechanical **PROBLEM 10-22**

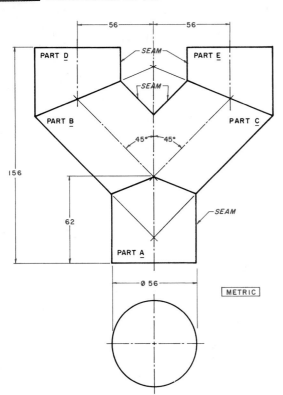

Problems 10-24 through 10-26

Develop the objects by using the given dimensions. Using bend allowance charts, design layouts to include material for bends.

Extra assignment(s): *Cut out the developments and glue or tape them together to prove their accuracy.*

Mechanical **PROBLEM 10-24**

MATERIAL, 1.5 THICK

ALL BENDS,
MIN. RADIUS

METRIC

Mechanical **PROBLEM 10-23**

Mechanical **PROBLEM 10-25**

MATERIAL 16 GAGE

ALL BENDS , MIN . RADIUS

Mechanical PROBLEM 10-26

Mechanical PROBLEM 10-28

Problems 10-27 through 10-30

Carefully lay out a full-size multiview drawing of the object; include all intersection lines. Make a full-size development of the object. Add tabs .125 (3) × 45° as required to hold parts together.
Extra assignment(s): *Cut out the developments and glue or tape them together to prove their accuracy.*

Mechanical PROBLEM 10-27

ALL DIMENSIONS ARE TO OUTSIDE
LIMITS OF PARTS

Mechanical PROBLEM 10-29

METRIC

Mechanical   PROBLEM 10-30

Mechanical   PROBLEM 10-32

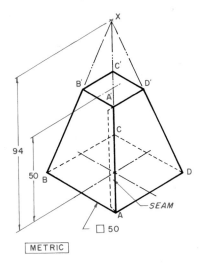

METRIC

Problems 10-31 through 10-42

Using the radial line development method, develop each object starting from the given seam. Label each point clockwise; add ends and/or auxiliary views as necessary to develop a complete object. Add tabs .125 (3) × 45° to suit. Use phantom lines for all fold lines.
Extra assignment(s): *Cut out the developments and glue or tape them together to prove their accuracy.*

Mechanical   PROBLEM 10-33

Mechanical   PROBLEM 10-31

Mechanical ## PROBLEM 10-34

Mechanical ## PROBLEM 10-35

Mechanical ## PROBLEM 10-36

Mechanical ## PROBLEM 10-37

Mechanical ## PROBLEM 10-38

Mechanical ## PROBLEM 10-39

PROBLEM 10-40

SECTION A-A

PROBLEM 10-41

PROBLEM 10-42

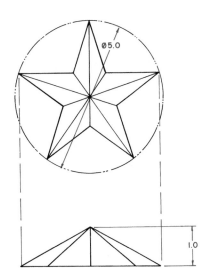

PROBLEMS

Using the triangulation development method, develop each object starting from the given seam. Label each point clockwise; add tabs .125 (3) × 45° to suit. Use phantom lines for all fold lines.
Extra assignment(s): *Cut out the developments and glue or tape them together to prove their accuracy.*

Mechanical   PROBLEM 10-43

Mechanical   PROBLEM 10-44

Mechanical **PROBLEM 10-45**

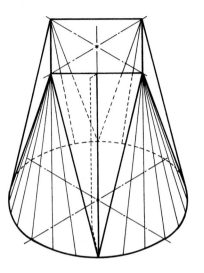

Problems 10-47 through 10-49

Using development charts, calculate the true developed length of each object. Round the answer to the nearest three places. Recheck all calculations.

Mechanical **PROBLEM 10-47**

Mechanical **PROBLEM 10-46**

Mechanical **PROBLEM 10-48**

Mechanical **PROBLEM 10-49**

PROBLEMS

Problems 10-50 through 10-54

Using the parallel line, the radial line, or the triangulation development method, develop each object. Be sure to leave a seam and to add all required tabs to suit as required. Use phantom lines for all fold lines.

Extra assignment(s): *Cut out the developments and glue or tape them together to prove their accuracy.*

Mechanical **PROBLEM 10-50**

METRIC

Mechanical **PROBLEM 10-51**

METRIC

Mechanical **PROBLEM 10-52**

Mechanical   PROBLEM 10-53

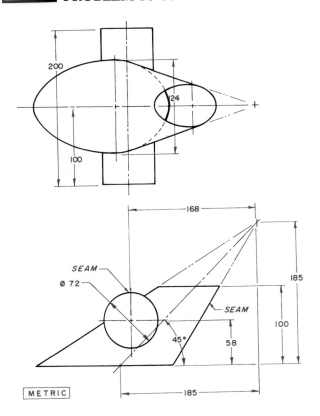

METRIC

Mechanical   PROBLEM 10-54

METRIC

11

Solid/3D Modeling: Computational Design and Analysis

CHAPTER OUTLINE

Factors in using visualization • Understanding data • Modeling techniques • Types of visualization images • Engineering visualization applications • Case study • Summary • Review questions

CHAPTER OBJECTIVES

Upon completion of this chapter, students should be able to do the following:

▶ Explain the advantages and disadvantages of using visualization.

▶ Describe the most widely used three-dimensional imaging techniques.

▶ Describe the Boolean operators used to generate solid models.

▶ Explain the various methods used to represent variables.

▶ Explain the various categories of engineering visualization.

KEY TERMS

Boolean operators

Computational fluid dynamics (CFD)

Finite-element analysis (FEA)

Finite-element model (FEM)

Gouraud

Holography

Modeling

Parametric solid modeling

Phong

Photorealism

Radiosity

Ray tracing

Rendering

Solid modeling

Texture mapping

Thermal analysis

Virtual reality

Visualization

Visualization is the pictorial representation of data for the purpose of a more clear and concise communication of information (**Figure 11-1**). The computer has made it possible to create pictorial representations of data that were virtually impossible with previous graphics techniques. The purpose of this chapter is to define and describe existing and emerging techniques for creating and analyzing data by means of computer-generated visualization. The old saying that a picture is worth a thousand words has never been more true. Today, however, the saying is more appropriately stated as: "A picture is worth 786,432 (1,024 × 768) pixels."

Visualization falls roughly into two categories. In some cases, visualization is used as an interactive design tool. An example of this is in mechanical CAD, where a clear image makes the design process easier. In other cases, visualization is used to represent complex data in a way that makes it easier to understand (**Figure 11-2**).

Increasingly complex engineering problems generate corresponding increases in the amount of information produced. The improvements in products made possible by more accurate data justify the large amounts of data to be utilized. In many cases, the amount of data to be analyzed is so great that it is almost impossible to comprehend it in numerical form. The extremely large amount of data is one of the major reasons for utilizing visualization techniques. Stacks of printouts consisting of numerical data are frequently impossible to comprehend without the benefit of visualization.

Factors in Using Visualization

Several factors are leading to the increased use of visualization techniques and output:

- More complex engineering problems are generating increasingly large amounts of information.

FIGURE 11-1   Visualization (© Swill Klitch/Stock Vector Illustration/Shutterstock.com).

FIGURE 11-2   Visualization in CAD (© Aquatic Creature/Stock Photo/Shutterstock.com).

- Increased computational speed makes it possible to perform visualization tasks on workstations and personal computers instead of being limited to supercomputers.

- Increasing the performance/price ratio makes visualization hardware and software affordable for more people and companies.

- Improved user interfaces make the operation of systems and the production of visualization output easier to perform. Rather than being so complex and unique that only a specialist can use a system to create visualization output, it has become increasingly easier for greater numbers of people to productively utilize visualization technology and techniques.

Even with tremendous improvements in performance and ease of use, major barriers to the adoption of visualization techniques still exist. Among the more critical factors are the following:

- In some cases, the complexity of visualization techniques and technology continues to make it difficult to learn and use. This is less true than it has been in the past, but it still takes substantial effort and training to produce high-quality visualization output.

- Visualization generally involves three-dimensional (3D) data. This represents a major increase in complexity from two-dimensional data and requires a solid understanding of three-dimensional information.

- An unlimited number of ways to represent any given data set presents complex choices. The decision of how to represent the data is subjective. Frequently, no standard rules exist to govern selection.

Understanding Data

Throughout this chapter, you will see a number of examples of powerful visualization techniques, from mechanical CAD to NASA satellite data to digitized biomedical data. In all cases, these examples use underlying 3D data. The difference among them is how the 3D data is obtained. In some cases, the 3D data is the product of complex calculations and is generated to represent something that would not be visible otherwise. In other cases, the data is the direct result of a designer creating a 3D model. The same data used for visualization can be used to create the tools to make a product.

Modeling Techniques

Modeling is the term used to describe the process of creating data that represents a real-world engineering

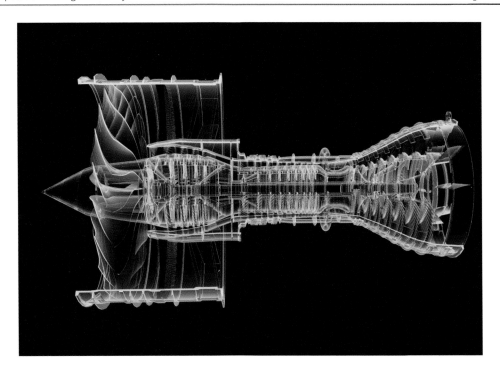

FIGURE 11-3   Computer model of a ships' propeller (© X-RAY Pictures/Stock Photo/Shutterstock.com).

problem. The term is used rather pervasively in a number of disciplines, so you should be comfortable with it. In art, the process of creating a sculpture is called modeling. In market research, the process of building data that represents how a demographic group responds is called modeling. And in CAD, the process of creating a 2D or 3D data set that represents a real object or system is likewise called modeling. In the case of 3D CAD, the process used to create the model limits the visualization techniques that can be used to represent it. The most primitive modeling technique is 3D wireframe (**Figure 11-3**). In this method, the geometry of 3D surfaces is represented, but the actual topology of those surfaces is not applied. (You can think of geometry as the edges and topology as the skins.)

Several years back, 3D wireframe modeling was more accepted because it required less in the way of software, hardware, and training. Today, CAD software and computers are powerful enough that there is no justification for using 3D wireframe modeling. The next logical step from 3D wireframe is surface modeling. In this method, skins are applied to the edges of surfaces. This modeling method has been popular for many years in the aerospace and automotive industries, as it can represent complex objects with great precision. It has a further advantage over 3D wireframe in that, because the surface topology is known, the computer can calculate which lines or edges will be obscured in any particular view.

From a designer's standpoint, building a surface model can be tedious, and there are no guarantees that all the surfaces in a model will fit together correctly when it is done. A surface model can be very accurate, but it takes great care on the part of the designer to ensure this (**Figures 11-4** and **11-5**).

Solid Modeling

A solid model (**Figures 11-6** and **11-7**) is simply an electronic model where the computer system has kept track of the surface topology, ensured that all the seams between surfaces match, and ensured that the entire model is closed. Sometimes, designers talk about a solid model "holding water." This means there are no gaps at the seams and no openings between surfaces. This makes solid models quite accurate.

There are a number of processes for creating solid models, but the end result is always the same—a topologically closed solid model. From a visualization standpoint, there is usually little difference between a surface model and a solid model. From a creation standpoint, solid models are simpler to create and require more computer horsepower.

Surface and solid models can be represented in 3D, with or without hidden lines removed. They can also be represented in a shaded form, where each surface is assigned a shading color. Depending on a computer's graphics power, it may be possible to

FIGURE 11-4 Surface modeling in CAD (© Yoshikazu Tsuno/Staff/AFP/Getty Images).

FIGURE 11-5 Surface model of a human head and neck (© Design36/Stock Vector Illustration/Shutterstock.com).

FIGURE 11-6 Surface model of an automobile frame (© Super3D/Stock Photo/Shutterstock.com).

create shaded views as fast or faster than hidden-line views. Some moderately priced graphics subsystems are able to manipulate shaded views in real time.

The use of 3D technology, particularly *solid modeling*, is more and more prevalent. A number of powerful yet easy-to-use solid modeling programs are available, and this has made a noticeable change in the acceptance of solid modeling for mainstream design tasks.

Solid models are formed by joining various geometric shapes such as cones, prisms, pyramids, boxes, wedges, and cylinders. It is as if the computer mathematically "welds" the shapes together to form the solid model. Examine the solid models in Figures 11-6 and 11-7 closely and identify as many geometric shapes as you can. The computer uses *Boolean operators* to join the various shapes that make up the solid model. The most commonly used Boolean operators are:

- Union (also called *addition*)
- Difference (also called *subtraction*)
- Intersection

Figure 11-8 illustrates how these Boolean operators work. In Figure 11-8A, solid X has been added to (union) solid Y to form a new solid. This new solid has a volume equal to the combined volume of solids X and Y before the union. In Figure 11-8B, solid X has been subtracted (difference) from the new solid formed in Figure 11-8A. Notice that the volume where solids X and Y overlapped has been removed from solid Y. That volume appears in Figure 11-8C as the intersection of solids X and Y. The volume of the solid in Figure 11-8C is equal to the overlap between solids X and Y.

A more advanced form of solid modeling is *parametric solid modeling*. With this advanced form of the concept, the model's geometry is controlled by design parameters—hence the name parametric solid modeling—and curved paths are controlled

mathematically rather than using coordinates. The benefit of parametric solid modeling is that as changes in the design are made, the model and corresponding drawings are updated automatically. This aspect of parametric solid modeling can shorten the design-prototyping-manufacturing process significantly.

Parametric Solid Modeling

Parametric modeling uses various *parameters*—hence the name—to define a model of an object such as a machine part, subassembly, or assembly. Commonly used parameters are dimensions. One of the major benefits of parametric modeling is that at any time the parameters that define the model are changed, the model is updated automatically in accordance with the change. This is critical because in the design process, an ongoing problem has always been that a change—even a seemingly minor change—to a part, subassembly, or assembly can cause other corresponding changes. Ensuring that these corresponding changes are made and that they

(A)

Union (addition)

(B)

Difference (subtraction)

(C)

Intersection

FIGURE 11-8   Boolean operators in solid modeling.

do not undermine the design is always a challenge. With parametric modeling, this problem is solved. In addition to using dimensions, other parameters may also be used to define an object. These other parameters include material density, formulas that describe swept features, and imported data.

In addition to parameters, parametric modeling can use *constraints*. Constraints consist of required relationships between features of an object or individual features that must be maintained, or the design is undermined. For example, two features in an object might have to remain parallel no matter what else is changed about the design. In such a case, parallelism would be considered a constraint. Perpendicularity is another example of a potential constraint. Roundness, flatness, straightness, circularity, concentricity, and symmetry are others.

Types of Visualization Images

Still Images

Single-frame, still images remain one of the most prevalent types of images used for visualization (**Figure 11-9**). The ability to separate product components demonstrates an additional benefit of 3D modeling. This same 3D database may be used to generate multiple views of the product by simply designating an alternate viewing location.

There are many types of still images. We've already spoken of wireframe and hidden-line images. Shaded images can be created with various algorithms to obtain differing levels of realism. The simplest shading method is flat shading, where any curved surfaces on a model are broken up into facets (a process called *tessellation*), and each surface is assigned a color based on its angle with respect to the viewer. This gives the model a decidedly chunky and faceted look. Flat shading is fairly rare these days, as the computation power required to generate a better image is negligible. The more common method is *Gouraud* shading, which improves on flat shading by interpolating the colors between adjacent facets. This gives rather a nice, smooth look— in fact, Gouraud shading is often called smooth shading. Most quality graphics boards in modern personal computers can Gouraud-shade images in real time. For very complex images, there are special 3D graphics boards that are optimized for displaying shaded images.

The next step up in shaded image quality is *Phong* shading, which adds specular reflection. This is a mirrorlike glint that makes objects look shiny. Phong shading is possible on higher-end graphics systems, and it is useful for simulations that must be very lifelike.

Beyond simple shading, there are a number of methods that can be used to enhance an image. It is possible, when assigning shading colors to a model, to specify a transparency. This makes it possible to view internal details of parts.

Photorealism is the term used to describe images that are created in such a fashion that they might fool the viewer into thinking that they are photographs. In general, photorealistic images include multiple light sources, shadow casting, transparency, reflections, and refractions (**Figure 11-10**). As a rule, they also include *texture mapping*, where patterns are applied to objects' surfaces. **Figure 11-11** shows a

FIGURE 11-9   Image of a satellite (© Science & Society Picture Library/Contributor/SSPL/Getty Images).

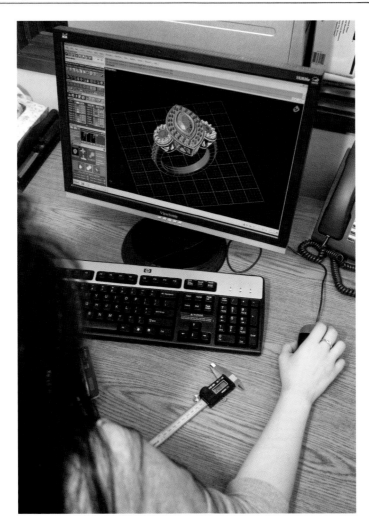

FIGURE 11-10   Photorealism (© Bloomberg/Contributor/Bloomberg/Getty Images).

FIGURE 11-11   Photorealism image (© Yoshikazu Tsuno/Staff/AFP/Getty Images).

typical photorealistic image. Photorealistic images are also often called rendered images, with the process of creating them being called *rendering*.

As a rule, every improvement in image quality with photorealistic images comes at the cost of substantial preparation work and processing time. Accurate shadow casting and reflection/refraction calculation require a rendering process called *ray tracing*, which literally involves tracing the rays of light from the eye of the viewer through the scene. Diffuse light calculations (involving, for example, the wash of light off a wall or through a stained glass window) require a rendering process called *radiosity*.

Animation

Animation provides a means of representing time-dependent variables and active visualization of objects or events. Products consisting of moving parts and active functions often cannot be clearly represented by still images. Animation is becoming increasingly useful for a variety of innovative applications. In addition to playing a programmed animation, you can now move through 3D models interactively. Advancements in visualization technology will continue to increase the reality with which 3D models can be viewed and toured electronically (**Figure 11-12**).

FIGURE 11-12   Computer animation (© Andrii Tkach/ Stock Photo/Shutterstock.com).

Holography

Holography produces images that simulate the effect of seeing the three-dimensionality of an object from multiple viewpoints. Holographic images have become fairly common for a variety of applications. Perhaps the most prevalent application has been the use of holographic images on credit cards. In the past, holographic images have been limited to physical objects that could be photographed in a manner that permits compositing multiple images into a holographic image.

It is now possible to produce holographic images from computer-generated 3D models. In their most basic form, these images are static images on sheet material. One advancement over a static image is the combination of computer and holographic technology to create holographic images in 3D space.

The complexity and computational requirements involved in generating a holographic image in this manner are extremely demanding. For this reason, the model is initially viewed as a wireframe and then rendered as a surface model when the desired viewpoint is established. As computational speeds increase, it will be possible to generate increasingly realistic images. Increased size and the addition of color now exist for computer-generated holograms. Faster computational speeds will also improve the capability to interact with the model in real time.

Virtual Reality

Virtual reality is a technology designed to simulate either a physical or hypothetical environment. While the emphasis in this chapter is on visualization, it is important to note that virtual reality is intended to involve all the natural senses. Virtual reality systems can incorporate stimuli for all the five basic senses, but most commonly they deal only with sight and sound.

Virtual reality has a wealth of applications. It can create an environment that permits a sense of how a particular engineering design may feel or function. By being able to experience such an environment, the researcher can test and analyze alternative designs before investing in physical prototypes. Where multiple senses are important to a design project, virtual reality provides a more complete and realistic simulation (**Figure 11-13**).

Virtual reality also has great potential for hypothetical and hazardous environments. These environments don't exist, are too dangerous for humans to enter physically, or may be beyond the reach of humans.

In the past few years, the Virtual Reality Modeling Language (VRML) standard has become popular.

FIGURE 11-13   Fluid flow researcher in a virtual reality environment (© Time Life Pictures/Contributor/ The LIFE Picture Collection/Getty Images).

Strictly speaking, it is not a virtual reality language. It is a language to model static scenes, which users can navigate through using whatever input and output devices they choose. It doesn't provide for object interaction or motion within a scene. Still, it is an effective tool for interactively viewing complex models. It doesn't provide for motion within a scene, but it does allow a user to navigate around a static 3D model.

Engineering Visualization Applications

Engineering visualization involves numerous types of design and research information. Each of these types of information requires unique and specific means of representation in order to most clearly and concisely convey selected information. The following examples represent major types of engineering visualization.

Design

The biggest application of engineering visualization is in design—whether of microchips or mile-long bridges. In the past, visualization was often seen as a marketing tool, primarily because the cost and difficulty involved couldn't be justified otherwise. As computers have become more powerful, visualization has become integrated into the design process. With this integration, it is possible not only to improve a product's time to market but also to optimize the product's cost and performance.

Finite-Element Analysis

Finite-element analysis (FEA) is a vital means of analyzing the structural response of objects. In this method, a model of a part is broken up into a finite number of polygonal elements, called a mesh. If the elements of the mesh are small enough, a structural analysis of this *finite-element model (FEM)* corresponds rather closely with the real-world response of the part. The truck in **Figure 11-14** has been created as a finite-element model. In this example, the model is being analyzed to determine the structural and impact characteristics of the catamaran.

Stress is a factor in engineering designs of all types. While mechanical engineers spend quite some time understanding stress, you can simply envision it as the internal forces in an object that result from an external load. It may be concentrated in a small area or distributed over a large span. A localized concentration of stress is called a stress riser and can lead to part failure. In designs where failure due to stress is a concern, FEA helps the designer reach an optimal solution much faster than building prototypes or models would.

FIGURE 11-14   Finite-element analysis of a truck. (© 3DDock/Stock Photo/Shutterstock.com).

In the past, FEA was the domain of stress specialists—typically PhDs in mechanical engineering. Recently, FEA programs have become easy enough to use that experienced designers can use them to get a quick readout on whether a part is likely to meet spec. Naturally, critical parts are still referred to the stress specialists. No sane person would want to fly in a plane in which the critical parts were not analyzed by an expert.

Thermal Analysis

Thermal analysis is a means of evaluating the thermal characteristics of a part or a product. Thermal analysis is vital to numerous types of design applications. A clear understanding of a product's thermal characteristics can improve both efficiency and longevity. It may, in fact, prevent product failure.

Thermal analysis may be used for applications where thermal characteristics are the only design variable. It may also be used for design projects where heat or cold is just one of many variables, and where each of the multiple variables has a direct impact on

related variables. The heat may be flowing through a conduit or around the exterior of an object. Other products may be solid objects involving critical thermal transfer through a specific material or materials.

One very common application for thermal analysis is the design of heat ducts. These same principles apply to a variety of similar types of heating and cooling systems. The cross-sectional shape and internal volume of a duct significantly impact thermal transfer characteristics.

Thermal analysis was critical to the success of NASA's Space Shuttle missions (**Figure 11-15**).

Thermal analysis systems employ visualization techniques that produce color-coded images, clearly illustrating temperature distributions. These simulations permit quick and easy evaluation of alternative conduit designs.

Computational Fluid Dynamics

Computational fluid dynamics (CFD) is another means of using computers to simulate physical events mathematically. This involves the study of

FIGURE 11-15   Accurate thermal analysis of the heat shield on NASA's Space Shuttle was critical to the mission's success (© Handout/Handout/Getty Images News/Getty Images).

fluids and their characteristics under a variety of conditions. Fluids include both gases and liquids. Computer programs and algorithms have evolved to a level that make them highly accurate for predicting fluid flow characteristics for mechanical and electrical products, architectural structures, medicine, and the automotive and aerospace industries.

CFD plays a vital role in the aerospace industry. The design of airplanes depends upon CFD to produce efficient planes that also perform optimally.

In addition to the aerospace industry, there are growing numbers of other industries discovering innovative applications for CFD visualization. Applications include sports equipment, architecture, automobiles, and even consumer products, where CFD can help designers improve internal airflow and cooling.

Pre-Assembly Analysis

On complex products, assembly is never simply a matter of bolting together a bunch of parts. In aircraft, for example, physically fitting systems and their components into an aircraft is a monumental task. Not only must the systems fit into limited space, they are also constrained by aerodynamic considerations for the shape of the aircraft. Visualization techniques make it possible to locate potential system interferences during the design stage and eliminate costly revisions. Boeing has been at the forefront of applying visualization technology to reduce prototyping costs.

Space allocation in the area beneath the flight deck is a complex task involving several groups within Boeing. Computer technology and visualization provide a common database that can be used interactively by multiple groups to resolve problems in the early stages of design. This concurrent engineering procedure saves a tremendous amount of time and resources. Equally as important, it results in optimum solutions to design criteria.

Kinematic Simulations

Kinematics is an excellent application for visualization and simulation. Linkages and mechanisms are present in numerous products that we use daily. Computers are ideal for solving the mathematical calculations necessary. Once the calculations have been performed, images may be generated that clearly illustrate resulting designs. Technology exists that makes it easy to computationally build kinematic systems. Once a system has been built, it may easily be animated and tested for performance. **Figures 11-16 and 11-17** illustrate applications of kinematic simulation.

Automotive Crash Simulations

The automotive industry is using visualization in computational simulations to better understand structural damage resulting from crashes. This crash information is vital to improving the safety of

FIGURE 11-16   Kinematic simulation application (© Time Life Pictures/Contributor/The LIFE Picture Collection/Getty Images).

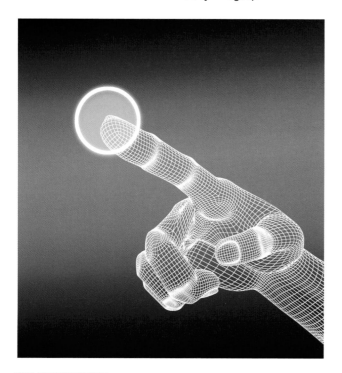

FIGURE 11-17   Kinematic simulation (© tawan/Stock Photo/Shutterstock.com).

vehicles. In the past, such information was available only from physically crashing real vehicles. The ability to interactively study computational crashes permits a better understanding of the crash results.

This information may then be used to optimize designs first with simulations. This reduces the need for crash testing real vehicles.

Case Study

Engineering visualization is one aspect of product design. In reality, there are multiple phases that occur between product conception and the final product. The following example illustrates how visualization is utilized to improve products, their production, and their cost effectiveness. The following example also consists of several types of visualization images and considerations that need to be addressed.

Product Design Example

Product design techniques are changing dramatically. Competition is demanding faster design cycles and improved design solutions. The following example describes how Xerox utilizes computer technology for product design. In addition to visualization techniques, methods of conducting real-time interactive design with variable controls on the computer screen are discussed. The goal is improved product quality and design time.

TO SOLID MODEL OR NOT? A DEBATE

The design team for Thompson Engineering International (TEI) had been debating the issue of visualization for over a week. All members of the team agreed that TEI's current 2D CAD capabilities were inadequate. The new space station project, the company's largest and most complex project ever, would be a particular challenge. The parts TEI will design for the joint U.S./European space station must be perfect; there is no room for error. This is why the topic of visualization is now being debated so intensely.

Several members of the team argue for a minimalist approach, combining 2D, 3D wireframe, and surface modeling where necessary. They argue that it will require less computing power and a shorter learning curve for design and drafting personnel. Plus, they won't have to throw away the investment they've made in their existing systems. Other members of the team favor moving completely to solid modeling; they counter the arguments of the wireframe advocates, claiming that approach won't yield sufficient information. In their words, "A wireframe image just is not descriptive enough to do us any good."

"Solid modeling," claims its advocates, "provides the most accurate, most comprehensive information possible, and the image is realistic." They add that "Revisions are particularly easy with solid modeling." The supporters of the minimalist approach counter by saying "We can't do solid modeling without an expensive hardware upgrade and it will take our team six months to develop modeling skills. This space station project doesn't have a six-month learning curve built into it." This type of debate happens frequently in engineering companies. Join the debate. What is your opinion?

The use of physical prototypes is both time-consuming and expensive. One means of significantly improving design solutions and time requirements is to perform initial design using computational prototypes. This permits ease of testing design alternatives and promotes investigating more designs than traditional physical prototyping methods. Testing more designs provides increased information from which to select the optimum design solution. **Figures 11-18** and **11-19** are examples of solid models being used in product design.

FIGURE 11-18   Computer model use in industrial design (© Mmaxer/Stock Photo/Shutterstock.com).

FIGURE 11-19   Solid model used in biomedical design
(© Ulrich Baumgarten/Contributor/
Ulrich Baumgarten/Getty Images).

SUMMARY

• Visualization is the pictorial representation of data for the purpose of a more clear and concise communication of information.

• The increased use of visualization can be attributed to the increasing complexity of engineering problems, the improved capabilities of computers, the relative affordability of computer hardware and software, and the relative ease of learning to create visualization output.

• Three-dimensional modeling techniques include wireframe, surface modeling, and solid modeling.

• Parametric solid modeling uses parameters of various kinds to define a model of an object. It also can use constraints such as parallelism that must be maintained as the model is revised.

• Types of visualization images include still images, transparency, photorealism, animation, stereo images, holography, and virtual reality.

• Engineering visualization categories include design, finite-element analysis, thermal analysis, computational fluid dynamics, kinematic simulation, and automotive crash simulations.

REVIEW QUESTIONS

Mark the following statements as either true or false.

1. Visualization is the pictorial representation of data for the purpose of a more clear and concise communication of information.

2. Visualization is beneficial because it allows extremely large amounts of data to be seen and therefore comprehended.

3. Computer technology and visualization have made it possible to resolve problems that were previously too complex to solve.

4. The wireframe method of modeling is the most time-consuming computationally.

5. A disadvantage of solid modeling is that it is not accurate.

6. Only costly problems, such as those faced by NASA, justify the effort involved in visualization.

7. A disadvantage of visualization is that the hardware and software are not affordable for most people and companies.

8. Normal personal computers are capable of displaying and manipulating shaded images in real time.

Answer the following questions by selecting the best answer.

1. Which of the following is **not** one of the basic methods of engineering modeling?
   a. Wireframe
   b. Surface
   c. Photorealism
   d. Solid modeling

2. Which of the following is **not** a type of visualization image?
   a. Animation
   b. Stereo images
   c. Transparency
   d. Intrinsic

3. Which of the following is the appropriate type of visualization image to represent an item in a physical or hypothetical environment?
   a. Virtual reality
   b. Holography
   c. Photorealism
   d. Animation

4. Which of the following is **not** a method for representing visualization variables?
   a. Color
   b. Motion
   c. Symbols
   d. Thermal analysis

5. Which of the following is **not** one of the five engineering visualization categories?
   a. Holography
   b. Finite-element analysis
   c. Thermal analysis
   d. Computational fluid dynamics

6. An example of an application using computational fluid dynamics is:
   a. The design of artificial turf
   b. The design of artificial intelligence
   c. The design of artificial knees
   d. The design of artificial hearts

7. Which of the following is **not** a factor in using visualization?
   a. More complex engineering problems are generating increasingly large amounts of information.
   b. Increased computational speed makes it possible to perform visualization tasks on workstations and personal computers instead of being limited to supercomputers.
   c. The decision of how to represent the data is subjective.
   d. Improved user interfaces make the operation of systems and the production of visualization output easier to perform.

8. Which of the following is **not** true regarding stress analysis?
   a. Stress is a critical factor in engineering designs of all types.
   b. Trained designers may be able to do basic stress analyses, but critical analyses should be left for specialists.
   c. Prior to being analyzed for stress, a model of a part must be broken up into a finite number of polygonal elements.
   d. Stress risers are desirable to ensure that parts will not fail.

9. Which of the following is **not** a constraint that might be used in solid modeling?
   a. Parallelism
   b. Flatness
   c. Visualization
   d. Symmetry

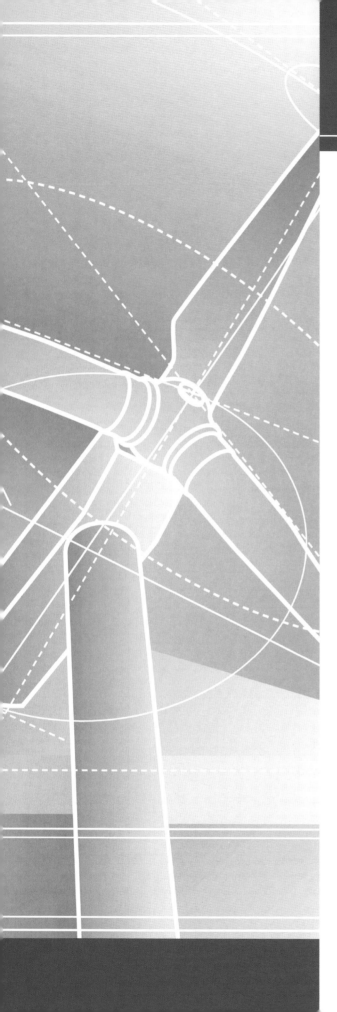

section 3

Design Drafting Applications

career profile

Mary Trahan

Mary started her career with drafting classes in high school and went on to a technical college, where she trained further on the subject. Coming out of the program, she joined the Consolidated Waterworks District 1 in Terrebonne Parish, Louisiana, as a drafter. At that time, her drawings were done with pen and paper. Things have changed a lot in 22 years!

Doing things on paper meant that when they had to replace valves or when a main line broke, they would have to redraft the map to account for the change. Thankfully, in the 1980s, the Waterworks started digitizing the maps, tracing aerial photographs electronically into computer-aided design (CAD) software. It was a big step up, but the program was not great for mapping. The roads, landmarks, and water mains are not merely straight lines, at which CAD programs are best. Often, the curves did not match. And technical specifications (like the depth at which a pipe is buried) had to be kept in a separate database.

By the 1990s, the Waterworks began working with other utilities, such as gas and electric, in order to fund and create a parish geographic information system (GIS). It became a functional way to view each other's information on intelligent maps. With the North American Datum coordinates, the maps can more closely approximate the spherical shape of the earth (versus the flatness of CAD). GIS also combined sophisticated graphics with the spatial integrity of tabular data. That means that Mary can see if a specific pipe has had several leaks in the past few years,

making it easier to prevent major problems before they happen. It also helps with safeguards: If there is a chemical leak at the plant, drafters can draw a polygon in the program and get customers' information for that area to notify them.

On a daily basis, Mary is responsible for maintaining all the water-related changes on the map, whether they are water lines or fire hydrants. For instance, in the case of a new subdivision, she will go out after it is completed to make a final inspection and to ensure that main lines and valves operate as designed and comply with her system. Once she and the board approve the job, the engineer will submit digital copies (usually in AutoCAD), and Mary will transfer those changes into the system map. Most of the drawings are flat, from a perspective above, but the benefit of the software is that she can also do other views. A profile, looking from the top of the earth into the depths, is especially necessary for new construction so engineers can ensure that all utilities cross within proper distances. Mary's insertions will make it easier for these other utilities to know where they can (and cannot) run lines.

12

Geometric Dimensioning and Tolerancing

CHAPTER OBJECTIVES

Upon completion of this chapter, students should be able to do the following:

▶ Describe what is meant by the term *general tolerancing*.

▶ Define the concept *geometric dimensioning and tolerancing*.

▶ Explain the purpose of a modifier.

▶ Distinguish between the concepts *maximum material condition (MMC)* and *regardless of feature size (RFS)*.

▶ Explain the concept *least material condition (LMC)*.

▶ Describe what is meant by *projected tolerance zone*.

▶ Make a sketch that illustrates the concept of datums.

▶ Demonstrate how to establish datums.

▶ Apply feature control symbols when dimensioning objects.

▶ Explain the concept of *true position*.

KEY TERMS

All around symbol

Angularity

Basic dimension

Between symbol

Bilateral tolerance

Circularity

Cylindricity

Datum

Datum feature

Datum feature simulator

Datum feature symbol

Datum plane

Datum reference frame

Datum surface

Datum target symbol

Feature control symbol

Flatness

Free-state variation

General tolerancing

Geometric dimensioning and tolerancing

Least material condition (LMC)

Limit dimensioning

Maximum material condition (MMC)

Modifiers

Parallelism

Perpendicularity

Positional tolerancing

Profile

Profile of a line

Profile of a surface

(continued)

Summary of Geometric Dimensioning and Tolerancing Terms

Actual Local Size. The value of any individual distance at any cross section of a feature.

Actual Mating Size. The dimensional value of the actual mating envelope.

Actual Size. The actual measured size of a feature.

Allowance. The difference between the larger shaft size limit and the smallest hole size limit.

Angularity. Tolerancing of a feature at any specified angle from a referenced datum.

Basic Dimension. A theoretically "perfect" dimension similar to a reference or nominal dimension. It is used to identify the exact location, size, shape, or orientation of a feature. Associated tolerances are applied by notes, feature control frames, or other methods, excluding tolerance within title blocks.

Bilateral Tolerances. Tolerances that are applied to a nominal dimension in the positive and negative directions.

Bonus Tolerance. The permitted allowable increase in tolerance as the feature departs from the material condition identified within the feature control frame.

Boundary, Inner. (external feature) A worst-case boundary generated by the least material condition minus the stated geometric tolerance and any additional applicable geometric tolerance resulting from the feature's departure from its specified material condition.

Boundary, Inner. (internal feature) A worst-case boundary generated by the maximum material condition minus the stated geometric tolerance and any additional applicable geometric tolerance resulting from the feature's departure from its specified material condition.

Boundary, Outer. (external feature) A worst-case boundary generated by the MMC plus the stated geometric tolerance and any additional applicable geometric tolerance resulting from the feature's departure from its specified material condition.

Boundary, Outer. (internal feature) A worst-case boundary generated by the LMC plus the stated geometric tolerance and any additional applicable geometric tolerance resulting from the feature's departure from its specified material condition.

Circular Runout. A tolerance that identifies an infinite number of single circular elements measured at cross sections on a feature when the feature is rotated 360° for each cross section.

(KEY TERMS continued)

Projected tolerance zone	Straightness
Regardless of feature size (RFS)	Tangent plane
	Tolerancing
Rule #1	True position
Runout	Unilateral tolerance
Size tolerance	Virtual condition
Statistical tolerancing symbol	

Circularity. A tolerance that controls the circular cross section of round features that is independent of other features. The tolerance zone boundary is formed by two concentric perfect circles.

Clearance Fit. A condition between mating parts in which the internal part is always smaller than the external parts it fits into.

Coaxiality. The condition of two or more features, or surfaces of revolution having coincident axes or coincident median points.

Compound Datum Features. Two datum features used to establish a datum axis or plane.

Concentricity. A tolerance in which the axis of a feature must be coaxial to a specified datum regardless of the datum's and the feature's size. The lack of concentricity is eccentricity.

Constraint. The condition of limiting one or more degrees of freedom.

Coplanarity. The condition of two or more features in the same plane.

Cylindricity. A tolerance that simultaneously controls a surface of revolution for straightness, parallelism, and circularity of a feature and is independent of any other features on a part. The tolerance zone boundary is composed of two concentric perfect cylinders.

Datum. Reference points, lines, planes, cylinders, and axes that are assumed to be exact. They are established from datum features.

Datum Axis. The axis of the true geometric counterpart of a referenced datum feature such as a hole or shaft.

Datum Feature. A feature that is used to establish a datum.

Datum Feature of Size.   A feature that has size, such as a shaft, which is used to establish a datum.

Datum Identification Symbol.   A special rectangular box that contains the datum reference letter and a dash on either side of the letter. It is used to identify datum features.

Datum: Feature Simulator.   A surface of adequately precise form (such as a surface plate, a gage surface, or a mandrel) contacting the datum feature(s) and used to establish the simulated datum(s).

Datum: Reference.   Entering a datum reference letter in a compartment of the feature control frame following the tolerance value.

Datum: Reference Frame.   Three mutually perpendicular planes that establish a coordinate system. It is created by datum references in a feature control frame or by a note.

Datum: Simulated.   A point, axis, or plane established by processing or inspection equipment such as the following: simulator, surface plate, gage surface, or mandrel.

Datum Simulation.   The use of a tool contacting a datum feature used to simulate a true geometric counterpart of the feature.

Datum Simulator.   A tool used to contact a datum feature.

Datum Target.   Specified points, lines, or areas on a feature used to establish datums.

Datum Target Area.   A specified area on a part that is contacted to establish a datum.

Datum Target Line.   A line on a surface that is contacted to establish a datum.

Datum Target Point.   A specified point on a surface used to establish a datum.

Datum Target Symbol.   A circle divided horizontally into halves containing a letter and number to identify datum targets.

Degrees of Freedom.   The allowed movement of a part along the three translational degrees of freedom, X, Y, and Z axes, and rotation around the X, Y, and Z axes. In the standard, the rotational degrees are labeled *u* for the X axis, *v* for the Y axis, and *w* for the Z axis.

Derived Median Line.   A straight or nonstraight line made of the midpoints between two opposed surface elements or by center points along the length of a cylinder.

Envelope, Actual Mating.   The term is defined according to the types of features as follows:

- For an *External Feature.* A similar perfect feature counterpart of smallest size that can be circumscribed about the features so that it just

contacts the surface at the highest points. For example, a smallest cylinder of perfect form or two parallel planes of perfect form at minimum separation that just contact(s) the highest points of the surface(s). For features controlled by orientation or positional tolerances, the actual mating envelope is oriented relative to the appropriate datum(s)—for example, perpendicular to a primary datum plane.

- For an *Internal Feature.* A similar perfect feature counterpart of largest size that can be inscribed within the feature so that it just contacts the surface at the highest points. For example, a largest cylinder of perfect form or two parallel planes of perfect form at maximum separation that just contact(s) the highest points of the surface(s). For features controlled by orientation or positional tolerance, the actual mating envelope is oriented relative to the appropriate datum(s).

Feature.   A component of a part such as a hole, slot, surface, pin, tab, or boss.

Feature, Axis of.   A straight line that coincides with the axis of the true geometric counterpart of the specified feature.

Feature, Center Plane of.   A plane that coincides with the center plane of the true geometric counterpart of the specified feature.

Feature, Derived Median Line of.   An imperfect line (abstract) that passes through the center points of all cross sections of the feature. These cross sections are normal to the axis of the actual mating envelope. The cross-section center points are determined as per ANSI B89.3.1.

Feature, Derived Median Plane of.   An imperfect plane (abstract) that passes through the center points of all line segments bounded by the feature. These line segments are normal to the actual mating envelope.

Feature of Size, Irregular.

- A directly tolerance feature or collection of features (sphere, cylinder, or parallel planes) that in combination establish a boundary.
- A directly tolerance feature or collection of features other than sphere, cylinder, or parallel planes that in combination establish a boundary.

Feature of Size, Regular.   One cylindrical or spherical surface, or a set of two opposed elements or opposed parallel surfaces, associated with a size dimension.

Feature-Relating Tolerance.   This is the tolerance value in the second line of a composite position tolerance. It controls the locations between features in a pattern.

Feature-Relating-Tolerance Zone Framework (FRTZF). This is the tolerance zone framework that controls the basic relationship between features in a pattern.

Fit. A term used to describe the range of assembly that results from tolerances on two mating parts.

Fixed Fastener Condition. In an assembly where two or more parts are stacked together with a fastener passing through them, and one of the holes fixes the location of the fastener.

Flatness. A tolerance that controls the amount of variation from the perfect plane on a feature independent of any other features on the part.

Floating Fastener Condition. In an assembly where two or more parts are stacked together with a fastener passing through them, and all of the holes produce a clearance with the fastener.

Form Tolerance. A tolerance that specifies the allowable variation of a feature from its perfect form.

Free-State Variation. The condition of a part that permits its dimensional limits to vary after removal from manufacturing or inspection equipment.

Geometric Counterpart, True. This is the exact opposite geometric shape of a feature. The shape's size may be a material boundary, virtual condition, or an actual mating envelope. It is the same shape as the geometric shape of a theoretically perfect tool or virtual simulator.

Least Material Boundary (LMB). The limits within a feature that it may extend to as a result of its stated tolerances or combination of tolerances. When MMC and LMC modifiers are used on datum features, their meaning includes maximum material boundary and least material boundary. Surfaces and features of size have material boundaries.

Least Material Condition (LMC). A condition of a feature in which it contains the least amount of material relative to the associated tolerances. Examples are maximum hole diameter and minimum shaft diameter.

Limit Dimensions. A tolerancing method showing only the maximum and minimum dimensions that establish the limits of a part size or location.

Limits. The maximum and minimum allowable sizes of a feature.

Location Tolerance. A tolerance that specifies the allowable variation from the perfect location of a feature relative to datums or other features.

Maximum Material Boundary (MMB). The limits outside a feature that it may extend to as a result of its stated tolerances or combination of tolerances. When MMC and LMC modifiers are used on datum features, their meaning includes MMB and LMB. Surfaces and features of size have material boundaries.

Maximum Material Condition (MMC). A condition in which the feature contains the maximum amount of material relative to the associated tolerances. Examples are maximum shaft diameter and minimum hole diameter.

Modifier. The application of MMC or LMC to alter the normally implied interpretation of a tolerance specification.

Parallelism. A tolerance that controls the orientation of interdependent surfaces and axes that must be of equal distance from a datum plane or axis.

Pattern-Locating Tolerance. This is the tolerance value in the first line of a composite position tolerance. It controls the locations and orientation of features relative to the referenced datums.

Pattern-Locating Tolerance Zone Framework (PLTZF). This is the tolerance zone framework that locates the true positions of controlled features relative to referenced datums.

Perpendicularity. A tolerance that controls surfaces and axes that must be at right angles with a referenced datum.

Position Tolerance. A tolerance that controls the position of a feature relative to the true position specified for the features, as related to a datum or datums.

Primary Datum. The first datum reference in a feature control frame. It is normally elected because it is most important to the design criteria and function of the part.

Profile of a Line. A tolerance that controls the allowable variation of line elements in only one direction on a surface along an elemental tolerance zone with regard to a basic profile.

Profile of a Surface. A tolerance that controls the allowable variation of a surface from a basic profile or configuration.

Profile Tolerance Zone. A tolerance zone that can control the form of an individual feature and provide for a composite control of form, orientation, and location.

Projected Tolerance Zone. A tolerance zone that applies to the location of an axis beyond the surface of the feature being controlled.

Reference Dimension. A nontolerance zone or location dimension used for information purposes only that does not govern production or inspection operations.

Regardless of Feature Size (RFS). A condition of a tolerance in which the tolerance must be met regardless of the produced size of the feature.

Regardless of Material Boundary (RMB). Indicates that a datum feature simulator must contact the extremities of a feature regardless of its physical size. Datum features references are understood to apply at regardless of material boundary unless MMC and LMC modifiers are used.

Related Actual Mating Envelope. As specified by the **Envelope, Actual Mating** definition, but constrained by any datum(s).

Rule #1. The dimension on a feature of size controls the size and form of the object within the specified limits of size. When the part is at maximum material condition, it must have perfect form. As the part moves toward least material condition, its form can vary equal to the movement from least material condition. When a part is at least material condition, it has the maximum amount of form tolerance available.

Rule #2. All geometric tolerances are understood to apply regardless of feature size unless a modifier is used to indicate otherwise.

Runout. The composite surface variation from the desired form of a part of revolution during full rotation of the part on a datum axis.

Secondary Datum. The second datum reference in a feature control frame. Established after the primary datum, it has less design influence and functionality.

Size, Virtual Condition. The actual value of the virtual condition boundary.

Stepped Datum. A datum established by two features offset in different planes.

Straightness. A tolerance that controls the allowable variation of a surface or an axis from a theoretically perfect line.

Symmetry. A condition for which a feature (or features) is equally disposed or shaped about the center plane of a datum feature.

Tangent Plane. A theoretically exact plane derived from the true geometric counterpart of the specified feature surface by contacting the high points on the surface.

Tertiary Datum. The third datum reference in a feature control frame. Established after the secondary datum, it has the least amount of design influence or functionality.

Tolerance. The acceptable dimensional variation or allowance of a part.

Total Runout. A tolerance that provides for a composite control of all surface elements as the part is rotated 360° about a datum axis.

Transition Fit. A condition in which the prescribed limits of mating parts produce either a clearance or an interference when the parts are assembled.

True Geometric Counterpart. The theoretically perfect boundary (virtual condition or actual mating envelope) or best-fit (tangent) plane of a specified datum feature.

True Position. The theoretically exact location of a feature.

Unilateral Tolerance. A tolerance that allows variations in only one direction.

Unrelated Actual Mating Envelope. As specified by the **Envelope, Actual Mating** definition, but not constrained by any datum(s).

Virtual Condition. A constant boundary produced by the combined effects of the maximum material condition size and geometric tolerance. It represents the worst-case condition of assembly at MMC.

Zero Tolerance at MMC or LMC. A tolerancing method where no tolerance is shown in the feature control frame. The tolerance allowed is totally dependent on the size of the feature departure from MMC or LMC.

General Tolerancing

The Industrial Revolution created a need for mass production, assembling interchangeable parts on an assembly line to turn out great quantities of a given finished product. Interchangeability of parts was the key. If a particular product was composed of 100 parts, each part could be produced in quantity, checked for accuracy, stored, and used as necessary.

Since it was humanly and technologically impossible to have every individual part produced exactly alike (it still is), the concept of geometric and positional tolerancing was introduced. *Tolerancing* means setting acceptable limits of deviation. For example, if a mass-produced part is to be 4″ in length under ideal conditions but is acceptable as long as it is not less than 3.99″ and not longer than 4.01″, there is a tolerance of plus or minus .01″ (**Figure 12-1**). This type of tolerance is called a *size tolerance*.

There are three different types of size tolerances: unilateral and bilateral, as shown in (**Figure 12-2**), and imit dimensioning. When a *unilateral tolerance* is

SIZE TOLERANCE

FIGURE 12-1 Size tolerance.

UNILATERAL TOLERANCE

$4.00 \begin{smallmatrix} +.01 \\ -.00 \end{smallmatrix}$

BILATERAL TOLERANCE
(Equally distributed)

$4.00 \pm .01$

BILATERAL TOLERANCE
(Unequally distributed)

$4.00 \begin{smallmatrix} +.01 \\ -.02 \end{smallmatrix}$

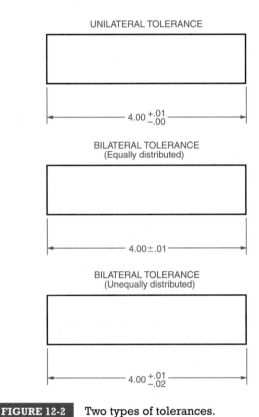

FIGURE 12-2 Two types of tolerances.

applied to a dimension, the tolerance applies in one direction only (for example, the object may be larger but not smaller, or it may be smaller but not larger). When a *bilateral tolerance* is applied to a dimension, the tolerance applies in both directions, but not necessarily evenly distributed. In *limit dimensioning,* the high limit, is placed above the low value. When placed in a single line, the low limit precedes the high limit and the two are separated by a dash.

Tolerancing size dimensions offers a number of advantages. It allows for acceptable error without compromises in design, cuts down on unacceptable parts, decreases manufacturing time, and makes the product less expensive to produce. However, it soon became apparent that in spite of advantages gained from size

tolerances, tolerancing only the size of an object was not enough. Other characteristics of objects also needed to be toleranced, such as the location of features, orientation, form, runout, and profile.

In order for parts to be acceptable, depending on their use, they need to be straight, round, cylindrical, flat, angular, and so forth. This concept is illustrated in (**Figure 12-3**). The object depicted is a shaft that is to be manufactured to within plus or minus .01 of 1.00″ in diameter. The finished product meets the size specifications but, since it is not straight, the part might be rejected.

The need to tolerance more than just the size of objects led to the development of a more precise system of tolerancing called *geometric dimensioning and tolerancing.* This new practice improved on conventional tolerancing significantly by allowing designers to tolerance size, form, orientation, profile, location, and runout (**Figure 12-4**). In turn, these are the characteristics that make it possible to achieve a high degree of interchangeability.

AS DIMENSIONED

1.01 MMC ENVELOPE

Ø1.00 ± .01

Ø1.00 ± .01

Ø.99 *AS MANUFACTURED*

FIGURE 12-3 Tolerance of form.

FOR INDIVIDUAL FEATURES	FORM
FOR INDIVIDUAL OR RELATED FEATURES	PROFILE
FOR RELATED FEATURES	ORIENTATION LOCATION RUNOUT

FIGURE 12-4 Types of tolerances.

Geometric Dimensioning and Tolerancing Defined

Geometric dimensioning and tolerancing is a dimensioning practice that allows designers to set tolerance limits not just for the size of an object but also for all of the various critical characteristics of a part. In applying geometric dimensioning and tolerancing to a part, the designer must examine it in terms of its function and its relationship to mating parts.

Figure 12-5 is an example of a drawing of an object that has been geometrically dimensioned and toleranced. It is taken from the dimensioning standards as defined by the American National Standards Institute (ANSI), written by the American Society of Mechanical Engineers (ASME) or ASME Y14.5—2009. This manual is a necessary reference for drafters and designers involved in geometric dimensioning and tolerancing.

The key to learning geometric dimensioning and tolerancing is to learn the various building blocks that make up the system and how to properly apply them. **Figure 12-6** contains a chart of the building blocks of the geometric dimensioning and tolerancing system. In addition to the standard building blocks shown in the figure, several modifying symbols are used when applying geometric tolerancing, as discussed in detail in upcoming paragraphs.

Another concept that must be understood in order to effectively apply geometric tolerancing is the concept of datums. For skilled, experienced designers, the geometric building blocks, modifiers, and datums

SYMBOL	CHARACTERISTIC	GEOMETRIC TOLERANCE
—	STRAIGHTNESS	FORM
▱	FLATNESS	
○	CIRCULARITY	
�	CYLINDRICITY	
⌒	PROFILE OF A LINE	PROFILE
⌓	PROFILE OF A SURFACE	
∠	ANGULARITY	ORIENTATION
⊥	PERPENDICULARITY	
//	PARALLELISM	
⊕	TRUE POSITION	LOCATION
◎	CONCENTRICITY	
≡	SYMMETRY	
* ↗	CIRCULAR RUNOUT	RUNOUT
* ↗↗	TOTAL RUNOUT	

* MAY BE FILLED IN

FIGURE 12-6   Building blocks.

blend together as a single concept. However, for the purpose of learning, they are dealt with separately and undertaken step-by-step as individual concepts. They are presented now in the following order: modifiers, datums, and geometric building blocks.

ANSI's dimensioning standards manual (Y14.5 series) changes from time to time as standards are updated. For example, the Y14.5 manual became Y14.5 in 1982 to accommodate metric dimensioning. Revised again in 1988 and 1994, it became Y14.5—1994. In the latest edition, the standard takes on the name of the developing agency, the American Society of Mechanical

FIGURE 12-5   Geometric dimensioning and tolerancing drawing example.

Engineers (ASME). ASME Y14.5—2009 is the latest edition in the ongoing revision process of the standard. This chapter helps students learn the basics of geometric dimensioning and tolerancing so they will be able to apply the latest standards set forth by ASME at any point in time and in accordance with any edition of the manual that is specified. Students should not use this chapter as a reference in place of the ASME standard. Always refer to the latest edition of the standard for specifics that go beyond the basics covered herein.

Modifiers

Modifiers are symbols that can be attached to the standard geometric building blocks to alter their application or interpretation. The proper use of modifiers is fundamental to effective geometric tolerancing. Various modifiers are often used: maximum material condition, least material condition, projected tolerance zone, free-state variation, tangent plane, all around, between symbol, and statistical tolerance (**Figures 12-7A** through **12-7C**).

Maximum Material Condition

Maximum material condition (MMC) is the condition of a characteristic when the most material exists. For example, the MMC of the external feature in (**Figure 12-8**) is .77″. This is the MMC because it represents the condition where the most material exists on the part being manufactured. The MMC of the internal feature in the figure is .73″. This is the MMC because the most material exists when the hole is produced at the smallest allowable size.

In using this concept, the designer must remember that the MMC of an internal feature is the smallest

allowable size. The MMC of an external feature is the largest allowable size within specified tolerance limits inclusive. A rule of thumb to remember is that MMC means most material.

Regardless of Feature Size

Regardless of feature size (RFS) tells machinists that a tolerance of form or position or any characteristic must be maintained regardless of the actual produced size of the object. Geometric tolerances are understood to apply regardless of feature size where the modifiers Ⓜ or Ⓛ are not used. It is no longer permissible to show the RFS modifier. The RFS concept is illustrated in (**Figure 12-9**). In the RFS example, the object is acceptable if produced in sizes from 1.002″ to .998″ inclusive. The form control is axis straightness to a tolerance of .002″ regardless of feature size. This means that the .002″ axis straightness tolerance must be adhered to, regardless of the produced size of the part.

Contrast this with the MMC example. In this case, the produced sizes are still 1.002″ to .998″. However, because of the MMC modifier, the .002″ axis straightness tolerance applies only at MMC or 1.002″.

If the produced size is smaller, the straightness tolerance can be increased proportionally. Of course, this makes the MMC modifier more popular with machinists for several reasons: (1) it allows them greater room for error without actually increasing the tolerance, (2) it decreases the number of parts rejected, (3) it cuts down on unacceptable parts, (4) it decreases the number of inspections required, and (5) it allows the use of functional gaging. All of these advantages translate into substantial financial savings while, at the same time, making it possible to produce interchangeable parts at minimum expense.

Least Material Condition

Least material condition (LMC) is the opposite of MMC. It refers to the condition in which the least material exists. This concept is illustrated in (**Figure 12-10A**).

In the top example, the external feature of the part is acceptable if produced in sizes ranging from .98″ to 1.02″ inclusive. The least material exists at .98″. Consequently, .98″ is the LMC.

In the bottom example, the internal feature (hole) is acceptable if produced in sizes ranging from .98″ to 1.02″ inclusive. The least material exists at 1.02″. Consequently, 1.02″ represents the LMC (**Figure 12-10B**).

Ⓜ	MAXIMUM MATERIAL CONDITION MAXIMUM MATERIAL BOUNDARY
Ⓛ	LEAST MATERIAL CONDITION LEAST MATERIAL BOUNDARY
Ⓟ	PROJECTED TOLERANCE ZONE
Ⓕ	FREE STATE
Ⓣ	TANGENT PLANE
⌓	ALL AROUND
↔	BETWEEN SYMBOL
ⓈⓉ	STATISTICAL TOLERANCE
ⒸⒻ	CONTINUOUS FEATURE
⌭	ALL OVER
▷	TRANSLATION
Ⓘ	INDEPENDENCY
Ⓤ	UNEQUALLY DISPOSED PROFILE

FIGURE 12-7A   Modifiers used with geometric tolerancing.

FIGURE 12-7B   Form and proportion of geometric tolerancing symbols.

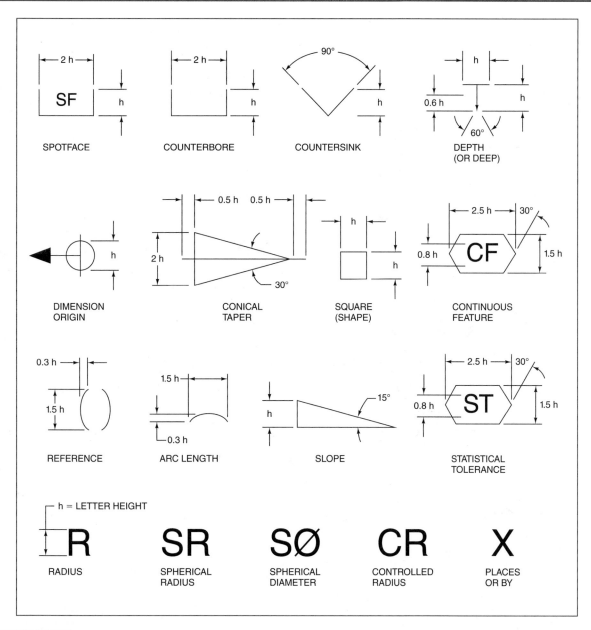

FIGURE 12-7C   Form and proportion of dimensioning symbols and letters.

FIGURE 12-8   MMC of an external and an internal feature.

Projected Tolerance Zone

Projected tolerance zone is a modifier that allows a tolerance zone established by a locational tolerance to be extended a specified distance beyond a given surface. This concept is discussed further later in this chapter under the heading "True Position."

Free-State Variation

Free-state variation is the concept that some parts cannot be expected to be contained within a boundary of perfect form. Some parts may vary in form beyond the MMC size limits after forces applied during manufacture are removed. For example, a

PRODUCED SIZE	TOLERANCE (form)
1.002	.002
1.001	.002
1.000	.002
.999	.002
.998	.002
RFS	

PRODUCED SIZE	TOLERANCE (form)
1.002	.002
1.001	.003
1.000	.004
.999	.005
.998	.006
MMC	

FIGURE 12-9   Regardless of feature size (RFS).

EXTERNAL FEATURE

INTERNAL FEATURE

FIGURE 12-10A   Least material condition (LMC).

thin-walled part shape may vary in its free state due to stresses being released in the part. This variation may require that the part meet its tolerance requirements while in its free state.

Parts that are subject to free-state variation do not have to meet the Rule #1 requirement of perfect form at MMC. These parts are standard stock such as bars, sheets, tubes, extrusions, structural shapes, or other items produced to established industry or government standards. The appropriate standard would govern the limits of form variation allowed after manufacture.

The free-state symbol specifies the maximum allowable free-state variation. It is placed within a feature control frame, following the tolerance and any modifiers (**Figure 12-11**).

PRODUCED SIZE	TOLERANCE (FORM)
1.020	.042
1.018	.040
1.016	.038
1.012	.034
1.008	.030
1.002	.024
.990	.012
.980	.002
LMC	

PRODUCED SIZE	TOLERANCE (FORM)
1.020	.002
1.018	.012
1.016	.024
1.012	.030
1.008	.034
1.002	.038
.990	.040
.980	.042
LMC	

FIGURE 12-10B   Least material condition (LMC).

FREE-STATE SYMBOL

FIGURE 12-11   Feature control frame with free-state symbol.

Tangent Plane

The *tangent plane* concept uses a modifying symbol with an orientation tolerance to modify the intended control of the surface. When an orientation tolerance is applied to a surface, the primary control is equivalent to the symbology used. An example of the primary control of a parallel callout is parallelism. However, when applied, the specified symbol controls not only parallelism but also other form variations such as concavity, convexity, waviness, flatness, and other imperfections as well.

If two such controlled surfaces are assembled, the abrupt variation in the surfaces can cause different mating effects and assembly conditions. There are several ways to control the effects of surface conditions when applying orientation tolerances. The obvious method is to refine the surface control with a form tolerance such as flatness. This is permissible because the orientation tolerance controls flatness to the extent of the specified tolerance value.

Another method is to modify the orientation tolerance to apply a tangent plane. When the modifier is applied, the orientation tolerance zone for a tangent plane is identical to any other orientation tolerance zones with one exception. The orientation tolerance no longer controls the form of the surface. The surface of the controlled feature must be within the specified limits of size, but it is not required to fall within the parallelism tolerance zone boundary. Only a plane tangent to the high points on the surface must be within the tolerance zone boundary. The symbol is placed within the feature control frame following the stated tolerance (**Figure 12-12**).

All Around Symbol

The *all around symbol* is the symbolic means of indicating that the specified tolerance applies all around the part. The normal tolerance zone of a geometric callout extends the length of the feature in question. If there is an abrupt change in surface condition, such as an offset, the tolerance zone would conclude at the beginning of the offset. Applying the all around symbol extends the tolerance zone all around the feature to include abrupt surface variations (**Figure 12-13**). This concept will be discussed further later in this chapter under the heading "Profile."

Between Symbol

The *between symbol* is a symbolic means of indicating that the stated tolerance applies to a specified segment

FIGURE 12-13   All around symbol.

of a surface between designated points. The normal tolerance zone of a geometric callout extends the length of the feature in question. Application of this symbol can be used to limit the tolerance zone to a specified area. It can also be used to clarify the extent of the profile tolerance when it is not clearly visible due to surface variations. **Figure 12-14** illustrates the use of this symbol.

Statistical Tolerancing Symbol

The *statistical tolerancing symbol* is a symbolic means of indicating that the stated tolerance is based on statistical process control (SPC). The symbol can be applied in one of two ways. When the tolerance is a statistical size tolerance, the symbol is placed next to the size dimension, as shown in (**Figure 12-15**). When the tolerance is a statistical geometric tolerance, the symbol is placed in the feature control frame, as shown in (**Figure 12-16**).

FIGURE 12-14   Between symbol.

FIGURE 12-15   Statistical tolerance symbol.

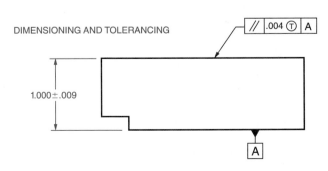

FIGURE 12-12   Specifying a tangent plane.

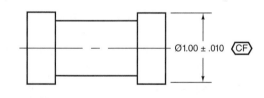

FIGURE 12-16   Symbol indicating the specified tolerance is a statistical geometric tolerance.

Continuous Feature Symbol

The *continuous feature symbol* is the symbolic means of indicating that the specified dimension or tolerance applies across interruptions in a feature so that it acts as a single continuous feature when considering the requirements for the feature (**Figure 12-17**).

All Over Symbol

The *all over symbol* is the symbolic means of indicating that the specified tolerance applies all over the surfaces of the part. Applying the all over symbol means that every surface feature gets the same tolerance. The all over symbol is two concentric circles located at the corner in the leader (**Figure 12-18**).

Translation Symbol

An equilateral triangle pointing to the right may be used as the *translation symbol*. This symbol is used with a datum feature reference to indicate that the datum feature simulator may translate. The symbol is placed in the feature control frame, as shown in (**Figure 12-19**).

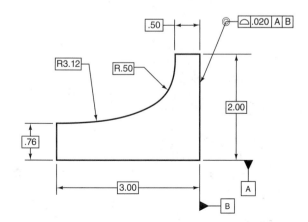

FIGURE 12-17   Continuous feature symbol.

FIGURE 12-19   Translation symbol.

Independency Symbol

Former practices allowed an exception to Rule #1 where application of a form tolerance larger than the size tolerance was desired. When this occurred, a note was included adjacent to the dimension that indicated "PERFECT FORM AT MMC NOT REQUIRED." Current practice introduces the *independency symbol* to replace the former required note. The independency symbol must be placed near the size dimension to invoke the exception to Rule #1. The letter "I" inside a circle is the independency symbol. This concept is illustrated in (**Figure 12-20**).

Unequally Disposed Profile Symbol

The letter "U" inside a circle is the unequally disposed tolerance symbol. This symbol is shown in a tolerance specification to indicate that a profile tolerance is unequally disposed, and it may be a unilateral zone or a zone that is unequally offset relative to the basic profile. This concept is illustrated in (**Figure 12-21**).

Datums

Datums are theoretically perfect points, lines, axes, surfaces, or planes used for referencing features of an object. They are established by the physical datum features that are identified on the drawing.

FIGURE 12-20   Independency symbol.

FIGURE 12-21   Unequally disposed profile symbol.

Identification of datum features is done by using a *datum feature symbol*. This symbol consists of a capital letter enclosed in a square frame. A leader line extends from the frame to the selected feature. A triangle is attached to the end of the leader and is applied in the appropriate way to indicate a datum feature. The symbols should only be applied to physical features. They should not be attached to center lines, axes, center planes, or other theoretical entities. **Figure 12-22** shows two ways in which datum feature symbols are placed on drawings. The datum symbol is attached to an extension line of the feature outline, clearly separated from the dimension line when the datum feature is a surface or placed on the visible outline of a feature surface.

In **Figures 12-23A** through **12-23C**, the datum feature symbol is placed on an extension of the

dimension line of a feature of size when the datum is an axis or center plane. In **Figures 12-23D** through **12-23F**, the datum is an axis. The symbol can be placed on the outline of a cylindrical surface or an extension line of the feature outline, separated from the size dimension. **Figure 12-23F** shows one arrow of the dimension line being replaced by the datum feature triangle when space is limited. If no feature control frame is used, the symbol is placed on a dimension leader line to the feature size dimension, as shown by the example of Datum B in (**Figure 12-24**). In (**Figure 12-25**), the symbol is attached to the feature control frame below (or above) when the feature(s) controlled is a datum center plane.

Establishing Datums

In establishing datums, designers must consider the function of the part, the manufacturing processes that will be used in producing the part, how the part will be inspected, and the part's relationship to other parts after assembly. Designers and drafters must also understand the difference between a datum, datum feature, datum feature simulator, datum surface, datum plane, and datum feature of size.

A datum is theoretical in nature and is located by the physical datum features identified on the drawing. A datum is considered to be the true geometric counterpart of the feature. It is the origin from which measurements are made or that provides geometrical

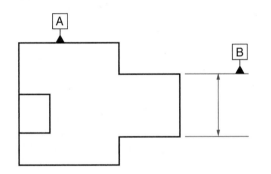

FIGURE 12-22   Datum feature symbols on a feature surface and an extension line.

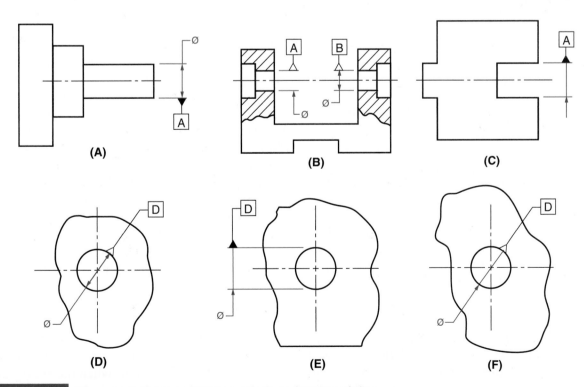

FIGURE 12-23   Placement of datum feature symbols on features of size.

NOTE: UNTOLERANCED DIMENSIONS LOCATING TRUE POSITION ARE BASIC

FIGURE 12-24   Datum reference on dimension leader line.

DIMENSIONING AND TOLERANCING

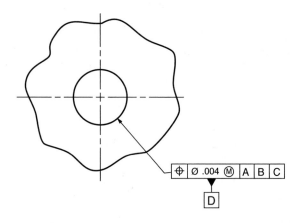

FIGURE 12-25   Placement of datum feature symbol in conjunction with a feature control frame.

references to which other features are established. A *datum feature* is the physical feature on a part used to establish a datum (**Figure 12-26**). It is identified on a drawing by use of a datum feature symbol (**Figure 12-27**).

A *datum feature simulator* is a surface, the form of which is of such precise accuracy (such as a surface plate, a gage surface, or a mandrel) that it is used to simulate the datum. The datum feature simulator contacts the datum feature(s) and simulates the

theoretical datum. Simulation is necessary because measurements cannot be made from the theoretical true geometric counterpart. It is therefore necessary to use high-quality geometric features to simulate datums. Although the features are not perfect, they are of such quality that they can be used for that purpose. **Figures 12-28** and **12-29** illustrate this concept with respect to a surface and a feature of size.

A *datum surface* (feature) is the inexact surface of the object used to establish a datum plane. A *datum plane* is a theoretically perfect plane from which measurements are made. Since inaccuracies and variations in the surface condition of the datum surface make it impractical to take measurements from, a theoretically perfect plane must be established from which measurements are made. To establish this datum plane, the high points of the datum surface are brought into contact with, in this case, a surface plane. which simulates the datum plane. This concept is illustrated in Figure 12-26.

Notice the irregularities on the datum surface. The high points on the datum surface actually establish the datum plane which, in this case, is the top of the manufacturing equipment. All measurements referencing Datum A are measured from the theoretically perfect datum plane. High point contact is used for establishing datums when the entire surface in question will be a machined surface.

THIS ON THE DRAWING

MEANS THIS

DATUM PLANE A
(TRUE GEOMETRIC COUNTERPART
OF DATUM FEATURE A)

DATUM FEATURE A

WORKPIECE

DATUM FEATURE SIMULATOR

SIMULATED DATUM PLANE A
(PLANE DERIVED FROM THE
DATUM FEATURE SIMULATOR)

SIMULATED DATUM FEATURE
(SURFACE OF MANUFACTURING OR
VERIFICATION EQUIPMENT)

(A) WORKPIECE AND DATUM FEATURE SIMULATOR PRIOR TO CONTACT

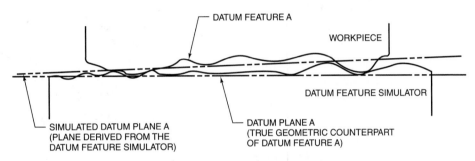

DATUM FEATURE A

WORKPIECE

DATUM FEATURE SIMULATOR

SIMULATED DATUM PLANE A
(PLANE DERIVED FROM THE
DATUM FEATURE SIMULATOR)

DATUM PLANE A
(TRUE GEOMETRIC COUNTERPART
OF DATUM FEATURE A)

(B) WORKPIECE AND DATUM FEATURE SIMULATOR IN CONTACT

FIGURE 12-26   Datum feature, simulated datum, and theoretical datum plane.

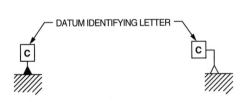

DATUM IDENTIFYING LETTER

DATUM FEATURE TRIANGLE MAY BE FILLED OR NOT FILLED.
LEADER MAY BE APPROPRIATELY DIRECTED TO A FEATURE.

FIGURE 12-27   Datum feature symbol.

A datum feature of size is established by associating the datum feature symbol with the size dimension of the selected feature size. When identified, the theoretical datum is the axis, center line, or center plane of the true geometric counterpart. It is simulated by the processing equipment (such as a chuck, vise, or centering device). The datum feature simulator establishes the datum axis, center line, or center plane from which

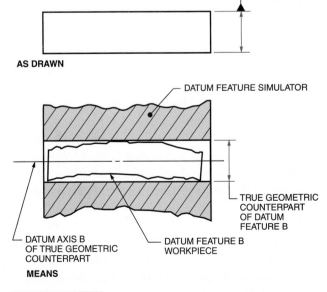

AS DRAWN

DATUM FEATURE SIMULATOR

TRUE GEOMETRIC
COUNTERPART
OF DATUM
FEATURE B

DATUM AXIS B
OF TRUE GEOMETRIC
COUNTERPART

DATUM FEATURE B
WORKPIECE

MEANS

FIGURE 12-28   Primary external datum diameter with datum feature simulator.

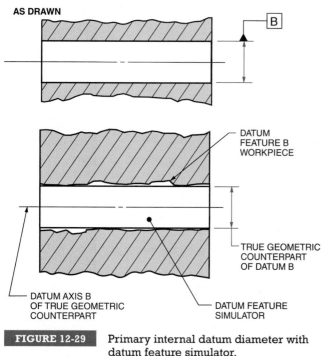

AS DRAWN

DATUM FEATURE B WORKPIECE

TRUE GEOMETRIC COUNTERPART OF DATUM B

DATUM AXIS B OF TRUE GEOMETRIC COUNTERPART

DATUM FEATURE SIMULATOR

FIGURE 12-29   Primary internal datum diameter with datum feature simulator.

measurements can be referenced. This concept is illustrated in Figures 12-28 and 12-29.

Datum Targets

On rougher, more irregular surfaces, such as those associated with castings, any specified points, lines, or area contacts are used for establishing datums. Datum targets designate specific points, lines, or areas of contact on a part that are used in establishing a datum. They are used when it is not always practical to identify an entire surface as a datum feature.

A *datum target symbol* is used to identify datum targets. It consists of a circle divided in half with a horizontal line. The lower portion contains the datum identifying letter followed by a datum target number. The numbers are sequential, starting with one for each datum. The letter and number establish a target label to identify planes or axes as datums. The upper half of the symbol is normally empty except when one uses a diameter symbol followed by a value to identify the shape and size of the target area (**Figure 12-30**). **Figure 12-31** shows a part using the datum's target areas to establish a datum plane.

Dimensions used to locate targets may be *basic dimensions* or toleranced dimensions. A basic dimension is a theoretically perfect dimension, much like a nominal or design dimension. The dimension is identified by enclosing the value in a rectangular box, as shown in (**Figure 12-32**). Tolerances placed in

TARGET AREA SIZE, WHERE APPLICABLE

DATUM IDENTIFYING LETTER

Ø.24
A1

TARGET NUMBER

OR

ØL.24

DATUM IDENTIFYING LETTER

A1

TARGET NUMBER

FIGURE 12-30   Datum target symbol.

B

.630   3.550   Ø.590
                     A3
.630   1.260
1.810

Ø.590 — A1   A2 — Ø.590

AS DRAWN

AREA CONTACT B1, B2, B3

DATUM FEATURE SURFACE

MEANS   DATUM PLANE B

FIGURE 12-31   Primary datum plane established by three datum target areas.

2.50

FIGURE 12-32   Basic dimension symbol.

general notes or within the title block do not apply to basic dimensions. In (**Figure 12-33**), the datum targets are located using basic dimensions. Points are located relative to one another and dimensioned to show the relationship between targets.

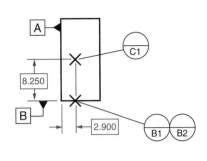

FIGURE 12-33   Dimensioning datum targets.

When specific datum target points are used for establishing datums, a minimum of three points, not in a straight line, are required for the primary datum, a minimum of two for the secondary, and a minimum of one for the tertiary, Figure 12-33. In Figure 12-33, primary datum plane A is the top of the object, and it is established by points A1, A2, and A3. Secondary datum plane B is the front of the object, and tertiary datum plane C is the right side. The datum feature symbol is placed on a drawing in the view where the surface in question appears as an edge.

Notice also that the secondary datum must be perpendicular to the first, and the tertiary datum must be perpendicular to both the primary and secondary datums. These three mutually perpendicular datum planes establish what is called the datum reference frame. The *datum reference frame* is a hypothetical, three-dimensional frame that establishes the three axes of an X, Y, and Z coordinate system into which the object being produced fits and from which measurements can be made. **Figure 12-34** shows an object located within a datum reference frame. For features that have sides (for example, rectangular and square objects), it takes three datums to establish a datum reference frame.

For cylindrical features, a complete reference frame is established with two datum references. **Figure 12-35** shows an object within a reference frame. Datum D is the primary datum feature and is used to establish datum plane K. Notice that datum feature E is established by two theoretical planes intersecting at right angles on the datum axis. The datum axis becomes the origin of measurements to locate other features on the object. Datum feature E uses the second and third planes to locate the datum axis. The reference frame is thus established using two datums.

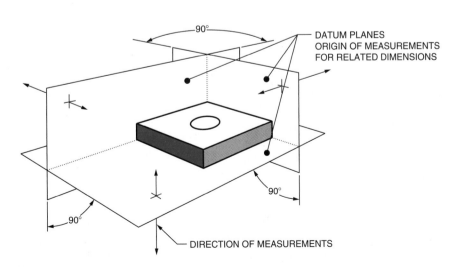

FIGURE 12-34   Datum reference frame.

DIMENSIONING AND TOLERANCING

(A)

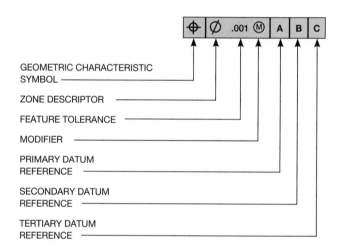

(B)

FIGURE 12-35   Part with cylindrical datum feature.

BASIC DIMENSION
EXACT DIMENSION

FIGURE 12-36   Basic dimensions.

Figure 12-36 is an example of a "basic dimension." A basic dimension is a theoretically perfect dimension, much like a nominal or design dimension, that is used to locate or specify the size of a feature. Basic dimensions are enclosed in rectangular boxes, as shown in Figure 12-36.

Feature Control Symbol

The *feature control symbol* is a rectangular box in which all data referring to the subject feature control is placed, including the symbol, datum references, the

feature control tolerance, and modifiers. These various feature control elements are separated by vertical lines. (Figure 12-5 contains a drawing showing how feature control symbols are actually composed.)

The order of the data contained in a feature control frame is important. The first element is the feature control symbol. Next is the zone descriptor, such as a diameter symbol where applicable. Then, there is the feature control tolerance, modifiers when used, and datum references listed in order from left to right (**Figure 12-37**).

Figures 12-38 through **12-42** illustrate how feature control symbols are developed for a variety of design situations. Figure 12-38 is a feature control symbol that specifies a .005 tolerance for flatness and no datum reference. Figure 12-39 specifies a tolerance of .005 for the true position of a feature relative to

GEOMETRIC CHARACTERISTIC
SYMBOL

ZONE DESCRIPTOR

FEATURE TOLERANCE

MODIFIER

PRIMARY DATUM
REFERENCE

SECONDARY DATUM
REFERENCE

TERTIARY DATUM
REFERENCE

FIGURE 12-37   Order of elements in a feature control symbol.

GEOMETRIC SYMBOL

FEATURE TOLERANCE

FIGURE 12-38   Feature control symbol with no datum reference.

GEOMETRIC SYMBOL

PRIMARY
DATUM
REFERENCE

FEATURE TOLERANCE

FIGURE 12-39   Feature control symbol with one datum reference.

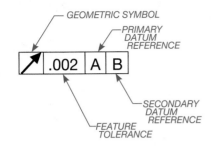

FIGURE 12-40   Feature control symbol with two datum references.

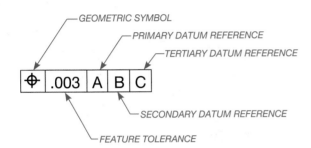

FIGURE 12-41   Feature control symbol with three datum references.

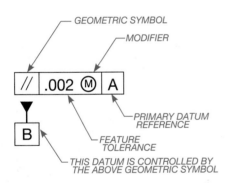

FIGURE 12-42   Feature control symbol with a modifier.

Datum A. Figures 12-40 and 12-41 show the proper methods for constructing feature control symbols with two and three datum references, respectively. Figure 12-42 illustrates a feature control symbol with a modifier and a controlled datum added.

True Position

True position is the theoretically exact location of the center line of a product feature such as a hole. The tolerance zone created by a position tolerance is an imaginary cylinder, the diameter of which is equal to the stated position tolerance. The dimensions used to locate a feature that is to have a position tolerance must be basic dimensions.

FIGURE 12-43   True position.

Figure 12-43 contains an example of a part with two holes drilled through it. The holes have a position tolerance relative to three datums: A, B, and C. The holes are located by basic dimensions. The feature control frame states that the positions of the center lines of the holes must fall within cylindrical tolerance zones having diameters of .030″ at MMC relative to Datums A, B, and C. The modifier indicates that the .030″ tolerance applies only at MMC. As the holes are produced larger than MMC, the diameter of the tolerance zones can be increased correspondingly.

Figure 12-44 illustrates the concept of the cylindrical tolerance zone from Figure 12-43. The feature control frame is repeated, showing a .030″ diameter tolerance zone. The broken-out section of the object from Figure 12-43 provides the interpretation. The cylindrical tolerance zone is shown in phantom lines. The center line of the hole is acceptable as long as it falls anywhere within the hypothetical cylinder.

FIGURE 12-44   Cylindrical tolerance zone.

Using the Projected Tolerance Zone Modifier

ASME recommends the use of the projected tolerance zone concept when the variation in perpendiculars of threaded or press-fit holes could cause fasteners, such as screws, studs, or pins, to interfere with mating parts.

The attitude of a threaded fastener is controlled by the inclination of the threaded hole into which it will assemble. There are instances where the inclination can be such that the fastener interferes with the mating feature. One method of overcoming this problem is to use a projected tolerance zone. When projected, the tolerance zone's intended outcome is to decrease the inclination of the fastener passing through the mating part. It is often thought that the tolerance zone extends through the feature being controlled to a point beyond the part equal to the projection, but this is not the case. Instead, the controlled feature has no internal tolerance; the zone is totally outside of the feature being controlled. The height of the zone is equal to the value specified within the feature control frame. **Figure 12-45** illustrates this concept.

The projected tolerance zone symbol is a capital P enclosed by a circle. It is placed within the feature control frame following the tolerance value or modifier where applicable. The projection height is placed after the projected tolerance zone symbol, as illustrated in Figure 12-45. When a projected tolerance zone modifier is used, the surface from which the

FIGURE 12-46　Projected tolerance zone using chain line.

tolerance is projected is identified as a datum, and the length of the projected tolerance zone is specified. In cases where it is not clear from which surface the projection extends, such as a through hole, a heavy chain line is used with a dimension applied to it, as illustrated in (**Figure 12-46**). The resultant tolerance zone lies totally outside the feature being controlled.

Flatness

Flatness is a feature control of a surface or median plane that requires all elements of the surface or median plane to lie within two hypothetical parallel planes. When flatness is the feature control, a datum reference is neither required nor proper.

Flatness is applied by means of a leader pointing to the surface or by an extension line of the surface. It can also be attached to a size dimension and control the derived median plane of a feature. This application is new to the current standard. The tolerance would apply at RFS unless the M or L modifier is used. For flatness of a surface, the tolerance is not additive and must be less than the tolerance of size of the part, unless the independency symbol is used, exempting it from Rule #1 requirements.

Flatness of a Surface

Figure 12-47 shows how flatness is called out in a drawing and the effect such a callout has on the produced part. The surface indicated must be flat within

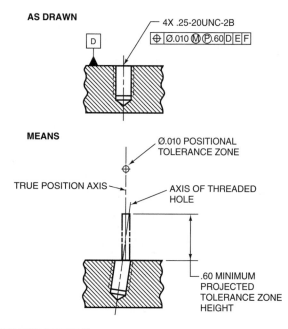

FIGURE 12-45　Specified projected tolerance zone.

FLATNESS

AS DRAWN

.010

.010 WIDE
TOLERANCE ZONE

MEANS   THE SURFACE MUST LIE BETWEEN TWO PARALLEL PLANES .010 APART.
IN ADDITION, THE LIMITS OF SIZE MUST NOT BE VIOLATED.

FIGURE 12-47   Flatness.

a tolerance zone of .010, regardless of the produced size of the part, as shown in Figure 12-47.

Flatness is specified when size tolerances alone are not sufficient to control the form and quality of the surface and when a surface must be flat enough to provide a stable base or a smooth interface with a mating part.

Flatness is inspected for a full indicator movement (FIM) using a dial indicator. FIM is the newer term that has replaced the older total indicator movement (TIR). FIM means that the swing of the indicator needle from one extreme to the other cannot exceed the amount of the specified tolerance. The dial indicator is set to run parallel to a surface table that is a theoretically perfect surface. The dial indicator is mounted on a stand or height gage. The machined surface is run under it, allowing the dial indicator to detect irregularities that fall outside of the tolerance zone.

Flatness of a Median Plane

In the current standard, flatness has been expanded to apply to a regular feature of size defined by parallel flat surfaces, meaning that the method in which the feature control frame is applied determines the intended control. If the feature control frame is attached to an extension line of the surface or attached to a leader pointing to the surface, the intended control is to the surface, Figure 12-47. However, if the feature control frame is attached to a dimension line

or adjacent to a dimension, the intended control is a median plane (center plane) (**Figure 12-48**). Thus, to locate the center plane of two parallel features, the distance between the features must be known. Moreover, to control the center plane, the feature control frame must be applied to the dimension of the part. This is an example of a feature of size. Like straightness, drastically different results are realized for flatness based on the application method.

In Figure 12-48, flatness is applied to a derived median plane by attaching the feature control in a position where it is associated with the size dimension of the feature. This controls the derived median plane (center plane) of the feature and indicates that the center plane must be straight in all directions within the tolerance specified.

Without the use of a modifier, the tolerance is understood to apply at RFS. When applied at RFS, a maximum outer boundary is created equal to the maximum part size and the combined effect of the form tolerance. Also, when applied to a feature of size, there is no perfect form boundary at MMC.

Use of the MMC modifier results in a boundary known as virtual condition. This boundary is a fixed boundary equal to the collective effects of the MMC and any applicable geometric tolerance. In this case, for an external feature, the virtual condition is equal to MMC plus the geometric tolerance. Thus, virtual condition is .753. With use of the MMC modifier, the tolerance in the feature control frame applies

FLATNESS

.740 ± .008

□ | .005

AS DRAWN

.005 BETWEEN
PARALLEL PLANES RFS

.753 MAXIMUM OUTER
BOUNDARY

.732 MIN   .748 MAX   DERIVED MEDIAN PLANE

FIGURE 12-48   Flatness of a derived median plane (RFS).

at MMC of the part. As the part moves from MMC, additional tolerance is allowed equal to the movement from MMC, as shown in (**Figure 12-49**).

Straightness

A *straightness* tolerance can be used to control surface elements or an axis. When used to control single elements for a flat surface, it is applied in the view where the element to be controlled is a straight line. When applied, it controls line elements in only one direction. It differs from flatness in that flatness covers an entire surface rather than just single elements on a surface. A straightness tolerance yields a tolerance zone of a specified width, within which all points on the line in question must lie. Straightness is generally applied to longitudinal elements.

A new similarity between straightness and flatness concerns the application of the feature control frame.

FLATNESS

.740 ± .008

□ | .005 Ⓜ

AS DRAWN

.005 AT MMC

.753 VIRTUAL
CONDITION

.748 MMC   DERIVED MEDIAN PLANE

ACTUAL LOCAL SIZE	.748	.746	.744	.743	.740	.735	.732
OUT-OF-FLATNESS	.005	.007	.009	.010	.013	.018	.021

FIGURE 12-49   Flatness of a derived median plane (MMC).

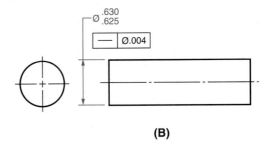

FIGURE 12-50   Dimensioning and tolerancing.

The method in which the feature control frame is applied determines the intended control. If the feature control frame is attached to an extension line of the surface or attached to a leader pointing to the surface, the intended control is to the surface (**Figure 12-50A**). However, if the feature control frame is attached to a dimension line or adjacent to a dimension, the intended control is an axis or center plane (**Figure 12-50B**). Drastically different results are realized based on the application method.

Straightness of a Flat Surface

Figure 12-51 shows how a straightness tolerance is applied on a drawing to the elements of a flat surface. The straightness tolerance applies only to the top surface. The bottom surface straightness error is controlled by the limits of size. In this case, the straightness tolerance is used as a refinement for the top surface only. The feature control frame states that any longitudinal element for the referenced surface, in the direction indicated, must lie between two parallel straight lines that are .002″ apart.

FIGURE 12-51   Straightness of a flat surface.

Straightness of a Cylindrical Surface

Straightness applied to the surface of a cylindrical feature is shown in (**Figure 12-52**). It is similar to that of a flat surface, with one exception. Since the surface is round, opposing surface line elements must also be considered when verifying straightness. The

AS DRAWN

MEANS

EACH LONGITUDINAL ELEMENT OF THE SURFACE MUST LIE BETWEEN TWO PARALLEL LINES (.004 APART) WHERE THE TWO LINES AND THE NOMINAL AXIS OF THE PART SHARE A COMMON PLANE. THE FEATURE MUST BE WITHIN THE SPECIFIED LIMITS OF SIZE AND THE BOUNDARY OF PERFECT FORM AT MMC (.650).

WAISTING (B) OR BARRELING (C) OF THE SURFACE MUST BE WITHIN THE STRAIGHTNESS TOLERANCE AND THE LIMITS OF SIZE OF THE FEATURE.

FIGURE 12-52   Straightness of surface elements.

full straightness tolerance may not be available for these elements due to conditions such as wasting or barreling of the surface. Additionally, the straightness tolerance is not additive to the size tolerance and must be contained within the limits of size. This means that if the part is made at MMC, no straightness tolerance is available because any variation in surface straightness would cause the part to exceed the MMC boundary of size. **Figure 12-53** illustrates the relationship between a straightness tolerance and a size tolerance of a part. Remember, each element of the surface must stay within the specified straightness tolerance zone and within the size tolerance envelope. Straightness is affected by running the single-line elements of a surface under a dial indicator for a full indicator movement (FIM).

Figures 12-54 through **12-57** further illustrate the concept of straightness. Figure 12-54 shows a part with a size tolerance but no feature control tolerance. In this example, the form of the feature is controlled by the size tolerance. The difference between maximum and minimum limits defines the maximum form variation that is allowed. ASME Y14.5M outlines the requirements of form control for individual features controlled only with a size dimension. This requirement is known as *Rule #1*. According to the standard, Rule #1 states: "Where only a tolerance size is specified, the limits of size of an individual feature define the extent to which variations in its geometric form, as well as size, are allowed." This means that the size limits of a part determine the maximum and minimum limits (boundaries) for that part. The MMC limit establishes a boundary limit of perfect form. If a part is at MMC, it must have perfect form. No variation in

ACTUAL LOCAL	.504	.503	.502	.501	.500
OUT-OF-STRAIGHTNESS	.000	.001	.002	.003	.004

FIGURE 12-54   Object with no feature control symbol (Rule #1 applies).

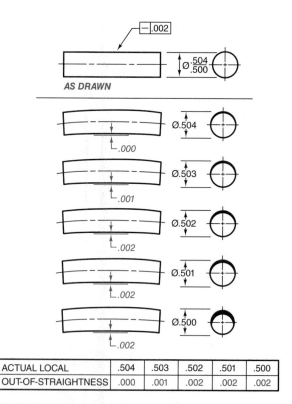

ACTUAL LOCAL	.504	.503	.502	.501	.500
OUT-OF-STRAIGHTNESS	.000	.001	.002	.002	.002

FIGURE 12-55   Straightness at RFS.

FIGURE 12-53   Straightness interpreted.

form is allowed. As the part varies in size toward LMC, the form of the part is allowed to vary equal to the variation in size from MMC. When the part is made at LMC, the form variation is equal to the

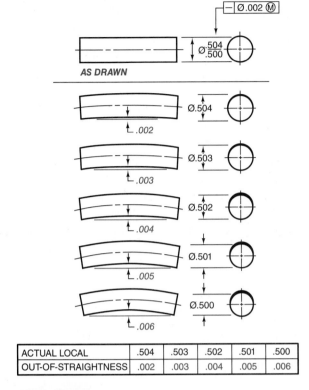

ACTUAL LOCAL	.504	.503	.502	.501	.500
OUT-OF-STRAIGHTNESS	.002	.003	.004	.005	.006

FIGURE 12-56   Straightness of an axis at MMC.

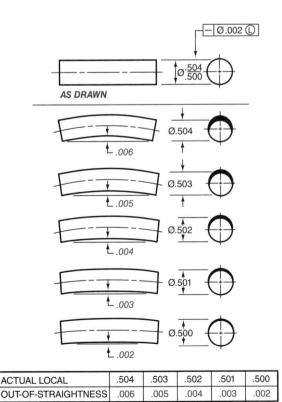

ACTUAL LOCAL	.504	.503	.502	.501	.500
OUT-OF-STRAIGHTNESS	.006	.005	.004	.003	.002

FIGURE 12-57   Straightness of an axis at LMC.

difference between the MMC and LMC sizes, as illustrated in Figure 12-54.

Figure 12-55 is the same part with a straightness tolerance of .002 regardless of feature size tolerance. The implied regardless of feature size tolerance limits the amount the surface can be out of straightness to a maximum of .002 regardless of the produced size of the part. However, because the straightness control is on a cylindrical surface, the .002 tolerance might not be available as the part approaches MMC. The drawing at the top of the figure illustrates how the part would be drawn. The five illustrations below the part as drawn illustrate the actual shape of the object with each corresponding produced size and the available tolerance.

Straightness of an Axis

To locate the axis of a part, the size of the part must be known. This is an example of what is known as a feature of size. Logically, then, to control the axis of a part, the feature control frame must be applied to the size dimension of that part (Figure 12-50). When straightness is applied to control the axis of the feature, the tolerance zone is cylindrical and extends the full length of the controlled feature.

Straightness applied to the axis of a feature creates a boundary condition known as *virtual condition*. Virtual condition in ASME Y14.5 is defined as follows: "A constant boundary generated by the collective effects of a size feature's specified MMC or LMC and the geometric tolerance for that material condition." This means that you are allowed to add the straightness tolerance to the MMC size for a shaft and subtract the straightness tolerance from the MMC size for a hole. The resultant boundary represents the extreme form variation allowed for the part. Although this boundary is theoretical, it represents the size boundary of mating features. Unlike straightness of a surface control, a straightness control of an axis allows for the availability of straightness tolerance even when the part is made at MMC. Axis control of a feature becomes more desirable because of the increased availability of tolerance and better control of mating features.

Figure 12-56 is the same part as in the previous examples with a straightness tolerance of .002 at MMC applied. The use of the MMC modifier is limited to tolerances controlling the axis of features. It specifies the tolerance allowed when the part is produced at MMC. The drawing at the top of the figure illustrates how the part would be drawn.

The five illustrations below the part as drawn illustrate the actual shape of the object with each corresponding produced size. A virtual condition boundary of .506 is created. When the part is at .504, the .506 virtual condition boundary allows for straightness of .002 at MMC. Since the .002 straightness tolerance applies at MMC, the amount that the part can be out of straightness increases correspondingly as the produced size decreases. The table at the bottom of Figure 12-56 summarizes the actual local sizes and the corresponding amounts that the part can be out of straightness for each size.

Figure 12-57 is an example of the same part with a .002 straightness tolerance at least material condition (LMC). It specifies the tolerance allowed when the part is produced at LMC. Notice that the .002 straightness tolerance applies at the LMC. As the actual produced size increases, the amount of out of straightness allowed increases correspondingly.

Figure 12-58 illustrates the same part from a .002 straightness tolerance regardless of feature size tolerance. Notice in this example that the .002 straightness tolerance applies regardless of the actual produced size of the part.

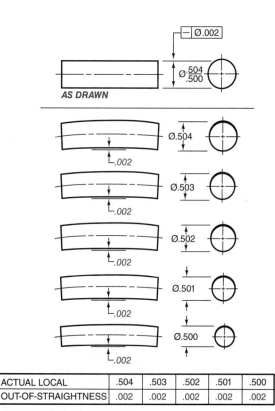

ACTUAL LOCAL	.504	.503	.502	.501	.500
OUT-OF-STRAIGHTNESS	.002	.002	.002	.002	.002

FIGURE 12-58   Straightness of an axis at RFS.

Circularity (Roundness)

Circularity, sometimes referred to as roundness, is a feature control for a surface of revolution (cylinder, sphere, cone, and so forth). It specifies that all points of a surface must be equidistant from the center line or axis of the object in question. The tolerance zone for circularity is formed by two concentric and coplanar circles between which all points on the surface of revolution must lie.

Figures 12-59 and **12-60** illustrate how circularity is called out on a drawing and provide an interpretation of what the circularity tolerance actually means. At any selected cross section of the part, all points on

MEANS   EACH CIRCULAR ELEMENT OF THE SURFACE IN A PLANE PERPENDICULAR TO AN AXIS MUST LIE BETWEEN TWO CONCENTRIC CIRCLES, ONE HAVING A RADIUS .007 LARGER THAN THE OTHER. EACH CIRCULAR ELEMENT OF THE SURFACE MUST BE WITHIN THE SPECIFIED LIMITS OF SIZE.

FIGURE 12-59   Circularity for a cylinder or cone.

EACH CIRCULAR ELEMENT OF THE SURFACE IN A PLANE PASSING THROUGH A COMMON CENTER MUST LIE BETWEEN TWO CONCENTRIC CIRCLES, ONE HAVING A RADIUS 0.25 LARGER THAN THE OTHER. EACH CIRCULAR ELEMENT OF THE SURFACE MUST BE WITHIN THE SPECIFIED LIMITS OF SIZE.

FIGURE 12-60   Circularity for a sphere.

the surface must fall within the zone created by the two concentric circles. At any point where circularity is measured, it must fall within the size tolerance. Notice that a circularity tolerance cannot specify a datum reference.

Circularity establishes elemental single-line tolerance zones that may be located anywhere along a surface. The tolerance zones are taken at any cross section of the feature. Therefore, the object may be spherical, cylindrical, tapered, or even hourglass-shaped, so long as the cross section for inspection is taken at 90° to the nominal axis of the object. A circularity tolerance is inspected using a dial indicator and making readings relative to the axis of the feature. In measuring a circularity tolerance, the full indicator movement (FIM) of the dial indicator should not be any larger than the size tolerance, and there should be several measurements made at different points along the surface of the diameter. All measurements taken must fall within the circularity tolerance.

Cylindricity

Cylindricity is a feature control in which all elements of a surface of revolution form a cylinder. It gives the effect of circularity extended the entire length of the object, rather than just a specified cross section. The tolerance zone is formed by two hypothetical concentric cylinders.

Figure 12-61 illustrates how cylindricity is called out on a drawing. Notice that a cylindricity tolerance does not require a datum reference.

Figure 12-61 also provides an illustration of what the cylindricity tolerance actually means. Two

hypothetical concentric cylinders form the tolerance zone. The outside cylinder is established by the outer limits of the object at its produced size within specified size limits. The inner cylinder is smaller (on radius) by a distance equal to the cylindricity tolerance.

Cylindricity requires that all elements on the surface fall within the size tolerance and the tolerance established by the feature control.

A cylindricity tolerance must be less than the size tolerance, and it is not additive to the MMC of the feature. Cylindricity is inspected using a dial indicator and making readings relative to the axis of the feature. In measuring cylindricity tolerance, the full indicator movement (FIM) of the dial indicator should not be any larger than the size tolerance. The part is rotated as the indicator is moved simultaneously along the surface of the diameter. At no point can the FIM exceed the tolerance specified.

Angularity

Prior to the current standard, *angularity* was defined as a feature control in which a given surface, axis, or center plane formed a specified angle other than 90° to a datum. However, under the ASME Y14.5—2009 standard, the definition of angularity includes the application to a 90° angle. As defined by the current standard, "angularity is the condition of a surface, feature's center plane, or feature's axis at any specified angle from a datum plane or datum axis." This begs the question: Why do you need a perpendicularity symbol if angularity can control perpendicularity? Although it might not be wrong to use an angularity callout to control perpendicularity, best practice might be better served if the perpendicularity symbol were only used for 90° angles and the angularity symbol were used for angles other than 90°.

Depending on the application, the tolerance zone formed by an angularity callout can consist of two hypothetical parallel planes or a cylindrical tolerance zone that form(s) the specified angle with the datum.

Angularity of a Surface

Figure 12-62 illustrates how an angularity tolerance on a surface is called out on a drawing. Notice that the specified angle is basic. This is required when applying an angularity tolerance. Figure 12-62 also provides an interpretation of what the angularity tolerance actually means. The surface must lie between two parallel

CYLINDRICITY /○/

Ø1.000 ±020
/○/ .005

.005 WIDE TOLERANCE ZONE

THE CYLINDRICAL SURFACE MUST LIE BETWEEN TWO
CONCENTRIC CYLINDERS, ONE HAVING A RADIUS .005
LARGER THAN THE OTHER. THE SURFACE MUST BE
WITHIN THE SPECIFIED LIMITS OF SIZE.

FIGURE 12-61   Specifying cylindricity.

AS DRAWN

MEANS

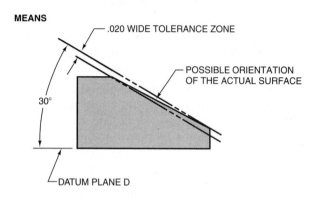

.020 WIDE TOLERANCE ZONE

POSSIBLE ORIENTATION OF THE ACTUAL SURFACE

DATUM PLANE D

THE SURFACE MUST LIE BETWEEN TWO PARALLEL PLANES .020 APART THAT ARE INCLINED AT 30° BASIC TO DATUM PLANE D. THE SURFACE MUST BE WITHIN THE SPECIFIED LIMITS OF SIZE.

FIGURE 12-62 Specifying angularity for a surface.

planes of .02 apart that are inclined at a 30° basic angle to datum plane A.

Angularity also controls the flatness of the surface to the same extent that it controls the angular orientation. When it is required that the flatness of the feature be less than the orientation, a flatness callout can be used as a refinement of the orientation callout. When using flatness as a refinement, the tolerance is less than the orientation tolerance. The feature control frame is normally placed on an extension line below the orientation control (**Figure 12-63**).

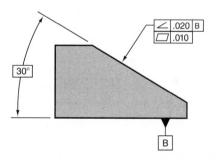

FIGURE 12-63 Angularity with flatness refinement.

AS DRAWN

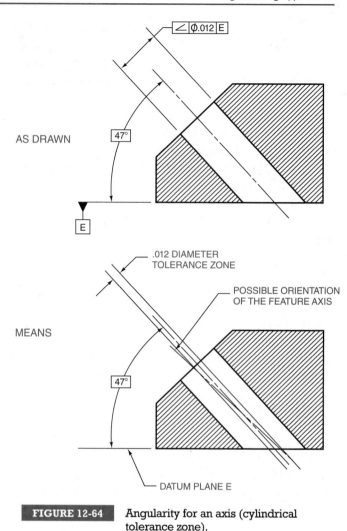

MEANS

.012 DIAMETER TOLERANCE ZONE

POSSIBLE ORIENTATION OF THE FEATURE AXIS

DATUM PLANE E

FIGURE 12-64 Angularity for an axis (cylindrical tolerance zone).

Angularity of an Axis or Center Plane

An angularity callout can also be used to control the axis or center plane of a feature. This is done by placing the feature control frame with the size dimension in an appropriate manner, as shown in (**Figure 12-64**). The tolerance zone for an axis control can be cylindrical in shape or two parallel planes. When the diameter symbol is used within the feature control frame, the tolerance zone is cylindrical. When no diameter is used, the tolerance zone shape is two parallel planes, (**Figure 12-65**).

Parallelism

Parallelism is a feature control that specifies that all points on a given surface, axis, or center plane must be equidistant from a datum. Consequently, a parallelism tolerance requires one or more datum references. A parallelism tolerance zone is formed by two hypothetical parallel planes that are parallel to a specified

FIGURE 12-65   Angularity for an axis (two parallel planes).

FIGURE 12-66   Parallelism for a surface to datum plane.

datum. They are spaced apart at a distance equal to the parallelism tolerance.

Parallelism of a Surface

Figure 12-66 illustrates how a parallelism is called out on a drawing and provides an interpretation of what the parallelism tolerance actually means. Notice that all elements of the toleranced surface must fall within the size limits.

Notice in Figure 12-66 that the .006 parallelism tolerance is called out relative to Datum A. You must specify a datum when calling out a parallelism tolerance. Parallelism should be specified when features such as surfaces, axes, and planes are required to lie in a common orientation. Parallelism is inspected by placing the part on an inspection table and running a dial indicator a full indicator movement across the surface of the part.

Parallelism also controls the flatness of the surface to the same extent that it controls parallel orientation. When it is required that the flatness of the feature be less than the orientation, a flatness callout can be used as a refinement of the orientation callout. When using flatness as a refinement, the tolerance is less than the orientation tolerance. The feature control frame is normally placed on an extension line below the orientation control (**Figure 12-67**).

Parallelism of an Axis or Center Plane

Parallelism can be used to control the orientation of an axis to a datum plane, an axis to an axis, or the center plane of noncylindrical parts. When applied to control the axis or center plane, the feature control frame must be placed with the size dimension in the appropriate fashion. When used to control an axis to a datum plane, the tolerance zone shape is two parallel planes separated by the amount of the stated tolerance. The tolerance control is only applicable relative to the specified datum surface (**Figure 12-68**). Virtual condition exists for the controlled feature, which allows for the availability of additional tolerance. The M and L modifiers can be used because we are controlling a feature of size.

FIGURE 12-67   Parallelism with flatness refinement.

AS DRAWN

MEANS

.006 WIDE
TOLERANCE
ZONE

POSSIBLE ORIENTATION
OF FEATURE AXIS

DATUM PLANE C

REGARDLESS OF FEATURE SIZE, THE FEATURE AXIS MUST
LIE BETWEEN TWO PARALLEL PLANES .006 APART THAT
ARE PARALLEL TO DATUM PLANE C. THE FEATURE AXIS
MUST BE WITHIN THE SPECIFIED TOLERANCE OF LOCATION.

FIGURE 12-68   Parallelism for an axis to datum plane.

When used to control an axis to a datum axis, the tolerance zone shape is cylindrical, and the diameter is equal to the amount of the stated tolerance. The tolerance control is three-dimensional, allowing the axis to float relative to the orientation of the datum, (**Figure 12-69**). Virtual condition exists for the controlled feature, which allows for the availability of additional tolerance. The M and L modifiers can be used because we are controlling a feature of size.

AS DRAWN

MEANS

POSSIBLE ORIENTATION
OF FEATURE AXIS

.010 CYLINDRICAL
TOLERANCE ZONE

DATUM AXIS F

REGARDLESS OF FEATURE SIZE, THE FEATURE AXIS MUST LIE
WITHIN A .010 DIAMETER CYLINDRICAL ZONE PARALLEL TO
DATUM AXIS A. THE FEATURE AXIS MUST BE WITHIN THE
SPECIFIED TOLERANCE OF LOCATION.

FIGURE 12-69   Parallelism for an axis to datum axis.

When used to control a center plane to a datum plane or a center plane to a center plane, the tolerance zone is similar to that of an axis to a surface or an axis to a datum axis. However, the tolerance zone shape is never cylindrical. The shape is two parallel planes separated by the amount of the stated tolerance. Virtual condition exists for the controlled feature, which allows for the availability of additional tolerance. The M and L modifiers can be used because we are controlling a feature of size.

Perpendicularity

Perpendicularity is a feature control that specifies that all elements of a surface, axis, or center plane form a 90° angle with a datum. Consequently, a perpendicularity tolerance requires a datum reference. A perpendicularity tolerance is formed by two hypothetical parallel planes that are at 90° to a specified datum. They are spaced apart at a distance equal to the perpendicularity tolerance.

Perpendicularity of a Surface

Figure 12-70 illustrates how a perpendicularity tolerance is called out on a drawing and provides an interpretation of what the perpendicularity tolerance actually means. The elements of the toleranced surface must fall within the size limits and between two hypothetical parallel planes that are a distance apart equal to the perpendicularity tolerance.

The perpendicularity of a part such as the one shown in Figure 12-70 could be inspected by clamping the part to an inspection angle. The datum surface should rest against the inspection angle. Then, a dial indicator should be passed over the entire surface for a full indicator movement to determine if the perpendicularity tolerance has been complied with.

Perpendicularity also controls the flatness of the surface to the same extent that it controls orientation. When it is required that the flatness of the feature be less than the orientation, a flatness callout can be used as a refinement of the orientation callout. When using flatness as a refinement, the tolerance is less than the orientation tolerance. The feature control frame is normally placed on an extension line below the orientation control (**Figure 12-71**).

AS DRAWN

THE SURFACE MUST LIE BETWEEN TWO PARALLEL PLANES .010 APART THAT ARE PERPENDICULAR TO DATUM PLANE F. THE SURFACE MUST BE WITHIN THE SPECIFIED LIMITS OF SIZE.

FIGURE 12-70   Perpendicularity for a surface to datum plane.

FIGURE 12-71   Perpendicularity with flatness refinement.

Perpendicularity of an Axis or Center Plane

Perpendicularity can be used to control the orientation of an axis to a datum plane, an axis to an axis, or the center plane of noncylindrical parts. When applied to control the axis or center plane, the feature control frame must be placed with the size dimension in the appropriate fashion. When used to control an axis to a datum plane, the tolerance zone shape is cylindrical, and its diameter equals the amount of the stated tolerance. The tolerance control is three-dimensional, allowing the axis to be at any orientation relative to the specified datum surface (**Figure 12-72**). Virtual condition exists for the controlled feature, which allows for the availability of additional tolerance. The M and L modifiers can be used because we are controlling a feature of size.

When used to control an axis to a datum axis, the tolerance zone shape is two parallel planes that are separated by a distance equal to the amount of the stated tolerance. The tolerance control is only applicable relative to the orientation of the datum (**Figure 12-73**). Virtual condition exists for the controlled feature, which allows for the availability of additional tolerance. The M and L modifiers can be used because we are controlling a feature of size.

When used to control a center plane to a datum plane or a center plane to a center plane, the tolerance is similar to that of an axis to a surface or an axis to a datum axis. However, the tolerance zone shape is never cylindrical. The shape is two parallel planes separated by the amount of the stated tolerance. Virtual condition exists for the controlled feature, which allows for the availability of additional tolerance. The M and L modifiers can be used because we are controlling a feature of size.

AS DRAWN

MEANS

REGARDLESS OF FEATURE SIZE, THE FEATURE AXIS MUST
LIE WITHIN A CYLINDRICAL ZONE .008 DIAMETER THAT IS
PERPENDICULAR TO AND PROJECTS FROM DATUM PLANE
G FOR THE FEATURE HEIGHT. THE FEATURE AXIS MUST BE
WITHIN THE SPECIFIED TOLERANCE OF LOCATION.

FIGURE 12-72   Perpendicularity for an axis to
datum plane.

AS DRAWN

MEANS

REGARDLESS OF FEATURE SIZE, THE FEATURE AXIS MUST
LIE BETWEEN TWO PARALLEL PLANES .020 APART THAT ARE
PERPENDICULAR TO DATUM AXIS F. THE FEATURE AXIS
MUST BE WITHIN THE SPECIFIED TOLERANCE OF LOCATION.

FIGURE 12-73   Perpendicularity for an axis to
datum axis.

Profile

Profile is a feature control that specifies the amount of allowable variance of a surface or line elements on a surface. There are three different variations of the profile tolerance: unilateral (inside), unilateral (outside), and bilateral (unequally disposed) (**Figure 12-74**). A profile tolerance is normally used for controlling arcs, curves, and other unusual profiles not covered by the other feature controls. It is a valuable feature control for use on objects that are so irregular that other feature controls do not easily apply.

When applying a profile tolerance, the symbol used indicates whether the designer intends profile of a line or profile of a surface (**Figures 12-75** and **12-76**). *Profile of a line* establishes a tolerance for a given single element of a surface. *Profile of a surface* applies to the entire surface. The difference between profile of a line and profile of a surface is similar to the difference between circularity and cylindricity.

When using a profile tolerance, drafters and designers should remember that the application of a profile tolerance is understood to create a bilateral tolerance that is equally disposed about the true profile, Figure 12-76. To indicate whether the tolerance application is unilateral or unequally disposed, the unequally disposed profile symbol is used. The previous practice of using the phantom line to indicate a unilateral tolerance is now an acceptable alternate practice. An ALL AROUND symbol should be placed on the leader line of the feature control frame to specify whether the tolerance applies all around or between specific points on the object (**Figure 12-77**). **Figure 12-78** provides an interpretation of what the BETWEEN A & B profile tolerance in Figure 12-77 actually means. The rounded top surface, and only the top surface, of the object must fall within the specified tolerance zone. **Figure 12-79** provides an interpretation of what the ALL AROUND profile tolerance is in Figure 12-77. An ALL OVER symbol should be placed on the leader line of the feature control frame to specify whether the tolerance applies all over (**Figure 12-80**). **Figure 12-81** provides an interpretation of what the ALL OVER profile tolerance in Figure 12-80 actually means. The entire surface of the object, all over the object, must fall within the specified tolerance zone.

Unilateral Profile

To indicate a unilateral tolerance, the use of the unequally disposed profile symbol is now the standard practice. The unequally disposed symbol is placed in

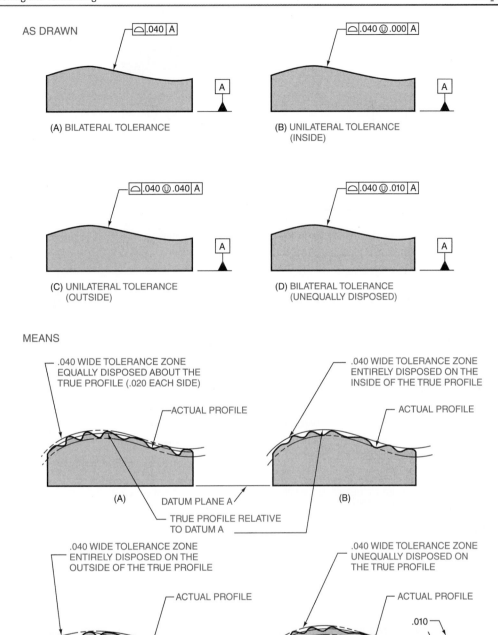

AS DRAWN

(A) BILATERAL TOLERANCE

(B) UNILATERAL TOLERANCE
(INSIDE)

(C) UNILATERAL TOLERANCE
(OUTSIDE)

(D) BILATERAL TOLERANCE
(UNEQUALLY DISPOSED)

MEANS

.040 WIDE TOLERANCE ZONE
EQUALLY DISPOSED ABOUT THE
TRUE PROFILE (.020 EACH SIDE)

ACTUAL PROFILE

(A)   DATUM PLANE A

TRUE PROFILE RELATIVE
TO DATUM A

.040 WIDE TOLERANCE ZONE
ENTIRELY DISPOSED ON THE
INSIDE OF THE TRUE PROFILE

ACTUAL PROFILE

(B)

.040 WIDE TOLERANCE ZONE
ENTIRELY DISPOSED ON THE
OUTSIDE OF THE TRUE PROFILE

ACTUAL PROFILE

(C)   DATUM PLANE A

TRUE PROFILE
RELATIVE TO
DATUM A

.040 WIDE TOLERANCE ZONE
UNEQUALLY DISPOSED ON
THE TRUE PROFILE

ACTUAL PROFILE

.010

.030

(D)

TRUE PROFILE
RELATIVE TO
DATUM A

FIGURE 12-74   Application of profile of a surface.

the feature control frame following the tolerance value. This first value represents the profile tolerance. A second value is then added following the unequally disposed symbol. The second value represents the value that will add material to the true profile or remove material from the true profile.

To create a unilateral tolerance where all the tolerance is outside the true profile (adding material), the value in front of the unequally disposed symbol and the value behind the unequally disposed symbol would be the same. In (**Figure 12-82**), the profile tolerance is .020. The value behind the unequally

FIGURE 12-75   Profile of a line with size control.

disposed symbol is .020. This indicates a profile tolerance of .020 (the value in front of the disposed symbol), and we want all of the .020 to add material to the outside of the part. Therefore, the .020 is disposed totally on the outside of the part.

To create a unilateral tolerance where all the tolerance is inside the true profile (removing material), the value in front of the unequally disposed symbol would represent the profile tolerance, and the value behind the unequally disposed symbol would be zero. In (**Figure 12-83**), the profile tolerance is .020. The value behind the unequally disposed symbol is .000. This indicates a profile tolerance of .020 (the value in front of the unequally disposed symbol), and we want zero amount of that profile tolerance to add material to the outside of the part. Therefore, the .020 is disposed totally on the inside of the part.

Unequally Disposed Profile Tolerance

To create an unequally disposed tolerance zone, the value in front of the unequally disposed symbol would represent the profile tolerance, and the value behind the unequally disposed symbol would be the amount of the profile tolerance that will add material to the part. In (**Figure 12-84**), the overall tolerance is .020, which is the first value. The second value behind the unequally disposed symbol is .008. This indicates a profile tolerance of .020 (the value in front of the unequally disposed symbol), and we want .008 of that .020 profile tolerance to add material to the outside of the part. This means that the .012 that remains will be the value that removes material from the part. The .020 tolerance is thus unequally disposed, where .008 is on the outside of the part and .012 is on the inside of the part.

DIMENSIONING AND TOLERANCING
AS DRAWN

MEANS

DATUM PLANE B

.020 WIDE TOLERANCE ZONE

90°

TRUE PROFILE
(INSIDE TOLERANCE
BOUNDARY)

THE SURFACES ALL AROUND MUST LIE
BETWEEN TWO PROFILE BOUNDARIES .020
APART, ONE COINCIDENT WITH AND THE
OTHER OUTSIDE THE TRUE PROFILE, AND
BOTH PERPENDICULAR TO DATUM PLANE B.
THE TOLERANCE ZONE EXTENDS TO THE
INTERSECTION OF THE BOUNDARY LINES.

OUTSIDE TOLERANCE BOUNDARY

FIGURE 12-76   Profile of a surface.

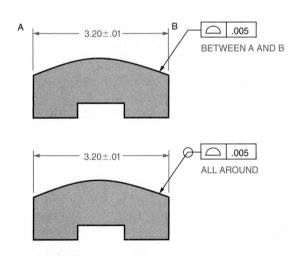

FIGURE 12-77   Profile "ALL AROUND."

FIGURE 12-79   Interpretation of "ALL AROUND."

FIGURE 12-78   Interpretation of "BETWEEN A AND B."

FIGURE 12-80   Profile "ALL OVER."

FIGURE 12-81   Interpretation of "ALL OVER."

FIGURE 12-82   Profile "Unilateral" material added.

FIGURE 12-83   Profile "Unilateral" material removed.

FIGURE 12-84   Profile "Unequally Disposed."

FIGURE 12-85   Profile "Unequally Disposed" internal part.

If the part were an internal part such as a hole or slot, the same would apply. To create an unequally disposed tolerance zone, the value in front of the unequally disposed symbol would represent the profile tolerance, and the value behind the unequally disposed symbol would be the amount of the profile tolerance that will add material to the part. In (**Figure 12-85**), the overall tolerance is .015, which is the first value to the left of the unequally disposed symbol. The second value is .008, which is the amount of the .015 on the outside of the true profile that will add material to the part. This means that the .007 that remains will be the value that removes material from the part. The .015 tolerance is thus unequally disposed, where .008 is on the outside of the part and .007 is on the inside of the part.

Profile tolerances may be inspected using a dial indicator. However, because the tolerance zone must be measured at right angles to the basic true profile and perpendicular to the datum, the dial indicator must be set up to move and read in both directions. Other methods of inspecting profile tolerances are becoming more popular, however. Optical comparators are becoming widely used for inspecting profile tolerances. An optical comparator magnifies the silhouette of the part and projects it onto a screen where it is compared to a calibrated grid or template so that the profile and size tolerances may be inspected visually.

Runout

Runout is a feature control that limits the amount of deviation from perfect form allowed on surfaces or rotation through one full rotation of the object about its axis. Revolution of the object is around a datum axis. Consequently, a runout tolerance does require a datum reference.

AS DRAWN

MEANS

AT ANY MEASURING POSITION, EACH CIRCULAR ELEMENT OF THESE SURFACES MUST BE WITHIN THE SPECIFIED RUNOUT TOLERANCE (.009 FULL INDICATOR MOVEMENT) WHEN THE PART IS ROTATED 360° ABOUT THE DATUM AXIS WITH THE INDICATOR FIXED IN A POSITION NORMAL TO THE TRUE GEOMETRIC SHAPE. THE FEATURE MUST BE WITHIN THE SPECIFIED LIMITS OF SIZE.

FIGURE 12-86   Specifying circular runout relative to a datum diameter.

Runout is most frequently used on objects consisting of a series of concentric cylinders and other shapes of revolution that have circular cross sections—usually the types of objects manufactured on lathes (**Figures 12-86** and **12-87**).

THE ENTIRE SURFACE MUST LIE WITHIN THE SPECIFIED
RUNOUT TOLERANCE ZONE (.009 FULL INDICATOR MOVEMENT)
WHEN THE PART IS ROTATED 360° ABOUT THE DATUM AXIS WITH
THE INDICATOR PLACED AT EVERY LOCATION ALONG THE
SURFACE IN A POSITION NORMAL TO THE TRUE GEOMETRIC
SHAPE WITHOUT RESET OF THE INDICATOR. THE FEATURE
MUST BE WITHIN THE SPECIFIED LIMITS OF SIZE.

FIGURE 12-87   Specifying total runout relative to a datum diameter.

Notice in Figures 12-86 and 12-87 that there are two types of runout: circular runout and total runout. The circular runout tolerance applies at any single-line element through which a section passes. The total runout tolerance applies along an entire surface, as illustrated in Figure 12-87. Runout is most frequently used when the actual produced size of the feature is not as important as the form, and the quality of the feature must be related to some other feature. Circular runout is inspected using a dial indicator along a single fixed position so that errors are read only along a single line. Total runout requires that the dial indicator move in both directions along the entire surface being toleranced.

Concentricity

It is not uncommon in manufacturing to have a part made up of several subparts all sharing the same center line or axis. Such a part is illustrated in (**Figure 12-88**). In such a part, it is critical that the center line for each

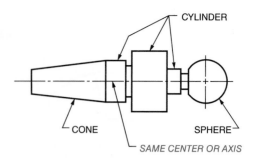

FIGURE 12-88   Part with concentric subparts.

subsequent subpart be concentric with the center lines of the other subparts. When this is the case, a concentricity tolerance is applied. A concentricity tolerance locates the axis of a feature relative to the axis of a datum. A concentricity tolerance deals only with the center line relationship. It does not affect the size, form, or surface quality of the part. Concentricity deals only with axial relationships. It is applied only on a regardless-of-feature-size basis. Regardless of how large or small the various subparts of an overall part are, only their axes are required to be concentric. A concentricity tolerance creates a cylindrical tolerance zone in which all center lines for each successive subpart of an overall part must fall. This concept is illustrated in (**Figure 12-89**). A concentricity tolerance is inspected by a full indicator movement of a dial indicator.

Symmetry

Parts that are symmetrically disposed about the center plane of a datum feature are common in manufacturing settings. If it is necessary that a feature be located symmetrically with regard to the center plane of a datum feature, a symmetry tolerance may be applied (**Figure 12-90**). The part in Figure 12-90 is symmetrical about a center plane. To ensure that the part is located symmetrically with respect to the center plane, a .030 symmetry tolerance is applied. This creates a .030 tolerance zone within which the center plane in question must fall, as illustrated in the bottom portion of Figure 12-90.

Positioning

Position tolerancing is used to locate features of parts that are to be assembled and mated. Position is symbolized by a circle overlaid by a large plus sign or cross. This symbol is followed by the tolerance, a modifier when appropriate, and a reference datum, (**Figure 12-91**). **Figures 12-92** and **12-93** illustrate the difference between conventional and position

AS DRAWN

MEANS

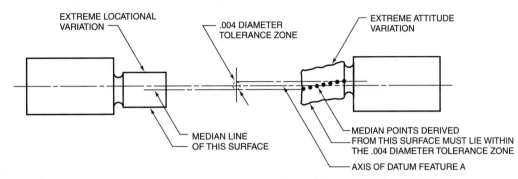

WITHIN THE LIMITS OF SIZE AND REGARDLESS OF FEATURE SIZE, ALL MEDIAN POINTS OF DIAMETRICALLY OPPOSED ELEMENTS OF THE FEATURE MUST LIE WITHIN A Ø.004 CYLINDRICAL TOLERANCE ZONE. THE AXIS OF THE TOLERANCE ZONE COINCIDES WITH THE AXIS OF DATUM FEATURE A. THE SPECIFIED TOLERANCE AND THE DATUM REFERENCE APPLY ONLY ON AN RFS BASIS.

FIGURE 12-89   Concentricity tolerancing.

AS DRAWN

MEANS

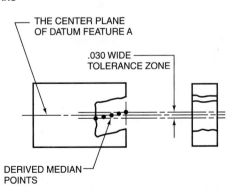

WITHIN THE LIMITS OF SIZE AND REGARDLESS OF FEATURE SIZE, ALL MEDIAN POINTS OF OPPOSED ELEMENTS OF THE SLOT MUST LIE BETWEEN TWO PARALLEL PLANES .030 APART, THE TWO PLANES BEING EQUALLY DISPOSED ABOUT DATUM PLANE A. THE SPECIFIED TOLERANCE AND THE DATUM REFERENCE CAN ONLY APPLY ON AN RFS BASIS.

FIGURE 12-90   Symmetry tolerancing.

FIGURE 12-91   Position symbology.

FIGURE 12-92   Conventional dimensioning.

dimensioning. The tolerance dimensions shown in Figure 12-92 create a square tolerance zone. This means that the zone within which the center line being located by the dimensions must fall takes the shape of a square. As shown in Figure 12-93, the tolerancing zone is round when position dimensioning

FIGURE 12-93   Position dimensioning.

FIGURE 12-94   Comparison of tolerance zones.

FIGURE 12-95   Positioning at MMC.

is used. The effect of this on manufacturing is that the round tolerancing zone with position dimensioning increases the size of the tolerance zone by 57%, (**Figure 12-94**). This means that for the same tolerance, the machinist has 57% more room for error without producing an out-of-tolerance part.

When using position dimensioning, the tolerance is assumed to apply regardless of the feature size unless modified otherwise. **Figure 12-95** illustrates the effect of modifying a position tolerance with a MMC modifier. In this example, a hole is to be drilled through a plate. The maximum diameter is .254, and the minimum diameter is .250. Therefore, the MMC of the part occurs when the hole is drilled to a diameter of .250. Notice from this example that as the hole increases, the positional tolerance increases. At MMC (.250 diameter), the tolerance zone has a diameter of .042. At LMC (.254 diameter), the tolerance zone increases to .046 diameter. The tolerance zone diameter increases correspondingly as the hole size increases.

Review of Datums

Fundamental to an understanding of geometric dimensioning and tolerancing is an understanding of datums. Since many engineering and drafting

students find the concept of datums difficult to understand, this section will review the concept in depth. It is important to understand datums because they represent the starting point for referencing dimensions to various features on parts and for making calculations relative to those dimensions. Datums are usually physical components. However, they can also be invisible lines, planes, axes, or points that are located by calculations or as they relate to other features. Features such as diameters, widths, holes, and slots are frequently specified as datum features.

Datums are classified as primary, secondary, or tertiary (**Figure 12-96**). Three points are required to

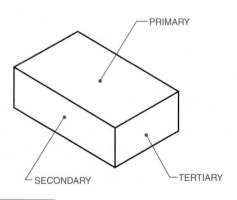

FIGURE 12-96   Datums.

establish a primary datum. Two points are required to establish a secondary datum. One point is required to establish a tertiary datum (**Figure 12-97**). Each point used to establish a datum is called off by a datum target symbol (**Figure 12-98**). The letter designation in the datum target symbol is the datum identifier. For example, the letter A in (**Figure 12-99**) is the datum designator for Datum A. The number 2 in (**Figure 12-100**) is the point designator for Point 2. Therefore, the complete designation of A2 means Datum A-Point 2.

Figure 12-101 illustrates how the points that establish datums should be dimensioned on a drawing. In this illustration, the three points that establish Datum A are dimensioned in the top view and labeled using the datum target symbol. The two points that establish Datum B are dimensioned in the front view. The one

FIGURE 12-101   Dimensioning datum points.

point that establishes Datum C is dimensioned in the right-side view. **Figure 12-102** illustrates the concept of datum plane and datum surface. The theoretically perfect plane is represented by the top of the machine table. The less perfect actual datum surface is the bottom surface of the part. **Figure 12-103** shows how the differences between the perfect datum plane and the actual datum surface are reconciled. The three points protruding from the machine table correspond with the three points that establish Datum A. Once this difference has been reconciled, inspections of the part can be carried out.

FIGURE 12-97   Establishing datums.

EACH POINT IS CALLED OFF BY A **DATUM TARGET SYMBOL**

FIGURE 12-98   Datum target symbol.

THE "A" INDICATES THE **DATUM**

FIGURE 12-99   Datum designation.

THE "2" INDICATES THE **POINT**

FIGURE 12-100   Point designator.

FIGURE 12-102   Datum plane versus datum surface.

FIGURE 12-103   Reconciling the datum surface to the datum plane.

SUMMARY

- General tolerancing involves setting acceptable limits of deviation for manufactured parts.

- Geometric dimensioning and tolerancing involves setting tolerance limits for all characteristics of a part.

- Modifiers are symbols that can be attached to the standard geometric building blocks to alter their application or interpretation.

- MMC is when the most material exists in the part. RFS means that a tolerance of form, position, or any characteristic must be maintained regardless of the actual produced size of the object.

- Projected tolerance zone is a modifier that allows a tolerance zone established by a locational tolerance to be extended a specified distance beyond a given surface.

- Datums are theoretically perfect points, lines, axes, surfaces, or planes used for referencing features of an object.

- True position is the theoretically exact location of the center line of a product feature such as a hole.

REVIEW QUESTIONS

Mark the following statements as either true or false.

1. Tolerancing means setting acceptable limits of deviation.

2. The three types of size tolerances are unilateral, location, and runout.

3. The need to tolerance more than just the size of an object led to the development of geometric dimensioning and tolerancing.

4. Geometric dimensioning specifies the allowable variation of a feature from perfect form.

5. The term *regardless of feature size* is a modifier that tells machinists that a tolerance of form, position, or any characteristic must be maintained regardless of the actual produced size of the object.

6. Datums are components of a part such as a hole, slot, surface, or boss.

7. A datum is established on a cast surface by a "flag" or symbol.

Answer the following questions by selecting the best answer.

1. Which of the following is the identification for the ASME standard on dimensioning?
   a. ASME Y14.5—2009
   b. ASME Y24.5—1992
   c. ASME Y34.5—1990
   d. ASME Y44.5—1988

2. Which of the following has the incorrect symbol?
   a. Flatness ▱
   b. Circularity O
   c. Straightness —
   d. True position ◆

3. Which of the following has the incorrect symbol?
   a. Perpendicularity =
   b. Straightness —
   c. Parallelism //
   d. Angularity ∠

4. Which of the following is **not** true regarding *flatness?*
   a. It differs from straightness.
   b. The term *flatness* is interchangeable with the term *straightness*.
   c. When flatness is the feature control, a datum reference is neither required nor proper.
   d. Flatness is specified when size tolerances alone are not sufficient to control the form and quality of the surface.

5. The term *least material condition* means:
   a. The opposite of MMC
   b. A condition of a feature in which it contains the least amount of material
   c. The theoretically exact location of a feature
   d. Both a and b

6. Which of the following is **not** true regarding feature control symbols?
   a. The order of data in a feature control frame is important.
   b. The first element is the feature control symbol.
   c. Various feature control elements are separated by //.
   d. Datum references are listed in order from left to right.

7. Which of the following feature controls must have a datum reference?
   a. Flatness   d. Parallelism
   b. Straightness   e. All of the above
   c. Cylindricity

CHAPTER 12 PROBLEMS

The following problems are intended to give beginning drafters practice in applying the principles of geometric dimensioning and tolerancing.

Conventional Instructions

The steps to follow in completing the problems are as follows:

STEP 1 Study the problem carefully.

STEP 2 Make a checklist of tasks you will need to complete.

STEP 3 Center the required view or views in the work area.

STEP 4 Include all dimensions according to ASME Y14.5—2009.

STEP 5 Recheck all work. If it's correct, neatly fill out the title block using light guidelines and freehand lettering.

NOTE: These problems do not follow current drafting standards. You are to use the information shown here to develop properly drawn, dimensioned, and toleranced drawings.

CAD INSTRUCTIONS

Students who plan to complete the following problems on a CAD system should begin with these general instructions.

1. Analyze the problem carefully.

2. Review the various CAD commands you will use to develop the drawing.

3. Read the specific instructions for the problem and begin.

Mechanical ## PROBLEM 12-1

Apply tolerances so that this part is straight to within .004 at MMC.

Mechanical ## PROBLEM 12-2

Apply tolerances so that the top surface of this part is flat to within .001 and the two sides of the slot are parallel to each other within .002 RFS.

Mechanical ## PROBLEM 12-3

Apply tolerances so that the smaller diameter has a cylindricity tolerance of .005 and the smaller diameter is concentric to the larger diameter to within .002. The shoulder must be perpendicular to the axis of the part to within .002.

Mechanical   ## PROBLEM 12-4

Apply tolerances to locate the holes using position and basic dimensions relative to Datums A-B-C.

Mechanical   ## PROBLEM 12-5

Apply angularity, true position, and parallelism tolerances of .001 to this part. Select the appropriate datums. The parallelism tolerances should be applied to the sides of the slot.

Mechanical   ## PROBLEM 12-6

Apply tolerances so that the outside diameter of the part is round to within .004 and the ends are parallel to within .001 at maximum material condition.

Mechanical   ## PROBLEM 12-7

Apply a line profile tolerance to the top of the part between points X and Y of .004. Apply position tolerances to the holes of .021 and parallelism tolerances of .001 to the two finished sides.

Mechanical   ## PROBLEM 12-8

Use the bottom of the part as Datum A and the right side of the part as Datum B. Apply surface profile tolerances of .001 to the top of the part between points X and Y.

Mechanical PROBLEM 12-9

Select datums and apply tolerances in such a way as to ensure that the slot is symmetrical to within .002 with the .625 diameter hole and the bottom surface is parallel to the top surface to within .004.

Mechanical PROBLEM 12-12

Select datums and apply a positional tolerance of .001 at MMC to the holes and a perpendicularity tolerance of .003 to the vertical leg of the angle.

Mechanical PROBLEM 12-10

Apply tolerances to this part so that the tapered end has a total runout of .002.

Problems 12-13 through 12-30

For each of the remaining geometric dimensioning and tolerancing problems, examine the problem closely with an eye to the purpose that will be served by the part. Then, select datums, tolerances, and feature controls as appropriate, and apply them properly to the parts. In this way, you will begin to develop the skills required of a mechanical designer. Do not overdesign. Remember: the closer the tolerances and the more feature controls applied, the more expensive the part. Try to use the rule of thumb that says: "Apply only as many feature controls and tolerances as absolutely necessary to ensure that the part will properly serve its purpose after assembly."

Mechanical PROBLEM 12-11

Apply tolerances to this part so that diameters X and Z have a total runout of .02 relative to Datum A (the large diameter of the part) and line runout of .004 to the two tapered surfaces.

Mechanical PROBLEM 12-13

Mechanical PROBLEM 12-14

.09 WIDE X .06 DEEP
UNDERCUT

125

Ø.75

.06 X 45°
CHAMFER

Ø.44

63

63

32

.98

.98

3.75

Ø1.25

.03 X 45°
CHAMFER

Mechanical PROBLEM 12-15

50

20

18

Ø16 THRU

82

18

50

25

20

56

METRIC

Mechanical PROBLEM 12-16

Ø76

Ø52

20

10

32

8

9

14

5

METRIC

Mechanical PROBLEM 12-17

1.50

.38

.38

.50

.50

1.50

.50

.50

2X
R.25

1.00

.38

.38

1.50

.38

1.50

.50

Mechanical PROBLEM 12-18

3X R24

24

64

14

14

25

38

36

METRIC

3X Ø20 THRU
(2 IN LINE)

Mechanical **PROBLEM 12-19**

Mechanical **PROBLEM 12-20**

Ø.31 THRU
⌴ Ø.38 ⤵.38

2.00
.62
2.50
1.50
1.0
.50
.31
.06
.12
.25
Ø.88 THRU
1.50
3.00
.25
.31
2X Ø.25 THRU

12
6
Ø62
16
6
3
12
44
Ø14 ⤵ 16
Ø40
METRIC

Mechanical **PROBLEM 12-21**

.50
.25
.50
.38
Ø2.00
Ø.38 ⤵.88
.75
.50
.75
3.75
1.50
Ø1.50 (BOTH ENDS)
Ø.495/.500

Mechanical **PROBLEM 12-22**

60°
SHARP
3 RIBS/120° APART
Ø3.14
RIB THICKNESS
.375, 3 REQ'D
1.50
.12
.06
Ø1.25
.12
.50
.63
3X Ø.50 THRU
120° APART ON
A Ø4.0 B.C.
R1.38
R1.00
Ø.56 THRU
60°
3.00
ALL UNMARKED RADIIS =
R.12

Mechanical   **PROBLEM 12-23**

Ø4 THRU
Ø6
Ø14
Ø26
Ø18
Ø28
Ø42
Ø22
Ø10
Ø6
64
42
14
38
116
134
9
76
24

METRIC

Mechanical   **PROBLEM 12-24**

4X R8
Ø18
Ø64
Ø46

SQUARE
KEYWAY FOR
Ø18 SHAFT

10
40
84

METRIC

Mechanical   **PROBLEM 12-25**

2.25
1.125
1.25
.625
(.25)
(.25)

SECTION THRU WEB

2.00   .50   R
.50   .50
.50
.50
3.00
.31
30°
.50
R
.50
2.00
.25
3.50   1.62   .50
.50
.50   1.12   2.50
.25

4X Ø.313 THRU
4X Ø.375 THRU

ALL UNMARKED RADII = R.06

Mechanical ## PROBLEM 12-26

(2.00)
.32
.84
R.50 (TYP.)
4X Ø.38, THRU
.38
1.38
R.18
(TYP.)
2.50
30°
Ø1.25
R.50
1.25
R.625
.50
1.5
.63
.50
.38
.38
.50
3.75
1.25
.50
.38
.50
45°
Ø.75 THRU
1.00
2.00

ALL UNMARKED RADII = R.09

Mechanical ## PROBLEM 12-27

.56
.50
(2.50)
R.44
2X Ø.44 THRU
.56
.44
R
Ø2.50
.56
2.50
Ø1.25 THRU
⌴ Ø1.88 ▼.50
2.00

ALL UNMARKED RADII = R.06

Mechanical ## PROBLEM 12-28

Ø1.75
Ø3.00 (O.D.)
Ø2.75 (I.D.)
.09
1.50
.25
.38
Ø2.25 (O.D.)
Ø1.00
3.25
4.06
2X Ø.50 THRU
.31

ALL UNMARKED RADII = R .06

PROBLEMS

PROBLEM 12-29

2X R10

60°

8

19

8

2X Ø6 THRU
⊔ Ø10 3

50

24

38

R23

2X R10

METRIC

ALL UNMARKED RADII = R.2

PROBLEM 12-30

Ø6

Ø14

Ø20

130

Ø40

58   110

68

Ø30

26

Ø20

METRIC

PROBLEM 12-31

This problem deals with feature control symbols. In Items 1–19, explain what each symbol means. In Items 20–30, draw the required symbols.

MEANS

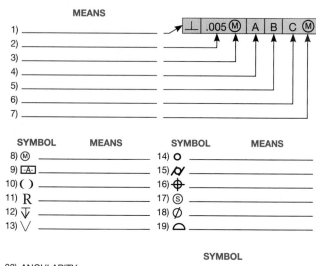

| ⊥ | .005 Ⓜ | A | B | C | Ⓜ |

1) _____
2) _____
3) _____
4) _____
5) _____
6) _____
7) _____

SYMBOL	MEANS		SYMBOL	MEANS
8) Ⓜ _____			14) ◯ _____	
9) ⊏A⊐ _____			15) ∠ _____	
10) ( ) _____			16) ⊕ _____	
11) R _____			17) Ⓢ _____	
12) ▽ _____			18) ⌀ _____	
13) ⋁ _____			19) ⌒ _____	

SYMBOL

20) ANGULARITY _____
21) POSITION _____
22) FLATNESS _____
23) PROFILE OF A SURFACE _____
24) PERPENDICULARITY _____
25) CIRCULAR RUNOUT _____
26) STRAIGHTNESS _____
27) TOTAL RUNOUT _____
28) PROFILE OF A LINE _____
29) CYLINDRICITY _____
30) CIRCULARITY _____

13

Fasteners

CHAPTER OUTLINE

Classifications of fasteners • Threads • Screw thread forms • Tap and die • Threads per inch (TPI) • Pitch • Single and multiple threads • Right-hand and left-hand threads • Thread relief (undercut) • Screw, bolt, and stud • Cap screws • Rivets • Keys and keyseats • Grooved fasteners • Spring pins • Fastening systems • Retaining rings • Summary • Review questions • Chapter 13 problems • CAD instructions

CHAPTER OBJECTIVES

Upon completion of this chapter, students should be able to do the following:

▶ Explain the classifications of fasteners.

▶ Explain the four uses of threads.

▶ List the various screw thread forms.

▶ Distinguish between the tap and die processes.

▶ Demonstrate two methods for measuring threads per inch.

▶ Define the term *pitch*.

▶ Distinguish between single and multiple threads.

▶ Demonstrate how to distinguish between right- and left-hand threads.

▶ Demonstrate the various methods of thread representation.

▶ Explain the concept of *thread relief*.

▶ Distinguish among the following: screw, bolt, and stud.

(continued)

(CHAPTER OBJECTIVES continued)

▶ Demonstrate the proper methods for drawing screws, bolts, and studs.

▶ Define the term *rivet*.

▶ Explain the concept of keys and keyseats.

▶ Explain how grooved fasteners are used.

▶ Describe the uses of spring pins.

▶ Explain the concept of fastening systems.

▶ Describe how retaining rings are used.

Classifications of Fasteners

As a new product is developed, determining how to fasten it together is a major consideration. The product must be assembled quickly using standard, easily available, low-cost fasteners. Some products are designed to be taken apart easily—others are designed to be permanently assembled. Many considerations are required as to what kind, type, and material of fastener is to be used. Sometimes the stress load upon a joint must be considered. There are two major classifications of fasteners: permanent and temporary. *Permanent fasteners* are used when parts will not be disassembled. *Temporary fasteners* are used when the parts will be disassembled at some future time.

Permanent fastening methods include welding, brazing, stapling, nailing, gluing, and riveting. Temporary fasteners include screws, bolts, keys, and pins.

Many temporary fasteners include threads in their design. In the early days, there was no such thing as standardization. Nuts and bolts from one company would not fit nuts and bolts from another company. In 1841, Sir Joseph Whitworth worked toward some kind of standardization throughout England. His efforts were finally accepted, and England came up with a standard thread form called Whitworth Threads.

In 1864, the United States tried to develop a standardization of its own but, because it would not interchange with the English Whitworth Threads, it was not adopted at that time. It was not until 1935 that the United States adopted the American Standard Thread. It was actually the same 60° V-thread form proposed back in 1864. Still, there was no standardization between countries. This created many problems, but nothing was done until World War II forced the issue of interchangeability on the Western Allies. As a result of problems experienced by the Allies, in 1948, the United States, Canada, and Great Britain developed the Unified Screw Thread. It was a compromise between the newer American Standard Thread and the old Whitworth Threads.

(KEY TERMS continued)

Self-locking rings	Stud
Series of thread	Temporary fasteners
Set screw	Thread relief
Single thread	Through hole
Spring pin	

Today, with the changeover to the metric system, new standards are being developed. The International Organization for Standardization (ISO) was formed to develop a single international system using metric screw threads. This new ISO standard will be united with the American National Standards Institute (ANSI) standards. At the present time, we are in a transitional period, and a combination of both systems is still being used.

Threads

Threads are used for four different applications:

1. To fasten parts together, such as a nut and a bolt.

2. For fine adjustment between parts in relation to each other, such as the fine adjusting screw on a surveyor's transit.

3. For fine measurement, such as a micrometer.

4. To transmit motion or power, such as an automatic screw threading attachment on a lathe or a house jack.

There are many types and sizes of fasteners, each designed for a particular function. Permanently fastening parts together by welding or brazing is discussed in Chapter 19. Although screw threads have other important uses, such as adjusting parts and measuring and transmitting power, only their use as fasteners and only the most-used kinds of fasteners are discussed in this chapter.

Thread Terms

Refer to **Figure 13-1** for the following terms.

• *External thread*—Threads located on the outside of a part, such as those on a bolt.

• *Internal thread*—Threads located on the inside of a part, such as those on a nut.

• *Axis*—A longitudinal center line of the thread.

• *Major diameter*—The largest diameter of a screw thread, both external and internal.

• *Minor diameter*—The smallest diameter of a screw thread, both external and internal.

FIGURE 13-1   Thread terms.

- *Pitch diameter*—The diameter of an imaginary diameter centrally located between the major diameter and the minor diameter.
- *Pitch*—The distance from a point on a screw thread to a corresponding point on the next thread, as measured parallel to the axis.
- *Root*—The bottom point joining the sides of a thread.
- *Crest*—The top point joining the sides of a thread.
- *Depth of thread*—The distance between the crest and the root of the thread, as measured at a right angle to the axis.
- *Angle of thread*—The included angle between the sides of the thread.
- *Series of thread*—A standard number of threads per inch (TPI) for each standard diameter.

Screw Thread Forms

The *form* of a screw thread is actually its profile shape. There are many kinds of screw thread forms. Seven major kinds are discussed next.

Unified National Thread Form

The Unified National thread form has been the standard thread used in the United States, Canada, and the Great Britain since 1948 (**Figure 13-2A**). This thread form is used mostly for fasteners and adjustments.

ISO Metric Thread Form

The ISO metric thread form is the new standard to be used throughout the world. Its form or profile is very similar to that of the Unified National thread, except that the thread depth is slightly less (**Figure 13-2B**). This thread form is used mostly for fasteners and adjustments.

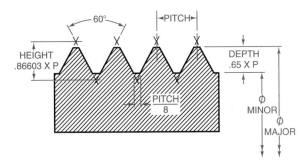

FIGURE 13-2A   Unified National (UN) thread form.

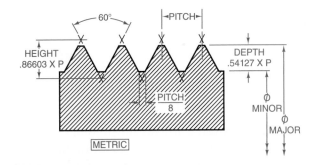

FIGURE 13-2B   ISO metric thread form.

Square Thread Form

The square thread's profile is exactly as its name implies—that is, square. The faces of the teeth are at right angles to the axis and, theoretically, this is the best thread to transmit power (**Figure 13-2C**). Because this thread is difficult to manufacture, it is being replaced by the Acme thread.

Acme Thread Form

The Acme thread is a slight modification of the square thread. It is easier to manufacture and is actually stronger than the square thread (**Figure 13-2D**). It, too, is used to transmit power.

FIGURE 13-2C   Square thread form.

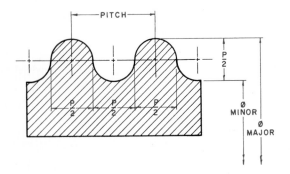

FIGURE 13-2F   Knuckle thread form.

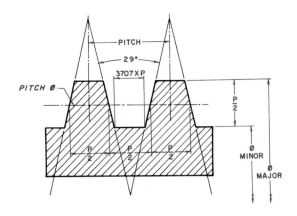

FIGURE 13-2D   Acme thread form.

FIGURE 13-2G   Buttress thread form.

Worm Thread Form

The worm thread is similar to the Acme thread, and it is used primarily to transmit power (**Figure 13-2E**).

Knuckle Thread Form

The knuckle thread is usually rolled from sheet metal and is used, slightly modified, in electric light bulbs, electric light sockets, and sometimes for bottle tops. The knuckle thread is sometimes cast (**Figure 13-2F**).

Buttress Thread Form

The buttress thread has certain advantages in applications involving exceptionally high stress along its axis in one direction only. Examples of applications are the breech assemblies of large guns, airplane propeller hubs, and columns for hydraulic presses (**Figure 13-2G**).

Tap and Die

Various methods are used to produce inside and outside threads. The simplest method uses thread-cutting tools called taps and dies. The tap cuts internal threads; the die cuts external threads (**Figure 13-3**).

FIGURE 13-2E   Worm thread form.

FIGURE 13-3   Tap and die.

In making an internal threaded hole, a tap-drilled hole must be drilled first. This hole is approximately the same diameter as the minor diameter of the threads.

Notice how the tap is tapered at the end; this taper allows the tap to start into the tap-drilled hole. This tapered area contains only partial threads.

Threads per Inch (TPI)

One method of measuring threads per inch (TPI) is to place a standard scale on the crests of the threads, parallel to the axis, and count the number of full threads within one inch of the scale (**Figure 13-4**). If only part of an inch of stock is threaded, count the number of full threads in one-half inch and multiply by two to determine TPI.

A simpler, more accurate method of determining TPI is to use a screw thread gage (**Figure 13-5**). By trial and error, the various fingers or leaves of the gage are placed over the threads until one is found that fits exactly into all the threads. Threads per inch are then read directly on each leaf of the gage (**Figure 13-6**).

Pitch

The pitch of any thread, regardless of its thread form or profile, is the distance from one point on a thread to the corresponding point on the adjacent thread as measured parallel to its axis (**Figure 13-7**). Pitch is found by dividing the TPI into one inch.

FIGURE 13-4   Use of a scale to calculate threads per inch (TPI).

FIGURE 13-5   Use of a screw pitch gage.

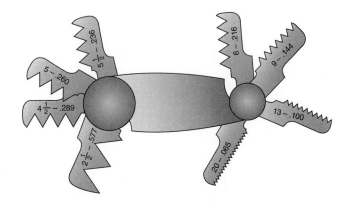

FIGURE 13-6   Reading with a screw pitch gage.

FIGURE 13-7   Coarse thread pitch.

FIGURE 13-8   Fine thread pitch.

In this example, a coarse thread pitch, there are 10 threads in one measured inch; 10 TPI divided into one inch equals a pitch of 10. In a fine thread of the same diameter, there are 20 threads in one measured inch; 20 TPI divided into one inch equals a pitch of 20 (**Figure 13-8**).

For metric threads, the pitch is specified in millimeters. Pitch for a metric thread is included in its call-off designation. For example: M10 × 1.5. The 1.5 indicates the pitch; therefore, it does not as a rule have to be calculated.

Single and Multiple Threads

A *single thread* is composed of one continuous ridge. The lead of a single thread is equal to the pitch. *Lead* is the distance a screw thread advances axially in one full turn. Most threads are single threads.

FIGURE 13-9   Single and multiple threads.

FIGURE 13-10   Right-hand and left-hand threads.

FIGURE 13-11   Thread representation.

Multiple threads are made up of two or more continuous ridges following side by side. The lead of a double thread is equal to twice the pitch. The lead of a triple thread is equal to three times the pitch (**Figure 13-9**).

Multiple threads are used when speed or travel distance is an important design factor. A good example of a double or triple thread is found in an inexpensive ballpoint pen. Take a ballpoint pen apart and study the end of the external threads. There will probably be two or three ridges starting at the end of the threads. Notice how fast the parts screw together. Speed, not power, is the characteristic of multiple threads.

Right-Hand and Left-Hand Threads

Threads can be either right-handed or left-handed. To distinguish between a right-hand and a left-hand thread, use this simple trick. A right-hand thread winding tends to lean toward the left. If the thread leans toward the left, the right-hand thumb points in the same direction. If the thread leans to the right (**Figure 13-10**), the left-hand thumb leans in that direction, indicating that it is a left-hand thread.

Thread Representation

The top illustration of **Figure 13-11** shows a normal view of an external thread. To draw a thread exactly as it will actually look takes too much drafting time. To help speed up the drawing of threads, one of two basic systems is used, and each is described and illustrated. The schematic system of representing threads was developed approximately in 1940, and it is still used somewhat today. The simplified system of representing threads was developed 15 years later, and it is actually quicker and in greater use today.

How to Draw Threads Using the Schematic System

STEP 1   Refer to **Figure 13-12**. Lightly draw the major diameter and locate the approximate length of full threads.

STEP 2   Lightly locate the minor diameter and draw the 45° chamfered ends as illustrated. Draw

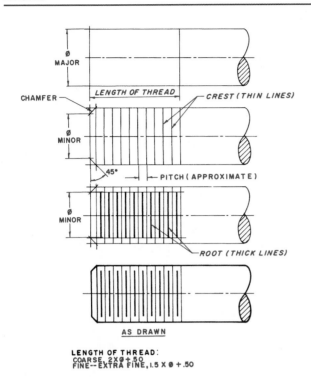

How to draw threads using the schematic system.

lines to represent the crest of the threads spaced approximately equal to the pitch.

STEP 3 Draw slightly thicker lines centered between the crest lines to the minor diameter. These lines represent the root of the threads.

STEP 4 Check all work and darken in. Notice that the crest lines are *thin* black lines and the root lines are *thick* black lines.

How to Draw Threads Using the Simplified System

STEP 1 Refer to **Figure 13-13**. Lightly draw the major diameter and locate the approximate length of full threads.

STEP 2 Locate the minor diameter and draw the 45° chamfered ends as illustrated. Draw dashed lines along the minor diameter. This represents the root of the threads.

STEP 3 Check all work and darken in. The dashed lines are thin black lines.

Standard External Thread Representation

The most recent standard to illustrate external threads using either the schematic or simplified system is illustrated in **Figure 13-14**. Note how section views are illustrated using the schematic and simplified systems

FIGURE 13-13 How to draw threads using the simplified system.

FIGURE 13-14 Standard external thread representation.

Standard Internal Thread Representation

There are two major kinds of interior holes: through holes and blind holes. A *through hole,* as its name implies, goes completely through an object. A *blind hole* is a hole that does **not** go completely through an object. In the manufacture of a blind hole, a tap drill must be drilled into the part first (**Figure 13-15**). To

FIGURE 13-15   Standard internal thread representation.

illustrate a tap drill, use the 30°–60° triangle. This is **not** the actual angle of a drill point, but it is close enough for illustration. The tap is now turned into the tap drill hole. Because of the taper on the tap, *full threads* do not extend to the bottom of the hole (refer back to Figure 13-3). The drafter illustrates the tap drill and the full threaded section, as shown to the right in Figure 13-15.

Using the schematic system to represent a through hole is illustrated in **Figure 13-16**. A blind hole and a section view are drawn as illustrated in **Figure 13-17**.

FIGURE 13-16   Standard internal thread representation for a through hole (schematic system).

FIGURE 13-17   Standard internal thread representation for a blind hole (schematic system).

Using the simplified system to represent a through hole is illustrated in **Figure 13-18**. A blind hole and a section view are drawn as illustrated in Figure 13-19.

Thread Relief (Undercut)

On exterior threads, it is impossible to make perfectly uniform threads up to a shoulder; thus, the threads tend to run out, as illustrated in **Figure 13-20**.

INTERIOR THREADS (THROUGH HOLE)

AS SEEN

SIMPLIFIED SYSTEM

SIMPLIFIED SYSTEM
(SECTION)

FIGURE 13-18   Standard internal thread representation for a through hole (simplified system).

INTERIOR THREADS (BLIND HOLE)

AS SEEN

SIMPLIFIED SYSTEM

FULL THD.

TAP HOLE

SIMPLIFIED SYSTEM (SECTION)

FIGURE 13-19   Standard internal thread representation for a blind hole (simplified system).

Where mating parts must be held tightly against the shoulder, the last one or two threads must be removed or *relieved*. This is usually done no farther than to the depth of the threads so as not to weaken the fastener. The simplified system of thread representation is illustrated at the bottom of Figure 13-20.

Full interior threads cannot be manufactured to the end of a blind hole. One way to eliminate this problem is to call off a thread relief or undercut, as illustrated in **Figure 13-21**. The bottom illustration is as it would be drawn by the drafter.

EXTERIOR THREAD RELIEF (UNDERCUT)

NOTE THREAD RUNOUT

DEPTH OF THREAD

THREAD RELIEF

AS SEEN

MINIMUM OF 1 OR 2 THREADS

THREAD RELIEF .06 X THD. DEPTH

AS DRAWN

FIGURE 13-20   External thread relief (undercut).

INTERIOR THREAD RELIEF (UNDERCUT)

NOTE THREAD RUNOUT

DEPTH OF THREAD

THREAD RELIEF

AS SEEN

MINIMUM OF 2 OR 3 THREADS

AS DRAWN

THREAD RELIEF .12 X THD. DEPTH

FIGURE 13-21   Internal thread relief (undercut).

Screw, Bolt, and Stud

Figure 13-22 illustrates and describes a screw, a bolt, and a stud. A *screw* is a threaded fastener that does not use a nut and is screwed directly into a part.

A *bolt* is a threaded fastener that passes directly through parts to hold them together and uses a nut to tighten or hold the parts together.

A *stud* is a fastener that is a steel rod with threads at both ends. It is screwed into a blind hole and holds other parts together by a nut on its free end. In general practice, a stud has either fine threads at one end and coarse threads at the other, or Class 3-fit threads at one end and Class 2-fit threads at the other end. Class of fit is fully explained later in this chapter under "Classes of Fit."

X = MINIMUM THREADS REQUIRED:
     STEEL, X = OUTSIDE DIAMETER
     CAST IRON/BRASS/BRONZE, X = 1.5 X OUTSIDE DIAMETER
     ALUMINUM/ZINC/PLASTIC, X = 2 X OUTSIDE DIAMETER

Y = MINIMUM SPACE = 2 X PITCH LENGTH

CLEARANCE HOLE: 0 TO .375 (9) = .03 (1) LARGER THAN OUTSIDE DIAMETER
                .375 (9) UP   = .06 (2) LARGER THAN OUTSIDE DIAMETER

FIGURE 13-22   Screw, bolt, and stud.

The minimum full thread length for a screw or a stud is:

In steel: equal to diameter.

In cast iron, brass, bronze: equal to 1.5 times the diameter.

In aluminum, zinc, plastic: equal to 2 times the diameter.

The clearance hole for holes up to .375 (9) diameter is approximately .03 oversize; for larger holes, .06 oversize.

Machine Screws

Machine screw sizes run from .021 (.3) to .750 (20) in diameter. There are eight standard head forms. Four major kinds are illustrated in **Figure 13-23**. *Machine screws* are used for screwing into thin materials. Most machine screws are threaded within a thread or two to the head. Although these are screws, machine screws sometimes incorporate a hex-head nut to fasten parts together.

The length of a machine screw is measured from the bottom of the head to the end of the screw (refer again to Figure 13-23).

Cap Screws

Cap screw sizes run from .250 (6) and up. There are five standard head forms (**Figure 13-24**). A *cap screw* is usually used as a true screw, and it passes through a clearance hole in one part and screws into another part.

How to Draw a Machine Screw or Cap Screw

The exact dimensions of machine screws and cap screws are given in Appendix A but, in actual practice, they are seldom used for drawing purposes. **Figures 13-25** and **13-26** show the various sizes as they are proportioned in regard to the diameter of the fastener. Various fastener templates are now available to further speed up drafting time.

Set Screws

A *set screw* is used to prevent motion between mating parts, such as the hub of a pulley on a shaft. The set screw is screwed into and through one part so that it applies pressure against another part, thus

FIGURE 13-23   Machine screws.

FIGURE 13-24   Cap screws.

MACHINE SCREW DIMENSIONS
APPROX. SIZES--FOR EXACT SIZES, SEE APPENDIX

FLAT HEAD   ROUND HEAD   OVAL HEAD   FILLISTER HEAD

FIGURE 13-25   Approximate sizes for machine screws or cap screws.

CAP SCREW DIMENSIONS
APPROX. SIZES--FOR EXACT SIZES, SEE APPENDIX

FLAT HEAD   ROUND HEAD   FILLISTER HEAD   HEX HEAD   HEX SOCKET HEAD

FIGURE 13-26   Approximate sizes for machine screws or cap screws.

preventing motion. Set screws are usually manufactured of steel, and they are hardened to make them stronger than the average fastener.

Set screws have various kinds of heads and many kinds of points. **Figure 13-27** illustrates a few of the more common kinds of set screws. Set screws are manufactured in many standard lengths of very small increments, so almost any required length is probably "standard." Exact sizes and lengths can be found in Appendix A. As with machine screws and cap screws, in actual practice, the drawing of set screws is done using their proportions in relationship to their diameters.

How to Draw Square- and Hex-Head Bolts

Exact dimensions for square- and hex-head bolts are given in Appendix A, but in actual practice, they are drawn using the proportions as given in **Figures 13-28** and **13-29**. Notice that the heads are shown in the profile so three surfaces are seen in

FIGURE 13-27   Approximate sizes for set screws and set screw points.

FIGURE 13-28   Approximate sizes for a square-head bolt.

FIGURE 13-29   Approximate sizes for a hex-head bolt.

FIGURE 13-30   Side view of hex- and square-head bolts.

the front view. In the event a square- or hex-head bolt must be illustrated 90°, the proportions as illustrated in **Figure 13-30** are used.

Nuts, Bolts, and Other Fasteners in Section

If the cutting plane passes through the axis of any fastener, the fastener is **not** sectioned. It is treated exactly as a shaft and drawn exactly as it is viewed. Refer to **Figure 13-31**. The illustration at the left is drawn correctly. The figure at the right is drawn incorrectly (notice how difficult it is to understand, especially the nut).

Thread Call-Offs

Although not all companies use the same call-offs for various fasteners, it is important that all drafters within one company use the same method. One method used to call off fasteners is illustrated in

FIGURE 13-31   Fasteners in section.

FIGURE 13-32   Thread call-off.

Figure 13-32. Regardless of which system is used, the first line contains the fastener's general identification, type of head, and classification. The second line contains all the exact detailed information. All threads are assumed to be right-hand (R.H.), unless otherwise noted. If a thread is to be left-hand (L.H.), it is noted at the end of the second line.

Various Kinds of Heads

Many different kinds of screw heads are used today. **Figure 13-33** illustrates a few of the standard heads.

Rivets

Rivets are permanent fasteners, usually used to hold sheet metal together. Most rivets are made of wrought iron or soft steel and, for aircraft and space missiles, copper, aluminum, alloy, or other exotic metals.

Riveted joints are classified by applications, such as pressure vessels, structural, and machine members. For data concerning joints for pressure vessels, refer to such sources as ASME boiler codes. For data concerning larger field structural rivets, such as bridges, buildings, and ships, see ANSI standards or a *Machinery's Handbook*. This chapter covers information for small-size rivets for machine-member riveted joints used for lighter mass-produced applications.

Two kinds of basic rivet joints are the lap joint and the butt joint (**Figure 13-34**). In the *lap joint,* the parts overlap each other and are held together by one or more rows of rivets. In the *butt joint,* the parts are butted and are held together by a cover plate or butt strap that is riveted to both parts.

Factors to consider are type of joint, pitch of rivets, type and diameter of rivet, rivet material, and size of clearance holes (**Figure 13-35**). The diameter of a rivet is calculated from the thickness of metal and commonly ranges between

$$d = 1.2 \sqrt{t} \text{ AND } d = 1.4 \sqrt{t}$$

where d is the diameter and t is the thickness of the plate.

FIGURE 13-34   Two basic rivet joints (lap joint and butt joint).

FIGURE 13-33   Kinds of screw heads.

FIGURE 13-35   Factors to consider in rivet joints.

Size and Type of Hole

Rivet holes must be punched, punched and then reamed or drilled. As a general rule, holes are usually made .06 (1.5 mm) larger in diameter than the nominal rivet diameter.

Rivet Symbols

Rivets applied in mass-produced applications are represented in **Figure 13-36A**, illustrating the kind of rivet, to which side it is applied, if it is to be countersunk, and so forth.

Kinds of Rivets

There are many different kinds of small-size rivets. The five major kinds are truss head, button head, pan head, countersunk head, and flat head.

The countersunk-head rivet is not as strong as the other kinds of rivets; therefore, more rivets must be used to gain strength equal to the other types.

Drawing of Rivets

American standard small solid rivets are shown in their approximate standard proportions in Figure 13-36A. These sizes are close enough to be used for drawing the rivet if necessary. For exact sizes, data must be obtained from other sources.

End-Points

The end of a standard rivet is usually cut off straight. If a point is required, a standard-size point is illustrated at the lower end of a countersunk head (refer back to Figure 13-36A).

(A)

FIGURE 13-36   (A) Approximate sizes for standard small rivets;

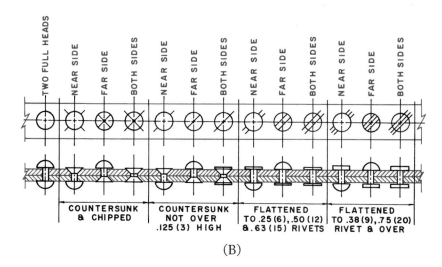

(B)

FIGURE 13-36 (*Continued*) (B) Illustrated rivet code.

The drafter must indicate what kind of rivet is to be used and on which side the head is to be positioned. To illustrate this, the drafter uses a code such as that illustrated in **Figure 13-36B.**

Keys and Keyseats

Key

A *key* is a demountable part that provides a positive means of transferring torque between a shaft and a hub.

Keys are used to prevent slippage and to transmit torque between a shaft and a hub. There are many kinds of keys. The five major kinds used in the industry today are illustrated in **Figure 13-37:** square key, flat key, gib-head key, Pratt & Whitney key, and Woodruff key. Where a lot of torque is present, a double key and keyseat are often used. In extreme conditions, a spline is machined into the shaft and into the hub. For spline information, refer to a *Machinery's Handbook.*

Keyseat

A *keyseat* is an auxiliary-located rectangular groove machined into the shaft and/or hub to receive the key.

Classes of Fit

There are three classifications of fit:

CLASS 1 A side surface clearance fit obtained by using bar stock key and keyseat tolerances. This is a relatively free fit.

FIGURE 13-37   Keys and keyseats.

CLASS 2 A possible side surface interference or side surface clearance fit obtained by using bar stock key and keyseat tolerances. This is a relatively tight fit.

CLASS 3 A side surface interference fit obtained by interference fit tolerances. This is a very tight fit and has not been generally standardized.

A FASTENER PROBLEM

The problem the design team at Reynolds Tool and Die Company faced at the beginning of the Anderson project had been fasteners. Specifically, the team had to decide how to anchor several springs that are critical components in a family of fixtures to be designed and manufactured for Anderson Aviation Company.

Using a small bolt was discussed and dismissed as being impractical because there would be a blind hole in the fixture. Without a through hole, a bolt could not be used. Various types of screws were discussed. One line of thought was that the head of the screw would anchor the end loop of the tension spring. In addition, a screw would work with a blind hole. The only negative brought up in the discussion of using a screw was the need for threads in the blind hole. The threading process would increase the cost of each fixture by 12%, a cost that had not been anticipated when the design team originally estimated the cost of the Anderson project.

The Reynolds design team was almost resigned to the 12% loss on the project when one member suggested using a grooved pin. After researching the recommendation, the team decided to use a Type G Number 2 pin 3/8″ in diameter. Such a pin has locking grooves on one end to fit into the blind hole and an angular groove at the other end to hold the end loop of a tension spring.

The decision to use a grooved pin turned out to be serendipitous. Not only did the groove perform as needed, it also decreased the anticipated cost of assembling the fixtures by 9%. Consequently, Reynolds Tool and Die Company earned more than anticipated on the Anderson project.

Key Sizes

As a general rule, the key width is about one-fourth the nominal diameter of the shaft. For exact recommended key sizes, refer to Appendix A or a *Machinery's Handbook*.

Dimensioning Keyseats

Methods of dimensioning a stock key are shown in **Figure 13-38**.

For dimensioning a Woodruff keyseat, the key number must be included (**Figure 13-39**). Refer to the key size in Appendix A to obtain the exact sizes. To dimension a Pratt & Whitney keyseat, see **Figure 13-40**.

FIGURE 13-39   Woodruff keyseat.

FIGURE 13-38   Dimensioning keyseats.

$$X = R - \sqrt{R^2 - (\tfrac{1}{2}Y)^2}$$

$$Z = 2R - X + \frac{Y}{2}$$

$$R = \frac{\varnothing}{2}$$

FIGURE 13-40   Pratt & Whitney keyseat.

Grooved Fasteners

Many types of fasteners are used in the industry, each with its own application. Threaded fasteners, such as nuts and bolts, are used to hold parts in tension.

FIGURE 13-41   Grooved fastener.

Grooved fasteners are used to solve metal-to-metal pinning needs with shear application (**Figure 13-41**).

Grooved fasteners have great holding power and are resistant to shock, vibration, and fatigue. They are available in a wide range of types, sizes, and materials. A grooved fastener often has a better appearance than most other methods of fastening. This can be important to the overall design if the fastener is visible.

Grooved fasteners have three parallel grooves, equally spaced, impressed longitudinally on their exterior surface. To make these grooves, a grooving tool is pressed *below* the surface to displace a carefully determined amount of material. Nothing is removed. The metal is displaced to each side, forming a raised portion or flute extending along each side of the groove (**Figure 13-42**). The crest of the flute constitutes the *expanded diameter* (Dx). The expanded diameter (Dx) is a few thousandths larger than the nominal diameter (D) of the stock.

Grooved fasteners can be custom-manufactured in order to meet most any application. For example, a custom-grooved pin can be made with a cross-drilled hole, a groove for a snap ring, or a threaded hole in one end. The groove length can also be varied and placed anywhere along the length of the pin.

Installation

The grooved fastener is forced into a drilled hole slightly larger than the nominal or specified diameter of the pin (**Figure 13-43**). The crest or flutes are forced back into the grooves when the fastener is driven into the hole. The resiliency of the metal

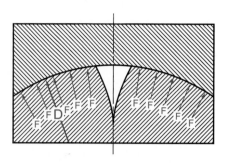

FIGURE 13-43   The same view after insertion.
A powerful radial thrust is obtained.

forced back into the grooves creates powerful radial forces against the hole wall.

In many cases, grooved fasteners are lower in cost than knurled pins, taper pins, pins with cotter pins, rivets, set screws, keys, or other methods used to fasten metal-to-metal parts together. The installation cost of grooved pins is invariably lower because of the required hole tolerances, and no special guides are required at assembly.

Material

Standard grooved fasteners are made of cold-drawn, low-carbon steel. The physical properties of this material are more than enough for ordinary applications. Alloy steel, hard brass, silicon bronze, stainless steel, and other exotic metals may also be specially ordered. These special materials are usually heat-treated for optimum physical qualities.

Finish

Standard grooved fasteners have a finish of zinc electroplate, deposited approximately .00015 inch deep. Chromate, brass, cadmium, and black oxide can also be specially ordered.

Standard Types

Study the various types of grooved fasteners and the related technical data in **Figures 13-44** through **13-48**. Note the various types of fasteners, how each functions, and what each replaces. For example, Type A is used in place of taper pins, rivets, set screws, and keys.

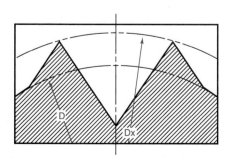

FIGURE 13-42   Enlarged sectional view of one of the grooves before inserting fastener.

TYPE A

The Type A DRIV-LOK pin has three **full-length tapered** grooves. This popular type is used in applications requiring excellent locking effect and ease of assembly. It is widely used in place of taper pins, rivets, set screws and keys. Typical applications include keying sprockets, gears, collars, knobs, handles, levers and wheels to shafts.

Locking Collar to Shaft

Lever and Shaft Assembly

TYPE A3

Type A3 pins have three **full-length parallel** grooves with a short pilot end to ensure easy starting. They are recommended for applications requiring maximum locking effect where severe vibration and/or shock loading are present.

Keying Gear to Shaft

Locking Gear to Shaft

TYPE U

Type U has three **parallel** grooves extending the **full length** of the pin. A short pilot at each end permits hopper or automatic feeding. Identical ends speed manual insertion as operator need not examine pin to determine proper end to start. Full-length parallel grooves provide maximum locking effect. Typical applications include keying gears, collars, knobs, handles, etc. to shafts.

Attaching Knob to Shaft

Pinning "V" Pulley to Shaft

FIGURE 13-44   Typical applications for grooved pin types A, A3, and U.

Standard Sizes

Refer to the standard size charts in **Figures 13-49** through **13-52**. Across the top of each chart are the nominal sizes from 1/16″ diameter to 1/2″ diameter. At the left side of each chart are listed the standard lengths from 1/4″ long to 4 1/2″ long. Various other technical information can be derived from the charts. For drilling procedure, hole tolerances, application data, and high-alloy pin applications, see **Figures 13-53** through **13-55**.

Grooved Studs

Grooved studs are widely used for fastening light metal or plastic parts to heavier members or assemblies (**Figures 13-56** and **13-57**). They replace screws, rivets, preened pins, and many other types of fasteners. The grooves function as any grooved fastener.

Spring Pins

Another type of fastener is the *spring pin* (**Figure 13-58**). Spring pins are manufactured by cold-forming strip metal in a progressive roll-forming operation. After forming, the pins are broken off and deburred to eliminate any sharp edges. They are then heat-treated to a R/C 46/53. This develops spring qualities or resiliency in the metal.

In the free state, the pins are larger in diameter than the hole into which they are to be inserted. The pins compress themselves as they are driven into the hole, thus exerting radial forces around the entire circumference of the hole.

Spring pins are made from a high-carbon steel—1074, stainless steel, either 420 heat-treated or 300 series cold-worked. Some spring pins are also made from brass or beryllium copper. Spring

TYPE

Type B pins have three **tapered** grooves extending **one-half** the length of the pin. This type is widely used as a hinge or pivot pin. Driven or pressed into a straight drilled hole, the grooved portion locks in one part, while the ungrooved portion will remain free. Also excellent for dowel and locating applications.

TYPE

The Type C Pin has three **parallel** grooves extending **one-quarter** its overall length. It is ideally suited for linkage or pivot applications, especially where a relatively short locking section and longer free length are required. Widely used in certain types of hinge applications. The long lead permits easy insertion.

Control Valve Hinge Assembly

Roller Pins

Hinge Pins

Linkage or Hinge Pin

| FIGURE 13-45 | Typical applications for grooved pin types B and C. |

TYPE

Type D has three **reverse** taper grooves extending **one-half** the length of the pin. Recommended for use in blind holes as a stop pin, roller pivot, dowel, or for certain hinge or linkage applications. Reverse taper grooves permit easy insertion in blind holes.

TYPE

Type E has three **parallel half-length** grooves located equidistant from each end. Widely used as T handle on valves and tools. Also used as a cross pin, cotter pin, pivot pin, etc. where center locking is required.

Roller End Pins

Linkage Pin

Stop Pins

T Handle for Valve

| FIGURE 13-46 | Typical applications for grooved pin types D and E. |

TYPE
G

Typical applications

Type G has three **parallel half-length** grooves including pilot. It is a very versatile pin, suitable for use in both blind and through holes as a spring anchor pin. The annular groove opposite the locking end is used to anchor the end loop of a tension spring. If snap or retainer rings are to be used, special section annular grooves can be machined to order.

Spring Anchor Pin
Used in Through
Hole

Spring Anchor in
Blind Hole

1 Straight pin with annular grooves at each end designed to accommodate snap or retainer rings. Available ³⁄₁₆″ through ¾″ diameters.

2 Pin with locking grooves on one end and annular groove on other end. Often used for roller or collar retention. Available ³⁄₁₆″ through ¾″ diameters.

3 Straight pin with both ends chamfered. Available ¹⁄₁₆″ through ¾″ diameters.

4 Grooved spring anchor pin with annular grooves at both ends. Available ³⁄₃₂″ through ¾″ diameters.

5 Special pin with grooves extending one-third length of pin. Groove length and position can be varied to suit specific applications. Available ¹⁄₁₆″ through ¾″ diameters.

6 Special straight pin with both ends chamfered and cross-drilled for cross pin or cotter. Available ¼″ through ¾″ diameters.

FIGURE 13-48   Special pins.

FIGURE 13-47   Typical applications for grooved pin type G.

pins meet the demand of many industrial applications. **Figures 13-59** and **13-60** list some suggested applications.

Fastening Systems

Fastening systems play a critical role in most product design. They often do more than position and secure components. In many cases, fastening systems have a direct effect upon the product's durability, reliability, size, and weight. They affect the speed with which the product may be assembled and disassembled, both during manufacturing and later in field service. Fastening systems also affect cost, not only for the fasteners but also for the machining and assembly operations they require.

Unless a drafter has had a great deal of experience in using different fastening devices and techniques, it is often difficult to choose a fastener that combines optimum function with maximum economy. A fastening system that is best for one product may not be desirable for another.

Retaining Rings

Retaining rings are precision-engineered fasteners. They provide removable shoulders for positioning or limiting the movement of parts in an assembly. Applications range from miniaturized electronic systems to massive earth-moving equipment. Retaining rings are used in automobiles, business machines, and complex components for guided missiles. They are found in such commonplace items as doorknobs and in sophisticated underwater seismic cable connectors.

Typical ring applications are shown in **Figure 13-61**. From these figures, the drafter may examine the design of various types of rings, determine their purpose and function, and decide how they may be used to the best advantage.

TYPE A | **TYPE A3** | **TYPE U**

STANDARD SIZES

Nominal diameter and recommended drill sizes	1/16	5/64	3/32	7/64	1/8	5/32	3/16	7/32	1/4	5/16	3/8	7/16	1/2
Dec. Equivalents	.0625	.0781	.0938	.1094	.1250	.1563	.1875	.2188	.2500	.3125	.3750	.4375	.5000
Crown Height, In.	.0065	.0087	.0091	.0110	.0130	.0170	.0180	.0220	.0260	.0340	.0390	.0470	.0520
Radius, In. ±.010	5/64	3/32	1/8	9/64	5/32	3/16	1/4	9/32	5/16	3/8	15/32	17/32	5/8
Pilot Length, In. (Ref.)	1/32	1/32	1/32	1/32	1/32	1/16	1/16	1/16	1/16	3/32	3/32	3/32	3/32
Chamfer Length, In. (Type U Only)	1/64	1/64	1/64	1/64	1/64	1/32	1/32	1/32	1/32	3/64	3/64	3/64	3/64

"Dx" EXPANDED DIAMETER—CAN BE DETERMINED ACCURATELY ONLY WITH RING GAGES

LENGTH OF PIN IN INCHES

		1/16	5/64	3/32	7/64	1/8	5/32	3/16	7/32	1/4	5/16	3/8	7/16	1/2
1/4	(.250)	.068	.084	.101	.117	.134								
3/8	(.375)	.068	.084	.101	.117	.134	.166	.198						
1/2	(.500)	.068	.084	.101	.117	.134	.166	.198	.230	.263				
5/8	(.625)	.068	.084	.101	.117	.134	.166	.198	.230	.263	.329			
3/4	(.750)	.068	.084	.101	.116	.134	.166	.198	.230	.263	.329	.394		
7/8	(.875)	.068	.084	.101	.116	.133	.165	.198	.230	.263	.329	.394	.459	
1	(1.000)	.068	.084	.101	.115	.133	.165	.198	.230	.263	.329	.394	.459	.525
1 1/4	(1.250)			.101	.115	.132	.164	.197	.230	.263	.329	.394	.459	.525
1 1/2	(1.500)					.132	.164	.197	.229	.262	.329	.394	.459	.525
1 3/4	(1.750)						.163	.197	.229	.262	.328	.393	.459	.525
2	(2.000)						.163	.196	.229	.262	.328	.393	.458	.525
2 1/4	(2.250)							.196	.229	.262	.328	.393	.458	.525
2 1/2	(2.500)								.228	.261	.327	.393	.458	.524
2 3/4	(2.750)								.228	.261	.327	.393	.458	.524
3	(3.000)								.227	.260	.327	.392	.457	.523
3 1/4	(3.250)									.260	.326	.392	.457	.523
3 1/2	(3.500)										.326	.391	.456	.522
3 3/4	(3.750)											.391	.456	.522
4	(4.000)											.390	.455	.521
4 1/4	(4.250)											.390	.455	.521
4 1/2	(4.500)												.454	.520

Nom. Diam. (D)	Exp. Diam. (Dx) reduced by
1/16	.002
5/64	.002
3/32	.002
7/64	.002
1/8	.002
5/32	.002
3/16	.003
7/32	.003
1/4	.003
5/16	.004
3/8	.005
7/16	.006
1/2	.006

TOLERANCES:
On Nominal Diameter "D"
+.000—.001 up to 7/64″ diameter
+.000—.002 7/64″ and above

On Expanded Diameter "Dx"
±.001 up to 7/64″ diameter
±.002 7/64″ and above
On over-all Length "L"
±.010 for all diameters

◄ For stainless steels and other special materials, the expanded diameters shown in above table are reduced by amounts shown at left.

Note: Intermediate pin lengths, pin diameters up to 3/4″, groove lengths, and groove positions to order as specials. When ordering, specify type, diameter, length, and special finishes.

FIGURE 13-49   Standard size chart of grooved pin types A, A3, and U.

STANDARD SIZES

Nominal diameter and recommended drill sizes	1/16	5/64	3/32	7/64	1/8	5/32	3/16	7/32	1/4	5/16	3/8	7/16	1/2
Dec. Equivalents	.0625	.0781	.0938	.1094	.1250	.1563	.1875	.2188	.2500	.3125	.3750	.4375	.5000
Crown Height, In.	.0065	.0087	.0091	.0110	.0130	.0170	.0180	.0220	.0260	.0340	.0390	.0470	.0520
Radius, In. ± .010	5/64	3/32	1/8	9/64	5/32	3/16	1/4	9/32	5/16	3/8	15/32	17/32	5/8

"Dx" EXPANDED DIAMETER—CAN BE DETERMINED ACCURATELY ONLY WITH RING GAGES

LENGTH OF PIN IN INCHES														
1/4	(.250)	.068	.084	.101	.117	.134								
3/8	(.375)	.068	.084	.101	.117	.134	.166	.198						
1/2	(.500)	.068	.084	.101	.117	.134	.166	.198	.230	.263				
5/8	(.625)	.068	.084	.101	.117	.134	.166	.198	.230	.263	.329			
3/4	(.750)	.068	.084	.101	.117	.134	.166	.198	.230	.263	.329	.394		
7/8	(.875)	.068	.084	.101	.117	.134	.166	.198	.230	.263	.329	.394	.459	
1	(1.000)	.068	.084	.101	.117	.134	.166	.198	.230	.263	.329	.394	.459	.525
1 1/4	(1.250)			.101	.117	.134	.166	.198	.230	.263	.329	.394	.459	.525
1 1/2	(1.500)					.134	.166	.198	.230	.263	.329	.394	.459	.525
1 3/4	(1.750)						.165	.198	.230	.263	.329	.394	.459	.525
2	(2.000)						.165	.198	.230	.263	.329	.394	.459	.525
2 1/4	(2.250)							.197	.230	.263	.329	.394	.459	.525
2 1/2	(2.500)								.230	.263	.329	.394	.459	.525
2 3/4	(2.750)								.229	.262	.329	.394	.459	.525
3	(3.000)								.229	.262	.329	.394	.459	.525
3 1/4	(3.250)									.262	.328	.393	.459	.525
3 1/2	(3.500)										.328	.393	.459	.525
3 3/4	(3.750)											.393	.458	.525
4	(4.000)											.393	.458	.525
4 1/4	(4.250)											.393	.458	.524
4 1/2	(4.500)												.458	.524

Nom. Diam. (D)	Exp. Diam. (Dx) reduced by
1/16	.002
5/64	.002
3/32	.002
7/64	.002
1/8	.002
5/32	.002
3/16	.003
7/32	.003
1/4	.003
5/16	.004
3/8	.005
7/16	.006
1/2	.006

TOLERANCES:

On Nominal Diameter "D"
+.000—.001 up to 7/64" diameter
+.000—.002 7/64" and above

On Expanded Diameter "Dx"
± .001 up to 7/64" diameter
± .002 7/64" and above
On over-all Length "L"
±.010 for all diameters

◄ For stainless steels and other special materials, the expanded diameters shown in above table are reduced by amounts shown at left.

Note: Intermediate pin lengths, pin diameters up to 1/4", groove lengths, and groove positions to order as specials. When ordering, specify type, diameter, length, and special finishes.

FIGURE 13-50   Standard size chart of grooved pin types B and C.

STANDARD SIZES

Nominal diameter and recommended drill sizes	1/16	5/64	3/32	7/64	1/8	5/32	3/16	7/32	1/4	5/16	3/8	7/16	1/2
Dec. Equivalents	.0625	.0781	.0938	.1094	.1250	.1563	.1875	.2188	.2500	.3125	.3750	.4375	.5000
Crown Height, In.	.0065	.0087	.0091	.0110	.0130	.0170	.0180	.0220	.0260	.0340	.0390	.0470	.0520
Radius, In. ±.010	5/64	3/32	1/8	9/64	5/32	3/16	1/4	9/32	5/16	3/8	15/32	17/32	5/8

"Dx" EXPANDED DIAMETER—CAN BE DETERMINED ACCURATELY ONLY WITH RING GAGES

LENGTH OF PIN IN INCHES

Length	(dec)	1/16	5/64	3/32	7/64	1/8	5/32	3/16	7/32	1/4	5/16	3/8	7/16	1/2
1/4	(.250)	.068	.084	.101	.117	.134								
3/8	(.375)	.068	.084	.101	.117	.134	.166	.198						
1/2	(.500)	.068	.084	.101	.117	.134	.166	.198	.230	.263				
5/8	(.625)	.068	.084	.101	.117	.134	.166	.198	.230	.263	.329			
3/4	(.750)	.068	.084	.101	.117	.134	.166	.198	.230	.263	.329	.394		
7/8	(.875)	.068	.084	.101	.117	.134	.166	.198	.230	.263	.329	.394	.459	
1	(1.000)	.068	.084	.101	.117	.134	.166	.198	.230	.263	.329	.394	.459	.525
1 1/4	(1.250)			.101	.117	.134	.166	.198	.230	.263	.329	.394	.459	.525
1 1/2	(1.500)					.134	.166	.198	.230	.263	.329	.394	.459	.525
1 3/4	(1.750)						.165	.198	.230	.263	.329	.394	.459	.525
2	(2.000)						.165	.198	.230	.263	.329	.394	.459	.525
2 1/4	(2.250)							.197	.230	.263	.329	.394	.459	.525
2 1/2	(2.500)								.230	.263	.329	.394	.459	.525
2 3/4	(2.750)								.229	.262	.329	.394	.459	.525
3	(3.000)								.229	.262	.329	.394	.459	.525
3 1/4	(3.250)									.262	.328	.393	.459	.525
3 1/2	(3.500)										.328	.393	.459	.525
3 3/4	(3.750)											.393	.458	.525
4	(4.000)											.393	.458	.525
4 1/4	(4.250)											.393	.458	.524
4 1/2	(4.500)												.458	.524

Nom. Diam. (D)	Exp. Diam. (Dx) reduced by
1/16	.002
5/64	.002
3/32	.002
7/64	.002
1/8	.002
5/32	.002
3/16	.003
7/32	.003
1/4	.003
5/16	.004
3/8	.005
7/16	.006
1/2	.006

TOLERANCES:

On Nominal Diameter "D"
+.000—.001 up to 7/64" diameter
+.000—.002 7/64" and above

On Expanded Diameter "Dx"
±.001 up to 7/64" diameter
±.002 7/64" and above

On over-all Length "L"
±.010 for all diameters

◄ For stainless steels and other special materials, the expanded diameters shown in above table are reduced by amounts shown at left.

Note: Intermediate pin lengths, pin diameters up to 3/4", groove lengths, and groove positions to order as specials. When ordering, specify type, diameter, length, and special finishes.

FIGURE 13-51   Standard size chart of grooved pin types D and E.

TYPE

(combined former F and G)

STANDARD SIZES

Nominal diameter and recommended drill sizes	3/32	7/64	1/8	5/32	3/16	7/32	1/4	5/16	3/8	7/16	1/2
Dec. Equivalents	.0938	.1094	.1250	.1563	.1875	.2188	.2500	.3125	.3750	.4375	.5000
Crown Height, In.	.0091	.0110	.0130	.0170	.0180	.0220	.0260	.0340	.0390	.0470	.0520
Radius, In., ±0.010	1/8	9/64	5/32	3/16	1/4	9/32	5/16	3/8	15/32	17/32	5/8
Pilot Length, In. (Ref.)	1/32	1/32	1/32	1/16	1/16	1/16	1/16	3/32	3/32	3/32	3/32
M Neck Radius (Ref.)	1/64	1/64	1/32	1/32	1/32	3/64	3/64	1/16	1/16	3/32	3/32
G Neck Width (Ref.)	1/32	1/32	1/16	1/16	1/16	3/32	3/32	1/8	1/8	3/16	3/16
Shoulder Width +.010—.000	1/32	1/32	1/32	3/64	3/64	1/16	1/16	3/32	1/8	1/8	1/8
K Neck Diameter ±.005	.062	.078	.083	.104	.125	.146	.167	.209	.250	.293	.312

"Dx" EXPANDED DIAMETER—CAN BE DETERMINED ACCURATELY ONLY WITH RING GAGES

LENGTH OF PIN IN INCHES

Length / Nominal Diam. (D)	3/32	7/64	1/8	5/32	3/16	7/32	1/4	5/16	3/8	7/16	1/2
3/8 (.375)	.101	.117	.134	.166	.198						
1/2 (.500)	.101	.117	.134	.166	.198	.230	.263				
5/8 (.625)	.101	.117	.134	.166	.198	.230	.263	.329			
3/4 (.750)	.101	.117	.134	.166	.198	.230	.263	.329	.394		
7/8 (.875)	.101	.117	.134	.166	.198	.230	.263	.329	.394	.459	
1 (1.000)	.101	.117	.134	.166	.198	.230	.263	.329	.394	.459	.525
1¼ (1.250)	.101	.117	.134	.166	.198	.230	.263	.329	.394	.459	.525
1½ (1.500)			.134	.166	.198	.230	.263	.329	.394	.459	.525
1¾ (1.750)				.165	.198	.230	.263	.329	.394	.459	.525
2 (2.000)				.165	.198	.230	.263	.329	.394	.459	.525
2¼ (2.250)					.197	.230	.263	.329	.394	.459	.525
2½ (2.500)						.230	.263	.329	.394	.459	.525
2¾ (2.750)						.229	.262	.329	.394	.459	.525
3 (3.000)						.229	.262	.329	.394	.459	.525
3¼ (3.250)							.262	.328	.393	.459	.525
3½ (3.500)								.328	.393	.459	.525
3¾ (3.750)									.393	.458	.525
4 (4.000)									.393	.458	.525
4¼ (4.250)									.393	.458	.524
4½ (4.500)										.458	.524

Inset (for stainless steels and other special materials):

Nom. Diam. (D)	Exp. Diam. (Dx) reduced by
3/32	.002
7/64	.002
1/8	.002
5/32	.002
3/16	.003
7/32	.003
1/4	.003
5/16	.004
3/8	.005
7/16	.006
1/2	.006

TOLERANCES:
On Nominal Diameter "D"
+.000—.001 up to 7/64" diameter
+.000—.002 7/64" and above
On Expanded Diameter "Dx"
±.001 up to 7/64" diameter
±.002 7/64" and above
On over-all Length "L"
±.010 for all diameters
For stainless steels and other special materials, the expanded diameters shown in above table are reduced by amounts shown at left.
Note: Intermediate pin lengths, pin diameters up to 3/4", groove lengths, and groove positions to order as specials. When ordering, specify type, diameter, length, and special finishes.

FIGURE 13-52   Standard size chart of grooved pin type G.

Most retaining rings are made of materials that have good spring properties. This permits the rings to be deformed elastically to a substantial degree, yet still spring back to their original shape during assembly and disassembly. This allows most rings to function in one of two ways: (1) they may be sprung into a groove or other recess in a part, or (2) they may be seated on a part in a deformed condition so that they grip the part by frictional means. In either case, the rings form a fixed shoulder against which other components may be abutted and prevented from moving.

Unlike wire-formed rings, which have a uniform section height, stamped rings have a tapered radial width. The width decreases symmetrically from the center section to the free ends. The tapered section permits the rings to remain circular after they have been compressed for insertion into a bore or expanded

Pin Diameter	Decimal Equivalent	Recommended Drill Size	Hole Tolerances ADD To Nominal Diameter
1/16″	.0625	1/16″	.002″
5/64	.0781	5/64	.002″
3/32	.0938	3/32	.003″
7/64	.1094	7/64	.003″
1/8	.1250	1/8	.003″
5/32	.1563	5/32	.003″
3/16	.1875	3/16	.004″
7/32	.2188	7/32	.004″
1/4	.2500	1/4	.004″
5/16	.3125	5/16	.005″
3/8	.3750	3/8	.005″
7/16	.4375	7/16	.006″
1/2	.5000	1/2	.006″

Tolerances for drilled holes shown in the table are based on depth-to-diameter ratio of approximately 5 to 1. Higher ratios may cause these figures to be exceeded, but in no case should they be exceeded by more than 10%. Specifications for holes having a depth-to-diameter ratio of 1 to 1 or less should be held extremely close; 60% of figures shown in the table is recommended.

Undersized drills should never be used to produce holes for Driv-Lok Pins. This malpractice results from the false assumption that the pins will ''hold better.'' Instead, they bend, damage the hole wall, crack castings, and peel their expanded flutes, thus reducing their retaining characteristics and preventing their reuse.

Holes made in hardened steel or cast iron are recommended to have a slight chamfer at the entrance. This eliminates shearing of the flutes as the pins are forced in.

Care should be exercised in all drilling for Driv-Lok Pins. Drills used should be new or properly ground with the aid of an approved grinding fixture. The drilling machine spindle must be in good condition and operated at correct speeds and feeds for the metal being drilled, and suitable coolant is always recommended. Drill jigs with accurate bushings always facilitate good drilling practice.

FIGURE 13-53   Drilling procedure and hole tolerance.

RECOMMENDED PIN DIAMETER FOR VARIOUS SHAFT SIZES AND TORQUE TRANSMITTED BY PIN IN DOUBLE SHEAR

Shaft Size	Pin Diameter	Torque Inch Lbs.	H.P. at 100 R.P.M.	Shaft Size	Pin Diameter	Torque Inch Lbs.	H.P. at 100 R.P.M.
3/16	1/16	4.6	.007	7/8	1/4	347	.555
7/32	5/64	8.4	.013	15/16	5/16	580	.927
1/4	3/32	13.7	.022	1	5/16	618	.990
5/16	7/64	23.6	.038	1 1/16	5/16	657	1.05
3/8	1/8	37.2	.060	1 1/8	3/8	1010	1.61
7/16	5/32	67.6	.108	1 3/16	3/8	1065	1.70
1/2	5/32	77.2	.124	1 1/4	3/8	1120	1.79
9/16	3/16	125.0	.200	1 5/16	7/16	1590	2.55
5/8	3/16	139.0	.222	1 3/8	7/16	1670	2.67
11/16	7/32	207.0	.332	1 7/16	7/16	1740	2.79
3/4	1/4	297.0	.476	1 1/2	1/2	2380	3.81
13/16	1/4	322	.516				

This table is a guide in selecting the proper size Driv-Lok Grooved Pin to use in keying machine members to shafts of given sizes and for specific load requirements. Torque and horsepower ratings are based on pins made of cold finished, low carbon steel and a safety factor of 8 is assumed.

MINIMUM SINGLE SHEAR VALUES (LBS.) OF DRIV-LOK PINS OF VARIOUS MATERIALS

DRIV-LOK PIN DIAM.	Cold Finished 1213 Steel	Shear-Proof® ALLOY STEEL R.C. 40 - 48	Brass	Silicon Bronze	Heat Treated Stainless Steel
1/16	200	363	124	186	308
5/64	312	562	192	288	478
3/32	442	798	272	408	680
7/64	605	1091	372	558	933
1/8	800	1443	492	738	1230
5/32	1240	2236	764	1145	1910
3/16	1790	3220	1100	1650	2750
7/32	2430	4386	1495	2240	3740
1/4	3190	5753	1960	2940	4910
5/16	4970	8974	3060	4580	7650
3/8	5810	12960	4420	6630	11050
7/16	7910	17580	6010	9010	15000
1/2	10300	23020	7850	11800	19640

FIGURE 13-54   Grooved pin application/engineering data.

MATERIAL HANDLING EQUIPMENT—The Type E pin provides positive locking with a half-length groove in the center of the pin. Extreme shear is exerted in this application, yet the Shear-Proof pin is used with complete safety for both men and materials. Type E is a special pin.

HEAVY-DUTY GEAR AND SHAFT ASSEMBLY—Type A Shear-Proof pin as specified for this application to give maximum locking power over the entire pin and gear hub area. The Type A pin, with grooves the full length of the pin, is the standard stock pin which meets most applications.

AUTOMATIC TRANSMISSION IN AUTOMOBILES—Special Type C was selected as a shaft in this transmission servo to replace a cross-drilled shaft with a cross pin for holding shaft in position. This eliminated a costly drilling operation and the cross pin.

UNIVERSAL JOINTS IN HAND TOOLS—Special Type E pin with center groove eliminates costly staking and grinding operations and improves product appearance. This pin is easily installed, fits flush, and permits plating before assembly.

EYE BOLT HINGE PIN—Type C pin, with quarter-length grooves, provides maximum ease of assembly. There is no interference until three-fourths of the pin is in position. The high safety factor inherent in Shear-Proof pins makes them practical and efficient for such constant-shear applications. Type C is a special pin.

HIGH-PRESSURE PISTON AND ROD ASSEMBLY—Type B pin was used here because the half-length grooves simplified the job of starting the pin into the hole. Ease of assembly was matched with sufficient locking power even when subjected to continuous, strong reciprocating forces. Type B is a special pin.

FIGURE 13-55   High-alloy shear-proof pins.

STANDARD SIZES and SPECIFICATIONS

Stud Number	Nominal Shank Diameter	Recommended Drill Size	Head Dia. Max.	Head Dia. Min.	Head Height Max.	Head Height Min.
0	.067	51	.130	.120	.050	.040
2	.086	44	.162	.146	.070	.059
4	.104	37	.211	.193	.086	.075
6	.120	31	.260	.240	.103	.091
7	.136	29	.309	.287	.119	.107
8	.144	27	.309	.287	.119	.107
10	.161	20	.359	.334	.136	.124
12	.196	9	.408	.382	.152	.140
14	.221	2	.457	.429	.169	.156
16	.250	¼"	.472	.443	.174	.161

Maximum Expansion—Standard Lengths

Stud No.	⅛"	3/16"	¼"	5/16"	⅜"	½"
0	.074	.074	.074			
2	.096	.096	.095			
4		.115	.113	.113		
6			.132	.130	.130	
7				.147	.147	.144
8					.155	.153
10					.173	.171
12						.206
14						.234
16						.263

TOLERANCES:
On length ± .010
On Exp. Diameter − .002
On Nominal Diameter +.000
                    −.002

Note: The expanded diameter can be determined accurately only with ring gages.

TYPICAL APPLICATIONS

Fastening knobs handles, etc.

Attaching nameplates, instruction panels

Widely used for fastening brackets

Fastening spring assemblies or control arms

FIGURE 13-56   Standard studs.

1. Flathead special stud with one-third length groove at lead end. Groove length can be varied.

2. Flathead special grooved stud with shoulder. Often hardened to provide wear surface in shoulder area.

3. Flathead grooved stud.

4. Round head reverse taper groove stud.

5. Stud with conical head and parallel grooves.

6. Round head stud with parallel grooves of special length.

7. Countersunk head grooved stud.

8. T head cotter used extensively in chain industry in place of cotter pins.

SPECIAL STUD APPLICATIONS

T Head Cotter in Chain   Linkage Assembly   Spring Anchor

FIGURE 13-57   Special studs.

FIGURE 13-58   Spring pin.

Dowel Application   Knob-to-Shaft   Stop Pin

Used as a Spacer   Hinge in Light Gage Metal   Keying Pulley to Shaft

To Prevent Shaft Rotation   Cotter Pin   T Handle

FIGURE 13-59   Spring pin application.

STANDARD SIZES

Nominal	A		B Max.	C		WALL THICK-NESS	RECOMMENDED HOLE SIZE		MINIMUM DOUBLE SHEAR STRENGTH POUNDS Carbon Steel and Stainless Steel
	Minimum ⅓(D₁+D₂+D₃)	Maximum (Go Ring Gage)		Min.	Max.		Min.	Max.	
.062	.066	.069	.059	.007	.028	.012	.062	.065	425
.078	.083	.086	.075	.008	.032	.018	.078	.081	650
.094	.099	.103	.091	.008	.038	.022	.094	.097	1,000
.125	.131	.135	.122	.008	.044	.028	.125	.129	2,100
.156	.162	.167	.151	.010	.048	.032	.156	.160	3,000
.187	.194	.199	.182	.011	.055	.040	.187	.192	4,400
.219	.226	.232	.214	.011	.065	.048	.219	.224	5,700
.250	.258	.264	.245	.012	.065	.048	.250	.256	7,700
.312	.321	.328	.306	.014	.080	.062	.312	.318	11,500
.375	.385	.392	.368	.016	.095	.077	.375	.382	17,600
.437	.448	.456	.430	.017	.095	.077	.437	.445	20,000
.500	.513	.521	.485	.025	.110	.094	.500	.510	25,800

All dimensions listed on this page are in accordance with National Standards. Wall thicknesses within the Spring Pin industry are standard.

SPRING PIN WEIGHT PER 1000 PIECES
Material—Steel · Nominal Diameter

LENGTH	.062	.078	.094	.125	.156	.187
0.187	.10	0.18				
0.250	.15	0.22	0.33			
0.312	.19	0.28	0.41			
0.375	.23	0.34	0.50	0.89		
0.437	.27	0.40	0.58	1.00	1.50	
0.500	.30	0.46	0.66	1.20	1.70	2.50
0.562	.34	0.51	0.75	1.30	1.90	2.90
0.625	.38	0.57	0.83	1.50	2.10	3.20
0.687	.42	0.63	0.92	1.60	2.40	3.50
0.750	.46	0.69	1.00	1.80	2.60	3.80
0.812	.49	0.74	1.10	1.90	2.80	4.10
0.875	.53	0.80	1.20	2.10	3.00	4.50
0.937	.57	0.86	1.30	2.20	3.20	4.80
1.000	.61	0.92	1.40	2.40	3.40	5.10
1.125		1.00	1.50	2.70	3.80	5.70
1.250		1.20	1.70	3.00	4.30	6.40
1.375		1.30	1.80	3.30	4.70	7.00
1.500		1.40	2.00	3.60	5.10	7.60
1.625				3.90	5.50	8.30
1.750				4.20	6.00	8.90
1.875				4.50	6.40	9.60
2.000				4.70	6.80	10.0
2.250					7.80	12.0
2.500					8.60	13.0

LENGTH	.219	.250	.312	.375	.437	.500
0.562	4.00					
0.625	4.50	5.30				
0.687	4.90	5.90				
0.750	5.30	6.30	9.90	15.0		
0.812	5.80	6.90	11.0			
0.875	6.20	7.40	12.0			
0.937	6.70	8.00	12.0			
1.000	7.00	8.50	13.0	20.0	24.0	
1.125	7.90	9.50	14.0			
1.250	8.80	11.0	16.0	24.0	30.0	41.0
1.375	9.70	12.0	18.0			
1.500	11.0	13.0	19.0	29.0	36.0	49.0
1.625	12.0	14.0	21.0			
1.750	12.0	15.0	22.0	34.0	42.0	57.0
1.875	13.0	16.0	24.0			
2.000	14.0	17.0	25.0	38.0	48.0	65.0
2.250	16.0	19.0	29.0	43.0	54.0	73.0
2.500	18.0	21.0	32.0	48.0	60.0	81.0
2.750	19.0	23.0	35.0	53.0	66.0	89.0
3.000	21.0	25.0	38.0	58.0	72.0	97.0
3.250		28.0	41.0	62.0	77.0	105.0
3.500		31.0	44.0	67.0	83.0	114.0
3.750			48.0	72.0	89.0	122.0
4.000			51.0	77.0	95.0	130.0

FIGURE 13-60   Spring pin data.

for assembly over a shaft. Most rings are designed to be seated in grooves. This constant circularity ensures maximum contact surface with the bottom of the groove. It is also an important factor in achieving high static and dynamic thrust load capacities.

Types of Retaining Rings

A great number of fastening requirements are involved in product design. This factor has led to the development of many different types of retaining rings.

Limited space prevents describing in detail all the different ring types available. In general, however, retaining rings can be grouped into two major categories.

Internal rings for axial assembly are compressed for insertion into a bore or housing. They generally have a large gap and holes in the lugs, located at the free ends, for pliers, which are used to grasp the rings securely during installation and removal.

External rings for axial assembly are expanded with pliers so they may be slipped over the end of a shaft, stud, or similar part. They have a small gap, and the lug position is reversed from that of the internal ring. Radially assembled external rings do not have holes for pliers. Instead, the rings have a large gap and are pushed into the shaft directly in the plane of the groove with a special application tool.

In addition to the tools shown here, retaining rings can be installed with automatic equipment. This equipment can be designed for specific, high-speed, automatic assembly lines. Retaining ring grooves serve two purposes: (1) they ensure precise seating of the ring in the assembly, and (2) they

INTERNAL	EXTERNAL	EXTERNAL	EXTERNAL
INTERNAL	EXTERNAL	EXTERNAL	EXTERNAL
INTERNAL	EXTERNAL	EXTERNAL	EXTERNAL
INTERNAL	EXTERNAL	EXTERNAL	EXTERNAL
INTERNAL	EXTERNAL	EXTERNAL	EXTERNAL
EXTERNAL	EXTERNAL	EXTERNAL	EXTERNAL

FIGURE 13-61   Ring applications.

permit the ring to withstand heavy thrust loads. The grooves must be located accurately and precut in the housing or shaft before the rings are assembled. Shaft grooves often can be made at no additional cost during the cut-off and chamfering operations.

Self-locking rings do not require any grooves because they exert a frictional hold against axial displacement. They are used mainly as positioning and locking devices where the ring will be subjected only to moderate or light loading.

Bowed rings differ from conventional types in that they are bowed around an axis perpendicular to the diameter bisecting the gap. The bowed construction permits the rings to function as springs and fasteners. This provides resilient end play take-up in the assembly.

Beveled rings have a 15° bevel on the groove-engaging edge. They are installed in grooves having a comparable bevel on the load-bearing wall. When the ring is seated in the groove, it acts as a wedge against the retained part. Sometimes play develops between the ring and retained part because of accumulated tolerances or wear in the assembly. If play develops, the spring action of the ring causes the

fastener to seat more deeply in its groove and move in an axial direction, automatically taking up the end play. (Because self-locking rings can be seated at any point on a shaft or in a bore, they too can be used to compensate for tolerances and eliminate end play.)

SUMMARY

- Threads are used for four applications: (1) to fasten parts together, (2) for fine adjustments between parts, (3) for fine measurements, and (4) to transmit motion or power.

- Screw thread forms include the Unified National, ISO, metric, square, Acme, worm, knuckle, and buttress.

- Inside and outside threads are produced using taps and dies. Taps are used to cut internal threads; dies are used to cut external threads.

- Pitch is the distance from one point on a thread to the corresponding point on the adjacent thread as measured parallel to its axis.

- A single thread is composed of one continuous ridge. Lead is the distance a screw thread advances axially in one full turn. Multiple threads are made up of two or more continuous ridges side by side.

- A right-hand thread rotates in a clockwise direction. A left-hand thread rotates in a counterclockwise direction.

- Threads may be represented graphically in the following ways: schematic or simplified.

- A screw is a threaded fastener that does not use a nut and is screwed directly into a part. A bolt is a threaded fastener that passes directly through parts to hold them together and uses a nut. A stud is a fastener that is a steel rod with threads at both ends.

- Rivets are permanent fasteners, usually used to hold sheet metal together.

- A key is a demountable part that provides a positive means of transferring torque between a shaft and a hub.

- Grooved fasteners are used to solve metal-to-metal pinning needs with shear application.

- Spring pins are manufactured by cold-forming strip metal in a progressive roll-forming operation. They are slightly larger in diameter than the hole into which they are inserted.

- Retaining rings are precision-engineered fasteners that provide removable shoulders for positioning or limiting movement in an assembly.

REVIEW QUESTIONS

Mark the following statements as either true or false.

1. Two other uses for threads besides fastening things together are adjusting parts and measuring and transmitting power.

2. If you have a 1/4 – 20 UNC threaded screw and rotate it 10 full turns, the end will travel 1/2″.

3. A 5115 series retaining ring would be recommended to provide resilient end play take-up.

4. The free diameter of an external 5100-775 retaining ring is 7.243.

5. The depth of a thread is figured by measuring the distance between the crest and the root of the thread as measured at a right angle to the axis.

Answer the following questions by selecting the best answer.

1. What #5100 series retaining ring would be recommended for a shaft diameter of 203.2 mm?
   a. 5100-343
   b. 5100-437
   c. 5100-675
   d. 5100-800

2. What is **not** a factor to be considered before selecting any type of retaining ring, other than load capacity?
   a. Lug size
   b. Adequate clearance
   c. The cost of installing the ring
   d. Whether or not a groove is required

3. What does 5.8 – 11 UNC – 2A × 21.2 LG mean?
   a. 5.8″ long, right-hand thread, average fit
   b. 5.8″ nominal size, 11 TPI, Unified National Series-Coarse, average fit, external thread, 21.2″ long
   c. 5.8″ diameter, 11″ long, coarse series
   d. 11 fit, coarse ×21.2″ long ×5.8″ diameter

4. What is the recommended hole size for a .312″ (8) diameter spring pin?
   a. Min .312, Max .318
   b. Min .321, Max .328
   c. Min .322, Max .325
   d. Min .315, Max .318

5. A screw that is 3″ long, 3.8″ in diameter, 15 threads per inch, Unified Coarse, right-hand thread, average fit, and rounded-head-style cap screw is called out by:
   a. 3.8 × 15 UNC − RDH Cap Screw × 3.0 LG
   b. RDH Cap Screw − 15 UNC 3.8 × 3.0 LG
   c. 3.8 − 15 UNC − 2A × 3.0 LG RDH
   d. 3″ Long Round Head Cap Screw (3.8″ diameter, coarse)

6. The housing diameter for an internal series N5000-25 ring is:
   a. .688
   b. .453
   c. .250
   d. .875

CHAPTER 13 PROBLEMS

The following problems are intended to give the beginning drafter practice using various size charts of fasteners and, by using these charts, practice in laying out and drawing the many fasteners used.

Conventional Instructions

The steps to follow in laying out fasteners are:

STEP 1 Study all required specifications.

STEP 2 Using the appropriate size chart for size dimension, lightly lay out fastener in place.

STEP 3 Check each dimension.

STEP 4 Darken in the fastener, starting with diameters and arcs first to the correct line thickness.

Use simplified thread representation.

STEP 5 Neatly add all required call-off specifications using the latest drafting standard.

CAD INSTRUCTIONS

Students who plan to complete the following problems on a CAD system should begin with these general instructions.

1. Analyze the problem carefully.

2. Plan your drawing sequence. In what order will you develop the drawing?

3. Read the specific instructions for the problem and begin.

Mechanical PROBLEM 13-1

On an A-size, 8.5 × 11 sheet of paper, lay out the center lines per the given dimensions.

Calculate standard thread lengths.

Draw the following fasteners on the center lines as illustrated.

A. Flathead cap screw—3.8 − 16 UNC − 2A × 2.0 lg.

B. Round-head cap screw—1/2 − 13 UNC − 2A × 2.25 lg.

C. Fillister-head cap screw—5.8 − 11 UNC − 2A × 2.5 lg.

D. Oval-head cap screw—3.8 − 24 UNF − 2A × 1.75 lg.

E. Hex-head cap screw with hex nut—1 − 8 UNC − 2A × 2.5 lg.

F. Socket-head cap screw—1.2 − 20 UNF − 2A × 2.0 lg.

G. Cotter pin—3.8 dia. × 3.5 lg.

Mechanical   **PROBLEM 13-2**

On an A-size, 8.5 × 11 sheet of paper, lay out the center lines per the given dimensions.

Calculate standard thread lengths.

Draw the following fasteners on the center lines as illustrated.

A. Hex-head bolt—13.8 − UNC × 3.0 lg.
B. Lock washer (for a 7.8 dia. cap screw).
C. Square nut—1 − 12 UNF 2B.
D. Square-head cap screw—3.4 − 10 UNC × 3.0 lg.

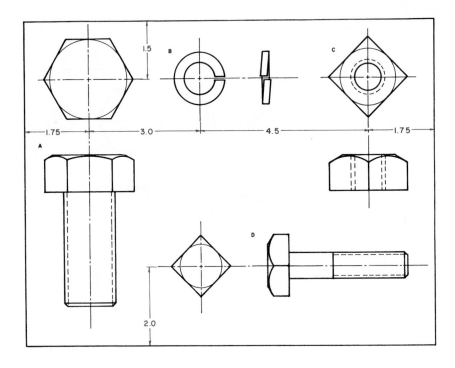

Mechanical   **PROBLEM 13-3**

On an A-size, 8.5 × 11 sheet of paper, lay out the center lines per the given dimensions.

A. Calculate the required size of square key needed for a 2.25 ID collar. Draw the collar and keyway, and dimension per drafting standards.

B. Calculate the required size of square key needed for the matching shaft of the collar. Draw the end view of the shaft and dimension per drafting standards.

C. Calculate the **minimum** length hex-head cap screw, 3.4 − 10 UNC − 2A, required to safely secure part X to part Y (material: aluminum).

Illustrate and call off the required pilot drill size and recommended depth into part Y.

Illustrate and call off the required size and depth of **full** threads for the hex-head cap screw.

Illustrate and call off the diameter for the clearance hole in part X.

D. Draw a square-head set screw with a full dog point, 5.8 − 18 UNF × 3.0 lg.

Mechanical   ## PROBLEM 13-4

On an A-size, 8.5 × 11 sheet of paper, lay out the center lines, part X, and part Y per the given dimensions.

Calculate the standard thread lengths.

Draw the following fasteners on the center lines as illustrated.

Give **all** specific hole call-off information for each fastener in the space above each as illustrated. Include clearance hole sizes, counterbore specifications, countersink specifications, spotface specifications, required tap drill size and depth for 75% thread, and tap size and required depth per the latest drafting standards.

A. Socket-head cap screw—7.8 − 9 UNC × 1.75 lg (calculate counterbore information).

B. Oval-head cap screw—3.4 − 10 UNC × 2.5 lg (calculate countersink information).

C. Hex-head cap screw—1 − 12 UNF × 2.5 lg (calculate spotface information).

D. Fillister-head cap screw with plain washer and hex-head nut—4 − 16 UNF × length to suit. (Calculate clearance hole required.)

Mechanical ## PROBLEM 13-5

Given: an illustration of an old design that uses a washer and cotter pin to hold the .3543 diameter shaft in place. Using this design, it was necessary to locate and drill a hole for the cotter pin. The new design incorporated the use of one retaining ring to achieve the same function at a much lower cost. Choose the correct type and size retaining ring and fill in the required dimensions at A, B, and C. Neatly letter all retaining ring data below the dimensioned shaft as illustrated.

Mechanical ## PROBLEM 13-6

Given: an illustration of an old design that uses a cover plate with four holes for four washers and four Round-head machine screws to hold the shaft in place. Using this design required drilling four blind tap-drill holes that had to be tapped for the four Round-head screws. The new design incorporated the use of one retaining ring to achieve the same function at a much lower cost. Choose the correct type and size retaining ring and fill in the required dimensions at A, B, and C. Neatly print all retaining ring data below the dimensioned collar as illustrated.

DATA:
RETAINING RING NUMBER _____
FREE DIAMETER _____
THICKNESS _____
LUG SIZE _____
LARGE SECTION _____
SMALL SECTION _____
PLIER HOLE SIZE _____
MAX. GAGING DIA. WHEN INSTALLED _____

DATA:
RETAINING RING NUMBER _____
FREE DIAMETER _____
THICKNESS _____
LUG SIZE _____
LARGE SECTION _____
SMALL SECTION _____
PLIER HOLE SIZE _____
MIN. GAGING DIA. WHEN INSTALLED _____

14

Springs

CHAPTER OUTLINE

Spring classification • Helical springs • Flat springs • Terminology of springs • Required spring data • Other spring design layout • Standard drafting practices • Section view of a spring • Summary • Review questions • Chapter 14 problems • CAD instructions

CHAPTER OBJECTIVES

Upon completion of this chapter, students should be able to do the following:

▶ Explain the classifications of springs.

▶ Describe the three different kinds of helical springs and their functions.

▶ Define the function of flat springs.

▶ Define the various terms associated with springs and their design.

▶ List the spring data that should accompany spring drawings.

▶ Demonstrate proficiency in drawing the various types of springs.

▶ Demonstrate proficiency in applying standard CAD practices when drawing springs.

▶ Demonstrate proficiency in drawing sectional views of springs.

▶ Demonstrate proficiency in drawing isometric views of springs.

KEY TERMS

Active coils	Inside diameter
Coil (turn)	Loaded length
Compression spring	Mean diameter
Direction of winding	Outside diameter
Extension spring	Solid length
Flat spring	Spring
Free length	Torsion spring
Helical spring	Wire size

A *spring* is a mechanical device that is used to store and apply mechanical energy. A spring can be designed and manufactured to apply a pushing action, a pulling action, a torque or twisting action, or a simple power action.

Product manufacturers usually use standard-size springs that are purchased from companies that specialize in making springs. These springs are usually mass-produced and are, therefore, relatively inexpensive. Occasionally, however, a special spring must be designed to perform a special function. To be able to design special-function springs, the drafter or designer must know the many terms associated with springs and how to design and construct special springs.

Spring Classification

A spring is generally classified as either a helical spring or a flat spring. *Helical springs* are usually cylindrical or conical; *flat springs,* as their name implies, are usually flat.

Helical Springs

Three types of helical springs are compression springs, extension springs, and torsion springs (**Figure 14-1**).

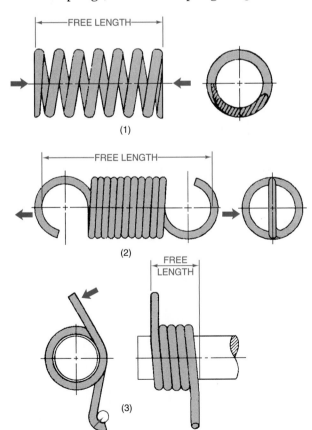

FIGURE 14-1   Types of helical springs: (1) Compression spring; (2) Extension spring; (3) Torsion spring.

Compression Spring

A *compression spring* offers resistance to a compressive force or applies a pushing action. In its free state, the coils of the compression spring do not touch. The types of ends of a compression spring are illustrated and described as follows:

Plain open-end—(Figure 14-2A). The plain open-end spring is very unstable, has only point contact, and has ends that are not perpendicular to the axis of the spring.

Plain closed-end—(Figure 14-2B). The plain closed-end spring is a little more stable and provides a round, parallel surface contact perpendicular to the axis of the spring.

Ground open-end—(Figure 14-2C). The ground open-end spring is much more stable and provides a flat surface contact perpendicular to the axis of the spring.

Ground closed-end—(Figure 14-2D). The ground closed-end spring is the most stable and the best design. This design provides a large, flat surface contact perpendicular to the axis of the spring.

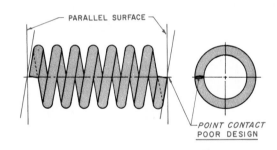

FIGURE 14-2A   Plain open-end compression spring.

FIGURE 14-2B   Plain closed-end compression spring.

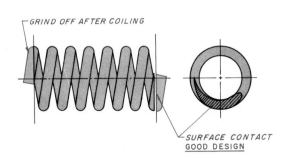

FIGURE 14-2C   Ground open-end compression spring.

FIGURE 14-2D   Ground closed-end compression spring.

A good example of a compression spring is the kind that is used in the front end of an automobile.

Extension Spring

An *extension spring* offers resistance to a pulling force. In its free state, the coils of the extension spring are usually either touching or very close. The ends of an extension spring are usually a hook or loop (**Figure 14-3**). Illustrated are a few of the many kinds of ends that are used. The loop or hook can be over the center or at the side. Each end is designed for a specific application or assembly. Regardless of the shape

SHORT LOOP OVER CENTER

LONG HOOK OVER CENTER

FULL LOOP OVER CENTER

FULL LOOP AT SIDE

'V' HOOK OVER CENTER

FIGURE 14-3   Ends of extension spring.

of the loop or hook, the overall length of an extension spring is measured from the inside of the loop or hook. A good example of an extension spring is the kind that is used to counterbalance a garage door.

Torsion Spring

A *torsion spring* offers resistance to a torque or twisting action. In its free state, the coils of a torsion spring are usually either touching or very close. The ends of a torsion spring are usually specially designed to fit a particular mechanical device (**Figure 14-4**). A good example of a torsion spring is the kind that is used to return a doorknob to its original position.

Flat Springs

The other classification or type of spring is the flat spring. The flat spring is made of spring steel and is designed to perform a special function. Flat springs are not standard and must be designed and manufactured to fit each particular need. The flat spring is considered a power spring, and it resists a pressure. A good example of a flat spring is the kind that is used for a door latch on a cabinet (**Figure 14-5**), or a leaf spring as used in a truck (**Figure 14-6**). A flat spiral spring used to power a windup clock or toy is another example (**Figure 14-7**).

FREE LENGTH

FIGURE 14-4   Torsion spring.

FIGURE 14-5   Flat spring.

FIGURE 14-6   Leaf spring.

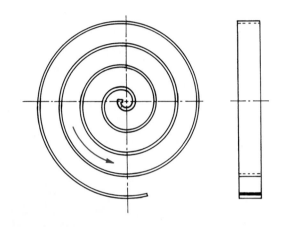

FIGURE 14-7   Special spring.

Terminology of Springs

It is important to know and fully understand the various terms associated with springs and their design (**Figure 14-8**).

Free Length

Free length is the overall length of the spring when it is in its free state of unloaded condition. Free length of a compression spring is measured from the extreme ends of the spring. In an extension spring, the free length is measured inside the loops or hooks.

FIGURE 14-8   Spring terminology.

Solid Length

Solid length of a compression spring is the overall length of the spring when **all** coils are compressed together so they touch. This can be mathematically calculated. If the wire diameter is .500, the overall solid length is 3.25.

Outside Diameter

Outside diameter (OD) is the outside diameter of the spring (2 × wire size **plus** inside diameter).

Wire Size

Wire size is the diameter of the wire used to make up the spring.

Inside Diameter

Inside diameter (ID) is the inside diameter of the spring (2 × wire size **minus** outside diameter).

Active Coils

Active coils in a compression spring are usually the total coils minus the two end coils. In a compression spring, the coil at each end is considered nonfunctional. In an extension spring, active coils include **all** coils.

Loaded Length

Loaded length is the overall length of the spring with a special given or designed load applied to it.

Mean Diameter

Mean diameter is the theoretical diameter of the spring measured to the center of the wire diameter. This theoretical diameter is used by the drafter to lay out and draw a spring. To mathematically calculate the mean diameter, subtract the wire diameter from the outside diameter.

Coil

Coil or *turn* is one full turn or 360° of the wire about the center axis (**Figure 14-9**).

FIGURE 14-9   Coils or turns.

FIGURE 14-10   Direction of winding.

Direction of Winding

Direction of winding is the direction in which the spring is wound. It can be wound right-hand or left-hand (**Figure 14-10**). The thumbs of your hands point toward the direction in which the coil windings are leaning. If the coil windings slant to the right, as the example to the left illustrates, it agrees with the direction of your left-hand thumb; thus, a left-hand winding. If the coil windings slant to the left, as the example to the right illustrates, it agrees with the direction of your right-hand thumb; thus, a right-hand winding. If the coil winding direction is not called off on the drawing, it will be manufactured with right-hand winding.

Required Spring Data

Each drawing must include complete dimensions and specific data. Besides the regular dimensions—that is, outside diameter (OD) and/or inside diameter (ID), wire size, free length, and solid length (compression spring)—the following data should be called off at the lower right side of the work area:

- Material
- Number of coils (including active and inactive coils)
- Direction of winding (if not noted, it is assumed to be right-hand)
- Torque data (if torsion spring)
- Finish
- Heat treatment specification
- Any other required data

How to Draw a Compression Spring

GIVEN: Plain open ends, 1.50 outside diameter, .25 diameter wire size, 4.00 free length, oil-tempered spring steel wire, eight total coils (six active coils), left-hand winding, heat treatment: heat to relieve coiling stresses, finish: black

paint. Make a rough sketch of the spring using the given specifications (**Figure 14-11A**). A rough sketch is extremely important in developing a new spring design.

STEP 1   Measure and construct two vertical lines the required overall length (front view). Divide the space between the two vertical lines into 17 equal spaces (front view) 2 × total coils plus 1; in this example, 2 × 8 + 1 = 17 (**Figure 14-11B**). Measure and construct the mean diameter. To calculate the mean diameter, subtract the wire diameter from the outside diameter (end view).

STEP 2   Determine the direction of windings; refer to the sketch and lightly construct the wire diameter accordingly (front view). The wire diameters are constructed on the mean diameter. Lightly draw the outside diameter and inside diameter (end view) (**Figure 14-11C**).

(A)

(B)

(C)

FIGURE 14-11   (A) How to draw a compression spring with plain open ends; (B) Step 1; (C) Step 2;

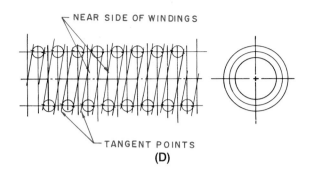

NEAR SIDE OF WINDINGS

TANGENT POINTS

(D)

REAR SIDE OF WINDINGS

END OF COIL

(E)

Ø.25

OPEN CLOSED ENDS

Ø1.50

2.38 SOLID LENGTH

4.00 FREE LENGTH

MATERIAL: TEMPERED SPRING STEEL

8 TOTAL COILS (6 ACTIVE)

LEFT-HAND COILS

HEAT TO RELIEVE COILING STRESSES

FINISH: BLACK

(F)

FIGURE 14-11   (*Continued*) (D) Step 3; (E) Step 4;
(F) Step 5.

STEP 3   Lightly construct the near side of the spring.
Add short, light tangent points (**Figure 14-11D**).

STEP 4   Lightly construct the rear side of the spring.
Add ends of spring; in this example, plain
open ends. Check all work against given
requirements (**Figure 14-11E**).

STEP 5   Recheck all work. Darken in spring using the
latest drafting standards. Add end of spring,
right end, dimensions, and all required spring
data (**Figure 14-11F**).

*NOTE: Outside diameter, wire size, free length,
and solid length are given as regular dimensions.*

*The material, number of coils, direction of winding,
heat treatment specifications, and finish requirements are listed below and to the right as a note. In
this example, the outside diameter is the controlling
dimension; that is, the outside dimension is more
important than the inside dimension. Inside and
outside dimensions should not both be added to the
same drawing.*

How to Draw a Compression Spring
(Plain Closed Ends)

GIVEN: Plain closed ends, 20 mm inside diameter,
5 mm diameter wire size, 100 mm free
length, material: hard drawn steel spring wire,
five active coils (total seven coils), left-hand
winding, heat treatment: heat to relieve
coiling stresses, finish: zinc plate. Make a
rough sketch of the spring using the given
specifications (**Figure 14-12A**).

STEP 1   Measure and construct two vertical lines the
required overall length (front view). Draw
a line **inside** the two vertical lines equal to
half the wire diameter from each end. Divide
the space between the two vertical lines into
13 equal spaces (front view) 2 × total coils
minus 1; in this example, 2 × 7 − 1 = 13
(**Figure 14-12B**). Measure and construct
the mean diameter. To calculate the mean

(A)

FREE LENGTH

MEAN DIAMETER

START
IN HALF WIRE DIA.

STOP
IN HALF WIRE DIA.

SCALE HELD AT ANGLE FOR 13
EQUAL PARTS

(B)

FIGURE 14-12   (A) How to draw a compression spring
with plain closed ends; (B) Step 1;

diameter, add the wire diameter to the inside diameter (end view).

STEP 2 Determine the direction of winding; refer to the sketch and lightly construct the wire diameter accordingly (front view). Lightly draw the outside diameter and inside diameter (end view) (**Figure 14-12C**).

STEP 3 Lightly construct the near side of the spring. Add short tangent points (**Figure 14-12D**).

STEP 4 Lightly construct the rear side of the spring. Add ends of springs. In this example, plan closed ends. Check all work against given requirements (**Figure 14-12E**).

STEP 5 Recheck all work. Darken in spring using the latest drafting standards. Add all required dimensions and spring data (**Figure 14-12F**).

How to Draw a Compression Spring (Ground Closed Ends)

(The following illustrations are for a ground closed-end compression spring. The same steps can be used for a ground open-end compression spring.)

GIVEN: Ground closed ends, 1.125 outside diameter, .188 diameter wire size, 3.00 free length, material: hard drawn steel spring wire, seven total coils (five active coils), right-hand winding, heat treatment: heat to relieve coiling stresses, finish: zinc plate. Make a rough sketch of the spring using the given specifications (**Figure 14-13A**).

STEP 1 Measure and construct two vertical lines the required free length (front view). Divide the space between the two vertical lines into 13 equal spaces (front view), 2 × total coils minus 1; in this example, 2 × 7 − 1 = 13 (**Figure 14-13B**). Measure and construct the mean diameter. To calculate the mean diameter, subtract the wire diameter from the outside diameter (end view).

STEP 2 Determine the direction of winding; refer to the sketch and lightly construct the wire diameter accordingly (front view). Lightly draw the outside diameter and inside diameter (end view) (**Figure 14-13C**).

STEP 3 Lightly construct the near side of the spring. Add short tangent points (**Figure 14-13D**).

STEP 4 Lightly construct the rear side of the spring. Add ends of spring; in this example, ground closed end. Check all work against given requirements (**Figure 14-13E**).

(C)

(D)

(E)

(F)

MATERIAL: HARD DRAWN SPRING STEEL WIRE
5 ACTIVE COILS (7 TOTAL COILS)
LEFT-HAND COILS
HEAT TO RELIEVE COILING STRESSES
FINISH : ZINC PLATE

FIGURE 14-12 (*Continued*) (C) Step 2; (D) Step 3; (E) Step 4; (F) Step 5.

STEP 5 Recheck all work. Darken in spring using the latest drafting standards. Add all required dimensions and spring data (**Figure 14-13F**).

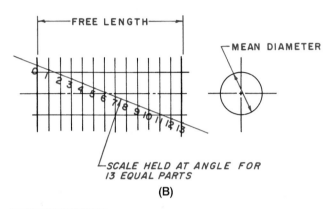

(A)

(B)

FIGURE 14-13   (A) How to draw a compression spring with ground closed ends; (B) Step 1;

How to Draw an Extension Spring

GIVEN: Full loop, over center each end 1.625 outside diameter, .188 diameter wire size, 4.750 approximate free length, material: hard drawn spring steel wire, heat treatment: heat to relieve coiling stresses, finish: black paint, windings as tight as possible. Make a rough sketch of the spring using the given specifications (**Figure 14-14A**).

STEP 1 Draw the outside diameter, inside diameter, and mean diameter (end view). Lightly draw the required free length. In the design of an extension spring, it is almost impossible to arrive at the **exact** free length. In designing an extension spring, it is best to design it slightly **shorter** than actually required. Construct the end loops with the **inside** diameter within the specified free length (**Figure 14-14B**). Note that in this example, the loops are the same exact diameter as the spring itself.

STEP 2 Starting at the intersection of the mean diameter of the right end loop and the mean diameter of the spring, lower side, locate and draw the first lower wire diameter (**Figure 14-14C**). Project directly up to the mean diameter of the spring, upper side, and draw the first upper wire diameter **over** to the left, half a wire diameter as illustrated. Lay out, side by side, wire

(C)

(D)

(E)

MATERIAL: HARD DRAWN SPRING STEEL WIRE

7 TOTAL COILS (5 ACTIVE COILS)

HEAT TO RELIEVE COILING STRESSES

FINISH: ZINC PLATE

(F)

FIGURE 14-13   (*Continued*) (C) Step 2; (D) Step 3; (E) Step 4; (F) Step 5.

(A)

(B)

FIGURE 14-14   (A) How to draw an extension spring with full loop ends; (B) Step 1;

(C)

(D)

(E)

(F)

FIGURE 14-14   (Continued) (C) Step 2; (D) Step 3; (E) Step 4; (F) Step 5.

diameters touching, as illustrated; in this example, 10 ½ coils. Note that the coils must end on the mean diameter, left end. This may mean slightly adjusting the location of the left end loop. It is best to bring it in slightly, if necessary.

STEP 3   Locate points X and Y, as illustrated in **Figure 14-14D**, and adjust the drafting machine at this angle and lock on this angle. Draw the wire loops.

STEP 4   Lightly draw the loop in the right-side view. To calculate just what the loop will actually look like, number various points along the loop in the front view; in this example, point 1, starting point; point 2, up; point 3, over; and point 4, around. Project these points to the end view, as illustrated in **Figure 14-14E**.

STEP 5   Recheck all work. Darken in spring using the latest drafting standards. Add all required dimensions and spring data (**Figure 14-14F**).

Other Spring Design Layout

When drawing any specially designed spring for a particular function, one should make a rough sketch of it, including all required specifications. A torsion spring is developed in a manner very similar to an extension spring with its tight coils. Usually, the required torque pressure is included in the given spring data. The actual designed deflection of the

torsion spring is illustrated by phantom lines, and the angle is noted.

Standard Drafting Practices

Some company standards specify that the drafter not draw the complete spring because of the time and cost involved. A shortcut method for drawing a spring is shown in **Figure 14-15**. At the top is the conventional method of representing a compression spring; below is the same spring drawn by the schematic drawing system.

SAVING DRAFTING TIME

John Dugan had been the outstanding drafting student at Minot Technical Institute. His work as a student had won him a closet full of awards and much recognition. Consequently, nobody was surprised when Dugan secured a job in the drafting department of Minot Machine Tool Company, an excellent employer.

John Dugan's talent was apparent from the outset, and within weeks he was assigned to a team that took on the company's most difficult projects. His first assignment on the "A" team was to draw the plans for several different extension springs. Springs had been Dugan's specialty at Minot Tech. In the words of his professor, "Dugan can draw springs that look so realistic you can lift them off the page and use them."

Consequently, Dugan was taken aback when his team leader stopped him after he had completed only one of the five spring drawings he was assigned. "What in the world are you doing?" asked the team leader, clearly flabbergasted. Dugan, unsure of what was wrong and shocked by the first negative reaction he had ever had to his work, could only stammer, "What do you mean? I'm drawing the springs you assigned me." "I asked you to draw them, not make them," said the team leader. "We don't do pictorial spring drawings here. This is the drafting department, not the art department. We use schematic representation with all the critical data called out in the notes. It takes time to create pictorial representations of springs, and time is money." "Besides," said the team leader, "there is no need to give the machining department anything but a schematic and some notes. That is all they need to make the springs, and it is our job to give them that much information and nothing more. Every minute you spend creating these beautiful pictorials is time that could have been put to better use on another project." John Dugan had just learned a valuable lesson that eventually every drafter must learn.

Another method of drawing a spring, especially a long spring, is illustrated in **Figure 14-16**. The uncompleted coils are illustrated by phantom lines. If the company does not have a standard method of drawing springs, it is the drafter's decision as to which method of representation is used. In most cases, it is best to draw the object, in this case the spring, in such a way that there is no question whatsoever as to exactly what is required. Of course, regardless of which system is used, full dimensions and specification data must be included.

Section View of a Spring

If a cutting-plane line passes through the axis of a spring, the spring is drawn in one of two ways. A small spring is drawn with the back coils showing, as illustrated in **Figure 14-17A**, with the section lining of the coils filled in solid. A larger spring is drawn the same way but uses standard section lining (**Figure 14-17B**).

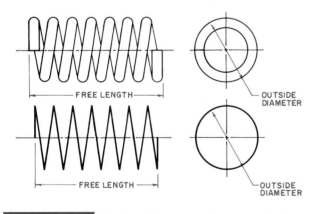

FIGURE 14-15 Schematic system of representing a compression spring.

FIGURE 14-16 Incomplete coils with phantom lines.

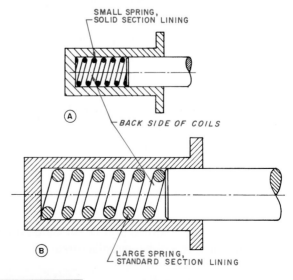

FIGURE 14-17 Section view of a spring.

PROBLEMS

Note that both illustrations are right-hand springs, but because the front half has been removed, only the rear half is seen, which appears as left-hand.

SUMMARY

- Springs can be classified as helical or flat. There are three types of helical springs: compression, extension, and torsion.

- Key terms associated with spring design include the following: free length, solid length, outside diameter, wire size, inside diameter, active coils, loaded length, wire diameter, coil, and direction of winding.

- Required spring data include the following: material, number of coils, direction of winding, torque data, finish, and heat treatment specifications.

- Some companies specify the use of schematic representation of springs as a way to save time and money. Another time-saving method is to draw only those coils that are at each end of the spring and connect them with phantom lines.

- If a cutting-plane line passes through the axis of a spring, the spring may be drawn in one of two ways. A small spring is drawn with the section lines filled in. A larger spring uses standard section lining.

REVIEW QUESTIONS

Mark the following statements as either true or false.

1. It is incorrect to dimension both the inside diameter and the outside diameter of a spring on the same drawing.

2. The two general classifications of springs are helical and flat.

3. The solid length of a compression spring is the length of the spring stretched out straight.

4. Section lining is illustrated in a large spring where the cutting-plane line passes through the center axis of the spring by filling the coils in solid.

5. The free length of a spring means the overall length of a spring when it is in its free state or unloaded condition.

Answer the following questions by selecting the best answer.

1. If the coil winding direction is not called off on the drawing, it will be manufactured:
   a. Left-hand winding
   b. Right-hand winding
   c. The same as other springs in the drawing
   d. None of the above

2. Which dimension should be given for the diameter of a spring on a drawing?
   a. Inside
   b. Outside
   c. Radial
   d. Both a and b

3. Other than the required dimensions, what spring data must be noted below and to the right?
   a. Material
   b. Number of coils
   c. Direction of winding
   d. All of the above

4. Which of the following is **not** a type of helical spring?
   a. Compression spring
   b. Flat spring
   c. Torsion spring
   d. Extension spring

5. Which of the following is **not** a common use for a torsion spring?
   a. A doorknob
   b. A clock
   c. A truck
   d. A windup toy

CHAPTER 14 PROBLEMS

The following problems are intended to give the beginning drafter practice in drawing various kinds of springs with special required specifications.

Conventional Instructions

The steps to follow in laying out springs are:

STEP 1 Study all required specifications.

STEP 2 Make a rough sketch of the spring, incorporating **all** required specifications and required dimensions.

STEP 3 Calculate the size of the basic shape of the front and end views.

STEP 4 Center the two views within the work area with correct spacing for dimensions between views.

STEP 5 Starting with the end view or the view that contains the diameter, lightly draw the spring, adhering to **all** specific required specifications. Take care to show correct direction of winding, correct type of ends, and total count of windings.

STEP 6 Check to see that both views are centered within the work area.

STEP 7 Check all dimensions in both views.

STEP 8 Darken in both views, starting first with all diameters and arcs.

STEP 9 Neatly add all dimensioning as required to fully describe the object using the latest drafting standards.

STEP 10 Neatly add all required specifications under the two views. The drawing must include the following: free length, wire size, total number of coils, type of ends, solid length, outside diameter, inside diameter, direction of windings, material (usually hard drawn steel spring wire), required heat-treating process

(usually heat to relieve coiling stresses), and any other special requirements.

STEP 11 For torsion-type springs, illustrate working location and travel with phantom lines and dimensions.

CAD INSTRUCTIONS

Students who plan to complete the following problems on a CAD system should begin with these general instructions.

1. Analyze the problem carefully.
2. Plan your drawing sequence. In what order will you develop the drawing?
3. Read the specific instructions for the problem and begin.

Problems 14-1 through 14-7

Construct a two view drawing of each spring using the listed steps. Center the views on an A-size, 8 ½ × 11″ sheet of

paper with a 1″ (25 mm) space between views. Use correct line thicknesses and all drafting standards. Add all required dimensions and specifications per the latest drafting standards.

Mechanical   ## PROBLEM 14-1

Compression-type spring. Free length, 7.00. Wire size, .25. 12 active coils (14 **total** coils). Plain open ends, OD 2.00/ID 1.50. R.H. winding. Calculate solid length.

Mechanical **PROBLEM 14-2**

Compression-type spring. Free length, 5.25. Wire size, .375. Four active coils (six **total** coils). Ground open ends, OD 2.75/ID 2.00. L.H. winding. Calculate solid length.

Mechanical **PROBLEM 14-3**

Compression-type spring. Free length, 6.25. Wire size, .375. Nine total coils (seven **active** coils). Plain closed ends, OD 2.00/ID 1.25. L.H. winding. Calculate solid length.

Mechanical   **PROBLEM 14-4**

(Metric) Compression-type spring. Free length, 90. Wire size, 12. Four total coils (two **active** coils). Ground closed ends, OD 96/ID 72. R.H. winding. Calculate solid length.

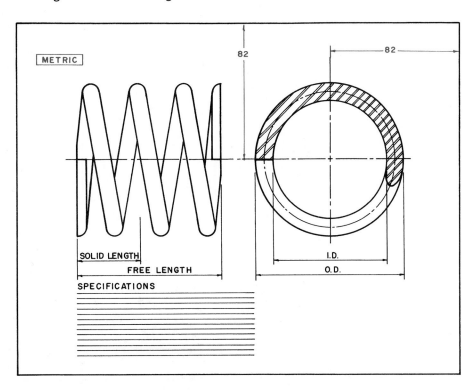

Mechanical   **PROBLEM 14-5**

Extension-type spring. Full loop over center, both ends. Approx. free length, 7.0. Wire size, .25. OD 2.0/ID 1.50. R.H. winding. (This is **standard** for all extension springs.)

PROBLEMS

PROBLEM 14-6

(Metric) Extension-type spring. Full loop over center (right end). Long hook over center (left end). Free length, 150. Wire size, 5 (12 total coils). 65 coil length, OD 42/ID 32. Standard R.H. winding.

PROBLEM 14-7

Torsion-type spring. Free length, 3.625. Wire size, .25. 13 total coils. OD 2.0/ID 1.50. R.H. standard winding, 90° maximum working flex (counterclockwise). Arm lengths 2.0 (as shown).

15

Cams

CHAPTER OBJECTIVES

Upon completion of this chapter, students should be able to do the following:

▶ Explain the cam principle.

▶ Describe the four basic types of followers.

▶ Distinguish between the two major kinds of cams used in the industry.

▶ Define the most important cam terms.

▶ Compare and contrast the four principal types of cam motion.

▶ Demonstrate proficiency at laying out a cam from a displacement diagram.

▶ Demonstrate proficiency in drawing various types of cams.

▶ Demonstrate proficiency in developing timing diagrams.

▶ Demonstrate proficiency in dimensioning cams.

KEY TERMS

Base circle	Parallel
Baseline	Perpendicular
Cam	Radial arm
Cylindrical	Rotation
Displacement diagram	Time interval
Dwell	Timing diagram
Harmonic motion	Uniform acceleration
Lowest position	Uniform velocity
Modified uniform velocity	Working circle

Cam Principle

A *cam* produces a simple means to obtain irregular or specified predictable, designed motions. These motions would be very difficult to obtain in any other way.

Figure 15-1 illustrates the basic principle and terms of a cam. In this example, a rotating shaft has an irregularly shaped disc attached to it. This disc is the cam. The follower, with a small roller attached to it, pushes against the cam. As the shaft is rotated, the roller follows the irregular surface of the cam, rising or falling according to the profile of the cam. The roller is held tightly against the cam either by gravity or a spring.

Figure 15-2 illustrates a simple cam in action. Notice how the rotation of the shaft converts into an up-and-down motion of the follower. A cam using a flat-faced follower is shown in **Figure 15-3**. This type of cam, for example, is used to raise and lower the valve in an automobile engine. A modified cam

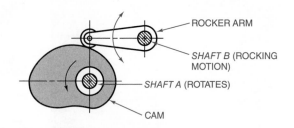

FIGURE 15-4   Rocker arm follower with wheel.

follower is used to change the rotary motion of shaft A into an up-and-down motion of the rocker arm and then into a rocking motion of shaft B (**Figure 15-4**).

Basic Types of Followers

Speed of rotation and the actual load applied upon the lifter determine the type of follower to be used. There are various basic types of followers: roller, pointed, flat-faced, and spherical-faced (**Figure 15-5**).

Cam Mechanism

Two major kinds of cams are used in the industry: radial arm design and cylindrical design. The *radial arm* design changes a rotary motion into either an up-and-down motion or a rocking action, as discussed previously. In the *cylindrical* design, a shaft rotates exactly as it does in the radial design, but the action or direction of the follower differs greatly. In the radial arm design, the follower operates *perpendicular* to the camshaft. In the cylindrical design, the follower operates *parallel* to the camshaft. See **Figure 15-6** for a simplified example of

FIGURE 15-1   Basic cam principle and terms.

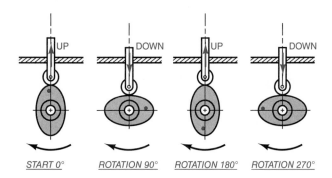

FIGURE 15-2   Simple cam in action.

FIGURE 15-5   Basic types of followers.

FIGURE 15-3   Flat-faced follower.

FIGURE 15-6   Cylindrical cam mechanism.

a cylindrical design cam. Shaft A has an irregular groove cut into it. As the cam rotates, the follower traces along the groove. The follower is directly attached to a shaft that moves to the left and right. Note that this shaft does **not** rotate; the motion is parallel to the axis of the rotating camshaft.

Cams produce one of three kinds of motion: uniform velocity, harmonic, or uniform acceleration. Each is discussed in full later.

Cam Terms

Working Circle

The *working circle* is considered to be a distance equal to the distance from the center of the camshaft to the highest point on the cam (**Figure 15-7**).

Displacement Diagram

The *displacement diagram* is a designed layout of the required motion of the cam. It is laid out on a grid, and its length represents one complete revolution of the cam. The length of the displacement diagram is usually drawn equal in length to the circumference of the working circle. This is not absolutely necessary, but it will give an in-scale idea of the cam's profile. The height of the displacement diagram is equal to the radius of a working circle (**Figure 15-8**). The bottom line of the displacement diagram is the *baseline*. All dimensions should be measured upward from the baseline. On the cam itself, think of the center of the cam as the baseline.

FIGURE 15-7   Cam terms and working circle.

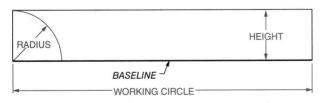

WORKING CIRCLE = CIRCUMFERENCE OF CIRCLE = 360°
BASELINE = CENTER OF WORKING CIRCLE

FIGURE 15-8   Displacement diagram (basic layout).

FIGURE 15-9   Displacement diagram with 30° increments.

The length of the displacement diagram is divided into equal lines or grids, each of which represents degrees around the cam. These divisions can be 30°, 15°, or even 10°. The finer the divisions, the more accurate the final cam profile (**Figure 15-9**).

Dwell

Dwell is the period of time during which the follower does not move. This is shown on the displacement diagram by a straight horizontal line throughout the dwell angle. On the cam, the dwell is drawn by a radius.

Time Interval

The *time interval* is the time it takes the cam to move the follower to the designed height.

Rotation

Rotation of the cam is either clockwise or counterclockwise. The actual cam profile is laid out opposite of the rotation.

Base Circle

The *base circle* is used to lay out an offset follower that is illustrated in detail later. The base circle is a circle with a radius equal to the distance from the center of the shaft to the center of the follower wheel at its *lowest position*. On an offset follower, the base circle replaces the working circle.

Cam Motion

Four major types of curves are usually employed. Special irregular curves other than the four major types can be designed to meet a specific movement, if necessary.

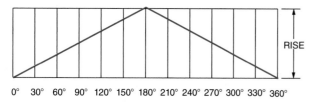

FIGURE 15-10   Uniform velocity cam.

FIGURE 15-12   Harmonic motion cam.

The four types are uniform velocity, modified uniform velocity, harmonic motion, and uniform acceleration. For comparison, the following four examples use the same working circle, same shaft, and same rise.

Uniform Velocity

In this type of motion, the cam follower moves with a *uniform velocity;* that is, it rises and falls at a constant speed, but the start and stop are very abrupt and rough (**Figure 15-10**).

Modified Uniform Velocity

Because of the abrupt and rough start and stop of the uniform velocity, it is modified slightly. *Modified uniform velocity* smoothes out the roughness slightly by adding a radius at the ends of the high and low points. This radius is equal to 1/3 the rise or fall (**Figure 15-11**). This radius smoothes out the start and stop somewhat and is good for slow speed.

Harmonic Motion

Harmonic motion is very smooth, but the speed is not uniform. Harmonic motion has a smooth start and stop and is good for fast speed.

To lay out a harmonic motion on the displacement diagram, draw a semicircle whose diameter is equal to the designed rise or fall. Divide the semicircle into equal divisions; in this example, 30°. Divide the overall horizontal distance of the rise or fall equally into increments of the same number as the semicircle; in this example, six equal divisions. Projection points on the semicircle are projected horizontally to the corresponding vertical lines. For example, the first point on the semicircle is

projected to the first vertical line, the second point on the semicircle is projected to the second vertical line, and so forth. Connect the curve with an irregular curve. This completes the harmonic curve (**Figure 15-12**).

Uniform Acceleration

Uniform acceleration is the smoothest motion of all cams, and its speed is constant throughout the cam travel. The uniform acceleration curve is actually a parabolic curve; the first half of the curve is exactly the reverse of the second half. This form is best for high speed.

To lay out a uniform acceleration motion on the displacement diagram, divide the designed rise or fall into 18 equal parts. To do this, place the edge of a scale with its "0" on the starting level of the rise or fall, and equally space 18 units of measure on the high or low elevation of the rise or fall (**Figure 15-13**). Mark off points 1, 4, 9, 4 (14) and 1 (17), and draw horizontal lines. Note that 14 is actually 4, and 17 is actually 1. Divide the overall horizontal distance of the rise or fall into 12 equal increments. From points 1, 4, 9, 4 (14) and 1 (17), project points to the corresponding vertical lines. For example, the first point is from point 1 on the division line to the first vertical line, and then from point 4 on the division line to the second vertical line, and so forth. Connect the curve with an irregular curve. This completes the uniform acceleration curve.

Combination of Motions

The illustrations thus far have covered the four major types of cams: uniform velocity, modified uniform velocity, harmonic motion, and uniform

FIGURE 15-11   Modified uniform velocity cam.

FIGURE 15-13   Uniform acceleration cam.

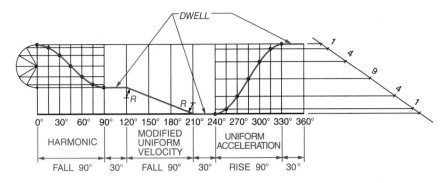

FIGURE 15-14   Combination of motions.

acceleration. Any combinations of these can be designed for a particular function or requirement. **Figure 15-14** illustrates a displacement diagram with a 90° fall using a harmonic motion followed by a 30° dwell. The cam then falls again 90° using a modified uniform velocity followed by a 30° dwell. The cam then rises using uniform acceleration motion followed by another 30° dwell.

Laying Out the Cam from the Displacement Diagram

Regardless of which type of motion is used, the layout process is exactly the same. The working circle is drawn equal to the height of the displacement diagram. The circle of the cam must be divided into the same number of equal divisions as the displacement diagram (**Figure 15-15A**). Label the increments on both the displacement diagram and the working circle of the cam; in this example, harmonic motion is illustrated.

Notice that the even increment on the cam layout is labeled opposite of the rotation of the cam. In this example, rotation is clockwise; therefore, the labeling of the radial lines should be counterclockwise. Transfer each distance from the displacement diagram to the corresponding radial line (**Figure 15-15B**). Using an irregular curve, complete the cam layout.

Offset Follower

The cam follower is usually in line with the same center line as the cam (see Figure 15-1). Occasionally, however, when position or space is a problem, the follower can be designed to be on another center line or offset, as illustrated in **Figure 15-16**. In order to lay out an offset follower, it is necessary to use slightly different steps than those used for a regular cam.

(A)

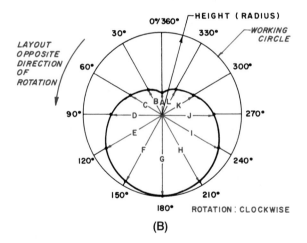

(B)

FIGURE 15-15   (A) Displacement diagram example; (B) Heights transferred from displacement diagram.

FIGURE 15-16   Associated cam terms.

How to Draw a Cam with an Offset Follower

GIVEN: Cam data: harmonic motion cam, follower offset .625, base circle .88 radius, rise of 1.50, shaft diameter .375, direction of rotation counterclockwise.

Follower data: follower width .250, roller .375 diameter.

STEP 1 Design and lay out a displacement diagram exactly as with any cam (**Figure 15-17A**).

STEP 2 From the center of the shaft of the cam, construct the offset circle. The offset circle has a radius equal to the required follower offset. From the same swing point, lay out the base circle. The radius of the base circle is equal to the distance from the center of the shaft to the center of the follower wheel at its *lowest* position (**Figure 15-17B**).

STEP 3 The point where it crosses a constructed vertical line from the center line of the follower is point A on the displacement diagram (**Figure 15-17C**). This represents the lowest position of the follower wheel. This vertical line is the measuring line and will be used to lay out the other points around the cam.

STEP 4 Divide the offset circle into the same equal divisions as used on the displacement diagram; in this example, 12 equal parts. Lightly and consecutively, letter or number the divisions on the offset circle. Start at the lowest point of the cam; in this example, point A. Letter or number in a direction opposite to that of the cam rotation. Note that point A is the exact location where the center line of the follower is tangent with the offset circle. Construct a layout projection line 90° from the tangent point of the circle (**Figure 15-17D**).

STEP 5 Transfer the distances from the displacement diagram to the measuring line, and lightly letter or number each point. With the compass point on the center of the shaft of the cam, swing these distances to the corresponding radial lines (refer to Figure 15-17D). Lightly letter or number each point. Do not forget to *lay out the cam opposite to the direction of rotation*. These points represent the path of the center of the follower wheel. Lightly connect these points together (**Figure 15-17E**).

STEP 6 With the compass set at the radius for the roller, lightly draw the roller around the path

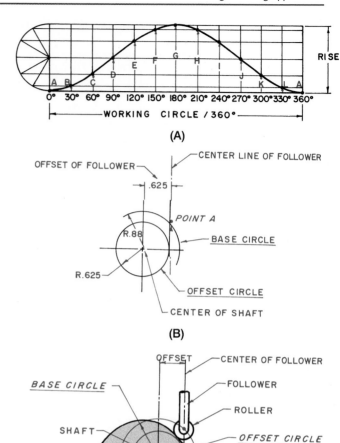

FIGURE 15-17   (A) Step 1 (drawing a cam with an offset follower design layout); (B) Step 2; (C) Step 3;

of the follower. The inside, high points of these arcs represent the actual profile of the cam (**Figure 15-17F**).

STEP 7 Check all construction work. If correct, darken in and complete the cam (**Figure 15-17G**).

How to Draw a Cam with a Flat-Faced Follower

GIVEN: Cam data: harmonic motion cam, base circle .75 radius, rise of 1.50, shaft diameter .375, direction of rotation clockwise.

Follower data: flat-faced follower, 1.25 diameter distance across the flat surface. Design and lay out a displacement diagram exactly as with any cam. In this example, the

(D)

(E)

(F)

(G)

FIGURE 15-17   (*Continued*) (D) Step 4; (E) Step 5;
(F) Step 6; (G) Step 7.

displacement diagram in Figure 15-17A will be used.

STEP 1   Construct the base circle. The base circle is drawn with a radius equal to the distance from the center of the shaft to the face of the follower in its lowest position. Divide the base circle into the same equal divisions as used on the displacement diagram. Lightly and consecutively letter or number the divisions on the base circle (**Figure 15-18A**).

STEP 2   Locate point A, which represents the location of the flat-faced follower at its lowest point. Construct a measuring line starting from the base circle, and transfer the distances from the displacement diagram to the measuring line. Letter or number each point (**Figure 15-18B**). Do not forget to lay out the cam *opposite to the direction of rotation.*

STEP 3   Swing these distances to the corresponding radial lines (**Figure 15-18C**).

STEP 4   From these distances, construct lines perpendicular from these radial lines at the point of intersection (**Figure 15-18D**).

STEP 5   Lightly construct the cam within these perpendicular lines (**Figure 15-18E**).

STEP 6   Check all construction work. If correct, darken in and complete the cam (**Figure 15-18F**).

(A)

(B)

FIGURE 15-18   (A) Step 1 (drawing a cam with a
flat-faced follower); (B) Step 2;

(C)

(D)

(E)

(F)

FIGURE 15-18 (*Continued*) (C) Step 3; (D) Step 4;
(E) Step 5; (F) Step 6.

Note the position of the flat-faced follower at
the 0°, 90°, and 150° positions.

Timing Diagram

Many times, more than one cam is attached to the
same shaft. In this case, each cam works independently but must function in relation to the other
cams. In order to be able to visualize graphically and
study the interaction of the cams on the same shaft,
place the various displacement diagrams in a column
with the starting points and ending points in line,
as illustrated in **Figure 15-19**. This represents one
full revolution and is referred to as a *timing diagram*.
For example, at 240°, it is easy to visualize the exact
position of each cam roller.

Dimensioning a Cam

A cam must be fully dimensioned. The cam
size dimensions are constructed radiating from
the center of the shaft by both radii length and
degrees from the starting point, as illustrated in
Figure 15-20. Other important dimensions pertain to the shaft hole size and keyway size and are
dimensioned according to the most recent drafting
standards.

Try not to place dimensions on the cam itself;
place them around the perimeter, if possible.

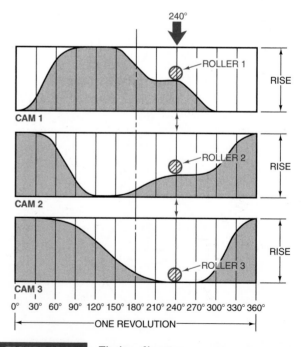

FIGURE 15-19   Timing diagram.

LEARNING ON THE JOB

"This new job has several cams in it," said Marge McCrory. "Does anybody know how to draw cams?" When she got nothing but blank stares from the members of her drafting team, McCrory said, "All right then, we'll just have to learn." She distributed copies of several different diagrams among team members.

"This is what we start with. These are displacement diagrams. Don't ask me how just yet, but we use these to lay out the cams. There is one displacement diagram for each cam." Then, McCrory told her team members to go to the company's Engineering Library and start researching cams. "I want to know what all of these terms on the diagrams mean, how we get a cam layout from a diagram, and how to completely detail a cam, including dimensions," said McCrory.

When the team gathered in the Engineering Department conference room the next day, McCrory could tell immediately that their research had been successful. Her team looked confident and anxious to get started. "Who wants to demonstrate how to lay out a cam from one of these displacement diagrams?" When several hands went up simultaneously, McCrory called on Doug Fulton, the team's senior drafter.

Fulton went to the marker board and quickly sketched his displacement diagram. Then, under it, he sketched the layout for the corresponding cam. He began the cam layout by drawing a circle, which he labeled "working circle." The radius of the circle was equal to the height of the displacement diagram. He divided the circle into 30° segments to match the 30° increments on the displacement diagram. He then transferred each displacement distance from the diagram onto the radial lines of the working circle, beginning at the center of the circle and working outward along each respective radial line.

Within just a few minutes, Fulton had laid out his cam. McCrory then asked another team member to use Fulton's diagram to illustrate cam terminology. Another team member put a transparency on an overhead projector to illustrate dimensioning techniques. McCrory was pleased with her team's work and told them so.

This case study from the workplace illustrates an important point. Students seldom graduate knowing everything they will need to know on the job. Even if they did, what must be known today will change tomorrow. Like Marge McCrory's team members, the modern drafting technician must be able to learn on the job.

ROTATION: CLOCKWISE

FIGURE 15-20   Dimensioning a cam.

SUMMARY

- A cam produces a simple means to obtain irregular or specified predictable, designed motions.

- Cam followers are of four types: roller, pointed, flat-face, and spherical-face.

- The two main types of cams are radial arm design and cylindrical design. Important cam-related terms are *working circle, displacement diagram, time interval, rotation,* and *base circle.*

- The four main types of cam motion are uniform velocity, modified uniform velocity, harmonic velocity, and uniform acceleration.

- By placing the displacement diagrams for different cams that are attached to the same shaft in a column, we can produce a timing diagram. A timing diagram allows you to visualize graphically the interaction of the cams.

REVIEW QUESTIONS

Mark the following statements as either true or false.

1. Speed of rotation and the actual load applied on the lifter determine the type of follower to be used.

2. The two major kinds of cams used in the industry are radial arm and parallel.

3. A working circle is considered a distance equal to the distance from the center of the camshaft to the highest point on the cam.

4. A displacement diagram is a designed layout of the required motion of the cam.

5. Dwell is the period of time during which the follower moves.

Answer the following questions by selecting the best answer.

1. Which of the following is **not** a type of motion produced by a cam?
   a. Harmonic
   b. Uniform acceleration
   c. Uniform velocity
   d. Harmonic displacement

2. Which of the following is **not** true regarding the design and layout of the cam displacement diagram?
   a. The bottom line of the displacement diagram is the baseline.
   b. All dimensions should be measured upward from the baseline.
   c. The length of the displacement diagram is usually drawn equal in length to the circumference of the working circle.
   d. A displacement diagram is a designed layout that measures the applied load.

3. Which of the following is **not** true regarding a timing diagram?
   a. It is used when more than one cam is attached to the same shaft.
   b. Each cam works independently.
   c. Each cam works dependently.
   d. Each cam must function in relation to the other cams.

4. Which of the following is **not** true regarding cam motion?
   a. With uniform velocity, there is a very rough stop and start.
   b. Harmonic motion is very smooth, but the speed is not uniform.
   c. Harmonic motion is very smooth, and the speed is uniform.
   d. Uniform acceleration is the smoothest motion.

5. Which of the following is **not** true regarding the length of the displacement diagram?
   a. The length of the displacement diagram represents one complete revolution of the cam.
   b. The length of the displacement diagram is usually drawn equal in length to the circumference of the working circle.
   c. The height of the displacement diagram is equal to the radius of the working circle.
   d. The height of the displacement diagram is equal to the circumference of the working circle.

CHAPTER 15 PROBLEMS

The following problems are intended to give the beginning drafter practice in the design and layout of cam displacement diagrams according to the required specifications to do a particular function. The cam displacement diagram will then be transferred to an actual cam profile layout. The drafter will practice laying out uniform velocity, harmonic motion, and uniform acceleration motions.

Conventional Instructions

The steps to follow in laying out cams are:

STEP 1 Study all required specifications.

STEP 2 Make a rough sketch of the cam displacement diagram, incorporating **all** required specifications and required dimensions.

STEP 3 Lightly lay out the displacement diagram. Space out horizontally as far as possible.

STEP 4 Break down the area into equal spaces that correspond to degrees around the cam. Label each line, indicating the degrees.

STEP 5 Following the required specifications, lightly develop the cam's travel on the displacement diagram. (Show all light layout work—do not erase.)

STEP 6 Check all work.

STEP 7 Darken in the displacement diagram.

STEP 8 To lay out the cam profile, locate the center of the cam.

STEP 9 Draw the shaft, hub, working circle, and key, if required, in place.

STEP 10 Divide the working circle into equal degrees as required.

STEP 11 Note direction of travel. Lightly letter the degrees around the working circle opposite of the direction of travel.

STEP 12 Locate the starting elevation and draw the cam follower from the starting point.

STEP 13 Lightly draw in the rest of the cam, following details as required.

STEP 14 Transfer all distances from the cam displacement diagram to the corresponding degree line around the cam layout.

STEP 15 Lightly connect all points using an irregular curve.

STEP 16 Check all work.

STEP 17 Darken in all work.

STEP 18 Neatly add all required specifications and dimensions to the drawing.

The drawing must include the following:
Shaft diameter
Hub size
Key or set screw size
Direction of rotation
Working circle size
All dimensions according to the most recent drafting standards. (For these problems, dimension only the front view.)

CAD INSTRUCTIONS

Students who plan to complete the following problems on a CAD system should begin with these general instructions.

1. Analyze the problem carefully.

2. Plan your drawing sequence. In what order will you develop the drawing?

3. Read the specific instructions for the problem and begin.

PROBLEMS 15-1A AND 15-1B

Construct a cam displacement diagram on an A-size, 8½ × 11″ sheet of paper per the dimensions given. Lay out a cam with the following specifications:

1. Working circle 5.0 dia.
2. Shaft dia. .75 counterclockwise
3. Hub dia. 1.25
4. Square key .18
5. Rotation: clockwise
6. Follower dia. 1.0
7. Start 1.0 from center

Required motion:

1. Rise 90°, modified uniform velocity 1.5
2. Dwell 60°
3. Fall 60°, modified uniform velocity .75
4. Dwell 30°
5. Fall 60°, modified uniform velocity .75
6. Dwell 60°

Construct the cam and follower on an A-size, 8½ × 11″ sheet of paper per the dimensions given, and transfer all dimensions from the displacement diagram to the cam layout.

Check all work and add all required dimensions and specifications according to the most recent drafting standards.

PROBLEMS

PROBLEMS

PROBLEM 15-1A

PROBLEM 15-1B

PROBLEMS 15-2A AND 15-2B

Construct a cam displacement diagram on an A-size, 8½ × 11″ sheet of paper per the dimensions given. Lay out a cam with the following specifications:

1. Working circle 5.0 dia.
2. Shaft dia. .75 clockwise
3. Hub dia. 1.25
4. Square key .18
5. Rotation: counterclockwise
6. Follower dia. 1.0
7. Start 1.0 from center

Required motion:

1. Rise 90°, harmonic motion 1.00
2. Dwell 30°

3. Rise 90°, harmonic motion .50
4. Dwell 30°
5. Fall 90°, harmonic motion 1.50
6. Dwell 30°

Construct the cam and follower on an A-size, 8½ × 11″ sheet of paper per the dimensions given, and transfer all dimensions from the displacement diagram to the cam layout.

Check all work and add all required dimensions and specifications according to the most recent drafting standards.

PROBLEM 15-2A

PROBLEM 15-2B

PROBLEMS 15-3A AND 15-3B

Construct a cam displacement diagram on an A-size, 8½ × 11″ sheet of paper per the dimensions given.

Lay out a cam with the following specifications:

1. Working circle 5.0
2. Shaft dia. .63
3. Hub dia. 1.0
4. Square key to suit
5. Rotation: clockwise
6. Follower dia. .75
7. Start 2.5 from center

Required motion:

1. Fall 180°, uniform acceleration 1.50
2. Dwell 45°
3. Rise 90°, uniform acceleration 1.50
4. Dwell 45°

Construct the cam and follower on an A-size, 8½ × 11″ sheet of paper per the dimensions given, and transfer all dimensions from the displacement diagram to the cam layout.

Check all work and add all required dimensions and specifications according to the most recent drafting standards.

Mechanical PROBLEM 15-3A

Mechanical PROBLEM 15-3B

PROBLEMS 15-4A AND 15-4B

Construct a cam displacement diagram on an A-size, 8½ × 11″ sheet of paper per the dimensions given. Lay out a cam with the following specifications (metric):

1. Working circle 245
2. Shaft dia. 28
3. Hub dia. 44
4. Square key to suit
5. Rotation: clockwise
6. Follower dia. 22
7. Start 32 from center

Required motion:

1. Rise 60°, harmonic motion 25
2. Dwell 15°

3. Rise 90°, harmonic motion 38
4. Dwell 15°
5. Fall 45°, harmonic motion 44
6. Dwell 15°
7. Rise 30°, harmonic motion 18
8. Dwell 15°
9. Fall to starting level 75°, harmonic motion 38

Construct the cam and follower on an A-size, 8½ × 11″ sheet of paper per the dimensions given, and transfer all dimensions from the displacement diagram to the cam layout.

Check all work and add all required dimensions and specifications according to the most recent drafting standards.

Mechanical **PROBLEM 15-4A**

Mechanical **PROBLEM 15-4B**

PROBLEMS 15-5A AND 15-5B

Construct a cam displacement diagram on an A-size, 8½ × 11″ sheet of paper per the dimensions given.

Lay out a cam with the following specifications:

1. Working circle 7.0
2. Shaft dia. .625
3. Hub dia. 1.0
4. Square key to suit
5. Rotation: counterclockwise
6. Follower dia. 1.0
7. Start 1.0 from center

Required motion:

1. Rise 120°, modified uniform velocity 2.5
2. Dwell 60°

3. Fall 30°, modified uniform velocity .75
4. Dwell 30°
5. Fall to starting level 90°, modified uniform velocity 1.75
6. Dwell 30°

Construct the cam and follower on an A-size, 8½ × 11″ sheet of paper per the dimensions given, and transfer all dimensions from the displacement diagram to the cam layout.

Check all work and add all required dimensions and specifications according to the most recent drafting standards.

Mechanical **PROBLEM 15-5A**

Mechanical **PROBLEM 15-5B**

PROBLEMS 15-6A AND 15-6B

Construct a cam displacement diagram on an A-size, 8½ × 11″ sheet of paper per the dimensions given.

Lay out a cam with the following specifications:

1. Working circle 7.75
2. Shaft dia. .56
3. Hub dia. .88
4. Square key to suit
5. Rotation: clockwise
6. Follower dia. .375
7. Start 1.625 from center

Required motion:

1. Fall 90°, uniform acceleration 3.0
2. Dwell 15°
3. Rise 60°, uniform acceleration 2.0
4. Dwell 105°
5. Rise 90°, uniform acceleration to starting level (1.0)

Construct the cam and follower on an A-size, 8½ × 11″ sheet of paper per the dimensions given, and transfer all dimensions from the displacement diagram to the cam layout.

Check all work and add all required dimensions and specifications according to the most recent drafting standards.

Mechanical **PROBLEM 15-6A**

Mechanical **PROBLEM 15-6B**

PROBLEMS 15-7A AND 15-7B

Construct a metric cam displacement diagram per the dimensions given.

Lay out a cam with the following specifications:

1. Rise/fall 56
2. Shaft dia. 16
3. Hub dia. 32
4. Square key to suit
5. Rotation: clockwise

6. Follower: flat-faced* (34 dia.)
7. Base circle 26 radius
8. Harmonic motion/one full turn

Construct the cam and flat-faced follower per the dimensions given.

Check all work and add required dimensions and specifications according to the most recent drafting standards and using the metric system.

*Design the flat-faced follower to suit.

Mechanical   **PROBLEM 15-7A**

Mechanical   **PROBLEM 15-7B**

PROBLEMS 15-8A AND 15-8B

Construct a cam displacement diagram per the dimensions given.

Lay out an offset cam with the following specifications:

1. Offset circle dia. 2.00
2. Shaft dia. .75
3. Hub dia. 1.25
4. Square key to suit
5. Rotation: clockwise
6. Follower dia. .75
7. Follower offset 1.00
8. Base circle 1.375 radius

Required motion:

1. Rise 75°, uniform acceleration 1.00
2. Dwell 10°

3. Rise 35°, harmonic motion .75
4. Dwell 5°
5. Fall 70°, modified uniform velocity 1.25
6. Dwell 15°
7. Rise 45°, harmonic motion .75
8. Dwell 15°
9. Fall 60°, uniform acceleration 1.26
10. Dwell 30°

Construct the cam and offset follower per the dimensions given, and transfer all dimensions from the displacement diagram to the cam layout.

Check all work and add all required dimensions and specifications according to the most recent drafting standards.

Mechanical   **PROBLEM 15-8A**

Mechanical   **PROBLEM 15-8B**

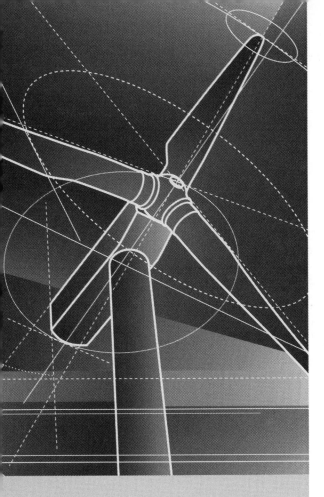

16

Gears

KEY TERMS

Addendum	Herringbone gear
Angle gear	Involute curve
Base circle	Miter gear
Bevel gear	Outside diameter
Center-to-center distance	Pinion gear
	Pitch diameter
Chain and sprockets	Rack gear
Chordal addendum	Ring gear
Chordal thickness	Root diameter
Circular pitch	Spiral bevel gear
Clearance	Spur gear
Dedendum	Train value
Diametral pitch	Whole depth
Gear tooth caliper	Working depth
Gear train	Worm
Helical gear	Worm gear

CHAPTER OUTLINE

Kinds of gears • Gear ratio • The involute curve • Pitch diameter and basic terminology • Pitch diameter (D) • Diametral pitch (P) • Gear blank • Pressure angle, base circle, and center-to-center distance • Rack • Gear train • Required tooth-cutting data • Measurements required to use a gear tooth caliper • Bevel gear • Worm and worm gear • Materials • Design and layout of gears • Summary • Review questions • Chapter 16 problems • CAD instructions

CHAPTER OBJECTIVES

Upon completion of this chapter, students should be able to do the following:

▶ Identify different kinds of gears.

▶ Discuss flexible drives such as chain and belt.

▶ Discuss terms related to the profiles and kinematics of gears.

▶ Determine the gear ratio between two gears.

▶ Calculate train values of gear trains.

▶ Construct an involute curve on a base circle.

▶ Calculate the center distance between two gears in contact.

▶ Prepare working drawings of gears.

Spur gears are the chief focus of this chapter; however, we plan to introduce you to various kinds of gears, gear nomenclature, technical drawing requirements for gears, the kinematics (study of motion without considering the forces) of gear trains, and related topics such as belt and chain drives. Brief consideration of forces on spur gear teeth is mentioned with respect to the pressure line between gears. Further inclusion of force topics appears in the problems at the end of the chapter.

Gear analysis and design is complex and beyond the scope of this text. For example, gear designers utilize equations that calculate stresses induced in gear teeth due to the forces transmitted and adverse operating conditions. Even more detailed analysis is required whenever wear on gear tooth surfaces is the determining design criterion.

Gears transfer rotary motion from one shaft to another shaft. Gears can change direction of rotation, speed up or slow down rotation, transmit rotational power (torque), and change rotary motion to straight-line motion. There are various kinds of gears, each with their own function (**Figure 16-1**).

Drafters must be able to prepare working drawings of the various gears using correct terminology.

Kinds of Gears

Gears are usually classified by the position or location of the shafts they connect. A spur gear, pinion gear, or helical gear is usually used to connect shafts that are parallel to each other. Intersecting shafts at 90° are usually connected by a beveled gear or angle gear. Shafts that are not parallel to each other and that do not intersect use a worm and worm gear. In order to connect rotary motion into a reciprocating or back-and-forth motion, a rack gear and pinion would be used. (Note that the rack gear and pinion are also known as a rack and pinion gear.)

Spur Gear

The *spur gear* is the most commonly used gear. It is cylindrical in form, with teeth that are cut straight across the face of the gear. All teeth are parallel to the axis of the shaft. The spur gear is usually considered the driven gear.

Pinion Gear

The *pinion gear* is exactly like a spur gear, but it is usually smaller and has fewer teeth. The pinion gear is normally considered the drive gear.

Rack Gear

The *rack gear* is a type of spur gear, but its teeth are in a straight line or flat instead of in a cylindrical form. The rack gear is used to transfer circular motion into straight-line motion.

Ring Gear

The *ring gear* is similar to the spur, pinion, and rack gears, except that the teeth are internal.

Bevel Gear

A bevel gear is another commonly used gear. A *bevel gear* is cone-shaped in form, with straight teeth that are on an angle to the axis of the shaft. Bevel gears are used to transmit power and motion between intersecting shafts that are at 90° to each other.

Angle Gear

The *angle gear* is similar to a bevel gear, except that the angles are at other than 90° to each other.

Miter Gear

The *miter gear* is exactly the same as a bevel gear, except that both mating gears have the same number of teeth. The shafts are at 90° to each other.

Spiral Bevel Ormiter Gears

Any bevel gears with curved teeth are called *spiral bevel gears*.

Worm Gear

Worm gears are used to transmit power and motion at a 90° angle between nonintersecting shafts. They are normally used to reduce speed. The worm gear is round like a wheel. The *worm* is shaped like a screw,

FIGURE 16-1   Various kinds of gears.

with threads (or teeth) wound around it. Because one full turn of the worm is required to advance the worm gear one tooth, a high-ratio speed reduction is achieved. The worm drives the larger worm gear.

Helical Gears

Helical gears operate on parallel shafts and transmit rotary motion in a manner identical to spur gears. The teeth have an involute shape but are not parallel to their rotational axis (**Figure 16-2**). Helical gear sets can transmit heavy loads at high speeds due to gradual engagement of the teeth and the smooth transfer of load from one tooth to another.

Herringbone Gears

Since helical gear teeth are cut at an angle to the axis of the gear, a force is generated along the axis. To compensate for this force, left-hand and right-hand teeth are combined (**Figure 16-3**). Large ships, for example, use *herringbone gears* in the drivetrain between the turbine and the propeller.

Chain and Sprockets

Chains and sprockets are used to transmit motion and power to shafts that are parallel to each other.

Sprockets are similar to spur and pinion gears, for the teeth are cut straight across the face of the sprocket and are parallel to the shaft. There are many types of chains and sprockets, but all use the same terms, formulas, ratios, and so forth.

Belt Drives

Belt drives, like the belts used on a typical automobile engine, are similar to chain-and-sprocket drives; however, slippage can occur between belts and pulleys. Belt drives are employed to start large masses such as ventilating fans so some slippage can occur if needed.

Gear Ratio

Friction wheels can be used to illustrate the basic kinetic relationship between a driver gear and a driven gear (**Figure 16-4**). Wheel A is on a shaft parallel to the shaft of wheel B, and it drives wheel B through friction at the point of contact. If wheel A rotates clockwise, then wheel B will rotate counterclockwise. Furthermore, if the circumference of wheel B is two times the circumference of wheel A, wheel A must make two complete revolutions to cause wheel B to make one (assuming no slippage).

The train value (TV) of this drivetrain is calculated by the equation:

$$TV = \frac{\text{driver circumference}}{\text{driven circumference}} = 0.5$$

For example, if the rotational speed of A is 30 RPM (revolutions per minute), then the rotational speed of B is calculated as follows:

$$B(RPM) = TV \times A(RPM) = 0.5 \times 30 = 15 \text{ RPM}$$

Assume that the drivetrain now is made up of gear A (20 teeth) rotating at 30 RPM and driving gear B (40 teeth). The train value (TV) is calculated by:

$$TV = \frac{\text{Number of Teeth on Driver}}{\text{Number of Teeth on Driven}} = \frac{20}{40} = 0.5$$

FIGURE 16-2   Helical gears.

FIGURE 16-3   Herringbone gears.

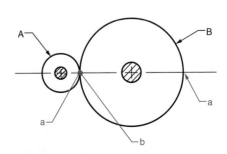

FIGURE 16-4   Relationship of two wheels of different diameters.

Then, the rotational speed of gear B = 0.5 × 30 = 15 RPM.

The Involute Curve

Spur gears use the involute curve as the profile of their gear teeth because of the involute's unique property of being able to transfer motion and forces in a uniform manner.

How to Construct an Involute Curve on a Base Circle

An *involute curve* may be considered to be the path traced by the end of a string as it is unwound from a circle called a *base circle*.

EXAMPLE:

GIVEN: Portion of a base circle m-n, which has a convenient radius of 3.2″ to fit on your paper (**Figure 16-5**).

STEP 1   Locate a point 6 to anchor a 3″ long theoretical string. Call the string 6-0, and label it 6, 5, 4, 3, 2, 1, and 0 at 0.5″ intervals (**Figure 16-5A**).

STEP 2   Beginning at point 1, construct a tangent to the 3.2″ curve (a base circle for the involute curve), lay out the distance 1-0 (0.5″) along the tangent from the point of tangency 1, and label it 1-0(1) (**Figure 16-5B**).

STEP 3   At point 2, construct a tangent to the base circle, and lay out a distance from the point of tangency 2 equal to two times the distance 1-0. Label it 2-0(2) (**Figure 16-5C**).

STEP 4   From points 3, 4, 5, and 6, lay off in a similar manner distances along the tangents, and label them 3-0(3), 4-0(4), 5-0(5), and 6-0(6), respectively (**Figure 16-5D**).

STEP 5   Finally, connect the points 0(1)–0(6) with a French curve to produce the involute curve (**Figure 16-5E**).

Pitch Diameter and Basic Terminology

The drafter must know and understand all major terms associated with various kinds of gears used today in the industry. Illustrated in **Figure 16-6** are the basic terms associated with most gears, regardless of type.

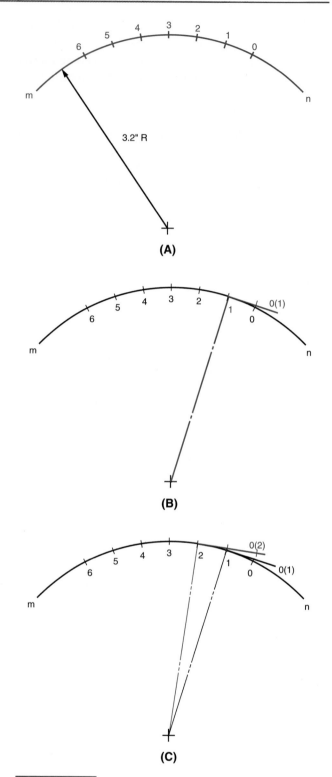

FIGURE 16-5   How to construct an involute curve on a base circle: (A) Step 1; (B) Step 2; (C) Step 3;

Pitch Diameter (D)

The *pitch diameters* (D) of two spur gears exist theoretically only when the two gears are in contact. When the point of contact between a tooth on gear A

(D)

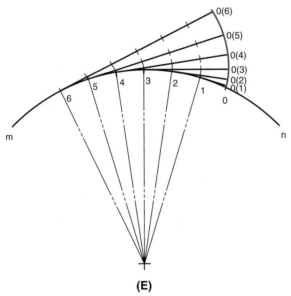

(E)

FIGURE 16-5   (*Continued*) (D) Step 4; (E) Step 5.

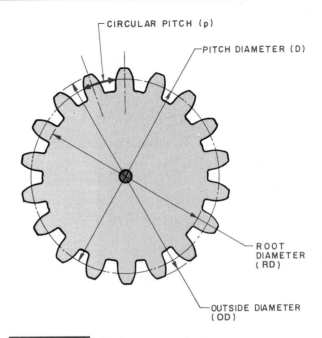

FIGURE 16-6   Basic gear terminology.

and a tooth on gear B falls on the line of centers between the two gears, then that point determines the radii of imaginary pitch circles of gear A and gear B, respectively (**Figure 16-7**).

Figure 16-8 illustrates additional terms that will help you to prepare working drawings of spur gears. Circular pitch (p) and the terms are explained in the paragraphs that follow.

Circular Pitch (p)

Circular pitch (p) is the distance between corresponding points of adjacent teeth measured on the circumference of the pitch diameter, Figure 16-8.

Diametral Pitch (P)

Diametral pitch (P) is a ratio equal to the number of teeth (N) per inch of pitch diameter (D). Follow the derivation that follows to better understand the term.

(p) $\times$ (N) = the circumference of the pitch circle

(D) $\times$ π = the circumference of the pitch circle

$$\text{so (p)} \times \text{(N)} = \text{(D)} \times \pi$$

$$\text{and (p)} \times \frac{\text{(N)}}{\text{(D)}} = \pi$$

$$\text{then (p)} \times \text{(P)} = \pi$$

EXAMPLE:

A gear with 48 teeth on a 3.0″ pitch diameter (D) would have a diametral pitch (P) of 16.

NOTE: All pairs of spur gears must have the same diametral pitch (P)!

Working Depth (HK)

The *working depth* is the distance that a tooth projects into the mating space. It is equal to gear addendum (A) plus pinion addendum (a).

Addendum (A)

The *addendum* is the radial distance from the pitch diameter to the top of the tooth. It is equal to 1/P.

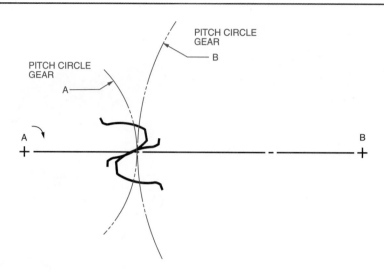

FIGURE 16-7   Pitch circles determined from point of contact on line of centers.

Dedendum (B)

The *dedendum* is the radial distance from the pitch diameter to the bottom of the tooth. It is equal to 1.157/P, 1.250/P, or 1.350/P; 1.250/P is preferred for 20° teeth.

Clearance (C)

Clearance is the space between the working depth and the whole depth. It is equal to dedendum (b) minus addendum (a).

Whole Depth (HT)

The *whole depth* is the total depth of a tooth space. It is equal to the addendum (a), plus the dedendum (b), plus the clearance (c).

NOTE: All notations in parentheses ( ) correspond to those used in the most recent edition of the Machinery's Handbook so that they agree if a more detailed formula is required by the drafter.

Outside Diameter (OD)

The *outside diameter* is the measurement over the extreme outer edge of teeth. It is equal to the pitch diameter plus 2 times the addendum (a).

Root Diameter (RD)

The *root diameter* is the measurement over the extreme inner edges of teeth. It is equal to the pitch diameter minus 2 times the dedendum (b).

How to Draw a Spur Gear Tooth

Draw a spur gear tooth for the experience and for learning spur gear nomenclature. Later you will discover that working drawings of gears and gear sets seldom require drawing the tooth profile.

GIVEN: Spur gear specifications of D = 6.0″, P = 4, T = 24, and ϕ = pressure angle of 20°.

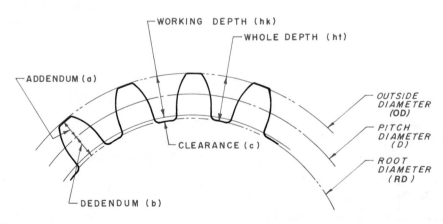

FIGURE 16-8   Major gear terms.

STEP 1   Construct an arc representing the basic circle (**Figure 16-9A**). The base circle radius = D cos φ/2 = 6.0 × cos 20°/2 = 2.819″.

STEP 2   At an arbitrary point B on the base circle, draw an involute curve one of several ways. Your choices are the basic method shown in Figure 16-5, the approximate method (D/8 = approximate involute curve), or the simplest method using a template. A fourth choice would be to use a CAD program. For this example, use the approximate method (**Figure 16-9B**). Then, D/8 = 6.0″/8 = 0.75″.

STEP 3   Draw an arc representing the pitch circle, D/2 = 6.0″/2 = 3.0″, and label point T where the pitch circle intersects the approximate involute curve (**Figure 16-9C**).

STEP 4   Locate the tooth width along the pitch circle (**Figure 16-9D**), and label the width T-S. The width T-S equals half the circular pitch p. The circular pitch p = π/P = π/4 = 0.785″. The tooth width then is 0.785″/2 = 0.392″. Construct an approximate involute curve from the base circle and going through point S.

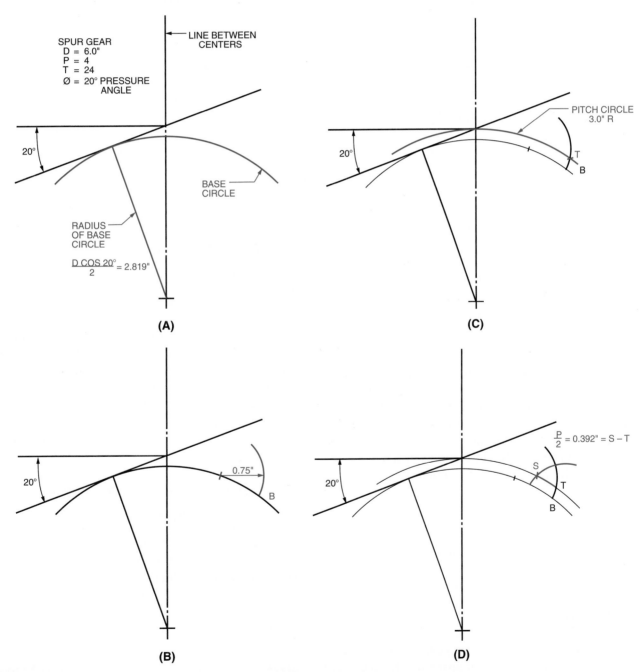

FIGURE 16-9   How to draw a spur gear tooth: (A) Step 1; (B) Step 2; (C) Step 3; (D) Step 4;

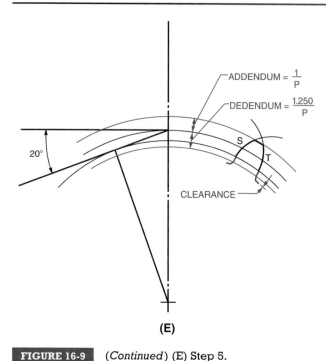

(E)

FIGURE 16-9   *(Continued)* (E) Step 5.

STEP 5   Finish the tooth profile by constructing the addendum and dedendum circles plus two radial lines (**Figure 16-9E**). Add the addendum $1/P = 1/4 = 0.25''$ to the pitch circle arc, and subtract the dedendum $1.250/P = 0.312''$ from the pitch circle arc. Finally, from where the base circle intersects the tooth profile, draw radial segments from the base circle to the dedendum arc. To provide a small amount of clearance, Figure 16-8, add small fillets.

Gear Template

A quick and efficient method of illustrating teeth is by using a gear template. When using a gear template, the drafter must first calculate the diametral pitch (P). The numbers next to each tooth indicate the diametral pitch. The lower part of the template opening is used to draw spur gear teeth. The top portion of the opening is used to draw teeth of a rack or worm.

Metric Spur Gears

Metric spur gears and inch spur gears are essentially identical. For a comparison of an inch gear with a metric one, a gear with a 20° pressure angle, 24 teeth, and a diametral pitch of P = 4 is shown in **Figure 16-10**.

Gear Blank

Gears are usually cut from a gear blank. The gear blank must allow sufficient space for the gears and a method to attach the gear to the shaft. Illustrated

SPUR GEAR TEETH Comparison

	Inch 20° 24T P = 4	Metric 20° 24T P = 4
Pitch Diameter	6.0"	152.4 mm
Circular Pitch	π/P = 0.785"	19.95 mm
Base Circle Diameter	D cos 20° = 5.638"	143.2 mm
Module (MDL) = (length of pitch diameter per tooth)		pitch dia/24 = 152.4/24 = 6.35 mm
Addendum	1/P = 0.25"	MDL = 6.35 mm
Dedendum	1.157/P = 0.289" or 1.250/P = 0.312	1.157 x MDL = 7.35 mm 1.25 x MDL = 7.94 mm

FIGURE 16-10   Comparison of metric and inch spur gears.

FIGURE 16-11   Gear blank.

in **Figure 16-11** is a common gear blank with an outside diameter and face width for the gears to be cut into, a hub, and a hole for the set screw to secure the gear blank to the shaft. A half-section is used to illustrate the gear blank in this example. Two complete sets of dimensions must be applied to the gear drawing: those relating to the gear blank, as illustrated, and those relating to the teeth. Gear dimensions do not hold tight tolerancing, but all gear tooth dimensions must hold very tight tolerancing.

Pressure Angle, Base Circle, and Center-to-Center Distance

The pressure angle determines the diameters of the base circles, which in turn determine the tooth forms. To illustrate this principle, the steps to prepare a drawing of two spur teeth in contact will be discussed.

How to Draw Two Spur Gear Teeth in Contact

GIVEN: A 20-tooth pinion gear to drive a 40-tooth gear, P = 5. The pressure angle ϕ is to be 20°.

STEP 1 Start the drawing by laying out the center-to-center distance (**Figure 16-12A**). The *center-to-center distance* equals the sum of the radius of the pinion plus the radius of the gear.

D (pinion) = N/P = 20/5 = 4″

R (pinion) = 4/2 = 2″

D (gear) = N/P = 40/5 = 8″

R (gear) = 8/2 = 4″

C-to-C   = 2 + 4 = 6″

Label the juncture of the two radii, T. Construct a perpendicular to the line between centers at point T.

STEP 2 Lay out the pressure line at 20° to the perpendicular at T, assuming that the pinion is rotating clockwise and driving the gear (**Figure 16-12B**). Next, construct perpendiculars from the pinion and gear centers to the pressure line.

STEP 3 Use the perpendiculars that represent the radii of the respective base circles to draw the base circles (**Figure 16-12C**).

NOTE: *The lengths of the radii of the base circles equal the radii of the pitch circles times cos 20°.*

STEP 4 To save time, use the approximate method (D/8, Figure 16-9) to draw the tooth profiles from each base circle, the profiles to go through point T (**Figure 16-12D**). Finish the two teeth and note that the base of the tooth on the gear is wider than the base on the pinion.

STEP 5 Assume that the pinion rotates approximately one tooth width clockwise (p/2), and construct a new pinion tooth (**Figure 16-12E**). Draw a gear tooth in contact and observe that the two teeth are in contact on the pressure line. The true involute profile ensures that the teeth will always be in contact along the pressure line—hence the smooth transfer of rotation and power.

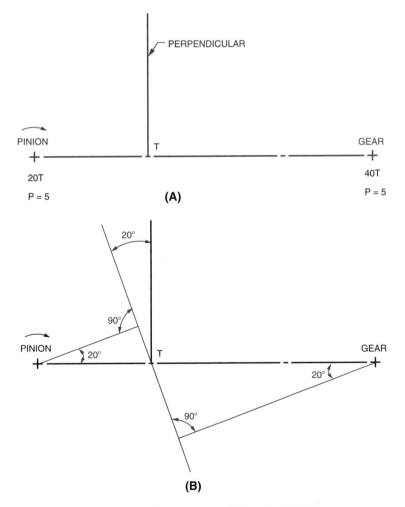

FIGURE 16-12   How to draw two spur gear teeth in contact: (A) Step 1; (B) Step 2;

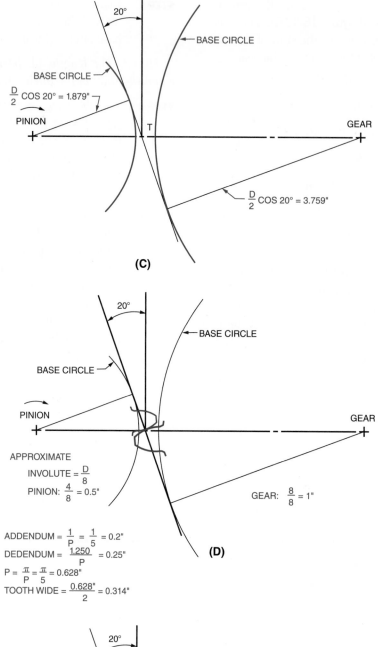

20°

BASE CIRCLE

BASE CIRCLE

$\frac{D}{2}$ COS 20° = 1.879"

PINION   T   GEAR

$\frac{D}{2}$ COS 20° = 3.759"

(C)

20°

BASE CIRCLE

BASE CIRCLE

PINION   GEAR

APPROXIMATE

INVOLUTE = $\frac{D}{8}$

PINION: $\frac{4}{8}$ = 0.5"   GEAR: $\frac{8}{8}$ = 1"

ADDENDUM = $\frac{1}{P}$ = $\frac{1}{5}$ = 0.2"

DEDENDUM = $\frac{1.250}{P}$ = 0.25"   **(D)**

P = $\frac{\pi}{P}$ = $\frac{\pi}{5}$ = 0.628"

TOOTH WIDE = $\frac{0.628"}{2}$ = 0.314"

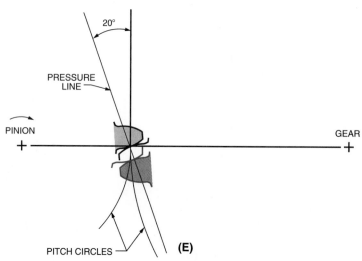

20°

PRESSURE
LINE

PINION   GEAR

PITCH CIRCLES   **(E)**

FIGURE 16-12   (*Continued*) (C) Step 3; (D) Step 4; (E) Step 5.

FIGURE 16-13 Gear terminology for a rack.

Rack

A rack is simply a gear with teeth formed on a flat surface (**Figure 16-13**). A rack changes rotary motion into reciprocating motion. All terms and formulas associated with spur and pinion gears apply to the rack. The sides of the teeth are straight, not involute as on a spur gear. The teeth are inclined at the same angle as the pressure angle of the mating pinion gear. Note that the circular pitch of the pinion gear is the same as the linear pitch of the rack. All dimensions for heights of depth, pitch, and addendum lines are calculated from a datum or reference line. The rack is usually manufactured out of rectangular stock but, occasionally, it is manufactured out of round stock to meet a specific design.

The required tooth-cutting data are for a rack:

- Pressure angle (same as mating pinion gear)
- Tooth thickness at pitch line (same as mating pinion gear)
- Whole depth (same as mating pinion gear)
- Maximum allowance pitch variation
- Accumulated pitch error max (over total length)

Gear Train

A *gear train* is two or more gears used to achieve a designed RPM. The ratio of the RPM of the last gear to the RPM of the first gear is called the *train value* of the gear train. Calculate the train value as follows:

$$\text{Train Value (TV)} = \frac{\text{product of number of teeth on driving gears}}{\text{product of number of teeth on driven gears}}$$

and

$$\text{RPM (last gear)} = \text{TV} \times \text{RPM first gear}$$

EXAMPLE:

Assume that a 150 RPM gear motor with a 15-tooth pinion drives a 40-tooth gear mounted on a conveyor driveshaft. The speed of the driveshaft would be:

$$\text{RPM (driveshaft)} = \frac{\substack{15 \text{ teeth} \\ \text{on driver}}}{\substack{40 \text{ teeth} \\ \text{on driven}}} \times \text{RPM (gear motor)}$$

$$\text{RPM (driveshaft)} = 0.375 \times 150 = 56.25 \text{ RPM}$$

Consider a more complicated gear train (**Figure 16-14A**):

- Shafts 1 and 2 are connected by pinion A and gear B.
- Shafts 2 and 3 are connected by pinion C and gear D.
- Shafts 3 and 4 are connected by pinion E and gear F.
- Pinion A has 12 teeth.
- Gear B has 45 teeth.
- Pinion C has 20 teeth.
- Gear D has 40 teeth.
- Pinion E has 15 teeth.
- Gear F has 60 teeth.

PROBLEM:

For an input speed of 200 RPM at shaft 1, what is the output speed of shaft 4 in RPM?

$$\text{Train Value (TV)} = \frac{12 \times 20 \times 15}{45 \times 40 \times 60} = 0.033$$

$$\text{Output speed shaft 4} = 0.033 \times 200 = 6.67 \text{ RPM}$$

Observe the use of arrows in Figure 16-14A to show the direction of gear rotation. Gear A is rotating clockwise; therefore, gear B is rotating counterclockwise.

FIGURE 16-14A A more complicated gear train.

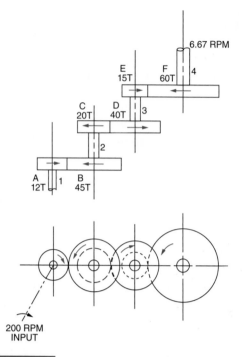

FIGURE 16-14B Optional methods of drawing gear trains.

REQUIRED CUTTING DATA			
ITEM	**TO FIND:**	**HAVING**	**FORMULA**
1	Number of teeth (N)	D & P	D x P
		DO & P	(DO x P) − 2
2	Diametral pitch (P)	p	$\dfrac{3.1416}{p}$
		D & N	$\dfrac{N}{D}$
		DO & N	$\dfrac{N+2}{DO}$
3	Pressure angle (ø)	−	20° STANDARD 14°-30' OLD STANDARD
4	Pitch diameter (D)	N & P	$\dfrac{N}{P}$
		N & DO	$\dfrac{N \times DO}{N+2}$
		DO & P	$DO - \dfrac{2}{P}$
5	Whole depth (ht)	a & b	a + b
		P	$\dfrac{2.157}{P}$
6	Outside diameter (OD)	N & P	$\dfrac{N+2}{P}$
		D & P	$D + \dfrac{2}{P}$
		D & N	$\dfrac{(N+2) \times D}{N}$
7	Addendum (a)	P	$\dfrac{1}{P}$
8	Working depth (hk)	a	2 x a
		P	$\dfrac{2}{P}$
9	Circular thickness (t)	P	$\dfrac{3.1416}{2 \times P}$
10	Chordal thickness	N, D & a	$\sin\left(\dfrac{90°}{N}\right) \times D$
11	Chordal addendum	N, D & a	$\left[1-\cos\left(\dfrac{90°}{N}\right)\right] \times \dfrac{D}{2} + a$
12	Dedendum (b)	P	$\dfrac{1.157}{P}$

FIGURE 16-15 Required tooth-cutting data for spur and pinion gears.

Occasionally, gear trains are drawn with their centers on a horizontal line (**Figure 16-14B**). The upper view clarifies the relationships of the gears and the directions of rotation.

The equation for calculating the train value of a spur gear train applies to other gear trains such as bevel, worm, and helical, plus combinations of all of them.

Required Tooth-Cutting Data

Overall gear blank dimensions are shown on the detail drawing (as illustrated in Figure 16-11) and can hold loose tolerancing. The actual gear dimensions must hold tight tolerancing.

Twelve essential items of information must be calculated and listed on the gear detail drawing. This list is usually located at the lower right-hand side of the drawing and generally consists of the data in **Figure 16-15**, but these items may vary slightly from company to company. The required material, heat-treating process, and other important information must also be included, usually in the title block.

In actual practice, the gear teeth are not drawn because it takes too much drawing time. A simplified method is used to illustrate the actual gear tooth, similar to that used to illustrate threads of a fastener (**Figure 16-16**). The outside diameter and the root diameter are illustrated by a thin phantom line. The

pitch diameter is illustrated by a thin center line. Actual teeth are not drawn, except to illustrate some special feature in relationship to a tooth, a spline, a keyway, or a locating point. In this example, all dimensions in color are related to the gear blank only. All gear dimensions are called off in the cutting data box underneath.

Recent practice of spur gear design favors 20° and 25° pressure angles; however, 14.5° full depth and 20° stub (shorter addendum) are not recommended for new designs according to the *Machinery's Handbook*.

Refer to the *Machinery's Handbook* for more extensive information on gearing and gearing standards, including ANSI B6.1-1968, R1974, and ANSI B6.7-1977.

CUTTING DATA	
NUMBER OF TEETH	24
DIAMETRAL PITCH	4
PRESSURE ANGLE	20°
PITCH DIAMETER	6.000
WHOLE DEPTH	0.5393
OUTSIDE DIAMETER	6.500
ADDENDUM	.250
WORKING DEPTH	.500
CIRCULAR THICKNESS	.3925
CHORDAL THICKNESS	.2566
CHORDAL ADDENDUM	.3924
DEDENDUM	.289

FIGURE 16-16 Example of a spur gear detail drawing.

Measurements Required to Use a Gear Tooth Caliper

A *gear tooth caliper* is used to check and measure an individual gear tooth. Gear tooth calipers measure two parts of the tooth profile: the chordal thickness and the chordal addendum (**Figure 16-17**). The center slide is set to the desired chordal addendum using the vertical vernier and scale. The tips of the two jaws are then used to measure the chordal thickness read on the horizontal vernier and scale.

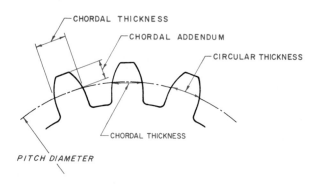

FIGURE 16-17 Measurements required to use a gear tooth caliper.

Chordal Thickness

The *chordal thickness* is the length of the chord measured straight across at the pitch diameter. It is measured straight across the tooth, not along the pitch diameter as is the circular thickness.

$$\text{Chordal thickness} = \sin\left(\frac{90°}{N}\right) \times \text{Pitch dia.}$$

Chordal Addendum

The *chordal addendum* is the distance from the top of the tooth to the point at which the chordal thickness is measured.

$$\text{Chordal addendum} = \left[1 - \cos\left(\frac{90°}{N}\right)\right] \times \frac{\text{Pitch dia.}}{2} + \text{addendum}$$

Bevel Gear

Bevel gears transmit power and motion between intersecting shafts at right angles to each other. **Figure 16-18** illustrates the various terms associated with bevel gears. The required tooth-cutting data for bevel gears are similar to spur/pinion gears. These must be calculated and listed on the detail drawing. **Figure 16-19** shows the order in which the data and formulas should be listed on the drawing. Bevel gears must be designed and drawn in pairs to ensure a perfect fit.

FIGURE 16-18 Gear terminology for bevel gears.

REQUIRED CUTTING DATA				
ITEM	TO FIND:	HAVING	FORMULA	
			SPUR	PINION
1	Number of teeth (N)	–	AS REQ'D.	
2	Diametral pitch (P)	p	$\dfrac{3.1416}{p}$	
3	Pressure angle (ø)	–	20° STANDARD 14°-30' OLD STANDARD	
4	Cone distance (A)	D & d	$\sin d \sqrt{\dfrac{D}{2}}$	
5	Pitch distance (D)	p	$\dfrac{N}{p}$	
6	Circular thickness (t)	p	$\dfrac{1.5708}{p}$	
7	Pitch angle (d)	N & d (of pinion)	90°–d(pinion)	$\tan d \dfrac{N \text{ pinion}}{N \text{ gear}}$
8	Root angle (γR)	d & δ	$d - \delta$	
9	Addendum (a)	p	$\dfrac{1}{p}$	
10	Whole depth (ht)	p	$\dfrac{2.188}{p} + .002$	
11	Chordal thickness (C)	D & d	$\dfrac{1}{2}\left(\dfrac{D}{\cos d}\right)\left[1 - \cos\left(\dfrac{90°}{\frac{N}{\cos d}}\right)\right] + a$	
12	Chordal addendum (aC)	d	$\sin\left(\dfrac{90°}{\frac{N}{\cos d}}\right)$	
13	Dedendum (bC)	P	$\dfrac{2.188}{P} - a(\text{pinion})$	$\dfrac{2.188}{P} - a(\text{gear})$
14	Outside diameter (OD)	D, a & d	D + (2 x a) x cos d	
15	Face	A	1/3 A (max.)	
16	Circular pitch (p)	p & N	$\dfrac{3.1416 \times p}{N}$	
17	Ratio	N gear & N pinion	$\dfrac{N \text{ gear}}{N \text{ pinion}}$	
18	Back angle (γO)	–	SAME AS PITCH ANGLE	
19	Angle of shafts	–	90°	
20	Part number of mating gear	–	AS REQ'D.	
21	Dedendum angle δ	A & b	$\dfrac{b}{A} = \tan \delta$	

FIGURE 16-19   Required tooth-cutting data for bevel gears.

Worm and Worm Gear

The worm and worm gear are used to transmit power between nonintersecting shafts at right angles to each other. With the worm and worm gear, a large speed ratio is possible because one revolution of a single-thread worm turns the worm gear only one tooth and one space.

The worm's thread is similar in form to a rack tooth. The worm gear is similar in form to a spur gear, except that the teeth are twisted slightly and curved to fit the curvature of the worm.

When drawing the worm and worm gear, an approximate representation is used (**Figure 16-20**). Cutting data for the worm and worm gear must be listed on the drawing in the lower right side in the order illustrated in **Figure 16-21** for the worm gear and in **Figure 16-22** for the worm.

NOTE: It is important that the mounting of a worm and gear set ensures that the central plane of the gear passes essentially through the axis of the worm. This may be accomplished by adjusting the gear axially at assembly by means of shims. When properly mounted and lubricated, worm gear sets will become more efficient after the initial breaking-in period.

Other information that must be on a worm/worm gear detail drawing includes:

- Gear blank information
- Tooth-cutting data
- Reference to mating part

Materials

Gears are made of many materials, such as brass, cast iron, steel, and plastic, to mention but a few. Many bevel gears are forged, some are stamped from thin material, some are cast as a blank and machined, and others are die-cast to the exact size and shape.

Each application must be carefully analyzed. Metal gears have been used for years, but if the load is not excessive, plastic gears have many advantages. Plastic runs quieter, has a self-lubricating effect, weighs much less, and costs much less. In addition, complicated multiple gears can be molded into a single piece, which further reduces costs.

Design and Layout of Gears

The initial design of gears starts with the nominal size of the pitch diameters. The required speed ratio, loading, space limitations, and center-to-center distances are also important factors to consider. Mating gears and pinions must have equal diametral pitch in order to correctly mesh; therefore, the diametral pitch should be one of the first considerations. A complete analysis and design of a gear or a complete

FIGURE 16-20   Examples of worm and worm gear data.

REQUIRED CUTTING DATA			
ITEM	**TO FIND:**	**HAVING**	**FORMULA**
1	Number of teeth (N)	–	AS REQ'D.
2	Pitch diameter (D)	N & p	$\dfrac{N \times p}{3.1416}$
3	Addendum (a)	p	p x .3181
		P	$\dfrac{1}{P}$
4	Whole depth (ht)	p	p x .6866
		P	$\dfrac{2.157}{P}$
5	Lead (L) Right-Left	p & N	p x N
6	Worm part no.	–	AS REQ'D.
7	Pressure angle ø	–	20° STANDARD 14°-30' OLD STANDARD
8	Outside diameter (OD)	Dt & Pa	Dt + .4775 x Pa
9	*Circular pitch (p)	P	$\dfrac{3.1416}{P}$
		L & N	$\dfrac{L}{N}$
10	Diametral pitch (P)	p	$\dfrac{3.1416}{p}$
11	Throat diameter (Dt)	D & Pa	D + .636 x Pa
12	Ratio of worm/ worm gear	N worm & N worm gear	$\dfrac{N \text{ worm gear}}{N \text{ gear}}$
13	Center-to-center distance between worm & worm gear	D worm & D worm gear	$\dfrac{D \text{ worm} + D \text{ worm gear}}{2}$

*Circular pitch (p) must be same as worm axial pitch (Pa)

FIGURE 16-21   Required tooth-cutting data for a worm gear.

REQUIRED CUTTING DATA			
ITEM	**TO FIND:**	**HAVING**	**FORMULA**
1	Number of teeth (N)	P	$\dfrac{3.1416}{P}$
2	Pitch diameter (D)	Pa	(2.4 x Pa) + 1.1
		DO & a	DO – (2 x a)
3	*Axial pitch (Pa)	–	Distance from a point on one tooth to same point on next tooth
4	Lead (L) Right or Left	p & N	p x N
5	Lead angle (La)	L & D	$\dfrac{L}{3.1416 \times D} = \tan La$
6	Pressure angle (ø)	–	20° STANDARD 14°-30' OLD STANDARD
7	Addendum (a)	p	p x .3183
		P	$\dfrac{1}{P}$
8	Whole depth (ht)	Pa	.686 x Pa
9	Chordal thickness	N, D & a	$\left[1 - \cos\left(\dfrac{90°}{N}\right)\right] \times \dfrac{D}{2} + a$
10	Chordal addendum	N, D & a	$\sin\left(\dfrac{90°}{N}\right) \times D$
11	Outside diameter (OD)	D & a	D + (2 x a)
12	Worm gear part no.	–	AS REQ'D.

*Axial pitch (Pa) must be same as worm gear circular pitch (p)

FIGURE 16-22   Required tooth-cutting data for worm.

INDUSTRY APPLICATION

GEARING UP FOR A JOB

The want ad read in part as follows: ***"Opening for entry-level mechanical drafting technician. Must have training in laying out and drawing gears."*** It was the sentence about gears that gave Peter Walton a problem. He had been an excellent student at Florida Community College (FCC). His field of study was mechanical drafting, and he had graduated with honors. But gears? Walton didn't remember studying gears. His program at FCC had focused on tool design. Walton knew a great deal about jigs and fixtures, but next to nothing about gears.

But Walton needed a drafting job and, at least for the time being, could not relocate. His mother was in poor health, and his father had died when Walton was a young child. Consequently, he needed to stay at home and look after his mother. There weren't many drafting jobs available in Walton's little town of Long Bend, Florida. Since graduating from FCC, he had been working at a convenience store. Walton had earned a degree in drafting, and he wanted to put it to work.

Knowing that he might not get another chance without relocating, Walton made up his mind. "I am going to get this job." Walton called and made an appointment for an interview. Regarding the gear issue, he told the company personnel manager, "I will bring several gear drawings that will verify my ability to lay out various types of gears." To himself he said, "I may not be an expert on gears now, but I will be when I interview for the job!" Walton had just one week to learn about gear design and to prepare several drawings. He didn't waste any time. Walton drove to FCC and met with his former drafting professor, who quickly pulled several books off the shelf and showed him what he would have to learn. Walton began his crash course immediately.

He read about spur, pinion, rack, ring, bevel, miter, and worm gears. He learned about gear ratio, pressure angle, and diametral pitch. He learned how to calculate the value of the gear train. Walton studied until his head hurt and his vision was blurry; then he studied some more. After three days, he was ready to start drawing. After two days of drawing, he had a comprehensive portfolio that included a spur gear, two bevel gears, and a worm gear.

Two days later, the interview went well, and Walton got the job. In fact, the company officials were so impressed with Walton's knowledge of gears that they increased the starting salary for the job!

gear chain are very complex and far beyond the scope of this text. Most designers try to use standard gears from a company that specializes in gears, gear chains, and gear design. Most gear manufacturing companies can be of assistance in designing gears for special applications. These companies can usually manufacture gears at a lower price than if they were made in-house.

Further analysis, in-depth study, and design data can be found in the most recent edition of *Machinery's Handbook*.

SUMMARY

- Gears are classified by the position or location of the shafts they connect. The most widely used types of gears are spur, pinion, rack, ring, bevel, angle, miter, worm, helical, and herringbone.
- Diametral pitch is a ratio equal to the number of teeth (N) per inch of pitch diameter (D). Circular

pitch is the distance between corresponding points of adjacent teeth measured on the circumference of the pitch diameter.

- A gear blank is the material from which the gear is cut. The gear blank must allow sufficient space for the gears and a method to attach the gear to the shaft.
- A rack is a gear with teeth formed on a flat surface. A rack changes rotary motion into reciprocating motion.
- A gear train is two or more gears used to achieve a designed RPM.
- A tooth gear caliper is used to check and measure an individual gear tooth.
- Bevel gears transmit power and motion between intersecting shafts at right angles to each other.
- The worm and worm gear are used to transmit power between nonintersecting shafts at right angles to each other.

• Gears are made of many materials such as brass, cast iron, steel, and plastic. Gears may be machined, forged, stamped, or cast.

REVIEW QUESTIONS

Mark the following statements as either true or false.

1. The outside diameter of a spur gear having a pitch diameter of 1.500 and 48 teeth is 1.68.

2. The pitch diameter of a spur gear having 40 teeth and a diametral pitch of 20 is 19.68.

3. The circular pitch of a bevel pinion with a 16 diametral pitch and 20 teeth is 1.43.

4. A ratio of 2:1 between gear A and gear B means that gear B rotates two times every time gear A rotates once.

5. The pitch diameter of a gear can be thought of as the outside diameter of the friction wheel.

6. The root diameter of a spur gear having a pitch diameter of 2.000 and 32 teeth is 1.75.

Answer the following questions by selecting the best answer.

1. Diametral pitch can be defined as:
   a. Diameter of the gear blank divided by two
   b. A ratio equal to the number of teeth on a gear per inch of pitch diameter
   c. The chordal thickness divided by the pressure angle
   d. Number of teeth divided by the dedendum

2. What is the dedendum of a spur gear having a pitch diameter of 3.000 and 48 teeth?
   a. 0.3857
   b. 0.2692
   c. 1.4617
   d. 0.3889

3. How many teeth are on a spur gear having a pitch diameter of 1.750, diametral pitch of 20, and outside diameter of 1.850?
   a. 3.24
   b. 37
   c. 16
   d. 35

4. What is the diametral pitch of a pinion gear having 75 teeth and 3.208 outside diameter? (Use the formula $P = N/D$.)
   a. 19.68
   b. 23.38
   c. 24.83
   d. 23.21

5. When cutting the teeth for bevel gears, how does one compute diametral pitch (P)?
   a. .5 (2.188) + sin d
   b. 1 ÷ p
   c. 3.1416 × sin d
   d. 3.1416 ÷ p

6. When cutting the teeth for a worm gear, how does one compute the pitch diameter?
   a. D + (2 × a)
   b. .686 × Pa
   c. P × N
   d. OD − (2 × a)

7. A gear train can best be described as:
   a. A series of gears linked together
   b. Two or more gears used to achieve a designed RPM
   c. A power transmission device for transferring motion
   d. A series of gear pinions

8. How does one compute the center-to-center distance between a worm and worm gear?
   a. P × .3181
   b. D + .636 × Pa
   c. Nworm gear ÷ Dworm
   d. Dworm + Dworm gear 2

CHAPTER 16 PROBLEMS

The following problems are intended to give the beginning drafter practice in using the many formulas associated with gears and practice in laying out finished professional detail drawings of major kinds of gears.

Conventional Instructions

The steps to follow in laying out gears are:

STEP 1   Study the problem carefully.

STEP 2   Make a sketch if necessary.

STEP 3   Do all math required for each problem. Keep all math work for rechecking.

STEP 4   Center the required views within the work area.

STEP 5   Include all dimensioning according to the most recent drafting standards.

STEP 6   Add all required gear-cutting specifications in the lower right side of the paper.

STEP 7   Recheck all work and, if correct, neatly fill out the title block using light guidelines and neat lettering.

CAD INSTRUCTIONS

Students who plan to complete the following problems on a CAD system should begin with these general instructions.

1. Analyze the problem carefully.
2. Plan your drawing sequence. In what order will you develop the drawing?
3. Read the specific instructions for the problem and begin.

PROBLEM 16-1

Use the following specifications to lay out a single-view drawing of a 2:1 ratio spur gear and pinion gear in mesh.

Spur gear:	Pinion gear:
88 teeth	hub diameter 1.625
pressure angle 20°	bore of 1.000 diameter
pitch diameter 5.500	pressure angle 20°
hub diameter 2.625	
bore of 2.000 diameter	

Use all standard drafting methods to illustrate the outside diameter, pitch diameter, and root diameter. Calculate and add the center-to-center distance between the shafts.

PROBLEM 16-2

Draw a spur gear having the following specifications: 56 teeth, pressure angle 20°, pitch diameter 3.500, hub diameter 1.00, gear blank with overall size (width) 1.00, face size .50, bore .5625; use a 1/8″ set screw.

Complete two views using the half-section method of representing gears. Do not show the gear teeth; use the conventional method of illustrating teeth. Dimension per all standard practices. Calculate and add all standard required cutting data to the lower right side of the work area. Material: S.A.E. #1320 cast steel, heat treatment: carburize .015 − .020 deep.

PROBLEM 16-3

Complete a two-view detail drawing of a pinion gear having the following specifications (calculate and add all standard cutting data): 30 teeth, pressure angle 20°, pitch diameter 6.000, hub diameter 2.000, heat treatment: carburize .050 deep, gear blank with overall size (width) 3.500, face size 1.250, bore 1.000, material: S.A.E. #4620 steel.

PROBLEM 16-4

Make a detail drawing, completely dimensioned, of a spur gear with the following specifications: diametral pitch 16, pressure angle 20°, pitch diameter 5.500, whole depth .1348, outside diameter 5.625, addendum .0625, working depth .125, circular thickness .098, chordal thickness .0633, chordal addendum .0985, whole depth .1348, dedendum .0723, material: cast brass.

Design a simple gear blank with a set screw to fasten the gear to a .75 dia. shaft.

PROBLEM 16-5

Design and draw a detail drawing, completely dimensioned, of a pair of bevel gears having teeth of a 4.0 diametral pitch, the gear with 25 teeth, the pinion gear with 13 teeth, face width of 1.00, gear shaft 1.25 diameter, and pinion shaft .875 diameter. (Calculate for FN-2 fits with the shaft.)

Design the hub diameter to be approximately twice the shaft diameter. Backing for the gear 1.375 and for the pinion .75. Design and dimension the remaining portion to suit using standard drafting practices. Add all cutting data to the lower right side of the work area. Material: S.A.E. #3120 cast steel, heat treatment: carburize .015/.020 deep.

PROBLEM 16-6

Make a design layout of a worm and worm gear with the following specifications: shaft diameter 1.25, single-thread worm, lead .75, worm gear with 28 teeth. Add all dimensions and cutting data as required.

17

Assembly and Detail Drawings for Design

KEY TERMS

Application engineering	ECR (engineering change request)
Assembly drawing	Invention agreement
Conceptual design	Mass production company
Contract	Parts list
Design development layout drawings	Patent drawings
Design procedure	Proposal
Design section	Purchased part
Detail drawings	Quotation department
DFA (design for assembly) procedure	Scientific design
DFM (design for manufacturing) procedure	Selling price
	Sketches
ECO (engineering change order)	Subassembly drawing
	Title block

CHAPTER OUTLINE

Technical drawings and the engineering department • The engineering department • Systems company • Design section • Working drawings: detail and assembly • Title block • Numbering system • Drawing revisions: ECRs and ECOs • Application engineering • Quotation department • Mass production company • Design for manufacturability (DFM) • DFM guidelines • Design for assembly (DFA) • DFA guidelines • Personal technical file • Invention agreement • Patent drawings • Summary • Review questions • Chapter 17 problems • CAD instructions

CHAPTER OBJECTIVES

Upon completion of this chapter, students should be able to do the following:

▶ Discuss the use of sketches, layout drawings, detail drawings, subassembly drawings, and final assembly drawings.

▶ Summarize the contents of a parts lists or a bill of materials, including purchased parts.

▶ Prepare sketches, layout drawings, detail drawings, subassembly drawings, and final assembly drawings.

▶ Describe a typical engineering department and the functions of a design section, an application department, and a quotation department.

(continued)

(CHAPTER OBJECTIVES continued)

▶ Describe the design procedure used in the industry and identify your own potential role in the procedure.

▶ Compare the similarities and differences between a proposal and a contract.

▶ Process ECRs and ECOs.

▶ Describe how a selling price is established for a product.

▶ Start your own technical file in a systematic manner.

▶ Explain DFM and DFA.

▶ Recognize invention agreements and patent drawings.

Technical Drawings and the Engineering Department

A representative engineering department and a flow-chart showing lines of communication (**Figure 17-1**), provide a framework for discussing technical drawings and related engineering practices. Technical drawings include sketches and layout, detail, and assembly drawings. Related engineering practices encompass the following: design procedure, parts lists, title blocks, engineering revisions, invention agreements, patents, personal technical files, designing for manufacturing and assembly, and quotation-proposals. In

the problems at the end of the chapter, you will have the opportunity to do basic design using the procedures, drawings, and practices discussed. You will be able to utilize topics from preceding chapters, such as dimensioning, standards, fasteners, tolerances, and machine elements.

The Engineering Department

Engineering departments vary from company to company, but most of them have functions in common. An example engineering department in a medium-sized engineering company, Figure 17-1, has a variety of sections usually directed by a chief engineer. Our example has three sections: design, application, and quotation. The design section is responsible for new products and minor design improvements for existing products. The application engineering section handles special-order modifications to existing products of the company. For example, if our company manufactured commercial scales for weighing trucks, the application section would modify scales to meet specific customer requirements. The quotation department estimates costs of new or modified existing products and prepares quotations of selling prices to present to potential customers. In larger engineering companies, design

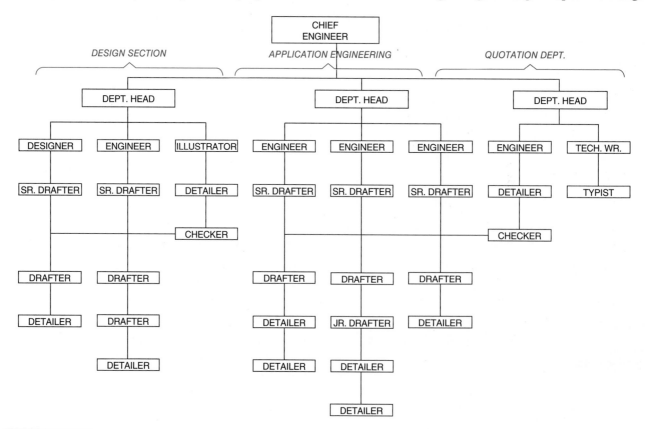

FIGURE 17-1   Chain of command in an engineering department.

and application sections may be divided into fields of engineering such as mechanical, civil, electrical, electronic, structural, materials, agricultural, petroleum, and so forth.

Directly under the chief engineer are the section heads, usually engineers or designers with extensive design-engineering experience plus an ability to lead teams of skilled personnel. The teams could consist of engineers, designers, drafters, detailers, checkers, technical writers, illustrators, and typists. Although most engineering departments have definite departmental structures, many companies move drafters, designers, and engineers back and forth among the various sections as needed. The moves help new personnel to learn the departmental structure and operation of their organization. Another exercise to gain an understanding of how an engineering organization functions is to follow paper documents along the chains of command as indicated in the flowchart, Figure 17-1.

Hence, for illustrative purposes in this chapter, two hypothetical engineering organizations are discussed. The first company produces one-of-a-kind systems to meet special needs of customers—in this case, a customer who produces electrical power. The second is a company that mass-produces a device used for positioning a small crosshair cursor or a pointer on a computer screen—the ubiquitous mouse.

Systems Company

Assume that a systems company has received a request for proposal (RFP) from a solid-waste-handling company that needs to convert burnable waste into electrical power. The proposal is to be submitted to the solid-waste-handling company in two months. The president of the systems company sends a memo to his chief engineer with directions to process the RFP. Consequently, the chief calls a meeting of the design, application, and quotation department heads. They decide that the design section will need to design a new furnace and one new conveyor for handling solid waste coming from a shredder. The application engineering group will need to modify several designs of existing conveyors. A steam turbine design and an electrical generator design will only require small changes. The quotation department has the task of preparing a proposal to be submitted to the waste-handling company before the two-month deadline. The proposal will include the selling price of the complete energy-generating system, the delivery dates, and descriptions of the system components. Then, if the waste-handling company accepts the proposal and

signs it, the proposal becomes a binding contract on both companies.

Design Section

The *design section* will be discussed first to illustrate the uses of the drawings and the related engineering practices listed earlier, which include preparing assembly, detail, and layout drawings; following design procedures; making comprehensive parts lists; executing engineering revisions; handling invention agreements; keeping personal technical files; and preparing quotations. Application engineering will be discussed later.

Design Procedure

A design process or *design procedure* used by the design section acts as a general guideline for a team of designers, engineers, and drafters (**Figure 17-2**). Their preliminary design tasks are to design a new conveyor and create a new furnace. Detailed information on these two system components needs to be

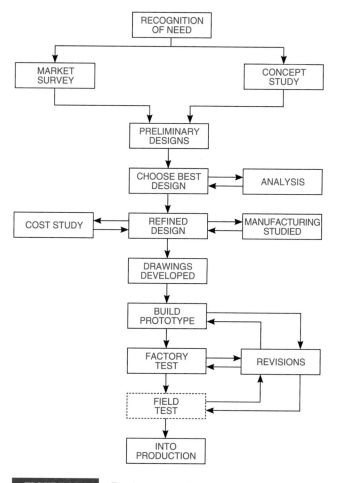

FIGURE 17-2   Design procedure.

developed so the quotation department can prepare cost estimates on them.

No one design process serves an individual designer, design team, or organization. Each individual, team, and organization develops a design process to suit their respective needs. (Similar-but-different design processes are also discussed in the Introduction and Chapters 1 and 21.)

The actual process of developing a new product from its established need (the RFP in this example) to its final production consumes many hours and involves a variety of highly trained, skilled personnel. The flowchart in Figure 17-2 shows the overall process to production; however, in this preliminary RFP phase, the design team only goes through to the cost study stage to prepare information for quoting purposes. Later, if the proposal is accepted by the customer, the design section team would develop the conveyor and furnace through to the production stage.

Nevertheless, to continue describing the design procedure followed by the design section team for the new conveyor and new furnace, concept-study *sketches* would be done first. (Note that a market survey would not be required for this project because the RFP stated the need.) Sketches would be made of conveyor ideas for moving the waste from a shredder to the furnace (**Figure 17-3**). Next, where and how

to burn the waste in the furnace and where to utilize the hot gases to generate steam would be considered. During these studies, many ideas are presented, usually in sketches. Designers are encouraged to generate and sketch ideas without restraint; that is, without being influenced by existing designs in order to arrive at a fresh and perhaps unique solution. The more ideas that are proposed in the time available, the greater the chances of developing the best solution to the problem. These brainstorming sessions are conducted in an atmosphere of "no criticism" so members will not feel inhibited. Analysis of the ideas is scheduled after the free-wheeling sessions.

The design section team employs two effective design approaches. In *conceptual design*, much use is made of known technical information from such sources as reference books, technical journals, handbooks, and manufacturers' catalogs. All members of the team should be aware of the latest technical information available. In *scientific design*, use is made of concepts from physics, mechanics, materials science, chemistry, and mathematics.

During the analysis phase, the best ideas are identified and developed further in sketches. Suitable engineering materials are investigated. Loads on the structures are analyzed, and energy balances (output versus input) are estimated.

FIGURE 17-3 Sketch for conveyor ideas for conveying waste to the furnace.

FIGURE 17-4   Sketch of preferred conveyor for conveying waste to the furnace.

Layout Drawings

The department head in the design section follows the team's progress and directs them to prepare *design development layout drawings* for a preliminary design review. The team draws the best sketches to scale either by pencil or a computer. For example **Figure 17-4** shows a sketch of a preferred drag-type conveyor for delivering burnable shredded waste to the furnace, and **Figure 17-5** shows portions of the preferred conveyor drawn to scale in a layout drawing. Note the informality of the design development layout drawing. Standard dimensioning practices are encouraged but not required. Clear handwritten notes are acceptable. Drafters, designers, and engineers do layout drawings as needed. The layouts may not be completely dimensioned, but they have overall dimensions and important center lines located. Informal notes specify tolerances, materials, special manufacturing processes, and related pertinent information. Title blocks are optional.

During the design review, the layout drawings help the team decide on the final designs to consider. The quotation department assists the team in preliminary cost analyses. Finally, the conveyor design and a new furnace concept that best meet the requirements are selected. Manufacturing personnel who also have participated in these discussions recommend appropriate manufacturing processes for the conveyor and furnace.

FIGURE 17-5   Layout drawing of preferred conveyor.

Assume now that the quotation department has received all the technical data and specifications they need from the design section and application engineering departments to prepare a cost estimate to satisfy the RFP. A proposal is written and presented to the waste-handling company just before the two-month deadline. In addition to the written proposal and several systems site drawings, three types of illustrative material may be presented for better visualization of the system. An illustrator may prepare a watercolor perspective drawing of the whole system or a computer-generated 3D picture. A third type that might be presented is an architectural-type scale model made of cardboard and balsa wood of the whole system showing dumping sites, chutes, conveyors, the furnace and boiler, and the electricity-generating equipment.

The waste-handling company representatives may ask for further information and explanations, such as brief descriptions of system components and when specific components will be delivered, installed, and operating. Finally, after further comments and questions, the proposal is accepted and signed. It is now a binding legal contract on both companies. **Figure 17-6** shows an example signature page from a proposal-contract form.

COMPANY

20—APPROVAL: There are no agreements or oral understandings outside of this proposal, which shall become a contract only when accepted by you,_____ _
_____, and approved in writing by a plant manager or an executive officer of
       Company.

Prepared and submitted to you for
Company

by _____

Accepted this _____ day of _____

19____ for

      _____

by _____

Title _____

Approved for   Company at

      _____

this _____ day of _____ 19 _____

by _____

Title _____

FIGURE 17-6   Proposal-contract signature page.

Working Drawings: Detail and Assembly

The systems company president notifies the chief engineer that the proposal is now a contract. The chief directs the designers and engineers to forward the design development layout drawings of the new conveyor and the furnace to drafters and detailers who use these drawings to create working drawings; that is, detail and assembly drawings for manufacturing. There are a number of different kinds of working drawings associated with the development and production of a product, so the drafter must be familiar with each kind, know the company standards, and be able to draw each kind. The major kinds of working drawings are:

- Assembly drawings
- Subassembly drawings
- Detail drawings
- Modified purchased drawings

Assembly Drawing

Any product that has more than one part must have an assembly drawing. The *assembly drawing* illustrates how a product is assembled when completed. The assembly drawing can have one, two, three, or more views as needed. Often, a full section view is used to show how the parts are assembled.

Assembly drawings usually are not dimensioned, except for general overall dimensions or to indicate the capabilities of the assembly. For example, a clamp assembly drawing might have a dimension to illustrate how wide it opens. (Refer to Problem 17-7.) Assembly drawings do not contain hidden lines unless they are absolutely necessary to illustrate some important feature that the reader might otherwise miss.

An assembly drawing must call off each part that is used to make up the assembly. Each part must be called off by its part number and title and the total number required to make up one complete assembly. Companies most often use one of two methods to call off the parts. In Method A, each part has a leader line with the part number, title, and number required (**Figure 17-7**). In Method B, each part has a leader line with a balloon call-off number inside (**Figure 17-8**). This method uses a tabulation chart to list the part number, title, and number required.

FIGURE 17-7   Assembly call-off (Method A).

MUST INCLUDE: **DRAWING NUMBER**
TITLE
NUMBER REQUIRED PER ASSEMBLY

NO.	DRAWING NO.	ITEM	REQ'D
–	–	1/4-20 UNC	–
5	PURCH	LOCK NUT-HEX	2
–	–	1/4-20 UNC X 1 1/2 LG.	–
4	PURCH	SCREW-CAP HEX HD	2
3	A 661982	HOOK	2
2	A 661981	SCREW-CENTER	1
1	A 661983	YOKE	1
NO.	DRAWING NO.	ITEM	REQ'D

FIGURE 17-8   Assembly call-off (Method B).

Subassembly Drawing

A subassembly is composed of two or more parts permanently fastened together. A *subassembly drawing* is very similar to an assembly drawing. All standard practices associated with assembly drawings apply to subassembly drawings. That is, full section views, few overall dimensions, hidden lines only if necessary, and call-offs using either Method A or B are employed.

FIGURE 17-9   Subassembly drawing.

Subassemblies sometimes require machining operations after assembly (**Figure 17-9**). For instance, a hole is to be drilled through the parts after assembly. Drilling each individual part and then aligning the holes at assembly would be much more difficult and costly. Thus, the required dimensions to drill the hole are added to the subassembly drawing.

There can be any number of subassemblies in the final assembly. For example, in an automobile, the engine is a subassembly, the alternator is a subassembly, the transmission is a subassembly, and so forth. These subassemblies, together with various single parts, make up the final assembly of the automobile. Note that subassemblies are often purchased, stocked, and stored for later final assemblies.

Detail Drawings

Every manufactured part must have its own fully dimensioned *detail drawing* with its own drawing number and title block (**Figure 17-10**). All information needed to manufacture the part is contained in the detail drawing, including as many views as needed to fully illustrate and dimension the parts.

The general practice in technical drawing is to draw only one part to one sheet of paper, regardless of how small the part is, because individual parts are often made on different machines or by different processes, and the bookkeeping plus costing is easier for one part per drawing.

Most companies use detail drawings as just described. However, some companies separate the detail drawings into various manufacturing processes, such as:

• Pattern detail drawings for castings

• Casting detail drawings

• Machining detail drawings

• Forging detail drawings

• Welding detail drawings

• Stamping detail drawings

Two observations regarding technical drawings are added here for your information.

1. Fold lines, construction lines, reference planes, and reference numbers and letters are almost never included on technical drawings. They may have been utilized to prepare views and sections, but they are omitted from the drawing.

2. A seldom-discussed drawing technique designed to clarify portions of drawings may be used effectively. Usually, a small portion of a large drawing is removed, enlarged, and labeled—e.g., DETAIL A.

Purchased Parts

Most manufacturing companies cannot make such purchased items as electric motors, bearings, shafting, bolts, nuts, screws, and washers as cheaply as companies that specialize in making these items. Therefore, designers should specify these standard items whenever possible. *Purchased parts* are drawn

FIGURE 17-10 Detail drawing.

FIGURE 17-11 Detail drawing of modified purchased part.

in assembly drawings, usually in outline form, to show their function and relationship to the product. All specific information for a purchased part should be listed on the drawing or parts list, including size, material, special features, vendor part number, performance, and so forth. For example, specifications for a DC motor might read as follows: "12 volt DC motor, GE no. 776113 or equal."

Modified Purchased Parts

A purchased part that needs to be modified, however slightly, must have the modification fully described and dimensioned as with any detailed part, except that the purchased part size and description are listed in the title block under Material. **Figure 17-11** shows a standard fastener, M20 × 2.5-6g × 70 long hex-head machine screw, which is to have a tip specially turned and a 4 mm diameter hole drilled through the tip. The standard fastener is drawn with the modifications added and dimensioned. Then, it would be called out as a standard metric fastener in the title block under Material. This directs the machinist to make the part from a standard fastener.

Title Block

Title blocks vary from company to company, although most companies follow the ANSI standard illustrated in Figure 17-10. The primary function of a *title block* is to provide information not normally given on the drawing, such as:

- Name and address of the company
- Drawing title
- Drawing or part number
- Scale of the drawing
- Name of the drafter and date completed
- Name of the checker and date checked

- Name of the person who approved the drawing and date
- Material the part is to be made of
- Tolerances/limits of dimensions
- Heat treatment requirements if not specified on the drawing
- Finish requirements if not specified on the drawing

Size of Lettering within the Title Block

- General information, .125 (3) freehand/.120 mechanical
- Drawing title, .250 (6) freehand/.240 mechanical
- Drawing number, .312 (8) freehand/.350 mechanical

Checking Drawings

The importance of absolute accuracy in engineering drawings cannot be stressed enough. The slightest error could cause large and unnecessary expenses. In some fields of engineering, such as in aircraft and space vehicles, an error could cause tremendous consequences—lost lives! Therefore, every precaution should be taken to avoid errors. In most engineering departments, checkers are employed to verify all dimensions and to see that standard materials and features (holes, threads, keyways, etc.) are used wherever possible. The checker's responsibility is to check the entire drawing and sign it before the drawing is finally approved (usually by a chief engineer) and released.

Conscientious drafters check their drawings before submitting them to the checkers to ensure accuracy and to save the checkers time. A partial checklist for drafters is as follows:

- Are all dimensions included?
- Are there unnecessary dimensions?
- Can the part be manufactured in the most economical way?

- Are dimensions and instructions clear and understandable?

- Will the part assemble with mating parts?

- Have all limits, tolerances, and allowances been properly analyzed for all moving parts?

- Have undesirable accumulations of tolerances been analyzed?

- Are all notes added?

- Are finish-texture symbols added?

- Are the material and treatment for each part specified?

- What is the drawing's general appearance? (legibility, neatness, arrangement of views)

- Does it follow all drawing and company standards?

- Is the title block complete? (title, part number, drafter's name, etc.)

Parts List

A *parts list* may contain the assembly drawing number, all subassembly drawing numbers, all detail drawing numbers, all purchased parts, material of each part, and the quantity of individual parts needed for one complete assembly.

A drafter usually prepares the parts list while following company guidelines. Some companies list the parts in order of size, some in the order of importance in the manufacturing process, and others in order of assembly. The assembly method is preferred due to the time savings later in the assembly area. An assembly method parts list is shown in **Figure 17-12** for a machine vise.

The parts are listed in the order used for assembling the machine vise. Note that the purchased parts do not have a drawing number; therefore, they are listed under the column titled Drawing No. as PURCH., indicating that these are standard parts to be purchased, not made, by the company. Additional typical purchases for most companies are motors, bearings, switches, gears, chains, belts, and controls. Complete specifications must be listed for purchased parts so that there is no question as to what must be purchased. The purchasing agent for the company may not have a technical background and therefore will rely on the parts list specifications for purchasing "buyouts." In this example, items numbered 7, 8, 17, 23, and 24 are purchased parts.

Some companies use a system of indents to list the parts, Figure 17-12. There are four indents, indicating the various kinds of drawings. The first indent indicates the main assembly drawing. The second indent

INDENT 1 = ASSEMBLY DRAWING
INDENT 2 = SUBASSEMBLY DRAWINGS
INDENT 3 = DETAIL DRAWINGS
INDENT 4 = PURCHASED PARTS

NO	DRAWING NO.				DESCRIPTION	MATERIAL	QUAN
					MASTER PARTS LIST		
1	D77942	VISE ASSEMBLY-MACHINE				AS NOTED	1
2							
3	C77947		BASE-VISE			AS NOTED	1
4	B77952			BASE-LOWER		C.I.	1
5	A77951			BASE-UPPER		C.I.	1
6	A77946			SPACER-BASE		STEEL	1
7	PURCH.				BOLT 1/2-13 UNC-2.0 LG.	STEEL	1
8	PURCH.				NUT 1/2-13 UNC	STEEL	1
9							
10	C77955		JAW-SLIDING			STEEL	1
11							
12	A77954		SCREW-VISE			STEEL	1
13	A77953			ROD-HANDLE		STEEL	1
14	A77956			BALL-HANDLE		STEEL	2
15							
16	A77961	PLATE-JAW				STEEL	2
17	PURCH.				SCREW 1/4-20 UNC-1.0 LG.	STEEL	4
18							
19	A77962	COLLAR				STEEL	1
20							
21	A77841	KEY-SPECIAL VISE				STEEL	2
22							
23	PURCH.				BOLT 1/2-13 UNC-4.0 LG.	STEEL	4
24	PURCH.				NUT 1/2-13 UNC	STEEL	4

JAN ENGINEERING PETERBOROUGH, NH	Model No. 160	Parts Lister NELSON	Date 6 AUG 98
Title VISE ASSEMBLY-MACHINE		PAGE 1 OF 1 PAGES	Drawing No. A1198891

FIGURE 17-12 Parts list.

indicates all subassembly drawings used to make up the assembly. The third indent indicates all detail drawings and modified purchase parts with drawing numbers to make up the assembly. The fourth indent indicates all purchased parts to make up the assembly.

Not all companies use this system, but using the indent system gives a quick indication of which parts are used to make up the various subassemblies. Note on the example list that item number 3 is indented one place, indicating a subassembly. Item numbers 4, 5, and 6 are detail drawings used to make up the subassembly item number 3. Item numbers 7 and 8 are purchased parts used to complete the subassembly item number 3.

Numbering System

The numbering system for identifying and recording technical drawings varies from company to company. Although no standard system is used by all companies, most companies assign a sequential number for each drawing as it is finished. Some also add a prefix letter to indicate the drawing size. A drawing on a C-size sheet with an assigned number of 114937 would be indicated on the drawing by C–114937. The drawing number could be done in freehand, 0.312 (8) high, or mechanically lettered, 350 (9) high.

Drawing Revisions: ECRs and ECOs

After any technical drawing has been signed off (drawn, checked, and approved) and incorporated into the company's system, changes in the drawing are usually accomplished through *ECRs* and *ECOs*. Anyone having a suggestion or need to improve a part or assembly within the product or an error to correct on the drawing must bring it up in an *engineering change request (ECR)* procedure meeting (**Figure 17-13**), and request an *engineering change order (ECO)*. No change can be made in a drawing after it has been released (signed off) to manufacturing, even by the drafter or designer who drew it. All departments concerned must agree to the change, Figure 17-13. The ECR and ECO explain what change is to be made, what it was before, why the change is to be made, whether it affects other parts, and who suggested the change. When the engineering change request has been approved, the committee issues the engineering change order. The ECR and ECO list the departments in the company that will be directly affected by the change.

Beginning drafters, detailers, designers, or newly hired engineering graduates may "get their feet wet" in the company by being assigned the task of incorporating ECOs in the company's technical drawings. These assignments are made primarily to introduce the employees, new to the company, to technical documents that flow through the company and to how the company functions. Introducing new employees to the flow of technical documents illustrates an admonition of a commencement speaker who said to the graduates, "Get dirt under your fingernails so you'll know what's going on!"

Minor changes on original drawings may be just erasures or additions made directly on the original, whether on paper or in a computer. If a dimension is not noticeably affected by the change, the dimension is simply corrected and underlined with a heavy black line. This indicates that the dimension is correct but slightly out of scale (**Figure 17-14**).

If the change is extensive, the drawing must be redrawn. The original drawing is not destroyed: it is stamped OBSOLETE and includes the note "SUPERSEDED BY (*the number of the corrected drawing*)." The new drawing must have a note referring back to the superseded drawing. The change on the drawing must be clearly marked where the change was made, and it must clearly describe what was changed. Each company has its own standards to indicate this information. The most common method is to place a letter or number in a small balloon near a changed dimension or dimensions or near the change made to the drawing. The letter or number must be listed in the revision block (**Figure 17-15**). The revision block, located either to the left of the title block or in the top right corner of the drawing sheet, shows the letter or number of the change, what was changed, the date of the change, and who made it. If the change is extensive, only the

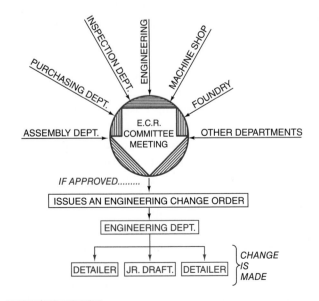

FIGURE 17-13   Engineering change request order procedure.

FIGURE 17-14   Balloon change call-off.

FIGURE 17-15   Revision block (Courtesy Bishop Graphics Accupress).

ECO number is listed. Anyone needing to know why the change was made can refer back to the ECO or the ECR.

Some companies add the last change letter to the drawing number which, in effect, actually changes the part number. For example, part number C–65498 would become C–65498–A for the first change. Then, if changed again at a later date, it would become C–65498–B, and so on.

In certain cases, appropriate administrative changes may be made without an engineering change order. Such simple changes as misspelled words or incorrect references are non-engineering changes and probably would not affect other departments.

The next section in the systems company's structure, application engineering, primarily processes existing designs.

Application Engineering

Preliminary activity in the *application engineering* section proceeds to the cost-study step in the design procedure, Figure 17-2, in a manner similar to the design section. The procedure acts as a guide for an application engineering team of engineers, designers, and drafters who consider designs of conveyors, steam turbines, and electrical generators the company has produced in the past. Also, the team considers concepts that already exist in the industry. The team's primary task is to modify and incorporate these existing concepts to meet the design requirements for the waste-burning system. For example, a scale is modified to monitor the amount of waste being delivered. A platform over the scale must be modified to accommodate large trucks. After being weighed, the trucks would dump the waste into a chute going to a flight conveyor, a standard product of the company, which would carry the waste to a hopper, where the waste would be stored until conveyed to a shredder. From the shredder, the waste would be conveyed to the furnace on the new drag conveyor created in the design section, Figures 17-4 and 17-5. Drawings of a steam turbine would be reviewed in an ECR

meeting, where ECOs would be written to modify the existing specifications and drawings of the turbine. Drawings of an electrical generator system also would be modified from ECOs. Drafters and detailers would process the ECOs.

Several new designs are developed in application engineering, but they are minor and are based on the designer's previous experience. For example, sketches and a layout drawing to determine overall dimensions of the chute previously mentioned are shown in **Figure 17-16**.

Finally, specifications and appropriate layout drawings are forwarded to the quotation department for their use in the cost study.

Quotation Department

During the design and development of the waste-burning furnace and during the processing of the conveyors, steam turbine, and electrical generator, the *quotation department* closely followed the design section's progress and the application engineering's activities. Thus, an estimate of the selling price for the entire system was nearly completed for the RFP by the time design and application finished their respective tasks. The selling price is the major dollar figure submitted to the customer, but subtotals for component systems, such as the furnace, the steam turbine, and the generator, may be included. The *selling price* is prepared by the quotation department in steps illustrated in **Figure 17-17**. Basic costs include labor and engineering materials. Basic labor (sometimes labeled direct labor) includes estimates of hours for engineering, drafting, manufacturing, assembling, and installing. Engineering materials include steel plates, structural steel shapes, fasteners, welding rods, switches, motors, controls, and so forth. Manufacturing expenses, such as depreciation of machines and maintenance, add to the basic costs to make up manufacturing costs. Sales salaries and expenses plus general overhead costs (electricity, water, office supplies, etc.) are added to manufacturing costs to arrive at the total cost to produce the energy system. Profit added to total cost establishes the selling price.

The selling price, specifications, delivery schedules, and the "boilerplate" make up the *proposal*. A typical boilerplate would be stored in a computer ready for use in any new proposal. It includes introductory comments and sections intended to promote the company to the prospective customer, such as descriptions of past systems designed and installed and summaries of the expertise of the

FIGURE 17-16   Sketches and layout drawing of chute.

ENGINEERING MATERIALS   DIRECT LABOR   MANUFACTURING EXPENSES

METAL PRODUCTS   ENGINEERING   DEPRECIATION
NONMETAL PRODUCTS   DRAFTING/   MAINTENANCE
STRUCTURAL STEEL   PLOTTING
WELDING RODS   MANUFACTURING
SWITCHES   ASSEMBLY
MOTORS   INSTALLATION
CONTROLS
ETC.

SALES AND OVERHEAD   PROFIT

OFFICES   PERCENTAGE ADDED
SALES SALARIES   ONTO TOTAL OF ALL
OFFICE PERSONNEL   EXPENSES
OFFICE SUPPLIES
ELECTRICITY & WATER
ETC.

XYZ ANN
MMM

SELLING PRICE = ENGINEERING + DIRECT + MANUFACTURING + SALES AND + PROFIT
               MATERIALS   LABOR   EXPENSES   OVERHEAD

FIGURE 17-17   Selling price components.

personnel who would handle the job if the proposal were accepted. Another section of the boilerplate protects the company with legal statements referring to late delivery due to strikes; acts of God (earthquakes, tornadoes, etc.); unavailability of materials; and so forth.

A cover letter with a bound copy of the proposal would be mailed to the prospective customer before the due date or delivered in person if an oral presentation was arranged. Oral presentations give the recipient an opportunity to ask questions, suggest changes, and correct any misunderstandings. When the customer accepts the proposal and signs it, it becomes a binding legal *contract* on both parties. It also becomes a document that immediately generates continuing work for the design and application sections of the systems company.

This continuing work for the design section was discussed earlier. Working drawings were prepared, checked, signed, and sent to manufacturing. In a similar manner, application engineering would produce working drawings and have them checked, signed, and sent on to manufacturing.

Mass Production Company

A *mass production company* that produces large quantities of a product would have an engineering departmental structure similar to the systems company discussed earlier, Figure 17-1; however, the mass production company would have a somewhat different approach to producing a product (**Figure 17-18**). The company's design, application-engineering, quotation, and manufacturing departments would follow this design procedure for mass-producing one product while working together more closely than departments in the company that produced a one-of-a-kind system. To illustrate these different procedures, assume that the mass production company primarily produces mouse units (**Figure 17-19**) for use with a computer. Essentially, a mouse provides a quick way for computer operators to position a small crosshair cursor or a small pointer on a video screen by moving the mouse across a special desk pad. Then, once the crosshair or pointer has been positioned, one key atop the mouse is clicked to initiate a basic computer operation.

Assume that the mass production company has received an order for two million mouse units and that each unit requires more functions than their standard mouse. The major challenge to the company is to design and mass-produce these units

FIGURE 17-18   DFM-DFA procedures.

FIGURE 17-19   Computer mouse.

as efficiently as possible and still have a reliable product. The company's design efforts are focused on two important areas of mass production. The first—DFM (design for manufacturability)—emphasizes design of the parts that make up a product. The second—DFA (design for assembly)—concentrates on assembly of the parts that make up a product.

Design for Manufacturability (DFM)

The *DFM procedure* is an iterative loop and acts as a guide for a design team, Figure 17-18. The primary goals of the DFM team are to develop the new mouse design for more efficient manufacturing, at the least cost, within a reasonable time, and to ensure its function for a reasonable life. The team made up of engineers, designers, drafters, and manufacturing personnel pursue these goals as illustrated by the iterative path shown. Start with optimizing (design, function, and cost); then go on to simplifying (DFM guidelines); ensuring (function, life, materials); and back to optimizing. One factor not shown in Figure 17-18 is the time constraint for making decisions. This can affect the success of the DFM process. Decisions must be made at each step in the loop based on information available at the time and the judgment of the team members.

DFM Guidelines

Guidelines for DFM evolved over a period of time, essentially due to an increasing need for collaboration on the design of products between a company's engineering department and manufacturing. To be competitive, companies needed to be more efficient and produce better products. Some DFM guidelines from DFM literature are:

- Minimize the number of parts.
- Minimize part variations.
- Design parts to have more than one function or use.
- Design parts for ease of manufacturing.
- Use fasteners that handle easily and can be readily installed.
- Arrange the assembly of parts so they assemble in layer fashion.
- Avoid flexible parts.

Minimize Number of Parts

One approach is to combine two parts into one, thus eliminating fasteners. An obvious example is shown in **Figure 17-20**.

Minimize Part Variations

If fasteners are required, use all the same size to save storage space and sorting-and-assembly costs. Use standard off-the-shelf parts where possible.

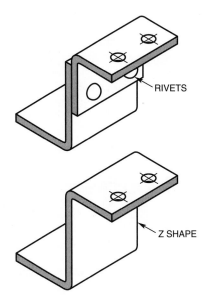

FIGURE 17-20   Combining two parts into one.

FIGURE 17-21   Post used for spacer and fastener.

Design Parts to be Multifunctional

For example, a leaf spring designed to exert a small force could also conduct an electrical current. A structural part could have ridges for guiding a second part into position. A spacer post could be used for a fastener (**Figure 17-21**). One shaft of a specific length and diameter could be used in a variety of products. One-size parts such as bearings, springs, or gears could be used in a variety of products in a company.

Design Parts for Ease of Manufacturing

Precision cast gears or stamped gears need little or no machining or surface finishing. Basic materials with relatively expensive finishes, such as anodized aluminum, may ultimately save painting, polishing, and machining later in the manufacturing process, thus keeping labor costs down.

Use Easily Handled and Installed Fasteners

Snap-in fasteners save assembly and screwdriver time (**Figure 17-22**).

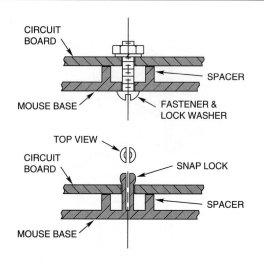

FIGURE 17-22   Snap-in fastener for circuit board.

FIGURE 17-24   Slotted disks inside a mouse.

FIGURE 17-23   Layering for mouse assembly.

Arrange Parts Assembly for Layering

Assembly of the mouse shown in **Figure 17-23** illustrates layering; that is, essentially, parts are stacked from the bottom up.

Avoid Flexible Parts

Use circuit boards instead of flexible, difficult-to-assemble wires for electrical circuits where possible. Of course, the mouse, Figure 17-19, needs one flexible connector for easy movement on a desk pad.

The next step in the loop, Figure 17-18, is to ensure function and life and to select the best materials. Most mouse units utilize a printed circuit on a board with electronic components to integrate the digital signals generated by light passing through rotating slotted disks inside the mouse (**Figure 17-24**). The shafts of the two slotted disks are driven by a small ball, mostly inside the body of the mouse, as it rolls on a special desk pad in a direction determined by the computer user, who observes a crosshair or pointer on the video screen. Since the disks are mounted in "X" and "Y" orientations, the sum of their rotations (digital signals) is used to position the crosshair or pointer on the video screen. The body of the mouse and a number of the inside parts are made of molded plastic. Resistors, capacitors, optical components, and switches are purchased buyouts. Stainless steel shafts support the disks; bearing supports for the shafts are part of the plastic body; thus, no lubrication is required. The keys for initiating basic operations are molded plastic with narrow portions acting as hinges. The only maintenance required for the mouse is to occasionally remove the ball (approximately 22 cm in diameter), wash it in soapy water, rinse, dry, and reinstall. The additional functions ordered by the customer for the new mouse design were incorporated by adding one key and one switch and expanding the printed circuit. These additions, of course, required engineering change orders (ECOs) to modify the company's existing mouse design.

After the team was satisfied that their designs for manufacturing parts of the mouse were optimum, they concentrated their efforts on design of the parts for assembly (DFA).

Design for Assembly (DFA)

The *DFA procedure* is an iterative loop that occurs simultaneously with the DFM loop, Figure 17-18, but the emphasis is on assembly of parts vis-à-vis the design and manufacturing of parts. The team has similar goals in the DFA loop—develop the best design of parts for assembly of the mouse, at the least cost, within a reasonable time, and ensure its function. They follow the DFA loop shown: first, consider which parts should be assembled automatically, and second, consider which parts need to be assembled manually.

DFA Guidelines

A partial list of DFA guidelines from DFA literature is as follows:

- Use pick-and-place robots for simple, routine placement of parts.
- Keep the number of parts to a minimum (DFM guideline also).
- Use gravity and vibration for moving parts into assembly areas.
- Avoid parts that tangle (hang up on each other).
- Use tapers and chamfers to guide parts into position.
- Avoid time-consuming manual operations such as screwdriving, soldering, and riveting.

Use Robots for Simple Operations

Robots can be programmed to pick electronic components off a conveyor or from a feeder and place them accurately into a circuit board for the mouse. A

FIGURE 17-25   Combine several parts into one.

mouse may require 25 to 50 electronic components, depending on the basic functions it is to perform.

Minimize the Number of Parts

Combine several components into one molded part (**Figure 17-25**).

Move and Orient Parts Using Gravity and Vibration

Figure 17-26 shows one method of orienting spacers and steps to avoid tangling of lock washers and tangling of rivets.

Use Tapers and Chamfers for Positioning

An example of tapered fasteners and a receptacle with a tapered hole, both for ease of assembly, is shown in **Figure 17-27**.

Avoid Time-Consuming Operations

The snap locks in the body shown in Figure 17-22 allow for quick assembly and accurate positioning of the circuit board.

FIGURE 17-26   Spacers handling and anti-tangling techniques.

INDUSTRY APPLICATION

UNDERSTANDING YOUR CUSTOMER'S NEEDS CAN PAY OFF

Mark and John are about the same age. They graduated from the same technical college at the same time, and both went to work for the same company. Both were excellent students, and both have done well on the job. But Mark just beat John in an internal competition for a team leader position. As a result, Mark now earns more than John, has a new office, and is on the fast track for future promotions.

When the time came to select a new team leader, the difference between Mark and John could be stated in just three words: "design for manufacturability." Mark was a proponent of DFM who had proven he was adept at applying its principles. In fact, he always referred to the manufacturing personnel who would use his drawings as his customers. Mark was often seen on the shop floor talking with manufacturing personnel and observing their processes. Over and over, Mark could be heard asking shop floor personnel the same questions: "What can I do to make our products easier to manufacture?" As a result, Mark had become a skilled practitioner of the art and science of design for manufacturability (DFM).

John, on the other hand, never visited the shop floor unless it was absolutely necessary. To him, the shop floor was a dirty place best avoided. John's attitude can be seen in one of his favorite statements: "I am a designer, not a machinist. If I wanted to know about machining, I would have majored in shop." John's mistake was to view design and drafting as an island that stands by itself rather than as a part of a larger process that includes design, manufacturing, and all of the other activities necessary to convert an idea into a product that can succeed in the marketplace.

Mark's first decision as a new team leader was to include manufacturing personnel on his team. He wanted to eliminate the "us-against-them" mentality that can grow between design and manufacturing so that his team's designs would be not just functional but also manufacturable.

FIGURE 17-27   Tapered fasteners and a tapered hole for ease of assembly.

Finally, after the DFA team members agree on the best designs for manufacturability and the best designs for assembly for the new mouse, and the DFM team concludes its optimization efforts, costs for producing the new mouse are checked. If the cost analysis results in further modifications, sketches and layout drawings may be prepared, and ECOs may be issued. To continue the design procedure, the new mouse design is formally released. Designers and drafters prepare working drawings. Eventually, all the drawings are released to manufacturing supervisors who have anticipated the drawings because

they had worked closely with the design teams during the development of the new mouse design.

As the new mouse goes into production, the personnel of the mass production company realize that even a small savings in cost resulting from following DFM and DFA guidelines is worthwhile when millions of units are involved. Moreover, the company finds that following the guidelines not only results in cost savings but also improves the design and function of products.

Personal Technical File

During the various design and development activities, each engineer, designer, detailer, checker, technical writer, and illustrator should develop and maintain a technical file to enhance his or her involvement in the company. The objective is to minimize the time spent locating technical information pertinent to their respective jobs. Usually, a company manufactures products and performs engineering in specific areas of technology, so files can be concentrated in these areas for efficient retrieval of information. An efficient technical filing system uses an alphabetical-numerical scheme. For example, the first part of a filing scheme for a drafter-designer in a company specializing in conveyors would be an alphabetical listing of technical topics with numerical references. This file should be kept in a three-ringed binder or a computer so that additional information can be added easily. Example entries would be:

Section A

Aluminum Products 112		Channel Sections	134
ANSI Standards	410	Conveyors	500
Section B		Codes and Standards	410
Bar Stock	132	Controls	210
Beams	133	*Section D*	
Belt Conveyors	510	Drives	600
Belt Materials	122	*Section E*	
Bucket Elevators	530	Electrical Components	200
Section C		Controls	210
Castings	140	Switches	230
Chain Conveyors	520	And so on . . .	

The numerical part of the file, kept in a file cabinet, would contain tab-numbered dividers, numbered folders, and numbered materials in the folders. Example tabs, folders, and materials would be numbered as follows. (The letters in parentheses are for discussion purposes only: T = tab number for a divider, F = folder number, and M = material labeled with a number.)

100 Engineering Materials	(T)
110 Metals	(F)
111 Ferrous Based	(M)
112 Aluminum	(M)
120 Nonmetals	(F)
121 Ceramics	(M)
122 Rubber (Belts)	(M)
130 Structural Shapes	(F)
131 Angles	(M)
132 Bar Stock	(M)
133 Beams	(M)
134 Channels	(M)
200 Electrical Components	(T)
210 Controls	(F)
220 Motors	(F)
221 AC Motors	(M)
222 DC Motors	(M)
230 Switches	(F)
300 Materials Handling	(T)
310 Solid Waste	(F)
400 Codes and Standards	(T)
410 ANSI Standards	(F)
411 ANSI Y14.5M–1982	(M)
500 Conveyors	(T)
510 Belt	(F)
520 Chain	(F)
530 Bucket	(F)
600 Drives	(T)
610 Gear Box	(F)
620 Roller Chain and Sprockets	(F)
630 V-Belt and Pulleys	(F)
And so on . . .	

Start a technical file now, as a student, and expand it later to suit your technical jobs in the industry.

In addition to your personal file, when you are working in a design section, the section usually has an extensive vendor catalog library and related references. There will most likely be a *Thomas Register* set (lists of vendors, their products, and addresses) and a set of ANSI (American National Standards Institute) standards related to your company's products. The vendor catalog collection would cover components such as the following:

- Bearings
- Belts
- Brakes
- Cams
- Clutches
- Controls
- Couplings
- Fasteners
- Gears
- Hydraulic power (liquid)
- Hydraulic power (pneumatic)
- Jigs and fixtures
- Keys and pins
- Lubricants
- Motors
- O-rings
- Piping lumber
- Plastics
- Programmable logic circuits
- Pulleys
- Pumps
- Reducers
- Retaining rings
- Roller chain
- Sprockets
- Structural steel and aluminum
- Shafts
- Switches
- Tools

Occasionally, a drafter, designer, or engineer may be required by an employer to keep a bound-page notebook for patent development. The pages are bound together, numbered, and each one signed by the individual inventor. Also, each page needs to be signed by at least two witnesses who understand the area of technology of the potential patent. Furthermore, the individual inventor may be required to sign an invention agreement.

Invention Agreement

Anyone who is creative and working for a company, such as a drafter, designer, or engineer, usually must sign an *invention agreement* form giving the company the right to any new invention designed while one is working for the company. The form specifies that the employee will not reveal any of the company's discoveries or projects. The invention agreement is binding on the employee from six months to two years after the employee leaves the company so that employees will not invent something, quit, and patent the invention themselves.

A company is not obligated to, so it usually does not give extra pay for an invention developed by an employee. This is what the employee is paid to do. However, companies recognize talent and will reward that talent with a promotion, stocks, bonds, or possibly a raise. Signing an employee/employer invention agreement is a normal request in any company. Of course, prospective employees should read the agreement carefully and understand it before signing. An applicant who refuses to sign the agreement may be asked to look elsewhere for employment.

Patent Drawings

All patent applications for a new idea or an invention must include a *patent drawing* to illustrate and explain its function. Patent drawings must be mechanically correct and must adhere to strict patent drawing regulations. For example, patent drawings are more pictorial and explanatory than regular detail or assembly drawings. Center lines, dimensions, and notes are omitted. All features and parts are identified by numbers that refer to written explanations and descriptions. Line shading is used to improve readability.

All patent drawings must be made so that they reproduce clearly, which essentially means that the drawings must be done in ink. Current requirements specify that India ink or its equivalent be used and that the drawings be done on strong, white paper, 8½ × 13″ (21.6 × 33.1 cm) or 8½ × 14″ (21.6 × 35.6 cm). A precise "sight" area (20.3 × 29.8 cm) must be used for the drawing itself on the strong, white paper. (For further requirements, refer to *37 CFR* Ch 1 7-1-91 edition.) The publication *A Guide for Patent Drawings* may be obtained from the Superintendent of Documents, U.S. Government Printing Office, Washington, DC 20402 or online at www.gpo.gov/.

SUMMARY

- A typical engineering department has three sections: design, application, and quotation. The design procedure used in the industry and the basis for establishing the selling price of a system or product should be understood by design and drafting students.

- Technical drawing types used in the industry include freehand sketches, layout drawings, detail drawings, and assembly drawings. Fold lines are seldom included on technical drawings in the industry. Procedures for handling ECRs and ECOs

on drawings should be understood by design and drafting students.

- Students should understand mass production and the role of DFM and DFA in the manufacturing process in a mass production company. Design for manufacturability (DFM) involves considering both function and production during the design process. Design for assembly (DFA) requires the consideration of both function and the assembly process.

REVIEW QUESTIONS

Mark the following statements as either true or false.

1. The application engineering section typically handles special-order modifications of existing products.

2. The only members of the design team should be designers and drafters.

3. Producing computer mouse units is an example of a one-of-a-kind system.

4. RFP stands for request for proposal.

5. In a layout drawing, a freehand sketch is redrawn to scale.

6. A design procedure acts as a general guideline for the design team.

7. Conceptual design is based on materials science, chemistry, and mathematics.

8. Any product that has more than two parts requires an assembly drawing.

9. A company uses purchased parts made by another company because it is cheaper.

10. The numbering system for identifying and recording technical drawings is standardized and used throughout the United States.

Answer the following questions by selecting the best answer.

1. The two types of illustrative material used by a systems company to help a customer visualize the overall system are:
   a. Assembly and detail
   b. Working and assembly
   c. Perspective and scale model
   d. Assembly and subassembly

2. Which of the following is **not** part of the information usually included in a title block?
   a. Material the part is to be made of
   b. Tolerances/limits of dimensions
   c. Part or drawing number
   d. Reference numbers of subassembly drawings

3. Which of the following is **not** true regarding checking drawings?
   a. Slight errors can cause large expenses.
   b. A good drafter thoroughly inspects his/her own drawings once completed.
   c. If the drafter does a thorough inspection of his/her own drawing, it is not necessary to send it to the checker.
   d. Every precaution should be taken to avoid errors.

4. Which of the following is **not** true regarding drawing revisions?
   a. A drawing can be modified once an ECR is signed and submitted.
   b. All departments concerned must agree to the change.
   c. A drafter can change his/her own drawing after it has been released and signed off.
   d. Minor changes to drawings could be simple erasures or additions made directly on the original.

5. Which of the following would be part of preparing a selling price?
   a. Engineering materials
   b. Basic labor (direct labor)
   c. Profit
   d. All of the above

6. Which of the following is **not** part of a typical boilerplate?
   a. Introductory comments
   b. Descriptions of past systems designed/installed
   c. An invention agreement for signature
   d. A legal section

7. Which of the following is **not** a guideline from DFM literature?
   a. Minimize number of parts.
   b. Minimize multifunctional parts.
   c. Minimize part variation.
   d. Minimize number of flexible parts.

8. Which of the following is **not** true regarding a mass production company?
   a. Its approach to producing a product would be different than a one-of-a-kind system.
   b. Its internal departments would work together more closely than those in a one-of-a-kind system.
   c. Its internal departments would work together less closely than those in a one-of-a-kind system.
   d. None of the above.

CHAPTER 17 PROBLEMS

The major drawings that were illustrated and emphasized in this chapter included freehand sketches, layout, detail, and assembly drawings—the types of drawings used in the industry to develop a product. Two completed sets of these drawings are included here for your information and guidance. Then, the problems for Chapter 17 follow.

EXAMPLE NUMBERS 17-1-1 THROUGH 17-1-6 (SUBASSEMBLY NOT REQUIRED)

Objective: Design a hold-down clamp that:

1. Bolts through a 3.4″ thick worktable and clamps materials 1″ to 1½″ thick.
2. Stays in the "open" (up) position when not clamping.

EX 17-1-1   Freehand sketch of hold-down clamp.

EX 17-1-2   Layout drawing of hold-down clamp.

PROBLEMS

DWG. NO.				SH	REV.
REVISIONS					
REV.	DESCRIPTION		DATE	APPROVED	

Ø.55
THRU

$1\frac{1}{2}$"

.75"

.75"   1"

1

.18"

3"

1	1		STEEL BAR 1/2" X 1-1/2" X 3"LG	AISI 1020	
QTY REQD	FSCM NO.	PART OR IDENTIFYING NO.	NOMENCLATURE OR DESCRIPTION	MATERIAL SPECIFICATION	
PARTS LIST					

UNLESS OTHERWISE SPECIFIED
DIMENSIONS ARE IN INCHES
TOLERANCES ARE:
FRACTIONS   DECIMALS   ANGLES
± 1/16" .XX ± .XXX ± ±.01°

CONTRACT NO.

		MATERIAL	APPROVALS	DATE		CLAMP BAR	
			DRAWN			HOLD DOWN CLAMP	
		FINISH	CHECKED				
NEXT ASSY	USED ON		ISSUED		SIZE A	FSCM NO.	DWG. NO. EX1-1	REV.
APPLICATION		DO NOT SCALE DRAWING			SCALE FULL		SHEET

EX 17-1-3   Detail drawing of clamp bar.

EX 17-1-4   Detail drawing of spring sleeve.

PROBLEMS

ASSEMBLY
HOLD DOWN CLAMP

QTY REQD	FSCM NO.	PART OR IDENTIFYING NO.	NOMENCLATURE OR DESCRIPTION		MATERIAL SPECIFICATION	

PARTS LIST

UNLESS OTHERWISE SPECIFIED DIMENSIONS ARE IN INCHES TOLERANCES ARE: FRACTIONS DECIMALS ANGLES ± .XX ± ± .XXX ±	CONTRACT NO.					
MATERIAL	APPROVALS	DATE		ASSEMBLY HOLD DOWN CLAMP		
	DRAWN					
	CHECKED					
FINISH	ISSUED		SIZE A	FSCM NO.	DWG. NO. EX1-3	REV.
DO NOT SCALE DRAWING			SCALE		SHEET 1 OF 2	

EX 17-1-5   Assembly drawing of hold-down clamp.

QTY REQD	FSCM NO.	PART OR IDENTIFYING NO.	NOMENCLATURE OR DESCRIPTION	MATERIAL SPECIFICATION
			.08" OD, FREE HT 1¹/₄"	
			.08" Ø5 COILS, PLAIN ENDS	
1	4		SPRING: HARD DRAWN SPRING STEEL WIRE RH	
1	3		SPRING SLEEVE	DWG EX1-2
1	2		CLAMP BAR	DWG EX1-1
1	1		HEX HEAD BOLT ¹/₂" DIA X 3¹/₄" LG. HEX NUT	

PARTS LIST

UNLESS OTHERWISE SPECIFIED DIMENSIONS ARE IN INCHES TOLERANCES ARE: FRACTIONS ± DECIMALS .XX ± .XXX ± ANGLES ±	CONTRACT NO.			
	APPROVALS	DATE	**ASSEMBLY** HOLD DOWN CLAMP	
MATERIAL	DRAWN			
	CHECKED			
FINISH	ISSUED		SIZE **A**	FSCM NO.	DWG. NO. EX1-3	REV.
DO NOT SCALE DRAWING			SCALE		SHEET 2 OF 2

EX 17-1-6 Parts list for hold-down clamp.

EXAMPLE NUMBERS 17-2-1 THROUGH 17-2-8 (SUBASSEMBLY REQUIRED)

Objective: Design an articulated parallelogram, approximately 7″ long and 2″ wide, made of brass, which:

1. Will maintain parallelism between two end members of the device and provide space in the end members for the customer's machining.

2. Uses ball bearings at the pivot points and uses retaining rings for the pivot pins.

   NOTE: Articulated parallelograms are used in pick-and-place robots and in goniometers (devices to measure angular displacements of knee joints).

EX 17-2-1 Freehand sketch of articulated parallelogram.

PROBLEMS

EX 17-2-2   Layout drawing of articulated parallelogram.

EX 17-2-3   Subassembly detail drawing of linkage.

DWG. NO.		SH	REV.
REVISIONS			
REV	DESCRIPTION	DATE	APPROVED

GROOVE   .029" +.003 / −.000   Ø.210" +.003 / −.000
GROOVE

.13"   .55"   ①

.8"

Ø.2497" +.0000 / −.0002
AS PURCHASED
PIC

1		1		PIN ¼"Ø X Ø.8" LG	PIC SS303	
QTY REQD	FSCM NO.	PART OR IDENTIFYING NO.		NOMENCLATURE OR DESCRIPTION	MATERIAL SPECIFICATION	
			PARTS LIST			

UNLESS OTHERWISE SPECIFIED
DIMENSIONS ARE IN INCHES
TOLERANCES ARE:
FRACTIONS DECIMALS ANGLES
± .XX ± .XXX ±.01" ±

CONTRACT NO.

MATERIAL

APPROVALS   DATE

DRAWN

CHECKED

FINISH

ISSUED

NEXT ASSY   USED ON

APPLICATION   DO NOT SCALE DRAWING

PIVOT PIN
PARALLELOGRAM

SIZE A   FSCM NO.   DWG. NO. EX2-2   REV.

SCALE   SHEET

EX 17-2-4   Detail drawing of pivot pin.

DWG. NO.		SH	REV.
REVISIONS			
REV	DESCRIPTION	DATE	APPROVED

2"

1.25"   3/8"   ①

1"/2

11"/64   11"/64   1"/2

3"/8   1"/2

1 3/4"

1		1		BLOCK ½ X 2 X 1¾ LG	BRASS	
QTY REQD	FSCM NO.	PART OR IDENTIFYING NO.		NOMENCLATURE OR DESCRIPTION	MATERIAL SPECIFICATION	
			PARTS LIST			

UNLESS OTHERWISE SPECIFIED
DIMENSIONS ARE IN INCHES
TOLERANCES ARE:
FRACTIONS DECIMALS ANGLES
± 1/64" .XXX ± .01" ±

CONTRACT NO.

MATERIAL

APPROVALS   DATE

DRAWN

CHECKED

FINISH

ISSUED

NEXT ASSY   USED ON

APPLICATION   DO NOT SCALE DRAWING

END BLOCK
PARALLELOGRAM

SIZE A   FSCM NO.   DWG. NO. EX2-3   REV.

SCALE   SHEET

EX 17-2-5   Detail drawing of end block.

1.25"

2"

$1\frac{3}{8}''$ 4.00" $1\frac{3}{8}''$

4

2   1   3

$\frac{1}{2}''$

CUSTOMER TO MODIFY END BLOCKS

QTY REQD	FSCM NO.	PART OR IDENTIFYING NO.	NOMENCLATURE OR DESCRIPTION		MATERIAL SPECIFICATION	
			PARTS LIST			

UNLESS OTHERWISE SPECIFIED
DIMENSIONS ARE IN INCHES
TOLERANCES ARE:

FRACTIONS   DECIMALS   ANGLES
±   .XX ±   ±
         .XXX ±

CONTRACT NO.

MATERIAL

FINISH

APPROVALS | DATE

DRAWN

CHECKED

ISSUED

DO NOT SCALE DRAWING

ASSEMBLY
PARALLELOGRAM

SIZE	FSCM NO.		DWG. NO.		REV.
A			*EX2-4*		
SCALE			SHEET *1 OF 2*		

EX 17-2-6   Assembly drawing of articulated parallelogram.

PROBLEMS

8	4	RETAINING RING	PIC Z3-3	
4	3	PIVOT PIN	DWG EX2-2	
2	2	END BLOCK	DWG EX2-3	
2	1	LINKAGE	DWG EX2-1	
QTY REQD	FSCM NO.	PART OR IDENTIFYING NO.	NOMENCLATURE OR DESCRIPTION	MATERIAL SPECIFICATION

PARTS LIST

UNLESS OTHERWISE SPECIFIED
DIMENSIONS ARE IN INCHES
TOLERANCES ARE:
FRACTIONS   DECIMALS   ANGLES
±   .XX ±   ±
                .XXX ±

CONTRACT NO.

	APPROVALS	DATE
MATERIAL	DRAWN	
	CHECKED	
FINISH	ISSUED	

ASSEMBLY
PARALLELOGRAM

| SIZE **A** | FSCM NO. | DWG. NO. EX2-4 | REV. |

DO NOT SCALE DRAWING   SCALE   SHEET   2 OF 2

EX 17-2-7   Parts list for articulated parallelogram.

| QTY REQD | FSCM NO. | PART OR IDENTIFYING NO. | NOMENCLATURE OR DESCRIPTION | MATERIAL SPECIFICATION |

PARTS LIST

UNLESS OTHERWISE SPECIFIED
DIMENSIONS ARE IN INCHES
TOLERANCES ARE:
FRACTIONS   DECIMALS   ANGLES
±   .XX ±   ±
                .XXX ±

CONTRACT NO.

	APPROVALS	DATE
MATERIAL	DRAWN	
	CHECKED	
FINISH	ISSUED	

ASSEMBLY
PARALLELOGRAM

| SIZE **A** | FSCM NO. | DWG. NO. EX2-4A | REV. |

DO NOT SCALE DRAWING   SCALE   SHEET   1 OF 1

EX 17-2-8   Exploded assembly drawing of articulated parallelogram (an alternative type of assembly drawing).

PROBLEMS

Unless otherwise directed, do the problems following the steps indicated using traditional technical drawing instruments or a CAD system as assigned.

Conventional Instructions

STEP 1   Prepare a freehand sketch(es) of the device or component to be designed or changed. Refer to Figure 17-4, Example 17-1-1, and Example 17-2-2 for example freehand sketches.

STEP 2   Prepare a layout drawing(s) of the freehand sketch(es) from Step 1. Refer to Figure 17-5, Example 17-2-1, and Example 17-2-2 for example layout drawings.

STEP 3   Prepare the necessary detail drawings of each component of the device, one component per drawing. Refer to Figure 17-10, Figure 17-11, Example 17-1-3, Example 17-1-4, Example 17-2-4, and Example 17-2-5 for examples. Include required fits and tolerances as required.

STEP 4   Prepare a subassembly drawing if needed. Refer to Figure 17-9 and Example 17-2-3 for example subassembly drawings.

STEP 5   Prepare the final assembly drawing with the appropriate parts list. Refer to Figure 17-7 for a Method A assembly drawing and to Figure 17-8 for a Method B drawing. Also refer to Example 17-1-5, Example 17-1-6, Example 17-2-6, and Example 17-2-7 for final assembly drawings with parts lists on separate sheets. Example 17-2-8 is an example of an exploded assembly drawing.

CAD INSTRUCTIONS

Students who plan to complete the following problems on a CAD system should begin with these general instructions.

1. Analyze the problem carefully.

2. Plan your drawing sequence. In what order will you develop the drawing?

3. Read the specific instructions for the problem and begin.

Mechanical   **PROBLEM 17-1**

Prepare a freehand sketch of a step stand, two levels, for general use in homes; institutions (libraries, schools, etc.); and business offices. Use black anodized aluminum sheet and structural angles. Use any combination of $1 \times 1 \times \frac{1}{8}$, $1\frac{1}{2} \times 1\frac{1}{2} \times \frac{1}{8}$, or $2 \times 2 \times \frac{1}{8}$ angles and assume bolted construction.

Mechanical   ## PROBLEM 17-2

The winch shown is to be used to pull boats up a ramp, and it needs a support base so it can be mounted on a horizontal wooden platform 1″ thick and about 4′ high.

a. Prepare a freehand sketch of the support base.
b. Prepare a layout drawing of the support base.

Mechanical

PROBLEM 17-3

Prepare a freehand sketch of a dolly approximately 2′ × 3′ using two fixed and two swivel casters. Fasteners should not project above the top surface. A convenient means to hand-carry the dolly should be provided.

Prepare layout, detail, and assembly drawings as assigned.

Mechanical

PROBLEM 17-4

The torque wrench shown in the freehand sketch needs to be developed further and sent on to be manufactured in the shop.

a. Prepare a layout drawing of the torque wrench.
b. Prepare detail drawings of the torque wrench.
c. Prepare an assembly drawing and parts list for the torque wrench.

Mechanical ## PROBLEM 17-5

The clamp shown in the freehand sketch needs to be developed further and sent on to be manufactured in the shop.

a. Prepare a layout drawing of the clamp.
b. Prepare detail drawings of the clamp.
c. Prepare an assembly drawing and a parts list for the clamp.

Mechanical ## PROBLEM 17-6

GIVEN: A layout drawing of adjustable dividers. Prepare detail and assembly drawings as assigned.

Mechanical **PROBLEM 17-7**

GIVEN: A layout drawing of a wood shop clamp. Prepare detail and assembly drawings as assigned.

Mechanical **PROBLEM 17-8**

Conceptual views of a toggle clamp are shown in the freehand sketches. When pivot B on the toggle clamp arm ABT passes below the line of centers between A and C, the pressure at P causes the elastic head to compress. Then, B is "locked" in position, and the object under the elastic head at P is clamped securely. A typical compression distance is approximately 1.5 mm. The adjustments at E accommodate different sizes of objects.

a. From the information given, prepare a layout drawing to establish the remaining dimensions of the components, the pivot point locations, and the bolt hole locations.

b. Prepare detail drawings of the components.

c. Prepare an assembly drawing and parts list for the toggle clamp.

Mechanical PROBLEM 17-9

GIVEN: A layout drawing of a jack. Prepare detail and assembly drawings as assigned.

Mechanical PROBLEM 17-10

GIVEN: A layout drawing of an adjustable compass. Prepare detail and assembly drawings as assigned.

Mechanical | **PROBLEM 17-11**

Limit switches, which are basically electrical on-off switches, limit the travel of machine parts, which act as interlocks, and indicate positions of devices. For example, the machines in the photographs in Chapter 20 use limit switches to limit travel. Limit switches activated by open doors prevent an elevator from moving. Positions of remote devices in hostile environments, such as control rods in nuclear reactors, are indicated safely by limit switches. Rotating antennas can be stopped at preset positions by limit switches to keep wires from tangling.

Use information in the figure to do a layout drawing to determine the dimensions of an actuating shoe that would cause the limit switch to operate and stop the rotation of the antenna. Also, do a layout drawing of the support plate for the actuating shoe. Include a hub and set screw. Hub diameters may average approximately 2 times shaft diameters and are about 2 times shaft diameters long. The hub is to be welded to the support plate. Use low-carbon steel.

Mechanical | **PROBLEM 17-12**

An air-actuated cylinder is to be used to bend a specimen in a fatigue test apparatus.

The air cylinder and the specimen are shown in the partially completed layout drawing. You are to devise a means to support the cylinder and the specimen and to provide adjustments as indicated.

a. Complete the layout drawing with a design that provides the support and adjustments.

b. Prepare detail drawings of your design.
c. Prepare an assembly drawing with a parts list of your design. When you draw the cylinder in the assembly drawing, use phantom lines. Purchased parts such as motors, reducers, generators, and cylinders often are drawn with phantom lines.

Mechanical   **PROBLEM 17-13**

GIVEN: A layout drawing of an adjustable lifting clamp. Prepare detail and assembly drawings as assigned.

HANDLE

2.63

LARGER THAN Ø

Ø.18

LOCK

SPRING

Ø.32 / UNF

RC FIT

RETAINING RING

3.25

RETAINING RINGS TO SUIT (.32 NOM. SIZE)

.125 NOM. SIZE

3.25

.56

PEEN OVER (3 PLACES)

OPEN POSITION

1.5

CLOSED POSITION

.88

JAWS

.31

120°

.56

5.5 OPEN POSITION

Mechanical ## PROBLEM 17-14

GIVEN: A layout drawing of a lever-controlled valve. Prepare detail and assembly drawings as assigned.

6X Ø.375-THRU ON A 3.00 B.C.

RC FIT

Ø3.75

SPRING PIN

Ø.38-UNC

.75

SPRING PIN

Ø.375 NOM SIZE

HANDLE   RETAINING RING

RC FIT   Ø7.50   *FN FIT*

LINK   RETAINING RING

WELD   FASTENER 1.0 LONG

TUBING Ø2.0/.13 WALL

TUBING Ø2.5/.375 WALL

WELD   3.75

SPRING Ø2.5 O.D. Ø.18 WIRE/4.0 FREE LENGTH

OPEN POSITION

RC FIT   1.5 **TRAVEL**   SHARP POINT

CLOSED POSITION

RETAINING RING   *WELD*

WELD

FLANGE FREE TO TURN

6X Ø.31- THRU ON A 3.00 B.C.

5.88

2.75

Ø3.75

3.00

Mechanical   ## PROBLEM 17-15

A hand-powered winch is shown in the isometric sketch. Using the information in the freehand sketches of the components of the winch, 17-15A through 17-15H:

a. Prepare detail drawings of each component. (Note that layout drawings are omitted in this exercise due to the completeness of the freehand sketches. Discrepancies, if any, will be found while doing the final assembly drawing.)

b. Prepare subassembly drawings of the drum 17-15B and the handle 17-15H.

c. Prepare a final assembly drawing with a parts list for the winch.

Drill and CSK
for ⅜"⌀ Flat
Head Cap Screws
4X on
5.75"⌀ B.C.

Spur Gear 20°
48T DP=8
¼" steel plate
1035

Drum Gear

1 1/16 ⌀

5.125

¼

.375

1/16" BOSS
1¾" OD

125

¼ R

1⅜⌀ RC-3

2¾⌀   1¾⌀

Drill and Tap for
⅜ –UNF-2B 4X
5.25 ⌀ B.C.

Cast Al Drum

RC-3 Fit
1" ID   1⅜ OO FN-1 Fit

Bronze

1⅛

Bushing for Drum
2 Reqd

PROBLEM 17-15 (continued)

1050 Steel

Drum Shaft

(C)

20° Spur gear
12T DP=8
Steel 1035

Pinion Gear

(D)

Cast Aluminum Base
6061 T6

(G)

Pawl 5/16 Plate
(Refer ch 4 for Geom.)

(E)

Use Truarc 5100-50
Retaining Ring

Winch Handle

Pawl Shoulder Bolt
Steel

(F)

Aluminum Handle
6061-T6

(H)

Mechanical   PROBLEM 17-16

The corrosion bath shown is rocked back and forth by an arm that is coupled to a variable-speed drive mechanism. Corrosive liquid is sloshed over the test specimens. The equipment was originally built by a skilled mechanic from sketches that have been misplaced or destroyed. Detail and assembly drawings of this testing equipment are needed by another company to build their own bath. From the information given in the sketches, do the following:

a. Prepare a layout drawing of the bath equipment.
b. Prepare detail drawings of the components.
c. Prepare subassembly drawings of the shaft A supports and the container for the plastic bath.
d. Prepare an assembly drawing with a parts list for the bath equipment.

PROBLEMS

PROBLEMS

Mechanical **PROBLEM 17-17**

Using detail drawings 17-17A through 17-17O, prepare a three-view assembly drawing and a parts list. Also, the following is a summary of the FITS specified for the various components. (Calculate these and show your work.)

(1) FN-1/1.062 Parts 1, 12
(2) FN-2/.688 Parts 1, 2
(3) LC-4/1.5 Parts 1, 11
(4) RC-5/.562 Parts 3, 4

(5) RC-5/.375 Parts 4, 5
(6) LC-2/.125 Parts 4, 8
(7) FN-2/.250 Parts 9, 14
(8) RC-6/.938 Parts 13, 14, 15

NO. 1 CRANKCASE
ALUMINUM ALLOY

(A)

NO. 2 BUSHING CRANKCASE
LEADED BRONZE

(B)

NO. 3 CRANKCASE SUBASSEMBLY
MAT'L AS NOTED

(C)

PROBLEM 17-17 (continued)

(D)

NO. 4 CRANKSHAFT
AISI 1040 STEEL

SECTION A-A

NO. 5 ROD-CONNECTING
ALUMINUM ALLOY 7075-T6

(E)

NO. 7 SPINNER
ALUMINUM ALLOY

(G)

NO. 8 PIN-DRIVE WASHER
1010 STEEL

(H)

NO. 9 PIN PISTON
1010 STEEL

(I)

NO. 6 DRIVE WASHER
AISI C1018 STEEL

(F)

NO. 10 SPACER-PISTON
BRASS

(J)

PROBLEM 17-17 (continued)

NO. 11 BACKPLATE
CAST ALUMINUM

(K)

NO. 12 FINS-CYLINDER
ALUMINUM ALLOY

(L)

NO. 14 PISTON
DUCTILE IRON

(N)

NO.13 SLEEVE-CYLINDER
DUCTILE IRON

(M)

NO. 15 HEAD CYLINDER
ALUMINUM ALLOY

(O)

PROBLEM 17-18: REVERSE ENGINEERING

Mechanical

Reverse engineering basically involves:

1. "Taking apart" a manufactured device.
2. Analyzing each component of the device for design features, size, materials, function, manufacturing processes, and any other factors such as aesthetics, color, and surface finishes.

Valuable insights can be learned from the exercise of "working backwards" as compared to "working forward."

1. Select or be assigned one of the following to be "taken apart" and analyzed as described above:

a. Component(s) from an electronic product such as a hand calculator, a printing calculator, a VCR, a TV, a light switch, or a timer. (Note that electronic repair shops are excellent sources of discarded products.)
b. Automotive devices
c. Appliances
d. Exercise and sports equipment
e. Hardware

2. Prepare a freehand sketch of the component selected or assigned, and list the design features, size, material(s), function, and other properties.

PROBLEM 17-19

Mechanical

Assume that you are a designer-drafter employed by a large company that designs and produces yard tools for average homeowners. Your company has received a request for a bid on 100,000 wheelbarrows of approximately a 5 cubic foot capacity. You are given a freehand sketch of a wheelbarrow to more accurately determine dimensions, sizes, and the buyout requirements (fasteners and the wheel).

Assume that the time constraints require that you must establish a selling price for the wheelbarrow because your company's quotation department must meet a major deadline on another project. An approximate method of quickly setting a selling price is to calculate a "price per pound" of existing wheelbarrows

already on the market. Then, multiply the weight of your wheelbarrow design times the price per pound for an approximate selling price.

You may need to visit a hardware store to obtain data or find an appropriate catalog. One of your colleagues said that his wheelbarrow was approximately $2 per pound.

Time is a constraint, so:

a. Prepare a layout drawing of the wheelbarrow and estimate its weight.
b. Calculate a price per pound of existing wheelbarrows and establish an approximate selling price.
c. Prepare an assembly drawing with a parts list for the wheelbarrow.

Mechanical | **PROBLEM 17-20**

Attribute listing is similar to reverse engineering but not as comprehensive. Problem 17-20 combines the attribute listing procedure with technical drawings.

Find a music stand (or an object assigned by your instructor) and do the following:

a. Measure the components of the stand, note the materials used, determine the manufacturing processes, list the function(s) of the stand, and comment on any other facets such as color, weight, and stability. Prepare freehand sketches of the components. Incorporate any design changes you deem appropriate.

b. Prepare layout drawings of your freehand sketches, especially if you recommend changes.

c. Prepare detail drawings of the components.

d. Prepare subassembly and assembly drawings as needed.

Mechanical | **PROBLEM 17-21**

The angle stop shown in the drawing is to be changed according to ECO Number C159, originated by GNC on Sept. 11, and justified because of a new model of anchorbolt puller to be manufactured. The description of change on a typical ECO form indicates that the 210 mm dimension changes to 250 and the holes change from a diameter of 13 to 19. An ECO committee determined that the change was necessary and that the change would not affect performance, reliability, or safety. Each member of the committee signed off on the ECO (see Figure 17-13). Make a machine copy of the drawing as it is shown. Make the changes. (Refer to Figure 17-14.) Record the changes in the revisions block (upper right-hand corner).

Mechanical   PROBLEM 17-22

An engineering change request (ECR) involving the articulated parallelogram in Example 17-2-1 has evolved into an engineering change order (ECO). The device is to be modified as follows:

1. Replace the ball bearings in the linkages with bronze bushings.

2. Use aluminum 6061-T6 instead of brass for the linkages and the end blocks.
   a. Prepare a layout drawing for the changes.
   b. Prepare detail drawings of the new design.
   c. Prepare an assembly drawing with a parts list for the new design.

Mechanical   PROBLEM 17-23

Use the five steps listed in the "Conventional Instructions" section earlier in this chapter to incorporate an ECO that requires that ball bearings replace the sliding bearings, Part #4.

| Mechanical | ## PROBLEM 17-24 |

Follow the five steps listed in the "Conventional Instructions" section earlier in this chapter to develop the puller shown. For this problem, use the information from Figures 17-7 and 17-8 to prepare freehand sketches for each part. Assume an overall length of 9″ for the assembly of the yoke, the screw center, and the hooks. Make the yoke and hooks of medium-carbon steel, AISI 1045. Make the hooks at least ¼″ thick, and use ½″ 20UNF threads or larger for the screw center.

MUST INCLUDE: **DRAWING NUMBER**
TITLE
NUMBER REQUIRED PER ASSEMBLY

NO.	DRAWING NO.	ITEM	REQ'D
–	–	1/4-20 UNC	–
5	PURCH	LOCK NUT-HEX	2
–	–	1/4-20 UNC X 1 1/2 LG.	–
4	PURCH	SCREW-CAP HEX HD	2
3	A 661982	HOOK	2
2	A 661981	SCREW-CENTER	1
1	A 661983	YOKE	1
NO.	DRAWING NO.	ITEM	REQ'D

| Mechanical | ## PROBLEM 17-25 |

The pressure transducer shown indicates pressure by the distortion of the piezoelectric crystal that generates a voltage proportional to the distortion. A digital voltmeter is used to read the voltage. Assembly of the transducer could be simplified if the tubular spacer were part of the hex-shaped fitting with the ½″ UNF threads. Prepare a freehand sketch of your idea to simplify the assembly. The crystal is a purchased item. Use stainless steel.

Mechanical PROBLEM 17-26

Manufacturing and assembly of the adjustable roller support shown could be improved upon. Prepare a freehand sketch of your idea for improvement and then follow the five steps listed in the "Conventional Instructions" section earlier in this chapter to develop the adjustable roller support. Note that slotted holes are used frequently for adjustments.

PROBLEM 17-27

Start your own technical file by selecting manufacturing catalogs that list products in areas of your interests. Label the catalogs with "100" numbers like the listings in this chapter in the section on Technical Files. Select at least 10 catalogs as a beginning.

As a suggestion, visit one or more of the following locations where you would most likely find a library of catalogs to help start your list: your own drafting classroom, an engineering library, maintenance shops on your campus, or local manufacturing firms.

PROBLEM 17-28

Drawings for patents of mechanical devices must clearly depict the assembly and function of the devices. Peruse a collection of patents or patent gazettes and list the patents with comments on the different types of drawings used to communicate information. List at least 10 patents (patent number, date, and a brief description of the device). Suggestion: librarians cannot only help you find the patents but can also help you find patents in your areas of technical interest.

18

Pictorial Drawings

CHAPTER OUTLINE

Purpose of pictorial drawings • Oblique drawings • Isometric drawings • Axonometric: drawing and projection • Other oblique drawings • Perspective drawings • Perspective view • Summary • Review questions • Chapter 18 problems • CAD instructions

CHAPTER OBJECTIVES

Upon completion of this chapter, students should be able to do the following:

▶ Explain the purpose of pictorial drawings.

▶ Demonstrate proficiency in developing oblique drawings, including cavalier and cabinet.

▶ Demonstrate proficiency in developing isometric drawings, including diametric and trimetric.

▶ Demonstrate proficiency in drawing circles, curves, and irregular shapes in isometric and on flat and inclined surfaces.

▶ Demonstrate proficiency in developing one-, two-, and three-point perspective drawings.

Purpose of Pictorial Drawings

Pictorial drawings facilitate the communication of technical information during all phases of the design process—from early sketching of product ideas for future uses to talking through concepts with colleagues and, finally, to the more formal pictorial drawings (for selling products, maintenance manuals, catalogs, assembly instructions, etc.). This chapter begins with the three most common types of pictorial drawings used by drafters, designers, and engineers—oblique, isometric, and perspective. Sketching these three types was introduced in Chapter 3.

After the discussion of oblique, isometric, and perspective drawings, more complete discussions will follow on additional types of pictorial drawings and projections.

Oblique Drawings

Oblique drawing is the easiest of the three to draw. In oblique drawing, one surface of the object, usually the front view, is drawn exactly as it would appear in a multiview drawing. Circles are true circles, and rectangles parallel to the frontal plane are in true size (**Figure 18-1**). Circles are drawn with a compass or circle template, and rectangles have square corners.

Constructing Oblique Drawings

Oblique drawings use three axes: two at right angles (for example, X and Y in Figure 18-1) and one, the Z axis, usually at an angle of the convenient 30° or 45° standard triangle. Measurements are taken from the multiview drawing; X and Y dimensions match those in front views, and Z dimensions, depth, are laid out parallel to the third axis.

FIGURE 18-1   Axes for oblique drawings.

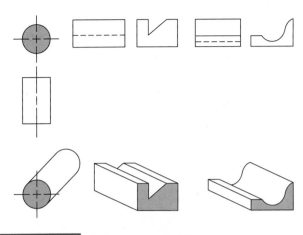

FIGURE 18-2   Advantages of oblique.

Cylinders, Angles, and Curves in Oblique Drawings

Figure 18-2 has further illustrations of the primary advantages of oblique drawings, especially starting with the front view. Cylinders are round, angles are in true size, and irregular curves are not distorted.

Isometric Drawings

Isometric drawings are the most used of the three common pictorials. They are easier to draw than perspective drawings, and they illustrate objects almost as well as perspective drawings.

Isometric drawings use three axes, sometimes called principal edges, at 120° to each other (**Figure 18-3**).

How to Make an Isometric Drawing

GIVEN: A multiview drawing of an object (**Figure 18-4A**).

STEP 1   Locate the starting point, and lightly draw the three principal edges at 120° using a 30°–60° triangle (**Figure 18-4B**).

STEP 2   Transfer the depth, height, and width directly from the multiview drawing (**Figure 18-4C**), to the three principal edges. Measure full size directly along each edge. Lightly construct an isometric, basic outline, or "box," of the object, Figure 18-4C.

STEP 3   Transfer true-size lengths, lengths parallel to the principal axes, from the various features of the object (**Figure 18-4D**).

STEP 4   Check all lines, correct if necessary, and darken in the object using the correct line thickness (**Figure 18-4E**).

FIGURE 18-3   Isometric principles.

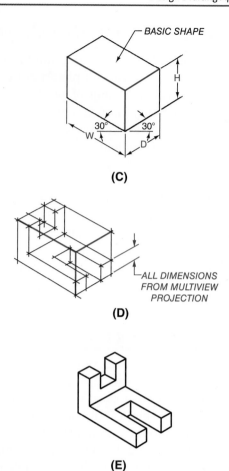

FIGURE 18-4   (*Continued*)(C) Step 2; (D) Step 3; (E) Completed isometric view.

GIVEN: MULTIVIEW PROJECTION

(A)

(B)

FIGURE 18-4   (A) How to draw an isometric drawing; (B) Step 1;

How to Center an Isometric Object

Save time in locating isometric drawings in or near the center of a drawing space by following these steps.

GIVEN: A multiview of an object (**Figure 18-5A**).

STEP 1 Locate the center point of the work area and do the following: go North ½ H, go S60° W a distance of ½ D, and go S60° E, ½ W to establish a starting point (**Figure 18-5B**).

STEP 2 From the starting point, draw the basic shape of the object (**Figure 18-5C**).

STEP 3 Construct the object within the basic shape (**Figure 18-5D**).

Offset Measurements

Offset measurements are used to locate one feature in relationship to another. Measure offsets parallel or perpendicular to one of the three standard planes—frontal, top, or profile—from the multiview of the object being drawn (**Figure 18-6A**).

GIVEN: MULTIVIEW DRAWING

(A)

THE WORK AREA

WORK AREA

—1/2 'D'
—1/2 'H'
—CENTER OF WORK AREA
—STARTING POINT

②①
③

—1/2 'W'

(B)

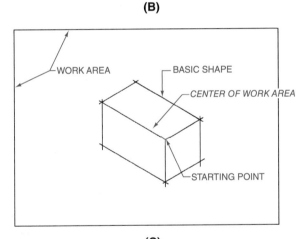

WORK AREA   BASIC SHAPE

CENTER OF WORK AREA

STARTING POINT

(C)

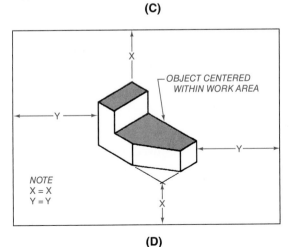

OBJECT CENTERED
WITHIN WORK AREA

X

Y

Y

X

NOTE
X = X
Y = Y

(D)

FIGURE 18-5   (A) Centering an isometric view within the work area; (B) Step 1; (C) Step 2; (D) Completed isometric view.

GIVEN: MULTIVIEW PROJECTION

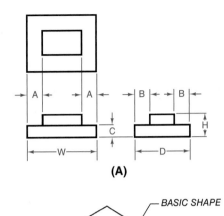

A   A   B   B

C

W   H

D

(A)

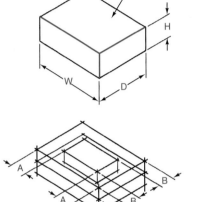

BASIC SHAPE

H

W   D

A   B
A   B
C

OFFSET DIMENSIONS
ISOMETRIC PROJECTION

(B)

FIGURE 18-6   (A) Multiview projection; (B) Offset measurements.

Develop the isometric drawing as before; first, the basic shape or "box" with dimensions W, D, and H, and then the offset measurements A, B, and C parallel or perpendicular to the frontal, top, or profile planes (**Figure 18-6B**). Note that lines that are truly parallel in the multiview drawing must be parallel in the isometric drawing.

Nonisometric Lines

If a line is not parallel or perpendicular to any of the three principal edges in an isometric drawing, it is a *nonisometric line.* **Figure 18-7A** shows a multiview drawing of an object with a 30° inclined surface, labeled as shown with a, b, a′, and b′.

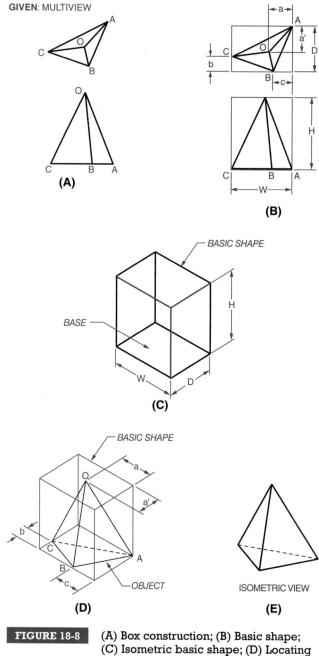

FIGURE 18-7   (A) Nonisometric lines; (B) Locating points.

FIGURE 18-8   (A) Box construction; (B) Basic shape; (C) Isometric basic shape; (D) Locating points of the object; (E) Isometric view completed.

Construct the isometric drawing as before; "box" first, then the feature using the X and Y dimensions and, finally, darken in the lines of the object (**Figure 18-7B**). Line a-b is not the true length, and the 30° angle is distorted.

Box Construction for Irregular Objects

Some objects (**Figure 18-8A**) have lines that do not line up with the isometric axes, so use a box or basic shape to establish offset dimensions for drawing the object (**Figure 18-8B**).

Next, draw an isometric box (**Figure 18-8C**). Locate key points of the object using offset measurements (**Figure 18-8D**). Finish the object (**Figure 18-8E**).

Hidden Lines

Hidden lines are not used in isometric drawings unless needed to show an important feature that otherwise would not be seen. Try to orient objects so hidden lines are not needed, not only in isometric drawings but also in all pictorials.

Isometric Curves

Construct isometric curves in isometric drawings using offsets or a grid. For a series of offset distances,

GIVEN: MULTIVIEW

GIVEN: MULTIVIEW

FIGURE 18-9A Drawing an isometric curve (offset method).

FIGURE 18-10A Drawing an isometric curve (grid method).

FIGURE 18-9B Drawing an isometric curve (offset distances and depth).

FIGURE 18-10B Isometric grid layout.

use a number of evenly spaced lines, all parallel to one of the principal edges (**Figure 18-9A**). Transfer these lines to the isometric layout and transfer each distance, respectively, to locate points on the curve (**Figure 18-9B**). Give depth to the drawing by projecting D distances at 30°, which is parallel to an isometric axis, from each point on the curve. Use a French curve to connect the points.

For transferring a curve using a grid, choose a grid spacing that will give a sufficient number of points along the curve to give as much detail of the curve as needed (**Figure 18-10A**).

Construct the basic shape of the object using the W, H, and D dimensions. Transfer the grid (**Figure 18-10B**). Finally, transfer the irregular curve, square by square. Use a French curve to connect the points on the curve.

Irregularly Shaped Objects

Transfer and draw irregularly shaped objects by dividing the object into a series of convenient sections. Draw each section as an isometric (**Figure 18-11A**). The two-view drawing of the boat has been simplified for this example. Divide the object into a series of sections, A through H. Use a logical baseline (in this case, a center line). Locate and draw each section (**Figure 18-11B**). Connect all the sections and complete the boat (**Figure 18-11C**).

Isometric Circles and Arcs

Drawing isometric circles and arcs, without using templates, takes too much time for isometric drawing. Use an isometric ellipse template (**Figure 18-12**). The template has eight hash marks printed around

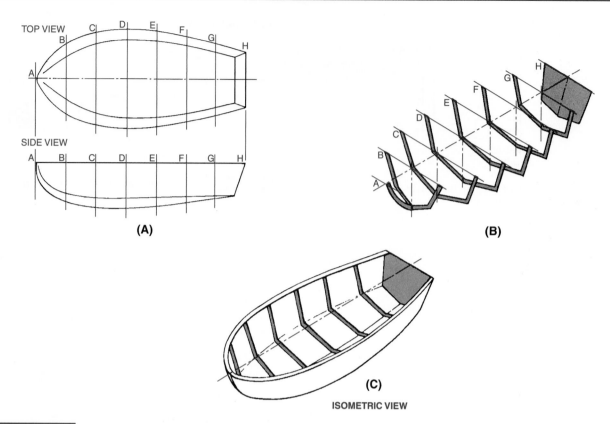

(A)

(B)

(C)

ISOMETRIC VIEW

FIGURE 18-11   (A) Irregularly shaped object; (B) Drawing each section; (C) Completed object.

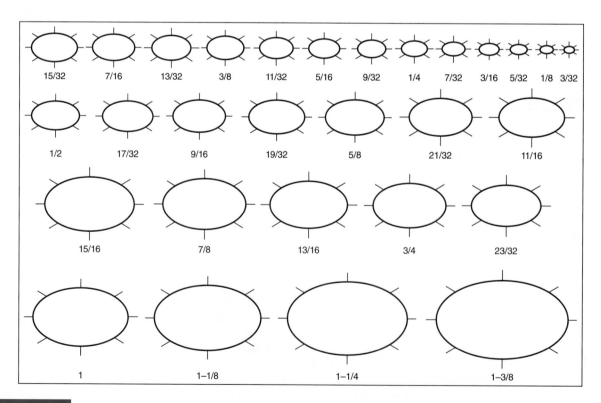

FIGURE 18-12   Isometric ellipse template.

the perimeter of each ellipse. Four indicate the isometric center lines of the circle or arc. For example, an ellipse for a one-inch circle would be labeled number –1– and would measure 1″ across the ellipse along the isometric center line. These same hash marks also indicate the four tangent points of the isometric circle with the isometric basic shape of the circle (**Figures 18-13A** and **18-13B**). The other four hash marks line up with the axes of the circle; two are coincident with the axis of the circle and are lined up across the narrow part of the ellipse, and two are perpendicular to the axis and are lined up across the longest part of the ellipse (**Figure 18-13C**). The longest distance, across the one-inch ellipse, actually measures more than one inch due to the way the basic shape is constructed for isometric drawing.

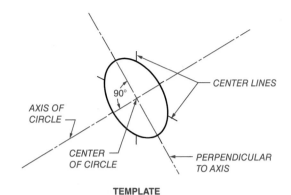

TEMPLATE

FIGURE 18-13C   Isometric ellipse template aligned on axis of circle 1 2 3 4.

How to Use Isometric Ellipse Templates for Isometric Drawings

STEP 1   For the object given (**Figure 18-14A**), locate the center of the circle or arc to be drawn using offset measurements, and construct the basic shape for the circle so that the sides of the basic shape are parallel to the two appropriate principal edges.

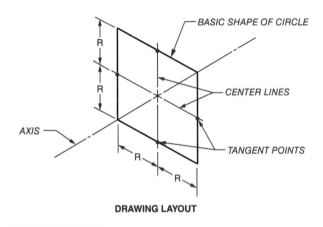

DRAWING LAYOUT

FIGURE 18-13A   Basic shape of the circle.

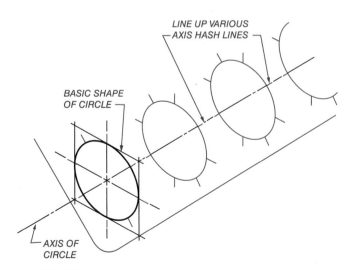

FIGURE 18-13B   Draw the ellipse within the basic shape.

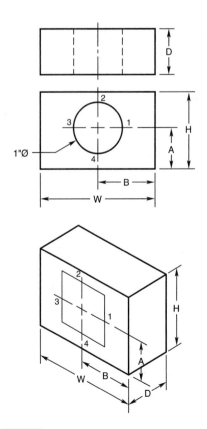

FIGURE 18-14A   How to use an isometric ellipse template.

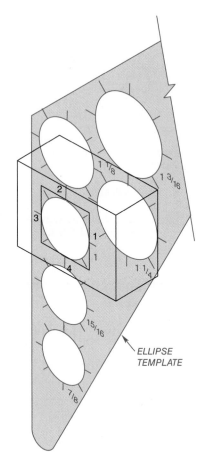

FIGURE 18-14B   Line up hash marks on tangent points.

STEP 2   Line up the 1″ ellipse template so that the isometric center line hash marks touch the tangent points 1, 2, 3, and 4 of the basic shape (**Figure 18-14B**). Trace the ellipse. If some doubt exists as to which hash marks to line up, refer to **Figure 18-15** for guidance.

FIGURE 18-15   Isometric circles in various planes.

How to Draw True Isometric Circles

Assume that a relatively accurate isometric drawing of a circle is required, but a template is not available for the size needed.

GIVEN: A regular circle in its basic shape and the same circle as it would be drawn in an isometric drawing in its basic shape (**Figure 18-16A**). In both drawings, the tangent points are given "clock" numbers.

STEP 1   Use the circle in the basic shape, a square, with the tangent points 3, 6, 9, and 12. Use the remaining clock numbers and divide the circle into the rectangles shown (**Figure 18-16B**).

STEP 2   Draw the basic shape of the circle, and use the offset method to locate points 1, 5, 7, and 11 using the X and Y dimensions (**Figure 18-16C**).

STEP 3   Locate points 2, 4, 8, and 10 in a similar manner (**Figure 18-16D**).

STEP 4   Connect the 12 points with a French curve, and finish the isometric drawing of the circle (**Figure 18-16E**).

To draw an arc of a circle, use these steps and the portion of the circle needed (**Figure 18-17**).

How to Draw Approximate Isometric Circles

Assume that an approximate isometric circle is needed in an isometric drawing, but a template is not available with the size required. The four-center method described in the following steps can be used.

GIVEN: A regular circle with a radius X, Figure 18-16A.

STEP 1   Draw the basic shape of the circle in the appropriate location in the drawing. Label the four tangent points 1, 2, 3, and 4 (**Figure 18-18A**).

STEP 2   From each of the four tangent points, draw a line at 90° to the opposite sides of the basic shape (**Figure 18-18B**). These lines intersect at four locations called the swing points or centers. Label them A, B, C, and D (**Figure 18-18C**). A line can be used to check the location centers A and B.

STEP 3   Construct a radius at each of the four centers, A, B, C, and D, as shown in **Figures 18-18D** and **18-18E**.

STEP 4   Darken in the completed isometric circle and center lines (**Figure 18-18F**).

Construct arcs by using the four-center method and the radius X discussed in these four steps (**Figure 18-18G**).

GIVEN: REGULAR CIRCLE   ISOMETRIC CIRCLE

(A)

(B)

(C)

(D)

ISOMETRIC CIRCLE

(E)

FIGURE 18-16   (A) Regular circle and isometric circle;
(B) Step 1; (C) Step 2; (D) Step 3;
(E) Completed isometric circle.

How to Draw an Ellipse on an Inclined Plane

GIVEN: A two-view drawing of a cylinder cut at an angle (**Figure 18-19A**).

STEP 1 Divide the view of the cylinder, showing it as a circle, in 12 equally spaced arcs (1 through 12). Project the points into the view showing

FIGURE 18-17   How to draw an isometric arc (offset method).

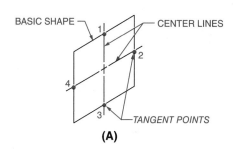

BASIC SHAPE   CENTER LINES

TANGENT POINTS

(A)

DRAW FOUR LINES 90° FROM EACH OF THE 4 TANGENT POINTS

(B)

SWING POINTS A, B, C, D

CHECK LINE

(C)

FIGURE 18-18   (A) Step 1 (how to draw an approximate isometric circle); (B) and (C) Step 2;

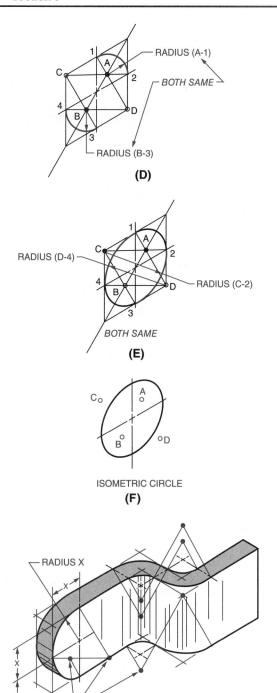

FIGURE 18-18   (*Continued*) (D) and (E) Step 3;
(F) Completed isometric circle;
(G) Drawing isometric arcs
(four-center method).

the base of the cylinder as an edge. Locate the true-length line segments from each point, Figure 18-19A.

STEP 2   Construct an isometric circle of the base, and locate each of the 12 points using the offset technique (**Figure 18-19B**).

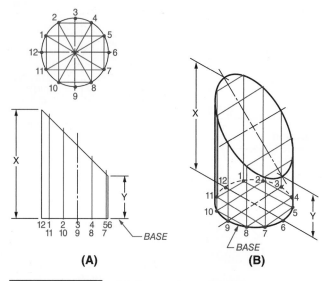

FIGURE 18-19   (A) How to draw an ellipse on an inclined plane. Given: two views; (B) An isometric ellipse on an inclined plane.

STEP 3   Project each true-length line segment up from the base, and connect the end-points to form the boundary of the elliptically shaped surface. Use a French curve.

Isometric Intersections

An isometric drawing can be constructed from a completed multiview drawing, showing an intersection(s), by following the steps outlined in this chapter. First, construct a basic shape, box, and then transfer a sufficient number of points from the multiview drawing, using the offset method, to complete the isometric drawing.

Isometric Dimensioning

Use the same dimensioning standards, rules, and methods discussed in Chapter 6 to dimension isometric drawings. Furthermore, use unidirectional dimensioning for clarity and ease of placing the dimensions (**Figure 18-20**). Note that extension lines and dimension lines are parallel to one of the three axes.

Isometric Rounds and Fillets

Rounds and fillets in isometric drawings can be shown three ways (**Figures 18-21A** through **18-21C**). Use whichever way, or combinations of ways, that best shows the fillets or rounds and where they are located.

Isometric Knurls, Screw Threads, and Spheres

Isometric knurls are usually drawn with straight lines (**Figures 18-22A and 18-22B**). Part A shows the

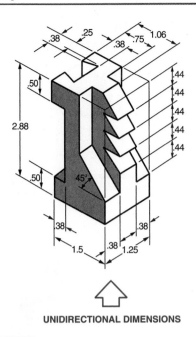

UNIDIRECTIONAL DIMENSIONS

FIGURE 18-20   Isometric dimensioning.

FIGURE 18-22   Isometric knurl.

location of a proposed knurl and how to indicate a chamfer. Part B shows the finished drawing.

Isometric threads are readily represented by a series of ellipses drawn along the center line of the fastener, parallel to each other, and approximating the thread spacing (**Figure 18-23**).

Isometric spheres require three ellipses to be drawn (**Figure 18-24**)—one along each axis and one circumscribing the first two. If only a portion of a sphere is wanted, then the first two ellipses can be used to outline a half-sphere (**Figure 18-25**), a quarter-sphere (**Figure 18-26**), or a three-quarter sphere (**Figure 18-27**).

Based on the two-view drawing in **Figure 18-28**, draw a flat surface in a sphere by following the implied steps in **Figure 18-29**.

FIGURE 18-23   Isometric threads.

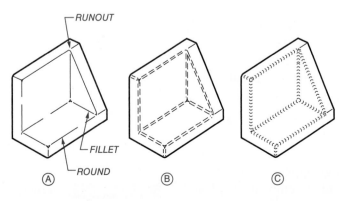

FIGURE 18-21   Isometric rounds and fillets.

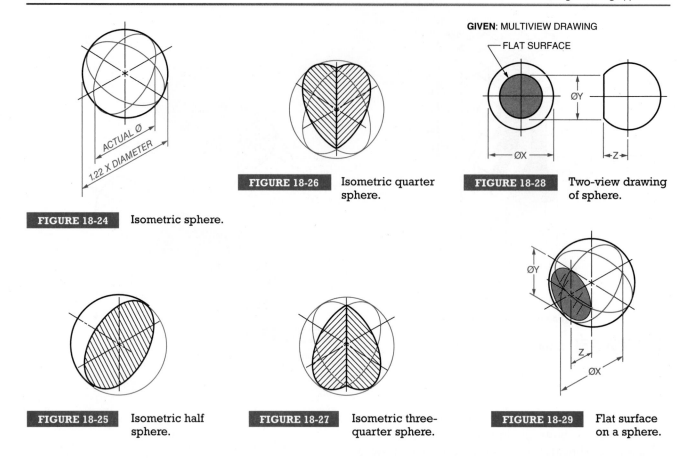

FIGURE 18-24   Isometric sphere.

FIGURE 18-26   Isometric quarter sphere.

GIVEN: MULTIVIEW DRAWING

FIGURE 18-28   Two-view drawing of sphere.

FIGURE 18-25   Isometric half sphere.

FIGURE 18-27   Isometric three-quarter sphere.

FIGURE 18-29   Flat surface on a sphere.

Axonometric: Drawing and Projection

Drafters, designers, and engineers use the pictorial drawings—oblique, isometric, and perspective—discussed in the first part of this chapter because approximations of these drawings are easy to sketch, and relatively accurate drawings of these types are easy to construct. Additional pictorial drawings and several projections will be considered next.

Isometric Projection

Isometric projection causes dimensions along the three principal edges to be foreshortened to approximately 82% of their true length (**Figure 18-30**). Note that the diagonal of the 1″ cube is seen as a point in the A view; the three edges 3-2, 3-4, and 3-7 equal 0.816″; and the angles between edges 3-2, 3-4, and 3-7 are all equal to 120°.

Isometric Assemblies

Isometric assembly drawings appear in two popular formats: exploded and assembled. *Exploded*

isometric drawings show all the components of an assembly as individual isometric drawings, including fasteners; however, the components are aligned in space in the same way they are to be assembled. See Problem 18-52 for an exploded-assembly sketch.

Exploded-assembly instructional drawings usually accompany products that are to be assembled

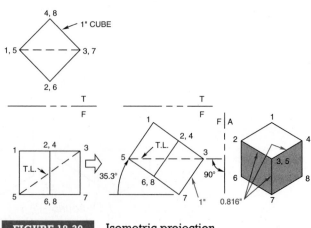

FIGURE 18-30   Isometric projection.

after the buyer purchases the product. Exploded-assembly drawings are used extensively in automotive maintenance manuals. Assembled isometric assembly drawings show how a single product looks when assembled.

Dimetric Drawing

Dimetric drawing differs from isometric drawing in that only two angles are equal. **Figure 18-31** shows

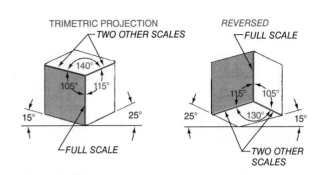

FIGURE 18-32   Trimetric drawing.

a number of orientations for drawing dimetric views; lines of an object that are parallel to the first two axes are drawn full size, lines parallel to the third are drawn to a different scale. Because two different scales are used, less distortion is apparent and an object looks more realistic. One disadvantage, however, is that circles are difficult to draw in dimetric views.

Trimetric Drawing

Trimetric drawing uses axes that are rotated so that each axis is drawn at a different angle, and each axis uses a different scale (**Figure 18-32**). Trimetric drawings can appear more realistic than any of the other types, but they are the most time-consuming to draw, so they aren't used by drafters.

Other Oblique Drawings

Cavalier Drawing

Cavalier drawings have a receding axis drawn between 30° and 60° to the horizon, and the receding distances are drawn full size (**Figures 18-33** and **18-34**).

FIGURE 18-31   Dimetric drawing.

FIGURE 18-33   Oblique drawing.

FIGURE 18-34   Cavalier drawing.

Cabinet Drawing

Cabinet drawings have the same range of receding axes as cavalier drawings, but the receding distances are drawn to half size (**Figure 18-35**).

General Oblique Drawing

General oblique drawing is similar to cabinet drawing except that the lines drawn along the direction of the receding distances may be at any scale between half and full (**Figure 18-36**).

FIGURE 18-35   Cabinet drawing.

FIGURE 18-36   General oblique.

Perspective Drawings

A *perspective drawing* is a three-dimensional drawing of an object or scene that shows visible lines almost as a person's eyes would perceive them from one particular viewing location. A perspective drawing is somewhat like a photograph (**Figure 18-37**). The upper drawing shows the lines of a building from the outside; the lower one shows an inside view of a room. Both views are drawn in two-point perspective like several of the sketches in Chapter 3.

Perspective Drawing Terms

Learn the terms used in perspective drawing while studying the figures and procedures that follow.

Station Point (SP). The exact location from which the observer views the object (**Figure 18-38**).

Ground Line (GL). The edge view of the surface upon which the object to be viewed rests, Figure 18-38.

Horizon (H). A line at the level of the viewer's eyes, Figure 18-38, and the line where the vanishing points are located.

Vanishing Point (VP). A point on the horizon line to which all other lines are projected (**Figure 18-39**). Six cubes, A through F, are used to show which sides of an object would be seen for various left and right positions of the cubes in perspective drawings. Note that if the dimension X shown is too great, a two-point perspective drawing would be needed.

Picture Plane (PP). Like the projection plane in multiview drawings. An imaginary plane perpendicular to the ground line, Figure 18-38.

Visible Rays (VRs). Lines from the eyes of the viewer to each and every point on the object to be viewed. The visible rays intersecting the picture plane create the perspective drawing, Figure 18-38.

Measuring Line (M). Actually used as a vertical line, but it appears horizontal because of the way it is established. The line originates at the projection plane and is 90° to it. Referring ahead, Figures 18-40D, G, and H show the measuring line as a vertical line used to measure true heights.

Presketching for Perspective Drawings

Make simple, quick sketches of an object to be drawn in two-point perspective to determine the best arrangement of the horizon, station point, ground line, and vanishing points (**Figures 18-40A** and **18-40B**). Two of the multiviews are required for the two-point perspective (usually a top view and a front or side view).

OUTSIDE VIEW IN TWO-POINT PERSPECTIVE

INSIDE VIEW IN TWO-POINT PERSPECTIVE

FIGURE 18-37   Outside and inside perspective views.

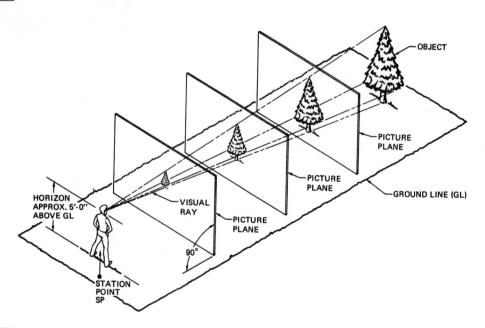

FIGURE 18-38   Perspective terms.

How to Draw a Two-Point Perspective

GIVEN: Multiview drawing Figure 18-40A and the sketches of the object in Figure 18-40B. Sketch C is selected as the best position to show the features of the object.

STEP 1   On the upper portion of a drawing sheet, draw the top view at any orientation, but attempt to match the best position from the sketches (**Figure 18-40C**). Locate the front edge of the object tangent to the picture plane.

STEP 2   From the front edge of the object, where it is tangent to the picture plane line, project the measuring line at 90° to the picture plane (**Figure 18-40D**). Locate the station point on the measuring line so the inclusive angle of view is 30° or less.

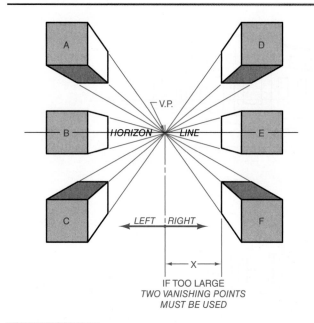

FIGURE 18-39   Vanishing point and horizon line.

STEP 3 From the station point, construct lines parallel to the two edges of the top view (**Figure 18-40E**), and label their intersections with the picture plane, 1 and 2.

STEP 4 Draw the horizontal line at a convenient location, preferably below the station point to avoid a cluttered drawing (**Figure 18-40F**). From points 1 and 2 on the picture plane, project downward to the horizon line to locate the left and right vanishing points. Locate a ground line and draw a side or front view on the line as shown.

STEP 5 From the view selected for the ground line—in this example, the right-side view—project the true height of the object to the measuring line (**Figure 18-40G**). From the true height, project lines to the left and right vanishing points, and then construct a perspective basic shape of the object. Note how the vertical boundaries of the basic shape are established by the top view.

STEP 6 Project the true heights of each feature from the right-side view to the measuring line and back toward the appropriate vanishing point to a point where the same feature intersects the picture plane line (**Figure 18-40H**). Develop the various features of the object in light lines until you are satisfied that the drawing is correct.

STEP 7 Finally, darken the lines of the object (**Figure 18-40I**). Hidden lines are not usually drawn except to show an important feature that otherwise would not be seen.

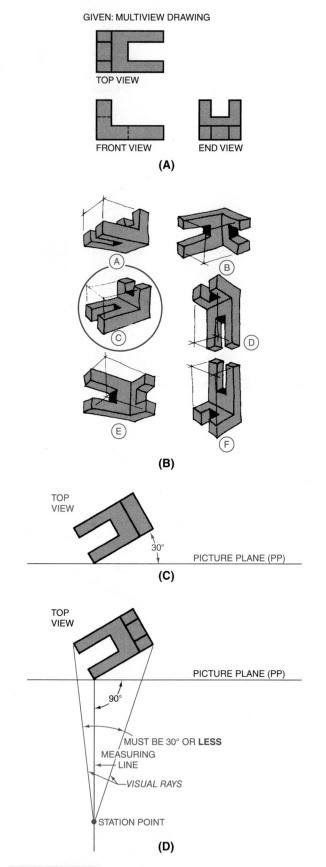

FIGURE 18-40   (A) How to draw a two-point perspective drawing; (B) Sketch of object in various positions; (C) Step 1; (D) Step 2;

(E)

(G)

(F)

(H)

(I)

FIGURE 18-40   (*Continued*) (E) Step 3; (F) Step 4; (G) Step 5; (H) Step 6; (I) Step 7.

Perspective View

Types of Perspective Views

Three types of perspective views are shown in **Figure 18-41**: one-, two-, and three-point, respectively. Only two-point perspective is discussed in this chapter because it is the type most often used in the industry.

Perspective Irregular Curves

Use a multiview drawing and a grid as was done for the isometric drawing. Then, follow the steps in the previous example to construct a perspective drawing (**Figure 18-42**).

Perspective Circles and Arcs

Divide the circle (or arc) into equal segments (**Figure 18-43**). Follow the steps from the previous example to construct the perspective drawing.

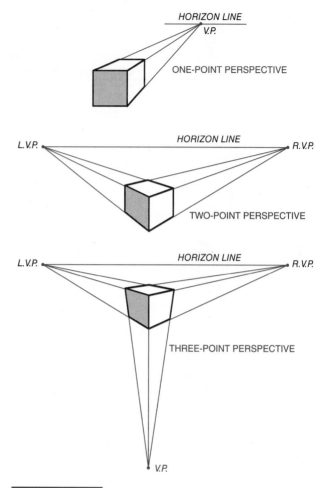

FIGURE 18-41   Kinds of perspective drawings.

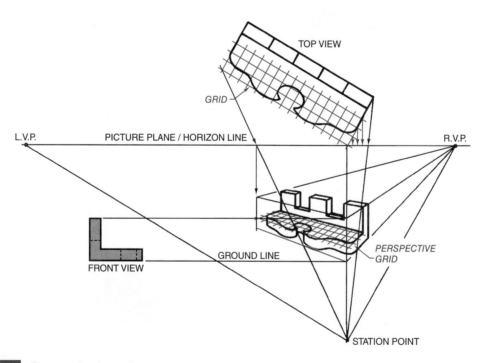

FIGURE 18-42   Perspective irregular curves.

DRAFTING STUDENTS ARE A VALUABLE RESOURCE

"Who remembers how to make isometric drawings?" asked Mark Malone—Chief Drafter for Pensacola Prestressed Concrete Company (PPCC). Malone had been asked by the foreman of PPCC's construction crew to develop pictorial drawings of several complicated connection joints for a bridge the company was building. The joints in question were so complicated that PPCC's construction crew was having trouble reading the plans. These joints had to be fitted together properly. The structural integrity of the bridge would depend on them.

Malone had agreed to provide pictorials without considering who would draw them. He could not remember the last time pictorial drawings had been called for. Prestressed concrete plans were almost always orthographic. In fact, he had not made a pictorial drawing himself since coming to PPCC 12 years ago. But he did remember being good at it as a drafting student at Pensacola Junior College.

That was when the idea came to him. "I'll ask the drafting class at PJC to take on this project." In the past, he had given the class projects when PPCC's drafting team got overloaded and deadlines were fast approaching. The students had always done an excellent job, and they appreciated the real-world experience.

Two days later, Malone had several outstanding isometric joint details, PJC's drafting students had additional real-world experience, the construction crew had drawings they could read, and Malone had found two talented students he could hire in the near future.

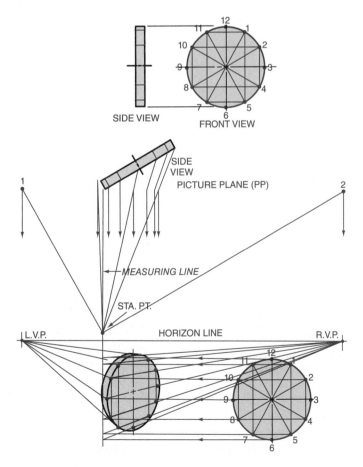

FIGURE 18-43   Perspective circles or arcs.

SUMMARY

- The most common types of pictorial drawings used by drafters, designers, and engineers are oblique, isometric, and perspective.

- Oblique drawings are the easiest type of pictorial drawing to make. The surface of the object is drawn exactly as it would appear in a multiview drawing. The depth axis is drawn at either 30° or 45°.

- Isometric drawings use three axes, sometimes called principal edges, at 120° to each other. They are the most frequently used pictorial drawings because they depict an object as well as a perspective drawing, but they are easier to make.

- A perspective drawing is a three-dimensional drawing of an object or scene that shows visible lines almost as a person's eye would perceive them from one particular viewing location. Key perspective concepts are the station point, ground line, horizon, vanishing point, picture plane, visible rays, and measuring line. There are one-, two-, and three-point perspective drawings.

- Dimetric drawings are similar to isometric except that only two angles are equal.

- Trimetric drawings are similar to isometric except that all three angles are different.

- A cavalier drawing is a type of oblique drawing in which depth is measured at full scale.

- A cabinet drawing is a type of oblique drawing in which depth is measured at half scale.

- General oblique drawings allow any appropriate scale to be used for the depth.

REVIEW QUESTIONS

Mark the following statements as either true or false.

1. The true isometric circle method is preferred over the four-center isometric circle method when time is an important factor.

2. An isometric drawing is dimensioned by using unidirectional dimensioning.

3. Trimetric drawings use axes that are rotated so that each axis is drawn at a different angle, and each axis uses a different scale.

4. Isometric, dimetric, and trimetric drawings are classified as perspective drawings.

5. Hidden lines are only used in pictorial drawings to show an important feature that otherwise would not be seen.

6. Isometric drawings use four axes that are sometimes called principal edges.

7. An isometric projection causes dimensions along the three principal edges to be foreshortened to approximately 87% of their true length.

8. Offset measurements are used to locate the center point of the work area.

Answer the following questions by selecting the best answer.

1. Which of the following is a type of pictorial drawing not frequently used in the industry?
   a. Cabinet
   b. Trimetric
   c. Isometric
   d. Two-point perspective

2. Which of the following is the perspective view most used in the industry?
   a. One-point
   b. Two-point
   c. Three-point
   d. Four-point

3. Which type of pictorial drawing best illustrates the object as it would be viewed by the eye?
   a. Oblique
   b. Isometric
   c. Perspective
   d. Dimetric

4. Which of the following is the correct definition of a nonisometric line?
   a. A line that is not parallel to any of the three principal edges in an isometric drawing
   b. A line that is not perpendicular to any of the three principal edges in an isometric drawing
   c. A line that is not parallel or perpendicular to any of the three principal edges in an isometric drawing
   d. None of the above

5. Which of the following is **not** true regarding an oblique drawing?
   a. It is easier to draw than isometric or perspective drawings.
   b. Usually, the front view is drawn exactly as it would appear in a multiview drawing.
   c. Oblique drawings use three axes.
   d. Oblique drawings are the most used of the three common pictorials.

CHAPTER 18 PROBLEMS

The following problems are intended to give the beginning drafter practice in laying out and developing isometric and perspective renderings of various objects.

Conventional Instructions

The steps to follow in laying out the drawings in this chapter are:

STEP 1 Study the problem carefully.

STEP 2 Position the object so it **best** illustrates the most features.

STEP 3 Make a pictorial sketch of the object.

STEP 4 Calculate the basic shape size.

STEP 5 Find the center of the work area.

STEP 6 Lightly center and lay out the basic shape within the work area.

STEP 7 Lightly develop the object within the basic shape.

STEP 8 Check to see that the completed object is centered within the work area.

STEP 9 Check all size dimensions, and check for accuracy of the object.

STEP 10 Darken in the object.

STEP 11 Recheck all work and, if correct, neatly fill out the title block using light guidelines and neat lettering.

CAD INSTRUCTIONS

Students who plan to complete the following problems on a CAD system should begin with these general instructions.

1. Analyze the problem carefully.

2. Plan your drawing sequence. In what order will you develop the drawing?

3. Read the specific instructions for the problem and begin.

Problems 18-1 through 18-27

Construct an isometric drawing of the object in Problems 18-1 through 18-27. (Note that these problems have only straight lines.)

PROBLEM 18-1

PROBLEM 18-2

PROBLEM 18-3

PROBLEMS

PROBLEM 18-4

METRIC

PROBLEM 18-7

METRIC

PROBLEM 18-5

PROBLEM 18-8

PROBLEM 18-6

PROBLEM 18-9

PROBLEM 18-10

PROBLEM 18-11

PROBLEM 18-12

PROBLEM 18-13

PROBLEMS

PROBLEM 18-14

METRIC

PROBLEM 18-15

PROBLEM 18-16

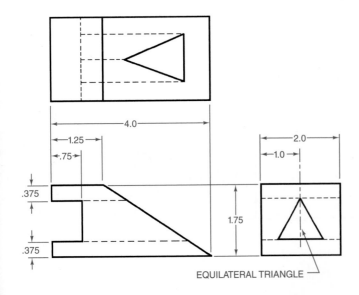

EQUILATERAL TRIANGLE

PROBLEM 18-17

PROBLEM 18-18

METRIC

PROBLEM 18-19

PROBLEM 18-21

PROBLEM 18-20

PROBLEM 18-22

PROBLEMS

PROBLEM 18-23

METRIC

PROBLEM 18-24

PROBLEM 18-25

PROBLEM 18-26

PROBLEM 18-27

.25 SQUARES

Problems 18-28 through 18-51

Construct an isometric drawing of the object in Problems 18-28 through 18-51. (Note that these problems have straight lines, arcs, and circles.)

PROBLEM 18-28

PROBLEM 18-29

PROBLEM 18-30

PROBLEM 18-31

PROBLEM 18-32

PROBLEM 18-33

METRIC

PROBLEM 18-34

PROBLEM 18-35

PROBLEM 18-36

PROBLEM 18-37

PROBLEM 18-38

METRIC

PROBLEM 18-40

METRIC

PROBLEM 18-39

PROBLEM 18-41

PROBLEM 18-42

ALL UNMARKED RADII = R.125

PROBLEM 18-43

PROBLEM 18-44

Ø2.75

Ø1.0 THRU
⌴ Ø2.0 ⫧ .50

R.75
(TYP.)

2X Ø.75,
THRU

R1.75

2.0

4.0

.50 ⊢(4 PLACES)
.25

(.50)

(2.0)

.75

PROBLEM 18-46

Ø18

S.Q. 75

REMOVE ONE
QUARTER

METRIC

PROBLEM 18-45

SR1.0

Ø2.0

Ø.75

3.0

Ø.50, THRU

PROBLEM 18-47

2.0

1.0

2.0

1.0

3X 1.5, THRU

1.0

2.0

PROBLEMS

PROBLEM 18-48

METRIC

PROBLEM 18-49

ALL UNMARKED RADII = R.09

Mechanical

PROBLEM 18-50

Mechanical

PROBLEM 18-51

Mechanical ## PROBLEM 18-52 (A TO I)

Construct an exploded isometric drawing of the assembly of a vise (Problem 18-52A). The vise consists of seven different parts and various fasteners (Problems 18-52B through 18-52I). Illustrate the exploded view with the parts in relationship to its assembly. Space parts with an approximate 1″ (25 mm) space between parts. Center the completed view within the work area.

(A)

PART NO. 1
1 REQ'D.

(B)

PROBLEM 18-52 (*continued*)

1/4-28 UNF-2B
V90° X Ø.44

2 X 1/4-20 UNC-2B

PART NO. 2
1 REQ'D.

(C)

.06 X 45° CHAMFER

Ø.190, THRU

1/2-28 UNF-2A THREAD

Ø.375

Ø.50

Ø.25

UNDERCUT TO DEPTH OF
THREAD

PART NO. 3
1 REQ'D.

(D)

Ø.188

Ø.125

PART NO. 4
1 REQ'D.

(E)

PROBLEM 18-52 (continued)

.06 X 45° CHAMFER $\varnothing\frac{.122}{.120}$ ↧ .28

$\varnothing.31$

.375

PART NO. 5
2 REQ'D.

(F)

$\varnothing.130$, THRU
V90° X .50

1.40

.70

.375   1.5   .188

2.25

(H)

1/4–20 UNC–2B

1.25

.625

.50   1.0   .188

2.25

(G)

TENTATIVE "BOX" ISOMETRIC
DRAWINGS. REARRANGE THE
DRAWINGS TO BEST SHOW
THE ASSEMBLY OF THE VISE.

(I)

PROBLEM 18-53

Using the given multiview drawing, develop a perspective drawing in the positions as illustrated by sketches A, B, C, D, E, and F.

SKETCHES

A   B   C

D   E   F

 PROBLEM 18-54

Using the given multiview drawing, develop a perspective drawing in the positions as illustrated by sketches A and B.

SKETCHES

A

B

PROBLEM 18-55

Choose various objects from Problems 18-1 through 18-51 and develop them into a perspective drawing. Compare the isometric view with the isometric drawing.

PROBLEM 18-56

Prepare an isometric drawing of the object shown in Problem 7-26.

PROBLEM 18-57

Prepare an isometric drawing of the object shown in Problem 10-49.

PROBLEM 18-58

Prepare an oblique drawing of the object shown in Problem 6-13.

PROBLEM 18-59

Prepare an oblique drawing of the object shown in Problem 6-40.

PROBLEM 18-60

Redesign parts 2 and 3 of the vise in Problem 18-52. Propose a more positive means for connecting part 2 to part 3. Show the redesign in any combination of the following formats: multiview drawings, section drawings, isometric drawings, oblique drawings, perspective drawings.

PROBLEM 18-61

Select an assembled product (one that can be disassembled easily or, if necessary, destroyed during disassembly) and draw an exploded, approximate isometric drawing of the components. The objective is to show how the product works. Suggestions: household items; automotive components; office devices (pencil sharpener, file drawer mechanism); sports equipment; weight-training apparatus; laboratory equipment; classroom demonstrations.

PROBLEM 18-62

Design a means, made of simple parts, to raise water one meter from one irrigation canal to another for a third-world country. Employ animal power or a windmill-type apparatus. Prepare an exploded isometric of the components that make up the design.

PROBLEM 18-63

Design a wooden toy (but metal for axles, fasteners, etc.) to be made out of simple shapes; for example, a train engine made of blocks, cylinders, circles, cones, etc. Prepare an exploded isometric showing how the pieces are to go together and how they are to be joined.

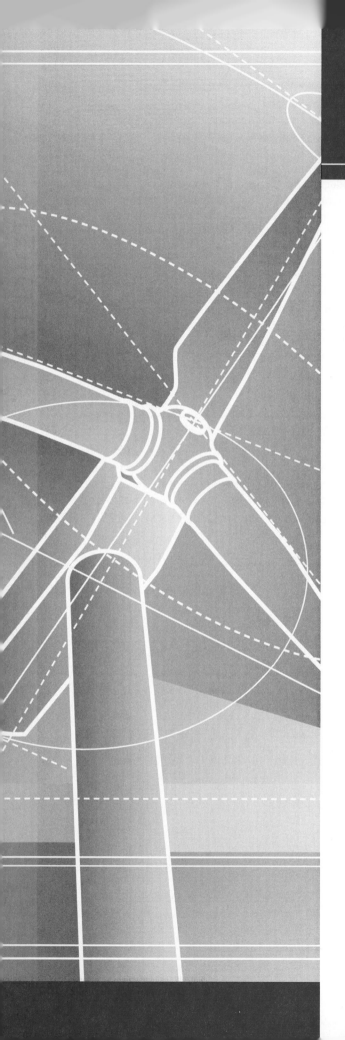

section 4

Related Technologies, Applications, and Processes

career profile

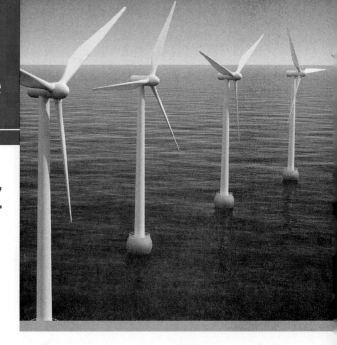

TERRY SCHULTZ

Terry Schultz came upon his career path late in life. At 40 years old, after many years managing restaurants, he was ready for a change. Remembering how much he enjoyed drafting classes in high school, Schultz went back to ITT to get a degree in drafting. After a few years working for others, he decided to go out on his own with his still-growing small business, Spokane CAD Consultants.

Primarily a structural drafter, Schultz works directly with an engineer on most projects. The engineer will give him sketches, which Schultz looks over to see if there's anything obviously missing. (He recognizes that engineers are not drafters and vice versa, but both need to serve as each other's check and balance.) Sometimes he does his own sketch to see if he understands before he lays a sketch out in CAD, paying particular attention to detail, scale, and ASME standards.

Often, Schultz and the engineer will work with the architect and the architect's drawings. But they also must do their own. Here's the difference: A structural drafter—and the engineer he or she is working with—is most interested in weights, stress loads, and pressures. He or she is thinking, how will I support this building? What beams and supports are necessary? While an architectural drafter might be looking at the facings of a building, a structural drafter will look at the column supports inside. The drafter also must work with the welder—where two pieces of a support will be seamlessly joined, there will be welding. Right now, Schultz and the engineer are creating a basin for a wastewater treatment plant. For this project, Schultz has to make sure his drawing outlines a setup that supports 36,000 gallons of water and contains the appropriate piping underneath. In this project, iron is welded to channel; he must know that it is a Phillips weld, what the depth of it is, and what the standard symbol for that is.

The engineer who brought Schultz into this project is someone he has worked with a number of times before. While owning his own business often requires a lot of legwork, he is starting to make contacts who will call him for projects. It is a good sign for the continued expansion of his business; Schultz hopes that within the next few years, he will be managing contract drafters as well. Until then, he can experience the day-to-day joy of what he is doing, especially when he gets to see something he drew turn into a sturdy and safe structure.

19

Welding

CHAPTER OBJECTIVES

**Upon completion of this chapter, students should be able to do
the following:**

▶ List the most widely used welding procedure.

▶ Demonstrate proficiency in the proper use of welding symbols.

▶ Demonstrate proficiency in the proper dimensioning of welds.

▶ Demonstrate proficiency in properly indicating weld placement.

▶ Demonstrate proficiency in properly indicating intermittent
welds.

▶ Explain the proper use of the contour symbols as part of the
welding symbol.

▶ Explain the proper use of the symbology for field welds.

▶ Describe the five basic types of welded joints.

▶ Describe the six major types of welds.

(continued)

(CHAPTER OBJECTIVES continued)

▶ Demonstrate proficiency in the proper use of multiple reference lines.

▶ Explain the concept of spot welding.

▶ Explain the differences between spot welds and projection welds.

▶ Explain the concept of the seam weld.

▶ Demonstrate proficiency in the use of welding templates.

Welding Processes

Pieces of metal can be fastened together with mechanical fasteners, as was discussed in Chapter 13, "Fasteners." They can also be held together by soldering or brazing, and some are fastened together by an adhesive. A permanent way to join pieces of metal together is by *welding*.

One method used to weld parts together is to heat the edges of the pieces to be joined until they melt and join or fuse together. When the pieces cool, they become one homogeneous mass, permanently joined together. A filler rod is sometimes used to mix with the molten metal to make the joint stronger. When a lot of heat is used to melt and fuse the joint, the weld is called a *fusion weld*. Heat for this process can come from a torch, burning gases, or with a high electric current. Because of the extremely high temperatures involved in welding, the part or parts may be distorted; thus, all machining is usually done after welding.

Another kind of weld used in years past was the pressure weld. In this process, the parts to be welded together are heated to a plastic state and forced together by pressure or by hammering. This was done by the local blacksmith. This old method was called *forge welding*. Today, faster and better methods are used.

Welding assemblies are usually built up from stock forms, such as plate steel, square bars, tubing, angle iron, and the like. These parts are cut to shape and welded together. The various welding processes are illustrated in **Figure 19-1**. Welded assemblies are much less expensive and more satisfactory than the casting process, especially in a case where only one or only a few identical parts are required.

Welding is used on large structures that would be difficult or impossible to build in a manufacturing plant. Large structures such as building frames, bridges, and ships are welded into one large assembly.

Major welding methods classified by the American Welding Society include brazing, gas welding, arc welding, resistance welding, and fusion welding. *Brazing* is the process of joining metals together with a nonferrous filler rod. A temperature above 1,000°F but just below the melting point is used to join the parts. The most commonly used *gas welding* is oxyacetylene welding. Resistance welding and fusion welding, the two major processes, are explained in this text. Detailed information regarding the other types of welding processes can be obtained by writing to the American Welding Society, P.O. Box 351040, Miami, Florida 33135.

Resistance welding is the process of passing an electric current through the exact location in which the parts are to be joined. This is usually done under pressure. The combination of the pressure and the heat generated by the electric current welds the parts together. This process is usually done on thin sheet metal parts.

FIGURE 19-1   Checklist of broad categories of welding, joining, and cutting processes.

| TYPE OF WELD | FILLET | GROOVE | | | | | | | BACK OR BACKING | PLUG OR SLOT | SURFAC- ING | FLANGE WELD | |
		SQUARE	V	BEVEL	U	J	FLARE V	FLARE BEVEL				EDGE	CORNER
SYMBOL	◿	‖	∨	⋁	∪	⊌	⟩⟨	⟨⟨	▽	▢	◠◠	⫼	⎮⎿
WELD ARROW SIDE													
WELD OPPOSITE SIDE													
WELD BOTH SIDES													
NO ARROW/ OPPOSITE SIDE SIGNIFICANCE													
EXAMPLE													

FIGURE 19-2   Fusion welding.

Fusion welding is usually used for large parts. The fusion welding process uses many standard types of welds (**Figure 19-2**). Each is explained in full.

The next few figures use the fillet welding symbol as an example, but the same method applies, regardless of which welding symbol is used.

Basic Welding Symbol

The *basic welding symbol* consists of a reference line, a leader line and arrow and, if needed, a tail (**Figure 19-3**). The tail is added for specific information or notes in regard to welding specifications, processes, or reference information. The reference line of the basic welding symbol is usually drawn horizontally. Any welding symbol placed on the upper side of the reference line indicates WELD OPPOSITE SIDE (**Figure 19-4**). Any welding symbol placed on the lower side of the reference line indicates WELD ARROW SIDE, Figure 19-4. The direction of the leader line and arrow has no significance whatsoever to the reference line.

A fillet weld symbol, Figure 19-2, is added above or below the reference line with the left leg always drawn vertical (**Figure 19-5**). A welding symbol placed above the reference line means to weld the OPPOSITE side, as illustrated in (**Figure 19-6**). A weld symbol placed below the reference line means to weld the ARROW side, as illustrated in Figure 19-6. The positions as drawn and as welded are shown in **Figures 19-7** and **19-8**. A weld symbol placed above and below the reference line means to weld both the OPPOSITE side and the ARROW side, as illustrated in (**Figure 19-9**).

FIGURE 19-3   Basic welding symbol.

FIGURE 19-4   Any welding symbol placed on the upper side of the reference line indicates weld OPPOSITE side.

FIGURE 19-5   Left leg of welding symbol is always drawn vertical.

FIGURE 19-6   Welding symbol placed above the reference line means to weld OPPOSITE side. Welding symbol placed below the reference line means to weld ARROW side.

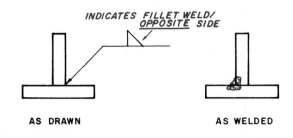

FIGURE 19-7   Position as drawn and as welded.

FIGURE 19-8   Position as drawn and as welded.

FIGURE 19-9   Welding symbol placed above and below the reference line means to weld both sides.

FIGURE 19-10   Dimension to the left of welding symbol indicates length of leg or side of weld.

FIGURE 19-11   Size of each weld is required.

FIGURE 19-12   Welds with legs or side of different sizes.

Size of Weld

The weld must be fully dimensioned so that there is no question whatsoever as to its intended and designed size. The size of a weld refers to the length of the leg or side of the weld. The size is placed directly to the left of the welding symbol. The two legs are assumed to be equal in size unless otherwise dimensioned (**Figure 19-10**). If a double weld is needed, the size of each weld must be included, as illustrated in **Figure 19-11**.

If the legs or sides of the weld are to be different, the dimensions of the sides must be indicated to the left of the welding symbol in parentheses as reference-only dimensions (**Figure 19-12**). Because the symbol does not indicate which is the .50 leg and which is the .25 leg, the detail drawing must include the dimensions.

Length of Weld

When a weld length is not specified, it is assumed to be continuous the full length (**Figure 19-13**). If a weld must be made to a special length, it must be indicated. This is done by a dimension directly to the right of the weld symbol (**Figure 19-14**).

When a weld is to be made continuously all around an object, it is indicated on the welding symbol by adding a small circle between the reference line and the leader line (**Figure 19-15**).

Placement of Weld

When a weld is not continuous and is needed in only a few areas, *section lining* is used to indicate where the weld is to be placed (**Figure 19-16**). If a weld is to

FIGURE 19-13   If weld is not specified, it is assumed to be continuous full length.

FIGURE 19-15   A small circle indicates weld continuously around an object.

FIGURE 19-14   Length of weld must be noted to the right of the welding symbol.

FIGURE 19-16   Section lining indicates location of weld.

be placed in a few areas, this is indicated by the use of multiple leader lines and arrows (**Figure 19-17**).

Intermittent Welds

Intermittent welds are a series of short welds. When this type of weld is needed, an extra dimension is added to the welding symbol (**Figure 19-18**). The usual size of the leg is noted to the left of the symbol, and the length of the weld is indicated to the right of the welding symbol, followed by a dash line and

the pitch of the intermittent welds (**Figure 19-19**). An *increment* is the length of the weld; the *pitch* is the center-to-center distance between increments.

Chain Intermittent Weld

A *chain intermittent weld* is a weld in which each weld is applied directly opposite to each other on

INDICATES FILLET WELD/
.125 LEGS -- THREE EDGES, BOTH SIDES

.125
.125

AS DRAWN

AS WELDED

FIGURE 19-17   Leader line and arrows indicate location of weld.

.125   1-2

PITCH
LENGTH OF WELD
WELD TYPE
SIZE OF WELD (LEG)

AS DRAWN

FIGURE 19-18   Intermittent welds are noted by dimensions to the right of the welding symbol.

INTERMITTENT WELDS

PITCH (2.0)

LENGTH OF WELD (1.0)

.125

.125

AS WELDED

FIGURE 19-19   Pitch is the center-to-center distance between welds.

opposite sides of the joint (**Figure 19-20**). Note that the welding symbol is applied above and below the reference line and that they are in line with each other.

INDICATES FILLET WELD/
.125 LEGS -- INTERMITTENT, IN LINE --
1.0 LONG/3.0 PITCH

.125   1-3
.125   1-3

WELDING SYMBOLS DRAWN IN LINE

AS DRAWN

LENGTH (1.0)   LENGTH (1.0)   LENGTH (1.0)

PITCH (3.0)   PITCH (3.0)

AS WELDED (TOP VIEW)

FIGURE 19-20   Chain intermittent welds are applied directly opposite each other.

Staggered Intermittent Weld

A *staggered intermittent weld* is similar to the chain intermittent weld except the welds are staggered directly opposite each other on opposite sides (**Figure 19-21**). Note that the welding symbols are applied above and below the reference line staggered to each other.

INDICATES FILLET WELD/
.125 LEGS -- INTERMITTENT, STAGGERED
1.0 LONG/2.5 PITCH

.125   1-3
.125   1-3

WELDING SYMBOLS DRAWN STAGGERED

AS DRAWN

PITCH (2.5)   PITCH (2.5)

LENGTH (1.0)   LENGTH (1.0)   LENGTH (1.0)

(1.0) LENGTH   (1.0) LENGTH   (1.0) LENGTH

PITCH (2.5)   PITCH (2.5)

AS WELDED (TOP VIEW)

FIGURE 19-21   Staggered intermittent welds are placed staggered opposite each other.

DESIGNATION OF WELDING AND ALLIED PROCESSES BY LETTERS

AAC ....air carbon arc cutting
AAW ....air acetylene welding
ABD...adhesive bonding
AB..arc brazing
AC...arc cutting
AHW ....atomic hydrogen welding
AOC...oxygen arc cutting
AW..arc welding
B..brazing
BB..block brazing
BMAW ....bare metal arc welding
CAC ....carbon arc cutting
CAW ....carbon arc welding
CAW-G ....gas carbon arc welding
CAW-S ....shielded carbon arc welding
CAW-T ....twin carbon arc welding
CW..cold welding
DB..dip brazing
DFB..diffusion brazing
DFW...diffusion welding
DS...dip soldering
EASP..electric arc spraying
EBC..electron beam cutting
EBW..electron beam welding
ESW...electroslag welding
EXW...explosion welding
FB..furnace brazing
FCAW..flux cored arc welding
FCAW-EG ....flux cored arc welding–electrogas
FLB...flow brazing
FLOW...flow welding
FLSP ....flame spraying
FOC ....chemical flux cutting
FOW..forge welding
FRW...friction welding
FS...furnace soldering
FW...flash welding

FIGURE 19-22   Designation of welding and allied processes by letters (abbreviations).

Process Reference

There are many welding processes developed by the American Welding Society. Each process has a standard abbreviation designation (**Figure 19-22**). Letter designations are used to specify a method used to obtain a particular contour of a weld (**Figure 19-23**). These abbreviations are taken from the latest American Welding Society's standard #AWS-2.4-79,71.

The standard process abbreviations or letter designations are usually added to the tail of the basic welding symbol as required (**Figure 19-24**). Sometimes they are called off in a note somewhere on the drawing (**Figure 19-25**).

Contour Symbol

If a special finish must be made to a weld, a contour symbol must be added to the welding symbol. There are three kinds of contour symbols used: flush, convex, or concave (**Figure 19-26**). The contour symbol

DESIGNATION OF WELDING AND ALLIED PROCESSES BY LETTERS

GMAC...gas metal arc cutting
GMAW...gas metal arc welding
GMAW-EG ....gas metal arc welding–electrogas
GMAW-P ....gas metal arc welding–pulsed arc
GMAW-S ....gas metal arc welding–short circuiting arc
GTAC...gas tungsten arc cutting
GTAW ....gas tungsten arc welding
GTAW-P................................gas tungsten arc welding–pulsed arc
HFRW ....high frequency resistance welding
HPW ....hot pressure welding
IB..induction brazing
INS...iron soldering
IRB ....infrared brazing
IRS...infrared soldering
IS...induction soldering
IW...induction welding
LBC ....laser beam cutting
LBW...laser beam welding
LOC ....oxygen lance cutting
MAC ....metal arc cutting
OAW ....oxyacetylene welding
OC..oxygen cutting
OFC ....oxyfuel gas cutting
OFC-A ....oxyacetylene cutting
OFC-H ....oxyhydrogen cutting
OFC-N ....oxynatural gas cutting
OFC-P ....oxypropane cutting
OFW..oxyfuel gas welding
OHW ....oxyhydrogen welding
PAC ....plasma arc cutting
PAW ....plasma arc welding
PEW ....percussion welding
PGW ....pressure gas welding
POC ....metal powder cutting
PSP ....plasma spraying
RB ....resistance brazing
RPW..projection welding
RS ....resistance soldering
RSEW ....resistance seam welding
RSW...resistance spot welding
ROW ....roll welding
RW..resistance welding
S ....soldering
SAW..submerged arc welding
SAW-S ....series submerged arc welding
SMAC..shielded metal arc cutting
SMAW...shielded metal arc welding
SSW..solid state welding
SW ....stud arc welding
TB..torch brazing
TCAB ....twin carbon arc brazing
TW..thermit welding
USW ....ultrasonic welding
UW ....upset welding

FIGURE 19-22   (*Continued*) Designation of welding and allied processes by letters (abbreviations).

LETTER	METHOD
C	CHIPPING
G	GRINDING
H	HAMMERING
R	ROLLING
M	MACHINING

FIGURE 19-23   Abbreviations of standard method used to obtain a particular contour.

FIGURE 19-24   Process abbreviations are placed in the tail of the basic welding symbol.

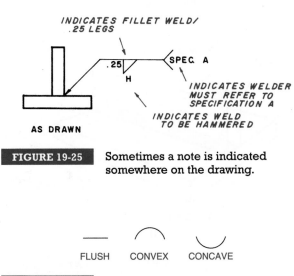

FIGURE 19-25   Sometimes a note is indicated somewhere on the drawing.

FLUSH   CONVEX   CONCAVE

FIGURE 19-26   Contour symbol.

FIGURE 19-27   Contour symbol is added to the welding symbol.

FIGURE 19-28   Finish number specifies finish required.

FIGURE 19-29   Finish weld is to be concave and ground to a 125-microinch surface finish.

is added directly above or below the welding symbol. If the welding symbol is above the reference line, the contour symbol is placed above the welding symbol. If the welding symbol is placed below the reference line, the contour symbol is placed below the welding symbol (**Figure 19-27**). The degree of finish is not usually included, but if a specific finish is desired, a finish number designation must be added (**Figure 19-28**).

Figure 19-29 represents that the finish weld must be concave and ground to a 125-microinch surface finish. **Figure 19-30** represents that the finish weld must be machined flat on the top surface to a surface finish of 125 microinches. The weld must be convex on the bottom surface with a hammered finish.

Field Welds

Any weld not made in the factory—that is, it is to be made at a later date, perhaps at final assembly on-site—is called a *field weld*. Its symbol is added to the basic weld symbol by a filled-in flag, located between the reference line and the leader line and drawn using a 30°–60° triangle (**Figure 19-31**).

Welding Joints

There are five basic kinds of welding joints, and they are classified according to the position of the parts that are being joined (**Figures 19-32A** through **19-32E**). There are many types of welds used to weld these joints together. Considerations as to which

AS DRAWN

AS WELDED

FINISH MACHINED, FLAT
TO 125 MICROINCHES,
ARROW SIDE

FINISH HAMMERED, CONVEX
OPPOSITE SIDE

AS FINISHED

FIGURE 19-30   Finish weld is to be machined flat on top surface to 125 microinches and convex on the bottom surface with a hammered finish.

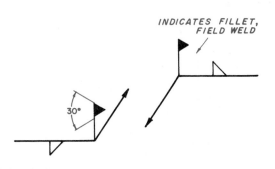

INDICATES FILLET,
FIELD WELD

30°

FIGURE 19-31   Field weld.

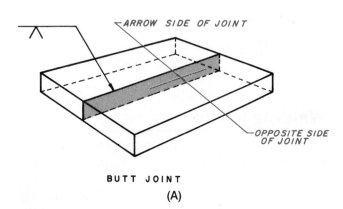

ARROW SIDE OF JOINT

OPPOSITE SIDE
OF JOINT

BUTT JOINT

(A)

FIGURE 19-32   (A) Butt joint;

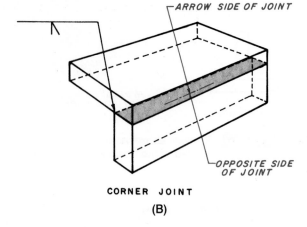

ARROW SIDE OF JOINT

OPPOSITE SIDE
OF JOINT

CORNER JOINT

(B)

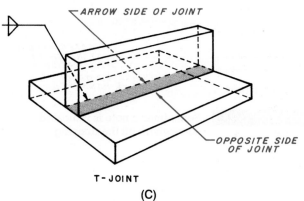

ARROW SIDE OF JOINT

OPPOSITE SIDE
OF JOINT

T-JOINT

(C)

OPPOSITE SIDE OF JOINT

ARROW SIDE OF JOINT

LAP JOINT

(D)

ARROW SIDE OF JOINT

OPPOSITE SIDE OF JOINT

EDGE JOINT

FROM 0° TO 30° MAX.

(E)

FIGURE 19-32   (*Continued*) (B) Corner joint; (C) T-joint; (D) Lap joint; (E) Edge joint.

type of weld to use are based on the particular application, thickness of material, required strength of the joint, and available welding equipment, among others.

Types of Welds

There are six major types of welds (refer to Figure 19-2): fillet, groove, back or backing, plug or slot, surface, and flange welds.

More than one type of weld may be applied to a single joint—usually for strength or appearance. For example, a groove V weld may be used on one side and a backing weld on the other side (**Figure 19-33**). This is known as a *multiweld.*

The same type of weld may be used on opposite sides of a single joint. Such a joint is called a *double weld;* for example, double V, double U, or double J weld (**Figure 19-34**). These, too, are used for strength and/or appearance.

Fillet Welds

Up to this point, the *fillet weld* and its symbol have been used as an example. The other five major types of welding symbols used are very similar.

Groove Weld

Seven types of welds are considered to be *groove welds:* square, V, bevel groove, J, U, flare V, and flare bevel. Each type has its own welding symbol (**Figure 19-35**). In the bevel groove and J welds, only one part is actually chamfered or grooved. To do this, the leader line and arrow point toward the part that has the bevel groove or J groove (**Figure 19-36**). Note that the welding symbol is placed above or below the reference line, as usual, to indicate on which side the chamfer or groove is located.

Size of Groove Weld. In the groove weld, the dimension refers to the actual depth of the chamfer or groove, **not** the size of the weld. With the fillet weld,

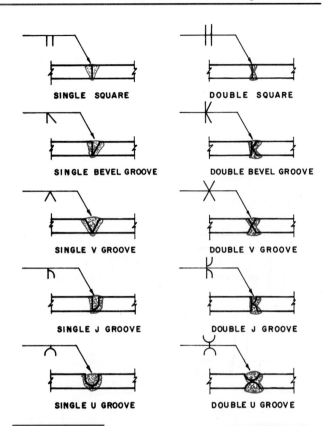

FIGURE 19-34   Double welds (same types of welds).

the size of the chamfer or groove is located directly to the left of the welding symbol and on the same side of the reference line (**Figure 19-37**). If a V groove weld symbol is shown without a size dimension, the

GROOVE WELDS		
TYPE	SYMBOL	AS SEEN
SQUARE	‖	
V	∨	
BEVEL GROOVE	⌶	
U	⌒	
J	Ρ	
FLARE V	⟩⟨	
FLARE BEVEL	⏊⟨	

FIGURE 19-33   Multiwelds (different types of welds).

FIGURE 19-35   Groove welds.

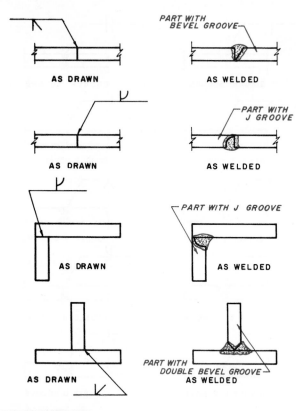

FIGURE 19-36   Groove weld illustrations.

size or depth is assumed to be equal to the thickness of the pieces.

Depth or Actual Size of Weld. The depth or actual size of the weld includes the penetration of the complete weld. This is illustrated in **Figure 19-38**. Note in this example that the depth or actual size of the weld is deeper than the depth of the chamfer. Occasionally, the depth or actual size of the weld is less than the size of the chamfer (**Figure 19-39**).

In order to call off the depth or actual size of the weld and the size of the chamfer or groove, a dual dimension is used, and both are added to the left of the welding symbol. In order to differentiate between the two, the size of the chamfer or groove is added to the extreme left of the welding symbol. The size of the depth or actual size of the weld is added directly to the left of the welding symbol, but in parentheses (**Figure 19-40**). Usually the weld is larger than the groove (**Figure 19-41**).

If the weld is less than the size of the chamfer, the dimensions are added to the left of the welding symbol in the same manner (**Figure 19-42**).

FIGURE 19-38   Depth or actual size of weld.

FIGURE 19-39   Size of weld only.

FIGURE 19-40   Size of weld is added to the left of the welding symbol in parentheses.

FIGURE 19-37   Groove welds with dimensions.

FIGURE 19-41 Example of effective throat-weld larger than groove.

FIGURE 19-45 Chamfered angle.

FIGURE 19-42 Example of effective throat-weld smaller than groove.

Root Opening. The allowed space between parts is called the *root opening* (**Figure 19-43**). The root opening is applied **inside** the welding symbol (**Figure 19-44**).

The root opening may be called off as a general note. For example, "unless otherwise noted, root opening for all groove welds is .06." If there is **no** opening between parts, a zero is added inside the welding symbol.

Chamfer Angle. In a groove weld, a V or bevel groove must be held to tight tolerances. The angle must also be included with the size dimensions in order to obtain the exact required geometric size and shape of all manufactured parts. The angle is also added

inside the welding symbol (**Figure 19-45**). Note that the leader line and arrow point toward the part with the chamfer. Because the welding symbol is placed above the reference line, the chamfer and weld are on the side opposite from the arrow.

Groove Radius. If a J or U groove weld must be held to tight tolerances, the angle and radius must be included with the size dimensions in order to obtain the exact required geometric size and shape of all manufactured parts. The angle is also added inside the welding symbol, and the required radius is called off as a note (**Figure 19-46**). Note that the leader line and arrow point toward the part with the groove. Because the welding symbol is placed below the reference line, the groove and weld are on the side of the arrow.

Backweld—Backing Weld

A *back weld* is a weld applied to the opposite side of the joint **after** the major weld has been applied. A *backing weld* is a weld applied first, followed by the major weld. Both use the same welding symbol, and both are used to strengthen a weld (refer to Figure 19-2). The back weld is also used for appearance.

If a welding melt-through is required, the same symbol is used, except it is filled in solid (**Figure 19-47**). When the back weld or backing weld symbol is used, it is placed on the opposite side of the reference line from the groove weld symbol (**Figures 19-48** and **19-49**).

FIGURE 19-43 Root opening.

FIGURE 19-44 Size of root opening.

FIGURE 19-46 Required radius.

BACK OR BACKING WELD   MELT-THRU WELD

FIGURE 19-47   Back weld symbol/melt-thru weld symbol.

BACK OR BACKING WELD
SYMBOL PLACED OPPOSITE SIDE OF
REFERENCE LINE FROM GROOVE WELD SYMBOL

WELD FIRST

AS DRAWN

WELD LAST
AS WELDED

FIGURE 19-48   Illustration of back weld symbol.

BACK OR BACKING WELD
SYMBOL PLACED OPPOSITE SIDE OF
REFERENCE LINE FROM GROOVE WELD SYMBOL

WELD LAST

AS DRAWN

WELD FIRST
AS WELDED

FIGURE 19-49   Illustration of back weld symbol.

AS DRAWN

AS WELDED

FIGURE 19-50   Plug and slot welds.

Plug and Slot Welds

The *plug weld* and *slot weld* both use the same welding symbol. The only difference between the two welds is the shape of the hole through which the weld is applied. A plug weld is made through a round hole. A slot weld is made through an elongated hole (refer to Figure 19-2).

The plug or slot welding symbol is applied to the basic welding symbol and interpreted in exactly the same way. If the welding symbol is applied above the reference line, the hole or slot is on the opposite side of the arrow. If below the reference line, the hole or slot is on the same side as the arrow (**Figure 19-50**).

Size of Plug or Slot Weld.   The given dimension for a plug or slot weld refers to the diameter of the plug or slot at the base of the weld. The required dimension is

placed directly to the left of the plug or slot weld symbol (**Figure 19-51**). The size of the slot weld is drawn and dimensioned directly on the detail drawing.

Depth of Weld.   Unless noted, the plug or slot weld hole is completely filled by the weld. If filling the hole is not necessary, the full depth of weld must be called off. The required depth of weld is added inside the welding symbol (**Figure 19-52**).

Tapered Plug or Slot.   If the plug or slot hole is tapered, the taper-inclusive angle is added directly above or below the welding symbol (**Figure 19-53**).

Surface Weld

If a surface must have material added to it or must be built up, a *surface weld* is added (refer to Figure 19-2). This welding symbol is **not** used to indicate the joining of parts together; therefore, the welding symbol is simply added below the reference line (**Figure 19-54**).

Size of Surface Weld.   If the surface is to be built up and a special height is **not** required, the welding symbol is drawn as illustrated in Figure 19-54. If the surface is to be built up and a specific height is required, the required height is added directly to the

FIGURE 19-51   Size of plug or slot weld.

FIGURE 19-53   Tapered plug or slot.

FIGURE 19-54   Surface weld.

FIGURE 19-55   Site of surface weld.

FIGURE 19-52   Depth of weld.

left of the welding symbol (**Figure 19-55**). If only a portion of the surface is to be built up, that portion must be illustrated and dimensioned on the detail drawing (**Figure 19-56**). The actual weld is represented by section lining.

Flange Weld

A *flange weld* is used to join thin metal parts together. Instead of welding to the surfaces of the parts to be joined, the weld is applied to the edges of thin material. A weld applied to thin metal could burn right through.

There are two distinct flange weld symbols (refer to Figure 19-2): the first is called an edge-flange weld (**Figure 19-57**); the other is called a corner-flange weld,

FIGURE 19-56   Illustration of a surface weld.

FIGURE 19-57   Edge-flange weld.

FIGURE 19-58   Corner-flange weld.

(Figure 19-58). The straight line of this welding symbol is always drawn to the left of the partially curved line, regardless of the actual joint being illustrated. The edge-flange or corner-flange weld symbol is placed above or below the reference line to indicate OPPOSITE side or ARROW side, but it is never on both sides (Figures 19-59 and 19-60).

FIGURE 19-59   Edge-flange weld symbol.

FIGURE 19-60   Corner-flange weld symbol.

Dimensioning a Flange Weld. It requires three dimensions to fully dimension a flange weld: the radius of the flange, the height above the point of tangency between parts, and the size of the weld itself (**Figures 19-61** and **19-62**). The three dimensions are added to the welding symbol as indicated previously. The welding symbol should appear at the peak of the joint as illustrated. If there is to be a root opening—that is, a space between parts—the required space must be indicated on the drawing, **not** on the welding symbol (**Figure 19-63**).

FIGURE 19-61   Dimensioning a flange weld.

FIGURE 19-62   Dimensioning a corner-flange weld.

FIGURE 19-63   Root opening.

Multiple Reference Line

If there is more than one operation on a particular joint, these operations are indicated on *multiple reference lines*. The first reference line closest to the arrow is for the first operation, the second reference line is for the second operation or supplemental data, and the third reference line is for the third operation or test information (**Figure 19-64**).

Spot Weld

Spot welding is a resistance welding process done by passing an electric current through the exact location where the parts are to be joined. Spot welding is usually done to hold thin sheet metal parts together. The resistance weld process uses three standard types of welds (**Figure 19-65**). Each is explained in full.

The symbol for a spot weld, as illustrated in **Figure 19-66**, is located tangent to or touching the reference line, and it uses the same added supplemental data symbols and dimensions as are illustrated with the fusion welding process (**Figure 19-67**). The tail is **always** included in order to indicate the process required to make the weld (refer to Figure 19-22).

Dimensioning the Spot Weld

The diameter size of the spot weld is added directly to the left of the weld symbol (**Figure 19-68**). Sometimes the diameter size is omitted, and a required shear strength is inserted in its place. In this condition, an explanation of the value is added to the drawing by a note (**Figure 19-69**).

FIGURE 19-65   Spot welds.

SPOT WELD SYMBOL

FIGURE 19-66   Spot weld symbol.

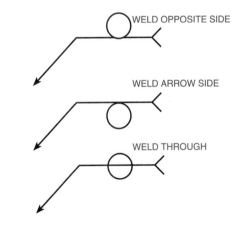

WELD OPPOSITE SIDE

WELD ARROW SIDE

WELD THROUGH

FIGURE 19-67   Spot weld symbol used with reference line.

FIGURE 19-64   Multiple reference line.

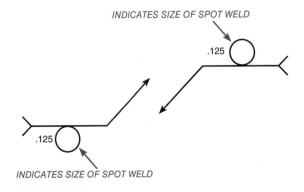

FIGURE 19-68   Diameter of spot weld.

ON-THE-JOB TRAINING

Fitness Equipment Corporation (FEC) is one of the largest manufacturers in the United States of heavy-gage fitness machines. Their equipment is used in some of the leading commercial health clubs and by several professional football teams. FEC's products are made primarily of steel tubing. Consequently, welding is one of the company's primary manufacturing processes.

Because welding is such an important process to the company, new hires in the drafting department spend their first six weeks of employment on the shop floor learning about welding. According to Jerry White, FEC's chief drafter, "We want our drafting and design personnel to know as much about our welding processes as our welders. Otherwise, how can they properly specify the various welds that might be needed on a given project? They don't have to develop the hand-eye coordination of a welder, but we do want them to know what kind of weld is called for in a given situation, how to properly indicate it using correct symbology, and how to properly dimension welds."

When asked about this practice, FEC's drafters are universally positive in their responses. Carla McDaniels summarized her reaction to six weeks on the shop floor as follows: "I was apprehensive at first. Having never been on the shop floor of a manufacturing plant, I had visions of a dark, dungeon-like atmosphere where men in hoods used torches on steel bars. However, my mentor was a woman, and the welding shop at FEC is cleaner than my house. It turned out to be a great six weeks. I learned a lot about welding, but I also learned what the welders need from me when I prepare a drawing. I also learned why my instructors back at MacArthur Technical Institute had stressed the need to put myself in the shoes of the people who would use my drawings. There are things I would have put in my drawings that are not needed and other things I would have left out that are needed."

Another drafter said, "Let me put it this way: Based on how much I learned on that shop floor in six weeks, if I ever decide to design houses for a living, I'm going to first spend six weeks working with a carpenter."

INDICATES 150 LB MIN ACCEPTABLE SHEAR
STRENGTH PER SPOT WELD

FIGURE 19-69   Shear strength of spot weld.

AS DRAWN

AS WELDED

FIGURE 19-70   Pitch of spot welds.

The spacing of the spot weld is called off by a number directly to the right of the weld symbol. This number is the pitch and refers to the center-to-center distance between spots (**Figure 19-70**). The number of welds for a particular joint is added above or below the welding symbol in parentheses.

Contour and Finish Symbols

Contour and finish symbols are added to the welding symbol exactly as is done in the fusion welding symbol (**Figure 19-71**).

AS DRAWN

AS WELDED

FINISH BY MACHINING TO A
SURFACE FINISH OF 125
MICROINCHES

FIGURE 19-71   Contour and finish symbols.

FIGURE 19-73   Projection weld symbol—dimple on
bottom.

WELD SYMBOL INDICATES DIMPLE
ON TOP PART

DIMPLE

FIGURE 19-74   Projection weld symbol—dimple on top.

Projection Weld

A *projection weld* is similar to a spot weld and uses the same welding symbol. One of the two parts has a series of evenly spaced dimples stamped into it. Each stamped dimple is the location of an individual spot weld (**Figure 19-72**). A projection weld allows

more penetration and, as a result, is a much better weld. Note that the dimples are **not** illustrated on the detail drawing.

The reference line (OPPOSITE side-ARROW side significance) is changed slightly as used for a projection weld. The OPPOSITE side-ARROW side indicates which of the two parts to be joined has the dimples on it (**Figures 19-73** and **19-74**).

Dimensioning, contour, and finish symbols are applied to projection welds exactly as they are for spot welds.

Seam Weld

A *seam weld* is like a spot weld except that the weld is actually continuous from start to finish.

The welding symbol is modified slightly to indicate a seam weld (**Figure 19-75**). Figure 19-76 indicates how the seam weld symbol is applied to the reference line.

SIDE VIEW

FIGURE 19-72   Projection weld.

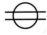

FIGURE 19-75   Seam weld symbol.

FIGURE 19-76   Seam weld symbol used with reference line.

Welding Template

In order to simplify and speed up the drawing time for welding symbols, welding templates can be used. The welding template standardizes the size of each welding symbol and provides a quick, ready reference for welding symbols (**Figure 19-77**).

Design of Weldments

Welded items are held together by fixtures during the actual welding process. This maintains the proper relationships between the parts. However, once the parts are released from the fixture, the potential for warping and other problems is introduced. This problem can be solved for the most part through proper design practices. Designers should keep the following objectives in mind when designing weldments.

1. Design for the convenience of the welder or welding machine/robot.

2. Design for proper heat control in the area of the weld.

3. Design in appropriate clamping mechanisms to prevent warping.

4. Design in outlets for the welding atmosphere.

5. Design in the appropriate clearances to accommodate for filler material.

6. Consider ease of operation after the weld is completed.

FIGURE 19-77   Welding template.

SUMMARY

- Welding is a permanent way to join metal pieces. Welding is less expensive and more satisfactory than the casting process.
- The basic welding symbol consists of a reference line, a leader line and arrow and, if needed, a tail. The tail is added for specific information or notes in regard to welding specifications, processes, or reference information.
- The size of a weld must be fully dimensioned so that there is no question whatsoever as to its intended and designed size. When a weld length is not specified, it is assumed to be continuous for the full length.
- When a weld is not continuous and is needed in only a few areas, section lining is used to indicate where the weld should be placed.
- Intermittent welds are a series of short welds. An increment is the length of the weld; the pitch is the center-to-center distance between increments.
- If a special finish must be made to a weld, a contour symbol must be added to the welding symbol. Three kinds of contour symbols are used: flush, convex, and concave.
- Any weld not made in the factory is called a field weld.
- There are five basic kinds of welding joints: butt, corner, T-joint, lap, and edge.
- Spot welding is a resistance welding process done by passing an electric current through the exact location where the parts are to be joined.
- A projection weld is similar to a spot weld and uses the same welding symbol. Dimples (projections) are stamped into one of the metal pieces. Each dimple is a location for a spot weld. A projection weld allows for more penetration. A seam weld is also like a spot weld except that the weld is continuous from start to finish.

REVIEW QUESTIONS

Mark the following statements as either true or false.

1. A plug weld and a slot weld both use the same symbol.
2. A field weld is one of the six major types of welds.
3. An intermittent weld is a series of short welds.
4. A welding symbol placed below the reference line means to weld the opposite side.
5. A flange weld is used to join thin metal parts together.
6. A process reference is a number that is placed at the bottom of the drawing as a reference.
7. A back weld is a weld applied to the opposite side of the joint after the major weld has been applied.
8. A backing weld is a weld applied to the opposite side of the joint after the major weld has been applied.

Answer the following questions by selecting the best answer.

1. Which of the following is **not** one of the major types of welds?
   a. Fillet
   b. Groove
   c. Joint
   d. Back or backing

2. Which of the following is the correct definition for the term *root opening*?
   a. The allowed space between parts
   b. The actual depth of the chamfer or groove
   c. The actual resistance between two parts
   d. None of the above

3. Which of the following correctly identifies the major welding processes?
   a. Resistance and fusion
   b. Field and contour
   c. Fillet and groove
   d. Plug and slot

4. Which of the following is **not** one of the major types of welding joints?
   a. Lap joint
   b. Edge joint
   c. L-joint
   d. T-joint

5. Which of the following is **not** true regarding spot welding?
   a. It is usually done to hold sheet metal together.
   b. It is stronger than a projection weld.
   c. It is represented by the same symbol as a projection weld.
   d. It is a resistance welding process.

6. Which of the following is **not** one of the types of groove welds?
   a. Bevel groove
   b. Flare V
   c. Square
   d. Flare U

7. What indicates a field weld on the welding symbol?
   a. A filled-in flag
   b. A filled-in circle
   c. A filled-in square
   d. A filled-in triangle

8. What indicates that a weld is to be continuous around an object?
   a. Adding a small square between the reference line and the leader line
   b. Adding a small circle between the reference line and the leader line
   c. Adding a small rectangle between the reference line and the leader line
   d. Adding a small triangle between the reference line and the leader line

CHAPTER 19 PROBLEMS

The following problems are intended to give the beginning drafter practice in studying the many factors involved in dimensioning and calling off welding processes. The student will sharpen skills in centering the views within the work area, line weight, dimensioning, and speed.

Conventional Instructions

The steps to follow in laying out the following problems are:

STEP 1 Study the problem carefully.

STEP 2 Choose the view with the most detail as the front view.

STEP 3 Make a sketch of the required views.

STEP 4 Add required dimensions to the sketch per the latest drafting standards.

STEP 5 Add welding symbols as required to the sketch per the latest welding standards.

STEP 6 Lightly lay out all required views and check that they are centered within the work area.

STEP 7 Check that all dimensions are added using the latest drafting standards.

STEP 8 Darken in all views using the correct line thickness and add all dimensioning using light guidelines.

STEP 9 Recheck all work and, if correct, neatly fill out the title block using light guidelines and neat lettering.

STEP 10 Recheck that all required welding symbols are added correctly.

CAD INSTRUCTIONS

Students who plan to complete the following problems on a CAD system should begin with these general instructions.

1. Analyze the problem carefully.

2. Plan your drawing sequence. In what order will you develop the drawing?

3. Read the specific instructions for the problem and begin.

Problems 19-1 through 19-11

Follow the listed steps using the latest drafting standards.

PROBLEM 19-1

PROBLEM 19-2

DOUBLE J GROOVE WELD/
.125 LEGS/.03 ROOT OPENING

.88

4.0

2X Ø.75 THRU

.88

.50

.50

1.0

30°

2.5

1.25

1.75

PROBLEM 19-4

3.5

.25

7X Ø.125 SPOT
WELDS

.25

6X.50 (3.0)

.25

R.125 (TYP.)

2.375

.44

.125

3.5

2.375

PROBLEM 19-5

Ø 2.5

.09

1.75

.188

MIN. RADIUS

SEAM WELD
.125 WIDE/ FULL LG.

5.0

PROBLEM 19-3

2X CHAMFER 15 X 45°

32

Ø 36 T THRU

140

28

25

BEVEL GROOVE
WELD/ 6 LEG/
CONVEX
GRIND FINISH

FILLET WELD/ 3 PLACES
6 LEGS/ 12 LONG/
24 PITCH

30

10

42

12

BEVEL GROOVE WELD/
INSIDE SURFACE
1.5 ROOT OPENING

112

60

12

METRIC

PROBLEM 19-6

2X R.75

2X Ø.75 THRU
IN LINE

2.0

.375

.50

1.125

3.75

1.125

.375

J GROOVE WELD/
MACHINE SMOOTH,.125/
FLUSH/ 2 PLACES

2X Ø.50 THRU

.50

1.125　2.50

FILLET WELD/
.25 LEG/ 4 PLACES

PROBLEM 19-7

10

10

20

40

30

25

Ø26 THRU

100

50

60

75

10

2X Ø16 THRU

METRIC

WELD TOGETHER TO SUIT

PROBLEM 19-8

3.125

.188

.188

2XR.125
.25

.188

(1.75)

FLARE BEVEL WELD/
WELD SIZE .09/2 PLACES

3.0

2X R.88

2X Ø .88 THRU

PROBLEM 19-9

2.5

.38

.38

.38

2.0

.38

.88

4X Ø.375 THRU

3.0

.50

.75

.50

.75

.50

.25

3X Ø .313 X 40°
PLUG WELD/ 3
PLACES/FILL
WELD .125 DEEP

.50

88

V GROOVE
WELD/CONCAVE/
.25 CHIP FINISH

1.0
OVERLAP

2.75　.25

FILLET WELD/ INSIDE/
BEVEL GROOVE/OUTSIDE
.125 LEG/MACHINE TO
SURFACE FINISH 125

PROBLEM 19-10

2 X Ø 18 THRU, AFTER WELDING

38

100

Ø 12 ⊤ THRU

2X Ø 32

38

40

15

70

140   40   20

6

SQUARE GROOVE
WELD/BOTH SIDES/
MACHINE FLUSH
TO 1.6

FILLET WELD/
ALL AROUND/.125
LEG / CHIP FINISH

METRIC

PROBLEM 19-11

2.0

Ø 2.0

3.0

Ø1.25 THRU

.50

.25

25

4.0   .50

.50

1.75   1.0   2.5

Problems 19-12 through 19-15

Convert the following problems from Chapter 5 from casting drawings to weldment drawings.

PROBLEM 19-12
Refer to Problem 5-27.

PROBLEM 19-13
Refer to Problem 5-28.

PROBLEM 19-14
Refer to Problem 5-36.

PROBLEM 19-15
Refer to Problem 5-37.

PROBLEMS

20

Modern Manufacturing: Materials, Processes, and Automation

CHAPTER OUTLINE

Engineering materials • Traditional manufacturing processes • Casting • Forging • Extruding • Stamping • Machining • Special workholding devices • Heat treatment of steels • Nontraditional machining processes • Automation and integration (CAM and CIM/FMS) • Computer-aided manufacturing (CAM) • Industrial robots • CIM/FMS • Just-in-time manufacturing (JIT) • Manufacturing resource planning (MRPII) • Statistical process control (SPC) • Summary • Review questions

CHAPTER OBJECTIVES

Upon completion of this chapter, students should be able to do the following:

▶ Define the concept of engineering materials.

▶ Distinguish between the two broad categories of engineering materials.

▶ List the most widely used types of metals in modern manufacturing.

▶ List the most widely used nonmetals in modern manufacturing.

▶ Define the term *composites*.

▶ Describe the principal types of casting processes.

▶ Describe the most common types of forging processes.

▶ Define the term *machining*.

(continued)

KEY TERMS

Abrasive water jet machining (AWJM)

Alloy steel

Aluminum

Angle plate fixture

Angle plate jigs

Annealing

Arbor press

Band saw (or contour band machine)

Bending

Blanking

Boring mill

Box jig

Breakdown (or bender)

Broaches

Broaching machine

Built-up tool

Case hardening

Casting

Cast irons

Cast steels

Cast tool

Centrifugal casting

Ceramics

Channel jigs

Cheek

Chemical milling

Chuck

Clapper box

Closed jigs

Composites

Compound

Continuous path

(continued)

(CHAPTER OBJECTIVES continued)

▶ Describe the most common machining processes.

▶ Explain the concept of computer-numerically controlled machines.

▶ Explain the concept of jigs.

▶ Explain the concept of fixtures.

▶ Describe the most widely used nontraditional machining processes.

▶ Explain the concept of automation.

▶ Explain the concept of computer-aided manufacturing (CAM).

▶ Define the term *robot*.

▶ Explain the concept of computer-integrated manufacturing (CIM).

▶ Explain the concept of the flexible manufacturing system (FMS).

This chapter describes the processes used in a modern manufacturing setting to produce metal and nonmetal products, as well as the engineering materials from which these products can be made. It also describes how these processes have been automated and integrated in an effort to improve quality and increase productivity.

The ultimate purpose of any engineering drawing is to provide the information necessary to make or fabricate an object. To achieve this purpose, an engineering drawing must completely describe and detail the desired object. The drawing must show the specific size and geometric shape of the object, as well as furnish the related information concerning the material specifications, finish requirements, and any special treatments required.

To properly construct an engineering drawing, the designer must be thoroughly familiar with the materials and methods used to transform the drawn image into the actual object specified in the drawing. By knowing the materials and basic processes used to manufacture an object, the designer can design the object with the manufacturing processes in mind. This will not only make manufacturing easier but will also often reduce the cost of a manufactured product. Therefore, the purpose of this chapter is to acquaint the new designer with the basic processes used to fabricate objects, as well as the capabilities and limitations of the machine tools used to machine these objects to their final form.

Engineering Materials

Manufacturing is the enterprise through which raw materials are converted into finished products. This conversion adds value to the raw materials. In a

(KEY TERMS continued)

Cope

Copper

Core prints

Cutoff saw

Cutting off

Cylindrical grinder

Die

Die casting

Draft

Drag

Drawing

Drill bushings

Drill press

Drop forging

Duplex fixture

Elastomers

Electrical discharge grinding (EDG)

Electrical discharge machining (EDM)

Electrical discharge wire cutting (EDWC)

Electrochemical discharge grinding (ECDG)

Electrochemical grinding (ECG)

Electrochemical honing (ECH)

Electrochemical machining (ECM)

Electrochemical turning (ECT)

Electromechanical machining (EMM)

Electron beam machining (EBM)

Engineering materials

External broaching

Extruding

Finishing impression

Fixed-renewable bushing

Fixtures

Flask

Foolproofing

Forging

Forming

Fuller

Gate

Green sand

Hardening

Heat treatment

High-performance alloys

Horizontal boring mills

Hurry-up-and-wait

Hydrodynamic machining (HDM)

Indexing fixture

Indexing jig

Internal broaching

Investment casting

Jigs

Laser beam machining (LBM)

Lathe

Lead

Lead time

Leaf jig

Liner bushings

Lost wax casting

Machining

Magnesium

Manufacturing

Maraging steel

Milling machine

Modified angle plate fixture

(continued)

modern engineering and manufacturing setting, these materials are referred to as *engineering materials*. Engineering materials can be divided into two broad categories: metals and nonmetals. Metals include steels, alloy steels, and a variety of high-performance alloys. Nonmetals include plastics, composites, elastomers, and ceramics.

Metals

Metals are widely used in modern manufacturing. The most frequently used metals are carbon steels; alloy steels; high-strength, low-alloy steels; stainless steels; maraging steels; cast steels; cast irons; high-performance alloys; tungsten, molybdenum, and titanium; aluminum; copper; magnesium; lead; tin; zinc; and powdered metals. These various metals are described in the following paragraphs.

Carbon Steels

The main elements in steels are iron and carbon. Carbon steels are divided into three groups based on their carbon content. Low-carbon steels contain from .001 to .30% carbon. Medium-carbon steels contain from .70 to 1.30% carbon. High-carbon steels contain from .70 to 1.30% carbon. Certain grades of carbon steels also contain boron to improve hardenability and aluminum to control grain size and promote deoxidation. The following types of products are made of carbon steels: bar, sheet, strip, plate, wire, tubing, and structural shapes.

Alloy Steels

An *alloy steel* is any steel that contains one or more alloying elements added to produce characteristics not available in unalloyed carbon steels. Widely used alloying elements include manganese, silicon, copper, aluminum, chromium, cobalt, columbium, molybdenum, nickel, titanium, tungsten, vanadium, and zirconium. Such alloys are added to improve the mechanical and physical properties of steels.

High-Strength, Low-Alloy Steels

These are steels that are significantly stronger than carbon steels as a result of the addition of small amounts of alloying elements and special processing. These alloying elements include manganese, silicon, phosphorus, copper, aluminum, chromium, niobium, vanadium, titanium, molybdenum, nickel, zirconium, nitrogen, and calcium. These steels are categorized according to composition. The different categories vary as to strength, toughness, formability, weldability, and corrosion resistance.

Stainless Steels

Stainless steels are steels that contain 10.5% or more chromium. The special appearance of stainless steels is the result of a chromium-based oxide film that coats the metal's surface. Stainless steels also contain a number of other elements that are added to improve their machinability and corrosion resistance. These elements include nickel, molybdenum, copper, titanium, silicon, manganese, columbium, aluminum, nitrogen, and sulfur. There are almost 60 standard grades of stainless steel, but there are five main types. These types are austenitic, ferritic, martensitic, prescription-hardening, and duplex.

Maraging Steels

Maraging steels are a special class of high-strength steels in which nickel is the primary alloy. The name *maraging* is derived from the term *martensite age hardening*, which is a process of age hardening of a matrix composed of low-carbon, iron-nickel martensite. The process used to produce maraging steels is a double-vacuum melting process. The first process involves vacuum-induction melting. The second process involves vacuum arc remelting. This double-vacuum process maintains a high level of purity and reduces residual elements. Maraging steels are produced in bar, plate, and sheet form.

Cast Steels

Cast steels are steels that are produced in ingot form by melting all component elements together and pouring the molten metal into a mold. This is an intermediate step. Cast steel ingots are subsequently hot or cold worked into another shape. Alloying elements include nickel, chromium, vanadium, and copper. The main types of cast steels are carbon, low-alloy, corrosion-resistant, and heat-resistant.

Cast Irons

Cast irons are metals in a family of cast ferrous metals. Cast irons are distinguished from steels by the relative carbon content (i.e., less than 2% in steels and over 2% in cast irons). In addition, cast irons also contain from 1 to 3% silicon. These differences are the basis of the corresponding differences in how steels and cast irons can be used. The metallurgical properties of cast irons make them easy to melt and cast. These are the basic types of cast irons:

white, malleable, gray, ductile, and compacted-graphite iron.

High-Performance Alloys

High-performance alloys are a group of metals that have high strength and/or high corrosion resistance over a broad range of temperatures. High-performance alloys are primarily iron-nickel, nickel, cobalt, or iron-based. They typically have a complex composition with at least three and sometimes over 10 component elements. Also known as the super-alloys, their most common elements include chromium, molybdenum, tungsten, columbium, aluminum, cobalt, titanium, carbon, and nitrogen. The most common applications of high-performance alloys are in aircraft and aerospace settings.

Tungsten, Molybdenum, and Titanium

Tungsten and molybdenum are known as *refractory metals* because of their exceptional ability to withstand high temperatures. They also have the following additional characteristics: low coefficient of expansion, high modulus of elasticity, high resistance to thermal shock, high conductivity, and excellent hardness properties at high temperatures. Unlike tungsten and molybdenum, titanium is not a refractory metal, but its properties are similar to such metals. The processes used to produce these three metals are similar. They are extracted from ore, processed into chemicals, and processed further into powders or sponges. Typical applications of tungsten and molybdenum are the automotive, aerospace, mining, chemical, processing, electrical, nuclear, petroleum processes, ordnance, and metal-processing industries.

Titanium is used primarily in the aerospace, chemical, and medical industries.

Aluminum

Aluminum is a bright, silvery metal that is very light. A given amount of ferrous metal will weigh three times as much as the same amount of aluminum. Aluminum is an abundant metal, accounting for approximately 8% of the earth's outer surface. Pure aluminum is ductile, soft, and weak compared with other metals. However, with the addition of alloying elements such as copper, magnesium, manganese, and zinc, it can be made much harder and stronger. Aluminum is light, an excellent conductor, and able to resist corrosion. Consequently, it is widely used in both manufacturing and construction.

Copper

Copper is a heavy metal that is salmon pink in its pure form and abundant. It is nonmagnetic and has excellent conductivity. Only silver is a better conductor. There are over 300 different types of copper and copper alloys. They are produced in a variety of forms, including strip, plate, sheet, pipe, tube, rod, forgings, wire, foil, extrusions, and castings. Copper is widely used in both manufacturing and construction.

Magnesium

Magnesium is a soft, silvery-white metal that lacks the strength needed for structural uses. However, it is extremely light and machinable. To increase its strength, magnesium is alloyed with aluminum, manganese, lithium, thorium, zinc, and zirconium. The result is a metal that can be machined at high speeds, has good conductivity and a high level of fatigue, and is dent-resistant.

Lead

Lead is a dark gray metal that has the following characteristics: high density, low melting point, corrosion resistance, chemical stability, malleability, lubricity, and electrical properties. Lead can be easily produced in a variety of forms.

Tin

Tin is a silvery-white metal that is soft and ductile. Its high level of malleability allows tin to be drawn into wire or hammered into extremely thin sheets. Tin is too weak to be used in its pure form, but alloyed with antimony, copper, lead, zinc, or silver, tin becomes useful. Its most common use is in coating steel containers used to preserve food. Tin is also used in solder alloys, bronzes, fusible alloys, type metals, pewter, and dental amalgams. Tin chemical compounds are used in the electroplating, ceramics, plastics, pesticide, and antifungal industries.

Zinc

Zinc is a bluish-white metal that is widely used in manufacturing. Only iron, aluminum, and copper are more widely used than zinc. Zinc is produced in castings and is widely used for coating other materials. Zinc alloys offer several advantages over other die-cast metals, including ease of casting, strength, ductility, and the ability to hold close tolerances and to be cast in thin sections.

Powdered Metals

Powdered metals are exactly what the name says. They are produced by atomization, electrolysis, chemical reduction, or oxide reduction. Various powdered metals are mixed and sintered to produce alloys with an almost endless number of compositions. Sintering produces bonds among the various powdered particles. The properties of products made from powdered metals depend on which metals are mixed, in what amounts, and by what processes. These characteristics can be controlled by varying particle size, particle shape, size distribution, sieve analysis, bulk density of the powder, rate of flow into the die, and the compressibility of the powder in the die.

Nonmetals

Nonmetals are widely used in modern manufacturing. The most widely used nonmetals are plastics, composites, ceramics, and elastomers. These various nonmetals are described in the following paragraphs.

Plastics

Plastics are nonmetal materials that are made from either natural or synthetic resins. Most plastics used in modern manufacturing are made from synthetic resins. There are many different kinds of plastics, including polystyrene, high-impact polystyrene, acrylics, polycarbonate, ABS, acetal, nylon, polypropylene, polyethylene, epoxy, and phenolic. Most plastics used in an engineering or manufacturing setting are made from thermoplastic resins. Such plastics have the following characteristics: (a) thermal, mechanical, chemical, corrosion resistance, and fabricability; (b) ability to sustain high mechanical loads; and (c) predictable performance.

Thermosets. Thermosets are a special subset of engineering plastics that are used in applications where resistance to heat and environmental conditions is important. For example, thermosets have long been used for insulation in electrical and electronics applications. The recipes for thermoset molding compounds can be varied to create just the right characteristics for a given application. Distinctive properties of thermosets include dimensional stability, low-to-zero creep, low water absorption, maximum physical strength, good electrical properties, high heat deflection temperatures, high heat resistance, minimum thermal expansion, and specific gravities in the 1.35 to 2.00 range.

Thermoplastics. Thermoplastics are a special subset of engineering plastics that are very similar to thermosets. However, processing advantages have allowed thermoplastics to replace thermosets in many applications. The benefits of thermoplastics over thermosets are faster molding, lighter weights, ability to have thinner walls that allow for more complex designs, and greater impact resistance. However, thermosets typically experience less creep at higher temperatures.

Composites

Composites are materials created by combining two or more materials. Perhaps the most readily recognizable composite is plywood. Fiberglass is another example of a composite. Composite technology is divided into four broad areas: organ matrix composites (resins), metal matrix composites, carbon-carbon composites, and ceramic matrix composites. Composites can be divided into two categories in terms of how they are produced: laminates and sandwiches. Laminates consist of two or more layers bonded together. Sandwiches consist of thick, low-density layers (i.e., honeycomb or foam materials) pressed between thin layers of higher-strength, higher-density material. Applications of composites are extensive and varied. They include aircraft and aerospace applications, athletic equipment, automobiles, prosthetic devices, printed circuit boards, boats, and many others.

Ceramics

Ceramics are materials made by combining metallic and nonmetallic elements such as oxides, carbides, and nitrites. Glass, bricks, tile, and porcelain are all common ceramics. Refractory ceramics are important in modern manufacturing because of their usefulness in electronics applications. They are used in thermistors, rectifiers, capacitors, transducers, and for high-voltage insulation. Ceramics are hard, brittle, have a high melting point, have low conductivity, are chemically and thermally stable, have high compressive strength, and have high creep resistance.

Elastomers

The most recognizable form of elastomer is natural rubber. *Elastomers* are linear polymers that are able to stretch and return to their original shape and size (i.e., a rubber band). Widely used elastomers include polyacrylate, ethylene propylene, neoprene, polysulfide, silicone, and urethane. Common applications of elastomers include oil hoses, o-rings, electric insulation, belts, gaskets, seals, value disks, and foam padding.

Traditional Manufacturing Processes

The term *manufacturing processes* refers to the basic methods used by the shop to make the object described in the engineering drawing. The specific processes used to make the object depend on the object itself. In some cases, the part may be cast, while other parts may be forged, extruded, or stamped. In many instances, the part, once fabricated into its basic form, must also be machined to maintain a specific degree of accuracy or to produce a feature not possible with other manufacturing processes.

When the manufacturing shop receives the engineering drawing, in the form of a part print, it is first reviewed to ensure that all pertinent data and information necessary to make the object are contained in the drawing. During this review, the shop personnel will consider several factors necessary to determine how the part must be made. These factors include the type and condition of the material used to make the object, the overall size and shape of the part, the types of operations required, the required accuracy of the part, and the number of parts to be made.

The type and condition of the material used to make the part are important considerations. Some parts may be made from solid bar stock, while other parts must be extruded, cast, forged, or stamped. In most cases, parts made from bar stock require the least lead time. The *lead time* is the interval from the time the shop receives the drawing until production begins. Most shops maintain a sufficient supply of bar stock to begin production as soon as the drawing is received. However, when a part must be extruded, cast, forged, or stamped, a longer lead time is required to make the necessary molds or dies to fabricate the object.

The size and shape of the object must be considered to determine if the object is within the capabilities of the shop to make. In addition, the size and shape may also determine the size of the machine tools required and the datum, or reference, surfaces used to locate the part during manufacture.

The types of operations required determine the types of equipment and machine tools needed to make the part. If, for example, the part requires holes, a drill press, vertical milling machine, or other machine tool may be used. The next consideration, the required accuracy of the part, also determines how the operations are performed. A hole with a required accuracy of .002″ (0.05 mm), for example, would require reaming, while a hole with a required accuracy of .020″ (0.5 mm) could be drilled.

The number of parts to be made will frequently determine if the part should be cast, forged, extruded, or machined from solid stock. Likewise, the number of parts will also determine if any special workholders are to be made. Larger production runs will normally justify more sophisticated tools and processes since the cost can be spread over a larger number of parts. Smaller production runs normally demand that the parts be made at the lowest possible cost with little or no investment in special molds, dies, or workholders.

While each of these factors has been separated for the purpose of this discussion, in practice, each is considered as part of the other. These factors are so closely related in production that they frequently overlap, and one cannot be considered without the others.

The primary processes used to fabricate manufactured products are casting, forging, extruding, stamping, and machining. To use these processes to their best advantage, the designer must be familiar with the strengths and weaknesses of each process, as well as the fundamental aspects of each process.

Casting

Casting is a process of pouring molten metal into a mold that contains the desired shape in the form of a cavity. The principal types of casting used in manufacturing today are sand casting, investment casting, centrifugal casting, and die casting.

Sand Casting

Sand casting is the most common type of casting method. The major components of the molds used to make sand castings are shown in **Figure 20-1** and include the *flask, pattern,* and *green sand.* The flask is a two-part box, or frame, used to contain the sand. The top half of the flask is called the *cope,* and the bottom half is called the *drag.* Occasionally, a third section may be installed between the cope and drag. This section is called a *cheek* and is used where a deep or complex shape must be cast. In sand casting, the mold is prepared by ramming the green sand around a model of a part, called the pattern. The model is then removed to form the cavity for the molten metal. The plane of division between the cope and drag is called the *parting line.* In many castings, the parting line occurs at the approximate middle of the part. The parting line can be seen on most castings by a ragged line that is usually ground off.

FIGURE 20-1   Components of a sand casting mold (from Thode, *Materials Processing,* Delmar Publishers Inc.).

The molten metal enters the mold through the *sprue hole* and is directed to the cavity by one or more *gates.* The sprue hole is formed by installing a sprue peg in the cope. This peg is then removed after the final ramming of the cope. The *riser* is used to vent the mold and to allow gases to escape. The riser also acts as a small reservoir to keep the cavity full as the metal begins to shrink during the cooling process.

Once the cope and drag are rammed and the pattern is removed, a solid part can be poured in the mold. However, in some cases, a hollow part, or one that has large holes, must be poured. To reduce the amount of material needed to fill the cavity and to reduce the time necessary to machine the part, a *sand core* may be installed in the cavity. When sand cores are intended to be installed in a cavity, *core prints* must also be provided to locate and anchor the sand cores during the casting process. Core prints are normally a simple extension of the pattern.

Making Patterns.   When patterns are made for casting, two important factors must always be considered. The first is the draft. *Draft* is the slope or taper of the sides of a pattern that permits it to be removed from the cope and drag without disturbing the cavity. The draft also permits the cast part to be removed easily from some molds. The amount of draft necessary will normally depend on the part being cast but will normally be about 1°.

The second consideration is shrinkage. When metals are cast, a certain amount of *shrinkage* occurs as the metal cools. The specific amount of shrinkage depends on the metal being cast. Steel, for example, shrinks at a rate of approximately 3/16″ per foot, while cast iron shrinks about 1/8″ per foot. To allow for this shrinkage and to make sure that the final part is the correct size, the pattern must be made slightly larger than the final size of the desired part (**Figure 20-2**).

When making a pattern, the pattern maker will normally use a shrink rule to compensate for the shrinkage.

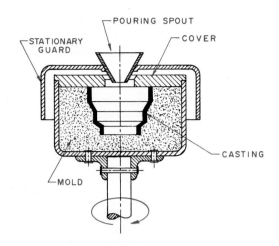

FIGURE 20-2   Shrinkage allowance for cast parts.

FIGURE 20-3   Centrifugal casting.

A *shrink rule* is a standard steel rule that has the graduations marked for a specific amount of shrinkage. A shrink rule used for cast iron has an extra 1/8″ added to each foot, while a shrink rule for steel has an additional 3/16″ added.

In most instances, when a pattern must be made, the pattern maker will receive the engineering drawing that contains all the final sizes of the part. The pattern maker will then make all the necessary calculations needed to make the oversized pattern. But, occasionally, the pattern maker will be given a drawing with all the calculations already made to produce the pattern. In either case, the patterns must be made to suit the material being cast.

Investment Casting

Investment casting produces parts with great detail and accuracy while at the same time allowing very thin cross sections to be cast. In this process, the pattern is made by casting wax in the desired form. The pattern is then placed in a sand mold, and the mold is fired to melt out the wax pattern. This process is also referred to as *lost wax casting*. The molten metal is then fed into the mold to produce the cast part.

Centrifugal Casting

Centrifugal casting (**Figure 20-3**) is a process of pouring a measured amount of molten metal into a rotating mold. This process can be used for a single mold or multiple molds. The centrifugal force created by the rotation forces the molten metal to fill the cavity. This same centrifugal force, along with the measured amount of material, also controls the wall thickness of the cast part and results in a less porous cast surface than is possible with sand casting. The molds used for this purpose are usually permanent molds made from metal rather than sand. These molds are used repeatedly and produce highly accurate parts requiring very little machining. However, due to their cost, permanent molds are normally used only for high-volume production.

Die Casting

Die casting (**Figure 20-4**) is a process in which molten metal is forced into metal dies under pressure. This process is very well-suited for materials such as zinc, aluminum, copper, and magnesium alloys.

FIGURE 20-4   Die casting (from Thode, *Materials Processing*, Delmar Publishers Inc.).

FIGURE 20-5   Powder metallurgy process.

Parts produced by die casting are superior in appearance and accuracy and require little or no machining to final size.

Powder Metallurgy

Powder metallurgy (**Figure 20-5**) while not an actual casting process, does have some similarities to casting. In the *powder metallurgy* process, metal particles, or powder, are blended and mixed to achieve the desired composition. The powder is then forced into a die of the desired form under pressure from 15,000 to 100,000 pounds per square inch. The resulting heat fuses the powder into a solid piece that can be machined.

Forging

Forging is the process of forming metals under pressure using a variety of different processes. The most common forms of forging are drop forging and press forging. Other variations of forging include rolling and upsetting.

Drop Forging

Drop forging is the process of forming a heated metal bar, or billet, in dies. In practice, the heated metal is placed on a lower portion of a forging die and struck repeatedly with the upper die portion. This forces the metal into the shape of the cavity of the dies. The pressure required to form the metal is produced by a drop hammer.

Most drop forge dies contain at least four different stations, or dies, to complete the part (**Figure 20-6**). The first station is called the *fuller*. This station is used to rough-form the part to fit the other cavities. The second station, when used, is called the *breakdown,* or *bender*. This station forms the part into any special contours required by the next station, the *roughing impression*. The roughing impression rough-forms the part to the desired form. The final station is called the *finishing impression* and is used to finish the part to the desired shape and size. Additional stations, such as a trimmer to remove the fins and a cutoff to sever the forged part from the bar, may also be included on the die if they are necessary.

Press Forging

Press forging is a single-step process in which the heated metal bar or billet is forced into a single die and is completed in a single press stroke. The press normally used for press forging is hydraulic.

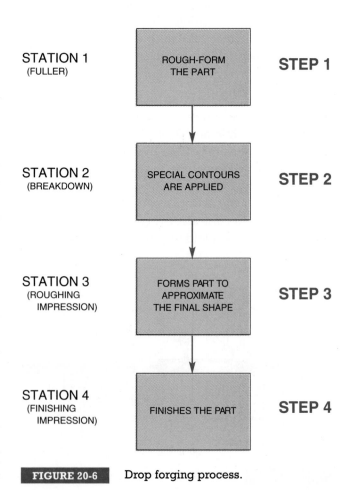

FIGURE 20-6   Drop forging process.

FIGURE 20-7   Upsetting.

Rolling

Rolling is a process in which the heated metal bar is formed by passing through rollers having the desired form impressed on their surfaces. Rolling is also frequently used to flatten and thin out thick sections.

Upsetting

Upsetting (**Figure 20-7**) is a process used to enlarge selected sections of a metal part. Bolt heads, for example, are normally produced by an upsetting process.

Extruding

Extruding is a process of forcing metal through a die of a desired form and cross section. The bars produced are then cut to the required lengths. Typical examples of extruded parts include parts for aluminum windows and doors. Extrusions provide a nearly final shape, as shown in **Figure 20-8** and, in many cases, only need to be cut off and machined slightly to complete a finished part.

Stamping

Stamping is a process of using dies to cut or form metal sheets or strips into a desired form. The main tool used for metal stamping is a die. The term *die*

FIGURE 20-9   Metal stamping die.

has a double meaning. It can be used to describe the entire assembled tool or the lower cutting part of the tool. The exact meaning of the term can usually be determined from the context of its use. The principal parts of a die are shown in **Figure 20-9**. The upper die shoe is used to mount the punch. The die is mounted on the lower die shoe. The guide pins and guide pin bushings are used to maintain the alignment between the punch and die. The stripper is designed to strip the stock material from the punch after the cutting stroke of the die.

When more than one operation takes place on a part during a single stroke (blanking and punching, for example), the operation is said to be *compound*. However, when several operations take place sequentially in a die, the die is called a *progressive die*.

The principal stamping operations normally performed are shearing, cutting off, parting, blanking, punching, piercing, perforating, trimming, slitting, shaving, bending, forming, drawing, coining, and embossing.

FIGURE 20-8   Extruding.

FIGURE 20-10   Operations that produce blanks.

FIGURE 20-11   (A) Punching operation to produce holes; (B) Piercing operation to produce holes.

Operations That Produce Blanks

The operations that produce blanks are shearing, cutting off, parting, and blanking (**Figure 20-10**). *Shearing* is a cutting action performed along a straight line. *Cutting off* is a cutting operation performed on a part to produce an edge other than a straightedge. In many cases, cutting off is used to finish the edges of a part. Like shearing, cutting off does not produce any scrap. *Parting,* on the other hand, is also an operation used to finish the edges of a part but, unlike the other operations, parting does produce scrap.

Blanking is an operation that produces a part cut completely by the punch and die. In blanking operations, the piece that falls through the die is the desired part. The scrap is the skeleton left on the stock strip.

Operations That Produce Holes

The operations that produce holes are punching, piercing, and perforating. *Punching* is an operation that cuts a hole in a metal sheet or strip (**Figure 20-11A**). In punching, the piece that falls through the die is scrap, and the area on the strip is the desired part. Punching is normally performed on parts that are to be blanked to provide holes for bolts, screws, or similar parts.

Piercing is an operation similar to punching, except that in piercing, no scrap is produced (**Figure 20-11B**). A pierced hole is sometimes used to increase the thickness of metal around a hole for tapping threads in a stamped part. *Perforating* is simply a punching operation performed on sheets to produce either a uniform hole pattern or a decorative form on the sheet.

Operations That Control Size

The operations used to control size are trimming, slitting, and shaving (**Figures 20-12A and 20-12B**). *Trimming* is an operation performed on formed or drawn parts to remove the ragged edge of the blank. *Slitting* is performed on large sheets to produce thin stock strips. Slitting may be performed by rollers or by straight blades. *Shaving* is an operation performed on a blanked part to produce an exact dimensional size or to square a blanked edge. Shaving produces a part with a close dimensional size and a cut edge that is almost perpendicular to the top and bottom surfaces of the blanked part.

FIGURE 20-12   (A) Operation that controls size-trimming; (B) Operation that controls size-shaving.

Operations That Bend or Form Parts

The operations used to bend or form parts are bending, forming, and drawing (**Figure 20-13**). *Bending* is an operation in which a metal part is simply bent to a desired angle. *Forming,* on the other hand, is an operation in which a part is bent or formed into a complex shape. Forming is also a catchall word frequently used to describe any bending operation that does not fall into one of the other categories. *Drawing* is a process of stretching and forming a metal sheet into a shape similar to a cup or top hat.

Machining

Machining is the process of removing metal with machine tools and cutters to achieve a desired form or feature. The principal types of machine tools used to perform these operations are lathes,

milling machines, drill presses, saws, and grinders. Other machines that are variations of the basic machines used in the machine shop include shapers and planers, boring mills, broaching machines, and numerically controlled machines. The specific type of machine used to perform a particular machining operation is determined by the type of operation and the degree of accuracy required.

Lathes

A lathe (**Figure 20-14**) is one of the most commonly used and versatile machine tools in the shop. Typically, a *lathe* performs such operations as straight and taper turning, facing, straight and taper boring, drilling, reaming, tapping, threading, and knurling (**Figure 20-15**).

In operation, the workpiece is held in the headstock and supported by the tailstock. The cutting

FIGURE 20-13   Operations that bend or form parts (from Thode, *Materials Processing*, Delmar Publishers Inc.).

FIGURE 20-15   Basic lathe operations (from Thode, *Materials Processing*, Delmar Publishers Inc.).

FIGURE 20-16   Lathe setup for straight turning (from Krar and Oswald, *Turning Technology*, Delmar Publishers Inc.).

tool is mounted in the tool post and is traversed past the rotating workpiece to machine the desired form (**Figure 20-16**). The most common method used to hold parts in a lathe is with a chuck. A *chuck* is a device much like a vise, having movable jaws that grip and hold the workpiece. The most popular type of chuck is the three-jaw chuck, but there are also two-jaw, four-jaw, and six-jaw chucks as well. Other types of devices used to hold and drive a workpiece in a lathe include collets, face plates, drive plates, and lathe dogs.

Milling Machines

Milling machines are used to machine workpieces by feeding the part into a rotating cutter (**Figure 20-17**). The two basic variations of the milling machine are the vertical milling machine and the horizontal milling machine.

In operation, the milling cutter is mounted in either the spindle or arbor, and the workpiece is held on the machine table. The most commonly used device to mount the workpiece for milling is the milling machine vise, but the workpiece may also be held directly on the machine table or in other work-holding devices. The operations most frequently performed on a milling machine include plain milling, face milling, end milling, straddle milling, gang milling, form milling, keyseat milling, and gear cutting

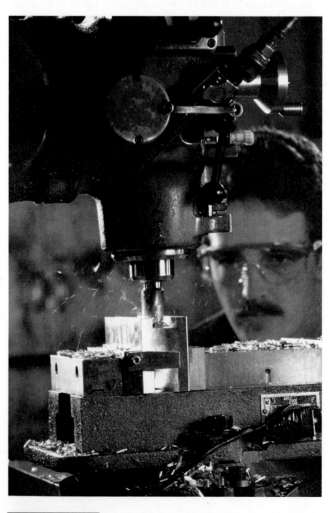

FIGURE 20-17   Milling process (Courtesy of Getty Images).

(**Figure 20-18**). **Figure 20-19** shows the types of cutters commonly used for milling operations.

Drill Presses

A *drill press* (**Figure 20-20**) is a machine that is used mainly for producing holes. The principal types of

CONVENTIONAL MILLING   CLIMB MILLING

FIGURE 20-18   Basic milling machine operations (from Thode, *Materials Processing*, Delmar Publishers Inc.).

FIGURE 20-20   Drill press (Courtesy of Fort Walton Machining).

FIGURE 20-19   Typical milling cutters (Courtesy of Fort Walton Machining).

drill presses used in the shop are the radial drill press and the sensitive drill press.

The operations normally performed on drill presses include drilling, reaming, tapping, chamfering, spotfacing, counterboring, countersinking, reverse countersinking, and reverse spotfacing (**Figure 20-21**).

When using a drill press, the workpiece is normally held in a vise or clamped directly to the machine table. The drill or other cutting tool is mounted in the machine spindle and is fed into the workpiece either by hand or with a mechanical feed unit.

Power Saws

The *power saws* used most often in the machine shop are the contour band machine or band saw and the cutoff saw (**Figure 20-22**). The contour band machine or *band saw* is used primarily for sawing intricate or detailed shapes; the *cutoff saw* is used to cut rough bar stock to lengths suitable for machining in other machine tools.

DRILL   REAM   TAP   CHAMFER   COUNTERBORE   COUNTERSINK   REVERSE COUNTERSINK   REVERSE SPOTFACE

FIGURE 20-21   Basic drill press operations.

FIGURE 20-22   Band saw (Courtesy of Getty Images).

Precision Grinders

Precision grinders are available in several styles and types to suit their many and varied applications. The principal types of precision grinders used in the machine shop are the surface grinder and the cylindrical grinder, among others.

The *surface grinder* (**Figure 20-23**) is mainly used to produce flat, angular, or special contours on flat workpieces. The most popular variation of this machine consists of a grinding wheel mounted on a horizontal spindle and a reciprocating table that traverses back and forth under the grinding wheel. One other variation of the surface grinder frequently found in the machine shop uses a vertical spindle and a round, rotating table (**Figure 20-24**). The workpiece is generally mounted and held on a magnetic chuck during the grinding operation on both styles of surface grinders.

FIGURE 20-23   Surface grinder (from Thode, *Materials Processing*, Delmar Publishers Inc.).

FIGURE 20-24   Vertical spindle-rotating table surface grinder (from Thode, *Materials Processing*, Delmar Publishers Inc.).

Cylindrical grinders are used to precisely grind cylindrical or conical workpieces. For grinding, the workpiece is mounted either between centers or in a precision chuck, much like a lathe. The grinding wheel is mounted behind the workpiece on a horizontal spindle. The table of the grinder traverses the workpiece past the grinding wheel in a reciprocating motion while the workpiece rotates at a preset speed.

Shapers and Planers

Shapers and *planers* are machine tools that use single-point cutting tools to perform their cutting operations. The shaper uses a reciprocating ram to drive the cutter. The cutting tool on this machine tool is mounted on the end of the ram in a unit called the *clapper box*. The clapper box can be positioned vertically or at an angle, depending on the work to be performed. The depth of cut is adjusted by lowering the vertical slide of the clapper box or by raising the position of the table. The workpiece is mounted on the table or vise and is fed past the reciprocating cutter by the table feed. The length and position of the stroke are regulated by the position of the ram and are easily adjusted to suit the size of the workpiece.

Planers operate in a manner similar to the shaper. The major differences between these two machines are their size and method of cutting. The planer is normally much larger than the shaper and the work, rather than the cutter, moves on the planer. In operation, the workpiece is mounted on the table of the planer, and the table reciprocates under the stationary tool. As the table reciprocates, the tool is moved across the workpiece by the feed unit. The depth of cut is determined by the height of the tool from the table and is adjusted by lowering or raising the cross beam. The length and position of the cut are determined by the position of the table.

These processes are seldom used today.

Boring Mills

A *boring mill* (**Figures 20-25** and **20-26**) is a machine tool normally used to machine large workpieces. The two common variations of these machines are horizontal and vertical. The distinction between the two is determined by the position of the spindle. *Horizontal boring mills* perform a wide variety of different machining tasks normally associated with a milling machine but on a much larger scale. *Vertical boring mills* are commonly used to turn, bore, and face large parts in much the same way as a lathe. Another variation of the horizontal boring mill sometimes found in the machine shop is the *vertical turret lathe*. This machine serves the same basic function as the vertical boring mill but, rather than using a single tool, a vertical turret lathe uses a turret arrangement to mount and position the cutting tools.

FIGURE 20-25   Milling machine (Courtesy of Fort Walton Machining).

FIGURE 20-26   CNC mill (Courtesy of Fort Walton Machining).

Broaching Machines

A *broaching machine* is used to modify the shape of a workpiece by pulling tools called *broaches* across or through the part. Both internal and external forms can be broached. *Internal broaching* can produce holes with a wide variety of different forms (**Figure 20-27**). *External broaching* is typically used to produce some gear teeth, plier jaws, and other similar details.

Another type of internal broaching operation is broached in an *arbor press* using a push-type rather than a pull-type broach. This type of broaching is often used in the top to produce keyways or other simple shapes.

Numerically Controlled Machines

Numerically controlled machine tools represent the norm in machine tool design in wide use today. These machine tools are operated by computers, and they can markedly reduce the errors caused by human operators.

The two basic variations of these machines in use today are the point-to-point and the continuous path machines. The principal difference between the two is in the movements of the tool with reference to the workpiece. *Point-to-point* machines operate on a series of preprogrammed coordinates to locate the position of the tool. When the tool finishes at one point, it automatically goes to the next point by the shortest route. This type of control is very useful for drilling machines, but machines such as milling machines require greater control of the tool movement. So, *continuous path* controls are used to control the movement of the tool throughout the cutting cycle. For example, if a circle or radius were to be milled, the operator or programmer would need only to program

FIGURE 20-27   Broached shapes.

a few points along the arc. The computer would compute the remaining points and guide the cutting tool throughout the complete cycle.

These controls are frequently used for a wide range of different machine tools, from drill presses and milling machines to lathes and grinders. Due to the advent of these controls, a whole new form of machine tool has evolved: the machining center (**Figure 20-28**). These machines can take a rough, unmachined part and completely machine the whole part without removing it from the machine or changing a setup.

Special Workholding Devices

Special workholding devices are frequently used to produce parts in high-volume production runs. The major categories of special workholding devices commonly used in the shop are jigs and fixtures. The primary function of jigs and fixtures is to transfer the required accuracy and precision from the operator to the tool. This permits duplicate parts to be produced within the specified limits of size without error. Both jigs and fixtures hold, support, and locate the workpiece; the principal difference between these tools is the method used to control the relationship of the tool to the workpiece. *Jigs* guide the cutting tool through a hardened drill bushing during the cutting cycle. *Fixtures,* on the other hand, reference the cutting tool by means of a set block (**Figure 20-29**).

Classification of Jigs

Jigs are normally classified by the type of operation they perform and their basic construction. Typically, jigs are used to drill, ream, tap, countersink, chamfer, counterbore, and spotface. Jigs are also divided into

FIGURE 20-29   Referencing the tool to the workpiece.

two general construction categories: open and closed. *Open jigs* are jigs that cover only one side of the part and are used for relatively simple operations. *Closed jigs* are jigs that enclose the part on more than one side and are intended to machine the part on several sides without removing the part from the jig.

The most common types of jigs include plate jigs, angle plate jigs, box jigs, and indexing jigs. While there are several other distinct styles of jigs, these represent the most common forms.

Plate jigs (**Figure 20-30**) are the most common form of jig. These jigs consist of a simple plate that contains

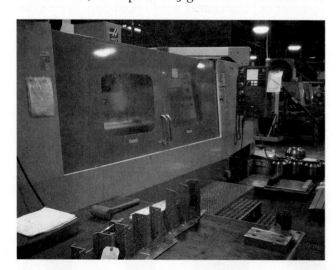

FIGURE 20-28   CNC machining center (Courtesy of Fort Walton Machining).

FIGURE 20-30   Plate jig.

the required drill bushings, locators, and clamping elements. Typical variations of the basic plate jig include the *table jig* (**Figure 20-31A**) and the *sandwich jig* (**Figure 20-31B**). Another and even simpler version of the plate jig is the *template jig* (**Figure 20-32**). These jigs are used where accuracy and not speed is the prime consideration. Template jigs may or may not have drill bushings and do not normally have a clamping device. This form of jig is frequently used for light machining or for layout work.

An *angle plate jig* (**Figure 20-33**) is a modified form of a plate jig in which the surface to be machined is not perpendicular to the locating surface. This type of jig is often used to machine pulleys, gears, hubs, or similar parts. Another variation of this type of jig is the *modified angle plate jig*. These jigs are used to machine parts at angles other than 90°.

FIGURE 20-32   Template jigs.

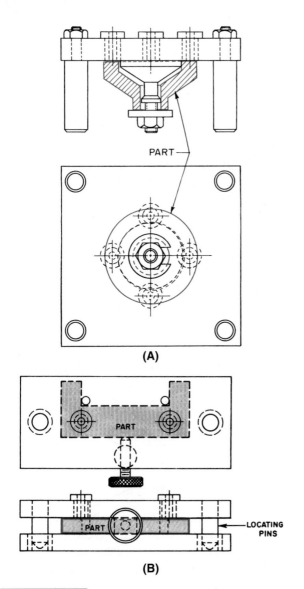

(A)

(B)

FIGURE 20-31   (A) Table jig; (B) Sandwich jig.

FIGURE 20-33   Angle plate jig.

A *box jig* (**Figure 20-34**) is designed to be used for parts that require machining on several sides. With these jigs, the part is mounted in the jig and clamped with a leaf or door. Other variations of the basic box jig include the *channel jig* (**Figure 20-35A**) and the *leaf jig* (**Figure 20-35B**). These jigs are similar in design to the box jig, but they machine the part on only two or three sides rather than all six sides.

FIGURE 20-34   Box jig.

FIGURE 20-35   (A) Channel jig; (B) Leaf jig.

FIGURE 20-36   Indexing jig.

An *indexing jig* (**Figure 20-36**) is used primarily to machine parts that have machined details at intervals around the part. Drilling four holes 90° apart is a typical example of the type of work this jig is best suited to perform. Another type of jig that uses an indexing arrangement to locate the jig, rather than the part, is the *multistation jig* (**Figure 20-37**). These jigs are used to machine several parts at one time.

Classification of Fixtures

Fixtures are classified by the type of machine they are used on, the type of operation performed, and their basic construction features. The principal types of fixtures normally used in the shop include plate fixtures, angle plate fixtures, vise jaw fixtures, and indexing fixtures. Typically, fixtures are used for milling, turning, sawing, grinding, inspecting, and several other varied operations.

FIGURE 20-37   Multistation jig.

FIGURE 20-38   Plate fixture.

A *plate fixture* (**Figure 20-38**) is the most common type of fixture. Like plate jigs, plate fixtures are simply a plate containing the locators, set blocks, and clamping devices necessary to locate and hold the workpiece and to reference the cutter. While similar in design to a plate jig, plate fixtures are normally made much heavier than plate jigs to resist the additional cutting forces.

An *angle plate fixture* (**Figure 20-39A**) and a *modified angle plate fixture* (**Figure 20-39B**) are simple modifications of the basic plate fixture design. These fixtures are used when the reference surface is at an angle to the surface to be machined.

A *vise jaw fixture* (**Figure 20-40**) is a useful modification to the standard milling machine vise. With this type of fixture, the standard jaws of a milling machine vise are replaced with specially shaped jaws to suit the part to be machined. The result is an accurate fixture that can be made at a minimal cost. Since the clamping device is contained within the vise, this fixture is very cost-effective. Likewise, one vise can be used for a countless number of different fixtures by simply changing the jaws.

FIGURE 20-40   Vise jaw fixture.

Indexing fixtures (**Figure 20-41**) are mainly used to machine parts that have a repeating part feature. Typical examples of the types of parts that are machined in an indexing fixture are shown in (**Figure 20-42**). Here again, the indexing feature may also be used to position the tool and the part. The *duplex fixture* shown in **Figure 20-43** shows a method of indexing the fixture to machine two parts. In use, the first part is machined while the second is loaded. The fixture is then rotated, the second

FIGURE 20-39   (A) Angle plate fixture; (B) Modified angle plate fixture.

FIGURE 20-41   Indexing fixture.

HEXAGON   SQUARE   GEAR   SPLINE   KEYWAYS

FIGURE 20-42   Typical parts machined in an indexing fixture.

FIGURE 20-43   Duplex fixture.

is unloaded, and a fresh part is loaded. This process is continued throughout the production run.

Locating Principles

To properly machine any part in a jig or fixture, the part must first be located correctly. The first rule of locating is repeatability. *Repeatability* is the feature of a jig or fixture that permits parts to be loaded in the tool in the same position part after part. When selecting locators for a part, the prime considerations must be position, foolproofing, tolerance, and elimination of duplicate or redundant locators.

Locators must be *positioned* as far apart as practical and should contact the workpiece on a reliable surface to ensure repeatability. The locators must also be designed to minimize the effect of chips or dirt. **Figure 20-44** shows a few methods generally used to relieve locators to prevent interference from chips.

Foolproofing is simply a method used to prevent the part from being loaded incorrectly. A simple pin (**Figure 20-45**) is normally enough to ensure proper loading of every part. The *tolerance* of a

A MID-CAREER CHANGE

Mark Roberts is the senior tool designer at Reynolds Manufacturing, Inc. (RMI). He is known in design and drafting circles as one of the best tool designers in the business, but it took a career change to help Roberts find his niche. He started work right out of high school as a machining trainee for RMI. In only half the time it normally takes, Roberts had earned his "ticket" as a full-fledged machinist.

Before long, he gained a reputation on the shop floor for making special workholding devices that improved the performance of machine tools, particularly on high-volume jobs. Over time, Roberts found himself spending more hours in the day making jigs and fixtures than doing his machining jobs. After discussing the situation with his supervisor, Roberts decided that he wanted to design jigs and fixtures full time.

Everyone agreed that this would benefit the company. The only problem was that Roberts actually knew very little about design. He knew machines and machining processes. He could visualize what was needed, and he could make excellent jigs and fixtures. But because he knew nothing about design, his workholding devices tended to be cumbersome. In addition, because he could not communicate his ideas on paper, he had to design, make, and modify every device himself. This slowed the process considerably.

Roberts and his supervisor agreed that if he could design workholding devices and develop understandable drawings, others could actually make them, thereby speeding the process. This realization caused Roberts to enroll in the night design and drafting program at Madison Technical Institute (MTI). He graduated with a certificate in mechanical design and drafting and was MTI's student of the year. Within six months of his graduation, Roberts was doing so well in the area of jig and fixture design that his company was able to open a tool design division that now designs and manufactures jigs and fixtures for other companies.

Roberts is still the senior tool designer at RMI, where the jig and fixture division is a commercial success. He also teaches tool design in the night program at MTI.

locator is determined by the specific size of the part to be machined. As a general rule, the tolerance of jigs and fixtures should be approximately 30 to 50% of the part tolerance. For example, if a hole were to be located within .010″ (.25 mm) from an edge, the locators in the jig or fixture should be positioned within .003 to .005″ (.08 to .13 mm) to make sure the part is properly positioned. An overly tight tolerance only adds cost, not quality, to a tool.

FIGURE 20-44   Relieving locators.

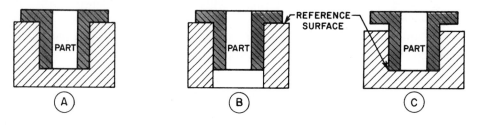

FIGURE 20-45   Foolproofing a workholder.

FIGURE 20-46   Duplicate locating.

Finally, locators should never duplicate any location. As shown in **Figure 20-46**, a part should be located on only one surface. Locating a part on two parallel surfaces only serves to reduce the effectiveness of the location and could improperly locate the part. First determine the reference surface, and use only that surface for locating.

Restricting Movement.   Every part is free to move in a limitless number of directions if left unrestricted. But, for the purpose of jig and fixture design, the number of directions in which a part can move has been limited to twelve: six axial and six radial (**Figure 20-47**). The methods used to restrict these twelve movements normally depend on the part itself, but the examples in **Figures 20-48A** and **20-48B** show two methods that are frequently used. In the first example, a three-pin base restricts five directions of movement. In the second example, a five-pin base restricts eight directions of movement. In **Figure 20-48C**, a six-pin base restricts nine directions of movement.

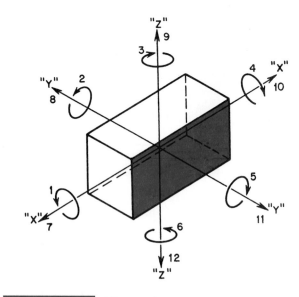

FIGURE 20-47   Planes of movement.

Types of Locators.   Locators are commercially available in many styles and types. **Figures 20-49A** through **20-49F** show several of the most common types of locators.

FIGURE 20-48   (A), (B), and (C) Restricting part movement.

FIGURE 20-49   (A), (B), and (C) Types of locators;

FIGURE 20-49 (*Continued*) (D), (E), and (F) Types of locators.

FIGURE 20-50 Always clamp a part at its most rigid point or a supported point.

FIGURE 20-51 Clamping forces.

Clamping Principles

In addition to locators, most jigs and fixtures use some type of clamping device to restrict the directions of movement not contained by the locators. In the design of a clamping arrangement, several factors should be considered. These include the position of the clamps, clamping forces, and tool forces.

Clamps should always be positioned to contact the part at either its most rigid point or a supported point. Clamping a part at any other point could bend or distort the part, as shown in **Figure 20-50**. The clamping forces used to hold a part should always be directed toward the most solid part of the tool. Clamping a part as shown in **Figure 20-51A** will normally result in an egg-shaped part because the clamping forces are directed only toward each other. However, by clamping the part as shown in **Figure 20-51B**, not only is the part held securely but the chance of distortion is also greatly reduced.

As a rule, use only enough clamping force to hold the part against the locators. The locators should always resist the bulk of the tool thrust, not the clamps.

When designing a jig or fixture, remember to direct the tool forces toward the locators, not the clamps. *Tool forces* are those forces generated by the cutting tool during the machining cycle. In most cases, the tool forces can be used to an advantage when holding any part. As shown in **Figure 20-52**, the downward tool forces are actually pushing the part into the tool. The rotational tool forces are contained by the locators. The only force the clamps need to hold are the forces generated by the drill as the point breaks out the opposite side of the part, and these forces are only a fraction of the cutting forces. So, when designing a jig or fixture, always direct the tool forces so that they act to hold the

FIGURE 20-52   Tool forces.

part in the tool. Use the clamps only to hold the part against the locators.

Types of Clamps.   Clamps, like locators, are commercially available in many styles and types. **Figures 20-53A** through **20-53D** show some of the

more common types of clamps used for jigs and fixtures. The specific type of clamp you should select for any workholding application will normally be determined by the part and the type of holding force desired.

Basic Construction Principles

When designing any jig or fixture, the first consideration is normally the tool body. Tool bodies are generally made in any of three ways: cast, welded, or built-up (**Figure 20-54**). *Cast tool* bodies are generally the most expensive and require the longest lead time to make. They do, however, offer such advantages as good material distribution, good stability, and good vibration-damping qualities. *Welded*

FIGURE 20-53   (A–D) Types of clamps.

FIGURE 20-54   Tool bodies.

tool bodies offer a faster lead time, but they must be machined frequently to remove any distortion caused by the heat of welding. The most popular and common type of tool body in use today is the built-up. *Built-up tool* bodies are made from pieces of preformed stock, such as precision-ground flat stock, ground rod, or plain cold-rolled sections that are pinned and bolted together. The principal advantages of using this type of construction are fast lead time, minimal machining, and easy modification.

Drill Bushings   *Drill bushings* are used to position and guide the cutting tools used in jigs. The three basic types of drill bushings used for jigs are press-fit bushings, renewable bushings, and liner bushings.

Press-fit bushings (**Figure 20-55**) as their name implies, are pressed directly into the jig plate. These bushings are useful for short-run jigs or in applications where the bushings are not likely to be replaced frequently. The two principal types of press-fit bushings are the head type and headless type.

Renewable bushings are used in high-volume applications or where the bushings must be changed to suit different cutting tools used in the same hole. The two types of renewable bushings that are commercially available are the slip-renewable type (**Figure 20-56A**) and the fixed-renewable type (**Figure 20-56B**).

Slip-renewable bushings (**Figure 20-56C**) are used where the bushing must be changed quickly. Typical applications include holes that must be drilled, countersunk, and tapped. The bushings are held in place with a lock screw and need only be turned counterclockwise and lifted out of the hole. The next

FIGURE 20-55   Press-fit drill bushings.

FIGURE 20-56   (A–E) Renewable drill bushings.

bushing is inserted in the hole and turned clockwise to lock it in place.

Fixed-renewable bushings are used for applications where the bushings do not need to be changed often, but where they are changed more often than press-fit bushings. High-volume production is a typical example where fixed-renewable bushings are often used.

Liner bushings (**Figure 20-56D**) are used in conjunction with renewable bushings and provide a hardened, wear-resistant mount for renewable bushings. Liner bushings are actually press-fit bushings with a large-diameter hole. Like press-fit bushings, liner bushings are available in both a head type and a headless style. The clamps normally used to hold the fixed-renewable bushing in the jig plate are shown in **Figure 20-56E**. The screw is used to mount slip-renewable bushings, whereas the other three styles are used to secure fixed-renewable bushings.

Several other variations of these basic bushing styles include special-purpose bushings, serrated and knurled bushings, and oil-groove bushings. Oil-groove bushings are shown in **Figure 20-57**.

Mounting Drill Bushings   When one is installing a drill bushing in the jig plate, the specific size of the jig plate and spacing from the part are important factors the designer must consider. As shown in **Figure 20-58A**, the jig plate should be one to two times the tool diameter. If a thinner jig plate must

FIGURE 20-57   Oil-groove bushings.

be used, a head-type bushing can be used to achieve the added thickness needed to support the cutting tool. The space between the jig plate and the workpiece should be one to one-and-one-half times the tool diameter for drilling (**Figure 20-58B**) and one-quarter to one-half the tool diameter for reaming (**Figure 20-58C**).

In those cases where both drilling and reaming operations are to be performed in the same hole, an arrangement similar to the one shown in **Figure 20-58D** may be used. Here, the jig plate is properly positioned for drilling, and the bushing used for reaming is made longer to achieve the desired distance for reaming. When special shapes or contours must be drilled, the ends of the bushings can be modified to suit the contour and maintain the proper support for the cutting tool (**Figures 20-58E** and **20-58F**).

Set Blocks   Set blocks are used along with thickness gages to properly position the cutting tool with a fixture. The specific design of the set block is determined by the part and shape to be machined. The basic set block designs shown in **Figure 20-59** are typical of the styles found on many fixtures.

Fastening Devices.   The most common fastening devices used with jigs and fixtures are dowel pins and socket-head cap screws. Other fasteners sometimes used with these tools include the nuts and washers shown in **Figure 20-60**.

FIGURE 20-58   (A–F) Mounting drill bushings.

FIGURE 20-59   Set blocks.

REGULAR   HEAVY DUTY   JAM   EXTENSION

FLANGE   SPHERICAL FLANGE   KNURLED JAM

NUTS

PLAIN   LOCK   C   SWINGING C

WASHERS

FIGURE 20-60   Nuts and washers frequently used with jigs and fixtures.

Heat Treatment of Steels

Heat treatment is a series of processes used to alter or modify the existing properties in a metal to obtain a specific condition required for a workpiece. The specific properties normally changed by heat treatment include hardness, toughness, brittleness, malleability, ductility, wear resistance, tensile strength, and yield strength. The five standard heat-treating operations normally performed on steel parts are hardening, tempering, annealing, normalizing, and case hardening.

Hardening is the process of heating a metal to a predetermined temperature, allowing the part to soak until thoroughly heated, and cooling rapidly in a cooling material called a *quench*. The most common quench media are air, oil, water, and a water and salt mixture called brine. Hardening is mainly used to increase the hardness, wear resistance, toughness, tensile strength, and yield strength of a workpiece.

Tempering is the process of heating a hardened workpiece to a temperature below the hardening

temperature, allowing the part to soak for a specific time period, and cooling. Tempering is mainly used to reduce the hardness so the toughness is increased and brittleness is decreased. Almost every hardened part is tempered to control the amount of hardness in the finished part.

Annealing is the process of heating a part, soaking it until it is thoroughly heated, and slowly cooling it by turning off the furnace. Annealing is mainly used to completely remove all hardness in a metal. Frequently, hardened parts are annealed to be remachined or modified and then rehardened.

Normalizing is the process used to remove the effects of machining or cold-working metals. In this process, the metal is heated, allowed to soak, and cooled in still air. This process produces a uniform grain size and eliminates almost all stresses in the metal.

Case hardening is the process of hardening the outside surface of a part to a preselected depth. In most case-hardening operations, carbon is added to the surface of the part by packing the part in a carbonous material and then heating it so that the carbon is transferred into the surface of the part. Once the carbon is added to the surface of the part, the part is then hardened to produce a hardened shell around a normalized core.

Nontraditional Machining Processes

The term *nontraditional machining* encompasses a number of advanced machining processes that have been developed in response to the demands of modern manufacturing for more effective, more efficient ways to machine the many new materials that have been developed and are still being developed. Although they are still referred to as "nontraditional" processes, they are now common and widely used. The most widely used nontraditional machining processes are explained in this section.

Hydrodynamic Machining (HDM)

Hydrodynamic machining (HDM) (**Figure 20-61**) removes material using a high-velocity, high-pressure stream of liquid solution. It is used primarily for slitting and contour-cutting many nonmetallic materials, such as wood and paper, asbestos, plastics, gypsum, leather, felt, rubber, nylon, fiberglass, and fiberglass-reinforced plastics. Some very thin workpieces of soft metal can be cut effectively by this process; steel sheet (0.005″ [0.13 mm] thick) and aluminum sheet (0.020″ [0.51 mm] thick) are processed,

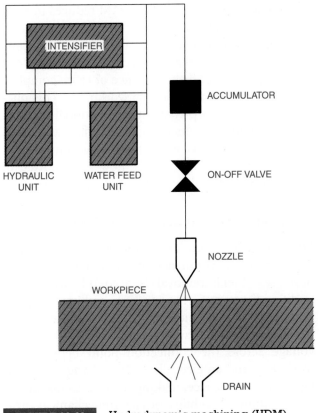

FIGURE 20-61   Hydrodynamic machining (HDM).

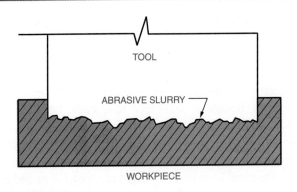

FIGURE 20-62   Ultrasonic machining (USM).

but water pressure in excess of 100 ksi (690 MPa) is usually required. Equipment that operates at these higher pressures is not commercially available. The cutting of hard metals using HDM remains mostly experimental.

Practical experience and experimental research have shown that, in some cases, brittle materials such as glass, acrylic, ceramics, and crystal do not appear suitable for cutting by the HDM process because these materials tend to develop severe cracks and may break under processing conditions.

Typically, soft materials are cut easily using HDM, and friable materials can be cut with good edge quality.

At its present stage of development, the industrial use of HDM is limited to the processing of relatively thin materials. When thicker materials are cut, stream lines increase significantly, causing poor edge quality.

Ultrasonic Machining (USM)

Ultrasonic machining (USM) (**Figure 20-62**) removes material using an abrasive slurry driven by a special tool vibrating at a high frequency along a longitudinal axis. It is used primarily for producing blind holes, through holes, slots, and irregular shapes. It is limited in complexity only by the configuration of the tooling. However, in some applications, tool

wear and/or taper in the cut may discount the process's effectiveness. The depth-to-width ratio of the cut is usually less than about 3:1. Current practice is limited to 3.5″ (89 mm) diameter tools machining cavities up to about 2.5″ (64 mm) deep.

Rotary Ultrasonic Machining (RUM)

Rotary ultrasonic machining (RUM) (**Figure 20-63**) is similar to USM except that: (1) the grinding motion is rotary instead of longitudinal, (2) RUM uses diamond tools instead of steel or Monel, and (3) RUM uses the diamond of the tool to remove material, while USM uses an abrasive slurry. RUM is used primarily in the development of prototypes. It is also used for machining ceramics, quartz, glass, ferrite computer parts, and composites.

FIGURE 20-63   Cutting internal threads using rotary ultrasonic machining (RUM).

The applications of RUM are currently limited by tool size. The horn/tool assembly must have a resonant (natural) frequency of about 20 kHz, so tool size is limited. Any variance in tool weight changes the natural frequency of the horn/tool assembly; too heavy a tool increases horn/tool assembly frequency beyond the resonant frequency of the transducer and power supply, causing the tool not to vibrate. Today, the largest vibrating horn feasible is about 1.5″ (38 mm) in diameter. Practical limit to tool weight is about 1.4 oz (40 g).

RUM is used widely in prototype and production applications. The process is effective in producing prototypes because it can make precise parts that can be used to make molds for large-volume production runs.

The process is particularly effective in the machining of sintered materials such as ceramics and ferrites. Conventionally, these materials are machined and drilled in the "green" state—prior to firing. When fired, the materials experience as much as 16% shrinkage, which destroys the accuracy created during machining. RUM is used to machine materials such as these after firing, and close tolerance relationships are maintained. Some applications include the machining of precision ceramic components; drilling small holes in ceramic printed circuit boards; and drilling small-diameter, deep, intersecting holes in quartz for laser development.

Ultrasonically Assisted Machining (UAM)

Ultrasonically assisted machining (UAM) involves assisting such traditional machining processes as drilling and turning by coupling the tool with an external source of vibrating energy. It is used in most drilling and turning applications and is effective in reducing the amount of subsurface tearing and plastic flow. It also evens out tool wear.

Under certain conditions, ultrasonic lathe turning has been shown to increase cutting rates by factors of four in aluminum and five in cutting steel. Nonmetallic materials have also shown marked increases in cutting rates when UAM is used. Alumina can be machined two times faster, and magnesium silicate can be machined up to four times faster in some cases, according to manufacturers of UAM equipment.

Some materials that are too brittle to machine traditionally can be machined effectively with UAM. For example, in one test, low-porosity mullite was machined with a good cut by applying ultrasonic vibrations to the tool post and carbide-tipped tool, but when ultrasonic power was turned off, the workpiece immediately shattered.

Some tests have shown that UAM reduces turning forces in some materials as much as 30–50% as compared to traditional turning methods. In the same tests, it was shown that surfaces produced by UAM exhibited a matte finish, evidence of more complete shearing of the chips from the workpiece. This phenomenon is in sharp contrast to traditional machining, which often produces a glossy surface as a result of tearing, materials enfoldment, and burnishing. Subsurface tearing and plastic flow are reported to be all but eliminated by UAM.

Electromechanical Machining (EMM)

Electromechanical machining (EMM) is a process that improves the performance of traditional machining processes. Metal removal is accomplished in the traditional manner (i.e., drilling, milling, turning, etc.) with the exception that the workpiece is electrochemically polarized by applying a controlled voltage across the connection point between the workpiece and an electrolyte. This changes the surface of the workpiece, making it easier to machine.

The limited feasibility of electromechanical turning (EMT) and electromechanical drilling (EMD) has been demonstrated in both laboratory and plant settings. The limited testing conducted to date indicates advantages of surface finish, tool wear, hole tolerance, and chip configuration. However, maximum improvement is tied closely to optimization of electrolytes, and it has been shown that the electrolytes should be modified through the use of inhibitors to minimize corrosion of machine tool components.

Electrochemical Discharge Grinding (ECDG)

Electrochemical discharge grinding (ECDG) is a combination of two processes: electrochemical grinding (ECG) and electrical discharge grinding (EDG). In ECDG, alternating current or pulsating direct current moves from a bonded graphite wheel (conductor) to a positively charged workpiece. No mechanical contact is made between the workpiece and the wheel. Instead, an electrolyte is pumped into the gap, and it serves as the connecting agent. Material removal occurs as follows: (1) the ECG process converts the surface of the metal into an oxide film, and (2) the EDG process removes the oxide film.

The production uses of ECDG are somewhat limited, although certain applications exist that are routinely performed, mostly in the grinding and sharpening of carbide tooling. Nearly any electrically conductive material can be processed, but careful comparison of the relative advantages and disadvantages of ECDG versus processes such as ECG and EDG should be made before specifying the process. For example, ECDG can remove material five times faster than EDG, but it uses up to 15 times the current used in EDG. Typical production tolerances of ±0.001″ (0.03 mm) are achieved with ECDG.

Current successful applications of the process include grinding and sharpening of carbide inserts, generation of delicate profiles using form grinding, grinding of honeycomb materials, and grinding of carbide thread chasers.

Electrochemical Grinding (ECG)

Electrochemical grinding (ECG) (**Figure 20-64**) combines an electrochemical reaction that occurs on the surface of the metal with abrasion. Subsequently, it is applied to the metal's surface to remove material. The process is typically used on tough, hard metals. The abrasion is applied by a rotary grinding wheel that grinds off oxidized surface metal.

In operations in which ECG can be applied, it produces results far beyond those that conventional grinding methods can provide. In many cases, it can reduce abrasive costs up to 90%. This reduction is most easily observed in connection with diamond wheels and carbide grinding. However, it is also significant with respect to steel and alloy steel grinding with nondiamond wheels.

Also, because it is a cool process, ECG can be used to grind any electrically conductive material without damage to it from heat. Therefore, ECG can simplify

fracture-inspection procedures or entirely eliminate scrap due to grinding-heat fractures. In addition, this process can grind steel or alloy steel parts without generating any burr. Thus, the costly operation of subsequent deburring is automatically eliminated.

Electrochemical Honing (ECH)

Electrochemical honing (ECH) is a process similar to ECG. It combines electrochemical oxidation of surface metal and abrasion. The primary difference is that honing is a process of removing material inside a drilled hole. Consequently, the abrasion is applied using nonconductive honing stones attached to a tool that rotates and reciprocates.

To be processed by ECH, workpieces must be conductive. The process is most effective when used to hone hard, tough metals and is well suited for the processing of parts that are susceptible to heat distortion. Electrochemical honing causes little heat buildup and no significant stresses and automatically deburrs the workpiece. The process is particularly effective for parts that require fast stock removal with good surface finish control.

Electrochemical Machining (ECM)

Electrochemical machining (ECM) (**Figure 20-65**) is a broad concept that applies to a number of process variations that all use an electrolytic action in removing metal. With ECM, electricity is combined with a chemical to form a reaction that is the opposite of metal plating. ECM can be used with any conductive workpiece but is used most often with hard metals. Its best application is external shaping of materials.

ECM is used in a wide variety of industries to machine many different metals. Experience has

FIGURE 20-64 Electrochemical grinding (ECG).

FIGURE 20-65 Electrochemical machining (ECM).

shown, however, that not all materials can be machined successfully (to acceptable metal removal rates and surface finish) using the ECM process. For example, some high-silicon aluminum alloys, such as cast alloys with substantial silicon content, sometimes cannot be machined with acceptable surface finish. Some troublesome experiences have been documented with SAE 332 aluminum pistons with 10% silicon, for example.

Electrochemical Turning (ECT)

Electrochemical turning (ECT) is a variation of ECM. The same basic principles apply except that the workpiece rotates. The primary application of ECT is the machining of large, disc-shaped forgings. In some cases, full-face electrodes are plunged into a rotating disc. Bearing races have been finished, with close tolerances and with surface roughness held to less than 5″. (0.13 μm) R_a. Another application, AISI 316 stainless steel workpieces (2.5″ [6.35 mm] diameter), are electrochemically turned, using an electrolyte of NaCl and NaNO₃ (3:3), to a surface finish of less than 10″. (0.25 μm) R_a with out-of-roundness of less than 0.0002″ (0.005 mm) TIR.

Electron Beam Machining (EBM)

Electron beam machining (EBM) is a thermal energy process. This means that it uses thermal energy to remove metal. The thermal energy comes in the form of a pulsating stream of high-speed electrons that are focused on a very small area of the workpiece by electrostatic and electromagnetic fields. Material removal occurs as the result of melting or evaporation. EBM is used for drilling small wire-drawing dies, metering holes, and holes for spinnerets used in textile manufacturing.

Any known material, metal or nonmetal, that will exist in a high vacuum can be cut, although experience has shown that diamonds do not cut well. Holes with depth-to-diameter ratios up to 100:1 can be cut. Limitations include high equipment costs and the need for a vacuum, which usually necessitates batch processing and restricts workpiece size. The process is generally economical only for small cuts in thin parts.

Electrical Discharge Machining (EDM)

Electrical discharge machining (EDM) (**Figure 20-66**) is a process that removes metal using a series of rapidly recurring electrical discharges between an electrically charged cutting tool and the workpiece submerged in a dielectric fluid.

FIGURE 20-66   Electrical discharge machining (EDM).

The application of EDM is limited to the machining of electrically conductive workpiece materials, but the process has the capability of cutting these materials regardless of their hardness or toughness. Nonconductors such as glass, ceramics, or plastics cannot be machined using EDM techniques. Machining of hardened steel using EDM eliminates the need for subsequent heat treatment with possible distortion. Complex shapes can be cut in hardened steel or carbide without costly sectional construction being necessary.

The production of stamping dies is a major application afforded by using EDM to match one portion of the die made conventionally. Extruding, heading, drawing, forging, and die casting dies, as well as molds for plastics, are also made using EDM techniques.

Electrical Discharge Wire Cutting (EDWC)

Electrical discharge wire cutting (EDWC) (**Figure 20-67**) is a process that is similar to the band-sawing process. Material removal occurs as the result of an electrical discharge as a wire electrode is drawn through the workpiece. EDWC can be used only with conductive workpieces. Primary applications of EDWC include the production of stamping dies, prototypes, molds, lathe tools, special form inserts, pot and wobble broaches, extrusion dies, and templates.

As is the case with any electrical discharge process, EDWC requires that the workpiece be electrically

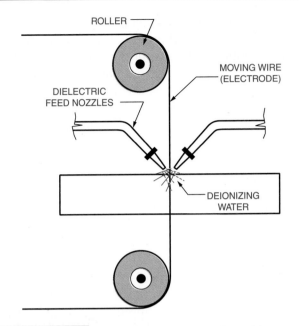

FIGURE 20-67   Electrical discharge wire cutting (EDWC).

conductive. The cut produced by the process is free of flaring and is controllable to produce small radii. Workpieces up to about 6″ (152 mm) in thickness can be processed using standard equipment; workpiece stacking effects greater productivity.

Normally produced from hardened metals such as tool steel, stamping dies are routinely cut using EDWC. The process facilitates cutting after heat treatment, thereby eliminating distortion. The use of EDWC in the manufacturing of dies affords significant savings. By conventional methods, dies are sometimes split into two or more sections to facilitate grinding with an optical projection form grinder. Usually, after the contour has been ground in the split sections, the sections are fitted in a holder or adapter. This time-consuming and costly process is eliminated with EDWC employed to produce the dies. Some experts claim that tool components such as dies can be produced in less than one-third the time required by conventional methods.

Electrical Discharge Grinding (EDG)

Electrical discharge grinding (EDG) is similar to EDM in that it employs rapidly recurring electrical discharges between an electrode and the workpiece in a dielectric fluid. The difference is that with EDG, the electrode is a rotating wheel made of graphite or brass. Primary applications include the grinding of steel and carbide at the same time, thin sections, and brittle and fragile parts.

EDG is generally used for operations such as the following: grinding steel and carbide at the same

time without wheel loading, grinding thin sections on which abrasive-wheel pressures might cause distortion, grinding brittle materials or fragile parts on which abrasive materials might cause fracturing, grinding through forms for which diamond-wheel costs would be excessive, and grinding circular forms in direct competition with abrasive-wheel methods.

Electrical discharge grinding is used to grind hard materials such as carbide form tools, hardened steel gear racks, or tungsten carbide inserts. The process is also used to grind hardened lamination dies. Cast iron workpieces are usually not processed using EDG because sand inclusions can damage the graphite grinding wheel.

Laser Beam Machining (LBM)

Laser beam machining (LBM) (**Figure 20-68**) is a process that uses lasers for performing operations that are traditionally performed with cutting tools. These operations include cutting, drilling, slotting, and scribing. The basic principle of LBM is the conversion of electrical energy into coherent light. This coherent light or laser beam is focused on the workpiece where it is converted into thermal energy. This thermal energy melts or vaporizes the material. LBM has broad applications, but it is particularly effective at cutting and drilling.

Any industrial application of the laser should be based on one of three criteria: (1) it can perform a superior job in terms of quality and cost over existing

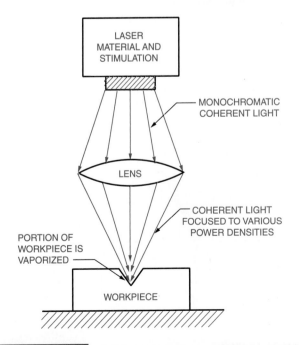

FIGURE 20-68   Solid state laser beam machining (LBM).

methods, (2) it is the only tool capable or available for the specific job, or (3) it allows restructuring of the manufacturing process, resulting in lower total cost. The cost of using laser systems falls into two categories: (1) the capital investment for the laser system, and (2) the operational cost. At present, both the equipment and the direct operational costs of a laser are higher than those of comparable conventional equipment and methods.

Because of the laser's ability to melt or vaporize any known metal and operate in any desired atmospheric environment, it is sometimes preferred over EBM (electron beam machining), which requires a vacuum chamber for certain applications. Other advantages include: (1) the ability to machine areas not readily accessible and extremely small holes, (2) the fact that no direct contact exists between the tool (laser) and the workpiece, (3) small heat-affected zones, and (4) easy control of beam configuration and size of exposed area.

Plasma Arc Machining (PAM)

Plasma arc machining (PAM) (**Figure 20-69**) is a process that makes use of a plasma arc to remove material. *Plasma* is a gas that has been heated to the point that it ionizes and becomes electrically conductive. With PAM, a stream of ionized particles directed through a nozzle against the workpiece removes material. The primary applications of PAM are hole piercing, stack cutting, gouging, grooving, and bevel cutting.

The plasma arc can be used for "machining" or removing the metal from the surface of a rotating

cylinder to simulate a conventional lathe or turning operation. As the workpiece is turned, the torch is moved parallel to the axis of the work. The torch is positioned so the arc will impinge tangentially on the workpiece and remove the outer layer of metal. Cutting can be accomplished with the workpiece rotating in either direction relative to the torch, but best results are obtained when the direction of rotation permits use of the shortest arc length for cutting. The flow of molten metal being removed must be in such a direction that it does not tend to adhere to the hot surface that has just been machined.

Chemical Milling

Chemical milling is a process that etches or shapes workpieces to close tolerances by chemically induced material removal. Primary applications include metal removal from irregularly shaped workpieces, reduction of web thicknesses, removal of the decarburized layer from low-alloy steel forgings, improved surface finish, and removal of defects.

In general, chemical milling is used to remove metal from a portion or the entire surface of formed or irregularly shaped parts such as forgings, castings, extrusions, or formed wrought stock; reduce web thicknesses below practical machining, forging, casting, or forming limits; taper sheets and preformed shapes; produce stepped webs, resulting in consolidation of several details into one integral piece; remove the decarburized layer from low-alloy steel forgings; remove up to 0.125″ (3.2 mm) per surface of metal to remove decarb and also create finished dimensions of die forgings; improve surface finish; remove surface cracks, laps, and other defects of forgings; remove alpha case from titanium forgings; and improve surface finish and control dimensions of aluminum forgings.

Photochemical Machining

Photochemical machining, also known as chemical blanking, involves producing parts by chemical action. It is accomplished by placing an exact image of the part to be produced on a sheet of metal (the prototype must be chemical-resistant) and immersing them both in a chemical. The chemical action dissolves all of the metal except the desired part. Most photochemically machined parts are thin and flat.

Photochemical machining has a number of applications wherein it provides unique advantages. Some of these include: working on extremely thin

FIGURE 20-69   Plasma arc machining (PAM).

materials when handling difficulties and die accuracies preclude the use of normal mechanical methods; working on hardened or brittle materials when mechanical action would cause breakage or stress-concentration points—chemical blanking works well on spring materials and hardened materials that are relatively difficult to punch; production of parts that must be absolutely burr-free; production of extremely complex parts for which die costs would be prohibitive; and producing short-run parts for which the relatively low setup costs and short time from print to production offer advantages. This is especially important in research and development projects and in model shops.

Water Jet Machining (WJM)

Unlike the nontraditional processes presented so far, *water jet machining (WJM)* is used with non-metal workpieces. It involves cutting plastics, nonmetal composites, ceramics, glass, and wood by directing a thin, pressurized jet of water against the workpiece.

Abrasive Water Jet Machining (AWJM)

Abrasive water jet machining (AWJM) (**Figure 20-70**) is similar to WJM except that an abrasive is added to the water jet after it exits the nozzle. This allows AWJM to be used with brittle and/or heat-sensitive materials such as glass, composites, and plastics.

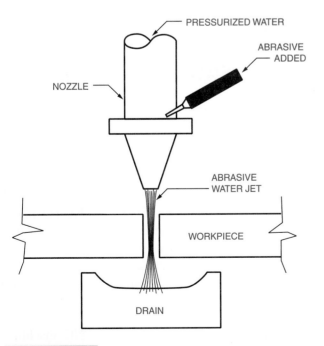

FIGURE 20-70   Abrasive water jet machining (AWJM).

Automation and Integration (CAM and CIM/FMS)

There have been four phases in the development of shop and manufacturing processes:

1. Manual phase
2. Mechanization phase
3. Automation phase (CAM)
4. Integration phase (CIM/FMS)

Of course, there has been and continues to be a good deal of overlap among these phases. Manual manufacturing did not suddenly stop when the age of mechanization began, nor did mechanized manufacturing stop when the automation phase began. Correspondingly, the beginning of the integration phase did not bring automated manufacturing to a sudden stop.

It takes many years for a given stage in the development of shop and manufacturing processes to phase out completely. For example, by the year 1900, mechanized manufacturing had replaced manual manufacturing as the dominant approach. However, manual manufacturing was still widely practiced at that time. Today, automated manufacturing has become the dominant approach; nevertheless, mechanized manufacturing is still widely practiced, and even manual manufacturing is still practiced in very isolated cases.

At present, we are seeing the integration phase emerge. It will eventually dominate but not completely replace—at least not for a long time—the other three approaches.

The manual phase preceded the Industrial Revolution. During these early times, manufactured products were produced by hand using a variety of manual tools. The Industrial Revolution ushered in the mechanization phase. The fundamental component of this phase was the machine. Early machines were steam-powered, while later machines, of course, were dependent on electricity. In this phase, machines such as mills, drills, saws, and lathes did the work, but people provided the control.

The automation phase has changed the way machines are controlled. It has also lessened the amount of human involvement in manufacturing. With the automation approach, manufacturing machines are controlled by computer. This has come to be known as CAM or computer-aided manufacturing. CAM machines are operated automatically according to computer programs. People still write the programs, set up the machines, monitor operations, and unload machines.

With CAM, individual operations such as milling, drilling, turning, and cutting are automated, but they still operate independently of each other. The most advanced manufacturing approach to date—computer-integrated manufacturing or CIM—solves this problem by linking automated manufacturing processes with other processes such as materials handling, warehousing, contracts and bids, scheduling, and design. CIM represents a major step toward the wholly automated factory. These two concepts—CAM and its eventual successor, CIM—deserve special attention.

Computer-Aided Manufacturing (CAM)

Through CAM, such manufacturing processes as machining and materials handling can be automated. CAM encompasses any manufacturing process that is automated through the use of the computer, but the best examples of CAM are computer numerical control (CNC) of machine tools and robotics.

Computer Numerical Control (CNC)

Numerical control is a method of controlling machine tools by using coded programs. These programs consist of numbers, letters, and special characters that define the path of the machine tool in accomplishing specific tasks. When the job changes, the program must be rewritten to accommodate the change. The programmability feature is the key to the growth of NC. Machine tools that are programmable are more flexible than their nonprogrammable counterparts.

NC is a broad term that encompasses the traditional approach to NC, as well as the more modern outgrowths of CNC and direct numerical control (DNC).

The first NC machines used punched cards as the programming medium. Later, punched tape was substituted, which was an improvement over manual control but did have some disadvantages. Paper tape was easy to damage and difficult to correct or edit.

Mylar tape, as a less fragile medium, began to replace paper tape. This solved the problem of frequent damage but did not make punched tape any easier to edit. Such problems, coupled with advances in microelectronic technology, led to the use of computers as the means for programming machine tools. Computers solved the damage and editing problems, and this gave birth to both CNC and DNC. CNC involves using a computer in writing, storing, and editing NC programs. DNC involves using a computer as the controller for one or more NC machines (**Figure 20-71**). A more advanced form

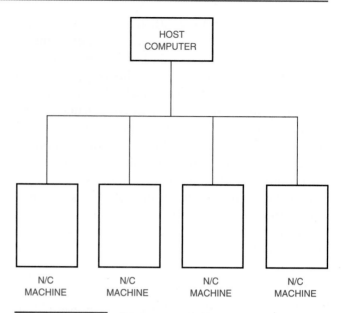

FIGURE 20-71   Direct numerical control (Adapted from Seames, *Computer Numerical Control*, Delmar Publishers Inc.).

FIGURE 20-72   Distributive numerical control (Adapted from Seames, *Computer Numerical Control*, Delmar Publishers Inc.).

of DNC is called "distributive numerical control" (**Figure 20-72**).

In this concept, there is a host coupler and several intermediate computers that are tied to NC machines, robots, and other NC manufacturing devices. The main host computer is the central repository for all programs for all jobs. Specific programs for specific jobs are "dumped" from the host computer to the intermediate computers. This cuts down on the amount of time required to get the programmed instructions from the computer to the machine tool. CNC machines are available for most machining processes, including cleaning and finishing, inspection and quality control, pressing and forming, and material removal.

Figure 20-73 is a CNC machining center.

The CNC machines shown in Figures 20-73 through 20-75 have controllers. There are as many different controllers as there are CNC machines (Figures 20-74 and 20-75).

CNC is the most modern and most technologically advanced method of controlling manufacturing machines. However, it is not the best control method in every case. Like any technological development, CNC has its advantages and disadvantages. There are times when the traditional manual approach is better. CNC is indicated when one or both of the following factors is important:

- Increased productivity
- Decreased labor and production costs

When CNC is indicated, there are advantages and disadvantages with which students of CAD/CAM should be familiar. The advantages include:

1. Better production and quality control
2. Increased productivity, flexibility, accuracy, and uniformity
3. Reduced labor, production, tool, and fixture costs
4. Less parts handling and tool storage

There are other advantages of CNC; however, these are the most important. There are also disadvantages associated with CNC. Those most frequently stated are:

1. High up-front costs
2. Higher operating costs
3. Retraining costs
4. Potential personnel problems

The advantages of CNC outweigh the disadvantages. But the list of disadvantages does make the point that it is not always the appropriate choice.

CNC Applications

CNC applications can be viewed at two levels. First, there is the industry level. At this level, all of the various manufacturing industries use CNC machines. These include:

- Aerospace manufacturers
- Electronics manufacturers
- Automobile, truck, and bus manufacturers
- Appliance manufacturers
- Tooling manufacturers
- Locomotive/train manufacturers

FIGURE 20-73 CNC machining center (Courtesy of Fort Walton Machining).

FIGURE 20-74 Typical CNC controller (Courtesy of Fort Walton Machining).

FIGURE 20-75 CNC controller (Courtesy of Fort Walton Machining).

The second application level to consider is the machine level. At this level, CNC applications include:

- Cleaning and finishing applications
- Material removal applications
- Presswork and forming applications
- Inspection/quality control applications

Cleaning and finishing machines perform such tasks as washing, degreasing, blasting, lapping, grinding, and deburring. There are currently several CNC cleaning and finishing machines on the market.

Material removal machines include mills, lathes, grinders, drills, and saws. All such machines may be CNC controlled. In fact, this is the machine-level application area most strongly associated with CNC.

Presswork and forming machines perform such tasks as pressing, stamping, blanking, punching, bending, and swagging. There are currently several CNC presswork and forming machines on the market.

Inspection and quality control machines perform such tasks as measuring, gaging, testing, and weighing. There are several CNC inspection and quality control machines on the market.

Industrial Robots

As is sometimes the case with new and emerging technological developments, there are a variety of definitions used for the term "robot." Depending on the definition used, the number of robot installations in this country and others will vary widely. There are a variety of single-purpose machines used in manufacturing plants that, to the layperson, would appear to be robots. These machines are hardwired to perform one function. They cannot be reprogrammed to perform a different function. These single-purpose machines do not fit the definition of industrial robots that is coming to be widely accepted. This definition is the one developed by the Robot Institute of America:

A robot is a reprogrammable multifunctional manipulator designed to move material, parts, tools, or specialized devices through variable programmed motions for the performance of a variety of tasks.

Note that the RIA's definition contains the words "reprogrammable" and "multifunctional." It is these two characteristics that separate the true industrial robot from the various single-purpose machines used in modern manufacturing firms. The term "reprogrammable" implies, first, that the robot operates according to a written program and, second, that this program can be rewritten to accommodate a variety of manufacturing tasks. The term "multifunctional" means that the robot is able, through reprogramming and the use of a variety of end effectors, to perform a number of different manufacturing tasks. Definitions written around these two critical characteristics are becoming the accepted definitions among manufacturing professionals. **Figures 20-76** and **20-77** are examples of modern industrial robots that fit this definition.

The microprocessor was the enabling device with regard to the wide-scale development and use of industrial robots. But major technological

FIGURE 20-76   G-Series SCARA robot (Courtesy of EPSON Robots).

FIGURE 20-77 S-5 Series six-axis robot (Courtesy of EPSON Robots).

developments do not take place simply because of a new capability. Something must provide the impetus for taking advantage of the new capability. In the case of industrial robots, the impetus was economics.

In the 1970s, it became imperative for the United States to produce better products at lower costs in order to be competitive with foreign manufacturers. Other factors, such as the need to find better ways of performing dangerous manufacturing tasks, contributed to the development of industrial robots. However, the principal rationale has always been, and is now, improved productivity.

Industrial robots offer a number of benefits that account for the rapid current and projected growth of industrial robot installations.

1. Increased productivity
2. Improved product quality
3. More consistent product quality
4. Reduced scrap and waste

5. Reduced reworking costs
6. Reduced raw goods inventory
7. Direct labor cost savings
8. Savings in related costs, such as lighting, heating, and cooling
9. Savings in safety-related costs
10. Savings from correctly forecasting production schedules

The Robot System

Work in a manufacturing setting is not accomplished by a robot. Rather, it is accomplished by a robot system. A robot system has four major components: the controller, the robot arm or manipulator, the end-of-arm tools, and the power sources (**Figure 20-78**). These components, coupled with the various other pieces of equipment and tools needed to perform the job for which a robot is programmed, are called the robot's work cell.

The contents of a robot's work cell will vary according to the application of the robot. However, the one constant in a robot's work cell is the robot system.

The Controller. A *robot* is a special-purpose device similar to a computer. As such, it has all the normal components of a computer, including the central processing unit—made up of a control section and an arithmetic/logic section—and a variety of input and output devices.

The controller for a robot system does not look like the microcomputer one is used to seeing on a

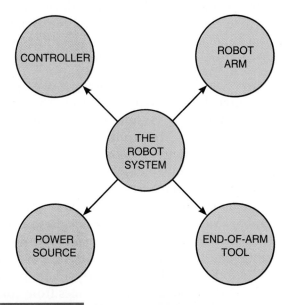

FIGURE 20-78   The robot system.

desk. It must be packaged differently so as to be able to withstand the rigors of a manufacturing environment. Typical input/output devices used in conjunction with a robot controller include teach stations, teach pendants, a display terminal, a controller front panel, and a permanent storage device.

Teach terminals, teach pendants, or front panels are used for interacting with the robot system. These devices allow humans to turn the robot on, write programs, and key in commands to the robot system. Display terminals give operators a soft copy output source. The permanent storage device is a special device on which reusable programs can be stored.

Mechanical Arm. The mechanical arm, or manipulator, is the part of a robot system with which most people are familiar. It is the part that actually performs the principal movements in doing manufacturing-oriented work. Mechanical arms are classified according to the types of motions of which they are capable. The basic categories of motion for mechanical arms are rectangular, cylindrical, and spherical. **Figure 20-79** shows the mechanical arm for the C3 compact robot.

End-of-Arm Tooling. The human arm by itself can perform no work. It must have a hand that, in turn, grips a tool. The mechanical arm of a robot, like the human arm, can perform no work. It, too, must have a hand, a tool, or a device that combines both functions. On industrial robots, the function of the wrist is performed by the tool plate. The tool plate is a special device to which end-of-arm tools are attached. The tools themselves vary according to the types of tasks the robot will perform. Any special device or tool attached to the tool plate to allow the robot to perform some specialized task is classified as an end-of-arm tool or an end effector.

The Power Source. A robot system can be powered by three different types of power. The power sources used in a robot system are electrical, pneumatic, or hydraulic. The controller, of course, is powered by electricity, as is any computer. The mechanical arm and end-of-arm tools may be powered by either pneumatic or hydraulic power. Some robot systems will use all three types of power. For example, a given robot might use electricity to power the controller, hydraulic power to manipulate the arm, and pneumatic power to manipulate the end-of-arm tool. Hydraulic power is fluid-based. Pneumatic power comes from compressed gas (**Figure 20-80**).

FIGURE 20-79   C3 compact six-axis robot (Courtesy of EPSON Robots).

CIM/FMS

Computer-integrated manufacturing is the most modern, most automated form of production. It involves tying different phases of production together into one wholly integrated system. The term "flexible manufacturing system" is sometimes used synonymously with computer-integrated manufacturing system. Actually, however, a flexible manufacturing system is one type of CIM system designed for medium-range production volumes and moderate flexibility. Other types of CIM systems are special systems and manufacturing cells. There are other terms that have been used to describe the same concept, but CIM and FMS are the two that are most widely used today.

FIGURE 20-80   RS-Series SCARA robot (Courtesy of EPSON Robots).

There are several different types of computer-integrated manufacturing systems. This is because each system is designed to meet the specific needs of the individual manufacturing setting where it will be used. However, in general, a CIM system is any computerized manufacturing system in which numerically controlled machines are joined together and connected by some form of automated material handling system.

In a CIM system, the computer is used in several ways: (1) it is used to control CNC machines,

(2) it is used to control the materials handling system, and (3) it is used for monitoring all production tasks accomplished by the system.

NOTE: Safety equipment may have been removed or opened to clearly illustrate products and must be in place prior to operation.

In spite of advances in manufacturing automation, there is human involvement with a CIM system. Typically, this involvement falls into six broad categories: (1) loading raw stock and materials onto the system for processing, (2) unloading processed workpieces from the system, (3) changing tools on machines within the system, (4) setting tools on machines within the system, (5) continuous maintenance of the system, and (6) occasional repair of the system when there is a breakdown or malfunction.

Stand-alone CNC machines are used in low-volume manufacturing applications that require a high degree of flexibility in order to produce a wide variety of parts. They represent one extreme of the manufacturing spectrum. At the other extreme are transfer lines. Transfer lines are used in high-volume manufacturing applications where all parts produced are identical. Transfer lines are not flexible; they cannot accommodate variety.

Transfer lines are a major component of what is sometimes referred to as "Detroit Automation." A transfer line consists of several workstations linked together by materials-handling devices that transfer workpieces from station to station.

The first station holds the raw material. The last station is a bin for collecting the finished parts. Each station in between performs some type of operation on the parts as they pass through. The transfer of parts from station to station, and the work performed on them, are automatic. Transfer lines are appropriate in situations that involve the high-volume production of identical parts.

There has always been a gap between these two extremes, and there has always been a need to fill this gap. The medium-volume, moderate-flexibility manufacturing situation has long been a problem in need of a solution. CIM systems are such a solution. CIM fills the void between stand-alone CNC machines and transfer lines (**Figure 20-81**).

FIGURE 20-81   Volume/flexibility continuum.

CIM offers a number of advantages that, when taken together, form the rationale for this modern approach to medium-volume, flexible production. The most important of these are:

1. Produces families of parts

2. Accommodates the random introduction of parts

3. Requires less lead time

4. Allows a closer relationship between parts to be produced and workpieces loaded onto the system

5. Allows better machine utilization

6. Requires less labor

CIM systems reduce the amount of direct and indirect labor costs associated with finished workpieces. This is because CIM systems require less human involvement in producing them. Ten to 12 traditional CNC machines require 10 to 12 operators. A CIM system with 10 to 12 CNC machines might require as few as four people. This means less direct labor. Indirect labor costs resulting from such tasks as materials handling are also reduced with CIM. This is because most materials handling in CIM systems is automated.

Types of CIM Systems

CIM systems fall between transfer lines and stand-alone CNC machines on a graph of production volume versus part variety flexibility. By applying these same criteria—production volume and part variety—CIM systems can be divided into three categories (**Figure 20-82**).

1. Special systems

2. Flexible manufacturing systems

3. Manufacturing cells

The preceding systems are listed in order from the least flexible (special systems) to the most flexible (manufacturing cells). Of all CIM systems, those classified as special systems are capable of the most volume production and the least flexibility. Special systems might produce as many as 15,000 parts per year, but they are limited to less than 10 different part types.

Flexible manufacturing systems represent the middle group in CIM systems. An FMS might have a production volume as high as 2,000 parts per year and a capability of handling as many as 200 different parts.

The most flexible of CIM systems are manufacturing cells, which in turn are capable of the lowest production volumes. A manufacturing cell might have a production volume as low as 500 parts per

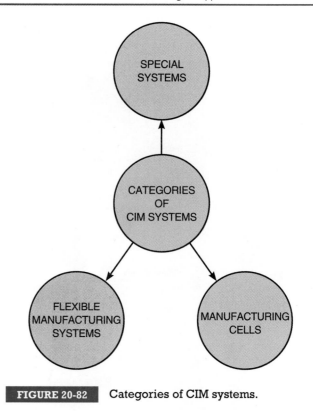

FIGURE 20-82   Categories of CIM systems.

year but a capability of handling as many as 500 different part types.

Just-in-Time Manufacturing (JIT)

Just-in-time manufacturing (JIT) is an approach to manufacturing that seeks to completely eliminate waste. The term *waste* is used here in the broadest sense and encompasses rejected parts, wasted personnel time, wasted machine time, excessive inventory, and any other situation that wastes time, material, or any other manufacturing resource. JIT has the following basic characteristics.

Holistic View

This characteristic means that JIT cannot be accomplished by any individual unit operating in isolation. The entire company—all units—must be included in the effort. All processes, from purchasing to accounting and production, must be synchronized in order for JIT to work properly.

Process Synchronization

This characteristic is what gives JIT its name. *Process synchronization* means that all processes involved in producing a product are in balance and synchronized so that production flows smoothly and continuously. Traditionally, production processes have been operated on a *hurry-up-and-wait* basis. While one process is hurrying to complete its task, the next one in succession is idle. Process synchronization smoothes this out so that each successive process receives its material just in time to be processed.

Simplicity

This characteristic is fundamental to JIT, the view being that the simpler things are, the easier JIT will be to accomplish. With JIT, a continuous effort is made to simplify processes so that they can be performed with a continually decreasing amount of resources.

Manufacturing Resource Planning (MRPII)

Manufacturing resource planning is often referred to as MRPII to distinguish it from its predecessor, which was material requirements planning (MRP). MRP was a way to reduce inventory and waste through systematic planning for material requirements. MRPII is a broader concept that allows companies to systematically plan for all resources needed in manufacturing (i.e., material, capacity of production systems, personnel, time, etc.).

An MRPII system is computer-based and typically consists of at least the following four components: (1) control database, (2) planning database, (3) functional modules, and (4) integrated modules. The control database contains inventory data and a master file on workcenters. The planning database contains a bill of materials file and a process routing file. Functional modules include production scheduling, capacity planning, material planning, and job cost reporting. Integrated modules or modules that might be integrated include such areas as sales, accounting, customer order processing, payroll, etc. Using MRPII properly simplifies the implementation of JIT because it helps synchronize all processes that are either directly or indirectly involved in producing the products in question.

Statistical Process Control (SPC)

Statistical process control (SPC) is a process that uses statistical control charts to monitor production processes in an attempt to reduce the amount of scrap and rework. Tolerances are transformed into upper and lower control limits that are placed on a chart. Periodically, points that correspond to the processing performance of the system are plotted on this chart. The plots are monitored visually or electronically.

As long as all points plotted fall within established limits on the chart, the process is under control. If a point falls outside of the limits, corrections should be made immediately to prevent further production of unacceptable parts.

SUMMARY

- Engineering materials are the raw materials that are converted into finished products in manufacturing. They can be divided into two broad categories: metals and nonmetals. Metals include steels, alloy steels, and a variety of high-performance alloys. Nonmetals include plastics, composites, elastomers, and ceramics.

- Casting is a process in which molten material is poured into a mold that contains the desired shape in the form of a cavity. The principal types of casting processes are sand casting, investment casting, centrifugal casting, and die casting.

- Forging is the process of forming metals under pressure using a variety of different processes. The most common forms of forging are drop forging and press forging. Other variations include rolling and upsetting.

- Extruding is a process of forcing metal through a die of a desired form and cross section.

- Stamping is a process of using dies to cut or form metal sheets or strips into a desired form.

- Machining is the process of removing metal with machine tools and cutters to achieve a desired form or feature. The principal types of machine tools are lathes, milling machines, drill presses, saws, and grinders.

- Special workholding devices are used to produce parts in high-volume production runs. These devices are called jigs and fixtures. Both jigs and fixtures hold, support, and locate the workpiece. Jigs guide the cutting tool through a hardened bushing during the cutting cycle. Fixtures reference the cutting tool by means of a set block.

- Heat treatment is a series of processes used to alter or modify the existing properties in a metal to obtain a specific condition required for a workpiece.

- Nontraditional machining encompasses a number of advanced machining processes that have been developed in response to the demands of modern manufacturing for more effective, more efficient ways to machine the many new materials that are now available.

- Computer-aided manufacturing (CAM) encompasses any manufacturing process that is automated through the use of computers. Examples of CAM are computer numerical control of machine tools and robotics.
- A robot is a reprogrammable, multifunctional manipulator designed to move materials, parts, tools, or specialized devices through variable programmed motions for the performance of a variety of tasks.
- Computer-integrated manufacturing (CIM) is the most modern form of production. It involves tying different phases of production together.
- A flexible manufacturing system is one type of CIM system designed for medium-range production volumes and moderate flexibility.
- Just-in-time manufacturing (JIT) is an approach to manufacturing that seeks to eliminate waste. Waste encompasses rejected parts, personnel time, machine time, excessive inventory, or any other resource. The key to JIT is process synchronization.
- Manufacturing resource planning or MRPII is a computer-based approach that allows companies to systematically plan for all resources needed in manufacturing (i.e., material, personnel, time, capacity of production systems, etc.).

REVIEW QUESTIONS

Mark the following statements as either true or false.

1. The term *lead time* means the interval of time from when the shop receives the drawing until production begins.
2. The cope and the cheek are terms used to describe the top and bottom sections of the flask used for casting.
3. The amount of shrinkage and the amount of expansion must be considered when making a pattern for casting.
4. The approximate percentage of shrinkage for steel is 3/16″ per foot.
5. Investment casting is also referred to as lost wax casting.
6. Centrifugal casting is the casting process in which molten metal is forced into metal dies.
7. There are usually six stations in a drop forge die.
8. Operations such as slitting, trimming, and shaving control size.
9. A drop hammer is the principal tool used in metal stamping.

Answer the following questions by selecting the best answer.

1. Which of the following is **not** one of the major factors to consider before any part is made?
   a. The material the part will be made of
   b. The overall size and shape of the finished object
   c. Types of operations required
   d. The cost of manufacturing the part
2. Which of the following is not one of the primary methods used to fabricate manufactured products?
   a. Welding
   b. Extruding
   c. Stamping
   d. Forging
3. Which of the following is **not** one of the common forging processes?
   a. Rough forging
   b. Drop forging
   c. Press forging
   d. Rolling
4. Which of the following is **not** one of the stamping operations that produces blanks?
   a. Cutting off
   b. Shearing
   c. Trimming
   d. Blanking
5. Which of the following stamping operations does **not** produce holes?
   a. Punching
   b. Piercing
   c. Perforating
   d. All of the above
6. Which of the following is the most common type of fixture?
   a. Plate fixture
   b. Angle plate fixture
   c. Vise jaw fixture
   d. Indexing fixture
7. Which of the following is **not** one of the four operations used to bend or form parts?
   a. Machining
   b. Drawing
   c. Forming
   d. Bending
8. Which of the following is **not** one of the four machine tools used for machining parts?
   a. Drill press
   b. Fixtures

c. Milling machine
d. Lathe

9. Which of the following is **not** true regarding the plate jig?
   a. It is the most common form of jig.
   b. It is the least common form of jig.
   c. A typical variation of the plate jig is the table jig.
   d. A typical variation of the sandwich jig is the table jig.

10. Which of the following is **not** one of the operations normally performed on a drill press?
    a. Cutting
    b. Tapping

c. Reaming
d. Drilling

11. Which of the following is **not** an element of JIT?
    a. To eliminate waste
    b. Process synchronization
    c. To build up inventory
    d. Each process receives its material just in time

12. Which of the following is **not** an element of MRPII?
    a. Control database
    b. Planning database
    c. Functional database
    d. Integrated database

21

The Design Process and Advanced Concepts

CHAPTER OUTLINE

Time • Learning the design process • The design process: phases and steps • Design projects: routine and non-routine • Modern design practices and standards • Summary • Review questions • Chapter 21 problems • CAD instructions

CHAPTER OBJECTIVES

Upon completion of this chapter, students should be able to do the following:

▶ Follow the phases and steps in the design process to do a routine design project as an individual effort.

▶ Follow the phases and steps in the design process to do a non-routine design project as a member of a team.

▶ Use various techniques such as the creative process, attribute listing, a decision table, and model building (actual and computer-generated) to enhance the steps in the design process.

▶ Use various information sources such as vendor catalogs, technical texts, technical periodicals, and technical handbooks to augment the steps in the design process.

▶ Identify the various types of technical drawings you would use in the design process, such as freehand sketches, design development layout drawings, kinematic layout drawings, and detail and assembly drawings.

(continued)

728

(CHAPTER OBJECTIVES continued)

▶ Discuss the objectives of DFM, DFA, DFR, and DFD.

▶ Recognize current trends in manufacturing.

▶ Identify the roles of drafters, designers, and engineers in the steps of the design process.

(KEY TERMS continued)

Steps (in the design process)

Trade-off

Vendor catalogs

Virtual prototyping

The *design process* is a plan of action for reaching a goal. The plan, sometimes labeled problem-solving strategy, is used by engineers, designers, drafters, scientists, technologists, and a multitude of professionals. This strategy, the design process, occurs in five *phases* during a product design, and each phase has a number of *steps*. The following phases occur sequentially with some overlapping, as shown in **Figure 21-1**.

Phase I	Recognize and Define the Problem
Phase II	Generate Alternative Solutions: SYNTHESIS
Phase III	Evaluate Alternative Solution Ideas: ANALYSIS
Phase IV	Document, Specify, and Communicate the Solution
Phase V	Develop, Manufacture, and Deliver

The steps, not shown in Figure 21-1, may occur sequentially; however, each phase has a different number of steps depending on the complexity of the problem to be solved and the time available.

Time

The available time, whether hours, days, months, or years, controls the starting and stopping of the phases during the design process. Therefore, successful problem solving requires careful planning. And since the phases must occur within the time span, they must be scheduled as needed to reach the required goal.

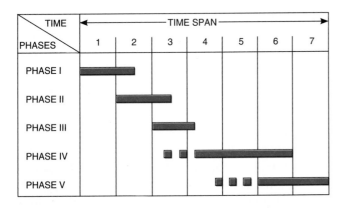

FIGURE 21-1   The phases in the design process.

Learning the Design Process

Learn the design process by doing it! That is, follow the phases and steps, as a guide, to do at least two design projects. First, individually, do a routine design project. Next, in a group, do a non-routine design project. Use these design experiences to augment and expand your design skills and to acquire an understanding that the design process can be modified to suit the level and complexity of the problem to be solved. This chapter provides the following to get you started:

1. An outline and discussion of the phases and steps in the design process (**Figure 21-2**)

2. A discussion of routine design projects (the type usually assigned to a neophyte drafter, designer, or engineer as their first task in a company)

3. A discussion of non-routine design projects (the type assigned to more experienced drafters, designers, and engineers as they assume more responsibility in their respective jobs)

4. A collection of design hints to enhance your problem-solving skills and design solutions

5. Examples from a non-routine design project (done by a student group)

Follow the phases and steps in Figure 21-2, generally in the sequence shown. Use the steps as a checklist. Samples from the student group's non-routine design project (#5 above) are included in the following discussion when appropriate.

The Design Process: Phases and Steps

Phase I: Recognize and Define the Problem

STEP 1   Establish Plans for the Phases, Steps, and Records

STEP 2   Gather Information
     a. Primary Sources
     b. Secondary Sources

DESIGN PROCESS: PHASES AND STEPS

**PHASE I
RECOGNIZE AND DEFINE THE PROBLEM**

STEP 1 Establish Plans for the Phases, Steps, and Records

STEP 2 Gather Information
　　　　a. Primary Sources
　　　　b. Secondary Sources

STEP 3 Combine Information into a Preliminary Definition
　　　　a. Constraints
　　　　b. Specifications
　　　　c. Preliminary Solution Ideas
　　　　d. Iterate

STEP 4 Prepare a Progress Report
　　　　a. Internal Clients
　　　　b. External Clients

**PHASE II
GENERATE ALTERNATIVE SOLUTIONS: SYNTHESIS**

STEP 5 Generate Solution Ideas
　　　　a. Use Systematic Techniques
　　　　b. Use Creative Techniques and Attitudes

**PHASE III
EVALUATE ALTERNATIVE SOLUTION IDEAS:
ANALYSIS**

STEP 6 Sift, Combine, and Prepare Solution Ideas for Analysis
　　　　a. Preliminary Preparation
　　　　b. Specific Preparation

STEP 7 Compare Solution Alternatives to Constraints and Specifications
　　　　a. Iterate
　　　　b. Persevere
　　　　c. Qualitative Evaluation
　　　　d. Quantitative Evaluation

STEP 8 Conduct a Design Review
　　　　a. In-House Design Reviews
　　　　b. Outside Design Reviews

STEP 9 Refine the Final Solution

**PHASE IV
DOCUMENT, SPECIFY,
AND COMMUNICATE THE SOLUTION**

STEP 10 Prepare Drawings and a Written Report

STEP 11 Give an Oral Presentation

**PHASE V
DEVELOP, MANUFACTURE, AND DELIVER**

STEP 12 Prepare Technical Detail Drawings of Each Component to Be Manufactured

STEP 13 Schedule Assembly, Final Testing, Packaging, and Delivery

STEP 14 Keep Records

FIGURE 21-2 Phases and steps in the design process.

STEP 3 Combine Information into a Preliminary Definition
　　　　a. Constraints
　　　　b. Specifications
　　　　c. Preliminary Solution Ideas
　　　　d. Iterate

STEP 4 Prepare a Progress Report
　　　　a. Internal Clients
　　　　b. External Clients

Step 1: Establish Plans for the Phases, Steps, and Records. Make a *planning chart*, which is an excellent visual display of when phases and steps are scheduled to begin and end (**Figure 21-3**). Note that over each bar is the name of the individual responsible for that activity, whether the individual does all of the work or not. See Figure 21-21 for the completed chart.

Start and keep a *log* of activities and the time spent. Each individual in the group should keep a log. Newspaper-style entries (who, what, where, when, why) are useful because they tell specifically what was done. Supervisors use the information for checking progress and for accounting purposes. But more importantly, the phone numbers, names, dates, and facts may be useful on future projects. Instructors use the log information for checking progress and for grading purposes. See **Figure 21-4** for a sample log from a student's notebook. See Figure 21-20 for a more complete log.

Some companies may require selected individuals to keep a more detailed log-design notebook for legal purposes or for patent development. For patent development, each page is bound, numbered, and then arranged to be signed by a knowledgeable witness.

Start a design file with every page dated to agree with the log, and file the pages chronologically for easy reference. Each individual keeps a *design file*, which includes freehand sketches, machine copies of technical information, copies of related drawings, summaries of telephone conversations and interviews, and vendor brochures. Samples from a student's design file are in **Figures 21-5A** and **21-5B**. A group log and file may be required by a supervisor in the industry.

Step 2: Gather Information.

a. *Primary Sources.* Contact the source of the problem, who will provide information to establish most of the project's constraints and specifications, verbally or in writing. Later, in Step 4, confirm any verbal decisions made by the source in a progress report.

PLANNING CHART									
TEMPOROMANDIBULAR JOINT FLEXOR									
K. Anderson	M. Bui		A. Hamamoto			W. Peterson			

ACTIVITY	WEEK								
	1	2	3	4	5	6	7	8	9
PHASE I Get Organized Client/Meeting Information Search Project Define Progress Report	KA	MB AH WP WP MB				⦿ MILESTONE			

FIGURE 21-3   Planning chart for student group.

TIME LOG	W. PETERSON		
TEMPOROMANDIBULAR JOINT FLEXOR			
DATE	ACTIVITY	TIME HR	CUM. TIME HR
10-2	Met with Anderson, Bui, Hamamoto- - Exchanged phone numbers - Copied Dr. P's proposal	½	½
10-8	MET - KA MB & AH - discussed proposal - made up questions for Dr. P Discussed planning chart	2 ¾	3 ¼

FIGURE 21-4   Log.

PRELIMINARY DESIGN CONSIDERATIONS

1. SIZE   Date 10-10
2. POWER
   –BELT POWER PACK
   –CHARGER
3. CONTROLS
   –FREQUENCY
   –OPENING
   –PATIENT OR DR.?
4. EXPECTED PRODUCTION
   –REUSABLE?
5. COST
6. MATERIALS
   –STAINLESS
   –PLASTIC
   –FDA RESTRICTIONS
7. NOISE LEVEL
8. FORCES REQUIRED
   –OPENING
   –CLOSING
9. SAFETY OVERDRIVE
10. JAW DIMENSIONS
11. ACCURACY

FIGURE 21-5A   File: Preliminary design considerations.

LATERAL VIEW

Date 10-10

a – MAXILLARY ACRYLIC SPLINT
b – MANDIBULAR ACRYLIC SPLINT
c – BUCCAL TUBES
d – UPPER MEMBER
e – LOWER MEMBER
f – PNEUMATIC PISTON
g – UNIVERSAL JOINTS

FIGURE 21-5B   File: Information from clients.

Next, review similar products and experiences for information because most designs are actually new combinations of existing ideas.

b. *Secondary Sources.* Investigate secondary sources, which include a variety of people and references.

The following checklist can be expanded to fit the needs of specific problems.

People: shop personnel (machine, structural, electronic, electrical, packaging, shipping, welding . . .)

Practicing professionals: engineers, lawyers, doctors, technologists, scientists . . .

City services: fire, police, waste, utility, library

Most helpful for students: instructors, librarians, and shop personnel

Patents: Note that most of the information on a patented device only appears in patent documents.

References:

Books and websites.

Handbooks: Each engineering discipline has a handbook—e.g., aeronautical, chemical, civil, electrical, mechanical, materials . . .

Standards and codes (a small sampling):

ANSI American National Standards Institute

ASTM American Society for Testing Metals

Mil Military Standard

UL Underwriters Laboratory ISO 9001 Catalogs: Vendor (products, hardware, electrical, electronic, structural shapes . . .)

Thomas Register (listing of companies, what they make, their address, sample catalogs . . .)

McMaster-Carr Supply Co. (Most complete hardware catalog found to date, P.O. Box 54960, Los Angeles, CA 90054, 213-945-2811)

Vendor Catalogs. Vendors provide customers with convenient guides to select products and materials from their catalogs. Many vendor catalogs are available online.

a. *Machine Elements.* V-belt drives, ball and roller bearings, roller chain drives, clutches, brakes, couplings, gearmotors (motor and gear reduction box as one unit), shafting, motors (AC and DC), and controls all have similar guidelines for selecting them. These guidelines include formulas, tables, and charts to convert most customer applications and environmental conditions to numbers or letters in their catalogs.

For example, assume that the belt conveyor shown in **Figure 21-6** needs a roller chain drive to transmit power from the gearmotor's 200 RPM output shaft to the headshaft of the conveyor. Calculations indicate 14 horsepower to operate the conveyor at 500 FPM (feet per minute).

The roller chain drive, sprockets, and chain can be selected from a vendor's catalog found online.

b. *Engineering materials and shapes* can be selected from *vendor catalogs,* most of which are available online.

(i) The catalogs published by vendors of materials list materials and their typical uses to aid in their selection. To illustrate, one catalog describes 6061 aluminum as being heat-treatable, having good formability (will not crack during bending), having good weldability, and being corrosion-resistant. Also,

FIGURE 21-6 Belt conveyor system schematic.

6061 aluminum is used for structural application, boats, and transportation equipment.

(ii) Some materials are selected primarily on the basis of their shape; for example, steel and aluminum are rolled into plates, bars, channels, beams, and angles and are extruded into pipe, square tubes, and rectangular tubes. Square and rectangular shapes are easier to join in fabricating than pipe sections. Pipe sections are equally strong in all directions.

(iii) Materials may be selected on the basis of their weight, rigidity, and coefficient of expansion due to temperature changes. INVAR is used in instrumentation because of its low coefficient of expansion of 1.6×10^{-6} inches per inch per degree Fahrenheit.

(iv) Additional bases for selecting materials are corrosion resistance, machinability, strength, and electrical properties.

(v) Combinations of all of the above, plus aesthetics and prior experiences, may be used for selecting materials.

Call or visit a local vendor of materials or machine elements and request one of their catalogs. Usually, they are eager to have their catalogs in the hands of future buyers, so they are happy to donate one or more.

Furthermore, since most vendors handle a variety of products, a call or a visit may result in a small library!

Step 3: Combine Information into a Preliminary Definition. Prepare a preliminary definition of the problem so that all parties concerned can see the direction and course of the project. The definition mainly consists of two types of statements: *constraints* and *specifications*. These are written as specifically as possible even at this early stage in the project. For example, a vague specification might say that the product is to be safe to use. A more specific specification would say that the product will be safe to use by a 12-year-old child who can read and understand simple directions, and that the directions will be affixed to the product.

a. *Constraints.* List the more restrictive constraints, such as cost, delivery date, availability of materials, codes, and standards. Then, list the less restrictive ones, such as shop capability, available personnel, product life, and aesthetics.

b. *Specifications.* Divide specifications into two main categories: performance and design. Performance specifications describe what the design is to do, what it is to accomplish, what it is to produce, who is to use it, what environment it will operate in, how it will be maintained, how safe it is, and how easy it is to use. Design specifications indicate what materials are to be used and how they are to be fabricated and finished. Purchased products also are included.

Well-written specifications are stated so that they can be tested or measured. For example, the chemistry and strength of a material can be measured and tested, respectively. Resistance, frequency, volume, temperature, and weight can all be measured.

c. *Preliminary Solution Ideas.* Collect preliminary solution ideas that occur while the definition, constraints, and specifications are being developed and written. File these for use in Phase II.

d. *Iterate.* Return to Step 2 to gather further information and to revise the definition. Returning and revising is an iterative process that occurs continually throughout the design process.

Step 4: Prepare a Progress Report. Prepare a *progress report* to the client, who is usually the primary contact during the project. By now, approximately one-fourth of the project time will have been used, so a report to the client should be made preferably in writing. Include the current definition—constraints and specifications—so that if there are any misunderstandings, they can be resolved early in the project. Decisions made verbally should be summarized. Indicate immediate plans and ask for confirmation as soon as possible. The client, a person or an organization, wants to know how the project is progressing and if any problems have arisen.

For the purposes of this discussion, clients are divided into two groups.

a. *Internal Clients* are instructors in classrooms and supervisory persons in the industry. And even with this close proximity to the client, important communications should be in writing.

b. *External Clients* comprise a large variety of people and organizations, both in classrooms and in the industry. In classroom situations, an external client could be another instructor, a student, a department, a design contest, a handicapped person, a rehabilitation department, a local government, or a national organization. In the industry, the external client could be another company; an individual; a department in a city, state, or national government; the military; or even another nation.

Phase II: Generate Alternative Solutions: Synthesis

STEP 5   Generate Solution Ideas

    a. Use Systematic Techniques

    b. Use Creative Techniques and Attitudes

Step 5: Generate Solution Ideas.   Review the solution ideas already collected and realize that additional effort and discipline need to be exerted to generate more. Frequently, an idea for a solution occurs early in the process, and there is a tendency—especially among students, because of their time constraints—to select the first idea as the final solution. Successful problem solvers know that additional ideas can be generated beyond the initial obvious ones by continuing to focus on the design problem and employing two types of idea-generating techniques: systematic techniques and creative techniques. Having a creative attitude also helps.

    a. *Use Systematic Techniques.* Verb lists provide a systematic checklist for generating new ideas, primarily building on existing ones. Ask the following questions regarding the idea being considered:

- Can it be modified?
- Can it be made larger? Smaller?
- Can it be used for something else?
- Can other materials be substituted?
- Can components be rearranged?
- Can it be combined with another idea?

Divide an idea into smaller parts so that each part can be systematically examined and developed.

Human Factors.   There are published tabulations of body dimensions of average males, females, and children. These are called the *human factors.* This statistical data may be found in textbooks and handbooks; however, for classroom designs, use a classmate or yourself for measurements, since most of the adult population is close to the average size. Use children if needed.

For a design file, observe or measure objects designed for average humans. Typical designs include chairs, tables, desks, computer stations, work counters, doorknobs, ladder rungs, switches, handles, and numerous objects encountered daily.

Also, consider and observe the various techniques that are used to interface machines with human operators. A variety of sounds such as voices speaking messages, a number of visual devices using a multitude of colors, numerous tactile devices such as clicks during dialing, and different shapes of knobs are all used.

Try a 180-degree approach. For instance, instead of breaking a walnut shell with a hammer, poke a hollow needle inside and explode the shell with air pressure.

List the obvious attributes of an object to uncover ideas.

- Describe the size using dimensioned sketches.
- List the intended functions of the object.
- Outline how it was probably manufactured.
- List the materials used to make it.
- Describe any other features, such as aesthetics, weight, and color.

Use attribute listing, not only to generate ideas but also as a checklist for the design of parts on technical drawings. That is, check for complete dimensions to describe the geometry of each part and dimensions to locate each feature, such as bolt holes, slots, special surfaces, and datums. Review appropriate dimensions to ensure the function of the part—for example, clearances and fits. Check the material specified for the part and whether or not the manufacturability of the part was considered during the design phase. Finally, check all remaining attributes, such as how easily the part can be assembled, surface finishes, heat treatments, and aesthetics. Also list the not-so-obvious attributes, such as other uses that were not intended functions. Developing the habit of proposing these "not-so-obvious" uses for objects will help to generate new ideas for future designs.

Figure 21-7 shows the obvious attributes of a paper clip, and the not-so-obvious ones are listed in **Figure 21-8**.

Browse through technical texts, periodicals, and handbooks that contain numerous ideas for components, mechanisms, and applications of existing products. The list that follows includes the three types.

- Texts:

Herkimer, Herbert, *The Engineer's Illustrated Thesaurus,* 1952, Chemical Publishing Co.

Schwartz, Otto B./Grafstein, Paul, *Pictorial Handbook of Technical Devices,* 1971, Chemical Publishing Co.

McCormick, Ernest J., *Human Factors Engineering,* 1970, McGraw-Hill, Inc.

Chapanis, A., *Man-Machine-Engineering,* 1965, Wadsworth Publishing Co.

- Periodicals (found in most libraries):

Machine Design

Product Engineering

EXAMPLE OF OBVIOUS ATTRIBUTES

SIZE

FUNCTION
Clip 2 or more papers
together with spring
action
Reusable

MATERIAL
Steel wire .034″ dia, ductile
Copper coated

MANUFACTURED
Coiled wire fed into automatic machine
——cut to length
——bent as shown
——copper coated

OTHER
Easy to attach
Low cost
Simple
Other shapes, but one
shown is most common

FIGURE 21-7 Attributes of a paper clip.

| EXAMPLES OF NOT-SO-OBVIOUS ATTRIBUTES | |
AS IS	CHANGE
–dig out ear wax	–fish hook
–make a chain	–cake tester
–balance for paper airplane	–electrical conductor
–clip other items	–chess pieces
–sinker for fishing	–lock pick
–checkers	–artistic shape
–poker chips	–therapy–bend for relaxing
–hanger for small picture	–melt many to make casting
–standard of weight	–fatigue failure demo

FIGURE 21-8 Not-so-obvious attributes.

Test and Measurement World

Control Engineering

Medical Product Manufacturing News

Hydraulics and Pneumatics

Research and Development

Engineer's Digest

- Handbooks:

Shigley, Joseph E., *Standard Handbook of Machine Design,* 1986, McGraw-Hill, Inc.

Laughner, V. H./Hargan, A. D., *Handbook of Fastening and Joining of Metal Parts,* 1956, McGraw-Hill, Inc.

Horger, A. J., *ASME Handbook Metals Engineering Design,* 1965, ASME.

b. *Use Creative Techniques and Attitudes.* A clear understanding of the *creative process* will help to generate ideas. The process—as described by technical people, composers, writers, builders, doctors, hobbyists, and many others—proceeds as follows:

(i) Accumulate all the information about the problem being attacked; that is, thoroughly and conscientiously get involved in the problem, even emotionally if necessary. This involvement, according to creative individuals, causes the subconscious portion of the brain to be alerted that a problem needs solving. Next, relax and let the information collected . . .

(ii) Incubate with other ideas that the subconscious brain retrieves from memory locations. Take a walk, jog. Plop both feet upon a desk and lean back while contemplating the ceiling. (Supervisors may need to be informed that the "process" is working.) Soon, one hopes that an idea will surface in the conscious portion of the brain; that is, the idea . . .

(iii) Illuminates its presence and causes exclamations like, "Eureka, I've got it!" Finally, the idea needs to be . . .

(iv) Verified. Does it actually solve the problem? Meet the constraints? Satisfy the specifications?

Develop a creative attitude and behavior by adopting characteristics of creative people. Several are listed here.

1. *Keep an open mind. Be willing to accept change and new ideas.*

2. *Look for similarities and unusual relationships.*

3. *Think positive. Avoid or ignore negative comments such as:* "It's been tried before." "It's not logical." "What a dumb idea." "It'll never work."

4. *Be persistent.*

Participate in exercises designed to revive creative talents, buried during years of being told what to do and to conform. For example, the late Professor John Arnold, M.I.T. and Stanford University, proposed a fictitious planet, Arcturus IV, populated with creatures who were unusually tall and slow-moving, had reaction times one-tenth as fast as humans, had three fingers on each hand, had three eyes, one of which had x-ray vision, and breathed methane. Professor Arnold asked his students to design an automobile-type vehicle for these inhabitants. Tackle this problem before reading on, and compare your design ideas with those of other members of your class. For example, what speed limit should be required due to the slow reaction time of the inhabitants?

Build a simple model related to the design problem being developed. Building a model, or a prototype, almost always produces new ideas because of the simultaneous involvement of the hands, eyes, and brain, plus the three-dimensional insight. Models can be inexpensive—cardboard, balsa wood, tape, etc. Computer graphics can be used.

Do exercises to develop imagination for the same reason that muscles are exercised: for growth and to avoid atrophying. One convenient source of creative exercises is the daily newspaper, which contains word puzzles, crossword puzzles, and picture puzzles.

Try the following exercises to help develop a creative attitude:

1. Use analogies. For example, pretend to be one of the components in a machine and imagine how each part would function if designed a certain way. What would happen if the design were changed?

2. Systematically propose that a product be changed in any one of the following ways: made bigger, smaller, longer, shorter, colorful, drab, nontraditional, expensive, or cheap.

Read books devoted to developing creativity and try some of their recommendations.

A person's creative potential can be enhanced by taking courses in areas outside of engineering and technology. Take courses to obtain different points of view in other disciplines, such as biology, geology, agriculture, environment, political science, art, philosophy, or oceanography.

Brainstorm according to the ideas of Alex Osborne in his *Applied Imagination*. The procedure is simple to follow. Meet with a group who is willing to abide by the following guidelines:

1. Have one person *record ideas as they are generated* on a surface visible to all participants. New ideas often stem from the visible ones.

2. Agree that *no one is to criticize* any ideas while the brainstorming is in progress, no matter how ridiculous an idea may seem. Frequently, viable ideas stem from less obvious ones.

3. Meet in as *pleasant surroundings* as practical, and include refreshments. Several one-hour sessions produce better results than one long session.

4. Finally, *meet to critique the ideas* for possible solutions.

Brainstorming can also be done individually.

Develop observation skills through freehand sketching by looking at a component while imagining how a sketch of the component would be made. Of course, do the sketch if possible! Looking at an object while imagining how a sketch would be made causes the viewer to see more about the object than he or she would if only a casual glance were given. Try sketching assemblies of components also.

Assume now that a number of design ideas have been generated. The group usually wants to continue in this creative mode because creating is a satisfying activity. However, since time controls the schedule, the group needs to move on. Students involved in the design process for the first time reluctantly leave this synthesis phase believing that if they had more time they could do a better job. Creative professionals know that they must do the best they can in each phase in the time available and then move on to complete their projects within the schedule.

So now is the time to focus on evaluating instead of generating.

Phase III: Evaluate Alternative Solution Ideas: Analysis

STEP 6 Sift, Combine, and Prepare Solution Ideas for Analysis
  a. Preliminary Preparation
  b. Specific Preparation

STEP 7 Compare Solution Alternatives to Constraints and Specifications
  a. Iterate
  b. Persevere
  c. Qualitative Evaluation
  d. Quantitative Evaluation

STEP 8 Conduct a Design Review
  a. In-House Design Reviews
  b. Outside Design Reviews

STEP 9 Refine the Final Solution

Step 6: Sift, Combine, and Prepare Solution Ideas for Analysis. Sift through all the collected ideas and keep those that seem to have potential as a partial solution or as a complete solution. File the remaining ideas for possible use in the future.

  a. *Preliminary Preparation.* Expand each potential idea in more detail. Clarify, if possible, vague statements and generalities. Ask each member of the group to develop preliminary statements and sketches of those of his or her ideas that survived the initial sifting and culling.

Do further research, if needed, to expand the ideas.

b. *Specific Preparation.* Prepare *layout development drawings* of the preliminary sketches. Layout drawings are essentially a designer's freehand sketch redrawn to scale with instruments or a computer. Designers use two types: design development and kinematic development (the latter is used primarily for mechanical products). Schematic layouts, using lines and symbols, but not to scale, are used for electronic circuits, electrical circuits, piping diagrams, routing paths, processing steps, chemical plants, and logic circuits.

A design development layout drawing is made from a freehand sketch of a design idea for an anchor bolt puller shown in **Figure 21-9A**. The design idea is drawn to scale in the design development layout drawing (**Figure 21-9B**). Notes and comments may be in writing or in casual lettering. Compare this informal approach to the formal technical detail and assembly drawings for shop use, which have standard lettering, lines, and symbols made on standard-size drawing sheets, complete with title blocks.

Use the design development layouts to establish the locations and sizes of parts to be fabricated so as to ensure the functions of the final product. Also use this layout to verify that purchased products will fit where required.

In industry, designers forward their design development layout drawings to drafters who prepare the formal technical detail and assembly drawings. Students in classrooms do both.

A kinematic development layout drawing is shown in **Figure 21-10**. This type of layout shows the relative motions of assembled components, verifies the function of the assembly, checks for interferences, and does velocity plus acceleration analyses. The kinematic analyses are usually done in a two-dimensional format even though the assemblies are, of course, three-dimensional.

ANCHOR BOLT PULLER FRAME

FIGURE 21-9A Freehand sketch of a design idea.

FIGURE 21-9B   Design development layout.

FIGURE 21-10   Kinematic development layout.

FIGURE 21-11   Kinematic model.

A useful complement to the kinematic development layout drawing is a simple cardboard-thumbtack model, as illustrated in **Figure 21-11**. CAD systems with programs for kinematic analyses are also useful and available.

Use cardboard-and-glue models (noted earlier) to further complement the layout drawings. Students, designers, drafters, and practicing engineers attest that two major phenomena happen during the process of making a model of a portion or all of a design from a design development layout. One, they discover that the components in the three-dimensional model may not work as planned, so adjustments and refinements are required. And two, new solution ideas surface, ideas no one had considered. They believe that the combination of using their hands and eyes while shaping and assembling materials triggers an untapped area in their subconscious minds. Furthermore, since ideas evolve as pictures in the mind, sketching, drawing, and model making seem to enhance the forming of these pictures.

Two additional advantages of cardboard models are the relative low cost (compared to prototypes) and the fact that the model may be used later in a design review.

The solution ideas that have been expanded and prepared for further analyses can now be considered as solution alternatives.

Step 7: Compare Solution Alternatives to Constraints and Specifications.

a. *Iterate.* Review Phases I and II for possible additions or deletions to the constraints and specifications. New information may have been discovered. The client may want changes made.

b. *Persevere.* Be cognizant of a morale problem that may occur at about this time in the design process. Imagine that a group, or an individual, has started a project with high hopes and confidence that a solution can be found. However, even though the first five or six steps have been conscientiously done, no real viable solution ideas have been generated. The individual or all of the participants find themselves at a low point in the morale curve shown in **Figure 21-12**. They discover that there is more to the problem than they thought, and useful information is difficult to find. They become discouraged. However, experienced problem solvers know that perseverance

will get them through the slump and that satisfactory results eventually will follow. They know that toward the end of Phase IV, their morale and self-confidence for attacking problems will be higher than when they started.

c. *Qualitative Evaluation.* Compare solution alternatives with what has been done before. If a solution alternative is similar to existing designs, either within the company or in industry, the feasibility of it is relatively easy to determine.

Use a range technique for evaluating preliminary alternatives. Intuition, experience, and judgment help to determine if the size or thickness of a member is within a reasonable range. For example, most students could evaluate design alternatives for home-use ladders based on their own experiences. Similar range approaches have names like back-of-the-envelope calculations and order-of-magnitude analyses. Interestingly, all of these approaches are useful in designing and in evaluating.

d. *Quantitative Evaluation.* Build a *prototype* of part or all of a solution alternative. For example, a student group had the problem of compressing a $2' \times 2' \times 4'$ bale of hay into a $2' \times 2' \times 2'$ cube so that their farmer client could ship more hay per shipload from Seattle, Washington, to a port in northern Alaska. The hay compressor was to be part of a complete hay-baler machine. One key factor that they needed for their design was the force required to compress a bale to half size. So they made a crude, but effective, set of rectangular welded-steel boxes, one to telescope within the other, to hold a bale in a testing machine that could generate up to 60,000 pounds of compressive force. See **Figure 21-13** for a conceptual

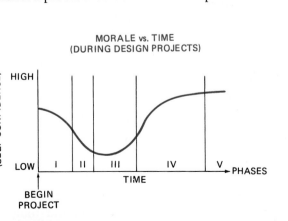

MORALE vs. TIME
(DURING DESIGN PROJECTS)

FIGURE 21-12   Morale curve.

BALE OF HAY AND
CONTAINERS SHOWN
IN CROSS SECTION

POUNDS OF
FORCE

FIGURE 21-13   Prototype test.

sketch showing the two containers and the bale of hay. The students learned that compressing the bale to half size required 14,000 pounds of force. A hydraulic cylinder was selected to provide the force.

Another type of prototype building is a process called stereolithography. Actual solid models are created layer by layer by an electronic-laser process in a special liquid solution. A focused beam causes the liquid to solidify over exact areas at accurate locations. It adheres to the previous layer until a complete component is built up.

Two alternatives to actually building a prototype of a proposed design can be done on a computer: wireframe drawing and solid modeling.

Perspective, wireframe drawings (line drawings in perspective of objects, no hidden lines) on expensive computer systems that can rotate views and zoom into or out of buildings or marine vessels are used by architects and marine engineers to visualize what the internal structure would look like.

Solid modeling on a computer screen can be helpful in evaluating a design for shape, structure, and interferences. Also, the solid model can be used as the basis for a finite-element analysis for checking the strength of an object.

Consider the feasibility of each alternative by asking questions such as: How much will it cost? Can the shop make it? Are engineering materials available? Are finished components available? Are sufficient labor hours available? Can the delivery date be met?

Manufacturing Process Costs. The following sample list is headed by the least costly and ends with the most costly manufacturing process. *Costs* are also related to the roughness of the surfaces of the materials. The roughness is a result of the manufacturing or production processes.

Sawing	Rough Surfaces
Cutting (torch)	(lower cost)
Drilling	
Turning	
Punching	
Milling	Average Surface Roughness
Reaming	(Average Cost)
Broaching	
Grinding	
Polishing	
Honing	
Lapping	Smooth Surface (Higher Cost)

Relative Costs of Materials. Use relative cost numbers as an aid to evaluating the feasibility of a design alternative and for establishing approximate costs. Mild steel is given a factor of 1.00; all other engineering materials are given relative numbers. Find the current cost of mild steel, and use the relative numbers to establish costs of the other materials. Revise the numbers as needed.

Material	Relative Cost (By Weight)
Mild steel	1.00
Spring steel	2.00
Tool steel	11.00
Cast iron	3.00
Aluminum	6.00
Stainless steel	7.00
Brass	7.00
Copper	7.50
Titanium	130.00
Teflon	67.00
Nylon	19.00
Rubber	4.00
Wood	1.00

Safe Designs. Use the following checklist to provide *safe designs*, safe installations, and protection against negligence suits.

Document the design to show evidence of the following:

a. Accurate calculations

b. Proper use of codes and standards

c. Use of state-of-the-art technology

d. Proper use of materials and hardware

e. Having run tests, if appropriate

f. Use of proper manufacturing processes, proper fasteners, and appropriate heat treatments

g. A quality control program

h. Preparation of instructions for safe use

i. Addition of warning signs to the product

j. Having told future owners of the proper safety devices and practices needed, such as guards around any exposed moving parts of machines and painted stripes to show danger areas

k. Thorough exploration of all possible uses of the product other than those for which it was designed

Access. Locate fasteners and components for installation and servicing so personnel can have access to them with their hands and tools. Also, provide access to weld areas for welders.

COMPONENTS SUBJECTED TO VARYING LOADS

FIGURE 21-14 Fatigue considerations.

DECISION TABLE					
Criteria	Weighting Factor	Walk	Bike	Bus	Car
$	10	10 100	7 70	6 60	4 40
Protection from weather	7	3 21	2 14	7 49	8 56
Time spent	6	2 12	5 30	4 24	6 36
Safety	3	7 21	6 18	9 27	8 24
Reliability	8	9 72	8 64	6 48	7 56
		226	**196**	**208**	**212**

FIGURE 21-15 Decision table.

Vibration. Use locking devices such as lock washers and lock nuts on threaded fasteners subjected to vibration. Some high-speed engines have wires passing through holes in the nuts to prevent the nuts from backing off because of vibrating conditions.

Fatigue. Avoid sharp changes of cross-section and small fillet radii on parts subjected to varying loads. See **Figure 21-14** for examples of good and poor practices. The sharp corners and radii (and scored surfaces) cause stress concentrations and therefore possible fatigue failure.

Cleaning/Corrosion. Provide sloping surfaces wherever standing fluids could cause corrosion. For example, tank bottoms should slope toward a drain hole so that chemicals and cleaning fluids can be completely drained. Upper surfaces should slope to prevent standing rainwater or melting snow.

Temperature. Avoid stringent tolerances on dimensions of parts subjected to extreme temperature changes. Furthermore, check the strength and ductility of materials at extreme temperatures. Also, the temperature of a part may need to be specified for verifying some critical dimensions at that temperature if tolerance on the dimension is less than ±0.0002″.

Prepare a *decision table* to systematically compare each solution alternative with every other alternative as to how well each one satisfies the constraints and specifications from the definition of the problem in Step 3.

An example decision table is shown in **Figure 21-15**. Although the example is simple and was made up for illustration purposes, the procedure is the same for more complex analyses. The example is based on the question to a class, "Which alternative for transportation to and from school for a distance of five miles (10 miles round trip) is the best one?" The class was asked to create a decision table for a quantitative evaluation of four transportation alternatives: walk, bike, bus, and car. First, a list of criteria was generated and tabulated in the left-hand column. Next, each criterion was compared to every other criterion and was given a weighting factor for its relative importance according to the students.

The four alternatives then were systematically compared against each other for each criterion. First, the expense criterion was considered. Walking was the least expensive, so it rated a score of 10 as compared to a car, which was given a 4. Biking and bussing scored 7 and 6, respectively.

After the four alternatives were compared in relation to each criterion, products of the weighting factors and the scores were totaled for each alternative. Walking scored the highest; therefore, walking would be the alternative to investigate and develop further. The bus and car were close contenders and were also looked over once more.

To evaluate solution alternatives in a design project, list the constraints and specifications in the left-hand column and give them weighting factors. Then, list the alternatives across the top. Compare and rate every alternative with the others for each constraint and specification.

Finally, compute the products of the weighting factors and the alterative scores, total the products for each alternative, and select the alternatives with the higher totals for further study.

Finally, select the best one, realizing that it is the best for the time available, the information collected, and the judgment of the group. Even though it is difficult to do, a decision must be made so that the project stays on schedule.

Schedule a design review next.

Step 8: Conduct a Design Review.

a. *In-House Design Reviews.* Prepare drawings, transparencies, slides, models, and appropriate documents to effectively and efficiently communicate the design solution to the attendees.

In a classroom, present sufficient information so that the students in other groups will ask thoughtful questions and make helpful comments. The goal of the review is to benefit from the collective expertise of classmates and to be sure that all facets of the design have been considered.

The instructor may be the only nonstudent present, as clients usually do not attend. Also, the instructor may schedule more than one design review throughout the project or may have conferences with each group.

During the design review, have someone record the questions and comments for modifying or reassessing the solution soon afterwards.

An in-house design review in industry is more comprehensive because, in addition to technical personnel, representatives from accounting, finance, sales, management, manufacturing, shipping, and installation may be present. Furthermore, the design solution discussed may involve a variety of products or an entire complex system.

b. *Outside Design Reviews.* An outside design review provides information to the client at a location selected by the client. Often, the main objective of an outside design review is to "sell" ideas to a client, whether the client is a private citizen, a private company, or a governmental organization. Companies and governmental organizations publish requests for bids or proposals on a variety of projects, such as aircraft, spacecraft, dams, highways, buildings, chemical processes, electronic devices, electrical systems, power plants, vehicles, military devices, ships, submarines, and numerous products. The bidding individual or proposing organization prepares an outside design review, and if the bid or proposal is accepted and a contract is signed, the bidder starts a new design process in more detail than the one leading to the outside review.

Students may give outside design reviews to a client as progress reports; however, most classroom projects require only in-house design reviews that lead to the final solution.

Step 9:   Refine the Final Solution. Review the questions, comments, and suggestions recorded during the in-house review and decide which ideas to incorporate in design changes or additions. Revise previous steps if necessary; however, time constraints dictate that the next phase should be started soon.

Phase IV: Document, Specify, and Communicate the Solution

STEP 10 Prepare Drawings and a Written Report

STEP 11 Give an Oral Presentation

In classroom projects, Phase IV requires about one-half of the time available. Moreover, most classroom projects end with Phase IV because of limited time, limited shop facilities, and limited funds.

In industry, Phase IV and Phase V overlap, and together they consume approximately one-half of the time available.

Step 10:   Prepare Drawings and a Written Report. Use the informal development layout drawings for preparing the formal technical detail and assembly drawings in standard views, lines, dimensions, and symbols. Only those parts that require shop processes are detailed in technical drawings. Parts—such as fasteners, keys, and pins—that do not require shop processes only appear in the parts list of the assembly drawings. Purchased components such as motors, switches, and bearings appear in assembly drawings in the parts list and are usually drawn in outline form.

Four examples of technical detail drawings of components to be made in a shop are shown in **Figure 21-16** and **Figures 21-17A** through **21-17C**. Figure 21-16 is a subassembly of structural steel parts, welded together for use in the anchor bolt puller.

FIGURE 21-16   Technical detail drawing of subassembly.

FIGURE 21-17A   Technical detail drawing of a spacer.

FIGURE 21-17B   Technical detail drawing of a support plate.

FIGURE 21-17C   Technical detail drawing of an angle stop.

FIGURE 21-18A   Assembly drawing of anchor bolt puller frame.

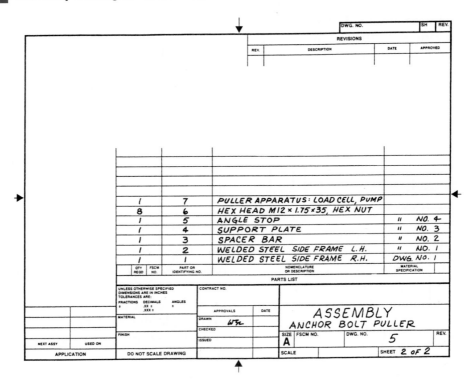

FIGURE 21-18B   Parts list for anchor bolt puller.

Figures 21-17A through 21-17C show several individual components. **Figure 21-18A** is an assembly drawing with a parts list (**Figure 21-18B**) for the complete anchor bolt puller. Note that the subassembly from Figure 21-16 is given only one part number in the assembly drawing parts list.

Logical Dimensioning.   Design components so they can be measured readily and logically—for two main reasons. One, shop personnel need to locate points, planes, and features such as holes from reference planes or surfaces. Two, the function of the component needs to be ensured.

Shop personnel use a large number of measuring devices, some very sophisticated, some relatively simple, but all measure from a reference plane or surface that is usually machined. Some of the more common measuring devices and their measuring capabilities are:

a. Flat, stainless steel scale: to 0.010″

b. Micrometer: inside and outside: to 0.0002″

c. Vernier calipers: inside and outside: to 0.001″

d. Lathes, milling machines, shapers, boring machines, and gear cutters with calibrated dials to indicate the movement of the cutting tool or the part being machined: to 0.0005″

e. Some newer milling machines and numerically controlled machines with electronic readouts: to 0.0001″

Engineers, designers, and drafters can facilitate these shop processes by dimensioning components from obvious reference planes or surfaces. The plane or surface may need to be created in the manufacturing process.

Ensure the function of a component or an assembly of components by dimensioning to common reference planes or surfaces. An example to illustrate reference planes is shown in **Figure 21-19**. The two identical L-shaped brackets are cast iron. Surface No. 1 is milled to establish a plane of reference to locate vertical surface No. 2 to be perpendicular to surface No. 1. Then,

the hole at No. 3 is machined perpendicular to surface No. 2 (and hence parallel to surface No. 1) so the shaft for the roller will be horizontal. The two holes in the bracket would probably be located with respect to vertical surface No. 2 and centered about the center line of the machined hole, surface No. 3.

A more exact dimensioning technique called positional tolerancing and form tolerancing (or geometrical tolerancing) establishes datum surfaces for referencing dimensions and features. The technique requires more skill and time to apply; however, time and money are saved because fewer parts are scrapped. They fit better and their function is ensured.

This tolerancing technique has grown as machines and systems have become more complex and expensive. Furthermore, when one company manufactures components that will be assembled together with components from another company, the dimensioning and features need to be accurately located so the components will go together. Refer to Chapter 12, Geometric Dimensioning and Tolerancing, for detailed information.

Technical detail and assembly drawings in industry are similar to the anchor bolt puller drawings, but with several differences. Complex drawing-numbering systems tell the user which ones are details, where individual components are used, which ones are assemblies, and which job or contract used the drawings. These systems are as varied as the number and size of industries and organizations. (Refer to Chapter 17.) Usually, the drawings are on larger sheets. Also, each individual part needing to be processed in a shop is detailed separately to accommodate accounting procedures and to merge into the flow of information within the company.

Design Hints

Use *design hints* during all phases of the design process as extensions to the design hints already included. These design hints help in initial designing and become checklists as the project progresses. Keep a file during design problem experiences to expand these lists, and remember: Carry a notebook at all times!

Attribute Listing for Checking Drawings. Use the technique of *attribute listing* as a systematic guide for checking the completeness of technical detail drawings. For example, ask the following questions:

a. What is the function? The function dictates the dimensioning plan.

b. What size is it? Are all geometrical shapes dimensioned? Are all the features located?

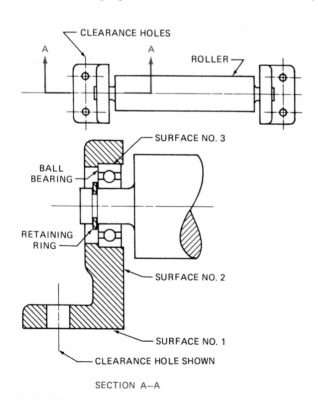

SECTION A–A

FIGURE 21-19   Reference plane for dimensioning.

c. What materials are used?

d. Can it be made? How will it be manufactured? What processes will be used?

e. What other facets need to be checked? Surface finish? Weight? Vibration? Safety? Aesthetics?

"Rules of Thumb." Use "rules of thumb," practical guides based on experience, to establish or check "ballpark" (within +1 or −10%) quantities. Use them as qualitative checks.

a. *Tolerance.* As a beginning problem solver in technical fields, keep in mind that the tighter the tolerance on the dimensions of an object, the more the object costs. Tolerances should only be as tight (small) as needed to ensure the function of the part. Use as a conceptual reference the thickness of the paper this sentence is written on: It is .003 to .004″ or 0.08 to 0.10 mm thick. Most machinists and machines can easily work to .001″, but .0001″ takes more time and costs more.

b. *Rigidity.* "When in doubt make it stout." This statement is just a reminder that not all designs need to be based on an extensive stress analysis and strength of a material. Rigidity and accurate locations may be more important than strength. For example, a surveyor's transit needs to be rigid to keep the lenses aligned. Instrument panels need to be rigid to support the dials and meters and to prevent vibration. Another practical application of the statement refers to sizing members when budgets are low. That is, a brief stress analysis producing a conservative size (stout) is less expensive than an extensive analysis and a more exact size.

c. *Murphy's Law.* "Anything that can go wrong will go wrong." Use this guide to evaluate how a designed part or assembly might behave. A more important application is to ask, "How could someone misuse the part or assembly?"

d. *Weld Size.* Fillet and butt welds constitute approximately 85% of all welds used in industry. And if the thinnest cross-section dimension of the weld is at least as thick as the thinnest member joined, the weld will be as strong as the members, assuming that the weld is done by a skilled welder.

e. *Threads.* The depth of threads for steel threaded fasteners should be equal to the diameter of the fastener. Use 1.5 to 2.0 times the diameter for the depth of threads for non-ferrous threaded fasteners or steel fasteners in non-ferrous metals.

f. *Intuitive Ranges.* Early in a design, establish sizes by asking what a reasonable range for this size would be. Establish performance, weights, and function in a similar manner. During an analysis, ask questions such as: Is this a reasonable size? Weight? Function? Performance?

Prepare a written report for the client. Although each client's needs in a report differ, the parts of reports are similar, and all the parts should be included. The order is arranged to suit the information to be communicated.

Use the following discussion as a guide for preparing professional, typed project reports to clients. Refer to samples from a student group's report later in the chapter.

a. *Letter of Transmittal.* A letter to the client, separate from but enclosed with the report, saying that the project has been completed, the report is enclosed, and that working with the client on future projects is anticipated. Fee statements may be a part of the letter. Include phone number(s) the client can call for questions and an address for mailing fees.

b. *Cover.* Enclose the report in a folder, unless otherwise directed.

c. *Title Page.* Include title, by whom written, to whom submitted, date, and organization (college or company).

d. *Abstract.* Include one or two sentences about the content of the report to capture a potential reader's interest. For product designs, describe the product, specify its performance, and give the cost or price.

e. *Table of Contents.* List the main headings and the subheadings with appropriate page numbers. List the illustrations, drawings, and tables, plus an appendix if used.

f. *Introduction.* Describe briefly—who, what, when, where, and why—the origin of the project/problem, because even though the client knows this information, other members of the client's organization—who may not—might read the report.

g. *Solution Description.* A written portion should use a liberal number of headings and subheadings and should describe the final solution, including the performance specifications and the design specifications. List the constraints, costs, materials, and special features. Also include the delivery date. Refer to illustrations, figures, charts, graphs, and drawings for efficient communication.

h. *Analyses.* Summarize the major assumptions and calculations. Place detailed *analyses* in an appendix.

i. *Costs.* For a classroom project, which usually ends with Phase IV, list the basic direct costs that the client would have to outlay to have the design made. These costs could include materials, shop labor, and a number of purchased components—fasteners, pins, shafts, switches, controls, electronic circuits, motors, and similar hardware. Even if the classroom project continues through Phase V and a prototype is built, the client will most likely be required to reimburse the students **only** for the actual expenses incurred. The excellent learning experience is payment for the students' labor!

In industry, only the selling price would be given in the report because the basis for arriving at the selling price (labor rates, material costs, overhead, sales, and profit margin) would be proprietary information.

j. *Drawings.* Include the technical assembly drawings in the main report and the technical detail drawings in an appendix. Pictorial drawings (isometric, oblique, or perspective) may also be appropriate, depending on the client's technical experience and background.

k. *Illustrations, Figures, Charts, and Graphs.* Check the illustrations, figures, charts, and graphs for complete information. Each one should have a legend. In addition, charts and graphs should have clearly marked and labeled axes.

l. *Alternatives.* Convince the client that a thorough investigation was done by the design group by including a brief summary of the *alternatives* that were considered but discarded. Use partially dimensioned sketches, drawings, and pictorials with appropriate comments as to why the alternatives were not selected.

m. *Recommendations.* Discuss briefly any recommended further developments or investigations that relate to the project. For example, one of the discarded alternatives may have some potential and should be investigated further.

n. *References.* List references and the specific pages that were used. Number the references, and use these numbers in the text of the report.

o. *Appendices.* Place information in the appendices that does not logically fit in the main report. Such information includes extensive calculations, computer printouts, vendor brochures, catalogs, machine copies of referenced material, and technical detail drawings.

In industry, when a company or organization wants to "sell" its expertise to clients, it includes a roster of individuals, indicating each individual's technical skills, education, and experience. A summary of projects completed by the company may be included.

Most organizations have an established format for their reports. A number of companies and organizations use computers to store boilerplate information that appears in all of their reports. Standard paragraphs and phrases appear at the beginning and at the end of their reports—hence, the wrap-around image of a boiler. New material fills the blanks.

Rewrite the report text at least two times, whether a classroom report is being prepared or one in industry! Check for completeness. Check grammar, punctuation, and style. Avoid clichés. Eliminate unnecessary words.

Step 11: Give an Oral Presentation. Prepare an *oral presentation* of the design solution to the client. The following guidelines for planning a presentation apply to student presentations and those done in industry.

a. For a simple but effective presentation, first tell the audience **what is** going to be presented. Then, present the material. Concluding, tell the audience **what was** presented.

b. Make projected words and numbers large enough so everyone can read them. A conservative guide is to make the letters on the screen as high as 1/250 of the distance to the person the farthest away from the screen. Thus, a 25-meter distance requires letters on the screen to be 0.1 meter high. Typed material reproduced on transparencies is usually unsatisfactory.

c. Restrict the number of different ideas on a visual to five or less. If more are on the transparency, distribute a copy to the audience beforehand.

d. Use a pointer.

e. Use models large enough to be seen easily.

f. PRACTICE at least once. One or more persons can do the presentation.

g. Allow time for questions.

In industry, a final oral presentation made to a client at the end of Phase IV could be done by a consulting-type firm whose final "product" consists of engineering drawings and a written report.

Oral Presentations. Public speaking is an important skill for engineers, designers, and drafting technicians.

During the design process discussed in this chapter, oral presentations are scheduled at least two times: the design review and the final presentation. As a guide for giving these two presentations and others, use the following three practical steps noted earlier.

1. Tell them what you're going to tell them.
2. Tell them.
3. Tell them what you've told them.

Further guidelines are in the exercises at the end of the chapter.

Phase V: Develop, Manufacture, and Deliver

STEP 12 Prepare Technical Detail Drawings of Each Component to Be Manufactured

STEP 13 Schedule Assembly, Final Testing, Packaging, and Delivery

STEP 14 Keep Records

Arrange for the necessary paperwork, drawings, materials manufacturing processes, personnel, and funds to develop the device or system to meet the client's requirements.

These requirements may range from modifying an existing device to producing a state-of-the-art system. Thousands of different combinations of requirements occur in industry, so describing them all here is impractical. Consequently, a more appropriate task for this chapter is to present the steps that are common to most industries.

Step 12: Prepare Technical Detail Drawings of Each Component to Be Manufactured. These drawings "get the wheels turning" within a company. A typical drawing enclosed in a clear plastic folder may follow the component through a manufacturing area from casting or forging to machining to heat treating to finishing to assembly and delivery. Additional areas could include checking, quality control, subassembly, and temporary storage.

Step 13: Schedule Assembly, Final Testing, Packaging, and Delivery. Plan the final processes prior to sending the product to the client. Compare appropriate subassemblies and complete assemblies with the specifications, particularly performance specifications. For example, an electronic device may be tested in a variety of environments before being released for shipment.

Design packaging and shipping containers so that the product will survive shipping without damage.

Some devices and assemblies must endure more severe loading en route than after being installed—for example, large frames, large cylindrical-shaped housings, and pressure vessels.

Write maintenance and service manuals to help prolong the life of the product. Provide instructions for proper and safe use of the product.

Step 14: Keep Records. Document and file all aspects of the product's development and use. Record the sources and quality of materials used. File the detail and assembly drawings. Summarize the manufacturing processes and prototype testing results. Add feedback comments from the client. Request cost, accounting, purchasing, and related records from appropriate personnel.

These records are useful for future use if the client wants replacement parts or another complete product. Also, other customers may purchase the product.

Drafters in Industry. Technical drawings were discussed in detail for Step 10 of Phase IV, primarily because student projects usually end with detail and assembly drawings. Drafters in industry would most likely do the detail and assembly drawings during Step 12 of Phase V. The difficulty of their drawing assignments would vary depending on their experience.

Newly hired drafters soon find that in industry, there are a number of sources of assistance. Colleagues in the drafting room can be very helpful by showing new members where to find information on previous jobs done by the company—things like where catalogs are filed; which CDs are available with vendor information; who to contact in the shop for fabricating and assembly practices; and where to have their drawings printed, plotted, and filed.

Experienced drafters not only do the technical drawings but they also become involved in designing. They develop ideas using layout drawings, solid modeling on computers, and frequently use cardboard, glue, and balsa wood to verify their ideas.

Two types of projects for students who are learning the design process are discussed next.

Design Projects: Routine and Non-Routine

Routine Design Problems

Routine design projects and problems are used to introduce new members of an organization to the various people in the company and to the routes design information needs to flow through. Beginning engineers, designers, technologists, and drafters need to know how jobs are initiated, documented, processed, and completed. An effective way to start these new employees is to assign routine projects and problems to them.

These first assignments usually are assigned to an individual and are characterized as follows:

a. The problem may be a modification to an existing product line, a change in the company's plant facilities, or just a drawing change.

b. The problem difficulty matches the individual's abilities and skills.

c. Information is communicated in a memorandum in a typical industrial format, such as an ECO (engineering change order) or an EDO (engineering design order).

d. Requirements are clearly stated and the information is nearly complete; however, some new data may be needed.

e. An analysis may be required.

f. For guidance, there are similar designs, knowledgeable colleagues, the company library, and vendor catalogs.

g. Technical drawings are required; they become part of the documentation of the modification or change.

h. Labor and materials costs are always present.

i. A condensed design process is followed; however, all the essential steps are followed, including planning, keeping a log, keeping a file, understanding the problem, synthesis, analysis, comparing alternatives, reviewing designs with supervisors, documenting the project in drawings and text, manufacturing, assembly, and installation.

An example first assignment that a newly graduated engineering student received in an EDO was to design a support for a tank containing a liquid for a paper-making process. But before resolving the technical parts of the support, the new engineer, escorted by the chief engineer, was introduced to personnel in drafting, purchasing, the shop, and the paper-making area where the tank was to be installed. Further introductions included individuals in general supplies, copy service, and even those who made coffee.

The technical challenges involved load calculation, sizing beams and columns and, finally, making a design development layout drawing. The chief engineer was consulted before the layout was handed to a drafter. The drafter questioned a few items and then produced technical detail drawings plus an assembly drawing.

The new engineer checked, approved, and signed the drawings. Then, he followed them through the shop and ultimately used the drawings at the installation site.

The last step was to document the project. The drawings and vendor information were filed. A short written report summarized the need for the tank support, and a copy of the EDO was included. Cost information from the accounting department completed the file.

Non-Routine Design Problems

Non-routine design problems are assigned to experienced engineers, designers, technologists, and drafters—their complexity and level of difficulty in direct proportion to the experience, skills, and ability of the recipients. Usually, a team of technical personnel attacks non-routine design problems.

Non-routine design problems are characterized as follows:

a. The problem originates outside the company, and it stems from the needs of a client who may be an individual, another company, or a governmental organization.

b. A team of individuals with diverse backgrounds or a group with similar backgrounds is formed, depending on the experience, skills, and abilities required.

c. The problem, not clearly defined, must be researched thoroughly so that constraints and specifications can be written.

d. A complete design process, with phases and steps, is required, from the initial planning to the final delivery and installation. More alternatives are generated than for routine problems and, therefore, more decisions are made.

e. The projects require a longer time span than routine projects.

f. Detailed planning is required for personnel assignments and hours, materials procuring, budget allocations, production processes, and delivery/installation.

g. More engineering trade-offs occur. A *trade-off* is a compromise, usually between costs and materials. For example, an expensive material (titanium) may cost more but weigh less than a less costly material (aluminum) that weighs more.

Since non-routine design problems vary as widely as the number and size of problem sources—individuals to organizations—there are no typical non-routine problems. However, the manner in which they are attacked—the design process—is typical. Therefore, follow the phases and steps presented in this chapter to obtain an insight into the efficacy of this process. Then, formulate your own design process as needed.

Examples from a student team member's log, the team's planning chart, and the team's project report are included here for your information. The challenge given to the student team from a doctor was to design a device that would exercise the muscles of a patient's jaw continuously, day and night, to accelerate healing of the jaw muscles after an operation.

The following lists some of the more important aspects of the step process just described. Use it to review the art that follows.

- Log (Step 1) (**Figure 21-20**)
- Planning Chart (Step 1) (**Figure 21-21**)
- Information from the Client (Step 1) (**Figures 21-22A through 21-22C**)
- Progress Report (Step 4) (**Figure 21-23**)
- Letter of Transmittal (Step 10) (**Figure 21-24**)
- Title Page and Abstract (Step 10) (**Figure 21-25**)
- Table of Contents (Step 10) (**Figure 21-26**)
- Introduction (Step 10) (**Figure 21-27**)
- Power Pack Assembly Sketch (Step 10) (**Figure 21-28**)
- Alternatives Discarded (Step 10) (**Figure 21-29**)
- Final Recommendations (Step 10) (**Figure 21-30**)
- Appendix: Materials, Electrical Schematics, and Costs (Step 10) (**Figures 21-31A through 21-31C**)

TIME LOG	W. PETERSON		
TEMPOROMANDIBULAR JOINT FLEXOR			
DATE	ACTIVITY	TIME HR	CUM. TIME HR

10-2	Met with Anderson, Bui, Hamamoto-
	  - Exchanged phone numbers
	  - Copied Dr. P's proposal
		½	½

10-8	MET - KA MB & AH - discussed
	  proposal - made up questions
	  for Dr. P
	  Discussed planning chart
		2 ¾	3 ¼

10-10	Questions for Dr. P MET at Dr. P's Office in MED SCHOOL	2 ½	5 ¾
10-15	Discussed electrical circuits w/ MB	¼	6

10-17	MET w/ group, checked drawings/
	  went to hobby shop for solenoids
	  Worked on progress report w/MB
	  Typed progress report
		2 ¼	8 ¼

10-20	Talked to Mike Wieggand re: valves Group meeting	2	10 ¼

10-30	Contacted vendors for bellows
	  information
	  Prepared transparencies for DESIGN
	  REVIEW - worked with Hamamoto
		3	13 ¼

11-1	Trip to surplus yard — found plastics	2 ½	15 ¾

FIGURE 21-20   Log.

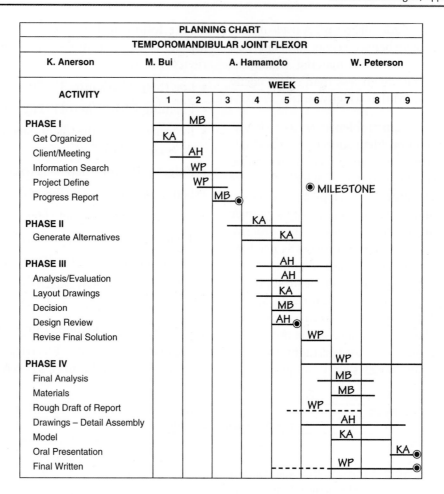

PLANNING CHART

TEMPOROMANDIBULAR JOINT FLEXOR

| K. Anerson | M. Bui | A. Hamamoto | W. Peterson |

ACTIVITY	WEEK								
	1	2	3	4	5	6	7	8	9
PHASE I		MB							
Get Organized	KA								
Client/Meeting		AH							
Information Search		WP							
Project Define		WP							
Progress Report			MB ⊙			⊙ MILESTONE			
PHASE II				KA					
Generate Alternatives					KA				
PHASE III					AH				
Analysis/Evaluation					AH				
Layout Drawings					KA				
Decision					MB				
Design Review					AH ⊙				
Revise Final Solution						WP			
PHASE IV							WP		
Final Analysis							MB		
Materials							MB		
Rough Draft of Report					WP				
Drawings – Detail Assembly							AH		
Model							KA		
Oral Presentation								KA ⊙	
Final Written							WP		⊙

FIGURE 21-21   Planning chart for student group.

PRELIMINARY DESIGN CONSIDERATIONS

1. SIZE   Date 10-10

2. POWER
   –BELT POWER PACK
   –CHARGER

3. CONTROLS
   –FREQUENCY
   –OPENING
   –PATIENT OR DR.?

4. EXPECTED PRODUCTION
   –REUSABLE?

5. COST

6. MATERIALS
   –STAINLESS
   –PLASTIC
   –FDA RESTRICTIONS

7. NOISE LEVEL

8. FORCES REQUIRED
   –OPENING
   –CLOSING

9. SAFETY OVERDRIVE

10. JAW DIMENSIONS

11. ACCURACY

LATERAL VIEW

Date 10-10

a — MAXILLARY ACRYLIC SPLINT
b — MANDIBULAR ACRYLIC SPLINT
c — BUCCAL TUBES
d — UPPER MEMBER
e — LOWER MEMBER
f — PNEUMATIC PISTON
g — UNIVERSAL JOINTS

FIGURE 21-22A   File: Preliminary design considerations.   **FIGURE 21-22B**   File: Information from clients.

FRONTAL VIEW

Date 10-10

FIGURE 21-22C   File: Information from clients.

Modern Design Practices and Standards

Drafters, designers, and engineers entering the technical industry today will find that the adjective "changing" describes current technology better than any other word. A number of these ongoing changes are described briefly in the following paragraphs.

One trend in industries that manufacture devices is to develop prototypes earlier in the design process than in previous years. Two terms describing this trend are *rapid prototyping* and *virtual prototyping*. Stereolithography (SL), mentioned earlier in this chapter, is a rapid prototype builder. SL involves building a solid model layer by layer by an electronic-laser process that uses data from three-view

Date 10-10

> Department of Mechanical Engineering
> (School)
> (City, State, Zip Code)

Dr.
University Hospital
(City, State, Zip Code)

Dear Dr.,

This letter is to inform you of our design team's progress and decisions regarding your CPM (continuous passive motion) device. We have discussed the design criteria set forth in your proposal and also examined the additional criteria we discussed with you at our last meeting. Presently, we are considering the feasibility of a pneumatic device with an external power pack. We are currently evaluating this concept and its ability to meet the following criteria:

1) Weight: less than 4 oz
2) Accurate and adjustable movement
3) Frequency in the range of 2–3 cycles per minute with a maximum range of 1–4 cycles per minute
4) Device must be safe for use by a 10-year-old and older for 24 hours at a time
5) Operation must be simple (few moving parts and adjustable by anyone who can read and understand simple directions)
6) Low cost (less than $500, including labor and materials)
7) Parts must be readily available (most from local sources)

Based on our pending evaluation of the above, we will pick what we feel is the best design, and prepare a report outlining our choice. Included in the report will be a cost analysis, detail and assembly drawings, and an overall evaluation. This report will be presented to you at a date and time decided by our instructor.

If you have any questions, please contact us through our instructor.

> Sincerely,
>
> W. Peterson
>
> K. Anderson
>
> M. Bui
>
> A. Hamamoto

FIGURE 21-23   Progress report.

Date 12-6

Department of Mechanical Engineering
(School)
(City, State, Zip Code)

Dr.
University Hospital
(City, State, Zip Code)

Dear Dr.,
This letter is to inform you that we have completed our design for a CPM device to be used in your research project. The enclosed report details the design, giving complete drawings, a cost analysis, and future design considerations.
The design is now ready for the prototype stage.

We are pleased to have had the opportunity to work on this project for you, and we hope we will be able to work with you more in the future.

Sincerely,

W. Peterson

K. Anderson

M. Bui

A. Hamamoto

FIGURE 21-24   Letter of transmittal.

DESIGN PROJECT: TEMPOROMANDIBULAR JOINT FLEXOR

SUBMITTED TO: DOCTOR

DEPARTMENT OF ORTHODONTICS
UNIVERSITY MEDICAL SCHOOL
(CITY, STATE, ZIP CODE)
Date 12-6

BY: W. PETERSON
     K. ANDERSON
     M. BUI
     A. HAMAMOTO

ABSTRACT

This report contains a description of a continuous passive motion (CPM) device to exercise the temporomandibular joint (TMJ). Briefly, the design incorporates the use of two compressible bellows that are connected by an air hose. While one of the bellows is compressed at a controlled rate, the other bellows expands at a controlled rate. The system is based on user feedback and provides for excellent emergency override, which is essential for safe use. The device costs less than $200.

FIGURE 21-25   Title page and abstract.

drawings. A more recent rapid process uses data from MRI (magnetic resonance imaging) examinations to generate plastic prototypes of human joints that need to be replaced. Doctors use the prototype models for guidance during operations; the data is used also for developing the metallic versions of the joint. This overall process is so accurate that the replacement joint needs little or no modification before installation.

An example of virtual prototyping requires a special head-mounted viewer and a hand-mounted positioning monitor for prototype development of

TABLE OF CONTENTS   PAGES

APPENDIX 1 Delrin Properties

APPENDIX 2 Analysis of Air Flow, Stress Analysis

APPENDIX 3 Vendor Catalogs

FIGURE 21-26   Table of contents.

INTRODUCTION

This project stems from a graduate research project in the department of orthodontics. Studies have shown that the continuous passive motion (CPM) on joints such as knees, elbows, and fingers have proven to be quite successful in alleviating post-operative pain, swelling, and adhesions. It is the Doctor's hopes as well as ours that a similar application of this research on the temporomandibular joint will prove to be equally successful. Our task was to develop a device that would work the jaw up and down to a controlled distance in a slow cyclic manner and at the same time provide a maximum in safety. It was also necessary to use feedback switches to control the inflating and deflating of a bellows that opened and closed the jaw of the patient so that any irregularities or overwhelming resistance to CPM by the individual (sneezing, coughing, yawning, etc.) will immediately be sensed by the system and the necessary override would be provided automatically.

FIGURE 21-27   Introduction.

a backhoe cab. The person wearing the viewer (probably the customer) can visualize the cab of the machine and the positions of the controls before the backhoe is built. Prior to virtual prototyping, expensive wood models of the cabs were constructed.

Additional rapid prototyping methods include:

- *Fusion Deposition Modeling* involves controlled placing of low-melting-point thermostatic filament to build up a model.
- *Laminated Object Manufacturing* builds models by fusing layers of plastic-coated papers cut out by a computer-controlled laser.

- *Selective Laser Sintering* prepares "slices" of an object using fused granules.
- *Solid Ground Curing* uses slices or layers of polymers to build models.

Another trend known as *parametric design* allows drafters and designers to change dimensions in three-dimensional computer models and have all affected dimensions change accordingly. Also, design parameters such as beam sizes, shaft diameters, the number of coils in a spring, and so on would be modified to suit the dimension changes.

COMPONENT NO.	DESCRIPTION
1	UPPER SUPPORT PLATE
2	LOWER SUPPORT PLATE
3	MOUTHPIECE BELLOWS
4	TRAVEL ADJUSTER
5	PRESSURE SWITCH
6	UPPER SPLINT RODS
7	LOWER SPLINT RODS
8	LOWER SPLINT
9	UPPER SPLINT

MOUTHPIECE SIDE VIEW
SCALE 1″ = 1″
DRAWING NO. 1.b.

1. RELAY
2. SOLENOID VALVE
3. MOTOR
4. MIN-E-PITCH CHAIN
5. SLIDER TRACK
6. DRIVE GEAR
7. BELLOWS
8. BELLOWS MOUNT
9. CONNECTING ROD
10. BATTERY
11. PRESSURE SWITCH

FIGURE 21-28   Power pack assembly sketch.

An international trend may affect your future job assignments. The desire of companies—both domestic and international—to sell their products in international markets created a need for international standards to ensure consistent quality and performance. To participate and be internationally competitive, your company must obtain certification based on ISO 9000 of the International Standardization Organization. Your involvement will most likely be in the technical

ALTERNATIVES CONSIDERED AND DISCARDED

1. Pneumatic Cylinders as Actuators
   Advantages:
                ..small in size
   Disadvantages:
                ..limited control of expansion rate
                ..higher working pressures, noise
                ..difficult to find stroke and size in standard products
                ..lubrication may be a problem
                ..allows minimal user resistance

2. Source of Compressed Air: Bottle and Intricate Valving
   Advantages:
                ..power pack simple and light-weight
                ..no batteries
                ..volume of air easily varied
   Disadvantages:
                ..complicated valving
                ..supplying air bottles cumbersome
                ..compressor needed to charge bottles
                ..noise
                ..difficult to get emergency override

3. Motor-Driven Cam and Linkage
   Advantages:
                ..simple to construct
   Disadvantages:
                ..difficult to interrupt
                ..excessive weight
                ..bulky, physically restrictive
                ..little adjustment
                ..little freedom of jaw movement (patients need rotation, translation,
                  and lateral movements)

FIGURE 21-29   Alternatives discarded.

drawing documentation for certification. Each company must document clearly what it plans to do and then keep copious records to show that it has carried out its plan.

Coupled with promoting the requirements of the ISO standards is another trend toward encouraging companies to develop a philosophy of TQM (Total Quality Management). Preparing a company for ISO 9000 certification and developing a TQM philosophy is somewhat similar to the phases in the design process. The planning, preparation, and execution phases in the Goetsch and Davis text involve defining the problems, generating solutions, evaluating them, selecting the best ones, and then implementing them.

An interesting trend in teaching technology has evolved from an industrial practice of thoroughly investigating competitive products by dissecting them and studying the components. Some schools call this procedure *reverse engineering*.

Several trends maturing in past decades involve companies that mass-produce products. Two of the trends, discussed in detail in Chapter 17, are outlined briefly in the next section. Then, two more recent trends are noted.

Design for Manufacturability (DFM) and Design for Assembly (DFA)

Design hints for DFM include goals such as the following:

- Minimize the number of parts.

- Minimize part variations.

- Design parts for more than one function or use.

- Design parts for ease of manufacturing.

- Use fasteners that handle easily and can be readily installed.

- Arrange assembly of parts so they assemble in layer fashion.

- Avoid flexible parts.

These goals are discussed early in the design process in meetings where not only designers, drafters,

FINAL RECOMMENDATIONS

ASSUMPTIONS

It is important to note that any state-of-the-art design is subject to change since so many variables, having no cut-and-dried answers, must be dealt with, and values must sometimes be assumed using good engineering judgement. Some of the variables in this design project include average mouth dimensions, forces to open and close the jaw, allowable weights for the mouthpiece and powerpack, influence of an oversize solenoid valve on battery life, and so on.

PROTOTYPE

The presence of these types of variables strongly implies a need for prototype testing to see if the actual design concept is valid. A second type of testing might be called human factor testing, which involves sizing and comfort factors. Are the dimensions on the mouthpiece satisfactory to accommodate an average-size person? Is the size of the powerpack cumbersome? Are weights unbearable? Are jaw movements restricted?

SOLENOID

It was previously mentioned that the specific solenoid valve was oversized. At this point in time the valve is the smallest one made. Because of its power consumption, which is excessive (7 watts max), the battery must be extraordinarily large. It is hoped that eventually a smaller solenoid valve could be purchased or even designed to be much more economical. This could drastically increase battery life and decrease battery size, ultimately decreasing the size of the powerpack.

CNC LATHE

From a manufacturing viewpoint it is recommended that the prototype have all its plastic parts machined on the CNC lathe, since molds are expensive as used in injection molding. However, if the device is ever mass produced, it is recommended that injection molding be the way to go, since production would be much faster, decreasing labor costs per unit.

FIGURE 21-30   Final recommendations.

APPENDIX NO. 1

DELRIN

The material chosen for the mouthpiece is Delrin. Delrin is a trade name for an acetal homopolymer. It was chosen because it has many desirable characteristics. Some of the advantages of Delrin are:

1) Good chemical resistance, especially to organic compounds

2) Desired regulatory status:
   a) FDA approved for intermittent contact with aqueous or alcohol-based foods
   b) USDA approved for meat and blood

3) Can be machined like soft brass

4) Can be welded or adhesively bonded

5) Forms good mechanical joints

6) Has good structural integrity

7) Tensile strength (at break) 9700 psi

8) Yield strength 9500 – 1200 psi

9) High impact resistance

10) Rockwell hardness 92 – 94

FIGURE 21-31A   Appendix: Material.

FIGURE 21-31B Appendix: Electrical schematic.

and engineers attend but manufacturing personnel participate also. Design hints for DFA include goals such as the following:

- Use "pick-and-place" robots for simple, routine placement of parts.
- Keep the number of parts to a minimum (DFM guideline also).
- Use gravity and vibration for moving parts into assembly areas.
- Avoid parts that may tangle (hang up on each other).
- Use tapers and chamfers to guide parts into position.
- Avoid time-consuming manual operations such as screwdriving, soldering, and riveting.

These goals are also discussed early in the design process in the same engineering and manufacturing meetings previously noted.

Designing for Recyclability (DFR) and Designing for Disassembly (DFD)

Recently, a large manufacturing company that produces automobiles, several companies that produce large appliances, and one large company that produces

QUANTITY	PART NO.	DESCRIPTION	COST
1	NP8-6	Yuasa Battery	$29.55
1	BC-3B	Yuasa Charger	$5.39
1	B11DK 040	Skinner Solenoid Valve	$33.00
1	1615 900-1	Mini Motor (9 rpm) 6v DC	$50.00
1	Rel CY6-5DC-SCO	Relay (Lafayette)	$4.88
2	80511-E	SPST Momentary switch (Lafayette)	$2.94
1	ZC-10B	Fuse Holder (Lafayette)	$1.20
1	3MP10A-10	10-Teeth pinion	$6.34
1	3MP10A-40	40-Teeth gear	$8.59
1	3CCF-50-E	Min-E-Pitch Chain 55 link	$3.24
1	E-62-10K	Unimax Switch	$3.08
1	4PTN-50K	Potentiometer 50 Kohm	$1.98
1	CH3-R10	Charging Jack	$2.99
		Miscellaneous Parts, Tubing, Fittings, Screws	$30.00
		TOTAL	$183.18

FIGURE 21-31C Appendix: Costs.

computers have initiated programs for DFR and DFD. Their goals are to recover as much material as feasible, mainly metals and plastic, to use again. Not only are their programs economically sound but they also acknowledge the fact that landfills are becoming less available, and some basic materials are becoming scarce. Both DFR and DFD programs are considered early in the design process, as are DFM and DFA. Design hints for DFR and DFD include the following:

- Mark plastic molded parts for easy identification.
- Reduce the number of different plastics used.
- Make assembly easy with snap-together parts (also easy to disassemble).
- Avoid permanent attachment (adhesives) of dissimilar materials.
- Use materials that can be recycled or used in other products.

The automobile manufacturer found that recycling engine blocks, alternators, starter motors, water pumps, and differentials cut expenses significantly. A manufacturer of large appliances marks all plastic parts (over 50 grams) with an SAE (Society of Automotive Engineers) identification, which gives the type and recyclability of the parts. The computer manufacturer recycles glass and phosphorus from CRTs, plastics from housings (ground up and made into shingles), and metals from parts (steel, aluminum, and copper).

Rapid Prototyping

Prototyping is the act of building a scale model or a life-size model of a part, object, device, assembly, or any other type of product that has been designed so that it can be tested before putting it into production. Prototyping ensures that the design works and everything about it is correct before beginning production runs. In this way, the company does not make the mistake of turning out a large production run of a faulty product.

Rapid prototyping speeds the process by using three-dimensional computer modeling. The rationale behind rapid prototyping is that it can 1) improve communication among all other personnel who have a role to play concerning the product in question (e.g., purchasing, design, manufacturing, quality, marketing, inventory, shipping, etc.), 2) decrease the amount of time required to develop the product in question, 3) decrease costs by helping all stakeholders avoid mistakes, 4)

minimize design changes; and 5) extend the product's life by showing designers features that should be added and unnecessary features that can be eliminated.

Rapid prototyping is accomplished in four steps: 1) a three-dimensional model is developed, 2) the model is *sliced* into layers, 3) the first layer of the physical model is developed, and the process is completed for each layer until the physical model is completed, and 4) the surface of the model is finished and cleaned.

Parametric Design

Parametric design is a concept enabled by three-dimensional modeling software in which objects can be described using algorithms, and certain parameters can be established so that the model can be altered while maintaining the parameters. For example, assume that no matter what changes are made to a certain part, the part is to have a hole of a specific size in a specific location, and the whole size and location can be established as an unchanging parameter. Other names for parametric design include computational design, generative design, digital design, associative design, and computer-aided design.

ISO 9000

ISO 9000 is a family of standards used to establish and maintain quality management systems that ensure that the needs of customers and other stakeholders are met. The basic standard is ISO 9001. It is supplemented by two additional standards in the ISO 9000 family: 1) ISO 9000—"Quality Management Systems. Fundamentals and Vocabulary"; and 2) ISO 9004—"Managing for the Sustained Success of an Organization. A Quality Management Approach." The ISO 9000 family of standards was developed and is kept up to date by the International Organization for Standardization. The standards are based on broad quality management principles. To be certified as an ISO 9000 registered organization, a company must establish and fully document the following procedures: 1) document control, 2) records control, 3) internal audits, 4) control of nonconforming products/services, 5) corrective action, and 6) preventive action.

The latest version of the ISO 9000 standard also requires organizations to establish and fully document any other procedures that are necessary for its effective operation. Further, the standard also requires that organizations do the following: 1) establish and issue a quality policy, 2) develop and maintain a quality manual, and 3) maintain all records specified in the standard.

Reverse Engineering

Reverse engineering is the process of starting with a finished product and disassembling it to discover the principles that guide its design, structure, function, and operation. It typically involves taking a product apart and analyzing its design so that the design can be replicated or even improved on. Reverse engineering can be used on most any product from mechanical parts to electronic devices to software. Reverse engineering is one way to develop a three-dimensional computer model of a product. This is done by breaking down all of the components of a product, making measurements, noting observations, and using the information gathered to construct a three-dimensional model.

For example, assume that an automobile company decides to bring out a new car model to compete with a specific model manufactured by a competitor. The company might purchase one of its competitor's cars and reverse engineer it to determine what will be necessary to develop a car that will outperform it. This would mean breaking the competitor's car down to the nuts and bolts level and analyzing everything about it (e.g., types of material used, how the car is put together, special features, attributes, etc.). This information would then be used to guide the designers responsible for developing a car that would outperform the competing model in quality, reliability, and cost.

SUMMARY

- The phases and steps in the design process are as follows: recognize and define the problem, generate alternative solutions, evaluate, document, and develop.

- Various techniques for complementing the design process are the creative process, attribute listing, decision tables, and model building.

- The technical drawings used in the design process are the same as those emphasized in Chapter 17: freehand sketches, layout drawings, and detail and assembly drawings.

- Design students should always carry a notebook to record design ideas.

- A well-written specification is written so that it can be tested or measured.

- Sources of design information include vendor catalogs, periodicals, patents, textbooks, and handbooks.

- Modern design practices and standards include DFM, DFA, DFR, DFD, rapid prototyping, parametric design, visual prototyping, ISO 9000 standards, and reverse engineering.

REVIEW QUESTIONS

Mark the following statements as either true or false.

1. A planning chart provides a visual display of when phases and steps are scheduled to begin and end.

2. One reason why a log is kept is because it could be useful on future projects.

3. Some sources of technical material are reference books, standards and codes, and catalogs.

4. Preliminary and design are the two major categories of specifications.

5. Iteration is the process of returning and revising, which occurs continually throughout the design process.

6. It is standard practice to give the first progress report when the project is 50% complete.

7. An attribute listing is a listing of characteristics.

8. Designers should carry a notebook at all times.

9. A design layout drawing is made from a freehand sketch of a design idea drawn to scale.

10. A decision table is used to determine the date the project will be completed and delivered to the client.

Answer the following questions by selecting the best answer.

1. Which of the following is **not** one of the five phases of the design process?
   a. Recognize and define the problem.
   b. Generate alternative solutions—synthesis.
   c. Evaluate solution ideas—analyze.
   d. Evaluate available time.

2. Which of the following is **not** typically a characteristic of a creative person?
   a. An open mind
   b. Thinks positive
   c. Pessimistic
   d. Accepts change

3. Which of the following is **not** an advantage of making a cardboard-glue model?
   a. No use for the model later
   b. Low cost
   c. Promotes new ideas
   d. Discover if the item will work as planned

4. Which of the following is **not** one of the practical steps for giving an oral presentation?
   a. Tell them what you are going to tell them.
   b. Tell them.
   c. Tell them again.
   d. Tell them what you've told them.

5. Which of the following is **not** a component of a brainstorming session?
   a. Record ideas.
   b. Criticize others' ideas.
   c. Have several sessions if necessary.
   d. Meet to critique ideas.

6. Which of the following is **not** true regarding non-routine design projects?
   a. They are usually given to experienced design project personnel.
   b. They usually require more time than routine design projects.
   c. Detailed planning is required.
   d. Fewer engineering trade-offs occur.

7. Which of the following acronyms has the **incorrect** definition?
   a. DFM—Design for Manufacturability
   b. DFA—Design for Assembly
   c. DFR—Design for Readability
   d. DFD—Design for Disassembly

8. Which of the following is **not** a reason for keeping a log of activities?

   a. For future projects
   b. For accounting purposes
   c. To track progress of the project
   d. A place to keep freehand sketches

CHAPTER 21 PROBLEMS

The following problems complement the design process and provide practice in understanding and learning the steps before attacking routine and non-routine design projects.

CAD INSTRUCTIONS

Students who plan to complete the following problems on a CAD system should begin with these general instructions. Before reading the specific instructions for each activity, go through each step in the following planning checklist. The checklist applies to any CAD system and will help ensure the optimum use of your time and resources.

1. Analyze the problem carefully. Decide exactly what you are being asked to do.

2. Determine what resources and references (if any) you will need in order to complete the problem. Collect those references and resources and have them readily available.

3. Decide if any particular standards apply to the project and have those standards available.

4. Determine what types of views will be required and how many of each.

5. Determine what the final plotted scale of the drawing will need to be, and select the appropriate paper size for plotting/printing.

6. Plan your drawing sequence. In what order will you develop the drawing (i.e., lines, features, dimension lines, leaders, dimensions)?

7. Review the various CAD commands you will have to use in order to develop the drawing.

8. Examine your CAD system to ensure that everything is in working order; then, begin the project.

Mechanical **PROBLEM 21-1 (STEP 2)**

Prepare a five-minute oral presentation on one of the following sources of information. Use transparencies for an overhead projector. Suggestions for transparencies (Tx):

Tx 1. Name of source, its location, and name of speaker.
Tx 2. Outline of the talk:
   a. Description of the source and its specific location
   b. Why source is useful to designers
   c. Example of use of source
   d. Conclusion
Tx 3. Description of source and its location.
Tx 4. Why source is useful to designers.
Tx 5. Example of use of source.
Tx 6. Conclusion.

Suggested sources:
1. ASTI (Applied Science and Technology Index)
2. ASTM (American Society for Testing Materials)
3. ANSI (American National Standards Institute)
4. Patents
5. UL (Underwriter's Laboratory)
6. *Machinery's Handbook*
7. Handbook (instructor's or your choice)
8. Vendor's catalog

Mechanical ## PROBLEM 21-2 (STEP 3)

Prepare performance and design specifications for one of the items in the following list. Well-written specifications can be verified by testing. For example, instead of specifying that a bumper jack will be easy to use, lightweight, and made of metal, well-written "specs" would be the following: easy to use by a licensed driver who is not handicapped, a maximum of 20 pounds of force to operate, weighs less than 10 pounds, and is made of SAE 1035 steel. Include a sketch with notes and dimensions.

1. Screwdriver for use in average household
2. Light switch in average household
3. Can opener for kitchen use
4. Knife for kitchen use
5. Pliers for electrician or average household
6. Hammer for carpenter or average household
7. Wood saw
8. Metal saw (hacksaw)

Mechanical ## PROBLEM 21-3 (STEP 7)

Prepare a decision table using criteria and weighting factors for the situations listed. Generate at least three alternatives for each situation.

1. Vehicle for town driving
2. Vehicle for country driving
3. Vehicle for backwoods driving
4. Vacation
5. College to attend
6. Job choice in one city
7. Job choice for different locations (West Coast, East Coast, South, North, etc.)
8. Your choice

Mechanical ## PROBLEM 21-4 (STEP 6)

Prepare a freehand sketch of a relatively uncomplicated design and then prepare a design development layout drawing of the sketched design. Draw the object to scale, but add notes and dimensions in an informal manner. No title block is needed. Complete information is required. Design suggestions:

1. Wall bracket to support object of your choice. Instructor's choice. Use welded assemblies.
2. Minor modification of a component from:
   a. This text (instructor's choice)
   b. Your choice
   c. Instructor's choice

Mechanical ## PROBLEM 21-5 (STEP 6)

Prepare a kinematic development layout drawing of one of the assembly drawings to follow or of an assembly assigned by your instructor (Problems 21-5A through 21-5E).

Suggestions for kinematic analyses:

1. Redraw the figures as single straight lines to represent the linkages.
2. Construct radii about fixed points of rotation.
3. Trace, on a separate sheet of paper, the lines representing the linkages whose motions are to be analyzed.
4. Use the lines as underlays to locate the new positions.

Mechanical ## PROBLEM 21-5A (STEP 6)

The figure below shows the linkages and hydraulic cylinder to operate a jet-blast deflector EF. The hydraulic cylinder AB causes linkage CD to rotate CCW (counterclockwise) about point C, which causes linkage DE to move the deflector EF in a CW (clockwise) direction. When EF is in the maximum CW position, it acts as a shield for personnel standing behind jet aircraft when the jets are warming up for launching from an aircraft carrier. Determine the maximum CW angle that EF rotates and the change in length (stroke) of cylinder AB to obtain the maximum angle. Draw lines EF, DE, CD, and AB when they are in position for the maximum CW rotation of EF.

ALL DIMENSIONS IN MM

Mechanical ## PROBLEM 21-5B (STEP 6)

One application of a four-bar linkage is to control the movement of the hood of an automobile: a pair of four-bar linkages supports the hood. Linkage GH is attached to the hood, as shown in the figure. When the hood closes, GH rotates and translates to a horizontal position. Determine the total CCW (counterclockwise) rotation of the hood from the open to the closed position. Also determine the change in length of the spring SP from the open to closed positions. Draw the lines representing the linkages and the spring in the closed position.

HOOD

ALL DIMENSIONS IN MM

Mechanical   ## PROBLEM 21-5C (STEP 6)

AB = 143 mm
CE = 286 mm
CD = 200 mm
CB = BE = AB
AB ROTATES +– 45°

Mechanical   # PROBLEM 21-5D (STEP 6)

FILM DRIVER

AB = 1.4"
CD = 0.9"

CD ROTATES 360° CW

Mechanical   # PROBLEM 21-5E (STEP 6)

AB = 4.9"
CD = 1.8"
BD = 2.2"

CD ROTATES +– 45°

MEMBERS SHOWN AS STRAIGHT LINES

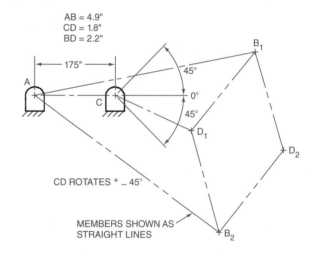

Mechanical   # PROBLEM 21-6 (STEPS 1–11)

The 19-step Kraft Mill Engineering Roadmap summarizes a design process from a company for improving or changing the company's plant facilities. A recent graduate from an engineering program sent the summary to one of her professors to show what she was doing during the first few months on the job.

Kraft Mill Engineering Roadmap

1. PROJECT SCOPE: Define the problem to be resolved, the changes that are required, and the objective to be achieved.

2. SCOPE REVIEW MEETING: Verify that both Maintenance and Operations approve of the project scope.

3. PRELIMINARY DESIGN: Select several options for consideration. Prepare rough cost estimates, sketch general layouts/schematics, and estimate R01/ payback for each option.

4. PRELIMINARY DESIGN REVIEW MEETING: Select one or two options to pursue. Obtain Management's approval to proceed. Determine if project is to be capitalized or expensed.

5. SUBMIT AR OR WO: Write a cost estimate. Use class $\pm 10\%$, $\pm 20\%$, or $\pm 40\%$, as required. Write and submit an AR (Authorization Request) or a WO (work order).

6. RECEIVE APPROVED AR OR WO: No money may be spent until approval is received.

7. DETAILED DESIGN: Make detailed layout drawings and flow schematics. Do energy balance. Develop schedule; make time-bar diagram for larger projects. Determine equipment requirements.

8. PROJECT DESIGN REVIEW MEETING: Obtain "buy off" from Operations and Maintenance of the design and layout. Verify schedule, shutdown periods, etc. Decide whether contractors or the mill will provide labor.

9. BIDS: Prepare bid package for material and for labor. Draw up a bidders list. Send to Purchasing Dept. and go out for bids.

10. REVIEW BIDS: Select the best one or two vendors for each portion of the project. Run background check on contractors. Check references of equipment manufacturers.

11. VENDOR APPROVAL MEETING: Obtain approval to purchase equipment and/or hire contractors.

12. PURCHASE ORDERS: Award bids. Write purchase orders (POs).

13. EQUIPMENT TRACKING: Track equipment as it comes into mill. Store in safe place until installation.

14. CONSTRUCTION: Install equipment.

15. PRE-STARTUP: Obtain and file vendor drawings, guarantees, and instruction files. Set up spare parts in the storeroom. Write an operating manual. Develop a startup plan.

16. PRE-STARTUP: Review startup plan. Set a startup date.

17. STARTUP: Put system into operation.

18. PROJECT REVIEW: Either hold meeting or circulate memo. Restate original project scope. Review the project. Determine that the desired objectives were (or were not) met. Discuss any problems. State final project cost.

19. PROJECT CLOSEOUT: Do necessary paperwork to close out project. Sign the WO as complete or fill out an AR Capital Job Closure Form.

Note: Some of these events overlap. Depending on the size and type of job, some events may be deleted, or some more may need to be added.

Study the 19 steps and do all or any one of the following as assigned by your instructor:

1. Prepare a planning chart assuming that a six-month period is required for a particular plant modification and that you are doing the job.

2. Relate the 19 steps to the design process discussion in this chapter.

3. List and comment on the different people most likely contacted during the 19 steps.

4. Comment on the apparent meaning of some of the terms and acronyms: for example, scope, ROI, capitalized, expensed, AR, WO, $\pm 10\%$ class, $\pm 20\%$ class, $\pm 40\%$ class, time-bar diagram, flow schematic, buy-off, bid package, background check, PO, tracking, and AR Capital Job Closure Form.

Mechanical **PROBLEM 21-7 (STEP 6)**

1. Contact local vendors of material to obtain current prices of common ones, such as steel shapes (angles, plates, bars, channels, and beams). Note that prices are usually given in dollars per 100 pounds and that price reductions are given for purchasing large quantities (e.g., a full railroad car).

2. Contact vendors of finished components—such as motors, switches, controls, bearings, chain, and similar products—to obtain costs to several categories of buyers.
   a. Private individuals usually pay the full retail price.
   b. Companies that include the finished components in a kit may get a 15% to 30% discount from the vendors because the companies sell the kits to customers.
   c. Companies that include the finished components as an integral part of an assembled machine are called original equipment manufacturers (OEMs). They may receive even larger discounts.

Mechanical **PROBLEM 21-8 (STEP 6)**

Frequently, professionals are asked to make quick estimates during conversations over lunch or while phoning a client. This type of quick thinking is also an excellent approach to use for checking final results, whether it is homework in an engineering science course or a system in industry. Ask questions such as does this look right, does this look reasonable, or does the output seem reasonable for the input?

For practicing "back-of-the-envelope" thinking and calculations, do the following in a time period assigned by your instructor and record the assumptions made:

1. Estimate the quantities of food and the cost of lunch for all of the people in your classroom.

2. Estimate the number of gallons of water used by your school in one day. (Perhaps maintenance personnel can verify the accuracy of the estimate.)

3. Estimate the cost of transporting a college football team 1,000 miles and expenses for a three-day road trip.

4. Estimate the selling price of a new automobile, truck, van, sports car, or motorcycle based on the price per pound of existing vehicles.

5. Estimate selling prices of other products using the approach in #4.

Mechanical ## PROBLEM 21-9 (STEP 12)

List the manufacturing processes required to make the problem parts shown in the following figures.

Mechanical ## PROBLEM 21-9A

Mechanical ## PROBLEM 21-9B

Mechanical ## PROBLEM 21-9C

Mechanical ## PROBLEM 21-9D

List the manufacturing processes required to make the part or parts assigned by your instructor.

Mechanical ## PROBLEM 21-10 (STEP 5)

From memory, make a freehand sketch of an object shown to you for a limited time by your instructor. Then, do a second freehand sketch of the same object while you have a much longer time to view and examine it.

This practice of viewing an object knowing you have to sketch it from memory will train you to see and remember more than you would at a casual glance.

Mechanical ## PROBLEM 21-11 (STEP 5)

List the obvious attributes of one of the following items. Include a freehand dimensioned sketch of the object. Then, list all the not-so-obvious uses of the object. Be creative.
1. Kitchen table
2. Clothespin
3. Bookshelves
   a. Wood
   b. Metal
4. Cardboard milk carton
5. Plastic container of 35 mm film

6. Screwdriver
7. Nail

Routine Design Problems (RDPs)

The routine design problems listed here are intended to give you experience in the design process while doing problems similar to those assigned to new employees in technical industries. Use the steps, as needed, in the design process to produce the following short report (printed or typed). The steps most needed are 2, 3, 5, 6, and 10.
1. A statement of the problem in your own words
2. A list of information sources used
3. Freehand sketch(es) or preliminary ideas of what needs to be done
4. Design development layout drawings of the ideas in the freehand sketches
5. Kinematic development layout drawings, if required
6. Complete detail and assembly drawings with parts lists
7. Cost analysis
8. Conclusions and recommendations
9. Appendices
   a. Machine copies of 3, 4, 5, and 6. (These copies will show how well pencil drawings reproduce as plotted or printed copies if CAD is used.)
   b. Vendor catalog information
   c. Any additional materials: work order, photographs, etc.
10. A title page and a folder for the report

PROBLEMS

PROBLEM 21-12 (RDP)

Refer to the torque wrench assembly in the accompanying figure and do the following:

1. Make the wrench just two times as large as shown for all dimensions.
2. Select steel for the handles and the fasteners.
3. Determine the bending stress in the necked-down portion of the handle. Specify as large a radius as practical at the necked-down location to avoid stress concentration. The bending stress will depend on the assumed force F and the length L.

F = FORCE BY OPERATOR IN POUNDS
(NEED TO ASSUME)
L = ASSUME DISTANCE FROM F TO d IN INCHES
d = DIAMETER OF NECKED–DOWN PORTION IN INCHES

STRESS DUE TO MOMENT F X L AT d IS

$$\text{STRESS} = \frac{(F \times L)\,(d/2)}{(d^4)/64} \text{ IN POUNDS PER SQUARE INCH}$$

PROBLEM 21-13 (RDP)

Refer to the clamp shown in the accompanying figure and assume that parts No. 1 and No. 2 are to be aluminum and parts No. 3 and No. 4 are steel. When threads in aluminum may be too weak because of the softer metal, a threaded-steel insert is pressed into the aluminum. Use an insert with appropriate threads inside and a nominal outside diameter of 15 mm but dimensioned for an FN-2 medium-drive fit.

Increase the 18 mm dimension for parts No. 1 and No. 2 to 25 mm.

Mechanical ## PROBLEM 21-14 (RDP)

The clamp shown in the accompanying figure needs to be modified to clamp rectangular assemblies 50 × 300 mm between the two threaded members instead of between the tapered extensions. The thread size should be increased 100% in diameter. Make any other changes that seem appropriate. Select materials. (*Note:* Part No. 4 must function like part No. 3, and the tapered portions of parts No. 1 and No. 2 are no longer needed.)

Mechanical ## PROBLEM 21-15 (RDP)

Modify the screw jack shown in the accompanying figure to have an eight-inch vertical movement instead of the apparent four-inch vertical one.

Mechanical   ## PROBLEM 21-16 (RDP)

Modify the roller assembly shown as follows:

1. Make the shaft, part No. 3, 1″ diameter × 3″ long.
2. Change the roller, part No. 2, to have a bronze bushing 1″ inside diameter, 1.25″ outside diameter × 1.6888″ long; FN-2 Fit for the 1.25″ diameter; and RC-3 for the 1″ inside diameter.
3. Change the supports, part No. 1, to have set screws to keep the shaft, No. 3, from turning.

Mechanical   ## PROBLEM 21-17 (RDP)

Change the vise shown in the accompanying figure to be 50% longer and 75% wider. In a kinematic layout drawing, show the extreme open and closed positions.

Mechanical ## PROBLEM 22-18 (RDP)

Develop the concept of the beverage can crusher shown in the accompanying figure. Use standard materials and parts as indicated. Include a kinematic layout drawing showing extreme positions of the crusher head.

BEVERAGE CAN CRUSHER
ALL ALUMINUM EXCEPT FOR FASTENERS

0   5   10

1 = 5

$\varnothing\frac{3}{4}$ TUBE SPACER X 4" LG

1" ⊡ TUBE 0.125 WALL

USE $\frac{1}{16}$" TUBE SPACER WASHERS TYP.

USE $\frac{1}{16}$" WASHERS TYP

1" ⊡ BAR

1" ⊡ TUBE 0.125 WALL

CRUSHER BAR $1\frac{1}{4}$ X $1\frac{1}{8}$ X 2" LG

BAR $1\frac{1}{4}$ X $\frac{1}{4}$

BAR $1\frac{1}{4}$ X $\frac{1}{4}$

SPACER $\varnothing\frac{3}{4}$ TUBE X 0.083" WALL X $1\frac{1}{8}$" LG

$\frac{1}{4}$" ℄

A   A

SEE DETAIL A–1   ∠2 X 2 X $\frac{1}{4}$

CSK FOR FLATHEAD MACHINE SCREW 4 PLS

$\frac{1}{2}$"SPACER

|←1"→|

1 = 2

DETAIL A–1

$\frac{1}{4}$" ℄ $\varnothing\frac{3}{8}$ BOLT

1" ⊡ X 0.125" WALL

SPACER $\varnothing\frac{3}{4}$ TUBE X 0.083" WALL X $1\frac{1}{8}$" LG

1 = 2

A–A

PROBLEMS

 PROBLEM 21-19 (RDP)

Use press-fit drill bushings with heads to make a leaf-type jig for the hole pattern shown in the part to be drilled.

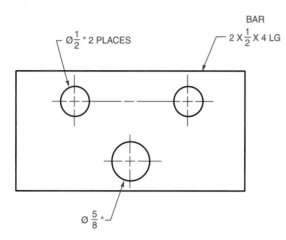

Non-Routine Design Problems

These non-routine problems are designed to give you experience in the overall design process, but mostly in Phases I–IV, Steps 1–11. When it is feasible, you are encouraged to build prototypes and discuss Steps 12 and 13 in Phase V.

Your instructor will most likely form student design teams to resolve the non-routine design problems, Problems 21-20 through 21-40, and the "real" problems from the suggested sources. Effective communication between members seems to be the major contributing factor to a team's success. What do you think?

The following groups of non-routine problems fall into three time-related categories: short, medium, and long projects. Suggestions for hours to allot for the three categories are included. Submit all reports in a folder with an appropriate title.

Short Non-Routine Design Problems

Suggestions: Allot 10–15 hours for one person to complete them; allot 5–7 hours for a team of two, etc.

Develop the problem only through Phase I of the design process. That is, prepare a plan of action for one or more students (Step 1 and Step 2), define the problem (Step 3), and prepare a progress report to a real or an assumed client (Step 4). Start a log and design file as assigned.

 PROBLEM 21-20

Develop a toy for a ( )-year-old child confined to a bed or a wheelchair. *Note:* Select or be assigned the child's age.

 PROBLEM 21-21

Design a small-capacity trailer that can be pulled by an automobile and that can be dismantled and stored in a relatively small space.

 PROBLEM 21-22

Design a means of mounting posts on automobiles or pick-up trucks for games such as badminton or volleyball at remote picnic sites.

 PROBLEM 21-23

Design a novel or practical use for discarded tires, cars, cartons, cans, or bottles.

 PROBLEM 21-24

Design a portable can crusher for aluminum or steel for use in areas where numerous cans are deposited or tossed: for example, on roadsides.

 PROBLEM 21-25

Design a teaching aid for one of the courses in your school, in a local high school, or in a local elementary school.

 PROBLEM 21-26

Design a portable solar cooker.

Medium Non-Routine Design Problems

Suggestions: Allot 20–30 hours for one person to complete them; allot 10–15 hours for a team of two, etc. Develop the following problems through Steps 1–8 of the design process. Present your progress report to your classmates, with all members taking part in the presentation. Start a log, start a design file, and prepare a planning chart.

 PROBLEM 21-27

Do any one of the short non-routine design problems listed, but pursue the problem through Step 8.

 PROBLEM 21-28

Design a device to assist an individual who has no legs but full use of two arms to get into a wheelchair from the floor and back to the floor.

PROBLEM 21-29

Design a device to lift a 150-pound handicapped person out of a swimming pool.

PROBLEM 21-30

Design a study or work area for an individual with little income who is confined to a wheelchair.

PROBLEM 21-31

Design a means to use "pedal power" from a typical one-speed bicycle for performing a useful function such as sawing, sewing, mixing, pumping, etc.

Long Non-Routine Design Problems

Suggestions: Allot 40–60 hours for one person to complete them; allot 20–30 hours for a team of two, etc. Develop the following problems through Phase IV of the design process. Note that Phase IV usually requires at least half the total time allotted to a project whenever complete detail and assembly drawings are needed.

 PROBLEM 21-32

Do any one of the short or medium non-routine design problems previously listed, but develop the problem through Phase IV of the design process.

PROBLEM 21-33

Design a means to rid aluminum beverage cans of moisture. The cans arrive in truckloads at a recycling center, and the driver wants to be paid a realistic price, based on the actual metal turned in, within half an hour.

PROBLEM 21-34

Design an inexpensive fruit dryer.

PROBLEM 21-35

Design a powered vehicle to move one or more employees throughout a huge manufacturing plant or warehouse. The vehicle should be cost-effective; that is, enough time (money) should be saved to justify its use.

 PROBLEM 21-36

Modify a tricycle to accommodate a child with minimal leg use.

PROBLEM 21-37

Design a safe vehicle for coasting on a track, and the track, for children weighing up to 100 pounds.

PROBLEM 21-38

Design a pool cue for a one-armed person.

PROBLEM 21-39

Design a manually powered device for lifting 25 kilograms and heavier weights, vertically, at least one meter.

PROBLEM 21-40

Design a device that utilizes a typical automobile jack to perform another function(s) such as pressing, clamping, splitting wood, operating a simple "jaws of life" mechanism, etc.

Other Sources

The following lists describe a number of general sources for finding real, non-routine design problems. These problems are usually arranged for by instructors. After bringing together the source of the real problem (the client) with the design team (or an individual student), the instructor can step back and become an advisor-coach. Real, non-routine design problems are usually longer, often due to the extra time spent visiting and corresponding with clients.

1. School classroom demonstrators of basic principles of chemistry, physics, statics, dynamics, mechanics of materials, engineering sciences, and manufacturing processes.
2. Handicapped
   a. Designs to help individuals confined to a wheelchair to:
      (i) Reach objects and prepare meals in a kitchen
      (ii) Obtain exercise
      (iii) Obtain access to heights higher than curbs
      (iv) Have access to classroom work surfaces
      (v) Have work surfaces on the wheelchair
   b. Designs to continually flex injured elbows and knee joints to accelerate healing.
   c. Designs to aid communications for individuals who have speech difficulties.
3. Hospitals
   a. Games or devices for children who are confined to a bed and have limited use of their arms.
   b. Devices to help injured individuals perform simple functions such as dress, read, play games, and eat.

4. Civic organizations
   a. Designs for YWCA or YMCA
   b. Designs to assist scouting programs such as neighborhood improvement projects
5. Local industry
   a. Projects for the industry
   b. Projects for the industry's client
6. Private individuals (designs for)
   a. Home improvements
   b. Hobbies
   c. Inventions
   d. Invalid care
7. City (designs for)
   a. Better parking meters
   b. Playgrounds

c. Park facilities
   d. Signs
   e. Junior Achievement
8. State (designs for)
   a. Signs
   b. Parks
   c. Recreational facilities
9. National (designs for)
   a. Design contests (Alcoa, Lincoln Arc Welding, engineering societies . . .)
   b. Environment
10. International (designs for)
    a. Third world countries
    b. Peace Corps volunteers

appendix A
tables

Contents

INCHES TO MILLIMETERS

in.	mm	in.	mm	in.	mm	in.	mm
1	25.4	26	660.4	51	1295.4	76	1930.4
2	50.8	27	685.8	52	1320.8	77	1955.8
3	76.2	28	711.2	53	1346.2	78	1981.2
4	101.6	29	736.6	54	1371.6	79	2006.6
5	127.0	30	762.0	55	1397.0	80	2032.0
6	152.4	31	787.4	56	1422.4	81	2057.4
7	177.8	32	812.8	57	1447.8	82	2082.8
8	203.2	33	838.2	58	1473.2	83	2108.2
9	228.6	34	863.6	59	1498.6	84	2133.6
10	254.0	35	889.0	60	1524.0	85	2159.0
11	279.4	36	914.4	61	1549.4	86	2184.4
12	304.8	37	939.8	62	1574.8	87	2209.8
13	330.2	38	965.2	63	1600.2	88	2235.2
14	355.6	39	990.6	64	1625.6	89	2260.6
15	381.0	40	1016.0	65	1651.0	90	2286.0
16	406.4	41	1041.4	66	1676.4	91	2311.4
17	431.8	42	1066.8	67	1701.8	92	2336.8
18	457.2	43	1092.2	68	1727.2	93	2362.2
19	482.6	44	1117.6	69	1752.6	94	2387.6
20	508.0	45	1143.0	70	1778.0	95	2413.0
21	533.4	46	1168.4	71	1803.4	96	2438.4
22	558.8	47	1193.8	72	1828.8	97	2463.8
23	584.2	48	1219.2	73	1854.2	98	2489.2
24	609.6	49	1244.6	74	1879.6	99	2514.6
25	635.0	50	1270.0	75	1905.0	100	2540.0

The above table is exact on the basis: 1 in. = 25.4 mm

MILLIMETERS TO INCHES

mm	in.	mm	in.	mm	in.	mm	in.
1	0.039370	26	1.023622	51	2.007874	76	2.992126
2	0.078740	27	1.062992	52	2.047244	77	3.031496
3	0.118110	28	1.102362	53	2.086614	78	3.070866
4	0.157480	29	1.141732	54	2.125984	79	3.110236
5	0.196850	30	1.181102	55	2.165354	80	3.149606
6	0.236220	31	1.220472	56	2.204724	81	3.188976
7	0.275591	32	1.259843	57	2.244094	82	3.228346
8	0.314961	33	1.299213	58	2.283465	83	3.267717
9	0.354331	34	1.338583	59	2.322835	84	3.307087
10	0.393701	35	1.377953	60	2.362205	85	3.364457
11	0.433071	36	1.417323	61	2.401575	86	3.385827
12	0.472441	37	1.456693	62	2.440945	87	3.425197
13	0.511811	38	1.496063	63	2.480315	88	3.464567
14	0.551181	39	1.535433	64	2.519685	89	3.503937
15	0.590551	40	1.574803	65	2.559055	90	3.543307
16	0.629921	41	1.614173	66	2.598425	91	3.582677
17	0.669291	42	1.653543	67	2.637795	92	3.622047
18	0.708661	43	1.692913	68	2.677165	93	3.661417
19	0.748031	44	1.732283	69	2.716535	94	3.700787
20	0.787402	45	1.771654	70	2.755906	95	3.740157
21	0.826772	46	1.811024	71	2.795276	96	3.779528
22	0.866142	47	1.850394	72	2.834646	97	3.818898
23	0.905512	48	1.889764	73	2.874016	98	3.858268
24	0.944882	49	1.929134	74	2.913386	99	3.897638
25	0.984252	50	1.968504	75	2.952756	100	3.937008

The above table is approximate on the basis: 1 in. = 25.4 mm, 1/25.4 = 0.039370078740+

TABLE 1

INCH/METRIC—EQUIVALENTS

Fraction	Decimal Equivalent Customary (in.)	Metric (mm)	Fraction	Decimal Equivalent Customary (in.)	Metric (mm)
1/64	.015625	0.3969	33/64	.515625	13.0969
1/32	.03125	0.7938	17/32	.53125	13.4938
3/64	.046875	1.1906	35/64	.546875	13.8906
1/16	.0625	1.5875	9/16	.5625	14.2875
5/64	.078125	1.9844	37/64	.578125	14.6844
3/32	.09375	2.3813	19/32	.59375	15.0813
7/64	.109375	2.7781	39/64	.609375	15.4781
1/8	.1250	3.1750	5/8	.6250	15.8750
9/64	.140625	3.5719	41/64	.640625	16.2719
5/32	.15625	3.9688	21/32	.65625	16.6688
11/64	.171875	4.3656	43/64	.671875	17.0656
3/16	.1875	4.7625	11/16	.6875	17.4625
13/64	.203125	5.1594	45/64	.703125	17.8594
7/32	.21875	5.5563	23/32	.71875	18.2563
15/64	.234375	5.9531	47/64	.734375	18.6531
1/4	.250	6.3500	3/4	.750	19.0500
17/64	.265625	6.7469	49/64	.765625	19.4469
9/32	.28125	7.1438	25/32	.78125	19.8438
19/64	.296875	7.5406	51/64	.796875	20.2406
5/16	.3125	7.9375	13/16	.8125	20.6375
21/64	.328125	8.3384	53/64	.828125	21.0344
11/32	.34375	8.7313	27/32	.84375	21.4313
23/64	.359375	9.1281	55/64	.859375	21.8281
3/8	.3750	9.5250	7/8	.8750	22.2250
25/64	.390625	9.9219	57/64	.890625	22.6219
13/32	.40625	10.3188	29/32	.90625	23.0188
27/64	.421875	10.7156	59/64	.921875	23.4156
7/16	.4375	11.1125	15/16	.9375	23.8125
29/64	.453125	11.5094	61/64	.953125	24.2094
15/32	.46875	11.9063	31/32	.96875	24.6063
31/64	.484375	12.3031	63/64	.984375	25.0031
1/2	.500	12.7000	1	1.000	25.4000

TABLE 2 From *Drafting for Trades and Industry—Basic Skills,* Nelson. Delmar Publishers Inc.

METRIC EQUIVALENTS

LENGTH

U.S. to Metric   *Metric to U.S.*

1 inch = 2.540 centimeters

1 foot = .305 meter

1 yard = .914 meter

1 mile = 1.609 kilometers

1 millimeter = .039 inch

1 centimeter = .394 inch

1 meter = 3.281 feet or 1.094 yards

1 kilometer = .621 mile

AREA

1 inch^2 = 6.451 centimeter^2

1 foot^2 = .093 meter^2

1 yard^2 = .836 meter^2

1 acre^2 = 4,046.873 meter^2

1 millimeter^2 = .00155 inch^2

1 centimeter^2 = .155 inch^2

1 meter^2 = 10.764 foot^2 or 1.196 yard^2

1 kilometer^2 = .386 mile^2 or 247.04 acre^2

VOLUME

1 inch^3 = 16.387 centimeter^3

1 foot^3 = .028 meter^3

1 yard^3 = .764 meter^3

1 quart = .946 liter

1 gallon = .003785 meter^3

1 centimeter^3 = 0.61 inch^3

1 meter^3 = 35.314 foot^3 or 1.308 yard^3

1 liter = .2642 gallons

1 liter = 1.057 quarts

1 meter^3 = 264.02 gallons

WEIGHT

1 ounce = 28.349 grams

1 pound = .454 kilogram

1 ton = .907 metric ton

1 gram = .035 ounce

1 kilogram = 2.205 pounds

1 metric ton = 1.102 tons

VELOCITY

1 foot/second = .305 meter/second

1 mile/hour = .447 meter/second

1 meter/second = 3.281 feet/second

1 kilometer/hour = .621 mile/second

ACCELERATION

1 inch/second^2 = .0254 meter/second^2

1 foot/second^2 = .305 meter/second^2

1 meter/second^2 = 3.278 feet/second^2

FORCE

N (newton) = basic unit of force, kg-m/s^2. A mass of one kilogram (1 kg) exerts a gravitational force of 9.8 N (theoretically 9.80665 N) at mean sea level.

TABLE 3

MULTIPLIERS FOR DRAFTERS

Multiply	By	To Obtain	Multiply	By	To Obtain
Acres	43,560	Square feet	Degrees/sec.	0.002778	Revolutions/sec.
Acres	4047	Square meters	Fathoms	6	Feet
Acres	1.562×10^{-3}	Square miles	Feet	30.48	Centimeters
Acres	4840	Square yards	Feet	12	Inches
Acre-feet	43,560	Cubic feet	Feet	0.3048	Meters
Atmospheres	76.0	Cms. of mercury	Foot-pounds	1.286×10^{-3}	British Thermal Units
Atmospheres	29.92	Inches of mercury	Foot-pounds	5.050×10^{-7}	Horsepower-hrs.
Atmospheres	33.90	Feet of water	Foot-pounds	3.241×10^{-4}	Kilogram-calories
Atmospheres	10,333	Kgs./sq. meter	Foot-pounds	0.1383	Kilogram-meters
Atmospheres	14.70	Lbs./sq. inch	Foot-pounds	3.766×10^{-7}	Kilowatt-hrs.
Atmospheres	1.058	Tons/sq. ft.	Foot-pounds/min.	1.286×10^{-3}	BTU/min.
Board feet	144 sq. in. × 1 in.	Cubic inches	Foot-pounds/min.	0.01667	Foot-pounds/sec.
British Thermal Units	0.2520	Kilogram-calories	Foot-pounds/min.	3.030×10^{-5}	Horsepower
British Thermal Units	777.5	Foot-lbs.	Foot-pounds/min.	3.241×10^{-4}	Kg-calories/min.
British Thermal Units	3.927×10^{-4}	Horsepower-hrs.	Foot-pounds/min.	2.260×10^{-5}	Kilowatts
British Thermal Units	107.5	Kilogram-meters	Foot-pounds/sec.	7.717×10^{-2}	BTU/min.
British Thermal Units	2.928×10^{-4}	Kilowatt-hrs.	Foot-pounds/sec.	1.818×10^{-3}	Horsepower
BTU/min.	12.96	Foot-lbs./sec.	Foot-pounds/sec.	1.945×10^{-2}	Kg-calories/min.
BTU/min.	0.02356	Horsepower	Foot-pounds/sec.	1.356×10^{-3}	Kilowatts
BTU/min.	0.01757	Kilowatts	Gallons	3785	Cubic centimeters
BTU/min.	17.57	Watts	Gallons	0.1337	Cubic feet
Cubic centimeters	3.531×10^{-5}	Cubic feet	Gallons	231	Cubic inches
Cubic centimeters	6.102×10^{-2}	Cubic inches	Gallons	3.785×10^{-3}	Cubic meters
Cubic centimeters	10^{-6}	Cubic meters	Gallons	4.951×10^{-3}	Cubic yards
Cubic centimeters	1.308×10^{-6}	Cubic yards	Gallons	3.785	Liters
Cubic centimeters	2.642×10^{-4}	Gallons	Gallons	8	Pints (liq.)
Cubic centimeters	10^{-3}	Liters	Gallons	4	Quarts (liq.)
Cubic centimeters	2.113×10^{-3}	Pints (liq.)	Gallons-Imperial	1.20095	U.S. gallons
Cubic centimeters	1.057×10^{-3}	Quarts (liq.)	Gallons-U.S.	0.83267	Imperial gallons
Cubic feet	2.832×10^{4}	Cubic cm	Gallons water	8.3453	Pounds of water
Cubic feet	1728	Cubic inches	Horsepower	42.44	BTU/min.
Cubic feet	0.02832	Cubic meters	Horsepower	33,000	Foot-lbs./min.
Cubic feet	0.03704	Cubic yards	Horsepower	550	Foot-lbs./sec.
Cubic feet	7.48052	Gallons	Horsepower	1.014	Horsepower (metric)
Cubic feet	28.32	Liters	Horsepower	10.70	Kg-calories/min.
Cubic feet	59.84	Pints (liq.)	Horsepower	0.7457	Kilowatts
Cubic feet	29.92	Quarts (liq.)	Horsepower	745.7	Watts
Cubic feet/min.	472.0	Cubic cm/sec.	Horsepower-hours	2547	BTU
Cubic feet/min.	0.1247	Gallons/sec.	Horsepower-hours	1.98×10^{6}	Foot-lbs.
Cubic feet/min.	0.4720	Liters/sec.	Horsepower-hours	641.7	Kilogram-calories
Cubic feet/min.	62.43	Pounds of water/min.	Horsepower-hours	2.737×10^{5}	Kilogram-meters
Cubic feet/sec.	0.646317	Millions gals./day	Horsepower-hours	0.7457	Kilowatt-hours
Cubic feet/sec.	448.831	Gallons/min.	Kilometers	10^{5}	Centimeters
Cubic inches	16.39	Cubic centimeters	Kilometers	3281	Feet
Cubic inches	5.787×10^{-4}	Cubic feet	Kilometers	10^{3}	Meters
Cubic inches	1.639×10^{-5}	Cubic meters	Kilometers	0.6214	Miles
Cubic inches	2.143×10^{-5}	Cubic yards	Kilometers	1094	Yards
Cubic inches	4.329×10^{-3}	Gallons	Kilowatts	56.92	BTU/min.
Cubic inches	1.639×10^{-2}	Liters	Kilowatts	4.425×10^{4}	Foot-lbs./min.
Cubic inches	0.03463	Pints (liq.)	Kilowatts	737.6	Foot-lbs./sec.
Cubic inches	0.01732	Quarts (liq.)	Kilowatts	1.341	Horsepower
Cubic meters	10^{6}	Cubic centimeters	Kilowatts	14.34	Kg.-calories/min.
Cubic meters	35.31	Cubic feet	Kilowatts	10^{3}	Watts
Cubic meters	61.023	Cubic inches	Kilowatt-hours	3415	BTU
Cubic meters	1.308	Cubic yards	Kilowatt-hours	2.655×10^{6}	Foot-lbs.
Cubic meters	264.2	Gallons	Kilowatt-hours	1.341	Horsepower-hrs.
Cubic meters	10^{3}	Liters	Kilowatt-hours	860.5	Kilogram-calories
Cubic meters	2113	Pints (liq.)	Kilowatt-hours	3.671×10^{5}	Kilogram-meters
Cubic meters	1057	Quarts (liq.)	Lumber Width (in.) ×		
Degrees (angle)	60	Minutes	$\dfrac{\text{Thickness (in.)}}{12}$	Length (ft.)	Board feet
Degrees (angle)	0.01745	Radians			
Degrees (angle)	3600	Seconds	Meters	100	Centimeters
Degrees/sec.	0.01745	Radians/sec.	Meters	3.281	Feet
Degrees/sec.	0.1667	Revolutions/min.	Meters	39.37	Inches

TABLE 4

MULTIPLIERS FOR DRAFTERS (Concluded)

Multiply	By	To Obtain	Multiply	By	To Obtain
Meters	10^{-3}	Kilometers	Pounds (troy)	373.24177	Grams
Meters	10^3	Millimeters	Pounds (troy)	0.822857	Pounds (avoir.)
Meters	1.094	Yards	Pounds (troy)	13.1657	Ounces (avoir.)
Meters/min.	1.667	Centimeters/sec.	Pounds (troy)	3.6735×10^{-4}	Tons (long)
Meters/min.	3.281	Feet/min.	Pounds (troy)	4.1143×10^{-4}	Tons (short)
Meters/min.	0.05468	Feet/sec.	Pounds (troy)	3.7324×10^{-4}	Tons (metric)
Meters/min.	0.06	Kilometers/hr.	Quadrants (angle)	90	Degrees
Meters/min.	0.03728	Miles/hr.	Quadrants (angle)	5400	Minutes
Meters/sec.	196.8	Feet/min.	Quadrants (angle)	1.571	Radians
Meters/sec.	3.281	Feet/sec.	Radians	57.30	Degrees
Meters/sec.	3.6	Kilometers/hr.	Radians	3438	Minutes
Meters/sec.	0.06	Kilometers/min.	Radians	0.637	Quadrants
Meters/sec.	2.237	Miles/hr.	Radians/sec.	57.30	Degrees/sec.
Meters/sec.	0.03728	Miles/min.	Radians/sec.	0.1592	Revolutions/sec.
Microns	10^{-6}	Meters	Radians/sec.	9.549	Revolutions/min.
Miles	5280	Feet	Radians/sec./sec.	573.0	Revs./min./min.
Miles	1.609	Kilometers	Radians/sec./sec.	0.1592	Revs./sec./sec.
Miles	1760	Yards	Reams	500	Sheets
Miles/hr.	1.609	Kilometers/hr.	Revolutions	360	Degrees
Miles/hr.	0.8684	Knots	Revolutions	4	Quadrants
Minutes (angle)	2.909×10^{-4}	Radians	Revolutions	6.283	Radians
Ounces	16	Drams	Revolutions/min.	6	Degrees/sec.
Ounces	437.5	Grains	Square yards	2.066×10^{-4}	Acres
Ounces	0.0625	Pounds	Square yards	9	Square feet
Ounces	28.349527	Grams	Square yards	0.8361	Square meters
Ounces	0.9115	Ounces (troy)	Square yards	3.228×10^{-7}	Square miles
Ounces	2.790×10^{-5}	Tons (long)	Temp. (°C.) + 273	1	Abs. temp. (°C.)
Ounces	2.835×10^{-5}	Tons (metric)	Temp. (°C.) + 17.78	1.8	Temp. (°F.)
Ounces (troy)	480	Grains	Temp. (°F.) + 460	1	Abs. temp. (°F.)
Ounces (troy)	20	Pennyweights (troy)	Temp. (°F.) – 32	5/9	Temp. (°C.)
Ounces (troy)	0.08333	Pounds (troy)	Watts	0.05692	BTU/min.
Ounces (troy)	31.103481	Grams	Watts	44.26	Foot-pounds/min.
Ounces (troy)	1.09714	Ounces (avoir.)	Watts	0.7376	Foot-pounds/sec.
Ounces (fluid)	1.805	Cubic Inches	Watts	1.341×10^{-3}	Horsepower
Ounces (fluid)	0.02957	Liters	Watts	0.01434	Kg.-calories/min.
Ounces/sq. inch	0.0625	Lbs./sq. inch	Watts	10^{-3}	Kilowatts
Pounds	16	Ounces	Watts-hours	3.415	BTU
Pounds	256	Drams	Watts-hours	2655	Foot-pounds
Pounds	7000	Grains	Watts-hours	1.341×10^{-3}	Horsepower-hrs.
Pounds	0.0005	Tons (short)	Watts-hours	0.8605	Kilogram-calories
Pounds	453.5924	Grams	Watts-hours	367.1	Kilogram-meters
Pounds	1.21528	Pounds (troy)	Watts-hours	10^{-3}	Kilowatt-hours
Pounds	14.5833	Ounces (troy)	Yards	91.44	Centimeters
Pounds (troy)	5760	Grains	Yards	3	Feet
Pounds (troy)	240	Pennyweights (troy)	Yards	36	Inches
Pounds (troy)	12	Ounces (troy)	Yards	0.9144	Meters

TABLE 4   *(Concluded)*

CIRCUMFERENCES AND AREAS OF CIRCLES
From 1/64 to 50, Diameter

Dia.	Circum.	Area	Dia.	Circum.	Area	Dia.	Circum.	Area	Dia.	Circum.	Area
1/64	.04909	.00019	8	25.1327	50.2655	17	53.4071	226.980	26	81.6814	530.929
1/32	.09818	.00077	8⅛	25.5254	51.8485	17⅛	53.7998	230.330	26⅛	82.0741	536.047
1/16	.19635	.00307	8¼	25.9181	53.4562	17¼	54.1925	233.705	26¼	82.4668	541.188
1/8	.39270	.01227	8⅜	26.3108	55.0883	17⅜	54.5852	237.104	26⅜	82.8595	546.355
3/16	.58905	.02761	8½	26.7035	56.7450	17½	54.9779	240.528	26½	83.2522	551.546
1/4	.78540	.04909	8⅝	27.0962	58.4262	17⅝	55.3706	243.977	26⅝	83.6449	556.761
5/16	.98175	.07670	8¾	27.4889	60.1321	17¾	55.7633	247.450	26¾	84.0376	562.002
3/8	1.1781	.11045	8⅞	27.8816	61.8624	17⅞	56.1560	250.947	26⅞	84.4303	567.266
7/16	1.3744	.15033	9	28.2743	63.6173	18	56.5487	254.469	27	84.8230	572.555
1/2	1.5708	.19635	9⅛	28.6670	65.3967	18⅛	56.9414	258.016	27⅛	85.2157	577.869
9/16	1.7671	.24850	9¼	29.0597	67.2007	18¼	57.3341	261.587	27¼	85.6084	583.207
5/8	1.9635	.30680	9⅜	29.4524	69.0292	18⅜	57.7268	265.182	27⅜	86.0011	588.570
11/16	2.1598	.37122	9½	29.8451	70.8822	18½	58.1195	268.803	27½	86.3938	593.957
3/4	2.3562	.44179	9⅝	30.2378	72.7597	18⅝	58.5122	272.447	27⅝	86.7865	599.369
13/16	2.5525	.51849	9¾	30.6305	74.6619	18¾	58.9049	276.117	27¾	87.1792	604.806
7/8	2.7489	.60132	9⅞	31.0232	76.5886	18⅞	59.2976	279.810	27⅞	87.5719	610.267
15/16	2.9452	.69029				19	59.6903	283.529	28	87.9646	615.752
1	3.1416	.78540	10	31.4159	78.5398	19⅛	60.0830	287.272	28⅛	88.3573	621.262
1⅛	3.5343	.99402	10⅛	31.8086	80.5156	19¼	60.4757	291.039	28¼	88.7500	626.797
1¼	3.9270	1.2272	10¼	32.2013	82.5159	19⅜	60.8684	294.831	28⅜	89.1427	632.356
1⅜	4.3197	1.4849	10⅜	32.5940	84.5408	19½	61.2611	298.648	28½	89.5354	637.940
1½	4.7124	1.7671	10½	32.9867	86.5902	19⅝	61.6538	302.489	28⅝	89.9281	643.548
1⅝	5.1051	2.0739	10⅝	33.3794	88.6641	19¾	62.0465	306.354	28¾	90.3208	649.181
1¾	5.4978	2.4053	10¾	33.7721	90.7626	19⅞	62.4392	310.245	28⅞	90.7135	654.838
1⅞	5.8905	2.7612	10⅞	34.1648	92.8856	20	62.8319	314.159	29	91.1062	660.520
2	6.2832	3.1416	11	34.5575	95.0332	20⅛	63.2246	318.099	29⅛	91.4989	666.226
2⅛	6.6759	3.5466	11⅛	34.9502	97.2053	20¼	63.6173	322.062	29¼	91.8916	671.957
2¼	7.0686	3.9761	11¼	35.3429	99.4020	20⅜	64.0100	326.051	29⅜	92.2843	677.713
2⅜	7.4613	4.4301	11⅜	35.7356	101.623	20½	64.4027	330.064	29½	92.6770	683.493
2½	7.8540	4.9087	11½	36.1283	103.869	20⅝	64.7954	334.101	29⅝	93.0697	689.297
2⅝	8.2467	5.4119	11⅝	36.5210	106.139	20¾	65.1881	338.163	29¾	93.4624	695.127
2¾	8.6394	5.9396	11¾	36.9137	108.434	20⅞	65.5808	342.250	29⅞	93.8551	700.980
2⅞	9.0321	6.4918	11⅞	37.3064	110.753				30	94.2478	706.858
3	9.4248	7.0686	12	37.6991	113.097	21	65.9735	346.361	30⅛	94.6405	712.761
3⅛	9.8175	7.6699	12⅛	38.0918	115.466	21⅛	66.3662	350.496	30¼	95.0332	718.689
3¼	10.2102	8.2958	12¼	38.4845	117.859	21¼	66.7539	354.656	30⅜	95.4259	724.640
3⅜	10.6029	8.9462	12⅜	38.8772	120.276	21⅜	67.1516	358.841	30½	95.8186	730.617
3½	10.9956	9.6211	12½	39.2699	122.718	21½	67.5442	363.050	30⅝	96.2113	736.618
3⅝	11.3883	10.3206	12⅝	39.6626	125.185	21⅝	67.9369	367.284	30¾	96.6040	742.643
3¾	11.7810	11.0447	12¾	40.0553	127.676	21¾	68.3296	371.542	30⅞	96.9967	748.693
3⅞	12.1737	11.7932	12⅞	40.4480	130.191	21⅞	68.7223	375.825	31	97.3894	754.768
4	12.5664	12.5664	13	40.8407	132.732	22	69.1150	380.133	31⅛	97.7821	760.867
4⅛	12.9591	13.3640	13⅛	41.2334	135.297	22⅛	69.5077	384.465	31¼	98.1748	766.990
4¼	13.3518	14.1863	13¼	41.6261	137.886	22¼	69.9004	388.821	31⅜	98.5675	773.139
4⅜	13.7445	15.0330	13⅜	42.0188	140.500	22⅜	70.2931	393.203	31½	98.9602	779.311
4½	14.1372	15.9043	13½	42.4115	143.139	22½	70.6858	397.608	31⅝	99.3529	785.509
4⅝	14.5299	16.8002	13⅝	42.8042	145.802	22⅝	71.0785	402.038	31¾	99.7456	791.731
4¾	14.9226	17.7206	13¾	43.1969	148.489	22¾	71.4712	406.493	31⅞	100.1383	797.977
4⅞	15.3153	18.6655	13⅞	43.5896	151.201	22⅞	71.8639	410.972	32	100.5310	804.248
5	15.7080	19.6350	14	43.9823	153.938	23	72.2566	415.476	32⅛	100.9237	810.543
5⅛	16.1007	20.6290	14⅛	44.3750	156.699	23⅛	72.6493	420.004	32¼	101.3164	816.863
5¼	16.4934	21.6476	14¼	44.7677	159.485	23¼	73.0420	424.557	32⅜	101.7091	823.208
5⅜	16.8861	22.6906	14⅜	45.1604	162.295	23⅜	73.4347	429.134	32½	102.1018	829.577
5½	17.2788	23.7583	14½	45.5531	165.130	23½	73.8274	433.736	32⅝	102.4945	835.971
5⅝	17.6715	24.8505	14⅝	45.9458	167.989	23⅝	74.2201	438.363	32¾	102.8872	842.389
5¾	18.0642	25.9672	14¾	46.3385	170.873	23¾	74.6128	443.014	32⅞	103.2799	848.831
5⅞	18.4569	27.1085	14⅞	46.7312	173.782	23⅞	75.0055	447.689	33	103.6726	855.299
6	18.8496	28.2743	15	47.1239	176.715	24	75.3982	452.389	33⅛	104.0653	861.791
6⅛	19.2423	29.4647	15⅛	47.5166	179.672	24⅛	75.7909	457.114	33¼	104.4580	868.307
6¼	19.6350	30.6796	15¼	47.9094	182.654	24¼	76.1836	461.863	33⅜	104.8507	874.848
6⅜	20.0277	31.9191	15⅜	48.3020	185.661	24⅜	76.5763	466.637	33½	105.2434	881.413
6½	20.4204	33.1831	15½	48.6947	188.692	24½	76.9690	471.435	33⅝	105.6361	888.003
6⅝	20.8131	34.4716	15⅝	49.0874	191.748	24⅝	77.3617	476.258	33¾	106.0288	894.618
6¾	21.2058	35.7847	15¾	49.4801	194.828	24¾	77.7544	481.106	33⅞	106.4215	901.257
6⅞	21.5985	37.1223	15⅞	49.8728	197.933	24⅞	78.1471	485.977	34	106.8142	907.920
7	21.9912	38.4845	16	50.2655	201.062	25	78.5398	490.874	34⅛	107.2069	914.609
7⅛	22.3839	39.8712	16⅛	50.6582	204.216	25⅛	78.9325	495.795	34¼	107.5996	921.321
7¼	22.7765	41.2825	16¼	51.0509	207.394	25¼	79.3252	500.740	34⅜	107.9923	928.058
7⅜	23.1692	42.7183	16⅜	51.4436	210.597	25⅜	79.7179	505.711	34½	108.3850	934.820
7½	23.5619	44.1787	16½	51.8363	213.825	25½	80.1106	510.705	34⅝	108.7777	941.607
7⅝	23.9546	45.6636	16⅝	52.2290	217.077	25⅝	80.5033	515.724	34¾	109.1704	948.417
7¾	24.3473	47.1730	16¾	52.6217	220.353	25¾	80.8960	520.768	34⅞	109.5631	955.253
7⅞	24.7400	48.7069	16⅞	53.0144	223.654	25⅞	81.2887	525.836			

TABLE 5

TRIGONOMETRIC FORMULAS

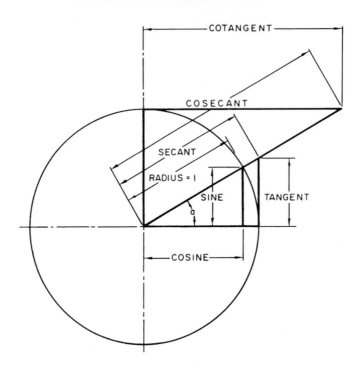

FORMULAS FOR FINDING FUNCTIONS OF ANGLES	
$\dfrac{\text{Side opposite}}{\text{Hypotenuse}}$ = SINE	
$\dfrac{\text{Side adjacent}}{\text{Hypotenuse}}$ = COSINE	
$\dfrac{\text{Side opposite}}{\text{Side adjacent}}$ = TANGENT	
$\dfrac{\text{Side adjacent}}{\text{Side opposite}}$ = COTANGENT	
$\dfrac{\text{Hypotenuse}}{\text{Side adjacent}}$ = SECANT	
$\dfrac{\text{Hypotenuse}}{\text{Side opposite}}$ = COSECANT	
FORMULAS FOR FINDING THE LENGTH OF SIDES FOR RIGHT-ANGLE TRIANGLES WHEN AN ANGLE AND SIDE ARE KNOWN	
Length of side opposite	Hypotenuse × Sine Hypotenuse ÷ Cosecant Side adjacent × Tangent Side adjacent ÷ Cotangent
Length of side adjacent	Hypotenuse × Cosine Hypotenuse ÷ Secant Side opposite × Cotangent Side opposite ÷ Tangent
Length of hypotenuse	Side opposite × Cosecant Side opposite ÷ Sine Side adjacent × Secant Side adjacent ÷ Cosine

TABLE 6

RIGHT-TRIANGLE FORMULAS

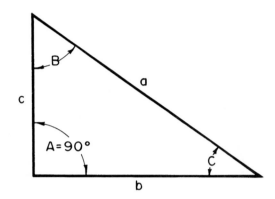

TO FIND ANGLES	FORMULAS	
C	$\dfrac{c}{a}$ = Sine C	90° − B
C	$\dfrac{b}{a}$ = Cosine C	90° − B
C	$\dfrac{c}{b}$ = Tan C	90° − B
C	$\dfrac{b}{c}$ = Cot C	90° − B
C	$\dfrac{a}{b}$ = Secant C	90° − B
C	$\dfrac{a}{c}$ = Csc C	90° − B
B	$\dfrac{b}{a}$ = Sine B	90° − C
B	$\dfrac{c}{a}$ = Cosine B	90° − C
B	$\dfrac{b}{c}$ = Tan B	90° − C
B	$\dfrac{c}{b}$ = Cot B	90° − C
B	$\dfrac{a}{c}$ = Secant B	90° − C
B	$\dfrac{a}{b}$ = Csc B	90° − C

TO FIND SIDES	FORMULAS	
a	$\sqrt{b^2 + c^2}$	
a	c × Csc C	$\dfrac{c}{\text{sine C}}$
a	c × Secant B	$\dfrac{c}{\text{Cosine B}}$
a	b × Csc B	$\dfrac{b}{\text{Sine B}}$
a	b × Secant C	$\dfrac{b}{\text{Cosine C}}$
b	$\sqrt{a^2 - c^2}$	
b	a × Sine B	$\dfrac{a}{\text{Cosecant B}}$
b	a × Cos C	$\dfrac{a}{\text{Secant C}}$
b	c × Tan B	$\dfrac{c}{\text{Cotangent B}}$
b	c × Cot C	$\dfrac{c}{\text{Tangent C}}$
c	$\sqrt{a^2 - b^2}$	
c	a × Cos B	$\dfrac{a}{\text{Secant B}}$
c	a × Sine C	$\dfrac{a}{\text{Cosecant C}}$
c	b × Cot B	$\dfrac{b}{\text{Tangent B}}$
c	b × Tan C	$\dfrac{b}{\text{Cotangent C}}$

TABLE 7

OBLIQUE-ANGLED TRIANGLE FORMULAS

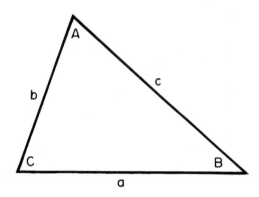

TO FIND	KNOWN	SOLUTION
C	A-B	$180° - (A + B)$
b	a-B-A	$\dfrac{a \times Sin\ B}{Sin\ A}$
c	a-A-C	$\dfrac{a \times Sin\ C}{Sin\ A}$
Tan A	a-C-b	$\dfrac{a \times Sin\ C}{b - (a \times Cos\ C)}$
B	A-C	$180° - (A + C)$
Sin B	b-A-a	$\dfrac{b \times Sin\ A}{a}$
A	B-C	$180° - (B + C)$
Cos A	a-b-c	$\dfrac{b^2 + C^2 - a^2}{2bc}$
Sin C	c-A-a	$\dfrac{c \times Sin\ A}{a}$
Cot B	a-C-b	$\dfrac{a \times Csc\ C}{b} - Cot\ C$
c	b-C-B	$b \times Sin\ C \times Csc\ B$

TABLE 8

RUNNING AND SLIDING FITS

VALUES IN THOUSANDTHS OF AN INCH

Nominal Size Range Inches		Class RC1 Precision Sliding			Class RC2 Sliding Fit			Class RC3 Precision Running			Class RC4 Close Running			Class RC5 Medium Running		
		Hole Tol. GR5	Minimum Clearance	Shaft Tol. GR4	Hole Tol. GR6	Minimum Clearance	Shaft Tol. GR5	Hole Tol. GR7	Minimum Clearance	Shaft Tol. GR6	Hole Tol. GR8	Minimum Clearance	Shaft Tol. GR7	Hole Tol. GR8	Minimum Clearance	Shaft Tol. GR7
Over	To	-0		+0	-0		+0	-0		+0	-0		+0	-0		+0
0	.12	+0.15	0.10	-0.12	+0.25	0.10	-0.15	+0.40	0.30	-0.25	+0.60	0.30	-0.40	+0.60	0.60	-0.40
.12	.24	+0.20	0.15	-0.15	+0.30	0.15	-0.20	+0.50	0.40	-0.30	+0.70	0.40	-0.50	+0.70	0.80	-0.50
.24	.40	+0.25	0.20	-0.15	+0.40	0.20	-0.25	+0.60	0.50	-0.40	+0.90	0.50	-0.60	+0.90	1.00	-0.60
.40	.71	+0.30	0.25	-0.20	+0.40	0.25	-0.30	+0.70	0.60	-0.40	+1.00	0.60	-0.70	+1.00	1.20	-0.70
.71	1.19	+0.40	0.30	-0.25	+0.50	0.30	-0.40	+0.80	0.80	-0.50	+1.20	0.80	-0.80	+1.20	1.60	-0.50
1.19	1.97	+0.40	0.40	-0.30	+0.60	0.40	-0.40	+1.00	1.00	-0.60	+1.60	1.00	-1.00	+1.60	2.00	-1.00
1.97	3.15	+0.50	0.40	-0.30	+0.70	0.40	-0.50	+1.20	1.20	-0.70	+1.80	1.20	-1.20	+1.80	2.50	-1.20
3.15	4.73	+0.60	0.50	-0.40	+0.90	0.50	-0.60	+1.40	1.40	-0.90	+2.20	1.40	-1.40	+2.20	3.00	-1.40
4.73	7.09	+0.70	0.60	-0.50	+1.00	0.60	-0.70	+1.60	1.60	-1.00	+2.50	1.60	-1.60	+2.50	3.50	-1.60
7.09	9.85	+0.80	0.60	-0.60	+1.20	0.60	-0.80	+1.80	2.00	-1.20	+2.80	2.00	-1.80	+2.80	4.50	-1.80
9.85	12.41	+0.90	0.80	-0.60	+1.20	0.80	-0.90	+2.00	2.50	-1.20	+3.00	2.50	-2.00	+3.00	5.00	-2.00
12.41	15.75	+1.00	1.00	-0.70	+1.40	1.00	-1.00	+2.20	3.00	-1.40	+3.50	3.00	-2.20	+3.50	6.00	-2.20

Nominal Size Range Inches		Class RC6 Medium Running			Class RC7 Free Running			Class RC8 Loose Running			Class RC9 Loose Running		
		Hole Tol. GR9	Minimum Clearance	Shaft Tol. GR8	Hole Tol. GR9	Minimum Clearance	Shaft Tol. GR8	Hole Tol. GR10	Minimum Clearance	Shaft Tol. GR9	Hole Tol. GR11	Minimum Clearance	Shaft Tol. GR10
Over	To	-0		+0	-0		+0	-0		+0	-0		+0
0	.12	+1.00	0.60	-0.60	+1.00	1.00	-0.60	+1.60	2.50	-1.00	+2.50	4.00	-1.60
.12	.24	+1.20	0.80	-0.70	+1.20	1.20	-0.70	+1.80	2.80	-1.20	+3.00	4.50	-1.80
.24	.40	+1.40	1.00	-0.90	+1.40	1.60	-0.90	+2.20	3.00	-1.40	+3.50	6.00	-2.20
.40	.71	+1.60	1.20	-1.00	+1.60	2.00	-1.00	+2.80	3.50	-1.60	+4.00	6.00	-2.80
.71	1.19	+2.00	1.60	-1.20	+2.00	2.50	-1.20	+3.50	4.50	-2.00	+5.00	7.00	-3.50
1.19	1.97	+2.50	2.00	-1.60	+2.50	3.00	-1.60	+4.00	5.00	-2.50	+6.00	8.00	-4.00
1.97	3.15	+3.00	2.50	-1.80	+3.00	4.00	-1.80	+4.50	6.00	-3.00	+7.00	9.00	-4.50
3.15	4.73	+3.50	3.00	-2.20	+3.50	5.00	-2.20	+5.00	7.00	-3.50	+9.00	10.00	-5.00
4.73	7.09	+4.00	3.50	-2.50	+4.00	6.00	-2.50	+6.00	8.00	-4.00	+10.00	12.00	-6.00
7.09	9.85	+4.50	4.00	-2.80	+4.50	7.00	-2.80	+7.00	10.00	-4.50	+12.00	15.00	-7.00
9.85	12.41	+5.00	5.00	-3.00	+5.00	8.00	-3.00	+8.00	12.00	-5.00	+12.00	18.00	-8.00
12.41	15.75	+6.00	6.00	-3.50	+6.00	10.00	-3.50	+9.00	14.00	-6.00	+14.00	22.00	-9.00

VALUES IN MILLIMETERS

Nominal Size Range Millimeters		Class RC1 Precision Sliding			Class RC2 Sliding Fit			Class RC3 Precision Running			Class RC4 Close Running			Class RC5 Medium Running		
		Hole Tol. H5	Minimum Clearance	Shaft Tol. g4	Hole Tol. H6	Minimum Clearance	Shaft Tol. g5	Hole Tol. H7	Minimum Clearance	Shaft Tol. f6	Hole Tol. H8	Minimum Clearance	Shaft Tol. f7	Hole Tol. H8	Minimum Clearance	Shaft Tol. e7
Over	To	-0		+0	-0		+0	-0		+0	-0		+0	-0		+0
0	3	+0.004	0.003	-0.003	+0.006	0.003	-0.004	+0.010	0.008	-0.006	+0.015	0.008	-0.010	+0.015	0.015	-0.010
3	6	+0.005	0.004	-0.004	+0.008	0.004	-0.005	+0.013	0.010	-0.008	+0.018	0.010	-0.013	+0.018	0.020	-0.013
6	10	+0.006	0.005	-0.004	+0.010	0.005	-0.006	+0.015	0.013	-0.010	+0.023	0.013	-0.015	+0.023	0.025	-0.015
10	18	+0.008	0.006	-0.005	+0.010	0.006	-0.008	+0.018	0.015	-0.010	+0.025	0.015	-0.018	+0.025	0.030	-0.018
18	30	+0.010	0.008	-0.006	+0.013	0.008	-0.010	+0.020	0.020	-0.013	+0.030	0.020	-0.020	+0.030	0.040	-0.020
30	50	+0.010	0.010	-0.008	+0.015	0.010	-0.010	+0.030	0.030	-0.015	+0.040	0.030	-0.030	+0.040	0.050	-0.030
50	80	+0.013	0.010	-0.008	+0.018	0.010	-0.013	+0.030	0.030	-0.020	+0.050	0.030	-0.030	+0.050	0.060	-0.030
80	120	+0.015	0.013	-0.010	+0.023	0.013	-0.015	+0.040	0.040	-0.020	+0.060	0.040	-0.040	+0.060	0.080	-0.040
120	180	+0.018	0.015	-0.013	+0.025	0.015	-0.018	+0.040	0.040	-0.030	+0.060	0.040	-0.040	+0.060	0.090	-0.040
180	250	+0.020	0.015	-0.015	+0.030	0.015	-0.020	+0.050	0.050	-0.030	+0.070	0.050	-0.050	+0.070	0.110	-0.050
250	315	+0.023	0.020	-0.015	+0.030	0.020	-0.023	+0.050	0.060	-0.030	+0.080	0.060	-0.050	+0.080	0.130	-0.050
315	400	+0.025	0.025	-0.018	+0.036	0.025	-0.025	+0.060	0.080	-0.040	+0.090	0.080	-0.060	+0.090	0.150	-0.060

Nominal Size Range Millimeters		Class RC6 Medium Running			Class RC7 Free Running			Class RC8 Loose Running			Class RC9 Loose Running		
		Hole Tol. H9	Minimum Clearance	Shaft Tol. e8	Hole Tol. H9	Minimum Clearance	Shaft Tol. d8	Hole Tol. H10	Minimum Clearance	Shaft Tol. e9	Hole Tol. GR11	Minimum Clearance	Shaft Tol. gr10
Over	To	-0		+0	-0		+0	-0		+0	-0		+0
0	3	+0.025	0.015	-0.015	+0.025	0.025	-0.015	+0.041	0.064	-0.025	+0.060	0.100	-0.040
3	6	+0.030	0.015	-0.018	+0.030	0.030	-0.018	+0.046	0.071	-0.030	+0.080	0.110	-0.050
6	10	+0.036	0.025	-0.023	+0.036	0.040	-0.023	+0.056	0.076	-0.036	+0.070	0.130	-0.060
10	18	+0.040	0.030	-0.025	+0.040	0.050	-0.025	+0.070	0.090	-0.040	+0.100	0.150	-0.070
18	30	+0.050	0.040	-0.030	+0.050	0.060	-0.030	+0.090	0.110	-0.050	+0.130	0.180	-0.090
30	50	+0.060	0.050	-0.040	+0.060	0.080	-0.040	+0.100	0.130	-0.060	+0.150	0.200	-0.100
50	80	+0.080	0.060	-0.050	+0.080	0.100	-0.050	+0.110	0.150	-0.080	+0.180	0.230	-0.120
80	120	+0.090	0.080	-0.060	+0.090	0.130	-0.060	+0.130	0.180	-0.090	+0.230	0.250	-0.130
120	180	+0.100	0.090	-0.060	+0.100	0.150	-0.060	+0.150	0.200	-0.100	+0.250	0.300	-0.150
180	250	+0.110	0.100	-0.070	+0.110	0.180	-0.070	+0.180	0.250	-0.110	+0.300	0.380	-0.180
250	315	+0.130	0.130	-0.080	+0.130	0.200	-0.080	+0.200	0.300	-0.130	+0.300	0.460	-0.200
315	400	+0.150	0.150	-0.090	+0.150	0.250	-0.090	+0.230	0.360	-0.150	+0.360	0.560	-0.230

TABLE 9   From *Drafting for Trades and Industry—Mechanical and Electronic*, Nelson. Delmar Publishers Inc.

LOCATIONAL CLEARANCE FITS

VALUES IN THOUSANDTHS OF AN INCH

Nominal Size Range Inches		Class LC1			Class LC2			Class LC3			Class LC4			Class LC5			Class LC6		
		Hole Tol. GR6	Minimum Clearance	Shaft Tol. GR5	Hole Tol. GR8	Minimum Clearance	Shaft Tol. GR7	Hole Tol. GR10	Minimum Clearance	Shaft Tol. GR9	Hole Tol. GR7	Minimum Clearance	Shaft Tol. GR6	Hole Tol. GR9	Minimum Clearance	Shaft Tol. GR8	Hole Tol. GR9	Minimum Clearance	Shaft Tol. GR8
Over	To	-0		+0	-0		+0	-0		+0	-0		+0	-0		+0	-0		+0
0	.12	+0.25	0	-0.15	+0.4	0	-0.25	+0.6	0	-0.4	+1.6	0	-1.0	+0.4	0.10	-0.25	+1.0	0.3	-0.6
.12	.24	+0.30	0	-0.20	+0.5	0	-0.30	+0.7	0	-0.5	+1.8	0	-1.2	+0.5	0.15	-0.30	+1.2	0.4	-0.7
.24	.40	+0.40	0	-0.25	+0.6	0	-0.40	+0.9	0	-0.6	+2.2	0	-1.4	+0.6	0.20	-0.40	+1.4	0.5	-0.9
.40	.71	+0.40	0	-0.30	+0.7	0	-0.40	+1.0	0	-0.7	+2.8	0	-1.6	+0.7	0.25	-0.40	+1.6	0.6	-1.0
.71	1.19	+0.50	0	-0.40	+0.8	0	-0.50	+1.2	0	-0.8	+3.5	0	-2.0	+0.8	0.30	-0.50	+2.0	0.8	-1.2
1.19	1.97	+0.60	0	-0.40	+1.0	0	-0.60	+1.6	0	-1.0	+4.0	0	-2.5	+1.0	0.40	-0.60	+2.5	1.0	-1.6
1.97	3.15	+0.70	0	-0.50	+1.2	0	-0.70	+1.8	0	-1.2	+4.5	0	-3.0	+1.2	0.40	-0.70	+3.0	1.2	-1.8
3.15	4.73	+0.90	0	-0.60	+1.4	0	-0.90	+2.7	0	-1.4	+5.0	0	-3.5	+1.4	0.50	-0.90	+3.5	1.4	-2.2
4.73	7.09	+1.00	0	-0.70	+1.6	0	-1.00	+2.5	0	-1.6	+6.0	0	-4.0	+1.6	0.60	-1.00	+4.0	1.6	-2.5
7.09	9.85	+1.20	0	-0.80	+1.8	0	-1.20	+2.8	0	-1.8	+7.0	0	-4.5	+1.8	0.60	-1.20	+4.5	2.0	-2.8
9.85	12.41	+1.20	0	-0.90	+2.0	0	-1.20	+3.0	0	-2.0	+8.0	0	-5.0	+2.0	0.70	-1.20	+5.0	2.2	-3.0
12.41	15.75	+1.40	0	-1.00	+2.2	0	-1.40	+3.5	0	-2.2	+9.0	0	-6.0	+2.2	0.70	-1.40	+6.0	2.5	-3.5

Nominal Size Range Inches		Class LC7			Class LC8			Class LC9			Class LC10			Class LC11		
		Hole Tol. GR10	Minimum Clearance	Shaft Tol. GR9	Hole Tol. GR10	Minimum Clearance	Shaft Tol. GR9	Hole Tol. GR11	Minimum Clearance	Shaft Tol. GR10	Hole Tol. GR12	Minimum Clearance	Shaft Tol. GR11	Hole Tol. GR13	Minimum Clearance	Shaft Tol. GR12
Over	To	-0		+0	-0		+0	-0		+0	-0		+0	-0		+0
0	.12	+1.6	0.6	-1.0	+1.6	1.0	-1.0	+2.5	2.5	-1.6	+1.0	4.0	-2.5	+6.0	5.0	-4.0
.12	.24	+1.8	0.8	-1.2	+1.8	1.2	-1.2	+3.0	2.8	-1.8	+5.0	4.5	-3.0	+7.0	6.0	-5.0
.24	.40	+2.2	1.0	-1.4	+2.2	1.6	-1.4	+3.5	3.0	-2.2	+6.0	5.0	-3.5	+9.0	7.0	-6.0
.40	.71	+2.8	1.2	-1.6	+2.8	2.0	-1.6	+4.0	3.5	-2.8	+7.0	6.0	-4.0	+10.0	8.0	-7.0
.71	1.19	+3.5	1.6	-2.0	+3.5	2.5	-2.0	+5.0	4.5	-3.5	+8.0	7.0	-5.0	+12.0	10.0	-8.0
1.19	1.97	+4.0	2.0	-2.5	+4.0	3.6	-2.5	+6.0	5.0	-4.0	+10.0	8.0	-6.0	+16.0	12.0	-10.0
1.97	3.15	+4.5	2.5	-3.0	+4.5	4.0	-3.0	+7.0	6.0	-4.5	+12.0	10.0	-7.0	+18.0	14.0	-12.0
3.15	4.73	+5.0	3.0	-3.5	+5.0	5.0	-3.5	+9.0	7.0	-5.0	+14.0	11.0	-9.0	+22.0	16.0	-14.0
4.73	7.09	+6.0	3.5	-4.0	+6.0	6.0	-4.0	+10.0	8.0	-6.0	+16.0	12.0	-10.0	+25.0	18.0	-16.0
7.09	9.85	+7.0	4.0	-4.5	+7.0	7.0	-4.5	+12.0	10.0	-7.0	+18.0	16.0	-12.0	+28.0	22.0	-18.0
9.85	12.41	+8.0	4.5	-5.0	+8.0	7.0	-5.0	+12.0	12.0	-8.0	+20.0	20.0	-12.0	+30.0	28.0	-20.0
12.41	15.75	+9.0	5.0	-6.0	+9.0	8.0	-6.0	+14.0	14.0	-9.0	+22.0	22.0	-14.0	+35.0	30.0	-22.0

VALUES IN MILLIMETERS

Nominal Size Range Millimeters		Class LC1			Class LC2			Class LC3			Class LC4			Class LC5			Class LC6		
		Hole Tol. H6	Minimum Clearance	Shaft Tol. h5	Hole Tol. H7	Minimum Clearance	Shaft Tol. h6	Hole Tol. H8	Minimum Clearance	Shaft Tol. h7	Hole Tol. H10	Minimum Clearance	Shaft Tol. h9	Hole Tol. H7	Minimum Clearance	Shaft Tol. g6	Hole Tol. H9	Minimum Clearance	Shaft Tol. f8
Over	To	-0		+0	-0		+0	-0		+0	-0		+0	-0		+0	-0		+0
0	3	+0.006	0	-0.004	+0.010	0	-0.006	+0.015	0	-0.010	+0.041	0	-0.025	+0.010	0.002	-0.006	+0.025	0.008	-0.015
3	6	+0.008	0	-0.005	+0.013	0	-0.008	+0.018	0	-0.013	+0.046	0	-0.030	+0.013	0.004	-0.008	+0.030	0.010	-0.018
6	10	+0.010	0	-0.006	+0.015	0	-0.010	+0.023	0	-0.015	+0.056	0	-0.036	+0.015	0.005	-0.010	+0.036	0.013	-0.023
10	18	+0.010	0	-0.008	+0.018	0	-0.010	+0.025	0	-0.018	+0.070	0	-0.040	+0.018	0.006	-0.010	+0.041	0.015	-0.025
18	30	+0.013	0	-0.010	+0.020	0	-0.013	+0.030	0	-0.020	+0.090	0	-0.050	+0.020	0.008	-0.013	+0.050	0.020	-0.030
30	50	+0.015	0	-0.010	+0.025	0	-0.015	+0.041	0	-0.025	+0.100	0	-0.060	+0.025	0.010	-0.015	+0.060	0.030	-0.040
50	80	+0.018	0	-0.013	+0.030	0	-0.018	+0.046	0	-0.030	+0.110	0	-0.080	+0.030	0.010	-0.018	+0.080	0.030	-0.050
80	120	+0.023	0	-0.015	+0.036	0	-0.023	+0.056	0	-0.036	+0.130	0	-0.080	+0.036	0.013	-0.023	+0.090	0.040	-0.060
120	180	+0.025	0	-0.018	+0.041	0	-0.025	+0.064	0	-0.041	+0.150	0	-0.100	+0.041	0.015	-0.025	+0.100	0.040	-0.060
180	250	+0.030	0	-0.020	+0.046	0	-0.030	+0.071	0	-0.046	+0.180	0	-0.110	+0.046	0.015	-0.030	+0.110	0.050	-0.070
250	315	+0.020	0	-0.023	+0.051	0	-0.030	+0.076	0	-0.051	+0.200	0	-0.130	+0.051	0.018	-0.030	+0.130	0.060	-0.080
315	400	+0.036	0	-0.025	+0.056	0	-0.036	+0.089	0	-0.056	+0.230	0	-0.150	+0.056	0.018	-0.036	+0.150	0.060	-0.090

Nominal Size Range Millimeters		Class LC7			Class LC8			Class LC9			Class LC10			Class LC11		
		Hole Tol. H10	Minimum Clearance	Shaft Tol. e9	Hole Tol. H10	Minimum Clearance	Shaft Tol. d9	Hole Tol. H11	Minimum Clearance	Shaft Tol. c10	Hole Tol. GR12	Minimum Clearance	Shaft Tol. gr11	Hole Tol. GR13	Minimum Clearance	Shaft Tol. gr12
Over	To	-0		+0	-0		+0	-0		+0	-0		+0	-0		+0
0	3	+0.041	0.015	-0.025	+0.041	0.025	-0.025	+0.064	0.06	-0.041	+0.10	0.10	-0.06	+0.15	0.13	-0.10
3	6	+0.046	0.020	-0.030	+0.046	0.030	-0.030	+0.076	0.07	-0.46	+0.13	0.11	-0.08	+0.18	0.15	-0.13
6	10	+0.056	0.025	-0.036	+0.056	0.041	-0.036	+0.089	0.08	-0.56	+0.15	0.13	-0.09	+0.23	0.18	-0.15
10	18	+0.070	0.030	-0.040	+0.070	0.050	-0.040	+0.100	0.09	-0.70	+0.18	0.15	-0.10	+0.25	0.20	-0.18
18	30	+0.090	0.040	-0.050	+0.090	0.060	-0.050	+0.130	0.11	-0.90	+0.20	0.18	-0.13	+0.31	0.25	-0.20
30	50	+0.100	0.050	-0.060	+0.100	0.090	-0.060	+0.150	0.13	-0.100	+0.25	0.20	-0.15	+0.41	0.31	-0.25
50	80	+0.110	0.060	-0.080	+0.110	0.100	-0.080	+0.180	0.15	-0.110	+0.31	0.25	-0.18	+0.46	0.36	-0.31
80	120	+0.130	0.080	-0.090	+0.130	0.130	-0.090	+0.230	0.18	-0.130	+0.36	0.28	-0.23	+0.56	0.41	-0.36
120	180	+0.150	0.090	-0.100	+0.150	0.150	-0.100	+0.250	0.20	-0.150	+0.41	0.31	-0.25	+0.64	0.46	-0.41
180	250	+0.180	0.100	-0.110	+0.180	0.180	-0.110	+0.310	0.25	-0.180	+0.46	0.41	-0.31	+0.71	0.56	-0.46
250	315	+0.200	0.110	-0.130	+0.200	0.180	-0.130	+0.310	0.31	-0.200	+0.51	0.51	-0.31	+0.76	0.71	-0.51
315	400	+0.230	0.130	-0.150	+0.230	0.200	-0.150	+0.360	0.36	-0.230	+0.56	0.56	-0.36	+0.89	0.76	-0.56

TABLE 10   From *Drafting for Trades and Industry—Mechanical and Electronic*, Nelson. Delmar Publishers Inc.

LOCATIONAL TRANSITION FITS

VALUES IN THOUSANDTHS OF AN INCH

Nominal Size Range Inches		Class LT1			Class LT2			Class LT3			Class LT4			Class LT5			Class LT6		
		Hole Tol. GR7	Maximum Interference	Shaft Tol. GR6	Hole Tol. GR8	Maximum Interference	Shaft Tol. GR7	Hole Tol. GR7	Maximum Interference	Shaft Tol. GR6	Hole Tol. GR8	Maximum Interference	Shaft Tol. GR7	Hole Tol. GR7	Maximum Interference	Shaft Tol. GR6	Hole Tol. GR8	Maximum Interference	Shaft Tol. GR7
Over	To	-0		+0	-0		+0	-0		+0	-0		+0	-0		+0	-0		+0
0	.12	+0.4	0.10	-0.25	+0.6	0.20	-0.4	+0.4	0.25	-0.25	+0.6	0.4	-0.4	+0.4	0.5	-0.25	+0.6	0.65	-0.4
.12	.24	+0.5	0.15	-0.30	+0.7	0.25	-0.5	+0.5	0.40	-0.30	+0.7	0.6	-0.5	+0.5	0.6	-0.30	+0.7	0.80	-0.5
.24	.40	+0.6	0.20	-0.40	+0.9	0.30	-0.6	+0.6	0.50	-0.40	+0.9	0.7	-0.6	+0.6	0.8	-0.40	+0.9	1.00	-0.6
.40	.71	+0.7	0.20	-0.40	+1.0	0.30	-0.7	+0.7	0.50	-0.40	+1.0	0.8	-0.7	+0.7	0.9	-0.40	+1.0	1.20	-0.7
.71	1.19	+0.8	0.25	-0.50	+1.2	0.40	-0.8	+0.8	0.60	-0.50	+1.2	0.9	-0.8	+0.8	1.1	-0.50	+1.2	1.40	-0.8
1.19	1.97	+1.0	0.30	-0.60	+1.6	0.50	-1.0	+1.0	0.70	-0.60	+1.6	1.1	-1.0	+1.0	1.3	-0.60	+1.6	1.70	-1.0
1.97	3.15	+1.2	0.30	-0.70	+1.8	0.60	-1.2	+1.2	0.80	-0.70	+1.8	1.3	-1.2	+1.2	1.5	-0.70	+1.8	2.00	-1.2
3.15	4.73	+1.4	0.40	-0.90	+2.2	0.70	-1.4	+1.4	1.00	-0.90	+2.2	1.5	-1.4	+1.4	1.9	-0.90	+2.2	2.40	-1.4
4.73	7.09	+1.6	0.50	-1.00	+2.5	0.80	-1.6	+1.6	1.10	-1.00	+2.5	1.7	-1.6	+1.6	2.2	-1.00	+2.5	2.80	-1.6
7.09	9.85	+1.8	0.60	-1.20	+2.8	0.90	-1.8	+1.8	1.40	-1.20	+2.8	2.0	-1.8	+1.8	2.6	-1.20	+2.8	3.20	-1.8
9.85	12.41	+2.0	0.60	-1.20	+3.0	1.00	-2.0	+2.0	1.40	-1.20	+3.0	2.2	-2.0	+2.0	2.6	-1.20	+3.0	3.40	-2.0
12.41	15.75	+2.2	0.70	-1.40	+3.5	1.00	-2.2	+2.2	1.60	-1.40	+3.5	2.4	-2.2	+2.2	3.0	-1.40	+3.5	3.80	-2.2

VALUES IN MILLIMETERS

Nominal Size Range Millimeters		Class LT1			Class LT2			Class LT3			Class LT4			Class LT5			Class LT6		
		Hole Tol. H7	Maximum Clearance	Shaft Tol. js6	Hole Tol. H8	Maximum Clearance	Shaft Tol. js7	Hole Tol. H7	Maximum Clearance	Shaft Tol. k6	Hole Tol. H8	Maximum Clearance	Shaft Tol. k7	Hole Tol. H7	Maximum Clearance	Shaft Tol. n6	Hole Tol. H8	Maximum Clearance	Shaft Tol. n7
Over	To	-0		+0	-0		+0	-0		+0	-0		+0	-0		+0	-0		+0
0	3	+0.010	0.002	-0.006	+0.015	0.005	-0.010	+0.010	0.006	-0.006	+0.015	0.010	-0.010	+0.010	0.013	-0.006	+0.015	0.016	-0.010
3	6	+0.013	0.004	-0.008	+0.018	0.006	-0.013	+0.013	0.010	-0.008	+0.018	0.015	-0.013	+0.013	0.015	-0.008	+0.018	0.020	-0.013
6	10	+0.015	0.005	-0.010	+0.023	0.008	-0.015	+0.015	0.013	-0.010	+0.023	0.018	-0.015	+0.015	0.020	-0.010	+0.023	0.025	-0.015
10	18	+0.018	0.005	-0.010	+0.025	0.008	-0.018	+0.018	0.013	-0.010	+0.025	0.020	-0.018	+0.018	0.023	-0.010	+0.025	0.030	-0.018
18	30	+0.020	0.006	-0.013	+0.030	0.010	-0.020	+0.020	0.015	-0.013	+0.030	0.023	-0.020	+0.020	0.028	-0.013	+0.030	0.036	-0.020
30	50	+0.025	0.008	-0.015	+0.041	0.013	-0.025	+0.025	0.018	-0.015	+0.041	0.028	-0.025	+0.025	0.033	-0.015	+0.041	0.044	-0.025
50	80	+0.030	0.008	-0.018	+0.046	0.015	-0.030	+0.030	0.020	-0.018	+0.046	0.033	-0.030	+0.030	0.038	-0.018	+0.046	0.051	-0.030
80	120	+0.036	0.010	-0.023	+0.056	0.018	-0.036	+0.036	0.025	-0.023	+0.056	0.038	-0.036	+0.036	0.048	-0.023	+0.056	0.062	-0.036
120	180	+0.041	0.013	-0.025	+0.064	0.020	-0.041	+0.041	0.028	-0.025	+0.064	0.044	-0.041	+0.041	0.056	-0.025	+0.064	0.071	-0.041
180	250	+0.046	0.015	-0.030	+0.071	0.023	-0.046	+0.046	0.036	-0.030	+0.071	0.051	-0.046	+0.046	0.066	-0.030	+0.071	0.081	-0.046
250	315	+0.051	0.015	-0.030	+0.076	0.025	-0.051	+0.051	0.036	-0.030	+0.076	0.056	-0.051	+0.051	0.066	-0.030	+0.076	0.086	-0.051
315	400	+0.056	0.018	-0.036	+0.089	0.025	-0.056	+0.056	0.041	-0.036	+0.089	0.062	-0.056	+0.056	0.076	-0.036	+0.089	0.096	-0.056

TABLE 11 From *Drafting for Trades and Industry—Mechanical and Electronic,* Nelson. Delmar Publishers Inc.

LOCATIONAL INTERFERENCE FITS

VALUES IN THOUSANDTHS OF AN INCH

Nominal Size Range Inches		Class LN1 Light Press Fit			Class LN2 Medium Press Fit			Class LN3 Heavy Press Fit			Class LN4			Class LN5			Class LN6		
		Hole Tol. GR6	Maximum Interference	Shaft Tol. GR5	Hole Tol. GR7	Maximum Interference	Shaft Tol. GR6	Hole Tol. GR7	Maximum Interference	Shaft Tol. GR6	Hole Tol. GR8	Maximum Interference	Shaft Tol. GR7	Hole Tol. GR9	Maximum Interference	Shaft Tol. GR8	Hole Tol. GR10	Maximum Interference	Shaft Tol. GR9
Over	To	-0		+0	-0		+0	-0		+0	-0		+0	-0		+0	-0		+0
0	.12	+0.25	0.40	-0.15	+0.4	0.65	-0.25	+0.4	0.75	-0.25	+0.6	1.2	-0.4	+1.0	1.8	-0.6	+1.6	3.0	-1.0
.12	.24	+0.30	0.50	-0.20	+0.5	0.80	-0.30	+0.5	0.90	-0.30	+0.7	1.5	-0.5	+1.2	2.3	-0.7	+1.8	3.6	-1.2
.24	.40	+0.40	0.65	-0.25	+0.6	1.00	-0.40	+0.6	1.20	-0.40	+0.9	1.8	-0.6	+1.4	2.8	-0.9	+2.2	4.4	-1.4
.40	.71	+0.40	0.70	-0.30	+0.7	1.10	-0.40	+0.7	1.40	-0.40	+1.0	2.2	-0.7	+1.6	3.4	-1.0	+2.8	5.6	-1.6
.71	1.19	+0.50	0.90	-0.40	+0.8	1.30	-0.50	+0.8	1.70	-0.50	+1.2	2.6	-0.8	+2.0	4.2	-1.2	+3.5	7.0	-2.0
1.19	1.97	+0.60	1.00	-0.40	+1.0	1.60	-0.60	+1.0	2.00	-0.60	+1.6	3.4	-1.0	+2.5	5.3	-1.6	+4.0	8.5	-2.5
1.97	3.15	+0.70	1.30	-0.50	+1.2	2.10	-0.70	+1.2	2.30	-0.70	+1.8	4.0	-1.2	+3.0	6.3	-1.8	+4.5	10.0	-3.0
3.15	4.73	+0.90	1.60	-0.60	+1.4	2.50	-0.90	+1.4	2.90	-0.90	+2.2	4.8	-1.4	+4.0	7.7	-2.2	+5.0	11.5	-3.5
4.73	7.09	+1.00	1.90	-0.70	+1.6	2.80	-1.00	+1.6	3.50	-1.00	+2.5	5.6	-1.6	+4.5	8.7	-2.5	+6.0	13.5	-4.0
7.09	9.85	+1.20	2.20	-0.80	+1.8	3.20	-1.20	+1.8	4.20	-1.20	+2.8	6.6	-1.8	+5.0	10.3	-2.8	+7.0	16.5	-4.5
9.85	12.41	+1.20	2.30	-0.90	+2.0	3.40	-1.20	+2.0	4.70	-1.20	+3.0	7.5	-2.0	+6.0	12.0	-3.0	+8.0	19.0	-5.0
12.41	15.75	+1.40	2.60	-1.00	+2.2	3.90	-1.40	+2.2	5.90	-1.40	+3.5	8.7	-2.2	+6.0	14.5	-3.5	+9.0	23.0	-6.0

VALUES IN MILLIMETERS

Nominal Size Range Millimeters		Class LN1 Light Press Fit			Class LN2 Medium Press Fit			Class LN3 Heavy Press Fit			Class LN4			Class LN5			Class LN6		
		Hole Tol. GR6	Maximum Interference	Shaft Tol. gr5	Hole Tol. H7	Maximum Interference	Shaft Tol. p6	Hole Tol. H7	Maximum Interference	Shaft Tol. t6	Hole Tol. GR8	Maximum Interference	Shaft Tol. gr7	Hole Tol. GR9	Maximum Interference	Shaft Tol. gr8	Hole Tol. GR10	Maximum Interference	Shaft Tol. gr9
Over	To	-0		+0	-0		+0	-0		+0	-0		+0	-0		+0	-0		+0
0	3	+0.006		-0.004	+0.010	0.016	-0.006	+0.010	0.019	-0.006	+0.015	0.030	-0.010	+0.025	0.046	-0.015	+0.041	0.076	-0.025
3	6	+0.008		-0.005	+0.013	0.020	-0.008	+0.013	0.023	-0.008	+0.018	0.038	-0.013	+0.030	0.059	-0.018	+0.046	0.091	-0.030
6	10	+0.010		-0.006	+0.015	0.025	-0.010	+0.015	0.030	-0.010	+0.023	0.046	-0.015	+0.036	0.071	-0.023	+0.056	0.112	-0.036
10	18	+0.010		-0.008	+0.018	0.028	-0.010	+0.018	0.036	-0.010	+0.025	0.056	-0.018	+0.041	0.086	-0.025	+0.071	0.142	-0.041
18	30	+0.013		-0.010	+0.020	0.033	-0.013	+0.020	0.044	-0.013	+0.030	0.066	-0.020	+0.051	0.107	-0.030	+0.089	0.178	-0.051
30	50	+0.015		-0.010	+0.025	0.041	-0.015	+0.025	0.051	-0.015	+0.041	0.086	-0.025	+0.064	0.135	-0.041	+0.102	0.216	-0.064
50	80	+0.018		-0.013	+0.030	0.054	-0.018	+0.030	0.059	-0.018	+0.046	0.102	-0.030	+0.076	0.160	-0.046	+0.114	0.254	-0.076
80	120	+0.023		-0.015	+0.036	0.064	-0.023	+0.036	0.074	-0.023	+0.056	0.122	-0.036	+0.102	0.196	-0.056	+0.127	0.292	-0.102
120	180	+0.025		-0.018	+0.041	0.071	-0.025	+0.041	0.089	-0.025	+0.064	0.142	-0.041	+0.114	0.221	-0.064	+0.152	0.343	-0.114
180	250	+0.030		-0.020	+0.046	0.081	-0.030	+0.046	0.107	-0.030	+0.071	0.168	-0.046	+0.127	0.262	-0.071	+0.178	0.419	-0.127
250	315	+0.030		-0.023	+0.051	0.086	-0.030	+0.051	0.119	-0.030	+0.076	0.191	-0.051	+0.152	0.305	-0.076	+0.203	0.483	-0.152
315	400	+0.036		-0.025	+0.056	0.099	-0.036	+0.056	0.150	-0.036	+0.089	0.221	-0.056	+0.152	0.368	-0.089	+0.229	0.584	-0.152

TABLE 12   From *Drafting for Trades and Industry—Mechanical and Electronic*, Nelson. Delmar Publishers Inc.

FORCE AND SHRINK FITS

VALUES IN THOUSANDTHS OF AN INCH

Nominal Size Range Inches		Class FN1 Light Drive Fit			Class FN2 Medium Drive Fit			Class FN3 Heavy Drive Fit			Class FN4 Shrink Fit			Class FN5 Heavy Shrink Fit		
		Hole Tol. GR6	Maximum Interference	Shaft Tol. GR5	Hole Tol. GR7	Maximum Interference	Shaft Tol. GR6	Hole Tol. GR7	Maximum Interference	Shaft Tol. GR6	Hole Tol. GR7	Maximum Interference	Shaft Tol. GR6	Hole Tol. GR8	Maximum Interference	Shaft Tol. GR7
Over	To	-0		+0	-0		+0	-0		+0	-0		+0	-0		+0
0	.12	+0.25	0.50	-0.15	+0.40	0.85	-0.25				+0.40	0.95	-0.25	+0.60	1.30	-0.40
.12	.24	+0.30	0.60	-0.20	+0.50	1.00	-0.30				+0.50	1.20	-0.30	+0.70	1.70	-0.50
.24	.40	+0.40	0.75	-0.25	+0.60	1.40	-0.40				+0.60	1.60	-0.40	+0.90	2.00	-0.60
.40	.56	+0.40	0.80	-0.30	+0.70	1.60	-0.40				+0.70	1.80	-0.40	+1.00	2.30	-0.70
.56	.71	+0.40	0.90	-0.30	+0.70	1.60	-0.40				+0.70	1.80	-0.40	+1.00	2.50	-0.70
.71	.95	+0.50	1.10	-0.40	+0.80	1.90	-0.50				+0.80	2.10	-0.50	+1.20	3.00	-0.80
.95	1.19	+0.50	1.20	-0.40	+0.80	1.90	-0.50	+0.80	2.10	-0.50	+0.80	2.30	-0.50	+1.20	3.30	-0.80
1.19	1.58	+0.60	1.30	-0.40	+1.00	2.40	-0.60	+1.00	2.60	-0.60	+1.00	3.10	-0.60	+1.60	4.00	-1.00
1.58	1.97	+0.60	1.40	-0.40	+1.00	2.40	-0.60	+1.00	2.80	-0.60	+1.00	3.40	-0.60	+1.60	5.00	-1.00
1.97	2.56	+0.70	1.80	-0.50	+1.20	2.70	-0.70	+1.20	3.20	-0.70	+1.20	4.20	-0.70	+1.80	6.20	-1.20
2.56	3.15	+0.70	1.90	-0.50	+1.20	2.90	-0.70	+1.20	3.70	-0.70	+1.20	4.70	-0.70	+1.80	7.20	-1.20
3.15	3.94	+0.90	2.40	-0.60	+1.40	3.70	-0.90	+1.40	4.40	-0.70	+1.40	5.90	-0.90	+2.20	8.40	-1.40

VALUES IN MILLIMETERS

Nominal Size Range Millimeters		Class FN1 Light Drive Fit			Class FN2 Medium Drive Fit			Class FN3 Heavy Drive Fit			Class FN4 Shrink Fit			Class FN5 Heavy Shrink Fit		
		Hole Tol. GR6	Maximum Interference	Shaft Tol. gr5	Hole Tol. H7	Maximum Interference	Shaft Tol. s6	Hole Tol. H7	Maximum Interference	Shaft Tol. t6	Hole Tol. GR8	Maximum Interference	Shaft Tol. gr7	Hole Tol. H8	Maximum Interference	Shaft Tol. t7
Over	To	-0		+0	-0		+0	-0		+0	-0		+0	-0		+0
0	3	+0.006	0.013	-0.004	+0.010	0.216	-0.006				+0.010	0.024	-0.006	+0.015	0.033	-0.010
3	6	+0.007	0.015	-0.005	+0.013	0.025	-0.007				+0.013	0.030	-0.007	+0.018	0.043	-0.013
6	10	+0.010	0.019	-0.006	+0.015	0.036	-0.010				+0.015	0.041	-0.010	+0.023	0.051	-0.015
10	14	+0.010	0.020	-0.008	+0.018	0.041	-0.010				+0.018	0.046	-0.010	+0.025	0.058	-0.018
14	18	+0.010	0.023	-0.008	+0.018	0.041	-0.010				+0.018	0.046	-0.010	+0.025	0.064	-0.018
18	24	+0.013	0.028	-0.010	+0.020	0.048	-0.013				+0.020	0.053	-0.013	+0.030	0.076	-0.020
24	30	+0.013	0.030	-0.010	+0.020	0.048	-0.013	+0.020	0.053	-0.013	+0.020	0.058	-0.013	+0.030	0.084	-0.020
30	40	+0.015	0.033	-0.010	+0.025	0.061	-0.015	+0.025	0.066	-0.015	+0.025	0.079	-0.015	+0.041	0.102	-0.025
40	50	+0.015	0.036	-0.010	+0.025	0.061	-0.015	+0.025	0.071	-0.015	+0.025	0.086	-0.015	+0.041	0.127	-0.025
50	65	+0.018	0.046	-0.013	+0.030	0.069	-0.018	+0.030	0.082	-0.018	+0.030	0.107	-0.018	+0.046	0.157	-0.030
65	80	+0.018	0.048	-0.013	+0.030	0.074	-0.018	+0.030	0.094	-0.018	+0.030	0.119	-0.018	+0.046	0.183	-0.030
80	100	+0.023	0.061	-0.015	+0.035	0.094	-0.023	+0.035	0.112	-0.023	+0.036	0.150	-0.023	+0.056	0.213	-0.036

TABLE 13 From *Drafting for Trades and Industry—Mechanical and Electronic*, Nelson. Delmar Publishers Inc.

UNIFIED STANDARD SCREW THREAD SERIES

Sizes		Basic Major Diameter	Series with graded pitches			Series with constant pitches								Sizes
						THREADS PER INCH								
Primary	Secondary		Coarse UNC	Fine UNF	Extra fine UNEF	4UN	6UN	8UN	12UN	16UN	20UN	28UN	32UN	
0		0.0600	—	80	—	—	—	—	—	—	—	—	—	0
	1	0.0730	64	72	—	—	—	—	—	—	—	—	—	1
2		0.0860	56	64	—	—	—	—	—	—	—	—	—	2
	3	0.0990	48	56	—	—	—	—	—	—	—	—	—	3
4		0.1120	40	48	—	—	—	—	—	—	—	—	—	4
5		0.1250	40	44	—	—	—	—	—	—	—	—	—	5
6		0.1380	32	40	—	—	—	—	—	—	—	—	UNC	6
8		0.1640	32	36	—	—	—	—	—	—	—	—	UNC	8
10		0.1900	24	32	—	—	—	—	—	—	—	—	UNF	10
	12	0.2160	24	28	32	—	—	—	—	—	—	UNF	UNEF	12
¼		0.2500	20	28	32	—	—	—	—	—	UNC	UNF	UNEF	¼
5⁄16		0.3125	18	24	32	—	—	—	—	—	20	28	UNEF	5⁄16
3⁄8		0.3750	16	24	32	—	—	—	—	UNC	20	28	UNEF	3⁄8
7⁄16		0.4375	14	20	28	—	—	—	—	16	UNF	UNEF	32	7⁄16
½		0.5000	13	20	28	—	—	—	—	16	UNF	UNEF	32	½
9⁄16		0.5625	12	18	24	—	—	—	UNC	16	20	28	32	9⁄16
5⁄8		0.6250	11	18	24	—	—	—	12	16	20	28	32	5⁄8
	11⁄16	0.6875	—	—	24	—	—	—	12	16	20	28	32	11⁄16
¾		0.7500	10	16	20	—	—	—	12	UNF	UNEF	28	32	¾
	13⁄16	0.8125	—	—	20	—	—	—	12	16	UNEF	28	32	13⁄16
7⁄8		0.8750	9	14	20	—	—	—	12	16	UNEF	28	32	7⁄8
	15⁄16	0.9375	—	—	20	—	—	—	12	16	UNEF	28	32	15⁄16
1		1.0000	8	12	20	—	—	UNC	UNF	16	UNEF	28	32	1
	1 1⁄16	1.0625	—	—	18	—	—	8	12	16	20	28	—	1 1⁄16
1 1⁄8		1.1250	7	12	18	—	—	8	UNF	16	20	28	—	1 1⁄8
	1 3⁄16	1.1875	—	—	18	—	—	8	12	16	20	28	—	1 3⁄16
1 ¼		1.2500	7	12	18	—	—	8	UNF	16	20	28	—	1 ¼
	1 5⁄16	1.3125	—	—	18	—	—	8	12	16	20	28	—	1 5⁄16
1 3⁄8		1.3750	6	12	18	—	UNC	8	UNF	16	20	28	—	1 3⁄8
	1 7⁄16	1.4375	—	—	18	—	6	8	12	16	20	28	—	1 7⁄16
1 ½		1.5000	6	12	18	—	UNC	8	UNF	16	20	28	—	1 ½
	1 9⁄16	1.5625	—	—	18	—	6	8	12	16	20	—	—	1 9⁄16
1 5⁄8		1.6250	—	—	18	—	6	8	12	16	20	—	—	1 5⁄8
	1 11⁄16	1.6875	—	—	18	—	6	8	12	16	20	—	—	1 11⁄16
1 ¾		1.7500	5	—	—	—	6	8	12	16	20	—	—	1 ¾
	1 13⁄16	1.8125	—	—	—	—	6	8	12	16	20	—	—	1 13⁄16
1 7⁄8		1.8750	—	—	—	—	6	8	12	16	20	—	—	1 7⁄8
	1 15⁄16	1.9375	—	—	—	—	6	8	12	16	20	—	—	1 15⁄16
2		2.0000	4½	—	—	—	6	8	12	16	20	—	—	2
	2 1⁄8	2.1250	—	—	—	—	6	8	12	16	20	—	—	2 1⁄8
2 ¼		2.2500	4½	—	—	—	6	8	12	16	20	—	—	2 ¼
	2 3⁄8	2.3750	—	—	—	—	6	8	12	16	20	—	—	2 3⁄8
2 ½		2.5000	4	—	—	UNC	6	8	12	16	20	—	—	2 ½
	2 5⁄8	2.6250	—	—	—	4	6	8	12	16	20	—	—	2 5⁄8
2 ¾		2.7500	4	—	—	UNC	6	8	12	16	20	—	—	2 ¾
	2 7⁄8	2.8750	—	—	—	4	6	8	12	16	20	—	—	2 7⁄8
3		3.0000	4	—	—	UNC	6	8	12	16	20	—	—	3
	3 1⁄8	3.1250	—	—	—	4	6	8	12	16	—	—	—	3 1⁄8
3 ¼		3.2500	4	—	—	UNC	6	8	12	16	—	—	—	3 ¼
	3 3⁄8	3.3750	—	—	—	4	6	8	12	16	—	—	—	3 3⁄8
3 ½		3.5000	4	—	—	UNC	6	8	12	16	—	—	—	3 ½
	3 5⁄8	3.6250	—	—	—	4	6	8	12	16	—	—	—	3 5⁄8
3 ¾		3.7500	4	—	—	UNC	6	8	12	16	—	—	—	3 ¾
	3 7⁄8	3.8750	—	—	—	4	6	8	12	16	—	—	—	3 7⁄8
4		4.0000	4	—	—	UNC	6	8	12	16	—	—	—	4
	4 1⁄8	4.1250	—	—	—	4	6	8	12	16	—	—	—	4 1⁄8
4 ¼		4.2500	—	—	—	4	6	8	12	16	—	—	—	4 ¼
	4 3⁄8	4.3750	—	—	—	4	6	8	12	16	—	—	—	4 3⁄8
4 ½		4.5000	—	—	—	4	6	8	12	16	—	—	—	4 ½
	4 5⁄8	4.6250	—	—	—	4	6	8	12	16	—	—	—	4 5⁄8
4 ¾		4.7500	—	—	—	4	6	8	12	16	—	—	—	4 ¾
	4 7⁄8	4.8750	—	—	—	4	6	8	12	16	—	—	—	4 7⁄8
5		5.0000	—	—	—	4	6	8	12	16	—	—	—	5
	5 1⁄8	5.1250	—	—	—	4	6	8	12	16	—	—	—	5 1⁄8
5 ¼		5.2500	—	—	—	4	6	8	12	16	—	—	—	5 ¼
	5 3⁄8	5.3750	—	—	—	4	6	8	12	16	—	—	—	5 3⁄8
5 ½		5.5000	—	—	—	4	6	8	12	16	—	—	—	5 ½
	5 5⁄8	5.6250	—	—	—	4	6	8	12	16	—	—	—	5 5⁄8
5 ¾		5.7500	—	—	—	4	6	8	12	16	—	—	—	5 ¾
	5 7⁄8	5.8750	—	—	—	4	6	8	12	16	—	—	—	5 7⁄8
6		6.0000	—	—	—	4	6	8	12	16	—	—	—	6

TABLE 14

DRILL AND TAP SIZES

DECIMAL EQUIVALENTS AND TAP DRILL SIZES OF WIRE GAGE LETTER AND FRACTIONAL SIZE DRILLS (TAP DRILL SIZES BASED ON 75% MAXIMUM THREAD)

FRACTIONAL SIZE DRILLS INCHES	WIRE GAGE DRILLS	DECIMAL EQUIVALENT INCHES	TAP SIZES — SIZE OF THREAD	TAP SIZES — THREADS PER INCH
1/64	80	.0135		
	79	.0145		
	....	.0156		
	78	.0160		
	77	.0180		
	76	.0200		
	75	.0210		
	74	.0225		
	73	.0240		
	72	.0250		
	71	.0260		
	70	.0280		
	69	.0292		
	68	.0310		
1/32	....	.0312		
	67	.0320		
	66	.0330		
	65	.0350		
	64	.0360		
	63	.0370		
	62	.0380		
	61	.0390		
	60	.0400		
	59	.0410		
	58	.0420		
	57	.0430		
	56	.0465		
3/64	....	.0469	0	80
	55	.0520		
	54	.0550		
	53	.0595	1	64
1/16	....	.0625		72
	52	.0635		
	51	.0670		
	50	.0700	2	56
	49	.0730		64
	48	.0760		
5/64	....	.0781		
	47	.0785	3	48
	46	.0810		
	45	.0820	3	56
	44	.0860		
	43	.0890	4	40
	42	.0935	4	48
3/32	....	.0937		
	41	.0960		
	40	.0980		
	39	.0995		
	38	.1015	5	40
	37	.1040	5	44
	36	.1065	6	32
7/64	....	.1094		
	35	.1100		
	34	.1110		
	33	.1130	6	40
	32	.1160		
1/8	31	.1200		
	....	.1250		
	30	.1285		
	29	.1360	8	32
	28	.1405		36

FRACTIONAL SIZE DRILLS INCHES	WIRE GAGE DRILLS	DECIMAL EQUIVALENT INCHES	TAP SIZES — SIZE OF THREAD	TAP SIZES — THREADS PER INCH
9/64	....	.1406		
	27	.1440		
	26	.1470		
	25	.1495	10	24
	24	.1520		
	23	.1540		
5/32	....	.1562		
	22	.1570		
	21	.1590	10	32
	20	.1610		
	19	.1660		
	18	.1695		
11/64	....	.1719		
	17	.1730		
	16	.1770	12	24
	15	.1800		
	14	.1820	12	28
	13	.1850		
3/16	....	.1875		
	12	.1890		
	11	.1910		
	10	.1935		
	9	.1960		
	8	.1990		
	7	.2010	1/4	20
13/64	....	.2031		
	6	.2040		
	5	.2055		
	4	.2090		
	3	.2130	1/4	28
7/32	....	.2187		
	2	.2210		
	1	.2280		
	A	.2340		
15/64	....	.2344		
	B	.2380		
	C	.2420		
	D	.2460		
1/4	E	.2500		
	F	.2570	5/16	18
	G	.2610		
17/64	....	.2656		
	H	.2660		
	I	.2720	5/16	24
	J	.2770		
	K	.2810		
9/32	....	.2812		
	L	.2900		
	M	.2950		
19/64	....	.2969		
	N	.3020		
5/16	....	.3125	3/8	16
	O	.3160		
	P	.3230		
21/64	....	.3281		
	Q	.3320	3/8	24
	R	.3390		
11/32	....	.3437		
	S	.3480		
	T	.3580		

FRACTIONAL SIZE DRILLS INCHES	WIRE GAGE DRILLS	DECIMAL EQUIVALENT INCHES	TAP SIZES — SIZE OF THREAD	TAP SIZES — THREADS PER INCH
23/64	....	.3594		
	U	.3680	7/16	14
3/8	....	.3750		
	V	.3770		
	W	.3860		
25/64	....	.3906	7/16	20
	X	.3970		
	Y	.4040		
13/32	....	.4062		
	Z	.4130		
27/64	....	.4219	1/2	13
7/16	....	.4375		
29/64	....	.4531	1/2	20
15/32	....	.4687		
31/64	....	.4844	9/16	12
1/2	....	.5000		
33/64	....	.5156	9/16	18
17/32	....	.5312	5/8	11
35/64	....	.5469		
9/16	....	.5625		
37/64	....	.5781	5/8	18
19/32	....	.5937		
39/64	....	.6094		
5/8	....	.6250		
41/64	....	.6406		
21/32	....	.6562	3/4	10
43/64	....	.6719		
11/16	....	.6875	3/4	16
45/64	....	.7031		
23/32	....	.7187		
47/64	....	.7344		
3/4	....	.7500		
49/64	....	.7656	7/8	9
25/32	....	.7812		
51/64	....	.7969		
13/16	....	.8125	7/8	14
53/64	....	.8281		
27/32	....	.8437		
55/64	....	.8594		
7/8	....	.8750	1	8
57/64	....	.8906		
29/32	....	.9062		
59/64	....	.9219		
15/16	....	.9375	1	12
61/64	....	.9531		
31/32	....	.9687		
63/64	....	.9844	1 1/8	7
1	....	1.0000		

TABLE 15

DRILLED HOLE TOLERANCE (UNDER NORMAL SHOP CONDITIONS)

STANDARD DRILL SIZE				TOLERANCE IN DECIMALS	
DRILL SIZE				PLUS	MINUS
Number	Fraction	Decimal	Metric (mm)		
80		0.0135	0.3412	0.0023	
79		0.0145	0.3788	0.0024	
—	1/64	0.0156	0.3969	0.0025	
78		0.0160	0.4064	0.0025	
77		0.0180	0.4572	0.0026	
76		0.0200	0.5080	0.0027	
75		0.0210	0.5334	0.0027	
74		0.0225	0.5631	0.0028	
73		0.0240	0.6096	0.0028	
72		0.0250	0.6350	0.0029	
71		0.0260	0.6604	0.0029	
70		0.0280	0.7112	0.0030	.0005
69		0.0292	0.7483	0.0030	
68		0.0310	0.7874	0.0031	
—	1/32	0.0312	0.7937	0.0031	
67		0.0320	0.8128	0.0031	
66		0.0330	0.8382	0.0032	
65		0.0350	0.8890	0.0032	
64		0.0360	0.9144	0.0033	
63		0.0370	0.9398	0.0033	
62		0.0380	0.9652	0.0033	
61		0.0390	0.9906	0.0033	
60		0.0400	1.0160	0.0034	
59		0.0410	1.0414	0.0034	
58		0.0420	1.0668	0.0034	
57		0.0430	1.0922	0.0035	
56		0.0465	1.1684	0.0035	
—	3/64	0.0469	1.1906	0.0036	
55		0.0520	1.3208	0.0037	
54		0.0550	1.3970	0.0038	
53		0.0595	1.5122	0.0039	
—	1/16	0.0625	1.5875	0.0039	
52		0.0635	1.6002	0.0039	
51		0.0670	1.7018	0.0040	
50		0.0700	1.7780	0.0041	
49		0.0730	1.8542	0.0041	.001
48		0.0760	1.9304	0.0042	
—	5/64	0.0781	1.9844	0.0042	
47		0.0785	2.0001	0.0042	
46		0.0810	2.0574	0.0043	
45		0.0820	2.0828	0.0043	
44		0.0860	2.1844	0.0044	
43		0.0890	2.2606	0.0044	
42		0.0935	2.3622	0.0045	
—	3/32	0.0937	2.3812	0.0045	
41		0.0960	2.4384	0.0045	
40		0.0980	2.4892	0.0046	
39		0.0995	2.5377	0.0046	
38		0.1015	2.5908	0.0046	
37		0.1040	2.6416	0.0047	
36		0.1065	2.6924	0.0047	
—	7/64	0.1094	2.7781	0.0047	

TABLE 16

DRILLED HOLE TOLERANCE (UNDER NORMAL SHOP CONDITIONS)

No./Letter	Fraction	Decimal	Metric (mm)	PLUS	MINUS
		STANDARD DRILL SIZE		TOLERANCE IN DECIMALS	
	DRILL SIZE				
35		0.1100	2.7490	0.0047	
34		0.1110	2.8194	0.0048	
33		0.1130	2.8702	0.0048	
32		0.1160	2.9464	0.0048	
31		0.1200	3.0480	0.0049	
—	1/8	0.1250	3.1750	0.0050	
30		0.1285	3.2766	0.0050	
29		0.1360	3.4544	0.0051	
28		0.1405	3.5560	0.0052	
—	9/64	0.1406	3.5719	0.0052	
27		0.1440	3.6576	0.0052	
26		0.1470	3.7338	0.0052	
25		0.1495	3.7886	0.0053	
24		0.1520	3.8608	0.0053	
23		0.1540	3.9116	0.0053	
—	5/32	0.1562	3.9687	0.0053	
22		0.1570	3.9878	0.0053	
21		0.1590	4.0386	0.0054	
20		0.1610	4.0894	0.0054	
19		0.1660	4.2164	0.0055	
18		0.1695	4.3180	0.0055	
—	11/64	0.1719	4.3656	0.0055	
17		0.1730	4.3942	0.0055	
16		0.1770	4.4958	0.0056	.001
15		0.1800	4.5720	0.0056	
14		0.1820	4.6228	0.0057	
13		0.1850	4.6990	0.0057	
—	3/16	0.1875	4.7625	0.0057	
12		0.1890	4.8006	0.0057	
11		0.1910	4.8514	0.0057	
10		0.1935	4.9276	0.0058	
9		0.1960	4.9784	0.0058	
8		0.1990	5.0800	0.0058	
7		0.2010	5.1054	0.0058	
—	13/64	0.2031	5.1594	0.0058	
6		0.2040	5.1816	0.0058	
5		0.2055	5.2070	0.0059	
4		0.2090	5.3086	0.0059	
3		0.2130	5.4102	0.0059	
—	7/32	0.2187	5.5562	0.0060	
2		0.2210	5.6134	0.0060	
1		0.2280	5.7912	0.0061	
A		0.2340	5.9436	0.0061	
—	15/64	0.2344	5.9531	0.0061	
B		0.2380	6.0452	0.0061	
C		0.2420	6.1468	0.0062	
D		0.2460	6.2484	0.0062	
E	1/4	0.2500	6.3500	0.0063	
F		0.2570	6.5278	0.0063	
G		0.2610	6.6294	0.0063	
—	17/64	0.2656	6.7469	0.0064	
H		0.2660	6.7564	0.0064	
I		0.2720	6.9088	0.0064	.002
J		0.2770	7.0358	0.0065	
K		0.2810	7.1374	0.0065	
—	9/32	0.2812	7.1437	0.0065	
L		0.2900	7.3660	0.0066	
M		0.2950	7.4930	0.0066	
—	19/64	0.2969	7.5406	0.0066	

TABLE 16   *(Continued)*

DRILLED HOLE TOLERANCE (UNDER NORMAL SHOP CONDITIONS)

STANDARD DRILL SIZE				TOLERANCE IN DECIMALS	
DRILL SIZE				PLUS	MINUS
Letter	Fraction	Decimal	Metric (mm)		
N		0.3020	7.6708	0.0067	
—	5/16	0.3125	7.9375	0.0067	
O		0.3160	8.0264	0.0068	
P		0.3230	8.2042	0.0068	
—	21/64	0.3281	8.3344	0.0068	
Q		0.3320	8.4328	0.0069	
R		0.3390	8.6106	0.0069	
—	11/32	0.3437	8.7312	0.0070	
S		0.3480	8.8392	0.0070	
T		0.3580	9.0932	0.0071	
—	23/64	0.3594	9.1281	0.0071	
U		0.3680	9.3472	0.0071	
—	3/8	0.3750	9.5250	0.0072	
V		0.3770	9.5758	0.0072	
W		0.3860	9.8044	0.0072	
—	25/64	0.3906	9.9219	0.0073	
X		0.3970	10.0838	0.0073	
Y		0.4040	10.2616	0.0073	
—	13/32	0.4062	10.3187	0.0074	
Z		0.4130	10.4902	0.0074	.002
	27/64	0.4219	10.7156	0.0075	
	7/16	0.4375	10.1125	0.0075	
	29/64	0.4531	11.5094	0.0076	
	15/32	0.4687	11.9062	0.0077	
	31/64	0.4844	12.3031	0.0078	
	1/2	0.5000	12.7000	0.0079	
	33/64	0.5156	13.0968	0.0080	
	17/32	0.5312	13.4937	0.0081	
	35/64	0.5469	13.8906	0.0081	
	9/16	0.5625	14.2875	0.0082	
	37/64	0.5781	14.6844	0.0083	
	19/32	0.5927	15.0812	0.0084	
	39/64	0.6094	15.4781	0.0084	
	5/8	0.6250	15.8750	0.0085	
	41/64	0.6406	16.2719	0.0086	
	21/32	0.6562	16.6687	0.0086	
	43/64	0.6719	17.0656	0.0087	
	11/16	0.6875	17.4625	0.0088	
	45/64	0.7031	17.8594	0.0088	
	23/32	0.7187	18.2562	0.0089	
	47/64	0.7344	18.6532	0.0090	
	3/4	0.7500	19.0500	0.0090	
	49/64	0.7656	19.4469	0.0091	
	25/32	0.7812	19.8433	0.0092	
	51/64	0.7969	20.2402	0.0092	
	13/16	0.8125	20.6375	0.0093	
	53/64	0.8281	21.0344	0.0093	
	27/32	0.8437	21.4312	0.0094	
	55/64	0.8594	21.8281	0.0095	
	7/8	0.8750	22.2250	0.0095	.003
	57/64	0.8906	22.6219	0.0096	
	29/32	0.9062	23.0187	0.0096	
	59/64	0.9219	23.4156	0.0097	
	15/16	0.9375	23.8125	0.0097	
	61/64	0.9531	24.2094	0.0098	
	31/32	0.9687	24.6062	0.0098	
	63/64	0.9844	25.0031	0.0099	
	1	1.0000	25.4000	0.0100	

From Drafting for Trades and Industry—Mechanical and Electronic. Nelson. Delmar Publishers. Inc.

TABLE 16   *(Concluded)* From *Drafting for Trades and Industry—Mechanical and Electronic,* Nelson. Delmar Publishers Inc.

INCH-METRIC THREAD COMPARISON

INCH SERIES			METRIC			
Size	Dia.(In.)	TPI	Size	Dia. (In.)	Pitch (mm)	TPI (Approx)
			M1.4	.055	.3 .2	85 127
#0	.060	80				
			M1.6	.063	.35 .2	74 127
#1	.073	64 72				
			M2	.079	.4 .25	64 101
#2	.086	56 64				
			M2.5	.098	.45 .35	56 74
#3	.099	48 56				
#4	.112	40 48				
			M3	.118	.5 .35	51 74
#5	.125	40 44				
#6	.138	32 40				
			M4	.157	.7 .5	36 51
#8	.164	32 36				
#10	.190	24 32				
			M5	.196	.8 .5	32 51
			M6	.236	1.0 .75	25 34
¼	.250	20 28				
5/16	.312	18 24				
			M8	.315	1.25 1.0	20 25
3/8	.375	16 24				
			M10	.393	1.5 1.25	17 20
7/16	.437	14 20				
			M12	.472	1.75 1.25	14.5 20
½	.500	13 20				
			M14	.551	2 1.5	12.5 17
5/8	.625	11 18				
			M16	.630	2 1.5	12.5 17
			M18	.709	2.5 1.5	10 17
¾	.750	10 16				
			M20	.787	2.5 1.5	10 17
			M22	.866	2.5 1.5	10 17
7/8	.875	9 14				
			M24	.945	3 2	8.5 12.5
1"	1.000	8 12				
			M27	1.063	3 2	8.5 12.5

TABLE 17

ISO BASIC METRIC THREAD INFORMATION

Basic Major DIA & Pitch	Tap Drill DIA	INTERNAL THREADS		EXTERNAL THREADS		Clearance Hole
		Minor DIA MAX	Minor DIA MIN	Major DIA MAX	Major DIA MIN	
M1.6 × 0.35	1.25	1.321	1.221	1.576	1.491	1.9
M2 × 0.4	1.60	1.679	1.567	1.976	1.881	2.4
M2.5 × 0.45	2.05	2.138	2.013	2.476	2.013	2.9
M3 × 0.5	2.50	2.599	2.459	2.976	2.870	3.4
M3.5 × 0.6	2.90	3.010	2.850	3.476	3.351	4.0
M4 × 0.7	3.30	3.422	3.242	3.976	3.836	4.5
M5 × 0.8	4.20	4.334	4.134	4.976	4.826	5.5
M6 × 1	5.00	5.153	4.917	5.974	5.794	6.6
M8 × 1.25	6.80	6.912	6.647	7.972	7.760	9.0
M10 × 1.5	8.50	8.676	8.376	9.968	9.732	11.0
M12 × 1.75	10.20	10.441	10.106	11.966	11.701	13.5
M14 × 2	12.00	12.210	11.835	13.962	13.682	15.5
M16 × 2	14.00	14.210	13.835	15.962	15.682	17.5
M20 × 2.5	17.50	17.744	17.294	19.958	19.623	22.0
M24 × 3	21.00	21.252	20.752	23.952	23.577	26.0
M30 × 3.5	26.50	26.771	26.211	29.947	29.522	33.0
M36 × 4	32.00	32.270	31.670	35.940	35.465	39.0
M42 × 4.5	37.50	37.799	37.129	41.937	41.437	45.0
M48 × 5	43.00	43.297	42.587	47.929	47.399	52.0
M56 × 5.5	50.50	50.796	50.046	55.925	55.365	62.0
M64 × 6	58.00	58.305	57.505	63.920	63.320	70.0
M72 × 6	66.00	66.305	65.505	71.920	71.320	78.0
M80 × 6	74.00	74.305	73.505	79.920	79.320	86.0
M90 × 6	84.00	84.305	83.505	89.920	89.320	96.0
M100 × 6	94.00	94.305	93.505	99.920	99.320	107.0

TABLE 18

KEY AND KEYWAY SIZES

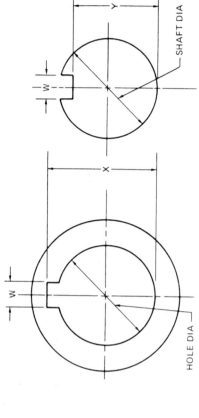

KEY SIZE = W x H
LENGTH (L) TO SUIT

Nom. Size (Inch)	– DIA. – (Shaft) Inch	mm	'X' (Collar) Inch	mm	'Y' (Shaft) Inch	mm
1/2	.500	12.700	.560	14.224	.430	10.922
9/16	.562	14.290	.623	15.824	.493	12.522
5/8	.625	15.875	.709	18.008	.517	13.132
11/16	.688	17.470	.773	18.618	.581	14.757
3/4	.750	19.050	.837	21.259	.644	16.357
13/16	.812	20.640	.900	22.860	.708	17.983
7/8	.875	22.225	.964	24.485	.771	19.583
15/16	.938	23.820	1.051	26.695	.791	20.091
1	1.000	25.400	1.114	28.295	.859	21.818
1 1/16	1.062	26.985	1.178	29.921	.923	23.444
1 1/8	1.125	28.575	1.241	31.521	.986	25.044
1 3/16	1.188	30.165	1.304	33.121	1.049	26.644
1 1/4	1.250	31.750	1.367	34.722	1.112	28.244
1 5/16	1.312	33.340	1.455	36.957	1.137	28.879
1 3/8	1.375	34.923	1.518	38.557	1.201	30.505

From *Drafting for Trades and Industry—Mechanical and Electronic*, Nelson. Delmar Publishers Inc.

Shaft Nom. Size – DIA. – From	To & Incl.	Square (W = H)	Type	Square Key From	To & Incl.	Tolerance
5/16 (8)	7/16 (11)	3/32 (2.38)	Bar Stock	--	3/4 (19.05)	+.000 – .002 (+.0000 – .0254)
7/16 (11)	9/16 (14)	1/8 (3.175)		--	1 1/2 (38.1)	+.000 – .003 (+.0000 – .0762)
9/16 (14)	7/8 (22)	3/16 (4.76)		3/4 (19.05)	2 1/2 (63.5)	+.000 – .004 (+.0000 – .1016)
7/8 (22)	1 1/4 (32)	1/4 (6.35)		1 1/2 (38.1)	3 1/2 (88.9)	+.000 – .006 (+.0000 – .1524)
1 1/4 (32)	1 3/8 (35)	5/16 (7.94)	Keystock	2 1/2 (63.5)	1 1/4 (31.75)	+.001 – .000 (+.0254 – .0000)
1 3/8 (35)	1 3/4 (44)	3/8 (9.53)		--	3 (76.2)	+.002 – .000 (+.0508 – .0000)
1 3/4 (44)	2 1/4 (57)	1/2 (12.7)		1 1/4 (31.75)	3 1/2 (88.9)	+.003 – .000 (+.0762 – .0000)
2 1/4 (57)	2 3/4 (70)	5/8 (15.88)		3 (76.2)		
2 3/4 (70)	3 1/4 (82)	3/4 (19.05)				
3 1/4 (82)	3 3/4 (95)	7/8 (22.23)				

(Figures in parenthesis = mm)

TABLE 19   From *Drafting for Trades and Industry—Mechanical and Electronic*, Nelson. Delmar Publishers Inc.

WOODRUFF KEY SIZES FOR DIFFERENT SHAFT DIAMETERS

Shaft Diameter	$\frac{5}{16}$ to $\frac{3}{8}$	$\frac{7}{16}$ to $\frac{1}{2}$	$\frac{9}{16}$ to $\frac{3}{4}$	$\frac{13}{16}$ to $\frac{15}{16}$	1 to $1\frac{3}{16}$	$1\frac{1}{4}$ to $1\frac{7}{16}$	$1\frac{1}{2}$ to $1\frac{3}{4}$	$1\frac{13}{16}$ to $2\frac{1}{8}$	$2\frac{3}{16}$ to $2\frac{1}{2}$
Key Numbers	204	304 305	404 405 406	505 506 507	606 607 608 609	807 808 809	810 811 812	1011 1012	1211 1212

TABLE 20

SHEET METAL AND WIRE GAGE DESIGNATION

GAGE NO.	AMERICAN OR BROWN & SHARPE'S A.W.G. OR B. & S.	UNITED STATES STANDARD	MANU-FACTURERS' STANDARD FOR SHEET STEEL	GAGE NO.
0000000	. . . .	.500	. . . .	0000000
000000	.5800	.469	. . . .	000000
00000	.5165	.438	. . . .	00000
0000	.4600	.406	. . . .	0000
000	.4096	.375	. . . .	000
00	.3648	.344	. . . .	00
0	.3249	.312	. . . .	0
1	.2893	.281	. . . .	1
2	.2576	.266	. . . .	2
3	.2294	.250	.2391	3
4	.2043	.234	.2242	4
5	.1819	.219	.2092	5
6	.1620	.203	.1943	6
7	.1443	.188	.1793	7
8	.1285	.172	.1644	8
9	.1144	.156	.1495	9
10	.1019	.141	.1345	10
11	.0907	.125	.1196	11
12	.0808	.109	.1046	12
13	.0720	.0938	.0897	13
14	.0642	.0781	.0747	14
15	.0571	.0703	.0673	15
16	.0508	.0625	.0598	16
17	.0453	.0562	.0538	17
18	.0403	.0500	.0478	18
19	.0359	.0438	.0418	19
20	.0320	.0375	.0359	20
21	.0285	.0344	.0329	21
22	.0253	.0312	.0299	22
23	.0226	.0281	.0269	23
24	.0201	.0250	.0239	24
25	.0179	.0219	.0209	25
26	.0159	.0188	.0179	26
27	.0142	.0172	.0164	27
28	.0126	.0156	.0149	28
29	.0113	.0141	.0135	29
30	.0100	.0125	.0120	30
31	.0089	.0109	.0105	31
32	.0080	.0102	.0097	32
33	.0071	.00938	.0090	33
34	.0063	.00859	.0082	34
35	.0056	.00781	.0075	35
36	.0050	.00703	.0067	36

TABLE 21

BEND ALLOWANCE FOR 90° BENDS (INCH)

Radii / Thickness	.031	.063	.094	.125	.156	.188	.219	.250	.281	.313	.344	.375	.438	.500	.531	.625
.013	.058	.108	.157	.205	.254	.304	.353	.402	.450	.501	.549	.598	.697	.794	.843	.991
.016	.060	.110	.159	.208	.256	.307	.355	.404	.453	.503	.552	.600	.699	.796	.845	.993
.020	.062	.113	.161	.210	.259	.309	.358	.406	.455	.505	.554	.603	.702	.799	.848	.995
.022	.064	.114	.163	.212	.260	.311	.359	.408	.457	.507	.556	.604	.703	.801	.849	.997
.025	.066	.116	.165	.214	.263	.313	.362	.410	.459	.509	.558	.607	.705	.803	.851	.999
.028	.068	.119	.167	.216	.265	.315	.364	.412	.461	.511	.560	.609	.708	.805	.854	1.001
.032	.071	.121	.170	.218	.267	.317	.366	.415	.463	.514	.562	.611	.710	.807	.856	1.004
.038	.075	.126	.174	.223	.272	.322	.371	.419	.468	.518	.567	.616	.715	.812	.861	1.008
.040	.077	.127	.176	.224	.273	.323	.372	.421	.469	.520	.568	.617	.716	.813	.862	1.010
.050		.134	.183	.232	.280	.331	.379	.428	.477	.527	.576	.624	.723	.821	.869	1.017
.064		.144	.192	.241	.290	.340	.389	.437	.486	.536	.585	.634	.732	.830	.878	1.026
.072			.198	.247	.296	.346	.394	.443	.492	.542	.591	.639	.738	.836	.885	1.032
.078			.202	.251	.300	.350	.399	.447	.496	.546	.595	.644	.743	.840	.889	1.036
.081			.204	.253	.302	.352	.401	.449	.498	.548	.598	.646	.745	.842	.891	1.038
.091				.212	.260	.309	.359	.408	.456	.505	.555	.604	.653	.752	.849	.898	1.045
.094				.214	.262	.311	.361	.410	.459	.507	.558	.606	.655	.754	.851	.900	1.048
.102				.268	.317	.367	.416	.464	.513	.563	.612	.661	.760	.857	.906	1.053
.109				.273	.321	.372	.420	.469	.518	.568	.617	.665	.764	.862	.910	1.058
.125				.284	.333	.383	.432	.480	.529	.579	.628	.677	.776	.873	.922	1.069
.156					.355	.405	.453	.502	.551	.601	.650	.698	.797	.895	.943	1.091
.188						.427	.476	.525	.573	.624	.672	.721	.820	.917	.966	1.114
.203								.535	.584	.634	.683	.731	.830	.928	.976	1.124
.218								.546	.594	.645	.693	.742	.841	.938	.987	1.135
.234								.557	.606	.656	.705	.753	.852	.950	.998	1.146
.250								.568	.617	.667	.716	.764	.863	.961	1.009	1.157

EXAMPLE: MATERIAL THICKNESS = 1/8", INSIDE RADII = 1/4" R. WHERE THESE CROSS = .480" BEND ALLOWANCE (B/A) PLUS TOTAL OF STRAIGHT LENGTHS = DEVELOPED LENGTH.

BEND ALLOWANCE FOR 90° BENDS (MILLIMETER)

Radii / Thickness	0.80	1.58	2.38	3.18	3.96	4.76	5.56	6.35	7.15	7.94	8.74	9.52	11.12	12.70	13.50	15.85
.330	1.46	2.74	3.98	5.22	6.45	7.73	8.96	10.20	11.44	12.71	13.95	15.19	17.70	20.18	21.42	25.16
.406	1.52	2.80	4.04	5.27	6.51	7.78	9.02	10.26	11.50	12.77	14.01	15.24	17.76	20.23	21.47	25.22
.508	1.59	2.87	4.11	5.34	6.58	7.86	9.09	10.33	11.57	12.84	14.08	15.32	17.83	20.30	21.54	25.29
.559	1.63	2.91	4.14	5.38	6.62	7.89	9.13	10.37	11.60	12.88	14.12	15.35	17.86	20.34	21.57	25.32
.635	1.68	2.96	4.20	5.43	6.67	7.95	9.18	10.42	11.66	12.93	14.17	15.40	17.92	20.39	21.63	25.38
.711	1.74	3.02	4.25	5.49	6.72	8.00	9.24	10.47	11.71	12.99	14.22	15.46	17.98	20.44	21.68	25.43
.813	1.81	3.08	4.32	5.56	6.79	8.07	9.30	10.54	11.78	13.05	14.29	15.53	18.04	20.52	21.75	25.50
.965	1.91	3.19	4.42	5.66	6.90	8.18	9.41	10.65	11.89	13.16	14.40	15.64	18.15	20.62	21.86	25.61
1.016	1.95	2.23	4.46	5.70	6.94	8.21	9.45	10.69	11.92	13.20	14.44	15.67	18.19	20.66	21.90	25.64
1.270		3.40	4.64	5.88	7.11	8.39	9.63	10.86	12.10	13.38	14.61	15.85	18.36	20.84	22.07	25.82
1.625		3.65	4.89	6.13	7.36	8.64	9.88	11.11	12.35	13.63	14.86	16.10	18.61	21.08	22.32	26.07
1.829			5.03	6.27	7.51	8.78	10.02	11.26	12.49	13.77	15.01	16.24	18.76	21.23	22.47	26.22
1.981			5.14	6.38	7.61	8.89	10.13	11.36	12.60	13.88	15.11	16.35	18.86	21.34	22.57	26.32
2.058			5.19	6.43	7.67	8.94	10.18	11.42	12.65	13.93	15.17	16.40	18.92	21.39	22.63	26.38
2.311			5.37	6.60	7.85	9.12	10.36	11.60	12.83	14.11	15.35	16.58	19.09	21.57	22.80	26.55
2.388			5.43	6.66	7.90	9.18	10.41	11.65	12.89	14.16	15.40	16.64	19.15	21.62	22.86	26.61
2.591				6.81	8.04	9.31	10.55	11.79	13.03	14.30	15.54	16.78	19.29	21.76	23.00	26.75
2.769				6.93	8.17	9.44	10.68	11.92	13.15	14.43	15.67	16.90	19.42	21.89	23.13	26.88
3.175				7.22	8.45	9.73	10.96	12.20	13.44	14.71	15.95	17.19	19.70	22.17	23.41	27.16
3.962					9.00	10.28	11.52	12.75	13.99	15.27	16.74	17.74	20.25	22.72	23.96	27.71
4.775					9.58	10.85	12.09	13.32	14.56	15.84	17.07	18.31	20.82	23.30	24.53	28.28
5.156						11.12	12.36	13.59	14.83	16.11	17.34	18.58	21.09	23.57	24.80	28.55
5.537								13.85	15.10	16.37	17.61	18.85	21.36	23.83	25.07	28.82
5.944								14.15	15.38	16.66	17.89	19.13	21.64	24.12	25.35	29.10
6.350								14.43	15.67	16.94	18.18	19.42	21.93	24.40	25.64	29.39

EXAMPLE: MATERIAL THICKNESS = 3.18 MM, INSIDE RADII = 6.35 MM R. WHERE THESE CROSS = 12.20 MM BEND ALLOWANCE (B/A) PLUS TOTAL OF STRAIGHT LENGTHS = DEVELOPED LENGTH.

TABLE 22   From *Drafting for Trades and Industry—Basic Skills*, Nelson. Delmar Publishers Inc.

BEND ALLOWANCE FOR EACH 1° OF BEND (INCH)

Radii Thickness	.031	.063	.094	.125	.156	.188	.219	.250	.281	.313	.344	.375	.438	.500	.531	.625
.013	.00064	.00120	.00174	.00228	.00282	.00338	.00392	.00446	.00500	.00556	.00610	.00664	.00774	.00883	.00937	.01101
.016	.00067	.00122	.00176	.00231	.00285	.00342	.00395	.00449	.00503	.00559	.00613	.00667	.00777	.00885	.00939	.01103
.020	.00069	.00125	.00179	.00233	.00287	.00343	.00397	.00452	.00506	.00561	.00616	.00670	.00780	.00888	.00942	.01106
.022	.00071	.00127	.00181	.00235	.00289	.00345	.00399	.00453	.00508	.00563	.00617	.00672	.00782	.00890	.00944	.01108
.025	.00074	.00129	.00184	.00238	.00292	.00348	.00402	.00456	.00510	.00566	.00610	.00674	.00784	.00892	.00946	.01110
.028	.00076	.00132	.00186	.00240	.00294	.00350	.00404	.00458	.00512	.00568	.00622	.00676	.00786	.00894	.00948	.01112
.032	.00079	.00134	.00189	.00243	.00297	.00353	.00407	.00461	.00515	.00571	.00625	.00679	.00789	.00897	.00951	.01115
.038	.00084	.00140	.00194	.00248	.00302	.00358	.00412	.00466	.00520	.00576	.00630	.00684	.00794	.00902	.00956	.01120
.040	.00085	.00141	.00195	.00249	.00303	.00359	.00413	.00468	.00522	.00577	.00632	.00686	.00796	.00904	.00958	.01122
.050		.00149	.00203	.00258	.00312	.00368	.00422	.00476	.00530	.00586	.00640	.00694	.00804	.00912	.00966	.01130
.064		.00160	.00214	.00268	.00322	.00378	.00432	.00486	.00540	.00596	.00650	.00704	.00814	.00922	.00976	.01140
.072			.00220	.00274	.00328	.00384	.00438	.00492	.00546	.00602	.00656	.00710	.00820	.00929	.00983	.01147
.078			.00225	.00279	.00333	.00389	.00443	.00497	.00551	.00607	.00661	.00715	.00825	.00933	.00987	.01152
.081			.00227	.00281	.00335	.00391	.00445	.00499	.00554	.00609	.00664	.00718	.00828	.00936	.00990	.01154
.091			.00235	.00289	.00343	.00399	.00453	.00507	.00561	.00617	.00671	.00725	.00835	.00944	.00998	.01162
.094			.00237	.00291	.00346	.00401	.00456	.00510	.00564	.00620	.00674	.00728	.00838	.00946	.00999	.01164
.102				.00298	.00352	.00408	.00462	.00516	.00570	.00626	.00680	.00734	.00844	.00952	.01006	.01170
.109				.00303	.00357	.00413	.00467	.00521	.00575	.00631	.00685	.00739	.00849	.00958	.01012	.01176
.125				.00316	.00370	.00426	.00480	.00534	.00588	.00644	.00698	.00752	.00862	.00970	.01024	.01188
.156					.00394	.00450	.00504	.00558	.00612	.00668	.00722	.00776	.00886	.00994	.01048	.01212
.188						.00475	.00529	.00583	.00637	.00693	.00747	.00802	.00911	.01019	.01073	.01237
.203								.00595	.00649	.00704	.00759	.00813	.00923	.01031	.01085	.01249
.218								.00606	.00660	.00716	.00770	.00824	.00934	.01042	.01097	.01261
.234								.00619	.00673	.00729	.00783	.00837	.00947	.01055	.01109	.01273
.250								.00631	.00685	.00741	.00795	.00849	.00959	.01068	.01122	.01286

EXAMPLE: MATERIAL THICKNESS = 1/8", INSIDE RADII = 1/4" R. WHERE THESE CROSS = .00534" IF THE INSIDE OF YOUR BEND IS 20° THE BEND ALLOWANCE (B/A) = .1068 (.00534 × 20°) BEND ALLOWANCE B/A PLUS TOTAL OF STRAIGHT LENGTHS = DEVELOPED LENGTH.

BEND ALLOWANCE FOR EACH 1° OF BEND (MILLIMETER)

Radii Thickness	0.80	1.58	2.38	3.18	3.96	4.76	5.56	6.35	7.15	7.94	8.74	9.52	11.12	12.70	13.50	15.85
.330	.0161	.0305	.0442	.0580	.0717	.0859	.0996	.1134	.1271	.1413	.1550	.1668	.1967	.2242	.2379	.2796
.406	.0169	.0311	.0448	.0586	.0723	.0865	.1002	.1140	.1277	.1419	.1556	.1694	.1973	.2248	.2385	.2802
.508	.0177	.0319	.0456	.0594	.0731	.0873	.1010	.1148	.1285	.1427	.1564	.1702	.1981	.2256	.2393	.2810
.559	.0181	.0323	.0460	.0598	.0735	.0877	.1014	.1152	.1289	.1431	.1568	.1706	.1985	.2260	.2397	.2814
.635	.0187	.0329	.0466	.0604	.0741	.0883	.1020	.1158	.1295	.1437	.1574	.1712	.1991	.2266	.2403	.2820
.711	.0193	.0335	.0472	.0610	.0747	.0889	.1026	.1164	.1301	.1443	.1580	.1718	.1997	.2272	.2409	.2826
.813	.0201	.0343	.0480	.0617	.0755	.0897	.1034	.1174	.1309	.1451	.1588	.1726	.2005	.2280	.2417	.2834
.965	.0213	.0355	.0492	.0629	.0767	.0909	.1046	.1183	.1321	.1463	.1600	.1737	.2017	.2291	.2429	.2845
1.016	.0217	.0358	.0496	.0633	.0771	.0913	.1050	.1187	.1325	.1467	.1604	.1741	.2021	.2295	.2433	.2849
1.270		.0378	.0516	.0653	.0790	.0932	.1070	.1207	.1345	.1486	.1624	.1761	.2040	.2315	.2453	.2869
1.625		.0406	.0543	.0681	.0818	.0960	.1097	.1235	.1372	.1514	.1652	.1789	.2066	.2343	.2480	.2897
1.829			.0559	.0697	.0834	.0976	.1113	.1251	.1388	.1530	.1667	.1805	.2084	.2359	.2496	.2913
1.981			.0571	.0709	.0846	.0988	.1125	.1263	.1400	.1542	.1679	.1817	.2096	.2371	.2508	.2925
2.058			.0577	.0715	.0852	.0994	.1131	.1269	.1406	.1548	.1685	.1823	.2102	.2377	.2514	.2931
2.311			.0597	.0734	.0872	.1014	.1151	.1288	.1426	.1568	.1705	.1842	.2122	.2396	.2534	.2950
2.388			.0603	.0740	.0878	.1020	.1157	.1294	.1432	.1574	.1711	.1848	.2128	.2402	.2540	.2956
2.591				.0756	.0894	.1035	.1173	.1310	.1448	.1589	.1727	.1864	.2143	.2418	.2556	.2972
2.769				.0770	.0907	.1049	.1187	.1324	.1461	.1603	.1741	.1878	.2157	.2432	.2570	.2986
3.175				.0802	.0939	.1081	.1218	.1356	.1493	.1635	.1772	.1910	.2189	.2464	.2601	.3018
3.962					.1001	.1142	.1280	.1417	.1555	.1696	.1834	.1971	.2250	.2525	.2663	.3079
4.775						.1206	.1343	.1481	.1618	.1760	.1897	.2035	.2314	.2589	.2726	.3143
5.156								.1510	.1648	.1789	.1927	.2064	.2344	.2618	.2756	.3172
5.537								.1540	.1677	.1819	.1957	.2094	.2373	.2648	.2785	.3202
5.944								.1572	.1709	.1851	.1988	.2126	.2405	.2680	.2817	.3234
6.350								.1603	.1741	.1883	.2020	.2157	.2437	.2711	.2849	.3265

EXAMPLE: MATERIAL THICKNESS = 3.175 MM, INSIDE RADII = 6.35 MM R. WHERE THESE CROSS = .1356 MM IF THE INSIDE OF YOUR BEND IS 20° THE BEND ALLOWANCE (B/A) = 2.712 MM (.1356 × 20°) BEND ALLOWANCE (B/A) PLUS TOTAL OF STRAIGHT LENGTHS = DEVELOPED LENGTH.

TABLE 23   From *Drafting for Trades and Industry—Basic Skills*, Nelson. Delmar Publishers Inc.

SPUR/PINION GEAR TOOTH PARTS

20 Degree Pressure Angles

Diametral Pitch	Circular Pitch	Circular Thickness	Addend.	Dedend.	Standard Fillet Radius
D.P.	C.P.	C.T.	A	D	
1	3.1416	1.5708	1.0000	1.2500	0.3000
2	1.5708	0.7854	0.5000	0.6250	0.1500
2.5	1.2566	0.6283	0.4000	0.5000	0.1200
3	1.0472	0.5236	0.3333	0.4167	0.1000
3.5	0.8976	0.4488	0.2857	0.3571	0.0857
4	0.7854	0.3927	0.2500	0.3125	0.0750
4.5	0.6981	0.3491	0.2222	0.2778	0.0667
5	0.6283	0.3142	0.2000	0.2500	0.0600
5.5	0.5712	0.2856	0.1818	0.2273	0.0545
6	0.5236	0.2618	0.1667	0.2083	0.0500
6.5	0.4833	0.2417	0.1538	0.1923	0.0462
7	0.4488	0.2244	0.1429	0.1786	0.0429
7.5	0.4189	0.2094	0.1333	0.1667	0.0400
8	0.3927	0.1963	0.1250	0.1563	0.0375
8.5	0.3696	0.1848	0.1176	0.1471	0.0353
9	0.3491	0.1745	0.1111	0.1389	0.0333
9.5	0.3307	0.1653	0.1053	0.1316	0.0316
10	0.3142	0.1571	0.1000	0.1250	0.0300
11	0.2856	0.1428	0.0909	0.1136	0.0273
12	0.2618	0.1309	0.0833	0.1042	0.0250
13	0.2417	0.1208	0.0769	0.0962	0.0231
14	0.2244	0.1122	0.0714	0.0893	0.0214
15	0.2094	0.1047	0.0667	0.0833	0.0200

TABLE 24

SURFACE TEXTURE ROUGHNESS

					Surface Roughness Average Obtainable by Common Production Methods							
				Roughness Height Rating in N Series of Roughness Grade, Microinches, µ in. AA								
Process	N12	N11	N10	N9	N8	N7	N6	N5	N4	N3	N2	N1
	2000	1000	500	250	125	63	32	16	8	4	2	1	.5

The ranges shown above are typical of the processes listed.

Higher or lower values may be obtained under special conditions.

KEY   █ Average Application   ▨ Less Frequent Application

From Interpreting Engineering Drawings. Jensen. Delmar Publishers Inc.

TABLE 25   From *Interpreting Engineering Drawings, Jensen.* Delmar Publishers Inc.

PROPERTIES, GRADE NUMBERS, AND USAGE OF STEEL ALLOYS

Class of Steel	*Grade Number	Properties	Uses
Carbon—Low 0.3% carbon	10xx	Tough—Less Strength	Rivet—Hooks—Chains – Shafts—Pressed Steel Products
Carbon—Medium 0.3% to 0.6% carbon	10xx	Tough & Strong	Gears—Shafts—Studs Various Machine Parts
Carbon—High 1.6% to 1.7%	10xx	Less Tough— Much Harder	Drills—Knives—Saws
Nickel	20xx	Tough & Strong	Axles—Connecting Rods—Crank Shafts
Nickel Chromium	30xx	Tough & Strong	Rings—Gears—Shafts – Piston Pins—Bolts – Studs—Screws
Molybdenum	40xx	Very Strong	Forgings—Shafts – Gears—Cams
Chromium	50xx	Hard W/Strength & Toughness	Ball Bearings—Roller Bearing—Springs – Gears—Shafts
Chromium Vanadium	60xx	Hard & Strong	Shafts—Axles—Gears – Dies—Punches—Drills
Chromium Nickel Stainless	60xx	Rust Resistance	Food Containers – Medical/Dental Surgical Instruments
Silicon—Manganese	90xx	Springiness	Large Springs

*The first two numbers indicate type of steel, the last two numbers indicate the approx. average carbon content—1010 steel indicates carbon steel w/approx. 0.10% carbon.

TABLE 26   From *Drafting for Trades and Industry—Mechanical and Electronic,* Nelson. Delmar Publishers Inc.

AMERICAN NATIONAL STANDARDS OF INTEREST TO DESIGNERS, ARCHITECTS, AND DRAFTERS

TITLE OF STANDARD

Abbreviations Y1.1-1989
American National Standard Drafting Practices
  Size and Format Y14.1M-1995
  Line Conventions and Lettering Y14.2M-1992
  Multi and Sectional View Drawings Y14.3M-1994
  Pictorial Drawing Y14.4M-1989 (R1994)
  Dimensioning and Tolerancing Y14.5M-1994
  Screw Threads Y14.6-1978 (R1993)
  Screw Threads (Metric Supplement) Y14.6aM-1981 (R1993)
  Gears and Splines
    Spur, Helical, and Racks Y14.7.1-1971 (R1993)
    Bevel and Hypoid Y14.7.2-1978 (R1994)
  Castings and Forgings Y14.8M-1989 (R1993)
  Springs Y14.13M-1981 (R1992)
  Electrical and Electronic Diagram Y14.15-1966 (R1988)
  Interconnection Diagrams Y14.15a-1971 (R1988)
  Information Sheet Y14.15b-1973 (R1988)
  Fluid Power Diagrams Y14.17-1966 (R1980)
  Digital Representation for Communication of Product Definition Data Y14.26M-1989
  Computer-Aided Preparation of Product Definition Data Dictionary of Terms Y14.26.3-1975
  Digital Representation of Physical Object Shapes .... Y14 Report
  Guideline—User Instructions .... Y14 Report No. 2
  Guideline—Design Requirements .... Y14 Report No. 3
  Ground Vehicle Drawing Practices .... In Preparation
  Chassis Frames Y14.32.1-1994
  Parts Lists, Data Lists, and Index Lists Y14.34M-1990
  Surface Texture Symbols Y14.36-1978 (R1993)
Illustrations for Publication and Projection Y15.1M-1979 (R1986)
Time Series Charts Y15.2M-1979 (R1986)
Process Charts Y15.3M-1979 (R1986)
Graphic Symbols for:
  Electrical and Electronics Diagrams .... with 315A-1986 (a supplement to Y32.2-1975) (R1994)
  Plumbing Y32.4-1977 (R1994)
  Use on Railroad Maps and Profiles Y32.7-1972 (R1994)
  Fluid Power Diagrams Y32.10-1967 (R1994)
  Process Flow Diagrams in Petroleum and Chemical Industries Y32.11-1961 (R1993)
  Mechanical and Acoustical Element as Used in Schematic Diagrams Y32.18-1972 (R1985)
  Pipe Fittings, Valves, and Piping Y32.2.3-1949 (R1994)
  Heating, Ventilating, and Air Conditioning Y32.2.4-1949 (R1993)
  Heat Power Apparatus Y32.2.6-1950 (R1993)
Letter Symbols for:
  Glossary of Terms Concerning Letter Symbols Y10.1-1972 (R1988)
  Quantities Used in Electrical Science and Electrical Engineering Y10.5-1968
  Structural Analysis Y10.8-1962
  Illuminating Engineering Y10.18-1967 (R1987)
  Mathematical Signs and Symbols for Use in Physical Sciences and Technology Y10.20-1990

TABLE 27

American National Standard Running and Sliding Fits (ANSI B4.1-1967, R1994)

Tolerance limits given in body of table are added or subtracted to basic size (as indicated by + or − sign) to obtain maximum and minimum sizes of mating parts.

Nominal Size Range, Inches	Class RC 1			Class RC 2			Class RC 3			Class RC 4		
	Clear-ance*	Standard Tolerance Limits		Clear-ance*	Standard Tolerance Limits		Clear-ance*	Standard Tolerance Limits		Clear-ance*	Standard Tolerance Limits	
		Hole H5	Shaft g4		Hole H6	Shaft g5		Hole H7	Shaft f6		Hole H8	Shaft f7
Over To	Values shown below are in thousandths of an inch											
0– 0.12	0.1 0.45	+ 0.2 0	− 0.1 − 0.25	0.1 0.55	+ 0.25 0	− 0.1 − 0.3	0.3 0.95	+ 0.4 0	− 0.3 − 0.55	0.3 1.3	+ 0.6 0	− 0.3 − 0.7
0.12– 0.24	0.15 0.5	+ 0.2 0	− 0.15 − 0.3	0.15 0.65	+ 0.3 0	− 0.15 − 0.35	0.4 1.12	+ 0.5 0	− 0.4 − 0.7	0.4 1.6	+ 0.7 0	− 0.4 − 0.9
0.24– 0.40	0.2 0.6	+ 0.25 0	− 0.2 − 0.35	0.2 0.85	+ 0.4 0	− 0.2 − 0.45	0.5 1.5	+ 0.6 0	− 0.5 − 0.9	0.5 2.0	+ 0.9 0	− 0.5 − 1.1
0.40– 0.71	0.25 0.75	+ 0.3 0	− 0.25 − 0.45	0.25 0.95	+ 0.4 0	− 0.25 − 0.55	0.6 1.7	+ 0.7 0	− 0.6 − 1.0	0.6 2.3	+ 1.0 0	− 0.6 − 1.3
0.71– 1.19	0.3 0.95	+ 0.4 0	− 0.3 − 0.55	0.3 1.2	+ 0.5 0	− 0.3 − 0.7	0.8 2.1	+ 0.8 0	− 0.8 − 1.3	0.8 2.8	+ 1.2 0	− 0.8 − 1.6
1.19– 1.97	0.4 1.1	+ 0.4 0	− 0.4 − 0.7	0.4 1.4	+ 0.6 0	− 0.4 − 0.8	1.0 2.6	+ 1.0 0	− 1.0 − 1.6	1.0 3.6	+ 1.6 0	− 1.0 − 2.0
1.97– 3.15	0.4 1.2	+ 0.5 0	− 0.4 − 0.7	0.4 1.6	+ 0.7 0	− 0.4 − 0.9	1.2 3.1	+ 1.2 0	− 1.2 − 1.9	1.2 4.2	+ 1.8 0	− 1.2 − 2.4
3.15– 4.73	0.5 1.5	+ 0.6 0	− 0.5 − 0.9	0.5 2.0	+ 0.9 0	− 0.5 − 1.1	1.4 3.7	+ 1.4 0	− 1.4 − 2.3	1.4 5.0	+ 2.2 0	− 1.4 − 2.8
4.73– 7.09	0.6 1.8	+ 0.7 0	− 0.6 − 1.1	0.6 2.3	+ 1.0 0	− 0.6 − 1.3	1.6 4.2	+ 1.6 0	− 1.6 − 2.6	1.6 5.7	+ 2.5 0	− 1.6 − 3.2
7.09– 9.85	0.6 2.0	+ 0.8 0	− 0.6 − 1.2	0.6 2.6	+ 1.2 0	− 0.6 − 1.4	2.0 5.0	+ 1.8 0	− 2.0 − 3.2	2.0 6.6	+ 2.8 0	− 2.0 − 3.8
9.85–12.41	0.8 2.3	+ 0.9 0	− 0.8 − 1.4	0.8 2.9	+ 1.2 0	− 0.8 − 1.7	2.5 5.7	+ 2.0 0	− 2.5 − 3.7	2.5 7.5	+ 3.0 0	− 2.5 − 4.5
12.41–15.75	1.0 2.7	+ 1.0 0	− 1.0 − 1.7	1.0 3.4	+ 1.4 0	− 1.0 − 2.0	3.0 6.6	+ 2.2 0	− 3.0 − 4.4	3.0 8.7	+ 3.5 0	− 3.0 − 5.2
15.75–19.69	1.2 3.0	+ 1.0 0	− 1.2 − 2.0	1.2 3.8	+ 1.6 0	− 1.2 − 2.2	4.0 8.1	+ 2.5 0	− 4.0 − 5.6	4.0 10.5	+ 4.0 0	− 4.0 − 6.5

See footnotes at end of table.

Nominal Size Range, Inches	Class RC 5			Class RC 6			Class RC 7			Class RC 8			Class RC 9		
	Clear-ance*	Standard Tolerance Limits		Clear-ance*	Standard Tolerance Limits		Clear-ance*	Standard Tolerance Limits		Clear-ance*	Standard Tolerance Limits		Clear-ance*	Standard Tolerance Limits	
		Hole H8	Shaft e7		Hole H9	Shaft e8		Hole H9	Shaft d8		Hole H10	Shaft c9		Hole H11	Shaft
Over To	Values shown below are in thousandths of an inch														
0– 0.12	0.6 1.6	+ 0.6 0	− 0.6 − 1.0	0.6 2.2	+ 1.0 0	− 0.6 − 1.2	1.0 2.6	+ 1.0 0	− 1.0 − 1.6	2.5 5.1	+ 1.6 0	− 2.5 − 3.5	4.0 8.1	+ 2.5 0	− 4.0 − 5.6
0.12– 0.24	0.8 2.0	+ 0.7 0	− 0.8 − 1.3	0.8 2.7	+ 1.2 0	− 0.8 − 1.5	1.2 3.1	+ 1.2 0	− 1.2 − 1.9	2.8 5.8	+ 1.8 0	− 2.8 − 4.0	4.5 9.0	+ 3.0 0	− 4.5 − 6.0
0.24– 0.40	1.0 2.5	+ 0.9 0	− 1.0 − 1.6	1.0 3.3	+ 1.4 0	− 1.0 − 1.9	1.6 3.9	+ 1.4 0	− 1.6 − 2.5	3.0 6.6	+ 2.2 0	− 3.0 − 4.4	5.0 10.7	+ 3.5 0	− 5.0 − 7.2
0.40– 0.71	1.2 2.9	+ 1.0 0	− 1.2 − 1.9	1.2 3.8	+ 1.6 0	− 1.2 − 2.2	2.0 4.6	+ 1.6 0	− 2.0 − 3.0	3.5 7.9	+ 2.8 0	− 3.5 − 5.1	6.0 12.8	+ 4.0 0	− 6.0 − 8.8
0.71– 1.19	1.6 3.6	+ 1.2 0	− 1.6 − 2.4	1.6 4.8	+ 2.0 0	− 1.6 − 2.8	2.5 5.7	+ 2.0 0	− 2.5 − 3.7	4.5 10.0	+ 3.5 0	− 4.5 − 6.5	7.0 15.5	+ 5.0 0	− 7.0 − 10.5
1.19– 1.97	2.0 4.6	+ 1.6 0	− 2.0 − 3.0	2.0 6.1	+ 2.5 0	− 2.0 − 3.6	3.0 7.1	+ 2.5 0	− 3.0 − 4.6	5.0 11.5	+ 4.0 0	− 5.0 − 7.5	8.0 18.0	+ 6.0 0	− 8.0 − 12.0
1.97– 3.15	2.5 5.5	+ 1.8 0	− 2.5 − 3.7	2.5 7.3	+ 3.0 0	− 2.5 − 4.3	4.0 8.8	+ 3.0 0	− 4.0 − 5.8	6.0 13.5	+ 4.5 0	− 6.0 − 9.0	9.0 20.5	+ 7.0 0	− 9.0 − 13.5
3.15– 4.73	3.0 6.6	+ 2.2 0	− 3.0 − 4.4	3.0 8.7	+ 3.5 0	− 3.0 − 5.2	5.0 10.7	+ 3.5 0	− 5.0 − 7.2	7.0 15.5	+ 5.0 0	− 7.0 − 10.5	10.0 24.0	+ 9.0 0	− 10.0 − 15.0
4.73– 7.09	3.5 7.6	+ 2.5 0	− 3.5 − 5.1	3.5 10.0	+ 4.0 0	− 3.5 − 6.0	6.0 12.5	+ 4.0 0	− 6.0 − 8.5	8.0 18.0	+ 6.0 0	− 8.0 − 12.0	12.0 28.0	+ 10.0 0	− 12.0 − 18.0
7.09– 9.85	4.0 8.6	+ 2.8 0	− 4.0 − 5.8	4.0 11.3	+ 4.5 0	− 4.0 − 6.8	7.0 14.3	+ 4.5 0	− 7.0 − 9.8	10.0 21.5	+ 7.0 0	− 10.0 − 14.5	15.0 34.0	+ 12.0 0	− 15.0 − 22.0
9.85–12.41	5.0 10.0	+ 3.0 0	− 5.0 − 7.0	5.0 13.0	+ 5.0 0	− 5.0 − 8.0	8.0 16.0	+ 5.0 0	− 8.0 − 11.0	12.0 25.0	+ 8.0 0	− 12.0 − 17.0	18.0 38.0	+ 12.0 0	− 18.0 − 26.0
12.41–15.75	6.0 11.7	+ 3.5 0	− 6.0 − 8.2	6.0 15.5	+ 6.0 0	− 6.0 − 9.5	10.0 19.5	+ 6.0 0	− 10.0 − 13.5	14.0 29.0	+ 9.0 0	− 14.0 − 20.0	22.0 45.0	+ 14.0 0	− 22.0 − 31.0
15.75–19.69	8.0 14.5	+ 4.0 0	− 8.0 − 10.5	8.0 18.0	+ 6.0 0	− 8.0 − 12.0	12.0 22.0	+ 6.0 0	− 12.0 − 16.0	16.0 32.0	+ 10.0 0	− 16.0 − 22.0	25.0 51.0	+ 16.0 0	− 25.0 − 35.0

All data above heavy lines are in accord with ABC agreements. Symbols H5, g4, etc. are hole and shaft designations in ABC system. Limits for sizes above 19.69 inches are also given in the ANSI Standard.

* Pairs of values shown represent minimum and maximum amounts of clearance resulting from application of standard tolerance limits.

TABLE 28   *ANSI standard running and sliding fits. (Courtesy ASME)*

American National Standard Clearance Locational Fits (ANSI B4.1-1967, R1994)

Tolerance limits given in body of table are added or subtracted to basic size (as indicated by + or – sign) to obtain maximum and minimum sizes of mating parts.

Nominal Size Range, Inches — Over / To	Class LC 1 Clearance*	Class LC 1 Hole H6	Class LC 1 Shaft h5	Class LC 2 Clearance*	Class LC 2 Hole H7	Class LC 2 Shaft h6	Class LC 3 Clearance*	Class LC 3 Hole H8	Class LC 3 Shaft h7	Class LC 4 Clearance*	Class LC 4 Hole H10	Class LC 4 Shaft h9	Class LC 5 Clearance*	Class LC 5 Hole H7	Class LC 5 Shaft g6
							Values shown below are in thousandths of an inch								
0– 0.12	0 / 0.45	+0.25 / 0	0 / −0.2	0 / 0.65	+0.4 / 0	0 / −0.25	0 / 1	+0.6 / 0	0 / −0.4	0 / 2.6	+1.6 / 0	0 / −1.0	0.1 / 0.75	+0.4 / 0	−0.1 / −0.35
0.12– 0.24	0 / 0.5	+0.3 / 0	0 / −0.2	0 / 0.8	+0.5 / 0	0 / −0.3	0 / 1.2	+0.7 / 0	0 / −0.5	0 / 3.0	+1.8 / 0	0 / −1.2	0.15 / 0.95	+0.5 / 0	−0.15 / −0.45
0.24– 0.40	0 / 0.65	+0.4 / 0	0 / −0.25	0 / 1.0	+0.6 / 0	0 / −0.4	0 / 1.5	+0.9 / 0	0 / −0.6	0 / 3.6	+2.2 / 0	0 / −1.4	0.2 / 1.2	+0.6 / 0	−0.2 / −0.6
0.40– 0.71	0 / 0.7	+0.4 / 0	0 / −0.3	0 / 1.1	+0.7 / 0	0 / −0.4	0 / 1.7	+1.0 / 0	0 / −0.7	0 / 4.4	+2.8 / 0	0 / −1.6	0.25 / 1.35	+0.7 / 0	−0.25 / −0.65
0.71– 1.19	0 / 0.9	+0.5 / 0	0 / −0.4	0 / 1.3	+0.8 / 0	0 / −0.5	0 / 2	+1.2 / 0	0 / −0.8	0 / 5.5	+3.5 / 0	0 / −2.0	0.3 / 1.6	+0.8 / 0	−0.3 / −0.8
1.19– 1.97	0 / 1.0	+0.6 / 0	0 / −0.4	0 / 1.6	+1.0 / 0	0 / −0.6	0 / 2.6	+1.6 / 0	0 / −1	0 / 6.5	+4.0 / 0	0 / −2.5	0.4 / 2.0	+1.0 / 0	−0.4 / −1.0
1.97– 3.15	0 / 1.2	+0.7 / 0	0 / −0.5	0 / 1.9	+1.2 / 0	0 / −0.7	0 / 3	+1.8 / 0	0 / −1.2	0 / 7.5	+4.5 / 0	0 / −3	0.4 / 2.3	+1.2 / 0	−0.4 / −1.1
3.15– 4.73	0 / 1.5	+0.9 / 0	0 / −0.6	0 / 2.3	+1.4 / 0	0 / −0.9	0 / 3.6	+2.2 / 0	0 / −1.4	0 / 8.5	+5.0 / 0	0 / −3.5	0.5 / 2.8	+1.4 / 0	−0.5 / −1.4
4.73– 7.09	0 / 1.7	+1.0 / 0	0 / −0.7	0 / 2.6	+1.6 / 0	0 / −1.0	0 / 4.1	+2.5 / 0	0 / −1.6	0 / 10.0	+6.0 / 0	0 / −4	0.6 / 3.2	+1.6 / 0	−0.6 / −1.6
7.09– 9.85	0 / 2.0	+1.2 / 0	0 / −0.8	0 / 3.0	+1.8 / 0	0 / −1.2	0 / 4.6	+2.8 / 0	0 / −1.8	0 / 11.5	+7.0 / 0	0 / −4.5	0.6 / 3.6	+1.8 / 0	−0.6 / −1.8
9.85–12.41	0 / 2.1	+1.2 / 0	0 / −0.9	0 / 3.2	+2.0 / 0	0 / −1.2	0 / 5	+3.0 / 0	0 / −2.0	0 / 13.0	+8.0 / 0	0 / −5	0.7 / 3.9	+2.0 / 0	−0.7 / −1.9
12.41–15.75	0 / 2.4	+1.4 / 0	0 / −1.0	0 / 3.6	+2.2 / 0	0 / −1.4	0 / 5.7	+3.5 / 0	0 / −2.2	0 / 15.0	+9.0 / 0	0 / −6	0.7 / 4.3	+2.2 / 0	−0.7 / −2.1
15.75–19.69	0 / 2.6	+1.6 / 0	0 / −1.0	0 / 4.1	+2.5 / 0	0 / −1.6	0 / 6.5	+4 / 0	0 / −2.5	0 / 16.0	+10.0 / 0	0 / −6	0.8 / 4.9	+2.5 / 0	−0.8 / −2.4

See footnotes at end of table.

Nominal Size Range, Inches — Over / To	Class LC 6 Clearance*	Class LC 6 Hole H9	Class LC 6 Shaft f8	Class LC 7 Clearance*	Class LC 7 Hole H10	Class LC 7 Shaft e9	Class LC 8 Clearance*	Class LC 8 Hole H10	Class LC 8 Shaft d9	Class LC 9 Clearance*	Class LC 9 Hole H11	Class LC 9 Shaft c10	Class LC 10 Clearance*	Class LC 10 Hole H12	Class LC 10 Shaft	Class LC 11 Clearance*	Class LC 11 Hole H13	Class LC 11 Shaft
								Values shown below are in thousandths of an inch										
0– 0.12	0.3 / 1.9	+1.0 / 0	−0.3 / −0.9	0.6 / 3.2	+1.6 / 0	−0.6 / −1.6	1.0 / 2.0	+1.6 / 0	−1.0 / −2.0	2.5 / 6.6	+2.5 / 0	−2.5 / −4.1	4 / 12	+4 / 0	−4 / −8	5 / 17	+6 / 0	−5 / −11
0.12– 0.24	0.4 / 2.3	+1.2 / 0	−0.4 / −1.1	0.8 / 3.8	+1.8 / 0	−0.8 / −2.0	1.2 / 4.2	+1.8 / 0	−1.2 / −2.4	2.8 / 7.6	+3.0 / 0	−2.8 / −4.6	4.5 / 14.5	+5 / 0	−4.5 / −9.5	6 / 20	+7 / 0	−6 / −13
0.24– 0.40	0.5 / 2.8	+1.4 / 0	−0.5 / −1.4	1.0 / 4.6	+2.2 / 0	−1.0 / −2.4	1.6 / 5.2	+2.2 / 0	−1.6 / −3.0	3.0 / 8.7	+3.5 / 0	−3.0 / −5.2	5 / 17	+6 / 0	−5 / −11	7 / 25	+9 / 0	−7 / −16
0.40– 0.71	0.6 / 3.2	+1.6 / 0	−0.6 / −1.6	1.2 / 5.6	+2.8 / 0	−1.2 / −2.8	2.0 / 6.4	+2.8 / 0	−2.0 / −3.6	3.5 / 10.3	+4.0 / 0	−3.5 / −6.3	6 / 20	+7 / 0	−6 / −13	8 / 28	+10 / 0	−8 / −18
0.71– 1.19	0.8 / 4.0	+2.0 / 0	−0.8 / −2.0	1.6 / 7.1	+3.5 / 0	−1.6 / −3.6	2.5 / 8.0	+3.5 / 0	−2.5 / −4.5	4.5 / 13.0	+5.0 / 0	−4.5 / −8.0	7 / 23	+8 / 0	−7 / −15	10 / 34	+12 / 0	−10 / −22
1.19– 1.97	1.0 / 5.1	+2.5 / 0	−1.0 / −2.6	2.0 / 8.5	+4.0 / 0	−2.0 / −4.5	3.0 / 9.5	+4.0 / 0	−3.0 / −5.5	5.0 / 15.0	+6 / 0	−5.0 / −9.0	8 / 28	+10 / 0	−8 / −18	12 / 44	+16 / 0	−12 / −28
1.97– 3.15	1.2 / 6.0	+3.0 / 0	−1.0 / −3.0	2.5 / 10.0	+4.5 / 0	−2.5 / −5.5	4.0 / 11.5	+4.5 / 0	−4.0 / −7.0	6.0 / 17.5	+7 / 0	−6.0 / −10.5	10 / 34	+12 / 0	−10 / −22	14 / 50	+18 / 0	−14 / −32
3.15– 4.73	1.4 / 7.1	+3.5 / 0	−1.4 / −3.6	3.0 / 11.5	+5.0 / 0	−3.0 / −6.5	5.0 / 13.5	+5.0 / 0	−5.0 / −8.5	7 / 21	+9 / 0	−7 / −12	11 / 39	+14 / 0	−11 / −25	16 / 60	+22 / 0	−16 / −38
4.73– 7.09	1.6 / 8.1	+4.0 / 0	−1.6 / −4.1	3.5 / 13.5	+6.0 / 0	−3.5 / −7.5	6 / 16	+6 / 0	−6 / −10	8 / 24	+10 / 0	−8 / −14	12 / 44	+16 / 0	−12 / −28	18 / 68	+25 / 0	−18 / −43
7.09– 9.85	2.0 / 9.3	+4.5 / 0	−2.0 / −4.8	4.0 / 15.5	+7.0 / 0	−4.0 / −8.5	7 / 18.5	+7 / 0	−7 / −11.5	10 / 29	+12 / 0	−10 / −17	16 / 52	+18 / 0	−16 / −34	22 / 78	+28 / 0	−22 / −50
9.85–12.41	2.2 / 10.2	+5.0 / 0	−2.2 / −5.2	4.5 / 17.5	+8.0 / 0	−4.5 / −9.5	7 / 20	+8 / 0	−7 / −12	12 / 32	+12 / 0	−12 / −20	20 / 60	+20 / 0	−20 / −40	28 / 88	+30 / 0	−28 / −58
12.41–15.75	2.5 / 12.0	+6.0 / 0	−2.5 / −6.0	5.0 / 20.0	+9.0 / 0	−5 / −11	8 / 23	+9 / 0	−8 / −14	14 / 37	+14 / 0	−14 / −23	22 / 66	+22 / 0	−22 / −44	30 / 100	+35 / 0	−30 / −65
15.75–19.69	2.8 / 12.8	+6.0 / 0	−2.8 / −6.8	5.0 / 21.0	+10.0 / 0	−5 / −11	9 / 25	+10 / 0	−9 / −15	16 / 42	+16 / 0	−16 / −26	25 / 75	+25 / 0	−25 / −50	35 / 115	+40 / 0	−35 / −75

All data above heavy lines are in accordance with American-British-Canadian (ABC) agreements. Symbols H6, H7, s6, etc. are hole and shaft designations in ABC system. Limits for sizes above 19.69 inches are not covered by ABC agreements but are given in the ANSI Standard.
* Pairs of values shown represent minimum and maximum amounts of interference resulting from application of standard tolerance limits.

TABLE 29   *ANSI standard clearance locational fits.* (Courtesy ASME)

American National Standard Transition Locational Fits (ANSI B4.1-1967, R1994)

Nominal Size Range, Inches Over To	Class LT 1 Fit*	Hole H7	Shaft js6	Class LT 2 Fit*	Hole H8	Shaft js7	Class LT 3 Fit*	Hole H7	Shaft k6	Class LT 4 Fit*	Hole H8	Shaft k7	Class LT 5 Fit*	Hole H7	Shaft n6	Class LT 6 Fit*	Hole H7	Shaft n7
0– 0.12	−0.12 / +0.52	+0.4 / 0	+0.12 / −0.12	−0.2 / +0.8	+0.6 / 0	+0.2 / −0.2							−0.5 / +0.15	+0.4 / 0	+0.5 / +0.25	−0.65 / +0.15	+0.4 / 0	+0.65 / +0.25
0.12– 0.24	−0.15 / +0.65	+0.5 / 0	+0.15 / −0.15	−0.25 / +0.95	+0.7 / 0	+0.25 / −0.25							−0.6 / +0.2	+0.5 / 0	+0.6 / +0.3	−0.8 / +0.2	+0.5 / 0	+0.8 / +0.3
0.24– 0.40	−0.2 / +0.8	+0.6 / 0	+0.2 / −0.2	−0.3 / +1.2	+0.9 / 0	+0.3 / −0.3	−0.5 / +0.5	+0.6 / 0	+0.5 / +0.1	−0.7 / +0.8	+0.9 / 0	+0.7 / +0.1	−0.8 / +0.2	+0.6 / 0	+0.8 / +0.4	−1.0 / +0.2	+0.6 / 0	+1.0 / +0.4
0.40– 0.71	−0.2 / +0.9	+0.7 / 0	+0.2 / −0.2	−0.35 / +1.35	+1.0 / 0	+0.35 / −0.35	−0.5 / +0.6	+0.7 / 0	+0.5 / +0.1	−0.8 / +0.9	+1.0 / 0	+0.8 / +0.1	−0.9 / +0.2	+0.7 / 0	+0.9 / +0.5	−1.2 / +0.2	+0.7 / 0	+1.2 / +0.5
0.71– 1.19	−0.25 / +1.05	+0.8 / 0	+0.25 / −0.25	−0.4 / +1.6	+1.2 / 0	+0.4 / −0.4	−0.6 / +0.7	+0.8 / 0	+0.6 / +0.1	−0.9 / +1.1	+1.2 / 0	+0.9 / +0.1	−1.1 / +0.2	+0.8 / 0	+1.1 / +0.6	−1.4 / +0.2	+0.8 / 0	+1.4 / +0.6
1.19– 1.97	−0.3 / +1.3	+1.0 / 0	+0.3 / −0.3	−0.5 / +2.1	+1.6 / 0	+0.5 / −0.5	−0.7 / +0.9	+1.0 / 0	+0.7 / +0.1	−1.1 / +1.5	+1.6 / 0	+1.1 / +0.1	−1.3 / +0.3	+1.0 / 0	+1.3 / +0.7	−1.7 / +0.3	+1.0 / 0	+1.7 / +0.7
1.97– 3.15	−0.3 / +1.5	+1.2 / 0	+0.3 / −0.3	−0.6 / +2.4	+1.8 / 0	+0.6 / −0.6	−0.8 / +1.1	+1.2 / 0	+0.8 / +0.1	−1.3 / +1.7	+1.8 / 0	+1.3 / +0.1	−1.5 / +0.4	+1.2 / 0	+1.5 / +0.8	−2.0 / +0.4	+1.2 / 0	+2.0 / +0.8
3.15– 4.73	−0.4 / +1.8	+1.4 / 0	+0.4 / −0.4	−0.7 / +2.9	+2.2 / 0	+0.7 / −0.7	−1.0 / +1.3	+1.4 / 0	+1.0 / +0.1	−1.5 / +2.1	+2.2 / 0	+1.5 / +0.1	−1.9 / +0.4	+1.4 / 0	+1.9 / +1.0	−2.4 / +0.4	+1.4 / 0	+2.4 / +1.0
4.73– 7.09	−0.5 / +2.1	+1.6 / 0	+0.5 / −0.5	−0.8 / +3.3	+2.5 / 0	+0.8 / −0.8	−1.1 / +1.5	+1.6 / 0	+1.1 / +0.1	−1.7 / +2.4	+2.5 / 0	+1.7 / +0.1	−2.2 / +0.4	+1.6 / 0	+2.2 / +1.2	−2.8 / +0.4	+1.6 / 0	+2.8 / +1.2
7.09– 9.85	−0.6 / +2.4	+1.8 / 0	+0.6 / −0.6	−0.9 / +3.7	+2.8 / 0	+0.9 / −0.9	−1.4 / +1.6	+1.8 / 0	+1.4 / +0.2	−2.0 / +2.6	+2.8 / 0	+2.0 / +0.2	−2.6 / +0.4	+1.8 / 0	+2.6 / +1.4	−3.2 / +0.4	+1.8 / 0	+3.2 / +1.4
9.85–12.41	−0.6 / +2.6	+2.0 / 0	+0.6 / −0.6	−1.0 / +4.0	+3.0 / 0	+1.0 / −1.0	−1.4 / +1.8	+2.0 / 0	+1.4 / +0.2	−2.2 / +2.8	+3.0 / 0	+2.2 / +0.2	−2.6 / +0.6	+2.0 / 0	+2.6 / +1.4	−3.4 / +0.6	+2.0 / 0	+3.4 / +1.4
12.41–15.75	−0.7 / +2.9	+2.2 / 0	+0.7 / −0.7	−1.0 / +4.5	+3.5 / 0	+1.0 / −1.0	−1.6 / +2.0	+2.2 / 0	+1.6 / +0.2	−2.4 / +3.3	+3.5 / 0	+2.4 / +0.2	−3.0 / +0.6	+2.2 / 0	+3.0 / +1.6	−3.8 / +0.6	+2.2 / 0	+3.8 / +1.6
15.75–19.69	−0.8 / +3.3	+2.5 / 0	+0.8 / −0.8	−1.2 / +5.2	+4.0 / 0	+1.2 / −1.2	−1.8 / +2.3	+2.5 / 0	+1.8 / +0.2	−2.7 / +3.8	+4.0 / 0	+2.7 / +0.2	−3.4 / +0.7	+2.5 / 0	+3.4 / +1.8	−4.3 / +0.7	+2.5 / 0	+4.3 / +1.8

Values shown below are in thousandths of an inch

All data above heavy lines are in accord with ABC agreements. Symbols H7, js6, etc. are hole and shaft designations in ABC system.
* Pairs of values shown represent maximum amount of interference (−) and maximum amount of clearance (+) resulting from application of standard tolerance limits.

TABLE 30   *ANSI Standard Transition Locational Fits.* (Courtesy ASME)

Tolerance limits given in body of table are added or subtracted to basic size (as indicated by + or − sign) to obtain maximum and minimum sizes of mating parts.

Nominal Size Range, Inches Over To	Class LN 1			Class LN 2			Class LN 3		
	Limits of Inter- ference	Standard Limits		Limits of Inter- ference	Standard Limits		Limits of Inter- ference	Standard Limits	
		Hole H6	Shaft n5		Hole H7	Shaft p6		Hole H7	Shaft r6
	Values shown below are given in thousandths of an inch								
0– 0.12	0 0.45	+0.25 0	+0.45 +0.25	0 0.65	+0.4 0	+0.65 +0.4	0.1 0.75	+0.4 0	+0.75 +0.5
0.12– 0.24	0 0.5	+0.3 0	+0.5 +0.3	0 0.8	+0.5 0	+0.8 +0.5	0.1 0.9	+0.5 0	+0.9 +0.6
0.24– 0.40	0 0.65	+0.4 0	+0.65 +0.4	0 1.0	+0.6 0	+1.0 +0.6	0.2 1.2	+0.6 0	+1.2 +0.8
0.40– 0.71	0 0.8	+0.4 0	+0.8 +0.4	0 1.1	+0.7 0	+1.1 +0.7	0.3 1.4	+0.7 0	+1.4 +1.0
0.71– 1.19	0 1.0	+0.5 0	+1.0 +0.5	0 1.3	+0.8 0	+1.3 +0.8	0.4 1.7	+0.8 0	+1.7 +1.2
1.19– 1.97	0 1.1	+0.6 0	+1.1 +0.6	0 1.6	+1.0 0	+1.6 +1.0	0.4 2.0	+1.0 0	+2.0 +1.4
1.97– 3.15	0.1 1.3	+0.7 0	+1.3 +0.8	0.2 2.1	+1.2 0	+2.1 +1.4	0.4 2.3	+1.2 0	+2.3 +1.6
3.15– 4.73	0.1 1.6	+0.9 0	+1.6 +1.0	0.2 2.5	+1.4 0	+2.5 +1.6	0.6 2.9	+1.4 0	+2.9 +2.0
4.73– 7.09	0.2 1.9	+1.0 0	+1.9 +1.2	0.2 2.8	+1.6 0	+2.8 +1.8	0.9 3.5	+1.6 0	+3.5 +2.5
7.09– 9.85	0.2 2.2	+1.2 0	+2.2 +1.4	0.2 3.2	+1.8 0	+3.2 +2.0	1.2 4.2	+1.8 0	+4.2 +3.0
9.85–12.41	0.2 2.3	+1.2 0	+2.3 +1.4	0.2 3.4	+2.0 0	+3.4 +2.2	1.5 4.7	+2.0 0	+4.7 +3.5
12.41–15.75	0.2 2.6	+1.4 0	+2.6 +1.6	0.3 3.9	+2.2 0	+3.9 +2.5	2.3 5.9	+2.2 0	+5.9 +4.5
15.75–19.69	0.2 2.8	+1.6 0	+2.8 +1.8	0.3 4.4	+2.5 0	+4.4 +2.8	2.5 6.6	+2.5 0	+6.6 +5.0

All data in this table are in accordance with American-British-Canadian (ABC) agreements. Limits for sizes above 19.69 inches are not covered by ABC agreements but are given in the ANSI Standard.

Symbols H7, p6, etc. are hole and shaft designations in ABC system.

* Pairs of values shown represent minimum and maximum amounts of interference resulting from application of standard tolerance limits.

TABLE 31   *ANSI Standard Interference Locational Fits.* (Courtesy ASME)

Nominal Size Range, Inches Over — To	Class FN 1 Inter-ference*	Standard Tolerance Limits Hole H6	Shaft	Class FN 2 Inter-ference*	Standard Tolerance Limits Hole H7	Shaft s6	Class FN 3 Inter-ference*	Standard Tolerance Limits Hole H7	Shaft t6	Class FN 4 Inter-ference*	Standard Tolerance Limits Hole H7	Shaft u6	Class FN 5 Inter-ference*	Standard Tolerance Limits Hole H8	Shaft x7
						Values shown below are in thousandths of an inch									
0–0.12	0.05 / 0.5	+0.25 / 0	+0.5 / +0.3	0.2 / 0.85	+0.4 / 0	+0.85 / +0.6				0.3 / 0.95	+0.4 / 0	+0.95 / +0.7	0.3 / 1.3	+0.6 / 0	+1.3 / +0.9
0.12–0.24	0.1 / 0.6	+0.3 / 0	+0.6 / +0.4	0.2 / 1.0	+0.5 / 0	+1.0 / +0.7				0.4 / 1.2	+0.5 / 0	+1.2 / +0.9	0.5 / 1.7	+0.7 / 0	+1.7 / +1.2
0.24–0.40	0.1 / 0.75	+0.4 / 0	+0.75 / +0.5	0.4 / 1.4	+0.6 / 0	+1.4 / +1.0				0.6 / 1.6	+0.6 / 0	+1.6 / +1.2	0.5 / 2.0	+0.9 / 0	+2.0 / +1.4
0.40–0.56	0.1 / 0.8	+0.4 / 0	+0.8 / +0.5	0.5 / 1.6	+0.7 / 0	+1.6 / +1.2				0.7 / 1.8	+0.7 / 0	+1.8 / +1.4	0.6 / 2.3	+1.0 / 0	+2.3 / +1.6
0.56–0.71	0.2 / 0.9	+0.4 / 0	+0.9 / +0.6	0.5 / 1.6	+0.7 / 0	+1.6 / +1.2				0.7 / 1.8	+0.7 / 0	+1.8 / +1.4	0.8 / 2.5	+1.0 / 0	+2.5 / +1.8
0.71–0.95	0.2 / 1.1	+0.5 / 0	+1.1 / +0.7	0.6 / 1.9	+0.8 / 0	+1.9 / +1.4				0.8 / 2.1	+0.8 / 0	+2.1 / +1.6	1.0 / 3.0	+1.2 / 0	+3.0 / +2.2
0.95–1.19	0.3 / 1.2	+0.5 / 0	+1.2 / +0.8	0.6 / 1.9	+0.8 / 0	+1.9 / +1.4	0.8 / 2.1	+0.8 / 0	+2.1 / +1.6	1.0 / 2.3	+0.8 / 0	+2.3 / +1.8	1.3 / 3.3	+1.2 / 0	+3.3 / +2.5
1.19–1.58	0.3 / 1.3	+0.6 / 0	+1.3 / +0.9	0.8 / 2.4	+1.0 / 0	+2.4 / +1.8	1.0 / 2.6	+1.0 / 0	+2.6 / +2.0	1.5 / 3.1	+1.0 / 0	+3.1 / +2.5	1.4 / 4.0	+1.6 / 0	+4.0 / +3.0
1.58–1.97	0.4 / 1.4	+0.6 / 0	+1.4 / +1.0	0.8 / 2.4	+1.0 / 0	+2.4 / +1.8	1.2 / 2.8	+1.0 / 0	+2.8 / +2.2	1.8 / 3.4	+1.0 / 0	+3.4 / +2.8	2.4 / 5.0	+1.6 / 0	+5.0 / +4.0
1.97–2.56	0.6 / 1.8	+0.7 / 0	+1.8 / +1.3	0.8 / 2.7	+1.2 / 0	+2.7 / +2.0	1.3 / 3.2	+1.2 / 0	+3.2 / +2.5	2.3 / 4.2	+1.2 / 0	+4.2 / +3.5	3.2 / 6.2	+1.8 / 0	+6.2 / +5.0
2.56–3.15	0.7 / 1.9	+0.7 / 0	+1.9 / +1.4	1.0 / 2.9	+1.2 / 0	+2.9 / +2.2	1.8 / 3.7	+1.2 / 0	+3.7 / +3.0	2.8 / 4.7	+1.2 / 0	+4.7 / +4.0	4.2 / 7.2	+1.8 / 0	+7.2 / +6.0
3.15–3.94	0.9 / 2.4	+0.9 / 0	+2.4 / +1.8	1.4 / 3.7	+1.4 / 0	+3.7 / +2.8	2.1 / 4.4	+1.4 / 0	+4.4 / +3.5	3.6 / 5.9	+1.4 / 0	+5.9 / +5.0	4.8 / 8.4	+2.2 / 0	+8.4 / +7.0
3.94–4.73	1.1 / 2.6	+0.9 / 0	+2.6 / +2.0	1.6 / 3.9	+1.4 / 0	+3.9 / +3.0	2.6 / 4.9	+1.4 / 0	+4.9 / +4.0	4.6 / 6.9	+1.4 / 0	+6.9 / +6.0	5.8 / 9.4	+2.2 / 0	+9.4 / +8.0

See footnotes at end of table.

Nominal Size Range, Inches Over — To	Class FN 1 Inter-ference*	Standard Tolerance Limits Hole H6	Shaft	Class FN 2 Inter-ference*	Standard Tolerance Limits Hole H7	Shaft s6	Class FN 3 Inter-ference*	Standard Tolerance Limits Hole H7	Shaft t6	Class FN 4 Inter-ference*	Standard Tolerance Limits Hole H7	Shaft u6	Class FN 5 Inter-ference*	Standard Tolerance Limits Hole H8	Shaft x7
						Values shown below are in thousandths of an inch									
4.73–5.52	1.2 / 2.9	+1.0 / 0	+2.9 / +2.2	1.9 / 4.5	+1.6 / 0	+4.5 / +3.5	3.4 / 6.0	+1.6 / 0	+6.0 / +5.0	5.4 / 8.0	+1.6 / 0	+8.0 / +7.0	7.5 / 11.6	+2.5 / 0	+11.6 / +10.0
5.52–6.30	1.5 / 3.2	+1.0 / 0	+3.2 / +2.5	2.4 / 5.0	+1.6 / 0	+5.0 / +4.0	3.4 / 6.0	+1.6 / 0	+6.0 / +5.0	5.4 / 8.0	+1.6 / 0	+8.0 / +7.0	9.5 / 13.6	+2.5 / 0	+13.6 / +12.0
6.30–7.09	1.8 / 3.5	+1.0 / 0	+3.5 / +2.8	2.9 / 5.5	+1.6 / 0	+5.5 / +4.5	4.4 / 7.0	+1.6 / 0	+7.0 / +6.0	6.4 / 9.0	+1.6 / 0	+9.0 / +8.0	9.5 / 13.6	+2.5 / 0	+13.6 / +12.0
7.09–7.88	1.8 / 3.8	+1.2 / 0	+3.8 / +3.0	3.2 / 6.2	+1.8 / 0	+6.2 / +5.0	5.2 / 8.2	+1.8 / 0	+8.2 / +7.0	7.2 / 10.2	+1.8 / 0	+10.2 / +9.0	11.2 / 15.8	+2.8 / 0	+15.8 / +14.0
7.88–8.86	2.3 / 4.3	+1.2 / 0	+4.3 / +3.5	3.2 / 6.2	+1.8 / 0	+6.2 / +5.0	5.2 / 8.2	+1.8 / 0	+8.2 / +7.0	8.2 / 11.2	+1.8 / 0	+11.2 / +10.0	13.2 / 17.8	+2.8 / 0	+17.8 / +16.0
8.86–9.85	2.3 / 4.3	+1.2 / 0	+4.3 / +3.5	4.2 / 7.2	+1.8 / 0	+7.2 / +6.0	6.2 / 9.2	+1.8 / 0	+9.2 / +8.0	10.2 / 13.2	+1.8 / 0	+13.2 / +12.0	13.2 / 17.8	+2.8 / 0	+17.8 / +16.0
9.85–11.03	2.8 / 4.9	+1.2 / 0	+4.9 / +4.0	4.0 / 7.2	+2.0 / 0	+7.2 / +6.0	7.0 / 10.2	+2.0 / 0	+10.2 / +9.0	10.0 / 13.2	+2.0 / 0	+13.2 / +12.0	15.0 / 20.0	+3.0 / 0	+20.0 / +18.0
11.03–12.41	2.8 / 4.9	+1.2 / 0	+4.9 / +4.0	5.0 / 8.2	+2.0 / 0	+8.2 / +7.0	7.0 / 10.2	+2.0 / 0	+10.2 / +9.0	12.0 / 15.2	+2.0 / 0	+15.2 / +14.0	17.0 / 22.0	+3.0 / 0	+22.0 / +20.0
12.41–13.98	3.1 / 5.5	+1.4 / 0	+5.5 / +4.5	5.8 / 9.4	+2.2 / 0	+9.4 / +8.0	7.8 / 11.4	+2.2 / 0	+11.4 / +10.0	13.8 / 17.4	+2.2 / 0	+17.4 / +16.0	18.5 / 24.2	+3.5 / 0	+24.2 / +22.0
13.98–15.75	3.6 / 6.1	+1.4 / 0	+6.1 / +5.0	5.8 / 9.4	+2.2 / 0	+9.4 / +8.0	9.8 / 13.4	+2.2 / 0	+13.4 / +12.0	15.8 / 19.4	+2.2 / 0	+19.4 / +18.0	21.5 / 27.2	+3.5 / 0	+27.2 / +25.0
15.75–17.72	4.4 / 7.0	+1.6 / 0	+7.0 / +6.0	6.5 / 10.6	+2.5 / 0	+10.6 / +9.0	9.5 / 13.6	+2.5 / 0	+13.6 / +12.0	17.5 / 21.6	+2.5 / 0	+21.6 / +20.0	24.0 / 30.5	+4.0 / 0	+30.5 / +28.0
17.72–19.69	4.4 / 7.0	+1.6 / 0	+7.0 / +6.0	7.5 / 11.6	+2.5 / 0	+11.6 / +10.0	11.5 / 15.6	+2.5 / 0	+15.6 / +14.0	19.5 / 23.6	+2.5 / 0	+23.6 / +22.0	26.0 / 32.5	+4.0 / 0	+32.5 / +30.0

All data above heavy lines are in accordance with American-British-Canadian (ABC) agreements. Symbols H6, H7, s6, etc. are hole and shaft designations in ABC system. Limits for sizes above 19.69 inches are not covered by ABC agreements but are given in the ANSI standard.

* Pairs of values shown represent minimum and maximum amounts of interference resulting from application of standard tolerance limits.

TABLE 32 *ANSI Standard Force and Shrink Fits. (Courtesy ASME)*

glossary

abrasive water jet machining (AWJM)   Similar to WJM, except that an abrasive is added to the water jet after it exits the nozzle.

labrado abrasivo con chorro de agua   Similar al labrado con chorro de agua, excepto que se le añade un abrasivo al chorro de agua luego que sale de la tobera.

absolute system   A type of numerical control system in which all coordinates are located from a fixed or absolute zero point.

sistema absoluto   Un tipo de sistema de control numérico en el cual todas las coordenadas se localizan a base de un punto fijo o de cero absoluto.

abstract   Part of the written report to the client, this includes one or two sentences about the content of the report to capture a potential reader's interest.

abstracto   Parte de un informe escrito al cliente, éste incluye una o dos oraciones acerca del contenido del informe para capturar el interés de un posible lector.

Acme   Screw thread form.

Acme   Tipo de rosca de tornillo.

active coils   The total coils minus the two end coils in a compression spring.

espirales activas   El número total de espirales menos las dos espirales de los extremos en un resorte de compresión.

actual size   Actual measured size of a feature.

tamaño real   El tamaño real de un rasgo.

acute angle   An angle less than 90°.

ángulo agudo   Un ángulo menor de 90°.

addendum   The radial distance from the pitch diameter to the top of the tooth. It is equal to 1/P.

altura de cabeza   La distancia radial desde el diámetro de paso hasta la altura del diente. Es igual a 1/P.

aligning of features   This occurs to correct the true projection concept. It is when the top hole is theoretically revolved to the cutting-plane line and projected to the sectional view.

alineación de rasgos   Esto ocurre para corregir el concepto de proyección verdadera. Es cuando un agujero teoréticamente se revuelve al plano cortante y se proyecta en la vista seccional.

alignment chart   A graphical representation of the relationship of two or more variables.

gráfica de alineamiento   Una representación gráfica de la relación entre dos o más variables.

Allen screw   Special set screw or cap screw with hexagon socket in head.

tornillo con cabeza hexagonal   Un tornillo de fijación o tornillo pasante especial con una cavidad hexagonal en la cabeza.

allowance   Minimum clearance between mating parts. Specifically, the amount of room between two main parts, usually considered the tightest possible fit.

concesión   La separación mínima entre piezas apareadas. Específicamente, la cantidad de espacio entre dos partes principales, generalmente considerado como el encaje más apretado posible.

alloy   Two or more metals in combination, usually a fine metal with a base metal.

aleación   Dos o más metales en combinación, generalmente un metal fino con un metal base.

alloy steel   This is any steel that contains one or more alloying elements added to produce characteristics not available in unalloyed carbon steels.

acero aleado   Esto es cualquier acero que contiene uno o más elementos de aleación que se le añaden para producir características que no están disponibles en los aceros con carbón sin alearse.

alphanumeric   Refers to the totality of characters that are either alphabetic or numeric.

alfanumérico Se refiere a la totalidad de los caracteres que son alfabéticos o numéricos.

alternate section lining Combination of true projection and correct projection.

revestimiento de sección alterna Una combinación de una proyección real y una proyección correcta.

alternatives Convincing the client that a thorough investigation was done by the design group by including a brief summary of the alternatives that were considered but discarded and why.

alternativas Convencer al cliente que se hizo una investigación exhaustiva por el grupo de diseño por medio de un resumen breve de las alternativas que se consideraron pero que se desecharon y por qué.

altitude (of a prism) The perpendicular distance between its end polygons (or bases).

altitud (de un prisma) La distancia perpendicular entre sus dos polígonos extremos o bases.

aluminum A bright silvery metal that is very light. In addition to being light, it is an excellent conductor and able to resist corrosion. It is usually alloyed with copper to increase hardness and strength.

aluminio Un metal plateado brilloso que es muy liviano. Además de ser liviano, es un excelente conductor eléctrico y es capaz de resistir la corrosión. Generalmente se alea con cobre para aumentar su dureza y resistencia.

ammonia A colorless gas used in the development process of diazo and sepia prints.

amoniaco Un gas incoloro que se usa en el desarrollo de las reproducciones de diazo y sepia.

analysis Summarizing the major assumptions and calculations. Also, the study and evaluation of an object's characteristics.

análisis Resumir las suposiciones principales y los cálculos. También, el estudio y evaluación de las características de un objeto.

angle gear Similar to a bevel gear, except that the angles are at other than 90° to each other.

engranaje angular Similar a un engranaje de bisel, excepto que los ángulos no son a 90° uno con respecto al otro.

angle iron A structural shape whose section is a right angle.

hierro angular Una forma estructural cuya sección transversal es un ángulo recto.

angle of thread The included angle between the sides of the thread.

ángulo de rosca El ángulo incluido entre los lados de una rosca.

angle plate jig A modified form of a plate jig in which the surface to be machined is perpendicular to the locating surface.

guía de placa angular Una forma modificada de una guía de placa en la cual la superficie a labrarse está perpendicular a la superficie de localización.

angularity A feature control in which a given surface, axis, or center plane must form a specified angle other than 90° with a datum.

angularidad Un rasgo de control en el cual una superficie, eje, o plano central dado debe pasar por un ángulo diferente de 90° respecto a una referencia.

annealing The process of heating a part, soaking it until it is thoroughly heated, and slowly cooling it by turning off the furnace.

recocer El proceso de calentar una pieza hasta que esté calentada completamente y luego enfriarla lentamente al apagar el horno.

anodizing The process of protecting aluminum by oxidizing in an acid bath using a direct current.

anodizar El proceso de proteger el aluminio oxidándolo en un baño ácido usando una corriente directa.

application A definable set of drafting tasks to be accomplished in a given drafting area; may be accomplished partly through manual procedures and partly through computerized procedures.

aplicación Un grupo definible de tareas de dibujo industrial que deben ser llevadas a cabo en un área de dibujo dada; pueden lograrse parcialmente por medio de procedimientos manuales y parcialmente por medio de procedimientos computarizados.

application engineering This type of engineering takes previous designs into consideration when creating a new design. It is generally a team effort whose primary task is to modify and incorporate these existing concepts to meet the design requirements for the new system.

ingeniería de aplicación Este tipo de ingeniería considera los diseños previos al crear diseños nuevos. Generalmente es un esfuerzo en equipo cuya tarea primaria es modificar e incorporar aquellos conceptos existentes para cumplir con los requisitos del nuevo sistema.

arbor press A machine that performs internal broaching operations using a push-type rather than a pull-type broach.

prensa de husillo Una máquina que lleva a cabo las operaciones de brochado internas usando un broche de tipo de empuje en vez de un tipo de halar.

arc A part of a circle.

arco Una parte de un círculo.

arc-weld To weld by electric arc; the work is usually the positive terminal.

soldadura de arco Soldar por medio de un arco eléctrico; la pieza de trabajo generalmente es el terminal positivo.

arrowhead The most commonly used termination symbol for dimension and leader lines.

flecha El símbolo de terminación más común para las líneas de dimensión y directrices.

artificial intelligence The ability of a machine to improve its own operation or the ability of a machine to perform functions that are normally associated with human intelligence, such as learning, adapting, reasoning, self-correction, and automatic improvement.

inteligencia artificial La habilidad de una máquina de mejorar su propia operación o la habilidad de una máquina para llevar a cabo funciones que normalmente están asociadas con la inteligencia humana, tales como el aprendizaje, adaptación, razonamiento, auto corrección, y mejoría automática.

assembled isometric assembly drawings This style shows how a single product looks when assembled.

dibujo de ensamblaje isométrico ensamblado Este estilo muestra cómo un solo producto se vería ensamblado.

assembly drawing Illustrates how a product is assembled when completed.

dibujo de ensamblaje Ilustra como un producto se ensambla cuando completado.

assembly section This section occurs when a sectional drawing is made up of two or more parts. It shows how the various parts go together.

sección de ensamblaje Esta sección ocurre cuando se crea un dibujo seccional de dos o más piezas. Muestra cómo se unen las varias piezas.

attribute listing A systematic guide for checking the completeness of technical detail drawings. It includes the questions: What is the function? What size is it? What materials are used? Can it be made—how and with what processes? What other factors need to be checked?

listado de atributos Una guía sistemática para verificar la plenitud de detalles técnicos en los dibujos. Incluye las preguntas: ¿Cuál es su función? ¿De que tamaño es? ¿Qué materiales se usan? ¿Puede hacerse – cómo y con qué procesos? ¿Qué otros factores necesitan verificarse?

attributes The characteristics of a part.

atributos Las características de una pieza.

automated assembly Assembly by means of operations performed automatically by machines that are controlled and monitored by computers.

ensamblaje automatizado Ensamblaje por medio de operaciones que se hacen automáticamente por máquinas que son controladas y monitoreadas por computadoras.

automated process planning Using a computer in developing a process plan.

planificación automática de proceso Usar una computadora al desarrollar un plan de proceso.

automation The conversion of a process or system to automatic operation.

automatización La conversión de un proceso o sistema a una operación automática.

auxiliary section An auxiliary view in section.

sección auxiliar Una vista auxiliar en sección.

auxiliary view An additional view of an object, usually of a surface inclined to the principal surfaces of the object to provide a true size and shape view.

vista auxiliar Una vista adicional de un objeto, generalmente de una superficie inclinada a las superficies principales de un objeto para proveer una vista del tamaño y forma real.

axes Plural of axis.

ejes El plural de eje.

axis A longitudinal center line of the thread. Also, an imaginary line around which parts rotate or are regularly arranged.

eje Una línea de centro longitudinal de una rosca. También, una línea imaginaria alrededor de la cual las piezas rotan o están ordenadas regularmente.

axonometric projection Three-dimensional drawings.

proyección axonométrica Dibujos tridimensionales.

axonometric sketching This type of sketching may be one of three types—isometric, dimetric, or trimetric. All relate to proportional scales and angle positions of length, width, and depth.

croquis axonométrico Este tipo de croquis puede ser de uno de tres tipos – isométrico, dimétrico o trimétrico. Todos se relacionan con las escalas de proporción y la posición angular del largo, ancho y profundidad.

azimuth bearing In this method, the total angle is used going clockwise from due north.

orientación de azimut En este método, el ángulo total se usa como si partiese en favor de las manecillas del reloj desde la dirección del norte.

Babbitt A soft alloy for bearings, mostly of tin with small amounts of copper and antimony.

Babbitt Una aleación suave de cojinetes, hecha mayormente de estaño con pequeñas cantidades de cobre y antimonio.

back weld A weld applied to the opposite side of a joint after the major weld has been applied.

soldadura trasera Una soldadura aplicada al lado posterior de la unión después de aplicar la soldadura principal.

backlash Lost motion between moving parts, such as threaded shaft and nut or the teeth of meshing gears.

contragolpe Perdida de movimiento entre partes móviles, tales como un eje de rosca y tuerca o los dientes de la malla de engranaje.

balloon Number within a circle used to identify parts.

globo Número dentro de un círculo utilizado para identificar las partes.

base circle Used to lay out an offset follower. It is a circle with a radius equal to the distance from the center of the shaft to the center of the follower at its lowest position.

círculo de base Utilizado para colocar un seguidor de desfase. Es un círculo con un radio igual a la distancia del centro del eje al centro del seguidor en su posición más baja.

baseline The bottom line of the displacement diagram.

Línea de base La línea inferior de un diagrama de desplazamiento.

baseline dimensioning A system of dimensioning where as many features of a part as are functionally practical are located from a common set of datums.

dimensionamiento de Línea de base Un sistema de dimensionamiento donde la mayoría de las características de una parte que sean funcionalmente prácticas están localizadas de un conjunto común de referencia.

basic dimension A theoretically "perfect" dimension similar to a reference or nominal dimension. It is used to identify the exact location, size, orientation, or shape of a feature.

dimensión básica Una dimensión teoréticamente perfecta similar a una referencia o una dimensión nominal. Esta es utilizada para identificar la localización exacta, tamaño, orientación o forma de una característica.

basic size The size from which the limits of size are derived by the application of allowances and tolerances.

tamaño Básico El tamaño del cual los limites de tamaño se derivan por la aplicación de provisiones y tolerancias.

batter The slope of the grade is used. For each unit of rise (R), there are four horizontal (H) units.

batter La pendiente de un grado es utilizada. Para cada unidad de elevación (R), hay cuatro unidades horizontales (H).

bearing A supporting member for a rotating shaft. Also, the bearing of a line is the direction of that line as it is drawn in the top view of a drawing.

cojinete Un miembro de apoyo para un eje que rota. También se refiere a la orientación o dirección de una línea según su trazado en la vista superior de un dibujo.

bend allowance The amount of sheet metal required to make a bend over a specific radius.

provisión de curvatura La cantidad de plancha de metal requerida para doblar sobre un radio específico.

bend allowance charts Two basic types of charts used to calculate bend allowances in both the English and metric systems. One is used for 90° bends, and the other is used for bends from 1° to 180°.

tablas de concesión de curvatura Son dos tipos básicos de tablas utilizadas para calcular la provisión de curvatura en el sistema métrico e inglés. Uno es utilizado para curvaturas de 90°, y el otro es utilizado para curvaturas entre 1° y 180°.

bending An operation in which a metal part is simply bent to a desired angle.

doblamiento Una operación en la cual una parte de metal es doblada a un ángulo deseado.

between symbol A symbolic means of indicating that the stated tolerance applies to a specified segment of a surface between designated points.

entre Símbolo Un medio simbólico para indicar que la tolerancia establecida aplica a un segmento específico de una superficie entre los puntos designados.

bevel An inclined edge; not a right angle to joining surface.

bisel Un borde inclinado; no en un ángulo recto a la superficie de unión.

bevel gear Cone-shaped in form with straight teeth that are on an angle to the axis of the shaft. It is used to transmit power and motion between intersecting shafts that are at 90° to each other.

engranaje biselado Conforme con dientes rectos que están en un ángulo con el eje. Este es utilizado para transmitir potencia y movimiento entre ejes que se intersecan que están a 90° el uno del otro.

beveled rings These have a 15° bevel on the groove-engaging edge. They are installed in grooves having a comparable bevel on the load-bearing wall.

anillos biselados Estos tiene 15° de bisel en el borde de ranura de encaje. Estos están instalados en las ranuras con un bisel comparable a una pared de carga.

bilateral tolerance When this tolerance is applied to a dimension, the tolerance applies in both directions but is not necessarily evenly distributed.

tolerancia bilateral Cuando esta tolerancia es aplicada a una dimensión, la tolerancia se aplica en ambas direcciones pero no es necesariamente distribuida equitativamente.

bisect To divide into two equal parts.

bisecar Dividir en dos partes iguales.

blanking   An operation that produces a part cut completely by the punch and die. Also, a stamping operation in which a press uses a die to cut blanks from flat sheets or strips of metal.

cortar   Una operación que produce una parte cortada completamente por el perforador y el troquel. Una operación de estampado en la cual una presa es utilizada como troquel para cortar blancos de planchas planas o tiras de metal.

blind hole   A hole that does not go completely through an object.

agujero ciego   Un agujero que no atraviesa completamente el objeto.

blueprint   A copy of a drawing.

fotocalco   Una copia de un dibujo.

boardroom sketching   This technique works better with larger groups than the conversational style. Since there are more people involved, a larger surface—for better viewing—is needed. It can be a chalkboard but more likely will be an enamel, glass, or dry-erase surface.

croquis de sala de conferencia   Esta técnica trabaja bien con grupos más grandes que los de estilo conversacional. Dado a que hay más personas involucradas, una superficie grande es necesaria para mejor visualización. Ésta puede ser una pizarra, pero más probable será una superficie de esmalte, cristal, o del borrar seco.

bolt   A threaded fastener that passes directly through parts to hold them together and uses a nut to tighten or hold the parts together. The tip of the nut is flat.

perno   Un sujetador de rosca que pasa directamente a través de las partes que sujeta juntas y utiliza una tuerca para sujetar estas partes. La punta de la tuerca es plana.

bolt circle   A circular center line on a drawing containing the centers of holes about a common center.

círculo del perno   Una línea circular central en el dibujo que contiene los centros de los agujeros alrededor de un centro común.

bore   To enlarge a hole with a boring bar or tool in a lathe, drill press, or boring mill.

taladrar   Agrandar un agujero con una barra taladrante o herramienta en un torno, presa taladrante, o fresadora taladrante.

boring mill   A machine tool normally used to machine large workpieces.

fresadora taladrante   Una maquinaria normalmente utilizada para fresar piezas de trabajo grandes.

boss   A cylindrical projection on a casting or a forging.

saliente cilíndrico   Una proyección cilíndrica en una fundición o en una pieza forjada.

bowed rings   They differ from conventional types in that they are bowed around an axis perpendicular to the diameter bisecting the gap. Their construction permits the rings to function as springs and fasteners.

anillos arqueados   Estos difieren de los tipos convencionales en que están arqueados alrededor del eje perpendicular al diámetro que biseca la apertura. La construcción de los anillos le permite funcionar como resortes al igual que como ajustadores.

box jig   Designed to be used for parts that require machining on several sides.

guía de cajón   Diseñada para ser utilizada en las partes que requieren trabajos en varios lados.

brass   An alloy of copper and zinc.

latón   Una aleación de cobre y zinc.

brazing   The process of joining metals together with a non-ferrous filler rod. A temperature above 1000°F, but just below the melting point, is used to join the parts.

soldadura de latón   El proceso de unir metales con una varilla de relleno no ferrosa. Se utiliza una temperatura sobre 1000°F, pero justo por debajo del punto de fusión, para unir estas partes.

breakdown (bender)   It is station 2 of the drop forging process—it forms the parts into special contours.

rotura (dobladora)   Es estación 2 del proceso de forjamiento de caída; ésta forma las partes en un contorno especial.

broach   A tool for removing metal by pulling or pushing it across the work; the most common use is producing irregular hole shapes such as squares, hexagons, ovals, or splines.

brocha   Una herramienta para remover metal cuando esta es halada o empujada a través de la pieza. La más utilizada comúnmente produce agujeros irregulares tales como cuadrados, hexágonos, óvalos, o estrías.

broaching machine   Used to modify the shape of a workpiece by pulling tools called broaches across or through the part.

brochadora   Utilizada para modificar la forma de una pieza halando herramientas llamadas brochas a través de la parte.

broken-out section   A portion removed from the drawing in order to make a particular feature easier to understand.

sección removida   Una porción removida de un dibujo para hacer que una característica en particular sea fácil de entender.

bronze   An alloy of eight or nine parts of copper and one part of tin.

bronce   Una aleación de ocho o nueve partes de cobre y una parte de estaño.

buff   To finish or polish on a buffing wheel composed of fabric with abrasive powders.

lustrar   Darle terminado o pulir en una rueda de pulido compuesta de tela con polvos abrasivos.

burr   The ragged edge or ridge left on metal after a cutting operation.

borde no terminado   Borde sin acabar o saliente dejado en el metal luego de la operación de cortar.

bushing   A metal lining that acts as a bearing between rotating parts such as a shaft and pulley; also used on jigs to guide cutting tools.

cojinete   Revestimiento de metal que actúa como un cojinete entre las partes que rotan tales como el eje y la polea; también usado para guiar herramientas cortantes.

butt joint   One of two kinds of basic rivet joints—the parts are butted and are held together by a cover plate or butt strap that is riveted to both parts.

union de tope   Una de dos tipos de uniones básicas de remache – las partes se unen y se sujetan por una placa o banda que se remacha a ambas partes.

cabinet drawings   Have the same range of receding axes as cavalier drawings, but the receding distances are drawn to half size.

dibujos de gabinete   Tienen el mismo rango de ejes que se alejan tales como en los dibujos cavalier, pero las distancias que se alejan se dibujan a media escala.

CAD (computer-aided drafting)   The use of computers and peripheral devices to aid in the documentation of design projects.

CAD (dibujo industrial asistido por computadora)   El uso de computadoras y aparatos periféricos para asistir en la documentación de los proyectos de diseño.

CAD software   The programs and ancillary materials that allow computers to perform CAD operations.

Software CAD   Programas y materiales auxiliares que permiten a las computadoras realizar operaciones CAD.

CAD system   The hardware, software, and user in computer-aided drafting. It is usually referred to as a configuration of computer hardware. This is a misuse of the term. CAD hardware is just one of the previously mentioned three components needed to have a CAD system.

sistema de CAD   El equipo físico, software, y el usuario en el dibujo asistido por computadora. Generalmente se conoce como la configuración del equipo físico de computadora. Esto es un uso inadecuado del término. El equipo físico para CAD es sólo uno de los tres componentes mencionados previamente necesarios para tener un sistema CAD.

CAE (computer-aided engineering)   Any use of a computer to assist in engineering tasks, analysis testing, calculating, design review, and experimentation.

ingeniería asistida por computadora   Cualquier uso de una computadora para asistir en las tareas de ingeniería, pruebas de análisis, cálculos, revisión de diseño, y experimentación.

calculator   Electro-mechanical device used for making mathematical computations.

Calculadora   Dispositivo electro mecánico usado para realizar cálculos matemáticos.

caliper   An instrument used for measuring diameters.

calibrador   Un instrumento usado para medir diámetros.

callout   A note on the drawing giving a dimension, specification, or machine process.

nota   Una nota en un dibujo que da una dimensión, especificación o proceso de labrado.

CAM (computer-aided manufacturing)   Refers to computer-controlled automation of the various manufacturing processes used in modern industry.

manufactura asistida por computadora   Se refiere a la automatización controlada por computadora de los varios procesos de manufactura usados en la industria moderna.

cam   A rotating shape for changing circular motion to reciprocating motion. Produces a simple means to obtain irregular or specified predictable, designed motion.

leva   Una figura que rota para cambiar el movimiento circular en un movimiento reciprocante. Produce una manera simple de obtener movimiento diseñado, irregular o predecible.

cap screw   Usually used as a true screw, it passes through a clearance hole in one part and screws into another part.

tornillo pasante   Generalmente se usa como un tornillo verdadero, pasa a través de un agujero pasante en una pieza y se atornilla en otra pieza.

carburize   To heat a low-carbon steel to approximately 2000°F in contact with material that adds carbon to the surface of the steel.

carbonizar   Calentar el acero bajo en carbón hasta aproximadamente 2000°F en contacto con material que le añade carbón a la superficie del acero.

case harden   To harden the outer surface of a carburized steel by heating and then quenching.

endurecer   Endurecer la superficie exterior del acero carbonizado calentándolo y luego enfriándolo rápido.

case hardening   The process of hardening the outside surface of a part to a preselected depth.

endurecimiento externo   El proceso de endurecer la superficie exterior de una pieza hasta una profundidad predeterminada.

cast irons   Metals in a family of cast ferrous metals.

hierros de fundición   Metales en una familia de metales férreos de fundición.

cast steels Steels that are produced in ingot form by melting all component elements together and pouring the molten metal into a mold.

aceros de fundición Aceros que se producen en forma de lingotes derritiendo todos los elementos componentes juntos y vertiendo el metal derretido en un molde.

casting A process of pouring molten metal into a mold that contains the desired shape in the form of a cavity.

fundición de molde Un proceso de verter metal derretido en un molde que contiene la forma deseada en la cavidad.

cavalier drawing A type of drawing in which a receding axis is drawn between 30° and 60° to the horizon, and the receding distances are drawn full size.

dibujo cavalier Un tipo de dibujo en el cual el eje que se aleja se dibuja entre 30° y 60° del horizonte, y las distancias que se alejan se dibujan a escala completa.

CD-ROM Compact Disc, Read-Only Memory; a high-volume storage medium used to store drawing files electronically. Can be used to store large projects for permanent archiving.

CD-ROM Disco compacto, sólo para lecturas; un medio de almacenamiento de alta capacidad usado para almacenar archivos de dibujo electrónicamente. Puede usarse para almacenar grandes proyectos para archivo permanente.

center drill A special drill to produce bearing holes in the ends of a workpiece to be mounted between centers.

taladro de centro Un taladro especial para producir agujeros para cojinetes en los extremos de una pieza de trabajo montada entre centros.

center line A type of line used in drafting to indicate the center of an object; its main characteristic is that it is a broken line with one short dash in the center.

línea de centro Un tipo de línea que se usa en el dibujo industrial para indicar el centro de un objeto; su característica principal es que es una línea entrecortada con un guión corto en el centro.

central angle An angle formed by two radial lines from the center of the circle.

ángulo central Un ángulo formado por dos líneas radiales desde el centro del círculo.

central processing unit (CPU) A unit of a computer that includes circuits controlling the interpretation and execution of instructions.

unidad de procesamiento central Una unidad de una computadora que incluye circuitos que controlan la interpretación y ejecución de instrucciones.

center wheel bow compass Manual drafting tool used for drawing circles. The desired diameter is accomplished by adjusting the wheel in the center of the device.

compás ajustable con rueda central Instrumento manual de dibujo que se usa para trazar círculos. El diámetro deseado se logra ajustando la rueda central.

centrifugal casting A process of pouring a measured amount of molten metal into a rotating mold.

fundición centrífuga El proceso de verter una cantidad medida de metal derretido en un molde que rota.

ceramics Materials made by combining metallic and nonmetallic elements. They have a high melting point, low conductivity, and are chemically and thermally stable. They are mainly used in electronics applications.

cerámicas Materiales hechos combinando elementos metálicos y no metálicos. Tienen un punto de fusión alto, conductividad eléctrica baja y son estables química y termalmente. Se usan mayormente para aplicaciones electrónicas.

chain and sprockets Used to transmit motion and power to shafts that are parallel to each other.

cadena y rueda dentada Se usa para transmitir movimiento y poder a los ejes que están paralelos unos con otros.

chain dimensioning Successive dimensions that extend from one feature to another, rather than each originating at a datum.

dimensionamiento de cadena Dimensiones sucesivas que extienden de un rasgo al otro, en lugar de originarse cada uno de un conjunto común de referencia.

chain intermittent weld A weld where each weld is applied directly opposite to each other on opposite sides of the joint.

cadena de soldadura intermitente Una soldadura donde cada soldadura se aplica directamente opuesta una frente a la otra en lados opuestos de una unión.

chamfer A beveled edge normally applied to cylindrical parts and a variety of different types of fasteners.

chaflán Un borde biselado que normalmente se aplica a las partes cilíndricas y a una variedad de tipos de sujetadores.

channel jig A variation of the basic box jig (see also *box jig*).

guía de canal Una variación de la guía de cajón básica (vea también *guía de cajón*).

character A letter, digit, or other symbol that is used as part of the organization, control, or representation of data.

carácter Una letra, dígito, u otro símbolo que se usa como parte de la organización, control o representación de datos.

chase To cut threads with an external cutting tool.

cincelar Cortar las roscas con una herramienta cortante externa.

check valves Used to restrict the flow of liquids through a pipe to one direction only.

válvulas reguladora Se usan para restringir el flujo de líquidos a través de una tubería hacia una dirección solamente.

cheek In sand casting, this is a section that might be installed between the cope and the drag.

mejilla En la fundición de molde en arena, esta es la sección que puede instalarse entre el armazón superior e inferior.

chemical milling A process that etches or shapes workpieces to close tolerances by chemically induced material removal.

fresado químico Un proceso que graba o forma las piezas de trabajo a tolerancias pequeñas por remoción de material químico.

chill To harden the outer surface of cast iron by quick cooling, as in a metal mold.

enfriar Endurecer la superficie externa del hierro fundido enfriándolo rápidamente, como en un molde de metal.

chord Any straight line whose opposite ends terminate on the circumference of the circle.

cuerda Cualquiera línea recta cuyos fines terminen en la circunferencia del círculo.

chordal addendum The distance from the top of the tooth to the point at which the chordal thickness is measured.

altura de cabeza cordal La distancia desde la parte superior del diente al punto donde se mide el espesor cordal.

chordal thickness The length of the chord measured straight across at the pitch diameter. It is measured straight across the tooth, not along the pitch diameter as is the circular thickness.

espesor cordal El largo de la cuerda medido directamente a través del diámetro de paso. Se mide directamente a través del diente, no a lo largo del diámetro de paso como es su espesor circular.

chuck A device much like a vise having movable jaws that grip and hold a rotating tool or workpiece.

prensa de sujeción Un aparato muy parecido a una prensa que tiene mandíbulas movibles que aguantan una pieza o herramienta en rotación.

circle A closed curve with all points at the same distance from the center point. Its major components are the diameter, the radius, and the circumference.

Círculo Una curva cerrada con todos los puntos a la misma distancia del punto central. Sus componentes principales son el diámetro, el radio y la circunferencia.

circle template Manual drafting tool used for making circles of various diameters. It consists of a thin plate (usually plastic) with holes of various sizes punched in it.

plantilla para círculos Instrumento para dibujar a mano usado para trazar círculos de diferentes diámetros. Consiste de una placa pequeña (generalmente de plástico) con perforaciones de diversos tamaños.

circular pitch The distance between corresponding points of adjacent teeth measured on the circumference of the pitch diameter. Also, the length of the arc along the pitch circle between the center of one gear tooth and the center of the next.

largo de paso circular La distancia entre puntos correspondientes de dientes adyacentes medida en la circunferencia del diámetro de paso. También, el largo del arco a lo largo del círculo de paso, entre el centro de un diente de engranaje y el centro del próximo.

circular runout A tolerance that identifies an infinite number of elemental tolerance zones on a feature when the feature is rotated 360° for each element.

excentricidad circular Una tolerancia que identifica un número infinito de zonas de tolerancia elemental en un rasgo cuando el rasgo se rota 360° para cada elemento.

circularity Sometimes referred to as roundness, it is a feature control for a surface of revolution (cylinder, sphere, cone, etc.). It specifies that all points of a surface must be equidistant from the center line or axis of the object in question.

circularidad A veces se conoce como redondez, es un control de rasgo para una superficie de revolución (cilindro, esfera, cono, etc.). Especifica que todos los puntos en una superficie tienen que estar equidistantes de la línea de centro o eje del objeto en cuestión.

circumference The distance around the outer surface of a circle.

circunferencia La distancia alrededor de la superficie exterior de un objeto.

civil engineer's scale Also called a decimal-inch scale; measurements are read directly from the scale. Each graduation is equal to 1/10 of an inch if the number in the corner is 10.

escala de ingeniero civil También se conoce como la escala de pulgada decimal; las medidas se miden directamente de la escala. Cada gradación es equivalente a 1/10 de una pulgada si el número en la esquina es 10.

clapper box This is where the cutting tool is mounted on the end of the ram in this unit.

caja de badajo Esto es cuando la herramienta cortante se monta sobre el extremo del pistón en esta unidad.

clearance fit A condition between mating parts in which there is always a clearance assembly. This is when the least material condition of the shaft is subtracted from the least material condition of the hole.

encaje de separación Una condición entre partes apareadas en la cual siempre hay un ensamblaje de separación. Esto es cuando la condición de material mínimo de un eje se resta de la condición de material mínimo de un agujero.

clearly defined ground rules   In a team situation, these are rules that are thoroughly understood and subscribed to by all team members. They state clearly what behaviors are expected of team members, including what will be tolerated and what will not.

normas Básicas claramente definidas   En una situación de equipo, éstas son las reglas que se entienden claramente y que son acogidas por todos los miembros del equipo. Declaran claramente qué comportamiento se espera de cada miembro del equipo, incluyendo qué se tolerará y qué no.

clockwise   Rotation in the same direction as the hands of a clock.

en favor de las manecillas de reloj   Rotación en el mismo sentido que las manecillas del reloj.

closed jigs   Jigs that enclose the part on more than one side and are intended to machine the part on several sides without moving the part from the jig.

guías cerradas   Guías que encierran la pieza en uno o más lados y su intención es poder labrar la pieza en varios lados sin mover la pieza de la guía.

coil (turn)   One full turn or 360° of the wire about the corner axis.

espiral   Un giro complete o 360° de un alambre alrededor del eje de esquina.

coin   To form a part in one stamping operation.

moneda   Formar una pieza en una operación de estampado.

cold rolled steel   Bessemer steel containing .12% to .20% carbon that has been rolled while cold to produce a smooth, quite accurate stock.

acero rodado en frío   Acero Bessemer que contiene .12% a .20% de carbón que se rueda mientras está frío para producir piezas en bruto que son lisas y bastante precisas.

collar   A round flange or ring fitted on a shaft to prevent sliding.

collar   Un reborde o anillo redondo que encaja en un eje para evitar deslizamiento.

communication   Transfer of information that is received and understood from one source to another.

Comunicación   Transferencia de información que se realiza y es recibida y comprendida de una fuente a otra.

compatibility of files   The ability of various CAD files to interact electronically with each other.

compatibilidad de archivos   La habilidad de diversos archivos CAD de interactuar entre sí electrónicamente.

composites   Materials created by combining two or more materials.

materiales compuestos   Materiales creados combinando dos o más materiales.

compound   This occurs when more than one operation takes place on a part during a single stroke.

combinado   Esto ocurre cuando ocurre más de una operación sobre una parte durante un solo tiro.

compression spring   Offers resistance to a compressive force or applies a pushing action. In its free state, the coils of the spring do not touch.

resorte de compresión   Ofrece resistencia a una fuerza de compresión o aplica una acción de empuje. En su estado libre, los espirales del resorte no se tocan.

compressive force   A force that "goes into" the joint.

fuerza de compresión   Una fuerza que actúa "hacia" la unión.

computational fluid dynamics (CFD)   A means of using computers to mathematically simulate physical events. This involves the study of fluids and their characteristics under a variety of conditions.

dinámica de fluido computacional   Una manera de usar computadoras para simular matemáticamente los eventos físicos. Esto consiste de estudiar los fluidos y sus características bajo una variedad de condiciones.

computational prototype   A three-dimensional computer-generated model that possesses the characteristics of a comparable physical object.

prototipo computacional   Un modelo tridimensional generado a computadora que posee las características de un objeto físico comparable.

computer graphics   A term that is sometimes misused a great deal. CG means using the computer and peripheral devices for producing any type of graphic image, technical or artistic.

gráficas computarizadas   Un término que frecuentemente se usa mal. Gráficas computarizadas significa el uso de computadoras y los aparatos periféricos para producir cualquier tipo de imagen gráfica, técnica o artística.

concentric   Having a common center, as with circles or diameters.

concéntrico   Que tiene un centro común, como en los círculos o diámetros.

concentric circle   Two or more circles with a common center point.

círculo concéntrico   Dos o más círculos con un punto central común.

concentricity   A tolerance in which the axis of a feature must be coaxial to a specified datum regardless of the datum's and the feature's size.

concentricidad   Una tolerancia en la cual el eje de un rasgo debe ser coaxial a una referencia independientemente de la referencia y del tamaño del rasgo.

conceptual design Much use is made of known technical information from such sources as reference books, technical journals, and handbooks.

diseño conceptual Hace mucho uso de la información técnica conocida de fuentes tales como libros de referencia, revistas técnicas y manuales.

cone A pyramid with a central axis and an infinite number of sides, which form a continuous curved lateral surface.

cono Una pirámide con un eje central y un número infinito de lados, los cuales forman una superficie lateral curva y continua.

configuration A group of machines, devices, parts, etc., that makes up a system.

configuración Un grupo de máquinas, aparatos, partes, etc., que forman un sistema.

conic section A section cut by a plane passing through a cone.

sección Cónica Una sección cortada por un plano que pasa a través de un cono.

continuous improvement Improvement to the performance of people and processes that never ends.

adelanto continuo Mejora en la actuación de personas y desarrollo de procesos que nunca termina.

continuous path Machine controls that are used to control the movement of the tool throughout the cutting cycle.

paso continuo Controles de máquina que se usan para controlar el movimiento de la herramienta a través del ciclo de corte.

contour interval The difference in elevation between neighboring contour lines.

intervalo de contorno La diferencia en elevación entre líneas de contorno adyacentes.

contour lines Used on topographical maps to connect points of elevation. They are a constant elevation on the surface of the earth, above sea level.

líneas de contorno Se usan en los mapas topográficos para conectar puntos de elevación. Muestran elevación constante en la superficie del terreno, por encima del nivel del mar.

contract A binding legal document.

contrato Un documento legal vinculante.

conventional break An effective way of drawing long objects of constant cross section on a relatively small sheet of drawing paper.

corte convencional Un método efectivo de dibujar los objetos largos de sección transversal constante en una hoja relativamente pequeña de papel.

conversational sketching This occurs when two or more people are huddled around a drafting board or a cafeteria table. The person doing the talking combines several types of sketches in one drawing as the idea is talked through.

croquis conversacional Esto ocurre cuando dos o más personas están reunidas alrededor de una mesa de dibujo o mesa de cafetería. La persona que habla combina distintos croquis en un dibujo a medida que se discute la idea.

coordinate An ordered set of data values, either absolute or relative, that specifies a location.

coordenada Un grupo ordenado de valores de datos, ya sean absolutos o relativos, que especifican una localización.

coordinate dimensioning A type of rectangular datum dimensioning in which all dimensions are measured from two or three mutually perpendicular datum planes; all dimensions originate at a datum and include regular extension and dimension lines and arrowheads.

dimensionamiento coordinado Un tipo de dimensionamiento de referencia rectangular en el cual las dimensiones se miden desde dos o tres planos de referencia mutuamente perpendiculares; todas las dimensiones se originan en una referencia e incluyen líneas de dimensión y de extensión regulares y flechas.

cope In sand casting, this is the top half of the flask.

armazón superior En la fundición de molde en arena, ésta es la mitad superior del armazón.

copper A heavy metal that is salmon pink in its pure form. It is nonmagnetic and has excellent conductivity.

cobre Un metal pesado que es de color rosado como el salmón en su estado puro. No es magnético y tiene conductividad eléctrica excelente.

core To form a hollow portion in a casting by using a dry-sand core or a green-sand core in a mold.

poner Núcleo Formar una porción hueca en una fundición de molde usando un centro de arena seca o un centro de arena verde en el molde.

core prints When sand cores are intended to be installed in a cavity, core prints must also be provided to locate and anchor the sand cores during the casting process. They are a simple extension of the pattern.

impresos de centro Cuando se tienen que instalar los centros de arena en una cavidad, también deben proveerse impresos de centro para localizar y anclar los centros de arena durante el proceso de fundición. Son simples extensiones del patrón.

costs This item is part of the written report to the client. It details the basic direct costs that the client will have to lay out to have the design made.

costos Este artículo es parte del informe escrito al cliente. Detalla los costos básicos directos que el cliente tiene que pagar para hacer el diseño.

counterbore The enlargement of the end of a hole to a specified diameter and depth.

avellanar Engrandecer un extremo de un agujero a un diámetro y profundidad específica.

counterbored hole Usually a plain through hole with an upper portion enlarged cylindrically to a specified diameter and depth.

agujero avellanado Generalmente un agujero pasante simple con la porción superior engrandecida cilíndricamente a un diámetro y profundidad específica.

countersink To chamfer the end of a hole to receive a flathead screw.

escariar Biselar un extremo de un agujero para recibir un tornillo de cabeza plana.

countersunk hole Usually a plain through hole with an upper portion enlarged conically, most often to 82°, to accommodate standard flathead machine screws.

agujero escariado Generalmente un simple agujero pasante con la porción superior engrandecida cónicamente, frecuentemente a 82°, para acomodar la cabeza estándar de los tornillos de máquina.

cover letter Brief letter sent to a potential employer explaining who the applicant is and what job he or she is interested in.

carta de presentación Breve que se envía a un eventual empleador y en la que se explica quién es el candidato o candidata y en qué puesto está interesado/a.

CPU Central Processing Unit (see also *central processing unit*).

CPU Unidad Central de Procesamiento (vea también *unidad central de procesamiento*).

creative process A process designed to produce a finished design. The elements include: accumulating the necessary information about the problem; "incubating" this information with other ideas that the brain retrieves from memory; illuminating its presence until a potential/probable solution is found; verifying its workability.

proceso creativo Un proceso diseñado para producir un diseño terminado. Los elementos incluyen el acumular la información necesaria acerca del problema; "incubar" esta información con otras ideas que el cerebro tiene en memoria; iluminar su presencia hasta que se encuentre una solución potencial/probable; verificar su viabilidad.

crest The top point joining the sides of a thread.

cresta El punto superior que une los lados de una rosca.

critical thinking and problem solving Thinking based on careful, thorough analysis of pertinent information rather than assumptions. Critical thinking is an important aspect of problem solving.

pensamiento crítico y resolución de problemas Pensamiento basado en un análisis cuidadoso y detallado de la información pertinente, más que en suposiciones. El pensamiento crítico es un aspect importante de la resolución de problemas.

cursor (1) On adisplay monitor, a movable marker that is visible on the viewing surface and is used to indicate a position at which an action is to take place or the position on which the next device operation would normally be directed. (2) On digitizers, a movable reference, usually optical crosshairs used by an operator to indicate manually the position of a reference point or line where an action is to take place.

cursor (1) En un monitor de despliegue, un marcador movible que se puede ver en la pantalla y que se usa para indicar la posición donde se llevará a cabo una acción o la posición adonde normalmente se dirigirá la próxima operación. (2) En los digitalizadores, una referencia movible, generalmente un "hilo de cabello" óptico usado por un operador para posicionar manualmente un punto o línea de referencia donde se llevará a cabo la acción.

cutoff saw Used to cut rough bar stock to lengths suitable for machining in other machine tools.

sierra tronzadora Se usa para cortar en bruto las barras de materiales en largos apropiados para labrarse con otras herramientas de labrado.

cuts and fills Leveling out a surface by excavating the earth of some of the higher points and using it to fill in the low points.

cortar y rellenar Aplanar una superficie excavando el terreno en los puntos más altos y usándola para rellenar los puntos bajos.

cutting off A cutting operation that is performed on a part to produce an edge other than a straightedge.

cortar Una operación de corte que se hace en una pieza para producir un borde diferente a un borde recto.

cutting-plane line Indicates the path an imaginary cutting plane follows to slice through an object.

línea plano cortante Indica el camino que sigue el plano cortante imaginario para cortar el objeto.

cylindrical A design where a shaft rotates exactly as it does in the radial design, but the action or direction of the follower differs greatly.

cilíndrico Un diseño en el cual el eje rota exactamente como en el diseño radial, pero la acción o dirección del seguidor varía grandemente.

cylindrical grinder Used to precisely grind cylindrical or conical workpieces.

moledora cilíndrica Se usa para moler piezas de trabajo cilíndricas o cónicas.

cylindricity A tolerance that simultaneously controls the surface of revolution for straightness, parallelism, and circularity of a feature and is independent of any other features on a part. Also a feature control in which all elements of a surface of revolution form a cylinder. It gives the effect of circularity extended the entire length of the object, rather than just a specified cross section.

cilindricidad Una tolerancia que simultáneamente controla la superficie de revolución para rectitud, paralelismo y circularidad de un rasgo y que es independiente de cualquier otro rasgo en la pieza. También un rasgo de control al en el cual todos los elementos de una superficie de revolución forman un cilindro. Proporciona el efecto de circularidad extendida a lo largo del largo completo de un objeto en lugar de sólo en la sección transversal.

data A representation of facts, concepts, or instructions in a formalized manner suitable for communication, interpretation, or processing by human or automatic means.

datos Una representación de los hechos, conceptos, o instrucciones en un formato adecuado para la comunicación, interpretación o procesamiento por medios automáticos o humanos.

datum Points, lines, planes, cylinders, and the like, assumed to be exact for purposes of computation, from which the location or geometric relationship (form) of features of a part may be established.

referencia Puntos, líneas, planos, cilindros, y así por estilo, que se asume que son exactos para propósitos de computación, desde los cuales se puede establecer la localización o forma geométrica de los rasgos de una parte.

datum feature The actual physical feature on a part used to establish a datum.

rasgo de referencia El rasgo físico real de una pieza que se usa para establecer una referencia.

datum feature simulator A surface, the form of which is of such precise accuracy that it is used to simulate the datum.

simulador de rasgo de referencia Una superficie, cuya forma es de tanta precisión que se usa para simular la referencia.

datum feature symbol This symbol identifies datum features. It consists of a capital letter enclosed in a square frame. A leader line extends from the frame to the selected feature.

símbolo de rasgo de referencia Este símbolo identifica los rasgos de referencia. Consiste de una letra mayúscula en un cuadro cuadrado. Una directriz se extiende desde el cuadro hasta el rasgo seleccionado.

datum identification symbol A special rectangular box that contains the datum reference letter.

símbolo de identificación de referencia Un cuadro rectangular especial que contiene la letra de la referencia.

datum plane A theoretically perfect plane from which measurements are made.

plano de referencia Un plano teóricamente perfecto desde el cual se hacen las medidas.

datum reference frame A hypothetical three-dimensional frame that establishes the three axes of an X-Y-Z coordinate system into which the object being produced fits and from which measurements can be made.

marco de referencia Un marco de referencia tridimensional e hipotético, que establece los tres ejes de un sistema de coordenadas X-Y-Z sobre el cual encaja el objeto en producción y desde el cual se hacen las medidas.

datum surface (feature) A surface, the form of which is of such precise accuracy that it is used to simulate the datum.

superficie de referencia (rasgo) Una superficie, cuya forma es de tal precisión que se usa para simular la referencia.

datum target symbol Identifies datum targets. It consists of a circle divided in half with a horizontal line.

símbolo de blanco de referencia Identifica los blancos de referencia. Consiste de un círculo dividido en dos con una línea horizontal.

decimal-inch dimensioning Frequently used in the dimensioning of technical drawings. It is much less cumbersome than fractional and still used more than metric.

dimensionamiento de pulgada decimal Se usa frecuentemente en el dimensionamiento de los dibujos técnicos. Es mucho menos incómodo que el dimensionamiento fraccional y se usa más que el métrico.

decision table A table that systematically compares solution alternatives with every other alternative as to how well each one satisfies the constraints and specifications from the definition of the problem.

tabla de decisiones Una tabla que sistemáticamente compara las alternativas de soluciones con otras alternativas acerca de cuán bien cada una satisface las restricciones y especificaciones de la definición del problema.

dedendum The radial distance from the pitch diameter to the bottom of the tooth.

altura de cabeza baja La distancia radial desde el diámetro de paso a la parte inferior del diente.

delineation Pictorial representations; a chart, a diagram, or a sketch.

delineación Representaciones pictóricas; una gráfica, un diagrama, o un croquis.

depth of thread The distance between the crest and the root of the thread, as measured at a right angle to the axis.

profundidad de rosca La distancia entre la cresta y la raíz de la rosca, medida a ángulos rectos del eje.

design file A file that includes freehand sketches, machine copies of technical information, copies of related drawings, summaries of telephone conversations, interviews, and vendor brochures.

archivo de diseño Un archivo que incluye dibujos a pulso, copias a máquina de información técnica, copias de dibujos relacionados y resúmenes de conversaciones telefónicas, entrevistas, y folletos de vendedores.

design hints Simple ideas used to help create the finished product. They help in initial designing and become checklists as the project progresses.

sugerencias de diseño Ideas simples usadas para ayudar a crear el producto final. Ayudan en el diseño inicial y se convierten en listas de comprobación a medida que el proyecto progresa.

design intent The ultimate goal when designing anything—the intended outcome. When changes are made to an object, the design intent remains the same and does not change.

propósito del diseño El propósito fundamental cuando se realiza un diseño – el resultado deseado. Aunque se realicen cambios en un objeto, el propósito del diseño se mantiene y no varía.

design procedure Acts as a general guideline for a team of designers, engineers, and drafters on the best way to accomplish the procedure sequentially.

procedimiento de diseño Sirve como una guía general para un equipo de diseñadores, ingenieros y delineantes acerca de la mejor manera de lograr el procedimiento secuencialmente.

design section This section illustrates the uses of the drawing and the related engineering practices. It includes preparing assembly, detail, and layout drawings; following design procedures; making comprehensive parts lists; executing engineering revisions; handling invention agreements; keeping personal technical files; and preparing quotations.

sección de diseño Esta sección ilustra los usos del dibujo y las prácticas de ingeniería relacionas. Incluye el preparar dibujos del ensamblaje, de detalles, y del arreglo; seguir los procedimientos de diseño; hacer listas de partes comprensivas; ejecutar revisiones de ingeniería; manejar los acuerdos de invención; mantener archivos técnicos personales; y el preparar cotizaciones.

design size The size of a feature after an allowance for clearance has been applied and tolerances have been assigned.

tamaño de diseño El tamaño de un rasgo después de aplicarle una concesión para separación y asignar las tolerancias.

design tools and methods Capabilities built into CAD software that allow specific functions to be performed.

Herramientas y Métodos de Diseño Capacidades incorporadas al software CAD que lo habilitan para realizar funciones específicas.

detail drawing A drawing of a single part that provides all the information necessary in the production of that part.

dibujo detallado Un dibujo de una sola pieza que provee toda la información necesaria en la producción de esa pieza.

detail notes Specific notes that pertain to one particular element or characteristic of a part.

notas de detalles Notas específicas relacionadas a un elemento o característica particular de una pieza.

developed length The total length of a pattern.

largo de desarrollo El largo total de un patrón.

development Drawing of the surface of an object unfolded on a plane. Also, the pattern or template of a shape that is laid out in a single flat plane in preparation for the bending or folding of a material to a required shape.

desarrollo Dibujo de la superficie de un objeto desdoblado en un plano. También, el patrón de una figura que se traza en un solo plano liso en preparación para el doblez de un material a una forma deseada.

development surface The exterior and/or interior of the sheet material used to form an object.

superficie de desarrollo El exterior y/o interior del material en hoja usado para formar un objeto.

deviation The variance from a specified dimension or design requirement.

desviación La variación de una dimensión o requisitos de dimensión especificada.

DFA (design for assembly) procedure An iterative loop that occurs simultaneously with the DFM loop. The emphasis is on assembly of parts. The DFA loop first considers which parts should be assembled automatically and next considers which parts need to be assembled manually.

procedimiento de diseño para ensamblaje Un ciclo iterativo que ocurre simultáneamente con el ciclo DFM. El énfasis está en el ensamblaje de las piezas. El ciclo DFA considera primero cuáles partes deberían ser ensambladas manualmente y después considera cuáles partes deberían ser ensambladas por mano.

DFD (design for disassembly) A design program with the goal of designing products that can be disassembled so that the greatest amount of reusable or recyclable materials, such as metal and plastic, can be recovered.

diseño para despiezado Un programa de diseño con el propósito de diseñar productos que pueden desensamblarse de manera que se pueda recuperar la cantidad más grande de materiales reciclables o reusables, tales como metales o plásticos.

DFM (design for manufacturability) A procedure in the form of an iterative loop that acts as a guide for a design team.

diseño para manufacturabilidad Un procedimiento en forma de un ciclo iterativo que actúa como una guía para el equipo de diseño.

DFR (design for recyclability) A design program with the goal of designing products made from materials that can be recycled for a variety of purposes.

diseño para reciclabilidad Un programa de diseño con el propósito de diseñar productos hechos de materiales que pueden reciclarse para una variedad de propósitos.

diagram A figure or drawing that is marked out by lines; a chart or outline.

diagrama Una figura o dibujo que está marcado por líneas; una gráfica o bosquejo.

diameter The length of a straight line passing through the center of a circle and terminating at the circumference on each end.

diámetro El largo de una línea recta que pasa a través del centro de un círculo y que termina en la circunferencia en cada extremo.

diametral pitch A ratio equal to the number of teeth (N) on a gear per inch of pitch diameter (D).

largo de paso diametral Una razón equivalente al número de dientes (N) en un engranaje por pulgada del diámetro de paso (D).

diazo Material that is either a film or paper sensitized by means of azo dyes used for photocopying.

diazo Material que es o una película o un papel hecho sensitivo por medio de tintes azo usados para fotocopias.

die The term has a double meaning. It can be used to describe the entire assembled tool or the lower cutting part of the tool. Also, a hardened metal piece shaped to cut or form a required shape in a sheet of metal by pressing it against a mating die.

troquel El término tiene significado doble. Puede usarse para describir una entera herramienta ensamblada o la parte inferior de la herramienta que corta. También, es un pedazo de metal endurecido formado para cortar la figura necesaria en una hoja de metal presionándolo contra otro troquel.

die casting A process in which molten metal is forced into metal dies under pressure, producing very accurate and smooth castings.

fundición a presión Un proceso en el cual el metal derretido se fuerza en discos metálicos bajo presión, produciendo piezas de fundición muy precisas y suaves.

die stamping The process of cutting or forming a piece of sheet metal with a die.

estampado de troquel El proceso de cortar y formar un pedazo de hojalata con un troquel.

digitize To use numeric characters to express or represent data.

digitalizar Usar caracteres numéricos para expresar o representar data.

digitizer A device for converting positional information into digital signals; typically, a drawing or other graphic is placed on the measuring surface of the digitizer and traced by the operator using a cursor.

digitalizador Un aparato para convertir la información posicional en señales digitales; típicamente, se coloca un dibujo u otra gráfica en la superficie de medida del digitalizador y se traza por operador usando un cursor.

dimension Measurements given on a drawing, such as size and location.

dimensión Medidas dadas en un dibujo, tales como tamaño y localización.

dimension line This line is drawn as a solid, thin line of approximately the same width as a hidden line. It should be broken for dimensions and have arrowheads for terminations. It is used to indicate graphically the linear distance being dimensioned.

línea de dimensión Esta línea se dibuja como una línea sólida y fina de aproximadamente el mismo espesor que una línea escondida. Debería ser entrecortada para dimensiones y debería tener flechas para las terminaciones. Se usa para indicar gráficamente la distancia lineal que se está dimensionando.

dimensioning The process whereby size and location data for the subject of a technical drawing are provided.

dimensionamiento El proceso en el cual se provee información de tamaño y localización para el sujeto de un dibujo técnico.

dimetric drawing Differs from isometric drawing in that only two angles are equal.

dibujo dimétrico Difiere de los dibujos isométricos en que sólo dos ángulos son iguales.

dip The downward angle of the vein with respect to a horizontal plane at the site of the ore vein.

depresión El ángulo hacia abajo de una vena con respecto al plano horizontal en el lugar de la vena de mineral.

direct numerical control (DNC) The use of a common host computer for the distribution of part programs to remote machine tools.

control numérico directo El uso de una computadora común para la distribución de programas de parte para las herramientas de labrado remotas.

direction   This is originally handed down from higher management. It outlines where the team should be going, what they are hoping to accomplish, and specific as well as general guidelines as to how to get there.

dirección   Esto originalmente se envía desde la administración superior. Delinea adónde debería dirigirse un equipo, lo qué esperan lograr, y guías específicas y generales de cómo llegar a ese punto.

direction of winding   The direction in which a spring is wound. It can be wound right- or left-hand.

dirección de enrollado   La dirección en la cual está enrollado un resorte. Puede estar enrollado a mano derecha o a mano izquierda.

displacement diagram   A designated layout of the required motion of a cam. It is laid out on a grid, and its length represents one complete revolution of the cam.

diagrama de desplazamiento   Una disposición designada que enseña la moción requisita de una leva. Se presenta sobre una rejilla, y su largo representa una revolución completa de la leva.

display monitor   An output device resembling a television that displays the information on which the CAD technician is working.

monitor de despliegue   Un aparato de salida parecido a un televisor que despliega la información en la cual el técnico de CAD está trabajando.

double weld   Similar to a multiweld, the same type of weld may be used on opposite sides of a single joint; for example, double V, double U, or double J welds. Used for strength and appearance.

soldadura doble   Similar a una soldadura múltiple, el mismo tipo de soldadura se puede usar en lados opuestos de una unión; por ejemplo, soldaduras V doble, U doble, o J doble. Se usa para fortaleza y apariencia.

double-curved surface   An object fully formed by curved lines with no straight lines.

superficie de doble curva   Un objeto completamente formado por líneas curvas sin ninguna línea recta.

dowel   A cylindrical pin, commonly used to prevent sliding between two contacting flat surfaces.

clavija   Un perno cilíndrico, comúnmente usado para evitar el deslizamiento entre dos superficies planas en contacto.

draft   The slope or taper of the sides of a pattern, which permits it to be removed from the cope and drag without disturbing the cavity.

calada   La pendiente o declive de los lados de un patrón, lo cual le permite que se remueva el armazón superior e inferior sin perturbar la cavidad.

drafters   People who do drafting.

delineantes   Personas que hacen dibujos industriales.

drafting   Creating technical drawings to support the design process.

dibujo industrial   El crear dibujos técnicos para apoyar al proceso de diseño.

drafting brush   Brush used to remove eraser and other types of residue from drawing prepared by hand.

brocha de dibujo   Brocha usada para remover residuos de goma de borrar y otros similares de un dibujo industrial hecho a mano.

drafting dots or tape   Semi-sticky tape cut into circular dots or regular form used to secure drafting paper to the drawing board in manual drafting.

redondeles o cinta para dibujo   Papel semi adhesivo cortado en forma circular o de otro tipo usado para fijar el papel de dibujo al tablero cuando se hacen dibujos a mano.

drafting machine or parallel bar   Tools attached to manual drafting boards that are used as guides in making lines of any angle.

máquina para dibujo o barra paralela   Herramientas que se adosan a los tableros de dibujo manual y que se usan como guías para trazar líneas de todo tipo de ángulo.

drag   In sand casting, this is the bottom half of the flask.

armazón inferior   En la fundición de molde en arena, ésta es la mitad inferior del armazón.

draw   To temper steel.

estirar   Templar el acero.

drawing   A process of stretching and forming a metal sheet into a shape similar to a cup or top hat.

estirado   El proceso de estirar y formar una hoja de metal en una forma similar a una taza o un sombrero de copa.

drill   To cut a cylindrical hole with a drill; a blind hole does not go through the piece.

taladrar   Cortar un agujero cilíndrico con una taladradora; un agujero ciego no atraviesa completamente la pieza.

drill bushings   Used to position and guide the cutting tools used in jigs. The three basic types are press-fit bushings, renewable bushings, and liner bushings.

cojinetes de taladro   Se usan para posicionar y guiar las herramientas de corte que se usan en guías. Los tres tipos básicos son los cojinetes de encaje a presión, los renovables y los de revestimiento.

drill press   A machine that is used mainly for producing holes.

taladradora de columna   Una máquina que se usa mayormente para producir agujeros.

drop forging   The process of forming a heated metal bar, or billet, into dies by using a drop hammer or with great pressure.

forjado   El proceso de formar una barra de metal calentada, o lingote, en matrices usando un martillo de forja o con gran presión.

dry cleaning pad   Bag-shaped manual drafting tool used for cleaning graphite from drafting paper so that it does not smear.

almohadilla seca de limpieza   Herramienta de dibujo manual usada para eliminar el grafito del papel de dibujo para evitar borrones y manchas.

duplex feature   A method of indexing the fixture to the machine in two parts. In use, the first part is machined while the second is loaded. The fixture is then rotated, and the first is unloaded while the second is worked on, while a third is loaded on the empty side.

rasgo doble   Un método de fijar un aditamento para labrar dos piezas. En uso, se labra la primera pieza mientras se carga la segunda. Luego se rota el aditamento, se descarga la primera pieza mientras se trabaja en la segunda y se carga una tercera pieza en el lado vacío.

dwell   The period of time during which the follower does not move.

tiempo de permanencia   El periodo de tiempo durante el cual el seguidor no se mueve.

eccentric   Not having the same center; off center.

excéntrico   Que no tiene el mismo centro; fuera de centro.

ECO (engineering change order)   An order for a change in a drawing to make improvements or correct an error. An ECO is issued by an Engineering Change Request (ECR) committee after all departments concerned have agreed to the change.

ECO (orden de cambio de ingeniería)   Una orden para cambiar un dibujo para hacer mejoras o para corregir un error. Se emite una Orden de Cambio de Ingeniería por medio de un comité de Petición de Cambio de Ingeniería después que todos los departamentos involucrados están de acuerdo con el cambio.

ECR (engineering change request)   A request to make a change in a drawing to make improvements or correct an error. The ECR explains what change is to be made, what it was before, why the change is to be made, whether it affects other parts, and who suggested the change. The request is made to an ECR committee, which issues an ECO after all departments concerned have agreed to the change.

ECR (petición de orden de cambio)   Una petición para hacer un cambio en un dibujo para hacer mejoras o para corregir un error. La Petición de Cambio de Ingeniería explica qué cambios se harán, cómo era antes, por qué se harán los cambios, si afecta a otras piezas, y quién sugirió el cambio. La petición se la hace ante un comité de Petición de Cambio de Ingeniería, el cual emite una Orden de Cambio de Ingeniería después que todos los departamentos involucrados están de acuerdo con el cambio.

elastomers   Linear polymers that are able to stretch and return to their original shape and size (like an elastic band).

elastómero   Polímeros lineales que pueden estirarse y que regresan a su forma y tamaño original (como una banda elástica).

electrical discharge grinding (EDG)   A nontraditional machining process similar to EDM in that it employs rapidly recurring electrical discharges between an electrode and the workpiece in a dielectric fluid. The difference is that with EDG, the electrode is a rotating wheel made of graphite or brass.

amoladura por descarga eléctrica   Un proceso de labrado no tradicional similar al labrado por descarga eléctrica (EDM) en que utiliza descargas eléctricas rápidas y recurrentes entre un electrodo y la pieza de trabajo en un líquido dieléctrico. La diferencia es que con la amoladura por descarga eléctrica, el electrodo es una rueda rotativa hecha de grafito o azófar.

electrical discharge machining (EDM)   A nontraditional machining process that removes metal using a series of rapidly recurring electrical discharges between an electrically charged cutting tool and the workpiece submerged in a dielectric fluid.

labrado por descarga eléctrica   Un proceso de labrado no tradicional que remueve metal usando una serie de descargas eléctricas rápidas y recurrentes entre una herramienta cortante cargada eléctricamente y la pieza de trabajo sumergida en un líquido dieléctrico.

electrical discharge wire cutting (EDWC)   A nontraditional machining process that is similar to the bandsawing process. Material removal occurs as the result of an electrical discharge as a wire electrode is drawn through the workpiece.

corte con alambre por descarga eléctrica   Un proceso de labrado no tradicional que es similar al proceso de corte con sierra de cinta. La remoción de material ocurre como resultado de una descarga eléctrica de un electrodo de alambre, a medida que se lo pasa a través de la pieza de trabajo.

electrical plan   This plan shows the house's electrical system superimposed on a floor plan without dimensions.

plano eléctrico   Este plano muestra el sistema eléctrico de la casa sobrepuesto en el plano de distribución sin dimensiones.

electrochemical discharge grinding (ECDG)   A combination of two nontraditional machining processes: electro-chemical grinding (ECG) and electrical discharge grinding (EDG). In ECDG, alternating current or pulsating direct current moves from a bonded graphite wheel to a positively charged workpiece.

amoladura por descarga electroquímica   Una combinación de dos procesos de labrado no tradicionales: amoladura electroquímica y amoladura por descarga eléctrica. En la amoladura por descarga electroquímica, una corriente alterna o corriente directa pulsante se mueve de una rueda de grafito a la pieza de trabajo cargada positivamente.

electrochemical grinding (ECG)   A nontraditional machining process that combines abrasion with an electrochemical reaction that occurs on the surface of the metal.

amoladora electroquímica   Un proceso de labrado no tradicional que combina la abrasión con una reacción electroquímica que ocurre en la superficie del metal.

electrochemical honing (ECH)   A nontraditional machining process similar to ECG. It combines electrochemical oxidation of surface metal and abrasion. The primary difference is that honing is a process of removing material inside a drilled hole.

afinación electroquímica   Un proceso de labrado no tradicional similar a la amoladura electroquímica. Combina la oxidación electroquímica de una superficie de metal y abrasión. La diferencia primaria es que la afinación es un proceso de remover material dentro de un agujero taladrado.

electrochemical machining (ECM)   The process, with many variations, of removing metal by combining electricity with a chemical to cause a reaction that is the opposite of metal plating.

labrado electroquímico   El proceso, con muchas variaciones, de remover metal combinando electricidad con un químico para causar una reacción que es lo opuesto del enchapado de metal.

electrochemical turning (ECT)   A variation of ECM. The primary difference is that the workpiece rotates.

torneado electroquímico   Una variación del labrado electroquímico. La diferencia principal es que la pieza de trabajo rota.

electromechanical machining (EMM)   A process that improves the performance of traditional machining processes. Metal removal is accomplished in the traditional manner, except that the workpiece is electrochemically polarized by applying a controlled voltage across the connection point between the workpiece and an electrolyte. This changes the surface of the workpiece, making it easier to machine.

labrado electromecánico   Un proceso que mejora el rendimiento de los procesos de labrado tradicional. La remoción de metal se logra de una manera tradicional, excepto que la pieza de trabajo se polariza electroquímicamente aplicándole un voltaje controlado a través de un punto de conexión entre la pieza de trabajo y un electrolito. Esto cambia la superficie de la pieza de trabajo, haciéndola más fácil de labrar.

electron beam machining (EBM)   A thermal energy process. It uses thermal energy to remove metal.

labrado con rayo de electrones   Un proceso de energía termal. Usa energía termal para remover metal.

elevations   Orthographic views of the front, back, and sides of a structure. Most are exterior views; however, interior elevations of the kitchen, bathrooms, and utility areas are frequently included in residential plans.

elevaciones   Vistas ortográficas del frente, la parte trasera y los lados de una estructura. La mayoría son vistas exteriores; sin embargo, elevaciones interiores de las cocinas, baños y áreas útiles se incluyen frecuentemente en los planos residenciales.

engineering materials   In modern manufacturing settings, this term refers to the raw materials.

materiales de ingeniería   En el ambiente de manufactura moderna, este término de refiere a los materiales brutos.

epicycloid   If the span of a cycloid is convex, an epicycloid is drawn.

epicicloide   Si la envergadura de un cicloide es convexa, se dibuja un epicicloide.

equilateral triangle   A triangle having three sides of equal length, as well as three equal interior angles.

triángulo equilátero   Un triángulo que tiene tres lados de largos iguales, al igual que tres ángulos interiores iguales.

eraser   Rubber-like manual drafting device used for eliminating graphite lines, symbols, and letters from drafting paper.

borrador   Implemento para dibujar a mano de tipo gomoso que se usa para eliminar líneas de grafito, símbolos y letras del papel de dibujo.

erasing shield   Manual drafting tool used in conjunction with an eraser to prevent erasing the wrong material. This is a thin plate (usually metal) with a variety of openings cut into it of various shapes. The plate covers what should not be erased, and the openings provide access to what should be erased.

bloqueador de borrado   Herramienta de dibujo manual que se usa junto con un borrador para prevenir que se borre material equivocado. Es una placa fina (generalmente de metal) con una diversidad de aberturas recortadas de distintas formas. La placa cubre lo que no se debe borrar y las aberturas proveen acceso a lo que debe ser borrado.

expanded diameter   When metal is displaced to each side, forming a raised portion or flute extending along each side of the groove, the crest of the flute creates this diameter. The diameter is a few thousandths larger than the nominal diameter of the stock.

diámetro expandido   Cuando se despliega el metal a cada lado, formando una porción levantada o canaladura que se extiende a cada lado de la ranura, la cresta de la canaladura crea este diámetro. El diámetro es unas cuantas milésimas más grande que el diámetro nominal del material.

exploded isometric drawing   This style shows all the components of an assembly as individual isometric drawings, including fasteners; however, the components are aligned in space the identical way they are to be assembled.

dibujo isométrico despiezado   Este estilo muestra todos los componentes de un ensamblaje como dibujos isométricos individuales, incluyendo los sujetadores; sin embargo, los componentes se alinean en el espacio idénticamente como serán ensamblados.

extension lines   Solid, thin lines of approximately the same width as hidden lines.

líneas de extensión   Líneas sólidas y finas de aproximadamente el mismo grueso que las líneas escondidas.

extension spring   A spring that offers resistance to a pulling force. In its free state, the coils are usually touching or very close.

resorte de extensión   Un resorte que ofrece resistencia a una fuerza de tirada. En su estado libre, las espirales están tocándose o están muy cerca.

external broaching   Typically used to produce some gear teeth, plier jaws, and other similar details.

brochado externo   Se usa típicamente para producir algunos dientes de engranajes, las mandíbulas de alicates y otros detalles similares.

external client   A large variety of people and organizations, both in classrooms and in the industry.

cliente externo   Una gran variedad de personas y organizaciones, tanto en los salones de clases y en la industria.

external rings   For axial assembly, rings that are expanded with pliers so they may be slipped over the end of a shaft, stud, or similar part.

anillos externos   Para los ensamblajes axiales, los anillos que se expanden con alicates de manera que puedan colocarse alrededor de un extremo de un eje, un perno o una parte similar.

external threads   Threads located on the outside of a part, such as those on a bolt.

roscas externas   Roscas que están localizadas en el exterior de la parte, tales como aquellos en los pernos.

extruding   A process of forcing metal through a die of a desired form and cross section.

formar por extrusión   Un proceso de forzar metal a través de un troquel de la forma y corte transversal deseado.

extrusion   Metal that has been shaped by forcing it in its hot or cold state through dies of the desired shape.

extrusión   Metal que se forma forzándolo en su estado frío o caliente a través de troqueles de la forma deseada.

fabrication   A term used to distinguish production operations from assembly operations.

fabricación   Un término que se usa para distinguir las operaciones de producción de las operaciones de ensamblaje.

face   To finish a surface at right angles, or nearly so, to the center line of rotation on a lathe.

encarar   Acabar en un torno una superficie a un ángulo recto, o casi a un ángulo recto, de la línea de rotación de centro.

FAO   Finish All Over.

acabar por todos lados   Acabar por Todos Lados.

fastener   A mechanical device for holding two or more bodies in definite positions with respect to each other.

sujetador   Un aparato mecánico para sujetar dos o más cuerpos en posiciones definidas, una con respecto a la otra.

feature   A portion of a part, such as a diameter, hole, keyway, or flat surface.

rasgo   Una porción de una parte, tal como un diámetro, agujero, muesca de posicionamiento o superficie plana.

features   In solid modeling, features are the building blocks of the part in question—shapes and operations that will be used to make the part electronically.

rasgos   En el modelado de sólidos, los rasgos son los elementos esenciales de la parte en cuestión – las figuras y operaciones que se usarán para crear la parte electrónicamente.

features and capabilities   Specific assets built into CAD software that determine the software's ability to perform certain tasks (e.g., drag-and-drop, cross hatching, textures, etc.).

rasgos y capacidades   Propiedades específicas integradas al software CAD que determinan la habilidad del mismo para realizar ciertas tareas (por ejemplo: arrastrar y soltar, sombreado, texturas, etc.)

feature control symbol   A rectangular box in which all data referring to the subject feature control are placed. These include the symbol, datum references, the feature control tolerance, and modifiers.

símbolo del control de rasgo   Una caja rectangular en que están colocados todos los datos que refieren al control del rasgo especificado. Puede incluir el símbolo, referencias del datum, la tolerancia del control de rasgo y los modificadores.

ferrous   A metal having iron as its base material.

ferroso   Un metal que tiene hierro como su material base.

field weld   Any weld not made in the factory; that is, a weld that is to be made at a later date, perhaps at final assembly on-site.

soldadura de campo   Cualquier soldadura que no se hace en la factoría; esto es, una soldadura que se hará en algún punto en el futuro, quizás en el lugar del ensamblaje final.

file   To finish or smooth with a file.

limar   Acabar o suavizar con una lima.

file compatibility   Ability of electronic files such as those used in CAD to interact with each other.

compatibilidad de archivos   Habilidad de los archivos electrónicos, como los usados en CAD, de interactuar entre sí.

fillet   An interior rounded intersection between two surfaces.

filete   Una intersección interior redondeada entre dos superficies.

fillet weld   A weld made at a concave junction formed where two metal parts meet.

soldadura de filete   Una soldadura que se hace en una unión cóncava donde se encuentran dos piezas de metal.

fin   A thin extrusion of metal at the intersection of dies or sand molds.

aleta   Una extrusión fina de metal en la intersección de los troqueles o moldes de arena.

finish   General finish requirements such as paint, chemical, or electroplating, rather than surface texture or roughness (see *surface texture*).

terminado   Requisitos generales de terminado tales como pintura, químico o galvanización, en lugar de una textura de superficie o aspereza (vea *textura de superficie*).

finishing impression   The final station on a drop forge die. It is used to finish the part to the desired shape and size.

impresión final   La estación final en un troquel de forja. Se usa para terminar la pieza a la forma y tamaño deseada.

finite-element analysis (FEA)   A virtual means of analyzing the structural response of objects. In this method, a model of a part is broken up into a finite number of polygonal elements called a mesh.

análisis de elementos finitos   Un método virtual de analizar la respuesta estructural de los objetos. En este método, un modelo de una pieza se divide en un número finito de elementos poligonales llamados una malla.

finite-element model (FEM)   A three-dimensional model of an object, with nodes containing specific information about the characteristics of the object at that point. The nodes are connected together to create an FEA mesh, which is then analyzed.

modelo de elementos finitos   Un modelo tridimensional de un objeto, con nodos que contienen información específica acerca de las características del objeto en ese punto. Los nodos están conectados para crear una malla de análisis de elementos finitos, que luego se analiza.

fit   The clearance or interference between two mating parts.

encaje   La separación o interferencia entre dos partes que encajan.

fixed-renewable bushing   Used for applications where the bushings do not need to be changed often, but where they are changed more often than press-fit bushings.

cojinete fijo renovable   Se usa para aplicaciones donde los cojinetes no necesitan cambiarse frecuentemente, pero que se cambian más frecuentemente que los cojinetes de encaje a presión.

fixture   A workholding tool that references the cutting tool to the workpiece by means of a set block.

aditamento   Una herramienta de trabajo que refiere la herramienta cortante a la pieza de trabajo por medio de un bloque.

flange   A relatively thin rim around a piece.

reborde   Un borde relativamente fino alrededor de una pieza.

flange weld   Used to join thin metal parts together. Instead of welding to the surfaces of the parts to be joined, the weld is applied to the edges of thin material.

soldadura de reborde   Se usa para unir piezas finas de metal. En lugar de soldar en las superficies a unirse, la soldadura se aplica a los bordes del material.

flask   In sand casting, this is a two-part box, or frame, used to contain the sand.

armazón   En los modelos de fundación de arena, ésta es la caja de dos partes, o armadura, que se usa para contener arena.

flat pattern   A layout showing true dimensions of a part before bending; may be an actual-size pattern on polyester film for shop use.

patrón plano   Un trazado que muestra las dimensiones verdaderas de una pieza antes de doblarse; puede ser un patrón de tamaño real en una película de poliéster para usarse en el taller.

flat spring   This spring is usually flat. Made of spring steel, it is designed to perform a special function. It is considered a power spring, and it resists a pressure.

resorte plano   Este resorte generalmente es plano. Está hecho de acero de resorte, está diseñado para llevar a cabo una función específica. Se considera un resorte de potencia y resiste a la presión.

flatbed plotter   A plotter that draws an image on a data medium such as paper or film mounted on a table.

trazador gráfico plano Un trazador gráfico que dibuja una imagen en un medio tal como papel o película sobre una mesa.

flatness A tolerance that controls the amount of variation that the perfect plane has on a feature independent of any other features on the part. Also, a feature control of a surface that requires all elements of the surface to lie within two hypothetical parallel planes.

llanura Una tolerancia que controla la cantidad de variación del plano perfecto en un rasgo independiente de otros rasgos en la pieza. También, un control de rasgo de una superficie que requiere que todos los elementos de la superficie estén dentro de dos planos paralelos hipotéticos.

flat-plane surface A surface where all points can be interconnected to form straight lines without exception.

superficie plana Una superficie donde todos los puntos pueden conectarse para formar líneas rectas sin excepción.

flexibility/adaptability Ability and willingness to change ideas, methods, and approaches in real time as circumstances dictate.

flexibilidad/adaptabilidad Habilidad y voluntad para cambiar de ideas, métodos y enfoques en tiempo real, según lo dicten las circunstancias.

floor plan A plan view of the structure in section with the ceiling and roof removed.

plano de distribución Una vista aérea de la estructura en sección con el techo removido.

flute Groove, as on twist drills, reamers, and taps.

acanaladura Ranura, como en las barrenas, escariadores y machos de rosca.

fold line The place where the imaginary perpendicular-orthographic viewing planes meet each other along the straight edge of intersection.

Línea de doblez El lugar dónde los planos de vista imaginarios perpendiculares y ortográficos se encuentran a lo largo de los bordes rectos de intersección.

foolproofing A method used to prevent parts from being loaded incorrectly. It involves a simple pin, which is normally enough to ensure proper loading of every part.

hacer infalible Un método usado para evitar que las partes sean cargadas incorrectamente. Consiste de un pasador simple, que normalmente es suficiente para asegurar que cada pieza esté cargada apropiadamente.

foreshortening A phenomenon wherein the lengths of edges are shorter in the image than their actual lengths.

escorzo Un fenómeno en el cual los largos de los bordes son más cortos en la imagen que sus largos actuales.

forge To force metal while it is hot to take on a new shape by hammering or pressing.

forjar Forzar el metal mientras está caliente para darle una nueva forma por medio de martilleo o prensado.

form The form of a screw thread is actually its profile shape.

forma La forma de una rosca de tornillo es en realidad su figura de perfil.

forming An operation in which a part is bent or formed into a complex shape.

moldeado Una operación en la cual una pieza se dobla o se forma en una figura compleja.

form tolerancing The permitted variation of a feature from the perfect form indicated on the drawing.

tolerancia de forma La variación permisible de un rasgo de su forma perfecta indicada en un dibujo.

foundation plan A plan view in section that shows the understructure of the building.

plano de cimiento Una vista aérea en sección que muestra la subestructura del edificio.

fractional dimensioning Generally not used on technical drawings, its primary use is on architectural or structural engineering drawings.

dimensionamiento fraccional Generalmente no se usa en los dibujos técnicos, su uso primario es en los dibujos arquitectónicos o estructurales.

free length The overall length of a spring when it is in its free state of unloaded condition.

largo libre El largo completo de un resorte cuando está en su estado libre de una condición sin carga.

free-state variation The concept that some parts cannot be expected to be contained.

variación de estado libre El concepto de que no se puede esperar que algunas partes estén contenidas.

front view The orthographic view showing the front elevation.

vista frontal La vista ortográfica que muestra la elevación frontal.

front view auxiliary An auxiliary view projected from the front view.

vista frontal auxiliar Una vista auxiliar proyectada desde la vista frontal.

full section A section of one of the regular multiviews that is sliced (cut) in two.

sección completa Una sección de una de las vistas múltiples regulares que fue cortada en dos.

full threads Threads that do not extend to the bottom of the hole.

rosca completa Las roscas que no se extienden hasta el fondo del agujero.

fuller The first station on a drop forge die. This station is used to rough-form the part to fit the other cavities.

batán   La primera estación en un troquel de una forja. Esta estación se usa para formar la parte en bruto para que quepa en las otras cavidades.

fusion weld   The intimate mating of molten metals producing the type of weld where a lot of heat is used to melt and fuse the joint.

soldadura de fusión   La unión íntima de metales fundidos produciendo un tipo de soldadura en la cual se usa mucho calor para derretir y fundir los metales.

gage   A system of numbers used to specify the thickness of sheet metal.

calibre   Un sistema de números que se usan para especificar el espesor de una hoja de metal.

galvanize   To cover a surface with a thin layer of molten alloy, composed mainly of zinc.

galvanizar   Cubrir una superficie con una capa fina de una aleación derretida, compuesta mayormente por zinc.

gasket   A thin piece of rubber, metal, or some other material placed between surfaces to make a tight joint.

junta   Un pedazo fino de goma, metal u otro material que se coloca entre superficies para formar una unión fuerte.

gate valves   Used to turn the flow of liquids through a pipe on or off without restricting the flow of the liquids through the valve or with as little restriction as possible.

válvulas de compuerta   Se usan para encender o detener el flujo de metal derretido a través de una tubería sin restringir el flujo de los líquidos a través de la válvula o con tan poca restricción como sea posible.

gear tooth caliper   Used to check and measure an individual gear tooth.

calibrador de dientes de engranaje   Se usa para verificar y medir un diente de engranaje individual.

gear train   Two or more gears used to achieve a designed RPM.

tren de engranajes   Dos o más engranajes que se usan para obtener un número de revoluciones por minuto designado.

general notes   Broad items of information that have a job- or project-wide application rather than relating to one element or part.

notas generales   Artículos amplios de información que tienen una aplicación a nivel del trabajo o de un proyecto en lugar de estar relacionadas con sólo un elemento o una parte.

general oblique drawing   Similar to cabinet drawing, except that the lines drawn along the direction of the receding distances may be at any scale the drafter chooses.

dibujo oblicuo general   Similar a un dibujo de gabinete, excepto que las líneas que se dibujan a lo largo de la dirección que se aleja pueden estar a cualquier escala escogida por el delineante.

generic CAD software skills   Skills CAD technicians need regardless of the specific software they might use (e.g., giving commands for basic geometric shapes, giving commands to control how the screen is viewed, and giving commands for modifying drawings).

habilidades genéricas del software cad   Habilidades que necesitan los técnicos en CAD independientemente del software específico que usen (por ejemplo: dar órdenes para trazar diversas figuras geométricas, para controlar cómo se ve la pantalla o para modificar dibujos).

geometric dimensioning and tolerancing   A means of dimensioning and tolerancing a drawing with respect to the actual function or relationship of part features that can be most economically produced; it includes positional and form dimensioning and tolerancing. It allows designers to set tolerance limits not just for the size of an object but also for all of the various critical characteristics of the part.

dimensionamiento y tolerancia geométrica   Un medio de dimensionamiento y de tolerar un dibujo con respecto a la función o relación actual de los rasgos de la pieza que pueden producirse más económicamente; incluye dimensionamiento y tolerancia de posición y de forma. Le permite a los diseñadores establecer límites de tolerancia no sólo para los tamaños de un objeto pero también para todas las características críticas de la parte.

geometric parameters   Defining characteristics in solid modeling that are relational in nature such as horizontal, vertical, concentric, parallel, tangent, and perpendicular.

parámetros geométricos   En el modelado de sólidos, la definición de características que se relacionan por naturaleza, tales como horizontal, vertical, paralelo, tangente y perpendicular.

globe valves   Used to turn the flow of liquids through a pipe on and off.

Válvulas de globo   Se usan para permitir y detener el flujo de líquidos a través de una tubería.

grade   The ratio of the vertical rise (R) to the horizontal run (H).

pendiente   La razón entre la elevación vertical (R) y el tiro horizontal (H).

graphic communication   Using visual material to relate ideas.

comunicación gráfica   El uso de materiales visuales para comunicar ideas.

graphical method   A method of approximating areas by dividing them into convenient shapes; e.g., trapezoids and triangles, scaling appropriate lengths, and calculating the areas.

método gráfico Un método de aproximar las áreas dividiéndolas en formas convenientes; por ejemplo, trapezoides y triángulos, ajustando los largos a la escala apropiada, y luego calculando las áreas.

green sand A type of sand used in sand casting. The mold is prepared by ramming the green sand around a model of a part.

arena verde Un tipo de arena que se usa en la fundición en moldes de arena. Los moldes se preparan compactando la arena verde alrededor de un modelo de una pieza.

grind To remove metal by means of an abrasive wheel, often made of carborundum; used where accuracy is required.

moler Remover material por medio de un material abrasivo, frecuentemente hecho de carborundo; se usa dónde se requiere precisión.

groove weld Seven types of welds are considered to be groove welds: square, V, bevel groove, J, U, flare V, and flare bevel.

soldadura de ranura Siete tipos de soldaduras se consideran soldaduras de ranuras: cuadrada, V, ranura bisel, J, U, V acampanado, y bisel acampanado.

grooved fasteners Used to solve metal-to-metal pinning needs with shear application.

sujetadores con ranura Se usan para resolver las necesidades de sujetar metales a metales en aplicaciones con esquilado.

grooved studs Widely used for fastening light metal or plastic parts to heavier members or assemblies.

clavijas ranuradas Usadas frecuentemente para sujetar partes de metal o plástico livianas a miembros de ensamblajes más pesados.

group technology Grouping of parts into families based on similarities in design or production processes.

tecnología de grupo Una agrupación de partes en familias basadas en similaridades de diseño o en el proceso de producción.

gusset A small plate used in reinforcing assemblies.

placa de refuerzo Una pequeña placa que se usa para reforzar los ensamblajes.

half section A type of section view in which the object is cut only halfway through and a quarter section is removed.

media sección Un tipo de vista seccional en la cual el objeto se corta sólo hasta la mitad y se remueve un cuarto de sección.

hard copy A copy of output in a visually readable form (e.g., printed reports, listings documents, summaries, and drawings).

copia impresa Una copia de la salida en una forma leíble (por ejemplo, informes impresos, documentos de listados, resúmenes y dibujos).

hard disk drive A storage device used in computers; it may be contained within the computer console or in an external cabinet that is attached to the processing unit. Key concerns relating to hard disk drives are storage capacity (typically measured in megabytes and gigabytes of memory) and access time.

unidad de disco duro Un aparato de almacenamiento usado en las computadoras; puede estar contenido dentro de la consola de la computadora o en un gabinete externo que está conectado a la unidad de procesamiento. Las preocupaciones principales relacionadas a las unidades de disco duro son su capacidad de almacenamiento (típicamente medida en megabytes y gigabytes de memoria) y el tiempo de acceso.

hardening The process of heating a metal to a predetermined temperature, allowing the part to soak until thoroughly heated, and cooling it rapidly in a cooling material.

endurecimiento El proceso de calentar el metal a una temperatura predeterminada, permitiéndole que absorba el calor hasta estar calentado completamente, y luego enfriándolo rápidamente en un material de enfriamiento.

hardness test Techniques used to measure the degree of hardness of heat-treated materials.

prueba de dureza Técnicas usadas para medir el grado de dureza de materiales tratados con calor.

harmonic motion A type of motion that is very smooth starting and stopping, but the speed is not uniform. It is good for fast speed.

movimiento armónico Un tipo de movimiento que es muy suave al comenzar y al detenerse, pero la velocidad no es uniforme. Es bueno para una velocidad rápida.

heading See *bearing*.

rumbo *Orientación*.

heat treatment A series of processes used to alter or modify the existing properties in a metal to obtain a specific condition required for a workpiece. Also, to change the properties of metals by heating and then cooling.

tratamiento de calor Una serie de procesos usados para alterar o modificar las propiedades existentes en un metal para obtener una condición específica requerida para una pieza de trabajo. También, para cambiar las propiedades de los metales calentándolos y luego enfriándolos.

heating/ventilating/air-conditioning plan This plan shows the ductwork for the house's heating and cooling system superimposed on the floor plan without dimensions.

plano de calefacción/ventilación/aire acondicionado Este plano muestra los conductos para el sistema de calefacción y de enfriamiento superpuestos sobre el plano de distribución sin dimensiones.

helical gear Operates on parallel shafts and transmits rotary motion in a manner identical to spur gears. The teeth have an involute shape but are not parallel to their rotational axis.

engranaje helicoidal Opera en ejes paralelos y transmite el movimiento rotativo en una manera idéntica a los engranajes de dientes rectos. Los dientes tienen una forma intricada pero no son paralelos a su eje de rotación.

helical spring Usually cyclindrical or conical in shape, there are three types: compression, extension, and torsion springs.

resorte helicoidal Generalmente de forma cilíndrica o cónica, hay tres tipos: resortes de compresión, de extensión y de torsión.

helix A form of spiral used in screw threads, worm gears, and spiral stairways, among other things.

hélice Una forma de espiral que se usa en las roscas de tornillos los engranajes de tornillo y las escaleras espirales, entre otras cosas.

hemisphere Hemispheres occur when a sphere is cut into two equal parts.

hemisferio Los hemisferios ocurren cuando una esfera se corta en dos partes iguales.

herringbone gear A type of gear in which left-hand and right-hand teeth are combined.

engranaje doble helicoidal Un tipo de engranaje en el cual se combinan los dientes de vuelta derecha y de vuelta izquierda.

hexagon A polygon having six angles and six sides.

hexágono Un polígono que tiene seis ángulos y seis lados.

hidden lines Lines not used in isometric drawings unless needed to show an important feature that otherwise would not be seen. Also, feature outlines (for example, intersections of plane surfaces, real but invisible); used in each viewing plane where appropriate.

líneas escondidas Líneas que no se usan en los dibujos isométricos a menos que sea necesario para mostrar rasgos importantes que no se verían de otra manera. También, los bordes de rasgos (por ejemplo, las intersecciones de planos, reales pero invisibles); usados en cada plano de vista dónde sea apropiado.

high-performance alloys A group of metals that have high strength and/or high corrosion resistance over a broad range of temperatures.

aleación de alto rendimiento Un grupo de metales que tienen alta resistencia y/o alta resistencia a la corrosión sobre un rango de temperaturas.

holography This process produces images that simulate the effect of seeing the three-dimensionality of an object from multiple viewpoints.

holografía Este proceso produce imágenes que simulan el efecto de ver la tridimensionalidad de un objeto desde múltiples puntos de vista.

hone A method of finishing a hole or other surface to a precise tolerance by using a spring-loaded abrasive block and rotary motion.

afinar Un método de terminar un agujero u otra superficie a una tolerancia precisa usando un bloque accionado por resorte y un movimiento giratorio.

horizontal Parallel to the horizon.

horizontal Paralelo al horizonte.

horizontal boring mills These machines perform a wide variety of different machining tasks normally associated with a milling machine but on a much larger scale.

fresadoras de taladrado horizontal Estas máquinas llevan a cabo una variedad de diferentes tareas de labrado normalmente asociadas con una fresadora pero a una escala mucho más grande.

human factors Published tabulations of body dimensions of average males, females, and children.

factores humanos Tablas publicadas de dimensiones corporales de los varones, mujeres y niños promedios.

hydrodynamic machining (HDM) A nontraditional machining process that removes material using a high-velocity, high-pressure stream of liquid solution. It is used primarily for slitting and cutting nonmetallic materials.

labrado hidrodinámico Un proceso de labra no tradicional que remueve material usando un chorro de una solución líquida de alta velocidad y alta presión. Se usa principalmente para cortar en tiras y para cortar materiales no metálicos.

icon A standardized graphical image designed to represent and communicate a computer application operation or function.

icono Una imagen gráfica estandarizada diseñada para representar y comunicar la operación o función de una aplicación de computadora.

inclined A line or plane at an angle to a horizontal line or plane.

inclinado Una línea o plano a cualquier ángulo de una línea o plano horizontal.

increment The length of a weld.

incremento El largo de una soldadura.

incremental system A system of numerically controlled machining that always refers to the preceding point when making the next movement; also known as continuous path or contouring method of NC machining.

sistema incremental Un sistema de labrado controlado numéricamente que siempre se refiere al punto anterior al hacer el próximo movimiento; también se conoce como el método de paso continuo o de contorno del labrado controlado numéricamente.

indexing jig Used primarily to machine parts that have machined details at intervals around the part.

guía de gradación Se usa principalmente para labrar partes que tienen detalles labrados a intervalos alrededor de una pieza.

information management Planning, organizing, and controlling the information that is needed by an organization to operate efficiently and effectively (typically associated with computer-based information systems).

administración de información El planeamiento, la organización y el control de ia información requeridos por una organización para funcionar eficiente y efectivamente (se asocia por lo general con sistemas de organización con base electrónica).

input The transfer of information into a computer or machine control unit (see *output*).

entrada La transferencia de información a una computadora o unidad de control de labrado (vea *salida*).

inside diameter The inside diameter of a spring.

diámetro interior El diámetro interior de un resorte.

intelligent robot A robot that can make decisions based on data received from sensing devices.

robot inteligente Un robot que puede tomar decisiones basadas en los datos que recibe de los aparatos de sensores.

interactive The capability of a computer application to permit an operator to change variables or test a design while the application is running. Virtual reality is an extreme example of an operator being able to interact with a computer application.

interactiva La capacidad de una aplicación de computadora para permitirle a un operador cambiar variables o probar un diseño mientras la aplicación está ejecutándose. Realidad virtual es un ejemplo extremo de un operador capaz de interactuar con una aplicación de computadora.

interchangeable Refers to a part made to limit dimensions so that it will fit any mating part similarly manufactured.

intercambiable Referente a una pieza que fue hecha a dimensiones límites de manera que pueda encajar en cualquier pieza de acoplamiento manufacturada de manera similar.

interface The means of interacting with a computer. Graphical user interfaces are becoming more prevalent and are designed to simplify applications by using graphical icons to guide the interaction with computer applications.

interfaz El medio de interactuar con una computadora. Los interfaces de uso gráfico están convirtiéndose más comunes y son diseñados para simplificar las aplicaciones usando iconos para guiar la interacción con las aplicaciones de computadora.

interference fit This represents the smallest amount of interference.

encaje de interferencia Esto representa la cantidad más pequeña de interferencia.

intermediate A translucent reproduction made on vellum, cloth, or film from an original drawing to serve in place of the original for making other prints.

intermedio Una reproducción transparente hecha en vitela, tela o película desde un dibujo original para funcionar en lugar del original para hacer otras reproducciones.

intermittent weld A series of short welds. When this type of weld is needed, an extra dimension is added to the welding symbol.

soldadura intermitente Una serie de soldaduras cortas. Cuando se necesita este tipo de soldadura, se añade una dimensión al símbolo de soldadura.

internal clients These clients can be instructors in classrooms and supervisory persons in industry.

clientes internos Estos clientes pueden ser instructores en los salones de clases y supervisores en la industria.

internal rings For axial assembly, these are compressed for insertion into a bore or housing.

anillos internos Para un ensamblaje axial, éstos se comprimen para insertarlos en un agujero o un cárter.

internal thread Threads located on the inside of a part, such as those on a nut.

rosca interna Roscas ubicadas en el interior de una pieza, tales como las de una tuerca.

interviewing tips Rules of thumb to help people prepare to do well during job interviews.

sugerencias para entrevistas de trabajo Normas generales para ayudar a los candidatos a un puesto a prepararse bien para una entrevista de trabajo.

invention agreement This form is signed by the drafter. It gives the company the right to any new invention designed while working with the company. It also (1) specifies that the employee will not reveal any of the company's discoveries or projects, and (2) is generally binding on the employee from six months to two years after leaving the company.

acuerdo de invención Este formulario lo firma el delineante y le provee a la compañía el derecho de cualquier invención nueva diseñada por éste mientras trabaja con la compañía. También (1) especifica que el empleado no revelará ninguno de los descubrimientos o proyectos de la compañía, y (2) es generalmente obligatorio en el empleado desde seis meses hasta dos años después de irse de la compañía.

investment casting A casting process in which a wax pattern is placed in a sand mold and the mold is fired to melt out the wax pattern, leaving a cavity into which molten metal is then poured.

fundición en molde de cera Un proceso de fundición en el cual se coloca un patrón de cera dentro de un molde de arena y se calienta el molde para derretir la cera, dejando una cavidad dentro de la cual se vierte el metal derretido.

involute curve May be considered the path traced by the end of a string as it is unwound from a circle called a base circle. Also a spiral curve generated by a point on a chord as it unwinds from a circle or a polygon.

curva intricada Puede considerarse como el camino trazado por el extremo de un hilo mientras se desenrolla de un círculo llamado un círculo base. También es una curva espiral generada por un punto en una cuerda a medida que se desenrolla de un círculo o un polígono.

isometric assembly These drawings appear in two popular formats: exploded and assembled (see *exploded isometric drawing* and *assembled isometric assembly drawings*).

ensamblaje isométrico Estos dibujos aparecen en dos formatos populares: despiezado y ensamblado (vea *dibujo isométrico despiezado* y *dibujo isométrico ensamblado*).

isometric drawing The most used of the three common pictorials. They are easier to draw than perspective drawings, and they illustrate objects almost as well as perspective drawings. The drawings are so positioned that all three axes make equal angles with the picture plane, and measurements on all three axes are made to the same scale.

dibujo isométrico El más usado de los tres tipos de dibujos pictóricos. Son más fáciles de dibujar que los dibujos de perspectiva e ilustran los objetos casi tan bien como los dibujos de perspectivas. Los dibujos se ubican de manera que los tres ejes forman ángulos iguales con el plano de dibujo, y las medidas en los tres ejes se hacen a la misma escala.

isometric projection Causes dimensions along the three principal edges to be foreshortened to approximately 82% of their true length.

proyección isométrica Causa que las dimensiones a lo largo de los tres ejes principales se acorten a aproximadamente un 82% de su largo verdadero.

isometric sketching Sketching in which the length and width lines recede at 30° to the horizontal and the height lines are vertical.

croquis isométrico Croquis en el cual las líneas de largo y ancho de las líneas retroceden a un ángulo de 30° de la horizontal y las líneas de alto son verticales.

isopleth The lines used to locate values in a nomogram.

isolínea Las líneas que se usan para localizar los valores en un nomograma.

iteration A procedure in which repetition of a sequence of operations yields results successively closer to a desired result.

iteración Un procedimiento en el cual la repetición de una secuencia de operaciones provee resultados sucesivamente más cercanos al resultado deseado.

jig A workholding tool that guides the cutting tool through a hardened drill bushing during the cutting cycle.

guía Una herramienta que agarra piezas y guía la herramienta cortante a través de los cojinetes de taladro endurecidos durante el ciclo de corte.

journal Portion of a rotating shaft supported by a bearing.

muñequilla Porción de un eje que rota que está apoyada por un cojinete.

just-in-time manufacturing (JIT) An approach to manufacturing that seeks to completely eliminate waste.

manufactura justo a tiempo Un método de manufactura que busca eliminar el desperdicio por completo.

kerf Groove or cut made by a saw.

corte Ranura o corte hecho por una sierra.

key A small piece of metal sunk partly into both shaft and hub to prevent rotation of mating parts. It is a demountable part that provides a positive means of transferring torque between a shaft and a hub.

llave Un pedazo pequeño de metal parcialmente empotrado en el eje y en el cubo para prevenir la rotación de partes que encajan. Es una parte removible que provee un medio positivo para transferir la fuerza de torsión entre un eje y un cubo.

keyboard The part of a CAD computer used for entering commands, text, annotation, and dimensions.

teclado La parte de una computadora de CAD que se usa para ingresar comandos, texto, anotaciones y dimensiones.

keying The process used for entering text on drawings, certain system commands, dimensions not entered automatically, and for logging on to the system.

tecleo El proceso usado para ingresar texto en los dibujos, ciertos comandos, dimensiones que no se ingresan automáticamente y para entrar al sistema.

keyseat An auxiliary-located rectangular groove machined into the shaft and/or hub to receive the key.

asiento de llave Una ranura auxiliar rectangular que se labra en un eje y/o cubo para recibir la llave.

keyslot The slot machined in a shaft for Woodruff-type keys.

ranura de llave Una ranura labrada en un eje para aceptar las llaves tipo Woodruff.

keyway   A slot in a hub or portion surrounding a shaft to receive a key.

muesca de posicionamiento   Una ranura en un cubo o la porción que rodea a un eje para recibir una llave.

knurl   A diamond-shaped or parallel pattern cut into cylindrical surfaces to improve gripping, the bonding between parts for permanent press fits and decoration.

estrías   Un patrón en forma de diamante o paralelo que se corta en las superficies cilíndricas para mejorar su agarre, la adherencia de partes para encajes de presión permanentes y para decoración.

laser beam machining (LBM)   A process that uses lasers for performing operations that are traditionally performed with cutting tools.

labrado con rayo Láser   Un proceso que usa láseres para llevar a cabo las operaciones que tradicionalmente se llevan a cabo con herramientas cortantes.

lathe   One of the most commonly used and versatile machine tools in the shop. It can perform such operations as straight and taper turning, facing, straight and taper boring, drilling, reaming, tapping, threading, and knurling.

torno   Una de las herramientas de labrado más frecuentemente usadas y más versátiles en el taller mecánico. El mismo puede llevar a cabo tales operaciones como el tornado, la cara, el taladrar derecho y gradualmente estrechado, el perforar, el escariar, el roscar, y el estriar.

layout   Lines drawn on material as guides for subsequent operations.

trazado   Líneas que se dibujan en un material como guías para las operaciones subsiguientes.

layout development drawing   A designer's freehand sketch redrawn to scale with instruments or a computer.

dibujo de desarrollo de trazado   Un croquis hecho a mano libre por un diseñador dibujado nuevamente a escala con instrumentos o con una computadora.

lead (noun, pronounced "led")   A dark gray metal with the following characteristics: high density, low melting point, corrosion resistance, chemical stability, malleability, lubricity, and electrical properties.

plomo   Un metal grisáceo oscuro con las siguientes características: densidad alta, punto de fundición bajo, resistencia a la corrosión, estabilidad química, maleabilidad, lubricidad y propiedades eléctricas.

lead pointer   Manual drafting device used to sharpen the lead in mechanical pencils.

sacapuntas de plomo   Instrumento de dibujo a mano que se usa para afinar la mina de los lápices mecánicos.

lead time   The interval from the time the shop receives the drawing until production begins.

tiempo de acción   El intervalo de tiempo desde que el taller recibe el dibujo hasta que comienza la producción.

leader line   A thin line used to direct the information contained in notes or dimensions to the appropriate surface, feature, or point on a drawing.

directriz   Una línea que se usa para dirigir la información contenida en notas o dimensiones a la superficie, rasgo o punto apropiado en un dibujo.

least material condition (LMC)   A condition that exists when the least material is available for the features being dimensioned. It is the lower limit for an external feature and the upper limit for an internal feature.

condición de material mínimo   Una condición que existe cuando está disponible la cantidad mínima de material para los rasgos que se dimensionan. Es el límite inferior para un rasgo externo y el límite superior para un rasgo interno.

letter of transmittal   A letter to a client, separate from but enclosed with a report, which states that the project has been completed, the report is enclosed, and that working with the client on future projects is anticipated.

carta de trámite   Una carta a un cliente, aparte de pero que incluye un informe que afirma que el proyecto fue completado, que se incluye el informe y que se anticipa trabajar con el cliente en proyectos futuros.

light pen   A handheld data-entry device used only with refresh displays. It consists of an optical lens and photocell, with associative circuitry mounted in a wand. Most light pens have a switch allowing the pen to be sensitive to light from the screen.

pluma de luz   Un aparato de entrada de datos manual que se usa solamente con las pantallas refrescadas. Consiste de un lente óptico y una fotocelda, con el circuito asociado colocado en el mango. La mayoría de las plumas de luz tienen un interruptor que le permite a las plumas a ser sensibles a la luz de la pantalla.

limit dimensioning   A tolerancing method that shows only the maximum and minimum dimensions that establish the limits.

dimensionamiento de Límite   Un método de tolerar que muestra sólo las dimensiones máximas y mínimas que establecen los límites.

limits   The maximum and minimum allowable sizes of a feature.

límites   Los tamaños máximos y mínimos permisibles para un rasgo.

line work   A generic term given to all of the various techniques used in creating the graphic data on technical drawings.

trabajo de Línea   Un término genérico que se le da a todas las varias técnicas usadas en la creación de datos gráficos en los dibujos técnicos.

liner bushings Used in conjunction with renewable bushings; they provide a hardened, wear-resistant mount for renewable bushings.

cojinetes de revestimiento Se usan en conjunto con los cojinetes renovables; proveen una montura endurecida, resistente al desgaste para los cojinetes renovables.

loaded length The overall length of a spring with a special given or designed load applied to it.

largo con carga El largo completo de un resorte con una carga dada o de diseño aplicada sobre él.

location tolerance A tolerance that specifies the allowable variation from the perfect location.

tolerancia de localización Una tolerancia que especifica la variación permisible de una localización perfecta.

log A record of activities completed and the time spent on each activity. It should take the who, where, what, when, and why style.

registro Un registro de actividades completadas y el tiempo usado en cada actividad. Debería tener el estilo de quién, dónde, qué, cuándo y por qué.

lowercase end points Each line in space is identified by two of these.

puntos de extremos minúsculos Cada línea en el espacio está identificada por dos de éstos.

lowercase letters Each point in space is called out by these.

letras minúsculas Cada punto en el espacio se llama por una de éstas.

lug An irregular projection of metal but not round as in the case of a boss, usually with a hole in it for a bolt or screw.

saliente Una proyección de metal irregular pero no redonda como en el caso de un saliente cilíndrico, generalmente con un agujero en él para aceptar un tornillo o un perno.

machine screw Used for screwing into thin materials. Most are threaded within a thread or two to the head. They sometimes incorporate a hex-head nut to fasten parts together.

tornillo de Máquina Se usa para atornillar materiales finos. La mayoría tiene una rosca dentro de una rosca hasta la cabeza. A veces incorporan una tuerca hexagonal para unir las piezas.

machining The process of removing metal with machine tools and cutters to achieve a desired form or feature.

labrar El proceso de remover material con herramientas de labrado y cortantes para lograr una forma o un rasgo deseado.

Magnaflux A nondestructive inspection technique that uses a magnetic field and magnetic particles to locate internal flaws in ferrous metal parts.

Magnaflux Una técnica de inspección no destructiva que usa campos magnéticos y partículas magnéticas para localizar los defectos internos en las piezas de metales ferrosos.

magnesium A soft, silvery white metal that is extremely light and machinable but lacks the strength needed for structural uses. To increase its strength, it can be alloyed with various metals.

magnesio Un metal blando, plateado blancuzco que es extremadamente liviano y fácil de labrar pero carece de la fortaleza necesaria para usos estructurales. Para aumentar su fortaleza, se puede alear con varios metales.

main storage The general-purpose storage of a computer; usually main storage can be accessed directly by the operating registers.

almacenamiento principal El almacenamiento de aplicación general en una computadora; generalmente el almacenamiento principal lo pueden acceder directamente los registros operacionales.

major diameter The largest diameter of a screw thread, both external and internal.

diámetro mayor El diámetro más grande de una rosca de tornillo, tanto externo como interno.

malleable casting A casting that has been made less brittle and tougher by annealing.

fundición maleable Una fundición que se hizo menos quebradiza y fuerte por medio del recocido.

Manual of Steel Construction This manual contains the dimensions and other design- and drafting-related information for structural steel shapes.

Manual de construcción en acero Este manual contiene las dimensiones y otra información relacionadas al dibujo industrial de las formas estructurales en acero.

manufacturing The enterprise through which raw materials are converted into finished products.

manufactura La empresa a través de la cual los materiales en bruto se convierten en productos acabados.

manufacturing resource planning (MRPII) A concept that allows companies to systematically plan for the resources needed in manufacturing (i.e., materials, capacity of production systems, personnel, time, etc.).

planificación de recursos de manufactura Un concepto que le permite a las compañías a planificar sistemáticamente para los recursos necesarios en la manufactura (por ejemplo materiales, capacidad de los sistemas de producción, personal, tiempo).

maraging steel A special class of high-strength steels in which nickel is the primary alloy.

acero martensítico envejecido Una clase especial de aceros de alta resistencia en los cuales el níquel es la aleación principal.

mass production company An organization that produces large quantities of a product.

compañía de producción en masa Una organización que produce grandes cantidades de un producto.

mass properties analysis A process involving performing analysis computations concerning mass properties such as weight, volume, surface area, moment of inertia, and center of gravity.

análisis de propiedades de masa Un proceso que consiste de hacer cálculos analíticos acerca de las propiedades de masa tales como peso, volumen, área de superficie, momento de inercia y centro de gravedad.

material handling The movement and handling of materials, either manually or through the use of powered equipment.

manejo de materiales El movimiento y manejo de materiales, ya sea manualmente o mediante el uso de equipo de potencia.

maximum material condition (MMC) When a feature contains the maximum amount of material; that is, minimum hole diameter and maximum shaft diameter. It is the upper limit for an external feature and the lower limit for an internal feature.

condición de material máximo Cuando un rasgo contiene la cantidad máxima de material; esto es, el diámetro de agujero mínimo y diámetro de eje máximo. Es el límite superior para un rasgo externo y el límite inferior para un rasgo interno.

mean diameter The theoretical diameter of the spring measured to the center of the wire diameter.

diámetro promedio El diámetro teórico de un resorte medido hasta el centro del diámetro del alambre.

mechanical method A method of approximating areas by using a planimeter.

método mecánico Un método de aproximar las áreas usando un planímetro.

mechanical pencils Pencils made of metal and plastic designed to hold lead of various degrees of hardness. Mechanical pencils have a built-in device for advancing the lead as it wears down from use.

lápices mecánicos Lápices hechos de metal y plástico diseñados para usarse con minas de distintos grados de dureza. Estos lápices tienen un mecanismo interno para hacer avanzar la mina a medida que se desgasta por el uso.

menu A list of options that are displayed either on the monitor or on a plastic or paper overlay.

menú Una lista de opciones que se despliegan en el monitor o en una cubierta de plástico o papel.

metes and bounds A system used to identify the perimeters of any property. A metes and bounds land survey begins at a selected known point of origin called the point-of-beginning or POB and then describes property boundaries in relation to this by using distance measurements, compass bearings, and boundaries such as streets, fences, and rivers.

medidas y límites Un sistema usado para identificar los perímetros de cualquier propiedad. Cualquier estudio de planimetría de medidas y límites comienza con un punto conocido de origen llamado el punto de comienzo y luego describe las relaciones de límites en relación a éste usando medidas de distancia, orientaciones de compás y demarcaciones tales como calles, cercas y ríos.

metric dimensioning One of three dimensioning systems used on technical drawings.

dimensionamiento Métrico Uno de tres sistemas de dimensionamiento que se usa en los dibujos técnicos.

microcomputer A computer that is constructed using a microprocessor as a basic element of the CPU; all electronic components are arranged on one printed circuit board.

microcomputadora Una computadora que se construyó usando un microprocesador como un elemento básico del CPU; todos los componentes electrónicos están puestos sobre una tarjeta de circuitos impresos.

micrometer A caliper used for making precise measurements of mechanical objects.

micrómetro Calibrador usado para tomar medidas precisas de objetos mecánicos.

microprocessor The central processing unit of a microcomputer. The microprocessor is contained on a single integrated circuit chip.

microprocesador La unidad de procesamiento central de una microcomputadora. El microprocesador está en un solo chip de circuito integrado.

mill To remove material by means of a rotating cutter on a milling machine.

fresar Remover material por medio de una cuchilla que rota en una fresadora.

milling machine Used to machine workpieces by feeding the part into a rotating cutter.

fresadora Se usa para labrar las piezas de trabajo insertando la pieza en una cortadora giratoria.

minor diameter The smallest diameter of a screw thread, both external and internal.

diámetro menor El diámetro más pequeño de una rosca de un tornillo, tanto externa como interna.

miter gear Exactly the same as a bevel gear, except that both mating gears have the same number of teeth. The shafts are at 90° to each other.

engranaje de inglete Exactamente igual al engranaje de bisel, excepto que los engranajes que se conectan tienen el mismo número de dientes. Los ejes están a 90° el uno del otro.

modified uniform velocity   It smoothes out the roughness slightly by adding a radius to the ends of the high and low points.

velocidad uniforme modificada   Suaviza la aspereza un poco añadiéndole un radio a los extremos de los puntos altos y bajos.

modifier   The application of MMC or RFS to alter the normally implied interpretation of a tolerance specification. There are symbols that can be attached to the standard geometric building block to alter their application or interpretation.

modificador   Una aplicación de la condición de material máximo o independiente de tamaño de rasgo para alterar la interpretación normal implicada de una especificación de tolerancia. Hay símbolos que se pueden adjuntar al bloque de construcción para alterar su aplicación o interpretación.

mold   The mass of sand or other material that forms the cavity into which molten metal is poured.

molde   La masa de arena u otro material que forma la cavidad en la cual se vierte el metal derretido.

mouse   A device that typically rolls on the desktop and directs the cursor on the display screen.

ratón   Un aparato que típicamente rueda sobre el escritorio y dirige el cursor en la pantalla del monitor.

multiple reference lines   Used when there will be more than one operation on a particular joint. The first operation goes on the first line, the second on the second, and so on.

líneas de referencias múltiples   Se usan donde habrá más de una operación un una unión particular. La primera operación se indica en la primera línea, la segunda en la segunda, y así sucesivamente.

multiple thread   A type of screw threading in which two or more continuous ridges follow side by side. Used when speed or travel distance is an important design factor.

rosca Múltiple   Un tipo de rosca de tornillo en que dos o más resaltos continuos siguen uno al lado del otro. Usado cuando velocidad o distancia de viaje es un factor importante del diseño.

multiweld   A weld where more than one type of weld may be applied to a single joint—for example, a groove V weld on one side and a backing weld on the other side. The weld is usually for strength or appearance.

soldadura múltiple   Una soldadura en la cual más de un tipo de soldadura se puede aplicar a una sola unión – por ejemplo, una soldadura de ranura V en un lado y una soldadura de espalda en el otro lado. Esta soldadura tiene usualmente el propósito de fuerza o apariencia.

neck   To cut a groove called a neck around a cylindrical piece.

collarín   Cortar una ranura llamada cuello alrededor una pieza cilíndrica.

nodes   Points of an object that contain specific information about the characteristics at that particular point.

nodos   Puntos en un objeto que contienen información específica acerca de las características de ese punto particular.

nominal dimension   A dimension that represents a general classification size given to a part or product.

dimensión nominal   Una dimensión que representa un tamaño de clasificación general que se le da a una pieza o producto.

non-ferrous   A description of metals not derived from an iron base or an iron alloy base; non-ferrous metals include aluminum, magnesium, and copper, among others.

no ferroso   Una descripción de los metales que no se derivan de una base de hierro o una base de una aleación de hierro; los metales no ferrosos incluyen aluminio, magnesio y cobre, entre otros.

nonisometric line   A line not parallel or perpendicular to any of the three principal edges in an isometric drawing.

línea no isométrica   Una línea que no es paralela o perpendicular a ninguno de los tres bordes principales de un dibujo isométrico.

nonroutine designs   Problems that are assigned to experienced engineers, designers, technologists, and drafters.

dibujos no de rutina   Problemas que se asignan a los ingenieros, diseñadores, tecnólogos y delineantes con experiencia.

nontraditional machining   Advanced machining processes that have been developed in response to the demands of modern manufacturing for more effective, more efficient ways to machine the many new materials that have been, and are still being, developed.

labrado no tradicional   Procesos de labrado avanzados que se han desarrollado en respuesta a las demandas de la manufacturación moderna para maneras más efectivas y eficientes para labrar los muchos materiales que se han desarrollado o que aún se están desarrollando.

normalize   To heat steel above its critical temperature and then cool it in air.

normalizar   Calentar el acero más allá de su temperatura crítica y luego dejarlo enfriar en aire.

normalizing   The process used to remove the effects of machining or cold-working metals. The process in which ferrous alloys are heated and then cooled in still air to room temperatures to restore the uniform grain structure free of strains caused by the cold working or welding.

normalizar   Un proceso usado para remover los efectos del labrado o del trabajado en frío en los metales. El proceso en el cual las aleaciones férreas se calientan y luego se enfrían en aire quieto a temperatura ambiente para restaurar la estructura granular uniforme libre de la tensión causada por el trabajo en frío o de soldadura.

north/south deviation A method of marking a drawing in which north is at the top of the page and south is at the bottom.

desviación norte/sur Un método de marcar un dibujo en el cual el norte está en la parte superior de la página y el sur en la parte inferior de la página.

notations Labeling that keeps track of all views and points in space and provides brief explanations/information on a drawing.

notación Rótulos que llevan cuenta de todas las vistas y puntos en el espacio y proveen explicaciones breves o información en un dibujo.

numerical control A system of controlling a machine or tool by means of numeric codes that direct commands to control devices attached or built into the machine or tool.

control numérico Un sistema para controlar una máquina o herramienta por medio de códigos numéricos que dirigen los comandos que controlan aparatos que están conectados a o que están dentro de la máquina o herramienta.

numeric parameters Defining characteristics in solid modeling that are numeric, such as line lengths and circle diameters.

parámetros numéricos En el modelado de sólidos, la definición de características numéricas, tales como el largo de las líneas y el diámetro de los círculos.

numerically controlled machine tools Machine tools that are controlled by means of numeric codes that direct commands to control devices attached or built into the machine or tool.

herramientas de labrado controladas numéricamente Herramientas de labrado que se controlan por medio de códigos numéricos que dirigen los comandos que controlan aparatos que están conectados a o que están dentro de la máquina o herramienta.

oblique drawing A style of pictorial drawing in which one surface of the object, usually the front view, is drawn exactly as it would appear in a multiview drawing. One of its principal faces is parallel to the plane of projection and is projected in its true size and shape; the third set of edges is oblique to the plane of projection at some convenient angle.

dibujo oblicuo Un estilo de dibujo pictórico en el cual una superficie del objeto, generalmente la vista frontal, se dibuja exactamente como aparecería en un dibujo de vistas múltiples. Una de las caras principales está paralela al plano proyección y se proyecta en tamaño y forma real; el tercer grupo de bordes está oblicuo al plano de proyección en un ángulo conveniente.

oblique projection It shows the full size of one view.

proyección oblicua Muestra el tamaño completo de una vista.

oblique sketching A type of sketching involving a combination of a flat, orthographic front with depth lines receding at a selected angle (usually 45°).

croquis oblicuo Un tipo de croquis que consiste de una combinación de un frente plano y ortográfico con líneas de profundidad que van disminuyendo en un ángulo seleccionado (generalmente 45°).

oblique surfaces Surfaces that are not perpendicular to any of the three principal projection planes.

superficies oblicuas Superficies que no son perpendiculares a ninguno de los tres planos de proyección.

obtuse angle An angle larger than 90°.

ángulo obtuso Un ángulo mayor de 90°.

obtuse angle triangle A triangle having an obtuse angle greater than 90° with no equal sides.

triángulo de Ángulo obtuso Un triángulo que tiene un ángulo obtuso mayor de 90° sin lados iguales.

octagon A polygon having eight angles and eight sides.

octágono Un polígono que tiene ocho ángulos y ocho lados.

offset measurements Used to locate one feature in relationship to another.

medida de compensación Se usa para localizar un rasgo en relación a otro.

offset section This section is accomplished by bending or offsetting the cutting-plane line. It is similar to a full section except that the cutting-plane line is not straight.

sección de compensación Esta sección se logra doblando o desplazando el plano cortante. Es similar a una sección completa, excepto que la línea del plano cortante no es recta.

ogee curve Used to join two parallel lines. It forms a gentle curve that reverses itself in a neat symmetrical geometric form.

curva de "S" Se usa para unir dos líneas paralelas. Forma una curva suave que vira en si misma en una forma geométrica muy simétrica.

open jigs Jigs that cover only one side of a part and are used for relatively simple operations.

guías abiertos Guías que cubren sólo un lado de una pieza y se usan para operaciones relativamente simples.

Opitz system One of the most widely used classification and coding systems. It uses characters in 13 places to code the attributes of parts and hence to classify them.

sistema Opitz Uno de los sistemas de clasificación y codificación más usado. Usa caracteres en 13 lugares para codificar los atributos de una pieza y por ende clasificarlos.

optimization   The process of improving a process or system to its maximum functionality.

optimización   El proceso de mejorar un proceso o sistema a su funcionalidad máxima.

ordinate   The Y coordinate of a point (i.e., its vertical distance from the X axis measured parallel to the Y axis or the vertical axis of a graph or chart).

ordenada   La coordenada Y de un punto (esto es, la distancia vertical desde el eje de X, medido paralelo al eje de Y o eje vertical de la gráfica).

orthographic projection   A projection on a picture plane formed by perpendicular projectors from the object to the picture plane; third-angle projection is used in the United States, while first-angle projection is used in most countries outside the United States.

proyección ortográfica   Una proyección en un plano de dibujo que está formado por proyecciones perpendiculares desde el objeto al plano de dibujo; en los Estados Unidos se usa la proyección de tercer ángulo, mientras que la proyección de primer ángulo se usa en la mayoría de los países fuera de Estados Unidos.

orthographic sketching   This type of sketching deals with flat, graphic facsimiles of a subject showing no depth. The views selected depend on the nature of the subject and the judgment of the sketcher.

croquis ortográfico   Este tipo de croquis trata con copias planas y gráficas de un sujeto que no muestra profundidad. Las vistas que se seleccionan dependen de la naturaleza del sujeto y el juicio de la persona que hace el croquis.

output   Data that have been processed from an internal storage to an external storage; opposite of input.

salida   Datos que se han procesado de un almacenamiento interno a un almacenamiento externo; lo opuesto a entrada.

outside diameter   The measurement over the extreme outer edge of teeth. It is equal to the pitch diameter plus two times the addendum.

diámetro externo   La medida alrededor del extremo externo de los dientes. Es igual al diámetro de paso más dos veces la altura de cabeza.

parallel   Having the same direction, such as two lines which, if extended indefinitely, would never meet.

paralelo   Que tiene la misma dirección, tales como dos líneas que, si se extienden infinitamente, nunca se encontrarían.

parallel alignment chart   The procedure for preparing a set of three parallel scales for a three-variable equation. The derivation aims at obtaining relationships from two of the scales so that the third can be located and graduated.

gráfica de alineación paralela   El procedimiento para preparar tres escalas paralelas para una ecuación de tres variables. La derivación aspira a obtener relaciones de dos de las escalas de manera que se pueda localizar y graduar la tercera.

parallel line development   Used for objects having parallel fold lines.

desarrollo de Líneas paralelas   Se usa para objetos que tienen líneas de doblez paralelas.

parallel projection   Is subdivided into the following three categories: orthographic, oblique, and axonometric projections.

proyección paralela   Se subdivide en las siguientes tres categorías: proyección ortográfica, oblicua y axonométrica.

parallel station lines   A baseline that must always be 90° from the fold lines. Used for objects having parallel fold lines.

líneas de estación paralelas   Una línea base que siempre tiene que estar a 90° de las líneas de doblez. Se usa para los objetos que tienen líneas de doblez paralelas.

parallelism   A feature control that specifies that all points on a given surface, axis, line, or center plane must be equidistant from a datum.

paralelismo   Un control de rasgo que especifica que todos los puntos de una superficie dada, eje, línea o plano central deben estar equidistantes de una referencia.

parallelogram   A quadrilateral whose opposite sides are parallel.

paralelogramo   Un cuadrilátero cuyos lados opuestos son paralelos.

parallelogram law   Two vectors, not parallel, drawn to scale, having the same units, and acting at the same point, can be added or subtracted.

ley del paralelogramo   Dos vectores, que no son paralelos, que están dibujados a escala, tienen las mismas unidades y actúan sobre un mismo punto, pueden sumarse o restarse.

parametric design   This allows drafters and designers to change dimensions in three-dimensional computer models and have all affected dimensions change accordingly.

diseño paramétrico   Éste le permite al delineante y a los diseñadores cambiar las dimensiones de los modelos tridimensionales de computadora y que como consecuencia se cambien todas las dimensiones afectadas.

part   Individual component of an assembly.

pieza   Un componente individual de un ensamblaje.

part family   A group of parts similar in design and/or the manufacturing process used to produce them.

familia de piezas   Un grupo de piezas de diseños y/o procesos de manufactura similares.

partial auxiliary view   A partial view makes it possible to eliminate one or more of the regular views which, in turn, saves drafting time and cost.

vista parcial auxiliar   Una vista parcial hace posible el eliminar una o más vistas regulares, lo que a su vez ahorra tiempo y gastos del proceso de dibujo industrial.

parting   Operation used to finish the edges of a part.

partir   Una operación que se usa para terminar los bordes de una pieza.

parting line   This is the plane of division between the cope and drag.

Línea divisoria   Es el plano de división entre la caja superior y la caja inferior del molde.

parts   This is a listing, identified first by number, part name, and quantity needed. It also contains a cross reference to the corresponding balloon number.

piezas   Este es un listado de piezas, identificadas primero por número, nombre de la pieza y cantidad necesaria. También tiene una referencia al número de globo correspondiente.

parts list   A bill of material that itemizes the parts needed to assemble one complete full assembly of a product.

lista de piezas   Una factura de materiales que desglosa las piezas necesarias para ensamblar un producto completo.

patent drawing   A drawing that must accompany a patent application for a new idea or invention. Its purpose is to illustrate and then explain the invention's function. It must be mechanically correct and must adhere to strict patent drawing regulations.

dibujos de patentes   Un dibujo que debe acompañar la aplicación para patente de una nueva idea o invención. Su propósito es ilustrar y explicar la función de la invención. Debe ser mecánicamente correcta y debe cumplir con los reglamentos estrictos para los dibujos de patentes.

pattern   A model, usually of wood, used in forming a mold for casting. In sand casting, this is a model of a part.

patrón   Un modelo, generalmente de madera, que se usa para formar el molde para la fundición. En la fundición en moldes de arena, éste es el modelo para una pieza.

peen   To hammer into shape with a ball peen hammer.

martillar   Formar mediante martillazos de un martillo de bola.

pentagon   A polygon having five angles and five sides.

pentágono   Un polígono que tiene cinco ángulos y cinco lados.

people skills   Sometimes referred to as "soft skills." People with good soft skills are able to interact effectively with others, work well on a team, and collaboratively solve problems. For example, communication with an emphasis on listening is a people skill.

habilidades sociales   También llamadas "don de gentes". Las personas con don de gentes pueden interactuar eficazmente con los demás, trabajar bien en equipo y resolver problemas en forma cooperativa. Por ejemplo, la comunicación con énfasis en la atención a lo que otros dicen, es un don de gentes.

perforating   A punching operation performed on sheets to produce either a uniform hole pattern or a decorative form in the sheet.

perforar   Una operación de perforado que se hace en hojas para producir un agujero de patrón uniforme o una forma decorativa en una hoja.

permanent fasteners   Used when parts will not be disassembled.

sujetadores permanentes   Se usan cuando las partes no serán desensambladas.

perpendicular   A type of motion of the follower specific to the radial arm cam design. Also, lines or planes at a right angle to a given line or plane.

perpendicular   Un tipo de movimiento del seguidor que es específico al diseño de levas de brazo radial. También, líneas o planos que están a ángulos rectos de una línea o un plano dado.

perpendicularity   A feature control that specifies that all elements of a surface, axis, center plane, or line form a 90° angle with a datum.

perpendicularidad   Un control de rasgo que especifica que todos los elementos de una superficie, eje, plano central o línea forman un ángulo de 90° con una referencia.

personal values   The sum of an individual's internal belief system concerning right and wrong, good and bad, and other aspects of life.

valores personales   El conjunto de las creencias fundamentales de un individuo en cuanto a correcto e incorrecto, bien y mal, y otros aspectos de la vida.

perspective drawing   A three-dimensional drawing of an object or scene that shows visible lines almost as a person's eyes would perceive them from one particular viewing location. Also a pictorial drawing in which receding lines converge at vanishing points on the horizon; the most natural of all pictorial drawings.

dibujo de perspectiva   Un dibujo tridimensional de un objeto o escena que muestra las líneas visibles casi como los ojos de una persona las percibiría desde un punto de vista particular. También es un dibujo pictórico en el cual todas las líneas que se alejan convergen en puntos en el horizonte; el más natural de todos los dibujos pictóricos.

perspective projection   A distortion phenomenon that occurs because the visual rays used to create the image in the viewer's eye are not parallel.

proyección de perspectiva   Un fenómeno de distorsión que ocurre debido a que los rayos visuales que se usan para generar la imagen no son paralelos.

perspective sketching   Involves creating a graphic facsimile of the subject in one-, two-, or three-point perspective using only a pencil.

croquis de perspectiva   Consiste en crear una copia gráfica del sujeto en perspectiva de uno, dos o tres puntos usando sólo un lápiz.

phantom line   A type of line that resembles a center line, except that it has two dashes.

línea fantasma   Un tipo de línea que se parece a la línea de centro, excepto que tiene dos guiones.

photochemical machining   Also known as chemical blanking, this involves producing parts by chemical action. The chemical dissolves all of the metal except the desired part.

labrado fotoquímico   También se conoce como troquelado químico, y consiste en producir piezas por acción química. El químico disuelve todo el metal excepto la parte deseada.

photorealism   A term used to describe images that are created in such a fashion that they might fool the viewer into thinking that they are photographs.

fotorealismo   Un término que se usa para describir las imágenes que se crean de tal manera que podrían engañar al espectador a pensar que están viendo fotografías.

physical prototype   A real, physical object that can be handled, analyzed, and tested.

prototipo Físico   Un objeto real y físico que puede manipularse, analizarse y probarse.

pickle   To clean forgings or castings in diluted sulfuric acid.

limpiar con Ácido   Limpiar las piezas forjadas o de fundición en ácido sulfúrico diluido.

pictorial drawing   A drawing that depicts the object in a three-dimensional format; these drawings are typically isometric or perspective but may also be dimetric, trimetric, or oblique.

dibujo pictórico   Un dibujo que muestra un objeto en tres dimensiones; estos dibujos típicamente son isométricos o de perspectiva pero también pueden ser dimétricos, trimétricos u oblicuos.

piercing   An operation similar to punching, except that no scrap is produced.

agujerar   Una operación similar al perforado, excepto que no produce material de desecho.

piercing point   This is the exact location of the intersection of a surface and a line.

punto penetrante   Ésta es la localización exacta de la intersección entre una superficie y una línea.

pilot   A piece that guides a tool or machine part.

guía   Una pieza que guía una herramienta o parte de una máquina.

pilot hole   A small hole used to guide a cutting tool for making a larger hole.

agujero guía   Un pequeño agujero que guía una herramienta cortante para hacer un agujero más grande.

pinion gear   Exactly like the spur gear but usually smaller and with fewer teeth.

engranaje de piñón   Exactamente como un engranaje de dientes rectos pero generalmente más pequeño y con menos dientes.

pitch   The distance from a point on a screw thread to a corresponding point on the next thread, as measured parallel to the axis. The center-to-center distance between increments.

paso   La distancia desde un punto en una rosca de un tornillo al punto correspondiente en la próxima rosca, medido paralelo al eje. La distancia centro a centro entre incrementos.

pitch circle   An imaginary circle corresponding to the circumference of the friction gear from which the spur gear is derived.

Círculo de paso   Un círculo imaginario que corresponde a la circunferencia del engranaje de fricción del cual se deriva el engranaje de dientes rectos.

pitch diameter   An imaginary diameter centrally located between the major diameter and the minor diameter.

diámetro de paso   Un diámetro imaginario localizado centralmente entre el diámetro mayor y el diámetro menor.

placing drawings   Shop drawings of concrete structural members that are poured on-site; they show the placement of reinforcing bars in the on-site form. Shop drawings and placing drawings are the same; the difference is in where the work occurs, in a shop or on site.

dibujos de colocación   Dibujos de taller de los miembros estructurales de concreto que se fundirán en el lugar de construcción; muestran la localización de las barras de refuerzo en el molde que está en el lugar. Los dibujos de taller y los dibujos de colocación son los mismos; la diferencia está en dónde se lleva a cabo el trabajo, en el taller o en el lugar de construcción.

plane   To remove material by means of a planer.

acepillar   Remover material por medio de una acepilladora.

plane surface   If any two points anywhere on a surface are connected to form a straight line, and that line rests upon the surface, it is a plane surface.

superficie plana   Si cualesquiera dos puntos en cualquier lugar en una superficie se conectan para formar una línea recta, y esa línea esta sobre la superficie, entonces es una superficie plana.

planer Single-edged cutting tool used to perform a cutting operation. It is normally much larger than a shaper, and the work, rather than the cutter, moves on the planer. Seldom used today.

acepilladora Una herramienta cortante de un solo filo que se usa para llevar a cabo una operación cortante. Normalmente es mucho más grande que una conformadora, y el trabajo en vez de la cortadora se mueve en la acepilladora. Muy poco usado hoy día.

planning chart A method of visually displaying when phases and steps are scheduled to begin and end. Some steps on the chart (within a phase) may overlap.

gráfica de planificación Un método de mostrar visualmente cuándo las fases y pasos están programadas para comenzar y terminar. Algunos pasos en la gráfica (dentro de una fase) pueden solaparse.

plasma A gas that has been heated to the point that it ionizes and becomes electrically conductive.

plasma Un gas que se ha calentado hasta el punto que se ioniza y conduce electricidad.

plasma arc machining (PAM) A process that makes use of a plasma arc to remove material.

labrado con arco de plasma Un proceso que usa un arco de plasma para remover material.

plastics Nonmetal materials that are made from either natural or synthetic resins. Their main characteristics are: thermal, mechanical, chemical, and corrosion resistance; fabricability; the ability to sustain high mechanical loads; and predictable performance.

plásticos Materiales no metálicos que están hechos de resinas naturales o sintéticas. Sus características principales son: resistencia termal, mecánica, química y a la corrosión, habilidad para fabricación; habilidad para soportar grandes cargas mecánicas; y funcionamiento predecible.

plate To coat a metal piece with another metal, such as chrome or nickel, by electrochemical methods.

enchapar Cubrir un pedazo de metal con otro metal, tal como cromo o níquel, por métodos electroquímicos.

plate jig The most common form of jig. It consists of a simple plate that contains the required drill bushings, locators, and clamping elements.

guía de placa La forma más común de guía. Consiste de una placa sencilla que contiene los cojinetes de taladro, localizadores, y elementos de fijación necesarios.

plot plan A plan view of the building site showing where the structure is located on the lot.

plano del terreno Una vista plana del lugar de construcción que muestra dónde está ubicada la estructura en el lote de terreno.

plotter An output device that makes a drawing by converting digital data into text and graphics.

trazador gráfico Un aparato de salida que crea un dibujo convirtiendo los datos digitales en texto y gráficos.

plug weld A weld made through a round hole. The symbols are the same. (See also *slot weld.*)

soldadura de tapón Una soldadura que se hace a través de un agujero. Los símbolos son los mismos. (Vea también *soldadura de ranura.*)

point An exact location in space or on a drawing surface.

punto Una localización exacta en el espacio o en la superficie de dibujo.

point-to-point A variation in the machine (along with continuous-path machines). In this type, the machine operates on a series of preprogrammed coordinates to locate the position of the tool. When the tool finishes at one point, it automatically goes to the next point by the shortest route.

punto a punto Una variación en la máquina (junto con las máquinas de camino continuo). En este tipo, la máquina opera con una serie de coordenadas preprogramadas para localizar la posición de la herramienta. Cuando la herramienta termina en un punto, automáticamente se mueve al próximo punto por el camino más corto.

poles Two reference measuring positions on the surface of the sphere on opposite sides of its center.

polos Dos posiciones de referencia en la superficie de la esfera en lados opuestos de su centro.

polish To produce a highly finished or polished surface by friction using a very fine abrasive.

pulir Producir una superficie bien acabada o pulida por fricción usando un abrasivo fino.

polygon A plane geometric figure with three or more sides.

polígono Una figura geométrica con tres o más lados.

polyhedron A solid object bounded by plane surfaces. Each surface is called a face.

poliedro Un objeto sólido definido por superficies planas. Cada superficie se llama una cara.

positional tolerance A tolerance that controls the position of a feature relative to the true dimension specified for the features, as related to a datum or datums.

tolerancia de posición Una tolerancia que controla la posición de un rasgo relativo a la dimensión real especificada para los rasgos, según se relacionan con una referencia o referencias.

powder metallurgy A process similar to casting in which metal particles, or powder, are blended and mixed. The powder is then forced into a die of the desired form under intense pressure. The resulting heat fuses the powder into a solid piece that can be machined.

metalurgia de polvo   Un proceso similar al proceso de moldes de fundición en el cual las partículas de metales o el polvo se mezclan. El polvo entonces se pasa a fuerza dentro de un troquel con la forma deseada bajo una presión intensa. El calor resultante funde el polvo en una pieza sólida que puede labrarse.

powdered metals   Metal powders that are produced by atomization, electrolysis, chemical reduction, or oxide reduction. They can be mixed or sintered to produce alloys with an almost endless number of compositions. The properties of products made from powdered metals depend on which metals are mixed, in what amounts, and by what processes.

metales en polvo   Polvos metálicos que se producen por atomización, electrólisis, reducción química, o reducción de óxido. Las propiedades de los productos que se hacen de los metales en polvo dependen de cuáles metales se mezclan, en qué cantidades, y por cuál proceso.

power saw   A generic term for all types of electric or gas-operated saws. The types of power saws used most often in the machine shop are the contour band machine, or band saw, and the cutoff saw.

sierra de potencia   Un término genérico para todo tipo de sierras eléctricas u operadas con gasolina. Los tipos de sierras de potencia que se usan con más frecuencia en un taller mecánico son la sierra de cinta de contorno o sierra de cinta y la sierra tronzadora.

precision   The quality or state of being precise or accurate; mechanical exactness.

precisión   La calidad o estado de ser preciso; exactitud mecánica.

precision grinder   Grinders that are used for precision work. The principal types used are surface and cylindrical grinders.

moledora de precisión   Moledoras que se usan para trabajo de precisión. Los tipos principales que se usan son las moledoras de superficie y cilíndricas.

press forging   A single-step process in which the heated metal bar or billet is forced into a single die and is completed in a single press stroke.

forjado a presión   Un proceso de un solo paso en el cual una barra de metal o un lingote caliente se presiona contra un troquel y se completa en un solo ciclo de presión.

press-fit bushings   A type of bushing used to position and guide the cutting tools used in jigs. They are pressed directly into the jig plate. They are used for short-run jigs or in applications where the bushings are not likely to be replaced frequently.

cojinete de encaje a presión   Un tipo de cojinete que se usa para posicionar y guiar las herramientas cortantes que se usan en las guías. Se encajan a presión directamente en la placa del guía. Se usan en las guías cortas o en las aplicaciones donde los cojinetes no serán reemplazados frecuentemente.

primary auxiliary view   Views that have been projected perpendicular from an inclined surface and viewed looking directly at that surface.

vista primaria auxiliar   Vista que se proyecta perpendicularmente desde una superficie inclinada y se ve directamente en la superficie.

principal viewing planes   The top, front, and right-side planes; they are all perpendicular to each other.

planos de vistas principales   Los planos superior, frontal, y del lado derecho; todos están perpendiculares unos a otros.

printer   A device that prints the output of a computer.

impresora   Un aparato que imprime el resultado de una computadora.

prism   A solid whose bases or ends are any congruent and parallel polygons and whose sides are parallelograms.

prisma   Un sólido cuyas bases o extremos son cualesquiera polígonos paralelos y congruentes y cuyos lados son paralelogramos.

prismatic   Pertaining to or like a prism.

prismático   Perteneciente a o parecido a un prisma.

process synchronization   The state that has been achieved when all processes involved in producing a product are in balance and synchronized so that production flows smoothly and continuously.

sincronización de proceso   El estado que se ha logrado cuando todos los procesos involucrados en producir un producto están balanceados y sincronizados de manera que su producción fluye suavemente y continuamente.

processor   The computer in a CAD system.

procesador   La computadora en un sistema de CAD.

profile   A feature control that specifies the amount of allowance variance of a surface or line elements on a surface.

perfil   Un control de rasgo que especifica la cantidad de variación permisible de una superficie o los elementos de línea en una superficie.

profile of a line   A tolerance that controls the allowable variation of a surface from a basic profile or configuration. It establishes a tolerance for a given single element of a surface.

perfil de una línea   Una tolerancia que controla la variación permisible en una superficie del perfil o configuración básica. Establece la tolerancia para un elemento de una superficie.

profile of a surface   A tolerance that controls the allowable variation of a surface from a basic profile or configuration. It applies to the entire surface. The difference between the profile of a line and surface is similar to the difference between circularity and cylindricity.

perfil de una superficie Una tolerancia que controla la variación permisible en una superficie del perfil o la configuración básica. Se aplica a toda la superficie. La diferencia entre el perfil de una línea y la superficie es similar a la diferencia entre la circularidad y la cilindricidad.

program A set of step-by-step instructions telling the computer to solve a problem with the information input to it or contained in memory.

programa Un grupo de instrucciones paso por paso que le dicen a una computadora que resuelva un problema con la información que le es dada o que está contenida en memoria.

progress report A written report, directed to the client, to inform him/her of the things that have been done to complete the job and the things that still need to be done. It can be issued at agreed-upon intervals, depending on the length of the project.

reportaje del estado del proyecto Un reportaje escrito, dirigido al cliente, para informarle sobre las cosas que habían hecho para completar el trabajo, y las cosas que todavía necesitan completarse. Puede publicarse a intervalos convenidos, dependiendo del largo del proyecto.

progressive die A die in which several operations take place sequentially.

troquel progresivo Un troquel en el cual se llevan a cabo varias operaciones en secuencia.

project work Work that is built around a specific and finite objective. Project is typically done by teams.

proyecto de trabajo Trabajo centrado en un objetivo específico y determinado. Normalmente se realiza en equipo.

projected tolerance zone A modifier that allows a tolerance zone established by a locational tolerance to be extended a specified distance beyond a given surface.

zona de tolerancia proyectada Un modificador que permite que se establezca una zona de tolerancia por medio de una tolerancia que se extenderá una distancia más allá de una superficie dada.

projection weld Similar to a spot weld and using the same welding symbol. This weld allows more penetration and, as a result, is a much better weld.

soldadura de proyección Similar a una soldadura de punto y usa el mismo símbolo de soldadura. Esta soldadura permite más penetración y, como resultado, es una soldadura mucho mejor.

projectors Imaginary lines of sight.
proyecciones Líneas de vista imaginarias.

proposal This is a combination of the selling price, specifications, delivery schedules, and the boilerplate.

propuesta Es la combinación del precio de venta, especificaciones, horarios de entregas y el texto modelo.

prototype An original model on which something is patterned.

prototipo Un modelo original según el cual se fabrica algo.

protractor A manual drafting tool used for measuring angles.

transportador Un dispositivo usado en el dibujo a mano para medir ángulos.

punch To cut an opening of a desired shape with a rigid tool having the same shape.

perforar Cortar una abertura de una forma deseada con una herramienta rígida de la misma forma.

punching An operation that cuts a hole in a metal sheet or strip.

perforado Una operación que corta un orificio en una hoja o tira de metal.

purchased part An item made elsewhere that is required to create a final product. Purchased parts are drawn on assembly drawings, usually in outline form, to show their function and relationship to the product.

pieza comprada Un artículo que fue hecho en otro lugar y que se requiere para crear el producto final. Las piezas compradas se dibujan en dibujos de ensamblaje, generalmente en silueta, para mostrar su función y relación con el producto.

pyramid A polyhedron having a polygon as its base.
pirámide Un poliedro que tiene un polígono como su base.

quadrant A sector with a central angle of 90° and usually with one of the radial lines oriented horizontally.

cuadrante Un sector con un ángulo central de 90° y generalmente con una de las líneas radiales horizontales.

quadrilateral A plane figure bounded by four straight lines.

cuadrilátero Una figura plana bordeada por cuatro líneas rectas.

quench The rapid cooling of a machined part in a cooling material. To immerse a heated piece of metal in water or oil in order to harden it.

enfriamiento rápido El enfriamiento rápido de una parte labrada en un material de enfriamiento. Sumergir un pedazo de metal calentado en agua o aceite para endurecerlo.

quotation department This department takes all factors, such as labor costs, material costs, manufacturing expenses, sales salaries and expenses, and general overhead costs, and uses them to arrive at the total cost to produce a product. It then adds the desired profit to the total cost to establish the selling price.

departamento de cotización   Este departamento toma todos los factores, tales como los costos de mano de obra, costos de materiales, gastos de manufactura, gastos y salarios de ventas, y los costos operativos generales, y los usa para llegar al costo total de producir un producto. Entonces le suma la ganancia deseada al costo total para establecer el precio de venta.

rack   A flat bar with gear teeth in a straight line to engage with teeth in a gear.

cremallera   Una barra plana con dientes de engrane alineados que encajan con los dientes de un engranaje.

rack gear   A type of spur gear whose teeth are in a straight line or flat.

engranaje de cremallera   Un tipo de engranaje de dientes rectos cuyos dientes están alineados o en un plano.

radial arm   Design that changes rotary motion into either an up-and-down motion or a rocking action.

brazo radial   Diseño que cambia el movimiento rotativo a un movimiento hacia arriba y hacia abajo o un movimiento de mecedora.

radial line development   A development in which the fold lines radiate from one point.

desarrollo de Línea radial   Un desarrollo en el cual las líneas de doblez parten desde un punto.

radius   The distance from the center point of a circle to the outside curved surface. It is half the diameter.

radio   La distancia desde el punto central de un círculo a la superficie exterior curva. Es la mitad del diámetro.

RAM   Random Access Memory.

RAM   Memoria de Acceso Aleatorio.

rapid prototyping   A prototyping process that builds the final product layer by layer by an electronic-laser process that uses data from three-dimensional drawings.

creación rápida de prototipo   Un proceso de creación de prototipo que construye el producto final capa por capa con un proceso electrónico con láser que usa datos de dibujos tridimensionales.

raster   The coordinate grid dividing the display area of a graphics display.

cuadrícula   La cuadrícula coordinada que divide el área de una pantalla.

real time   Response times in the interaction between humans and computers that are, for all intents and purposes, equal to real-world response times.

tiempo real   Tiempos de respuesta en la interacción entre humanos y computadoras que son, para todo propósito práctico, iguales a los tiempos de respuesta del mundo real.

rectangular system   A method of surveying in which a large area of land is surveyed beginning from one basic reference point and divided into a grid of six-mile-square blocks called townships. Townships are further divided into 36 one-mile-square blocks called sections; these are subsequently divided into smaller and smaller rectangular sections of land.

sistema rectangular   Un método de planimetría en el cual se mide un área grande de terreno, comenzando desde un punto básico de referencia y dividido en una cuadrícula de bloques de seis millas cuadradas llamados municipios. Los municipios se subdividen en 36 bloques de una milla cuadrada llamados secciones; éstos a su vez se subdividen en secciones de terreno cada vez más pequeñas.

reference dimension   A nontoleranced dimension used for information purposes only, which does not govern production or inspection operations.

dimensión de referencia   Una dimensión sin tolerar que se usa con propósitos informativos solamente y que no gobierna las operaciones de producción o inspección.

reference line   A line that is always placed through the center of any symmetrical object.

línea de referencia   Una línea que siempre se coloca a través del centro de cualquier objeto simétrico.

refractory metals   Metals with an exceptional ability to withstand high temperatures; they also have a low coefficient of expansion, high modulus of elasticity, high resistance to thermal shock, high conductivity, and excellent hardness properties at high temperatures. Tungsten and molybdenum are two of the best known.

metales refractarios   Metales con una habilidad excepcional para resistir altas temperaturas; también tienen un coeficiente de expansión termal bajo, un módulo de elasticidad alto, alta resistencia a shock termal, alta conductividad y propiedades excelentes de dureza a altas temperaturas. El tungsteno y el molibdeno son dos de los más conocidos.

refresh display   A CRT device that requires the refreshing of its screen presentation at a high rate so that the image will not fade or flicker.

pantalla refrescada   Un aparato de tubo de rayos catódicos (CRT) que requiere refrescar su presentación en pantalla a alta velocidad para que la imagen no se desvanezca o parpadee.

regardless of feature size (RFS)   The condition in which tolerance of position or form must be met irrespective of where the feature lies within its size tolerance. It tells machinists that the tolerance must be maintained regardless of the actual produced size of the object.

independiente de tamaño de rasgo   La condición en la cual la tolerancia de posición o forma debe de cumplirse independientemente de dónde caiga el rasgo dentro de la tolerancia de tamaño. Le dice a los labradores que las tolerancias deben mantenerse independiente del tamaño de producción del objeto.

relief   An offset of surfaces to provide clearance.

relieve   Un desbalance entre superficies para proveer separación.

removed section   This section is very similar to a rotated section, except that it is drawn removed or away from the regular views.

sección removida   Esta sección es muy similar a la sección rotada, excepto que se dibuja removida o separada de las vistas regulares.

rendering   Finishing a drawing to give it a realistic appearance; a representation.

rendición   Acabar un dibujo dándole una apariencia realista; una representación.

renewable bushing   A type of drill bushing that is used in high-volume applications or where the bushings must be changed to suit different cutting tools used in the same hole.

cojinete renovable   Un tipo de cojinete de taladro que se usa en las aplicaciones de alto volumen en el cual los cojinetes tienen que cambiarse para cumplir con las diferentes herramientas cortantes que se usan en el mismo agujero.

repeatability   The ability to redraw lines without creating a double image.

reproducibilidad   La habilidad de volver a dibujar líneas sin crear una imagen doble.

resistance welding   The process of passing an electric current through the exact location where the parts are to be joined. This is usually done under pressure. Also, the process of welding metals by using the resistance of the metals to the flow of electricity to produce the heat for fusion of the metals.

soldadura de resistencia   El proceso de pasar corriente eléctrica a través de una localización exacta dónde se unirán las partes. Esto generalmente se hace bajo presión. Además, el proceso de soldar los materiales usando la resistencia de los materiales al flujo de electricidad para producir calor para fundir los metales.

responsibility   The condition of being obliged to account for something to someone.

responsabilidad   El sentimiento de sentirse obligado a rendir cuentas de algo a alguien.

Résumé   A brief, well-structured summary of an individual's qualifications for employment.

currículum vítae   Un resumen breve y bien estructurado de las calificaciones de una persona para un trabajo.

retaining rings   Precision-engineered fasteners. They provide removable shoulders for positioning or limiting the movement of parts in an assembly.

anillos retenedores   Sujetadores diseñados con precisión. Éstos proveen bordes removibles para posicionar o limitar el movimientos de las partes de un ensamblaje.

revolved section   Sometimes referred to as the rotated section, this is used to illustrate the cross section of ribs, webs, bars, arms, spokes, or other similar features of an object.

sección revuelta   A veces llamada la sección girada, ésta es usada para ilustrar la sección transversal de costillas, entramados, barras, rayos o otros rasgos similares de un objeto.

rib   A relatively thin, flat member acting as a brace or support.

costilla   Un miembro plano y relativamente fino que sirve como refuerzo o apoyo.

right angle   An angle of 90°.

ángulo recto   Un ángulo de 90°.

right profile   The right-side view of a drawing.

perfil derecho   La vista del lado derecho de un dibujo.

right triangle   A triangle having a right angle or an angle of 90°.

triángulo recto   Un triángulo que tiene un ángulo recto o de 90°.

right-side view   The orthographic view that is projected to the right of the front view; the right profile.

vista lateral   La vista ortográfica que se proyecta al lado derecho de la vista frontal; el perfil derecho.

ring gear   Similar to the spur, pinion, and rack gears, except that the teeth are internal.

engranaje de anillo   Similar a un engranaje de dientes rectos, un piñón o un engranaje de cremallera, excepto que los dientes son internos.

riser   In sand casting, used to vent the mold and to allow gases to escape; also serves as a small reservoir for excess material so that the mold cavity will remain full as the metal begins to shrink during the cooling process.

tubo de subida   En la fundición en moldes de arena, se usan como respiradero del molde para permitir el escape de los gases; también sirve como un pequeño estanque para el material en exceso de manera que la cavidad del molde permanezca llena a medida que el metal comienza a contraerse durante el proceso de enfriamiento.

rivet (verb)   To connect with rivets or to clench over the end of a pin by hammering.

remachar   Unir con remaches o remachar el extremo de un pasador martillándolo.

rivets (noun)   Permanent fasteners, usually used to hold sheet metal together.

remaches   Sujetadores permanentes, generalmente se usan para sujetar hojalata.

robot A reprogrammable multifunction manipulator designed to move material parts, tools, or specialized devices through variable programmed motions for the performance of a variety of tasks.

robot Un manipulador multifuncional programable diseñado para mover partes de material, herramientas, o aparatos especializados a través de movimientos variables programados para el funcionamiento de una variedad de tareas.

rolling A process where the heated metal bar is formed by passing through rollers having the desired form impressed on their surfaces.

rodar Un proceso en el cual se le da forma a una barra de metal caliente pasándola a través de rodillos que tienen la forma deseada en su superficie.

ROM (read-only memory) Refers to computer chips that have programs built into them at the factory.

ROM (memoria Sólo lectura) Se refiere a un chip de computadora en el cual se almacenan los programas en la fábrica.

roof plan A plan view showing the location and spacing of rafters and/or trusses. It is superimposed on the floor plan without dimensions.

plano de techo Una vista de plano que muestra la localización y separación de las vigas y/o armaduras. Está sobrepuesta sobre el plano de distribución sin dimensiones.

root The bottom point joining the sides of a thread.

raíz El punto inferior que une los lados de una rosca.

root diameter The measurement over the extreme inner edges of teeth.

diámetro de raíz La medida sobre el extremo interior de los dientes.

root opening The allowed space between parts.

abertura de raíz El espacio permisible entre piezas.

rotary ultrasonic machining (RUM) A nontraditional machining process whereby the grinding motion is rotary and diamond tools remove material. It is used in the development of prototypes.

labrado ultrasónico rotativo Un proceso de labrado no tradicional en el cual el movimiento moledor es rotativo y las herramientas de diamante remueven el material. Se usa en el desarrollo de prototipos.

rotate To revolve through an angle relative to an origin.

rotar Girar por un ángulo relativo a un origen.

round An exterior rounded intersection of two surfaces.

redondo Una intersección exterior redondeada de dos superficies.

routine designs Relatively basic projects and problems used to introduce new members of an organization to the various people in the company and to the routes design information needs to flow through.

diseños de rutina Proyectos y problemas relativamente básicos que se usan para introducir a los nuevos miembros de la organización a las diferentes personas en la compañía y a las rutas por las cuales debe fluir la información de diseño.

runout A feature control that limits the amount of deviation from perfect form that is allowed on surfaces or rotation through one full rotation of the object about its axis.

excentricidad Un control de rasgo que limita la cantidad de desviación de una forma perfecta que es permisible en las superficies o en la rotación de un objeto alrededor de su eje a través de una rotación completa.

sandblast To blow sand at high velocity with compressed air against castings or forgings to clean them.

pulir con chorro de arena Soplar arena a alta velocidad con aire comprimido contra las piezas de moldes de fundición o forjadas para limpiarlas.

sandwich jig A variation of the plate jig in which the workpiece is held between two plates.

guía de Sándwich Una variación de la guía de placa en la cual la pieza de trabajo se aguanta entre dos placas.

scalars Magnitudes without directions, such as temperature and pressure.

escalares Magnitudes sin direcciones, tales como temperaturas y presiones.

scalene triangle A triangle having no equal sides or angles.

triangulo escaleno Un triángulo que no tiene lados o ángulos iguales.

scientific design A design process that uses concepts from physics, mechanics, materials science, chemistry, and mathematics.

diseño científico Un proceso de diseño que usa conceptos de física, mecánica, ciencia de materiales, química y matemática.

scrape To remove metal by scraping with a hand scraper, usually to fit a bearing.

raspar Remover metal raspándolo con raspador de mano, generalmente para que encaje en un cojinete.

screw A fastener that does not use a nut and is screwed directly into a part.

tornillo Un sujetador que no usa una tuerca y se atornilla directamente en una pieza.

seam weld Like a spot weld, except that the weld is actually continuous from start to finish.

soldadura continua Como una soldadura de punto, excepto que la soldadura es continua desde el principio hasta el fin.

seams The points at which two pieces of material are joined together by gluing, soldering, riveting, welding, and so on.

junta Los puntos dónde se unen dos pedazos de material por medio de pegamento, soldadura, remaches y así por el estilo.

secondary auxiliary view A view that is projected directly from the primary auxiliary view; used when a primary auxiliary view is not enough to fully illustrate an object.

vista auxiliar secundaria Una vista que se proyecta directamente desde la vista primaria auxiliar; se usa cundo la vista primaria auxiliar no es suficiente para ilustrar un objeto completamente.

section lining In welding, used to indicate where the weld is to be placed when a weld is not continuous and is needed in only a few areas. Otherwise, it shows where an object is sliced or cut by the cutting-plane line.

Línea de sección En soldadura, se usa para indicar dónde se debe colocar la soldadura cuando la soldadura no es continua y se necesita sólo en unas áreas. También, muestra dónde se corta el objeto por la línea del plano cortante.

sectional view A view of an object obtained by the imaginary cutting away of the front portion of the object to show the interior detail.

vista seccional Una vista de un objeto que se obtiene al hacer un corte imaginario y removiendo la porción frontal de un objeto para mostrar los detalles interiores.

sector The area of a circle lying between two radial lines and the circumference.

sector El área de un círculo que queda entre dos líneas radiales y la circunferencia.

segment The smaller portion of a circle separated by a chord.

segmento La porción más pequeña de un círculo separada por una cuerda.

self-locking rings A type of retaining ring; they do not require any grooves because they exert a frictional hold against axial displacement.

anillos de bloqueo automático Un tipo de anillo retenedor; no requieren ranuras porque ejercen un agarre de fricción contra el movimiento axial.

selling price Prepared by the quotation department. The result of adding basic costs such as labor, engineering materials, salaries, sales costs, general overhead, and so on to come up with the cost of manufacturing the product and then adding the desired profit to the cost.

precio de venta Es preparado por el departamento de cotización. Es el resultado de sumar los costos básicos tales como mano de obra, materiales de ingeniería, salarios, gastos de venta, costos operativos y cosas por el estilo para obtener el costo de manufacturar el producto y luego añadirle al costo la ganancia deseada.

semicircle Half of a circle.

semicírculo Mitad de un círculo.

set screw A screw used to prevent motion between mating parts, such as the hub of a pulley on a shaft. It is screwed into and through one part so that it applies pressure against another part, thus preventing motion.

tornillo de fijación Un tornillo para prevenir el movimiento entre piezas apareadas, tales como el centro de una polea en un eje. Se atornilla a través de una pieza de manera que aplica presión contra otra pieza, evitando de esta manera el movimiento.

shape To remove metal from a piece with a shaper.

conformar Remover metal de una pieza con una conformadora.

shaper A machine tool that uses a single-point cutting tool to perform the cutting operation. It is much smaller than the planer (see also *planer*). The workpiece is mounted on a table or vise, and the cutter is driven by a reciprocating ram. Seldom used today.

conformadora Una maquina de labrado que usa una herramienta cortante de un solo punto para la operación cortante. Es mucho más pequeña que la acepilladora (vea también *acepilladora*). La pieza de trabajo se monta sobre una mesa y el cortador lo maneja con un martinete reciprocante. Casi no se usa hoy día.

shaving An operation performed on a blanked part to produce an exact dimensional size or to square a blanked edge.

rasurar Una operación que se hace en una parte en una pieza bruta para producir un tamaño dimensional o para cuadrar un borde de una pieza bruta.

shearing A cutting action performed along a straight line. It uses two blades in sliding contact.

cizallar La acción cortante a lo largo de una línea recta. Usa dos cuchillas en contacto deslizante.

shim A thin piece of metal or other material used as a spacer in adjusting two parts.

calce Un pedazo fino de metal u otro material que se usa como espaciador para ajustar dos partes.

shrink rule A standard steel rule that has the graduations marked for a specific amount of shrinkage.

regla de contracción Una regla estándar de acero que tiene gradaciones marcadas para determinar la contracción.

shrinkage The percentage of size lost of a particular cast metal as it cools.

contracción El porcentaje del tamaño perdido en un metal fundido mientras se enfría.

side view auxiliary An auxiliary view that is projected from the right-side view.

vista lateral auxiliar Una vista auxiliar que es proyectada desde la vista del lado derecho.

simulation The computer-generated emulation of a system or process.

simulación Una emulación generada a computadora de un sistema o proceso.

single thread A screw thread type that is composed of one continuous ridge.

rosca sencilla Un tipo de rosca de tornillo que consiste de un resalto continuo.

single-curved surface A surface that can be unrolled to form a plane.

superficie de una curva Una superficie que puede desenrollarse para formar un plano.

size tolerance A tolerance where an ideal size is considered acceptable if it does not exceed certain parameters; for example, a size tolerance of $\pm.01''$ for a piece that is ideally $4''$ long means that it is acceptable if it is no longer than $4.01''$ and no shorter than $3.99''$.

tolerancia de tamaño Una tolerancia en la cual el tamaño ideal se considera aceptable si no excede ciertos parámetros; por ejemplo, una tolerancia de tamaño de $\pm.01''$ para una pieza que idealmente es de $4''$ de largo significa que es aceptable si no es más larga de $4.01''$ ni más corta de $3.99''$.

sketches Rough drawings of what will be the final drawing.

croquis Dibujos rudos de lo qué será el dibujo final.

slip-renewable bushings A type of renewable drill bushing used where the bushing must be changed quickly. It is held in place by a lock screw but can be removed by turning it counterclockwise to the release position and lifting it out of the hole.

cojinetes deslizables renovables Un tipo de cojinete de taladro renovable que se usa dónde se debe cambiar un cojinete rápidamente. Se aguanta en su lugar por un tornillo de seguridad, pero puede removerse por girarlo de derecha a izquierda hasta la posición de libertad y despúes levantarlo desde su agujo.

slitting Performed on large sheets to produce thin stock strips. It may be performed by rollers or straight blades.

cortar en tiras Se hace con hojas grandes para producir tiras finas de material. Puede hacerse por medio de rodillos o cuchillas rectas.

slope The angle, in degrees, that a line makes with a level plane.

pendiente El ángulo, en grados, que forma una línea con el plano de nivel.

slot weld (see also *plug weld*) A weld made through an elongated hole. The symbols are the same.

soldadura de ranura (vea también *soldadura de tapón*) Una soldadura hecha a través de un agujero alongado. Los símbolos son los mismos.

soft skills Also known as people skills. Soft skills are nontechnical skills that allow people to work effectively with others, including skills such as communication.

don de gentes También se le da el nombre de habilidades sociales. El don de gentes es una habilidad no técnica que permite a las personas trabajar eficazmente con otras e incluye habilidades tales como la comunicación.

software A set of programs, procedures, rules, and possibly associated documentation concerned with the operation of a CAD system.

software Un grupo de programas, procedimientos, reglas y la posible documentación asociada con la operación de un sistema de CAD.

solder To join with solder, a compound usually composed of lead and tin.

estañar Unir con soldadura, un compuesto generalmente compuesto por plomo y estaño.

solid length The overall length of the spring when all coils are compressed together so they touch.

largo Sólido El largo completo de un resorte cuando todas las espirales se comprimen de manera que estén tocándose.

solid modeling Electronically creating a solid representation of an object that represents not only length, width, and depth but also volume. A solid model is as much like the actual object in question as it is possible to be in an electronic replication.

modelado de Sólidos La creación por medios electrónicos de la representación sólida de un objeto que representa no sólo el largo, el ancho y la profundidad, sino también el volumen, Un modelo sólido es lo más parecido al objeto real que puede darse en una representación electrónica.

solution description Describes the final solution, including the performance specifications and the design specifications.

descripción de solución Describe la solución final, incluyendo las especificaciones de funcionamiento y de diseño.

specifications A detailed, precise presentation of something or a proposal for something. Usually divided into two main categories: performance and design. Performance specs describe what the design is supposed to accomplish; design specs describe how the product is made and the materials used.

especificaciones Una presentación detallada y precisa de algo o una propuesta para algo. Generalmente se divide en dos categorías principales: funcionamiento y diseño. Las especificaciones de funcionamiento describen qué se supone que logre el diseño; las especificaciones de diseño describen cómo se hará el producto y qué materiales se usarán.

sphere A closed surface, every point on which is equidistant from a common point or center.

esfera Una superficie cerrada, cada punto de la cual está equidistante de un punto común o centro.

spin To form a rotating piece of sheet metal into a desired shape by pressing it with a smooth tool against a rotating form.

rotar Formar una pieza de hojalata que rota en la forma deseada prensándola con una herramienta lisa contra un molde que rota.

spiral bevel gear Any bevel gear with curved teeth.

engranaje Cónico en espiral Cualquier engranaje cónico con los dientes curvos.

spot welding A resistance-type weld that joins pieces of metal by welding separate spots rather than using a continuous weld.

soldadura de punto Un tipo de soldadura de resistencia que une pedazos de metal soldando puntos separados en lugar de formar una soldadura continua.

spotface To produce a round spot or bearing surface around a hole, usually with a spotfacer; similar to a counterbore.

avellanar Producir una superficie redonda o de apoyo alrededor de un agujero, generalmente con un avellanador.

spotfaced hole A plain through hole with a shallow counterbore deep enough to provide a flat surface for the head of a bolt or machine screw or a nut for a threaded member.

agujero avellanado Un agujero pasante avellanado lo suficientemente para proveer una superficie plana a la cabeza de un tornillo o para la tuerca de un miembro con rosca.

spring A mechanical device that is used to store and apply mechanical energy.

resorte Un aparato mecánico que se usa para almacenar y aplicar energía mecánica.

spring pin A type of fastener; manufactured by cold forming strip metal in a progressive roll-forming operation. They are larger in diameter than the hole into which they are to be inserted and compress themselves as they are driven into the hole so that they exert radial forces around the entire circumference of the hole.

pasador de resorte Un tipo de sujetador; manufacturado por trabajo en frío en una operación de formación por rodillo. Son más grandes en diámetro que el agujero en el cual se van a insertar y se comprimen a medida que se meten a presión dentro del agujero de manera que ejercen fuerzas radiales a lo largo de toda la circunferencia del agujero.

sprue hole The hole in the top half of a mold flask through which the molten metal is poured to reach the gate or gates, which direct the metal to the cavity of the mold.

orificio de colada El orificio en la mitad superior de un molde a través del cual se vierte el metal derretido para que llegue a los canales que dirigen al metal a la cavidad del molde.

spur gear The most commonly used gear. It is cylindrical in form, with teeth that are cut straight across the face of the gear. All teeth are parallel to the axis of the shaft.

engranaje de dientes rectos El tipo de engranaje más común. Es de forma cilíndrica, con dientes que se cortan rectos en la cara del engranaje. Todos los dientes están paralelos al eje.

staggered intermittent weld A type of intermittent weld in which the welds are staggered directly opposite each other on opposite sides.

soldadura intermitente alterna Un tipo de soldadura intermitente en la cual las soldaduras están alternadas directamente opuestas una de otra en lados opuestos.

stainless steels Steels that contain 10.5% or more chromium.

aceros inoxidables Aceros que contienen 10.5% o más de cromo.

stamping A process of using dies to cut or form metal sheets or strips into a desired form.

acuñado Un proceso de usar troqueles para cortar o formar hojas o tiras de metal en una forma deseada.

station lines These line segments must be assumed or chosen on the lateral surface.

líneas de estación Estos segmentos de línea tienen que asumirse o escogerse en la superficie lateral.

statistical process control (SPC) A process that uses statistical control charts in monitoring production processes in an attempt to reduce the amount of scrap and rework.

control de proceso estadístico Un proceso que usa tablas de control estadístico en el monitoreo de los procesos de producción en un intento de reducir la cantidad de desperdicio y de trabajo repetido.

statistical tolerancing symbol A symbolic means of indicating that the stated tolerance is based on statistical process control (SPC).

Símbolo de tolerancia estadística Un medio simbólico de indicar que la tolerancia indicada está basada en un proceso de control estadístico.

steps (in the design process) A set of recognizable tasks in the design process that must be done in a specific order.

pasos (en el proceso de diseño) Un grupo de tareas reconocibles en el proceso de diseño que tienen que hacerse en un orden específico.

stereolithography   A process whereby solid models are created layer by layer by an electronic-laser process in a special liquid solution.

estereolitografía   Un proceso mediante el cual los modelos sólidos se crean capa por capa en un proceso electrónico por láser en una solución líquida especial.

straightness   A tolerance that controls the allowable variation of a surface or an axis from a theoretically perfect plane or line. It can be used to control surface elements, an axis, or a center plane.

rectitud   Una tolerancia que controla la variación permisible de una superficie o un eje de un plano o línea perfecta. Puede usarse para controlar los elementos de superficies, un eje o un plano central.

stress relieving   Heating a metal part to a suitable temperature and holding that temperature for a determined time, then gradually cooling it in air; this treatment reduces the internal stresses induced by casting, quenching, machining, cold working, or welding.

alivio de tensión   Calentar una pieza metálica a una temperatura adecuada y mantenerla a esa temperatura por un tiempo determinado y luego enfriarla gradualmente; este tratamiento reduce las tensiones inducidas por los procesos de fundición, enfriado, labrado, trabajado en frío o de soldadura.

stretchout   A flat pattern development for use in laying out, cutting, and folding lines on flat stock, such as paper or sheet metal, to be formed into an object.

patrón extendido   Un patrón de desarrollo plano que se usa para establecer las líneas de corte y de dobleces en material plano, tales como papel u hojalata, para formarlos en un objeto.

strike   A compass direction of a horizontal line lying in the plane of an ore vein.

dirección de descubrimiento   La dirección de compás de una línea horizontal que está en el plano de una vena de mineral.

structure   Anything constructed of parts.

estructura   Cualquier cosa que está formada de partes.

stud   A fastener that is a steel rod with threads at both ends. It is screwed into a blind hole and holds other parts together by a nut on its free end.

perno   Un sujetador que es una barra de metal con roscas en ambos extremos. Se atornilla en un agujero ciego y sujeta otras piezas por medio de una tuerca en su extremo libre.

subassembly drawing   A drawing of two or more parts permanently fastened together; very similar to an assembly drawing.

dibujo de subensamblaje   Un dibujo de dos o más partes que están permanentemente unidas; muy similar al dibujo de ensamblaje.

support available   After-purchase support for CAD software from the vendor (e.g., online help feature, telephone assistance, etc.).

apoyo disponible   Apoyo al software CAD que el vendedor da posterior a la compra (por ejemplo, ayuda en línea, asistencia telefónica, etc.)

surface grinder   A grinder mainly used to produce flat, angular, or special contours on flat workpieces.

moledora de superficie   Una moledora que se usa principalmente para producir contornos planos, angulares o especiales en las piezas de trabajos planas.

surface texture   The roughness, waviness, lay, and flaws of a surface.

textura de superficie   La aspereza, ondulación y fallas de una superficie.

surface weld   This weld is done when a surface must have material added to it or be built up.

soldadura de superficie   Esta soldadura se hace cuando hay que añadirle material a una superficie o ésta tiene que acrecentarse.

swage   To hammer metal into shape while it is held over a swage or die that fits into a hole in the swage block or anvil.

estampar   Martillar el metal para darle forma mientras se aguanta sobre un estampador o un troquel que encaja en un agujero en el bloque de estampar o yunque.

sweat   To fasten metal together by the use of solder between the pieces and the application of heat and pressure.

soldadura sudada   Unir metal por medio del uso de soldadura entre los pedazos y la aplicación de calor y presión.

symbol   A letter, character, or schematic design representing a unit or component.

símbolo   Una letra, carácter o diseño esquemático que representa una unidad o componente.

tabular dimension   A type of rectangular datum dimensioning in which dimensions from mutually perpendicular datum planes are listed in a table on the drawing instead of on the pictorial portion.

dimensión tabular   Un tipo de dimensionamiento de referencia rectangular en el cual las dimensiones de planos de referencia mutuamente perpendiculares se enumeran en una tabla en el dibujo en lugar de en la parte pictórica del dibujo.

tangent plane   This concept uses a modifying symbol with an orientation tolerance to modify the intended control of the surface.

plano tangente   Este concepto usa un símbolo modificado con una tolerancia de orientación para modificar el control intencionado de una superficie.

tap   A tool used to produce internal threads by hand or machine.

macho de rosca   Herramienta que se usa para producir las roscas internas a mano o por máquina.

taper pin   A small, tapered pin for fastening, usually to prevent a collar or hub from rotating on a shaft.

pasador Cónico   Un pasador pequeño y cónico que se usa para sujetar, generalmente para evitar que un collar o cubo rote sobre un eje.

taper reamer   A reaming tool that produces accurately tapered holes, as for a taper pin.

escariador Cónico   Una herramienta para escariar que produce agujeros cónicos precisos, como para un pasador cónico.

tapered hole   A hole that must first be drilled .015″ smaller in diameter than the smallest diameter of the tapered hole and then reamed.

agujero Cónico   Un agujero que tiene que taladrarse .015″ más pequeño que el diámetro más pequeño del agujero cónico y luego tiene que escariarse.

teamwork   A team is a group of people that works together toward a common purpose. Working in ways that ensure the efficiency and effectiveness of a team is known as teamwork.

trabajo de equipo   Un equipo es un grupo de personas que trabajan juntas para lograr un objetivo común. Se aplica el término trabajar en equipo a hacer un trabajo tal que asegura la eficacia y la efectividad del equipo.

tempering   The process of heating a hardened workpiece to a temperature below the hardening temperature, allowing the part to soak for a specific time period, and cooling.

templar   El proceso de calentar una pieza de trabajo endurecida a una temperatura por debajo de su temperatura de endurecimiento, permitiéndole a la pieza que se remoje por un tiempo específico y se enfríe.

template   A guide or pattern used to mark out the work.

plantilla   Una guía o patrón que se usa para marcar el trabajo.

temporary fasteners   Used when the parts will be disassembled at some future time.

sujetadores provisionales   Se usan cuando las piezas se desarmarán en algún momento futuro.

tensile force   A force that "goes away from" the joint.

fuerza en tensión   La fuerza que "es hacia fuera" de una unión.

tensile strength   The maximum load (pull) a piece supports without breaking or failing.

resistencia a la tensión   La carga máxima (en tensión) que una pieza aguanta sin romperse o fallar.

tesselation   Flat shading, where any curved surfaces on a model are broken up into facets.

mosaico   Sombrado plano donde cualquier superficie curva de un modelo se divide en facetas.

thermal analysis   A means of evaluating the thermal characteristics of a part or product.

análisis termal   Una manera de evaluar las características termales de una pieza o producto.

thickness gage   A standard gage used for determining what type of metal to use on any given project. It deals with the thickness—rather than the weight—of various metals used in construction. In current usage, it replaced the older but more confusing weight gage system.

calibre de espesor   Un calibre estándar que se usa para determinar qué tipo metal usar en cualquier proyecto. Utiliza el espesor – en lugar del peso – de varios metales que se usan en construcción. En su uso actual, reemplazó al sistema más viejo y más confuso del calibre de peso.

thinwall section   This section is any very thin object that is drawn in section. It should be filled in solid black.

sección de pared fina   Esta sección es cualquier objeto muy fino que se dibuja en secciones. Debería ser rellenado de negro sólido.

through hole   As its name implies, this hole goes completely through an object.

agujero pasante   Como su nombre implica, es un agujero que pasa completamente a través de un objeto.

time interval   The time it takes a cam to move the follower to the designed height.

intervalo de tiempo   El tiempo que le toma a una leva mover al palpador a la altura designada.

time-dependent variables   Variable characteristics that change over time; time-dependent variables are frequently best viewed as animations.

variables dependientes del tiempo   Características variables que cambian con el tiempo; las variables dependientes del tiempo se ven mejor como animaciones.

timing diagram   A diagram used to visualize graphically and study the interaction of cams on the same shaft by placing the different displacement diagrams in a column with the starting and ending points in line.

diagrama de sincronización   Un diagrama que se usa para visualizar gráficamente y estudiar la interacción de las levas en el mismo eje colocando los diferentes diagramas de desplazamiento en una columna con los puntos de comienzo y de fin en línea.

tin   A silvery white metal that is soft and ductile. It is too weak to be used in its pure form, so it is alloyed with other metals such as antimony, copper, lead, zinc, or silver. It is most commonly used to coat steel containers used to preserve food.

estaño Un metal plateado blancuzco que es blando y dúctil. Es muy débil para usarse en su estado puro, así que se alea con otros metales tales como antimonio, cobre, plomo, zinc o plata. Comúnmente se usa para recubrir los envases de acero que se usan para preservar alimentos.

title block A section on the drawing sheet whose primary function is to provide information not normally given on the drawing itself.

bloque titular Una sección en una hoja de dibujo cuya función principal es proveer información que normalmente no se provee en el dibujo mismo.

tolerance Regarding locators, acceptable deviation limits determined by the specific size of the part to be machined. As a general rule, the tolerance of jigs and fixtures should be approximately 30% to 50% of the part tolerance.

tolerancia Referente a los localizadores, son los límites aceptables de desviación determinados por un tamaño específico de la pieza a ser labrada. Como regla general, la tolerancia de las guías y aditamentos debería ser aproximadamente 30% a 50% de la tolerancia de la pieza.

tolerancing Setting acceptable limits of deviation.

tolerar Establecer los límites aceptables de desviación.

tool forces Those forces generated by the cutting tool during the machining cycle.

fuerzas de corte Aquellas fuerzas que son generadas por la herramienta de corte durante el ciclo de labrado.

tooling Standard and/or special tools for producing a particular part, including jigs, fixtures, gauges, cutting tools, and so on.

herramientas Herramientas estándares y/o especiales para producir una pieza particular, incluyendo guías, aditamentos, instrumentos de medición, herramientas cortantes y cosas por el estilo.

top auxiliary view An auxiliary view projected from the top view.

vista superior auxiliar Una vista auxiliar que es proyectada desde la vista superior.

top view A viewing plane that is usually above the object and perpendicular to the other two; the plan view.

vista superior Un plano de vista que generalmente está encima el objeto y perpendicular a los otros dos planos; vista aérea.

torque The rotational or twisting force in a turning shaft.

fuerza de torsión La fuerza rotacional o de torsión en un eje que rota.

torsion spring Offers resistance to a torque or twisting action. In its free state, the coils are usually either touching or very close.

resorte de torsión Ofrece resistencia a una fuerza de torsión o a una acción de torcer. En su estado libre, las espirales del resorte se están tocando o están muy cerca.

total runout A tolerance that simultaneously controls a surface related to a datum in all directions.

excentricidad total Una tolerancia que simultáneamente controla una superficie en relación a una referencia en todas las direcciones.

trade-off A compromise, usually between costs and materials.

concesión Un compromiso, generalmente entre costos y materiales.

trammel An instrument consisting of a straightedge with two adjustable fixed points for drawing curves and ellipses.

elipsógrafo Un instrumento que consiste de una regla con dos puntos fijos ajustables y se usa para dibujar curvas y elipses.

transition fit A condition in which two parts are toleranced so that either a clearance or an interference can result when the parts are assembled.

encaje de transición Una condición en la cual dos piezas se han tolerado de manera que resulte en una separación o una interferencia cuando se ensamblen las piezas.

translucent A quality of material that passes light but diffuses it so that objects are not identifiable.

translúcido La cualidad de un material que permite que la luz pase pero que la difunde de manera que los objetes no son identificables.

triangle A closed plane figure with three straight sides.

triángulo Una figura de plano cerrado con tres lados rectos.

triangular scales Various types of scales—mechanical engineering, architectural, civil engineering—that are triangular in cross section, a feature that allows each device to have six different scales or to contain multiple scales.

reglas triangulares Diversos tipos de reglas – para ingeniería mecánica, para arquitectura, para ingeniería civil – que son triangulares en cruce transversal, característica que permite que en cada herramienta se dispongan de seis o más reglas diferentes.

triangulation A method of dividing a surface into a number of triangles and then transferring each triangle's true size and shape to the development.

triangulación Un método de dividir una superficie en un número de triángulos y luego transferir el tamaño real y forma de cada triangulo al desarrollo.

triangulation development The development process of breaking up an object into a series of triangular plane surfaces.

desarrollo de triangulación El proceso de desarrollo de romper un objeto en una serie de superficies planas triangulares.

trimetric drawing A pictorial drawing technique that uses axes that are rotated so that each axis is drawn at a different angle and each axis uses a different scale.

dibujo trimétrico Una técnica de dibujo pictórico en el que se rotan los ejes de manera que cada eje se dibuja a un ángulo diferente y a una escala diferente.

trimming An operation performed on formed or drawn parts to remove the ragged edge of the blank.

desorillar Una operación que se hace en las piezas formadas para remover el borde irregular de una pieza bruta.

true distance The true actual straight distance between any two points.

distancia real La distancia recta real entre cualesquiera dos puntos.

true length True length of a line is the actual straight distance between its two ends.

largo real El largo real de una línea es la distancia recta real entre sus dos extremos.

true position The basic or theoretically exact position of a feature.

posición real La posición básica o teórica exacta de un rasgo.

true size True actual size of a surface or object.

tamaño real El tamaño real de una superficie u objeto.

true slope This slope can only be seen in the view that has a level line and where it is drawn at a true length.

pendiente real Esta pendiente sólo puede verse en una vista que tiene una línea nivelada y que se dibujó al largo real.

truncated Having the apex, vertex, or end cut off by a plane.

truncado Que tiene un ápice, vértice o un extremo cortado por un plano.

trusses Frames that are usually two-dimensional and are made of structural shapes, such as angles, tubes, rounds, and channels.

armadura Armazones que generalmente son de dos dimensiones y que están compuestas por formas estructurales como ángulos, tubos, círculos y canales.

turn To produce on a lathe a cylindrical surface parallel to the center line.

tornear Producir en un torno una superficie cilíndrica paralela a la línea de centro.

two-force members Members that have force at each end. The forces are equal in magnitude, in opposite directions, and collinear.

miembros de dos fuerzas Miembros que tienen una fuerza en cada extremo. Las fuerzas son iguales en magnitud, en direcciones opuestas, y son colineles.

typical This term, when associated with any dimension or feature, means the dimension or feature applies to the locations that appear to be identical in size and shape.

típico Este término, al asociarlo con algún rasgo o dimensión, significa que el rasgo o dimensión se aplica a los lugares que parecen ser idénticos en tamaño y forma.

ultrasonic machining (USM) A nontraditional machining process that removes material using an abrasive slurry driven by a special tool vibrating at a high frequency along a longitudinal axis.

labrado ultrasónico a Máquina Un proceso no tradicional de labrar a máquina que remueve material usando una lechada abrasiva que la guía una herramienta especial que vibra a altas frecuencias a lo largo del eje longitudinal.

ultrasonically assisted machining (UAM) A nontraditional machining process that involves assisting such traditional machining processes as drilling and turning by coupling the tool with an external source of vibrating energy.

labrado a máquina asistido por ultrasonido Un proceso no tradicional de labrar a máquina que consiste en asistir un proceso de labrado tradicional, tal como el taladrado o el torneado, acoplando la herramienta a una fuente externa de energía vibratoria.

uniform Having the same form or characteristic; unvarying.

uniforme Que tiene la misma forma o característica; invariante.

uniform acceleration The smoothest motion of all cams, and its speed is constant through the cam travel. The acceleration curve is actually a parabolic curve; the first half of the curve is exactly the reverse of the second half. This form is best for high speed.

aceleración uniforme El movimiento más suave de todos las levas, y su velocidad es constante a través del viaje de la leva. La curva de aceleración es en realidad una curva parabólica; la primera mitad de la curva es exactamente el reverso de la segunda mitad. Esta forma es la mejor para velocidad alta.

uniform velocity Regarding a cam, a type of motion. That is, it rises and falls at a constant speed, but the start and stop are very abrupt and rough.

velocidad uniforme Referente a una leva, un tipo de movimiento. Esto es, sube y baja a una velocidad constante, pero el comenzar y el detener son muy abruptos y ásperos.

unilateral tolerance When applied to a dimension, the tolerance applies in one direction (for example, the object may be larger but not smaller, or it may be smaller but not larger).

tolerancia unilateral   Cuando se aplica a una dimensión, la tolerancia aplica en una dirección (por ejemplo, el objeto puede ser más grande pero no más pequeño, o puede ser más pequeño pero no más grande).

unlimited plane   A term meaning that there is a limitless extension beyond the indicated boundaries.

plano ilimitado   Un término que significa que hay una extensión ilimitada más allá de los bordes indicados.

uppercase letters   A lettering system where all the letters are uppercase. It is used on the viewing plane.

letras mayúsculas   Un sistema de letras en el cual todas las letras son mayúsculas. Se usa en el plano de vista.

upset   To form a head or enlarged end on a bar by pressure or by hammering between dies.

recalcar   Formar una cabeza o un extremo engrandecido en una barra por medio de presión o martillando entre troqueles.

upsetting   A process used to enlarge selected sections of a metal part.

recalcadura   Un proceso usado para engrandecer secciones selectas de una parte metálica.

vendor catalogs   Catalogs published by vendors of materials that list materials and their typical uses to aid in their selection.

catálogos de vendedores   Catálogos publicados por los vendedores de materiales que listan materiales y sus usos típicos para ayudar en su selección.

verification   A technique used in preparing notes on drawings; the notes are first written down on a print of the drawing, and then another drafter checks the notes to ensure that they are accurate and not open to multiple interpretations.

verificación   Una técnica que se usa para preparar notas en los dibujos; las notas se escriben primero en una impresión del dibujo y luego otro redactor verifica las notas para asegurar que son precisas y no están abiertas a interpretaciones múltiples.

vernier caliper   A caliper with a Vernier scale for making fine measurements of the length, depth, and width of mechanical objects and their features.

calibrador vernier   Un calibradir con una escala Vernier para realizar medidas precisas de largo, profundidad y ancho de objetos mecánicos y de sus características.

vernier scale   A small, movable scale, attached to a larger fixed scale, for obtaining fractional subdivisions of the fixed scale.

escala vernier   Una escala pequeña y movible que está pegada a una escala más grande, y se usa para obtener subdivisiones fraccionales de la escala fija.

vertex   The highest point of something; the top; the summit. Plural: vertices.

Vértice   La posición más alta de algo; la cimbre; la cumbre. Plural: vértices.

vertical   Perpendicular to the horizon.

vertical   Perpendicular al horizonte.

vertical boring mill   A machine tool commonly used to turn, bore, and face large parts in much the same way as a lathe.

fresa de taladrado vertical   Una herramienta para labrar que se usa comúnmente para tornar, taladrar y encarar partes grandes de manera similar a un torno.

vertical turret lathe   A variation of the vertical boring mill. Rather than using a single tool, this machine uses a turret arrangement to mount and position the cutting tools.

torno de torre vertical   Una variación de la fresa de taladrado vertical. En lugar de usar una herramienta sola, esta máquina usa un arreglo de torre para montar y posicionar las herramientas cortantes.

virtual condition   A condition of a feature in which all tolerances of location, size, and form are considered. Also, where straightness applied to the axis or center line of a feature creates a boundary condition.

condición virtual   Una condición de algún rasgo en la cual se consideran todas las tolerancias de localización, tamaño y forma. También, dónde la rectitud aplicada al eje o la línea central de un rasgo crea una condición de frontera.

virtual reality   A technology designed to simulate either a physical or hypothetical environment. It is intended to involve all the natural senses.

realidad virtual   Una tecnología diseñada para simular un ambiente físico o hipotético. Su intención es involucrar a todos los sentidos.

visible line   One of 12 basic types of lines, it is thin and dark.

línea visible   Una de 12 tipos básicos de líneas. Es fina y oscura.

visualization   A pictorial representation of data for the purpose of more clearly and concisely communicating information.

visualización   Una representación pictórica de los datos con el propósito de comunicar información más clara y concisamente.

vorticity   The stage of a fluid in vortical (rotational) motion; the whirling of a fluid (like blood in the heart).

vorticidad   El estado de un fluido en movimiento voraginoso (rotacional); el arremolinar de un fluido (como la sangre en el corazón).

warped surface A surface that cannot be developed because it is neither a plane surface nor a curved surface.

superficie combada Una superficie que no puede desarrollarse porque no es una superficie plana ni es una superficie curva.

waste Used here in the broadest sense and encompasses rejected parts, wasted personnel time, wasted machine time, excessive inventory, and any other situation that wastes time, material, or any other manufacturing resource.

desperdicio Aquí se usa en el sentido más amplio y comprende las partes rechazadas, el tiempo perdido del personal, el tiempo perdido de las máquinas, el inventario en exceso, y cualquier otra situación que desperdicia tiempo, material o cualquier otro recurso manufacturero.

water jet machining (WJM) A nontraditional machining process used with nonmetal workpieces. It uses a thin, pressurized jet of water against the workpiece.

labrado con chorro de agua Un proceso no tradicional de labrado que se usa con piezas que no son de metal. Usa un chorro fino de agua presurizada contra la pieza de trabajo.

web A thin, flat part joining larger parts; also known as a rib.

entremado Una pieza plana y fina que une piezas más grandes; también se conoce como una costilla.

weight gage An obsolete system for determining metal usage. It was based on the weight of wrought iron. It has been replaced by the thickness gage.

calibre de peso Un sistema obsoleto para determinar el uso de metal. Estaba basado en el peso del hierro forjado. Se ha reemplazado por el calibre de espesor.

welded tool The bodies offer a faster lead time but must be machined frequently to remove any distortion caused by the heat of welding.

herramienta soldada Las piezas ofrecen un tiempo de procesamiento más rápido pero tienen que labrarse a máquina frecuentemente para remover cualquier distorsión causada por el calor de la soldadura.

welding The process of holding two pieces of metal together, permanently, by means of pressure or fusion-welding processes.

soldadura El proceso de unir dos pedazos de metal, permanentemente, por medio de procesos de soldadura a presión o por fusión.

welding symbol A symbol used to convey welding instructions on a drawing. It consists of a reference line, a leader line and arrow and, if needed, a tail.

símbolo de soldadura Un símbolo que se utiliza para comunicar las instrucciones de soldadura en un dibujo.

Consiste de una línea de referencia, una línea principal y una flecha y, si es necesario, una cola.

whiteprinter A machine that reproduces a drawing through a chemical process. An original drawing is placed on a sheet of coated whiteprint paper and exposed to light, and the exposed sheet is passed through ammonia vapor for developing. This causes the portions of the sheet that were shielded from the light by the lines on the original to turn blue or black.

impresora de blanco Máquina que reproduce los dibujos por medio de un proceso químico. El dibujo original se coloca sobre una hoja de papel de impresión en blanco y se expone a la luz, y la hoja expuesta se pasa por vapor de amoniaco para revelarse. Esto causa que las partes de la hoja que fueron bloqueadas de la luz por las líneas en el dibujo original se tornen azul o negras.

whole depth The total depth of a tooth space.

profundidad total La profundidad total de un espacio de engranaje.

wire size The diameter of the wire used to make up the spring.

tamaño de alambre El diámetro del alambre usado para formar un resorte.

Woodruff key A semicircular flat key.

llave Woodruff Una llave plana y semicircular.

working circle Considered a distance equal to the distance from the center of the camshaft to the highest point on the cam.

círculo de trabajo Se considera como una distancia igual a la distancia desde el centro del eje de levas al punto más alto de la leva.

working depth The distance that a tooth projects into the mating space. It is equal to gear addendum (a) plus pinion addendum (a).

profundidad de trabajo La distancia que se proyecta un diente dentro de su pieza de engrane. Es igual a la altura de cabeza del engranaje (a) más la altura de cabeza del piñón (a).

working drawings A set of drawings that provides details for the production of each part and information for the correct assembly of the finished product.

dibujos de trabajo Grupo de dibujos que provee detalles para la producción de cada pieza e información para el ensamblaje correcto del producto final.

workpiece A part in any stage of manufacturing before it becomes a finished part.

pieza de trabajo Un parte en cualquier paso de la manufactura antes de convertirse en una pieza acabada.

workplace safety Working in such a way as to ensure that the health and safety of personnel are not threatened by potential hazards in the workplace. To achieve peak performance and continual improvement, people need a safe work environment.

seguridad en el lugar de trabajo Trabajar con la seguridad de que la salud y la seguridad del personal no se hallan amenazadas por eventuales riesgos en el lugar de trabajo. Para lograr un desempeño tope y un mejoramiento continuo el personal necesita un ambiente de trabajo seguro.

workstation The assigned location where a worker performs his or her job (i.e., the keyboard and the system display).

estación de trabajo El lugar asignado dónde un trabajador hace su trabajo (esto es, el teclado y la pantalla del sistema).

worm gear Used to transmit power and motion at a 90° angle between nonintersecting shafts. It is normally used as a speed reducer.

engranaje de tornillo Se usa para transmitir potencia y movimiento a un ángulo de 90° entre los ejes que no se cruzan. Normalmente se usa como un reductor de velocidad.

written notes Brief, carefully worded statements placed on drawings to convey information not covered or not adequately explained using graphics.

notas escritas Declaraciones cortas, redactadas cuidadosamente que se colocan en los dibujos para comunicar información que no está cubierta o que no se explica adecuadamente usando gráficos.

wrought iron Iron of low carbon content; useful because of its toughness, ductility, and malleability.

hierro forjado Hierro con poco contenido de carbón; es útil debido a su dureza, ductilidad, y maleabilidad.

yards Volume in cubic yards; used in construction.

yardas Volumen en yardas cúbicas; utilizado en construcción.

zinc A bluish-white metal that is widely used in manufacturing. It is produced in castings and is widely used for coating other materials. Its advantages include ease of casting, strength, ductility, and the ability to hold close tolerances.

zinc Un metal azul-blancuzco que se usa extensamente en la manufactura. Se produce en moldes fundidos y se usa extensamente para recubrir otros materiales. Sus ventajas incluyen la facilidad de fundición, fortaleza, ductilidad y su habilidad para mantener tolerancias pequeñas.

index